Using Information Technology

Technology

Third
Edition

A Practical Introduction to Computers & Communications

Using Information Technology

Technology
Third Edition 3

A Practical Introduction to Computers & Communications

Brian K. Williams

Stacey C. Sawyer

Sarah E. Hutchinson

 **Irwin McGraw-Hill**

Boston Burr Ridge, IL Dubuque, IA Madison, WI New York San Francisco St. Louis Bangkok Bogotá Caracas
Lisbon London Madrid Mexico City Milan New Delhi Seoul Singapore Sydney Taipei Toronto

Irwin/McGraw-Hill

A Division of The McGraw-Hill Companies

USING INFORMATION TECHNOLOGY

This book is printed on acid-free paper.

international   2 3 4 5 6 7 8 9 0   VNH/VNH   9 3 2 1 0
domestic   3 4 5 6 7 8 9 0   VNH/VNH   9 3 2 1 0

ISBN 0-256-26146-6

Vice president/Editor-in-chief: *Michael W. Junior*
Senior sponsoring editor: *Garrett Glanz*
Developmental editor: *Burrston House, Ltd.*
Marketing manager: *Jodi McPherson*
Senior project manager: *Gladys True*
Senior production supervisor: *Madelyn S. Underwood*
Production management & page layout: *Stacey C. Sawyer, Sawyer & Williams*
Interior designer: *Laurie J. Entringer*
Cover designer: *Matthew Baldwin*
Cover illustrator: *Bob Commander © SIS*
Senior photo research coordinator: *Keri Johnson*
Photo research: *Monica Suder*
Supplement coordinator: *Nancy Martin*
Compositor: *GTS Graphics, Inc.*
Typeface: *10/12 Trump Mediaeval*
Printer: *Von Hoffman Press, Inc.*

Library of Congress Cataloging-in-Publication Data
Williams, Brian K.
    Using information technology: a practical introduction to
  computers & communications / Brian K. Williams, Stacey C. Sawyer,   Sara E.
Hutchinson. — 3rd ed.
        p.   cm.
    Includes index.
    ISBN 0-256-26146-6
    1. Computers.   2. Telecommunications systems.   3. Information
technology.   I. Sawyer, Stacey C.   II. Hutchinson, Sarah E.
    III. Title
    QA76.5W5332 1999
    004—dc21   98-21695

http://www.mhhe.com

Photo and other credits are listed in the back of the book.

Brief Contents

Preface to the Instructor

"Computer technology is the most powerful and the most flexible technology ever developed," says Terry Bynum, who chairs the American Philosophical Association's Committee on Philosophy and Computing. "Even though it's called a technical revolution, at heart it's a social and ethical revolution because it changes everything we value."

For evidence we need only look to the daily headlines: The boom in Web sites. The portal wars. Internet 2. Broader bandwidth connections. Falling PC prices. Greater microprocessor power. Intranets and extranets. Telephone and cable company mergers. GEO, MEO, and LEO satellite systems. And behind the headlines are the social and ethical changes: The mobile workplace. The blurring of work and leisure. The risk to privacy. The altering of photographs. Net gossip passing as truth. The technological distancing of the haves from the have-nots.

Information technology—the fusion of computing and communications— is creating far-reaching changes in the way we work, the way we live, and even in the way we think.

The Audience for & Promises of This Book

USING INFORMATION TECHNOLOGY: A Practical Introduction to Computers & Communications, THIRD EDITION, is intended for use as a concepts textbook to accompany a one-semester or one-quarter introductory course on computers or microcomputers. It is, we hope, a book that will make a difference in the lives of our readers.

The **key features** of *USING INFORMATION TECHNOLOGY,* THIRD EDITION, are as follows. We offer:

1. **Emphasis on unification of computer and communications systems.**
2. **Careful revision in response to extensive instructor and student feedback.**
3. **Commonsense illustration program.**
4. **Emphasis on practicality.**
5. **Emphasis throughout on ethics.**
6. **Use of techniques for reinforcing student learning.**
7. **Up-to-the-minute material—in the book and on our Web site.**

We elaborate on these features next.

Key Feature #1: Emphasis on Unification of Computers & Communications

The First Edition of this text broke new ground by emphasizing the technological merger of the computer, communications, consumer electronics, and media industries through the exchange of information in the digital format used by computers. This is the relatively new phenomenon known as **technological convergence.**

Since the First Edition, other texts have also added coverage of the Internet and the World Wide Web. However, we agree with analysts who say the revolution is far broader than this, and we continue to stress the unification of entire industries and technologies and their effects. Thus, the THIRD EDITION continues to embrace the theme of convergence by giving it in-depth treatment in six chapters—the introduction, systems software, telecommunications, communications technology, databases, and promises and challenges (Chapters 1, 3, 7, 8, 9, 12). Convergence is also brought out in examples throughout other chapters.

This theme covers much of the technology currently found under such phrases as *the Information Superhighway, the Multimedia Revolution,* and *the Digital Age: mobile computing, the Internet, Web search tools, online services, workgroup computing, the virtual office, video compression, PC/TVs, "intelligent agents,"* and so on.

Key Feature #2: Careful Revision in Response to Extensive Instructor & Student Feedback

Our publisher has told us that the First Edition of *USING INFORMATION TECHNOLOGY* was apparently the most successful new text in the field at that time, with over 300 schools adopting both comprehensive and brief versions. We were delighted to learn that the Second Edition reached an even wider audience. An important reason for this success, we believe, was all the valuable contributions of our reviewers, both instructors and students.

Both the printed version of the Second Edition and the manuscript and proofs of the THIRD EDITION underwent a highly disciplined and wide-ranging reviewing process. This process of expert appraisal drew on instructors who were both users and nonusers, who were from a variety of educational institutions, and who expressed their ideas in both written form and in focus groups.

We also received input from a number of student users and nonusers of the Second Edition. Many indicated their appreciation for the Experience Boxes, as well as such pedagogical devices as section "Previews & Reviews," our unique end-of-chapter Summary, the practical emphasis of the book, and the people-oriented writing.

We have sometimes been overwhelmed with the amount of feedback, but we have tried to respond to all consensus criticisms and countless individual suggestions. Every page of the THIRD EDITION has been influenced by instructor feedback. The result, we think, is **a book addressing the needs of most instructors and students.**

New to this edition! In particular, we have addressed the following matters:

- **Communications material separated into two chapters:** Because of the overwhelming amount of new material, and following the direction of our reviewers, we split the old "Communications" chapter into two chapters. Chapter 7, "Telecommunications," covers online resources, the Internet, and the World Wide Web. Chapter 8, "Communications Technology," covers communications hardware, channels, and networks. (The chapters may be assigned in reverse order without loss of continuity.)

- **Input and output material made one chapter:** The two chapters "Input" and "Output" are combined into a single chapter, which allows us to continue to offer a book of just 12 chapters, which instructors (particularly in quarter-system schools) have indicated they prefer.

- New Experience Boxes—on digital photography, Web use for term papers, and identity theft: Recent developments suggested a need for **new end-of-chapter Experience Boxes.** The arrival of digital photography has led us to create *"Photo Opportunities: Working with Digitized Photographs"* (Chapter 6). Also, in keeping with our ethics theme, the newly popular use of the Web by students for term-paper research—and for online plagiarism and other abuses—has resulted in *"Web Research, Term Papers, & Plagiarism"* (Chapter 7). The escalation of theft-of-identity crime has resulted in *"Preventing Your Identity from Getting Stolen"* (Chapter 9). Other Experience Boxes have been thoroughly updated to reflect the latest advances.

- New README Boxes throughout: The well-received README boxes that appear in every chapter are almost **completely new** but they still retain their practical orientation. Examples are "Batteries for Laptops," "Getting Real About Credit Cards," "Comparing Mobile Phones," "How Long Will Digitized Data Last?" and "Programmers Wanted—*Really* Wanted." See the list on the inside front cover.

In addition to these major structural and substantive changes, we have made hundreds of line-by-line and word-by-word adjustments to refine coverage and to conform with instructor's requests.

Key Feature #3: Commonsense Illustration Program

In an era of overillustrated introductory texts, more and more instructors have become concerned about texts that now seem to favor illustrations—and especially glitzy photos—over information. The THIRD EDITION of *USING INFORMATION TECHNOLOGY* addresses this concern in several ways. Artwork in the book is designed principally to be **didactic.** There are no unnecessary, space-filling photo "galleries," for instance. To support learning concepts, photographs are often coupled with additional information—an elaboration of the discussion in the text, some how-to advice, an interesting quotation, or a piece of line art. In general, then, we do not think the practical and pedagogical should be diminished in favor of glamorous artwork and photography.

Key Feature #4: Emphasis on Practicality

As with past editions, we are trying to make this book a "keeper" for students. Thus, we not only cover fundamental concepts but also offer a great deal of **practical advice.** This advice, of the sort found in computer magazines and general-interest computer books, is expressed principally in two kinds of boxes—Experience Boxes and README boxes:

- The Experience Box: Appearing at the end of each chapter, the Experience Box is **optional** material that may be assigned at the instructor's discretion. However, students will find the subjects covered are of immediate value.
  Some examples: "Becoming a Mobile Computer User"; "How to Buy Software"; "Preventing Your Identity from Getting Stolen." Five of the Experience Boxes show students how to benefit from going

online. They include "Online Résumés & Other Career Strategies for the Digital Age" and "Job Searching on the Internet & World Wide Web."

- **README boxes:** README boxes consist of optional material of two types—Practical Matters, and Case Studies:

  Practical Matters offer practical advice—such as tips for managing your e-mail or staying focused to avoid information overload.

  Case Studies offer behind-the-scenes looks at information technology—such as how many years digitized data in secondary storage will last or how virtual teams in business transcend the usual organizational hierarchy.

Key Feature #5: Emphasis Throughout on Ethics

Ethics

Many texts discuss ethics in isolation, usually in one of the final chapters. We believe this topic is too important to be treated last or lightly. Thus, **we cover ethical matters in 19 places** throughout the book, as indicated by the special logo shown here in the margin. For example, the all-important question of what kind of software can be legally copied is discussed in Chapter 2 ("Applications Software"), an appropriate place for students just starting software labs. Other ethical matters discussed are the manipulation of truth through digitizing of photographs, intellectual property rights, netiquette, censorship, privacy, and computer crime.

A list of pages with ethics coverage appears on the inside front cover. Instructors wishing to teach all ethical matters as a single unit may refer to this list.

Key Feature #6: Reinforcement for Learning

Having individually or together written nearly two dozen textbooks and scores of labs, the authors are vitally concerned with reinforcing students in acquiring knowledge and developing critical thinking. Accordingly, we offer the following to provide learning reinforcement:

- **Interesting writing:** Studies have found that textbooks **written in an imaginative style** significantly improve students' ability to retain information. Thus, the authors have employed a number of journalistic devices—such as the short biographical sketch, the colorful fact, the apt direct quote—to make the material as interesting as possible. We also use real anecdotes and examples rather than fictionalized ones.

- **Key terms and definitions in boldface: Each key term AND its definition is printed in boldface** within the text, in order to help readers avoid any confusion about which terms are important and what they actually mean.

- **"Preview & Review" presents abstracts of each section for learning reinforcement:** Each main section heading throughout the book is followed by **an abstract or précis entitled Preview & Review.** This enables the student to get a preview of the material before reading it and then to review it afterward, for maximum learning reinforcement.

- **Innovative chapter Summaries for learning reinforcement:** The end-of-chapter Summary is especially innovative—and especially helpful to students. In fact, research through student focus groups has shown that this format was clearly first among five different choices of summary formats. Each concept is discussed under **two columns, headed "What It Is/What It Does" and "Why It's Important."**

Each concept or term is also given a cross-reference page number that refers the reader to the main discussion within the chapter.

In addition, as we discuss next, the term or concept is also given a Key Question number (such as *KQ 2.1, KQ 2.2,* and so on) corresponding to the appropriate Key Question (learning objective) at the beginning of the chapter.

- **Key Questions to help students read with purpose:** ***New to this edition!*** Lists of learning objectives at the start of chapters are common in textbooks—and most students simply skip them. Because we believe learning objectives are excellent instruments for reinforcement, we have crafted ours to make them more helpful to students. We do this in two ways:

    (1) By **phrasing the learning objectives as Key Questions.** These Key Questions appear on the chapter-opening page and again at the start of each chapter section. By phrasing learning objectives as Key Questions we give students a tool to help them read with purpose.

    (2) By **tying terms and concepts in the end-of-chapter Summary to the Key Questions.** That is, in the Summary we have given "KQ" numbers to the terms and concepts that relate to the particular Key Question numbers in the text.

    For example, in Chapter 2, *Key Questions 2.11* ask "When is copying a violation of copyright laws, what is a software license agreement, and what types of agreements are there?" Terms and concepts appearing in the end-of-chapter Summary that relate to these questions—such as "copyright," "freeware," and "intellectual property"—are identified with the notation *KQ 2.11* and the page number in the chapter where they are discussed.

- **Cross-referencing system for key terms and concepts:** Wherever important key terms and concepts appear throughout the text that students might need to remind themselves about, we have added **"check the cross reference"** information, to indicate the first definition or usage of a key term or concept, as in: "use of machine language (✔ p. 111)." In student focus groups, this cross-reference device was found to rank *first* out of 20-plus study/learning aids.

- **Material in "bite-size" portions:** Major ideas are presented in **bite-size form,** with generous use of advance organizers, bulleted lists, and new paragraphing when a new idea is introduced.

- **Short sentences:** Most sentences have been kept short, the majority not exceeding **22–25 words** in length.

- **End-of-chapter exercises:** For practice purposes, students will benefit from several exercises at the end of each chapter: **fill-in-the-blank questions, short-answer questions, multiple-choice questions,** and **true-false questions.** Answers to selected exercises appear upside down at the end of the Exercises section.

    In addition, we present several "Knowledge in Action," end-of-chapter **projects/critical-thinking questions,** generally of a practical nature, to help students absorb the material. In a typical example, students are asked to identify the security threats to which their home computers are vulnerable.

Key Feature #7: Up-to-the-Minute Material—in the Text & on the Irwin/McGraw-Hill Web Site

Writing a text like this is a constant steeplechase of trying to keep up with changing technological developments. Every day seems to bring reports of something new and important. As we write this, our September 1998 publication date is only three months away. However, because our publisher has allowed us to do several steps concurrently (writing, reviewing, editing, production), our text includes coverage of the following material:

ActiveX. The bandwidth economy. Brain-wave input. Cable modems. Cyberspace job hunting. Data mining. Digital cameras. Digital TV. Divx. DSL. DVD. EPIC architecture. Extranets. GEO, MEO, and LEO satellite systems. Gigabit Ethernet. High-capacity bar codes. Identity theft. Internet 2. Map software. The Merced chip. Net addiction. NGI. Online secondary storage. Photonics. Plagiarism and online term papers. Portal sites. Presentation technology. R/3 software. Radio-frequency identification devices. Set-top boxes. Silicon germanium chips. Telephony. 3-D displays. Virtual teams. VRML. Web authoring tools. WebTV. Windows 98. XML. The Y2K problem . . . And more.

Still, we recognize that a Gutenberg-era lag exists between our last-minute scribbling and the book's publication date. And of course we also realize that fast-moving events will unquestionably overtake some of the facts in this book by the time it is in the student's hands. Accordingly, after publication we are periodically offering instructors updated material and other interaction on the Irwin/McGraw-Hill UIT Web Site: **http://www.mhhe.com/cit/ concepts/uit**.

Complete Course Solutions: Supplements That Work—Four Distinctive Offerings

It's less important how many supplements a textbook has than whether they are truly useful, accurate, and of high quality. Irwin/McGraw-Hill presents **four distinctive kinds of supplement offerings** to complement the text:

1. **Application-software tutorials—four types**
2. **McGraw-Hill Learning Architecture Web-based software**
3. **Classroom presentation software**
4. **Instructor support materials**

We elaborate on these below.

Supplement Offering #1: Application-Software Tutorials—Four Types

Our publisher, Irwin/McGraw-Hill, offers four different series of tutorials, which present four different hands-on approaches to learning various types of application software. An Irwin/McGraw-Hill sales representative can explain the specific software covered by each series.

- Advantage Series tutorials: Written by *Sarah E. Hutchinson* and *Glen J. Coulthard*, manuals in the **Advantage Series for Computer Education** average just over 200 pages each and cover a large number of popular software packages, including the latest versions of Microsoft Office. Each tutorial leads students through step-by-step instructions not only for the most common methods of executing commands but also for alternative methods.

Each session begins with a case scenario and concludes with case problems showing real-world application of the software. "Quick Reference" guides summarizing important functions and shortcuts appear throughout. Boxes introduce unusual functions that will enhance the user's productivity. Hands-on exercises and short-answer questions allow students to practice their skills.

- **Advantage Interactive CD-ROM tutorials:** Offered by Irwin/McGraw-Hill in partnership with *MindQ Publishing*, the **Advantage Interactive** CD-ROM tutorials are based on the printed *Advantage Series* texts described above. The CD-ROMs combine sight, sound, and motion into a truly interactive learning experience. Video clips, simulations, hands-on exercises, and quizzes reinforce every important concept. *Advantage Interactive* tutorials are available for latest versions of Microsoft Office and may be used independently or with corresponding manuals in the *Advantage Series*.

- **O'Leary Series print tutorials:** Written by *Linda* and *Timothy O'Leary*, the **O'Leary Series** manuals are designed for application-specific short courses. Each manual offers a project-based approach that gives students a sense of the real-world capabilities of software applications. Extensive screen captures provide easy-to-follow visual examples for each major textual step, while visual summaries reinforce the concepts, building on students' knowledge. Manuals are available for a wide variety of software applications, including latest versions of Microsoft Office.

- **Interactive Computing Skills CD-ROM tutorials:** Created by *Ken Laudon* and *Azimuth Multimedia*, the **Interactive Computing Skills** CD-ROM tutorials offer complete introductory coverage of software applications, including Microsoft Office 4.3 and 97. Each narrated and highly interactive lesson takes 45–60 minutes to complete. "SmartQuizzes" at the end of the lessons actively test software skills within a simulated software environment. With up to four lessons per disk, *Interactive Computing Skills* is a valuable addition to an instructor's courseware package or an excellent self-study tool for students.

Supplement Offering #2: McGraw-Hill Learning Architecture

New to this edition! The future of interactive, networked education is here today! This exciting Web-based software provides complete course administration, including content customization, authoring, and delivery. With the **McGraw-Hill Learning Architecture (MHLA)** and a standard Web browser, students can take online quizzes and tests, and their scores are automatically graded and recorded. *MHLA* also includes useful features such as e-mail, message boards, and chat rooms, and it easily links to other Internet resources. Your Irwin/McGraw-Hill sales representative can explain *MHLA* in detail.

Supplement Offering #3: Classroom Presentation Software

To help instructors enhance their lecture presentations, Irwin/McGraw-Hill makes available the **CIT Classroom Presentation Tool,** a graphics-intensive set of electronic slides. This CD-ROM-based software helps to clarify topics that may otherwise be difficult to present. Topics are organized to correspond with the text chapters. The *Presentation Tool* also includes electronic files for all of the graphics in the text, allowing instructors to customize their presentations.

Minimum system requirements: IBM PC or compatible with a Pentium processor, 4X CD-ROM drive, and at least 16 MB of RAM, running Windows 95 or later. An LCD panel is needed if the images are to be shown to a large audience.

Supplement Offering #4: Instructor Support Materials

We offer the instructor the following other kinds of supplements and support to complement the text:

- **Instructor's Resource Guide:** This complete guide supports instruction in any course environment. For each chapter, the **Instructor's Resource Guide** provides an overview, chapter outline, lecture notes, notes regarding the boxes (README boxes) from the text, solutions, and suggestions, and additional information to enhance the project and critical thinking sections.

- **Test bank:** The test bank contains over 1200 different questions, which are directly referenced to the text. Specifically, it contains *true/false, multiple-choice,* and *fill-in questions,* categorized by difficulty and by type; *short-essay questions; sample midterm exam; sample final exam;* and *answers to all questions.*

- **Diploma 97—computerized testing software:** Created by *Brownstone Research Group,* **Diploma 97** has been consistently ranked number one in evaluations over similar testing products. *Diploma 97* gives instructors simple ways to write sophisticated tests that can be administered on paper or posted over a campus local area network, an intranet, or the Internet.

    Test results can be merged into *Diploma 97's* gradebook program, which automates grading, curving, and reporting functions. Indeed, thousands of students and hundreds of assignments can be put into the same gradebook file. In addition, teaching programs can be attached to questions to create interactive study guides.

    System requirements: (a) IBM PC or compatible with at least 2 MB of RAM running Windows 3.1 or (b) Macintosh with at least 2 MB of RAM running System 6.01 or later; CD-ROM drive or 3.5-inch floppy-disk drives.

- **Videos:** A selection of 10 video segments of the acclaimed PBS television series, *Computer Chronicles,* is available to qualified adopters. Each video is approximately 30 minutes long. The videos cover topics ranging from computers and politics, to online financial services, to the latest developments in PC technologies.

- **Technical support services:** Irwin/McGraw-Hill's Technical Support is available to instructors on any of our software products, such as the McGraw-Hill Learning Architecture or the CIT Classroom Presentation Tool. Instructors can access the Online Helpdesk at **www.mhhe.com/helpdesk** or by calling toll free 1-800-331-5094.

- **UIT Web site:** It's appropriate that a text with a strong communications focus also find a way to employ the communications technology available. Accordingly, a text-specific Irwin/McGraw-Hill UIT Web is available, located at **http://www.mhhe.com/cit/concepts/uit.**

This Web site was developed as a place to go for periodic updates of text material, relevant links, downloads of supplements, an instructor's forum for sharing information with colleagues, and other value-added features.

Instructor Scenarios for Using the Text

USING INFORMATION TECHNOLOGY, THIRD EDITION, was carefully designed based on marketplace feedback. We have tried to write the kind of book that instructors asked for, and the materials are designed to serve a consensus kind of course.

Thus, to serve the new generation of students we are presenting a book that, we hope, reads like a magazine, offers interesting illustrations, and helps the reader learn through many extra pedagogical features—README boxes, Experience Boxes, Key Questions, section Preview & Reviews, innovative end-of-chapter Summaries ("What It is/What It Does," and "Why It's Important"), and end-of-chapter exercises. *Actual material on which the student is to be tested—the general text—constitutes only slightly more than half of each chapter,* as determined from representative chapters. In Chapter 3, for instance, general text constitutes about 24 of the 41 pages. (The rest consists of chapter opening, illustrations, boxes, section Previews & Reviews, end-of-chapter Summaries, and end-of-chapter exercises.)

Many instructors have told us that having the material presented in **just 12 chapters,** rather than the customary 14 or 15 or more, better suits their teaching approach. With 12 chapters, readings may be assigned at the rate of slightly over a chapter a week in a quarter system, less than a chapter a week in a semester system. Chapters are organized according to the topic outline of traditional introductory computer texts. Thus, most instructors can continue to follow their present course outlines.

NOTE: The text allows for **a good deal of instructor flexibility.** After Chapter 1, the remaining 11 chapters may be taught in any sequence, or selectively omitted, at the instructor's discretion. As mentioned, to make this possible, we offer **"check the cross-reference"** information to indicate the first definition or usage of a key term or concept, as in "(✔ p. 111)." For instructors whose courses are less than 3 units or who must teach students software labs in addition to computer concepts, there are other options. Any one or combination of the following scenarios will allow instructors to teach selectively from this book without loss of continuity:

Ethics

- Scenario 1—Teach all "ethics" segments as one component: Rather than discuss ethical matters just in one place, we have spread this topic around through the book, as indicated by the special sign shown here. All the pages of ethics coverage are indicated on the inside front cover. Instructors wishing to teach all ethical matters as a single component (as toward the end of the semester or quarter) may direct students to read the ethics material in the order shown on that list.

- Scenario 2—Skip the Experience Boxes: Some instructors may wish to assign all 12 chapters but not the end-of-chapter essays we call Experiences Boxes. All Experience Boxes are considered optional (not testable) material, but some instructors may wish to pick and choose which they assign, and some instructors may wish not to assign any.

- Scenario 3—Skip chapters on systems and software development: Some instructors may choose to forego Chapter 10, "Information Systems: Management & Development," and Chapter 11, "Software Development: Programming & Languages."

- Scenario 4—Skip the last chapter: Chapter 12, "Society & the Digital Age: Promises & Challenges," could be skipped. Instead, for a discussion of security and ergonomic issues, the instructor may choose to assign the Chapter 5 Experience Box: "Good Habits: Protecting Your Computer System, Your Data, & Your Health."

- Scenario 5—skip chapters on applications and systems software: Instructors whose courses include software labs may feel their students are already getting enough knowledge about applications and systems software that they do not need to read Chapters 2 and 3. (Chapter 2 is "Applications Software: Tools for Thinking & Working." Chapter 3 is "Systems Software: The Power Behind the Power.")

With these kinds of options, we feel sure that most instructors will be able to tailor the text to their particular course.

Finally, we should mention that a brief version of this text is also available: *USING INFORMATION TECHNOLOGY, BRIEF VERSION,* by Stacey C. Sawyer, Brian K. Williams, and Sarah E. Hutchinson. This book offers 10 rather than 12 chapters and four rather than 12 Experience Boxes, and in general the coverage has been selectively reduced.

Acknowledgments

Three names are on the front of this book, but a great many others are powerful contributors to its development.

First among the staff of Irwin/McGraw-Hill is our sponsoring editor, Garrett Glanz, our lifeline, who once again did a top-notch job of supporting us and of coordinating the many talented people whose efforts on development and supplementary materials help strengthen our own. Garrett, you've been great. We also want to welcome aboard our new sponsoring editor, Kyle Lewis, who joined us when this book was in the final stages of production.

We also appreciate the cheerfulness and efficiency of other people in editorial and marketing at Irwin/McGraw-Hill, specifically Jodi McPherson, Tony Noel, and Carrie Berkshire. Irwin's top management—the very supportive John Black, Mike Junior, David Littlehale, Merrily Mazza, Jerry Saykes, Kurt Strand, and Jeff Sund—actively backed our revision, and we are extremely grateful to them. Many others at Irwin have also closely assisted us, and we would like to single out designer Laurie Entringer, who designed the interior of the book and bore with us through all our picky changes, and project manager Gladys True who worked actively with us. We also appreciate having the help of Madelyn Underwood, production supervisor; Keri Johnson, photo research coordinator; and Nancy Martin, supplement coordinator.

Outside of Irwin/McGraw-Hill we were fortunate to find ourselves in a community of first-rate publishing professionals. We are ecstatic fans of the editorial development company of Burrston House, Ltd., and their active participants on this book, the highly experienced Glenn and Meg Turner and the very hard-working and always supportive Cathy Crow.

Two-thirds of the author team would like once again to thank the third, Stacey Sawyer, who not only co-wrote this book but also massaged in all the reviewers' comments, picked the photos, conceptualized much of the art, laid out the pages, and directed the production of the entire enterprise under frantic deadlines. Stacey, thanks for everything.

Stacey worked with freelance designer Matt Baldwin on the cover and with Monica Suder on the photo research, and we are grateful to both. Anita Wagner, top-notch copy editor, performed her usual careful scrutiny of the manuscript. Standing behind Anita was our able proofreader Martha Ghent, who saved us from ourselves by pointing out potential embarrassments. James Minkin did his customary unbeatable job of indexing. David Sweet did his always highly competent job of obtaining permissions.

GTS Graphics turned in their usual top-drawer performance in handling prepress production. We especially want to thank Elliott and Bennett Derman, Gloria Fontana, and their dynamite production coordination team of Ruth Sakata, Donna Machado, and Mary Zelinski.

Finally, the authors are grateful to a number of people for their superb work on the ancillary materials.

Acknowledgment of Focus Group Participants, Survey Respondents, & Reviewers

We are grateful to the following people for their participation in focus groups, response to surveys, or reviews on manuscript drafts or page proofs of all or part of the book. We cannot overstate their importance and contributions in helping us to make this the most market-driven book possible.

INSTRUCTOR FOCUS GROUP PARTICIPANTS

Russell Breslauer
Chabot College

Patrick Callan
Concordia University

Joe Chambers
Triton College

Hiram Crawford
Olive Harvey College

Edouard Desautels
University of Wisconsin—Madison

William Dorin
Indiana University—Northwest

Bonita Ellis
Wright City College

Pat Fenton
West Valley College

Bob Fulkerth
Golden Gate University

Charles Geigner
Illinois State University

Julie Giles
DeVry Institute of Technology

Dwight Graham
Prairie State College

Don Hoggan
Solano Community College

Stan Honacki
Moraine Valley Community College

Tom Hrubec
Waubonsee Community College

Alan Iliff
North Park College

Julie Jordahl
Rock Valley College

John Longstreet
Harold Washington College

Paul Lou
Diablo Valley College

Ed Mannion
California State University—Chico

Jim Potter
California State University— Hayward

Pattie Riden
Western Illinois University

Behrooz Saghafi
Chicago State University

Naj Shaik
Heartland Community College

Charlotte Thunen
Foothill College

James Van Tassel
Mission College

STUDENT FOCUS GROUP PARTICIPANTS

Virginia Amarna
Laney College

Roger Lyle
College of Marin

Kerry Bassett
California State University—Chico

Susan Malibiran
San Francisco City College

Jeff Ferreira
Chabot College

Karey Mathews
Chabot College

Jocelyn Lander
Chabot College

Teresa Taganat
San Francisco City College

Alfred Lepori
Mission College

Robin Torbet
College of Marin

SURVEY RESPONDENTS

Nancy Alderdice
Murray State University

Norman Muller
Greenfield Community College

Margaret Allison
University of Texas—Pan American

Paul Murphy
Massachusetts Bay Community College

Angela Amin
Great Lakes Junior College

Sonia Nayle
Los Angeles City College

Connie Aragon
Seattle Central Community College

Janet Olpert
Cameron University

Gigi Beaton
Tyler Junior College

Pat Ormond
Utah Valley State College

William C. Brough
University of Texas—Pan American

Marie Planchard
Massachusetts Bay Community College

Jeff Butterfield
University of Idaho

Helen Corrigan-McFadyen
Massachusetts Bay Community College

Fernando Rivera
University of Puerto Rico—Mayaguez Campus

James Frost
Idaho State University

Naj Shaik
Heartland Community College

Candace Gerrod
Red Rocks Community College

Jack Shorter
Texas A&M University

Julie Heine
Southern Oregon State College

Randy Stolze
Marist College

Jerry Humphrey
Tulsa Junior College

Ron Wallace
Blue Mountain Community College

Jan Karasz
Cameron University

Steve Wedwick
Heartland Community College

Alan Maples
Cedar Valley College

REVIEWERS

Nancy Alderdice
Murray State University

Bonnie Bailey
Morehead State University

Margaret Allison
University of Texas—Pan American

David Brent Bandy
University of Wisconsin—Oshkosh

Sharon Anderson
Western Iowa Tech Community College

Robert Barrett
Indiana University, Purdue University at Fort Wayne

Anthony Baxter
University of Kentucky

Virginia Bender
William Rainey Harper College

Warren Boe
University of Iowa

Randall Bower
Iowa State University

Phyllis Broughton
Pitt Community College

J. Wesley Cain
City University, Bellevue

Judy Cameron
Spokane Community College

Kris Chandler
Pikes Peak Community College

William Chandler
University of Southern Colorado

John Chenoweth
East Tennessee State University

Ashraful Chowdhury
Dekalb College

Erline Cocke
Northwest Mississippi Community College

Robert Coleman
Pima County Community College

Glen Coulthard
Okanagan University

Robert Crandall
Denver Business School

Thad Crews
Western Kentucky University

Jim Dartt
San Diego Mesa College

Patti Dreven
Community College of Southern Nevada

John Durham
Fort Hays State University

John Enomoto
East Los Angeles College

Ray Fanselau
American River College

Eleanor Flanigan
Montclair State University

Ken Frizane
Oakton Community College

James Frost
Idaho State University

JoAnn Garver
University of Akron

Jill Gebelt
Salt Lake Community College

Charles Geigner
Illinois State University

Frank Gillespie
University of Georgia

Myron Goldberg
Pace University

Sallyann Hanson
Mercer County Community College

Albert Harris
Appalachian State University

Jan Harris
Lewis & Clark Community College

Michael Hasset
Fort Hays State University

Martin Hochhauser
Dutchess Community College

James D. Holland
Okaloosa-Waltoon Community College

Wayne Horn
Pensacola Junior College

Christopher Hundhausen
University of Oregon

Jim Johnson
Valencia Community College

Jorene Kirkland
Amarillo College

Victor Lafrenz
Mohawk Valley Community College

Sheila Lancaster
Gadsden State Community College

Stephen Leach
Florida State University

Paul Leidig
Grand Valley State University

Chang-Yang Lin
Eastern Kentucky University

Paul Lou
Diablo Valley College

Deborah Ludford
Glendale Community College

Peter MacGregor
Estrella Mountain Community College

Donna Madsen
Kirkwood Community College

Kenneth E. Martin
University of North Florida

Curtis Meadow
University of Maine

Timothy Meyer
Edinboro University

Marty Murray
Portland Community College

Charles Nelson
Rock Valley College

Wanda Nolden
Delgado Community College

E. Gladys Norman
Linn-Benton Community College

George Novotny
Ferris State University

Pat Ormond
Utah Valley State College

John Panzica
Community College of Rhode Island

Rajesh Parekh
Iowa State University

Merrill Parker
*Chattanooga State Technical
  Community College*

Jim Potter
*California State University—
  Hayward*

Leonard Presby
William Patterson State College

Delores Pusins
Hillsborough Community College

Eugene Rathswohl
University of San Diego

Alan Rea
Western Michigan University

Jerry Reed
Valencia Community College

John Rezac
Johnson County Community College

Jane Ritter
University of Oregon

Stan Ross
Newbury College

Judy Scheeren
*Westmoreland County Community
  College*

Al Schroeder
Richland College

Earl Schweppe
University of Kansas

Tom Seymour
Minot State University

Elaine Shillito
Clark State Community College

Denis Titchenell
Los Angeles City College

Jack VanDeventer
Washington State University

Jim Vogel
Sanford Brown College

Dale Walikainen
Christopher Newport University

Reneva Walker
Valencia Community College

Patricia Lynn Wermers
North Shore Community College

Ron West
Umpqua Community College

Doug White
Western Michigan University

Edward Winter
Salem State College

Floyd Winters
Manatee Community College

Israel Yost
University of New Hampshire

Eileen Zisk
Community College of Rhode Island

Write to Us

We welcome your response to this book, for we are truly trying to make it as useful as possible. Write to us in care of Kyle Lewis, Sponsoring Editor, Irwin/McGraw-Hill, 1333 Burr Ridge Parkway, Burr Ridge, IL 60521 or via e-mail: **kyle_lewis@mcgraw-hill.com**

Brian K. Williams
Stacey C. Sawyer
Sarah E. Hutchinson

Detailed Contents

Chapter 2

APPLICATIONS SOFTWARE: TOOLS FOR THINKING & WORKING 47

Chapter 3

SYSTEMS SOFTWARE: THE POWER BEHIND THE POWER 103

Chapter 4

PROCESSORS: HARDWARE FOR POWER & PORTABILITY 145

Chapter 5

INPUT & OUTPUT: TAKING CHARGE OF COMPUTING & COMMUNICATIONS 193

Chapter 6

STORAGE: FOUNDATIONS FOR INTERACTIVITY, MULTIMEDIA, & KNOWLEDGE 263

Chapter 9

FILES & DATABASES: FROM DATA ORGANIZING TO DATA MINING 413

Chapter 10

INFORMATION SYSTEMS: INFORMATION MANAGEMENT & SYSTEMS DEVELOPMENT 453

Chapter 11

SOFTWARE DEVELOPMENT: PROGRAMMING & LANGUAGES 507

Chapter 12
SOCIETY & THE DIGITAL AGE: CHALLENGES & PROMISES 553

The Digital Age

An Overview of the Revolution in Computers & Communications

key questions

You should be able to answer the following questions:

1.1 **From the Analog to the Digital Age: The "New Story" of Computers & Communications** What are analog and digital signals, and what is "technological convergence"?

1.2 **Overview of a Computer-&-Communications System: System Elements 1 & 2—People & Procedures** What are the six elements of a computer-and-communications system?

1.3 **System Element 3: Data/Information** What is the difference between data and information, and what are the principal measurements of data?

1.4 **System Element 4: Hardware** What are the five basic operations of computing, and what are the corresponding categories of hardware devices?

1.5 **System Element 5: Software** What is software, and what are the two kinds of software?

1.6 **System Element 6: Communications** How is "communications" defined and how does digital communications present us with the possibility of having an Information Superhighway?

1.7 **Overview of Developments in Computer Technology** What are the three developments in computing, and what are the five types of computers?

1.8 **Overview of Developments in Communications Technology** What are three developments in communications?

1.9 **Computer & Communications Technology Combined: Connectivity & Interactivity** What are connectivity and interactivity?

1.10 **The Ethics of Information Technology** What are some ethical concerns in the field of information technology?

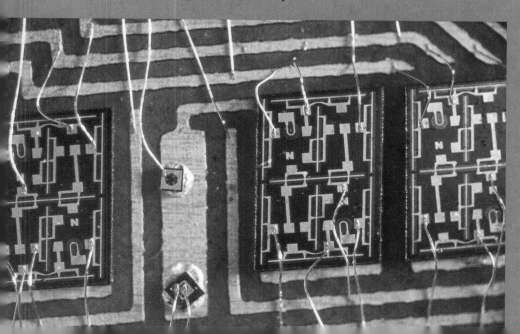

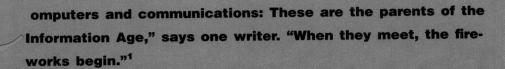

Computers and communications: These are the parents of the Information Age," says one writer. "When they meet, the fireworks begin."[1]

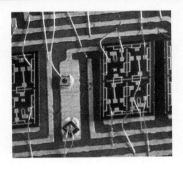

What sort of fireworks are we talking about? Here are some starbursts:

- In 1991, says this writer, "companies for the first time spent more on computing and communications gear . . . than on industrial, mining, farm, and construction machines." The significance? "Info tech is now as vital . . . as the air we breathe."

- "Info tech"—information technology—is changing conventional meanings of time, distance, and space. With cell phones, pagers, fax machines, and portable computers, says one expert, "the physical locations we traditionally associate with work, leisure, and similar pursuits are rapidly becoming meaningless."[2]

- The nature of work is changing, and changing fast. Among nonusers of computers, according to one survey, 70% reported they were struggling with serious employment problems—layoffs, low pay, dead-end jobs. Less than a third of those calling themselves "sophisticated" computer users reported any such problems.[3] Moreover, people who use a computer at work are estimated to make 20% higher wages than those who don't.[4]

- Entire industries are being redefined. The Internet and the World Wide Web, for instance, are not only changing traditional mass media—television, newspapers, recordings—they have themselves become mass media.

Computers and communications are bringing about a revolution that will make—indeed, is making now—profound changes in your life. This wrenching change in human history goes under many names: The Computer Revolution. The Information Revolution. The Communications Revolution. The Internet Revolution. The Multimedia Revolution. The Binary Age. The Information Age. The Information Society. The Information Superhighway—or "Infobahn" or I-way or Dataway. We prefer to call it the Digital Age, but whatever its name, it is happening in all parts of society and in all parts of the world, and its consequences will reverberate throughout our lifetime.

The technological systems and industries that the computer and communications revolution is bringing forth may seem overwhelmingly complex. However, the concept on which they are based is as simple as the flick of a light switch: *on* and *off*. Let us begin to see how this works.

1.1 From the Analog to the Digital Age: The "New Story" of Computers & Communications

KEY QUESTIONS

What are analog and digital signals, and what is "technological convergence"?

Preview & Review: Each "Preview & Review" in this book gives you a brief overview of the information discussed in the section that follows it. You can use it again as a review to test your knowledge. This is the first one.

Information technology is technology that merges computers and high-speed communications links. This merger of computer and communications technologies is producing "technological convergence"—the technological coming together of several

2   Chapter 1

industries through various devices that exchange information in the electronic format used by computers. The industries include computers, communications, consumer electronics, entertainment, and mass media.

Computers are based on digital, two-state (binary) signals—0 and 1. However, most phenomena in the world are analog, meaning they have the property of continuously varying in strength and/or quantity. Today the word "digital" is used almost interchangeably with "computer."

The essence of all revolution, stated philosopher Hannah Arendt, is the start of a *new story* in human experience. For us, the new story is the arrival of information technology. **Information technology is technology that merges computing with high-speed communications links carrying data, sound, and video.**[5] The most important consequence of information technology is that it is producing a gradual fusion of several important industries in a phenomenon that has been called *technological convergence.*

What Is "Technological Convergence"?

"I won't give up my WebTV," raves Annie D. Johnson, an 81-year-old retiree in Mountain View, California. "I've had the postman in to look at it, I've even had the Meals on Wheels man in to look at it, and they thought it was great, too."[6]

WebTV technology allows consumers to add $300 set-top boxes to their television sets, so that they can access and view the World Wide Web, the graphical part of the worldwide computer network known as the Internet. Because it's far cheaper than a personal computer and easier to use, WebTV has been enthusiastically received by senior citizens, latecomers to the Internet revolution, who use it to stay in touch electronically with friends and do research on investments and health matters.

Although WebTV has its drawbacks—television screens aren't designed for reading the text-filled pages from the Internet—it illustrates exactly what we mean by technological convergence.

Technological convergence, also known as digital convergence, is the technological merger of several industries through various devices that exchange information in the electronic, or digital, format used by computers. The industries are computers, communications, consumer electronics, entertainment, and mass media. In the recent past, it was not possible to use your television set as a computer or to use a personal computer to watch broadcast TV programs. Now, however, the technologies of television and computing are coming together in new devices such as WebTV.

Technological convergence has tremendous significance. It means that from a common electronic base, information can be communicated or delivered in all the ways we are accustomed to receiving it. These include the familiar media of newspapers, photographs, films, recordings, radio, and television. However, it can also be communicated through newer technology—satellite, fiber-optic cable, cellular phone, fax machine, or compact disk, for example. More important, as time goes on, *the same information will be exchanged among many kinds of equipment, using the language of computers.* For example, you'll be able to view movies on disks that will work with both the television set in your living room and with the computer on your desk.

What are the implications of this shift from single, isolated technologies to a unified digital technology? The effect of technological convergence on your life could be quite profound. Among other things, it means that you will have to become accustomed to:

- The stepped-up pace of technological change
- The increased need for continuous learning
- Being prepared to interact with people from other cultures and backgrounds
- Continually evaluating the usefulness and reliability of huge quantities of information

Is this consolidation of technologies an overnight phenomenon? Actually, it has been developing over several years, as we explain next.

The Merger of Computer & Communications Technologies

Technological convergence is derived from a combination of two recent technologies: *computers* and *communications.* (■ *See Panel 1.1.*)

■ PANEL 1.1

Fusion of computer and communications technology

Today's new information environment came about gradually from two separate streams of technological development.

- **Computer technology:** It's highly unlikely that anyone reading this book would not have seen a computer by now. Nevertheless, let's define what it is. **A *computer* is a programmable, multiuse machine that accepts data, raw facts, and figures, and processes, or manipulates, it into information we can use, such as summaries or totals.** Its purpose is to speed up problem solving and increase productivity.

  If you've actually touched a computer, it's probably been a personal computer, such as the widely advertised desktop or portable models

Computer Technology

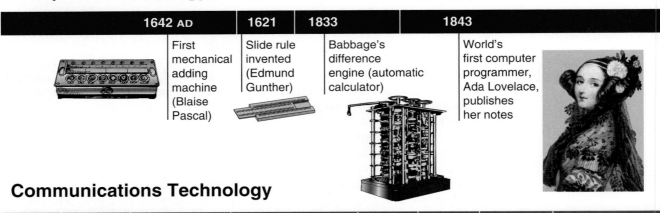

	1642 AD	1621	1833	1843
	First mechanical adding machine (Blaise Pascal)	Slide rule invented (Edmund Gunther)	Babbage's difference engine (automatic calculator)	World's first computer programmer, Ada Lovelace, publishes her notes

Communications Technology

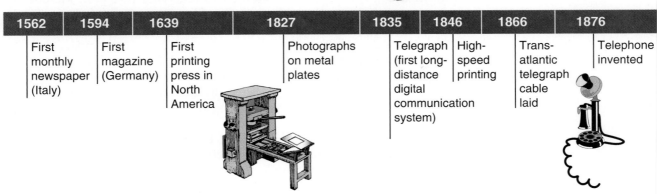

1562	1594	1639	1827	1835	1846	1866	1876
First monthly newspaper (Italy)	First magazine (Germany)	First printing press in North America	Photographs on metal plates	Telegraph (first long-distance digital communication system)	High-speed printing	Trans-atlantic telegraph cable laid	Telephone invented

from Apple, IBM, Compaq, Dell, Gateway, or Packard Bell NEC. However, many other machines, such as automobiles, microwave ovens, and portable phones, use miniature electronic processing devices (microprocessors, or microcontrollers) similar to those that control personal computers.

An example of how raw data is computer-processed into useful information is provided by the computer connected to an automated teller machine (ATM). The unseen computer processes your deposit and withdrawal data to give you the total in your account, printed on the ATM receipt.

● **Communications technology:** Unquestionably you've been using communications technology for years. ***Communications, or telecommunications, technology* consists of electromagnetic devices and systems for communicating over long distances.** The principal examples are telephone, radio, broadcast television, and cable TV.

Before the 1950s, computer technology and communications technology developed independently, like rails in a railroad track that never merge. Since then, however, they have gradually fused together, producing a new information environment.

Why have the worlds of computers and of telecommunications been so long in coming together? The answer is this: *Computers are digital, but most of the world is analog.* Let us explain what this means.

1890	1900		1930	1944	1946
Electricity used for first time in a data-processing project (punched cards)	Hollerith's automatic census-tabulating machine (used punched cards)		General theory of computers	First electro-mechanical computer (Mark I)	First programmable electronic computer in United States (ENIAC)

1888	1894	1895	1912	1915	1928	1939	1946	1947	1948
Radio waves identified	Edison makes a movie	Marconi develops radio; motion-picture camera invented	Motion pictures become a big business	AT&T long-distance service reaches San Francisco	First TV demonstrated; first sound movie	Commercial TV broad-casting	Color TV demon-strated	Transistor invented	Reel-to-reel tape recorder

The Digital Basis of Computers

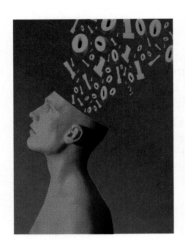

Computers may seem like incredibly complicated devices, but their underlying principle is simple. When you open up a personal computer, what you see is mainly electronic circuitry. And what is the most basic statement that can be made about electricity? Simply this: It can be either *turned on* or *turned off.*

With a two-state on/off arrangement, one state can represent a 1 digit, the other a 0 digit. Because computers are based on on/off or other two-state conditions, they use the *binary system,* which is able to represent any number using only two digits—0 and 1. (*Binary* means having two components, alternatives, or outcomes. It is derived from the Latin word for "double," *binarius.*) Today, **digital specifically refers to communications signals or information represented in a two-state (binary) way.** More generally, *digital* is usually synonymous with "computer-based."

In the binary system, each 0 (off) or 1 (on) is called a *bit,* short for *binary digit.* In turn, bits can be grouped in various combinations to represent characters of data, numbers, letters, punctuation marks, and so on. For example, the letter H could correspond to the electronic signal 01001000 (that is, off-on-off-off-on-off-off-off). (A group of eight bits is called a *byte.*)

Digital data, then, consists of data represented by on/off signals, symbolized as 0s and 1s. This is the method of data representation by which computers process and store data and communicate with each other.

Computer Technology

1952	1963	1964	1967	1969	1970	1971	1975	1977	
UNIVAC computer correctly predicts election of Eisenhower as U.S. President	BASIC developed at Dartmouth	IBM introduces 360 line of computers	Hand-held calcu-lator	ARPA-Net established, led to Internet	Micro-processor chips come into use; floppy disk introduced for storing data	First pocket calculator	First microcomputer (MITs Altair 8800)	Apple II computer (first personal computer sold in assembled form)	

Communications Technology

1950	1952	1957	1961	1968	1975	1976	1977
Cable TV	Direct-distance dialing (no need to go through operator); transistor radio introduced	First satellite launched (Russia's Sputnik)	Push-button telephones	Portable video recorders; video cassettes	Flat-screen TV	First wide-scale marketing of TV computer games (Atari)	First inter-active cable TV

The Analog Basis of Life

"The shades of a sunset, the flight of a bird, or the voice of a singer would seem to defy the black or white simplicity of binary representation," points out one writer.[7] Indeed, these and most other phenomena of the world are **analog, continuously varying in strength and/or quantity.** Sound, light, temperature, and pressure values, for instance, can fall anywhere along a continuum or range. The highs, lows, and in-between states have historically been represented with analog devices rather than digital ones. Examples of analog devices are a speedometer, a thermometer, and a pressure sensor, which can measure continuous fluctuations. In fact, a conventional watch with hour, minute, and second hands that sweep around the dial is an analog device (as opposed to the digital watch in which time is expressed in little windows with changing numbers).

Thus, *analog data* is transmitted in a continuous form that closely resembles the information it represents. The electrical signals on a telephone line are analog-data representations of the original voices. Telephone, radio, broadcast television, and cable-TV have traditionally transmitted analog data.

The differences between analog and digital transmission are apparent when you look at a drawing of an on/off digital signal and one of a wavy analog signal, such as a voice message appearing on a standard telephone line. In general, for your computer to receive communications signals transmitted over a telephone line, you need a *modem* to translate the telephone line's analog signals into the computer's digital signals. (■ *See Panel 1.2., p. 8.*)

The modem provides a means for computers to communicate with one another while the old-fashioned copper-wire telephone network, an analog system that was built to transmit the human voice, still exists. Our concern, however, goes far beyond telephone transmission. How can the analog

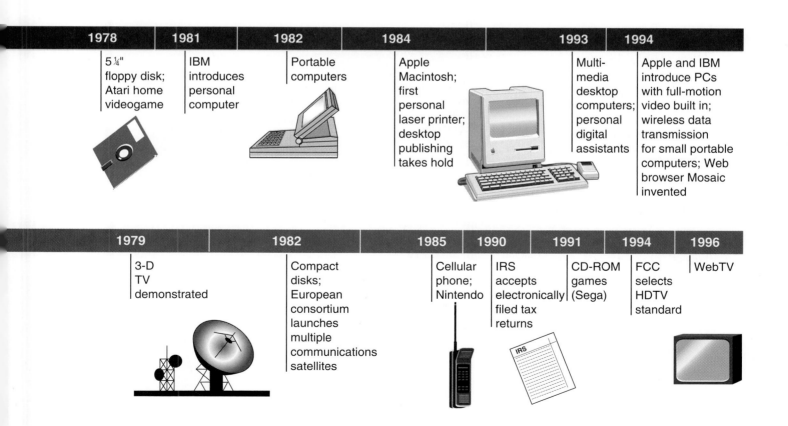

1978	1981	1982	1984	1993	1994
5¼" floppy disk; Atari home videogame	IBM introduces personal computer	Portable computers	Apple Macintosh; first personal laser printer; desktop publishing takes hold	Multi-media desktop computers; personal digital assistants	Apple and IBM introduce PCs with full-motion video built in; wireless data transmission for small portable computers; Web browser Mosaic invented

1979	1982	1985	1990	1991	1994	1996
3-D TV demonstrated	Compact disks; European consortium launches multiple communications satellites	Cellular phone; Nintendo	IRS accepts electronically filed tax returns	CD-ROM games (Sega)	FCC selects HDTV standard	WebTV

■ PANEL 1.2

Analog versus digital signals, and the modem
Note the wavy line for an analog signal and the on/off line for a digital signal. (The modem shown here is outside the computer; today most modems are inside the computer's cabinet.)

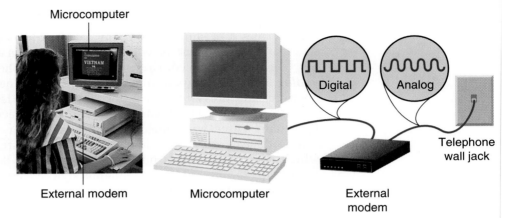

Microcomputer

External modem   Microcomputer   External modem

Digital   Analog

Telephone wall jack

realities of the world be expressed in digital form? How can light, sounds, colors, temperatures, and other dynamic values be represented so that they can be manipulated by a computer? Let us consider this.

Converting Reality to Digital Form

Suppose you are using an analog tape recorder to record a singer during a performance. The analog process will produce a near duplicate of the sounds. This will include distortions, such as buzzings and clicks, or electronic hums if an amplified guitar is used.

The digital recording process is different. The way in which music is captured for music CDs (compact disks) does not provide a duplicate of a musi-

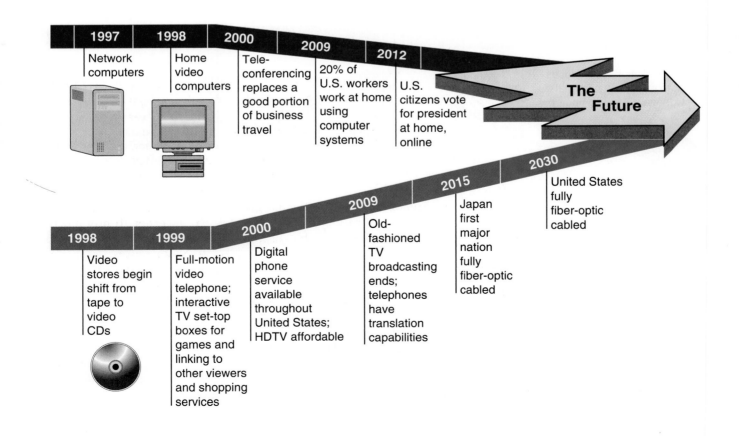

1997 Network computers

1998 Home video computers

2000 Tele-conferencing replaces a good portion of business travel

2009 20% of U.S. workers work at home using computer systems

2012 U.S. citizens vote for president at home, online

The Future

1998 Video stores begin shift from tape to video CDs

1999 Full-motion video telephone; interactive TV set-top boxes for games and linking to other viewers and shopping services

2000 Digital phone service available throughout United States; HDTV affordable

2009 Old-fashioned TV broadcasting ends; telephones have translation capabilities

2015 Japan first major nation fully fiber-optic cabled

2030 United States fully fiber-optic cabled

cal performance. Rather, the digital process uses *representative selections (samples)* to record the sounds and produce a copy that is virtually exact and free from distortion and noise. Computer-based equipment takes samples of sounds at regular intervals—nearly 44,100 times a second. The samples are then converted to numbers that the computer uses to express the sounds. The sample rate of 44,100 times per second and the high level of precision fool our ears into hearing a smooth, continuous sound. Similarly, for visual material a computer can take samples of values such as brightness and color. The same is true of other aspects of real-life experience, such as pressure, temperature, and motion.

Are we being cheated out of experiencing "reality" by allowing computers to sample sounds, images, and so on? Actually, people willingly made this compromise years ago, before computers were invented. Movies, for instance, carve up reality into 24 frames a second. Television frames are drawn at 30 lines per second. These processes happen so quickly that our eyes and brains easily jump the visual gaps. Digital processing of analog experience represents just one more degree of compromise.

Let us now look at how a digital computer-and-communications system works. The following sections present a brief overview that is important to an understanding of the rest of the book.

1.2 Overview of a Computer-&-Communications System: System Elements 1 & 2—People & Procedures

KEY QUESTION

What are the six elements of a computer-and-communications system?

Preview & Review: A computer-and-communications system has six elements: (1) people, (2) procedures, (3) data/information, (4) hardware, (5) software, and (6) communications.

People are the most important part—the creators and the beneficiaries—of a computer-and-communications system. Two types of people use information technology—computer professionals and end-users.

Procedures are steps for accomplishing a result. Procedures may be expressed in print-based manuals or online documentation.

Erie Benhamou, chief executive of 3Com Corporation in California's Silicon Valley, travels a lot—18 days a month in 1997. But the computer system at home allows him and his family to keep in touch via e-mail—electronic mail—from nearly anywhere in the world.

"I was in Mexico for spring break last year," says his son Ori, 17, "and I e-mailed him to make sure he signed me up for Bay-to-Breakers," an annual, somewhat goofy 7.5-mile footrace in San Francisco. "Dad's always forgetting stuff like that."

Benhamou's French-born wife, Illeana, grew up with a suspicion of technology. Now, she says, "I'm a convert." She uses the Internet to buy cheap airline tickets and to find people who share her interest in literature. It's also easier to keep in touch with family and friends. "It is so much faster and easier to write an e-mail and punch a button and send it," she says. "No stamps, no postman."[8]

What keeps the Benhamous in touch is a system using computers and communications. A *system* is a group of related components and operations that interact to perform a task. A system can be many things: the registration process at your college, the 52 bones in the foot, a weather storm front, the monarchy of Great Britain. Here we are concerned with a technological kind of system. **A *computer-and-communications system* is made up of six elements: (1) people, (2) procedures, (3) data/information, (4) hardware,**

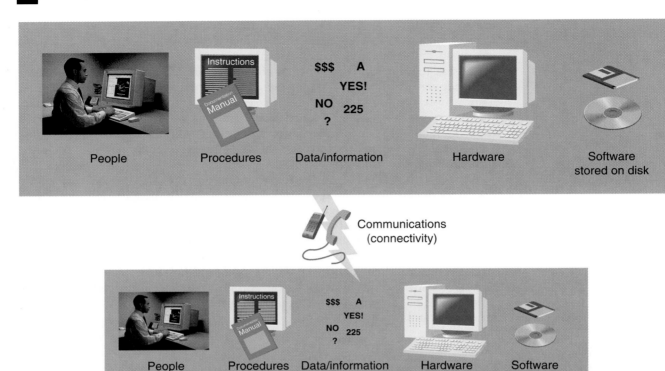

People   Procedures   Data/information   Hardware   Software stored on disk

Communications (connectivity)

People   Procedures   Data/information   Hardware   Software stored on disk

■ PANEL 1.3

A computer-and-communications system
The five basic elements of the system include people, procedures, data/information, hardware, and software. These five system elements are connected to other systems via the sixth element, communications.

(5) **software,** and (6) **communications.** (■ *See Panel 1.3.*) We briefly describe these elements in the next six sections and elaborate on them in subsequent chapters.

System Element 1: People

People are the most important part of a computer-and-communications system. People of all levels and skills, from novices to programmers, are the users and operators of the system. The whole point of the system, of course, is to benefit people. However, all the technology in the world won't benefit anyone if people improperly design, develop, implement, and manage computer systems.

Two types of people use information technology—*computer professionals* and *"end-users."*

- Computer professionals: **A *computer professional,* or an *information technology professional,* is a person who has had extensive education or considerable experience in the technical aspects of using a computer-and-communications system.** For example, a *computer programmer* creates the programs (software) that process the data in a computer system.

  Such professionals are in high demand. As one article points out, automation "in industries from food services to textiles means that even Ben & Jerry's in rural Vermont needs a systems engineer [designer/developer] or two to help the firm make ice cream."[9] Colette Michaud, 32, of San Rafael, California, works 90- and 100-hour work weeks as a project manager overseeing the creation of CD-ROM computer games, but the average pay for information technology workers in Silicon Valley, near where she works, is $70,000 a year versus $27,000 for the average U.S. worker. Moreover,

35% of computer-related businesses give stock options to employees, compared to 5% in other industries.[10]

- **End-users:** An end-user is a person probably much like yourself. **An end-user, or simply a *user*, is someone with moderate technical knowledge of information technology who uses computers for entertainment, education, or work-related tasks.** The user does not understand all the technical nuances of a system but instead usually reacts to the programs and procedures instituted by an information technology professional.

Lawyers, for example, may know how to use a computer to search a huge data bank of information for legal decisions relevant to their case, but they don't need to know how to actually create such a huge bank of electronically classified information. Another example of an end-user is a person who does not know exactly how the Internet works but who knows how to use it. For instance, one woman found the Internet's power as a research and education tool a boon in

README

Practical Matters: High-Tech Help Wanted—Computers & Higher Pay

HIGH-TECH HELP WANTED: NO EXPERIENCE NECESSARY

A STUDY LINKS COMPUTERS TO HIGHER PAY

STUDY WARNS OF A SHORTAGE OF PEOPLE TRAINED FOR JOBS IN TECHNOLOGY

THE PAYOFF FROM COMPUTER SKILLS: A BIG INCOME LIFT FOR COLLEGE GRADS

Behind the headlines are people like Carl Glaum and Debby Atkins.

Opera singer Glaum, 47, orchestrated a career change by taking six months of training in computer programming offered by a Milwaukee financial data processing company to anyone with a college degree. After training, his new job will pay about $35,000.[11]

Texas police officer Atkins, 38, quit her job to take a special seven-month programming course. Now she hunts glitches on a data network instead of criminals. Her starting pay: $45,000. This is equal to her last police salary—which she attained after 15 years. In addition, she says, "You learn something new every day."[12]

No doubt about it, computer users earn more. And it's not just computer professionals. A study by the Information Technology Association of America estimates that 190,000 information-technology jobs stand vacant in the U.S.[13] Half are for jobs such as high-level software developers and computer engineers. "But the bar is dropping for less demanding slots," says one report, "including net-work administrator, basic computer programmer, and Web page designer."[14]

Today, people using computers at work earn more than 20% higher wages than those who don't.[15] "With more than 70% of college-educated workers using computers . . . ," says one article, "researchers figure that as much as 30% to 50% of their relative wage gains in recent decades reflect the spread of computer technology."[16]

The real winners, however, are people like Bhuvanesh Abrol and Lisa Rogers. Recent college graduate Abrol was earning degrees in economics and electrical engineering at Rensselaer Polytechnical Institute in Troy, New York, and entertaining six job offers prior to graduation.[17] He decided to accept a signing bonus of $2000 to work for Advanced Micro Devices in California for starting pay of $46,000 a year.

Rogers, 37, of San Francisco, is a former banker who is now a Webmaster, though she doesn't call herself that.[18] She is the general manager for the corporate Web site for Visa International. The average pay in the United States for a job of this nature is $75,400 a year.[19] Salaries for Web site programmers, artists, and managers are continuing to rise as more companies open Web sites.

In addition, as in sports and entertainment, there are also the superstars. An executive with a California communications company reported he had lost a five-year programmer making $80,000 to a consultancy offering two years guaranteed at $300,000 per year.[20]

(We cover more on computer industry jobs in the README box in Chapter 11.)

helping her 12-year-old daughter complete a research project about soap. "In the old days, you'd have had to go to the library, write to the companies for information," she says. "But we sat down together and dialed into Colgate-Palmolive's Web page and Lever Brothers' Web page. . . . It was a blast."[21]

System Element 2: Procedures

***Procedures* are descriptions of how things are done—steps for accomplishing a result or rules and guidelines for what is acceptable.** Sometimes procedures are unstated, the result of tradition or common practice. You may find this out when you join a club or are a guest in someone's house for the first time. Sometimes procedures are laid out in great detail in manuals, as is true, say, of tax laws.

When you use a bank ATM—a form of computer system—the procedures for making a withdrawal or a deposit are given in on-screen messages. In other computer systems, procedures are spelled out in manuals. Manuals, called *documentation,* contain instructions, rules, or guidelines to follow when using hardware or software. When you buy a microcomputer or a software package, it comes with documentation, or procedures. Nowadays, in fact, many such procedures come not only in a book or pamphlet but also on a computer disk, which presents directions on your display screen. Many companies also offer documentation on the Internet.

1.3 System Element 3: Data/Information

KEY QUESTIONS

What is the difference between data and information, and what are the principal measurements of data?

Preview & Review: The distinction is made between raw data, which is unprocessed, and information, which is processed data. Units of measurement of data/information capacity include kilobytes, megabytes, gigabytes, and terabytes.

Though used loosely all the time, the word *data* has some precise and distinct meanings.

"Raw Data" Versus Information

Data can be considered the raw material—whether in paper, electronic, or other form—that is processed by the computer. In other words, **data consists of the raw facts and figures that are processed into information.**

***Information* is summarized data or otherwise manipulated data that is useful for decision making.** Thus, the raw data of employees' hours worked and wage rates is processed by a computer into the information of paychecks and payrolls. Some characteristics of useful information are that it is *relevant, timely, accurate, concise,* and *complete.*

Actually, in ordinary usage the words *data* and *information* are often used synonymously. After all, one person's information may be another person's data. The "information" of paychecks and payrolls may become the "data" that goes into someone's yearly financial projections or tax returns.

Units of Measurement for Capacity: From Bytes to Terabytes

A common concern of computer users is "How much data can this gadget hold?" The gadget might be a diskette, a hard disk, or a computer's main memory (all terms we'll explain shortly). The question is a crucial one. If you have too much data, the computer may not be able to handle it. Or if a software package takes up too much storage space, it cannot be run on a particular computer.

We mentioned that computers deal with "on" and "off" electrical states, which are represented as 0s and 1s, called *bits.* Bits are combined in groups of eight, called *bytes,* to hold the equivalent of a character. A *character* is a single letter, number, or special symbol (such as a punctuation mark or dollar sign). Examples of characters are A, 1, and ?.

A computer system's data/information storage capacity is represented by bytes, kilobytes, megabytes, gigabytes, and terabytes:

- **Kilobyte:** A *kilobyte,* abbreviated K or KB, is equivalent to approximately 1000 bytes (or characters). More precisely, 1 kilobyte is 1024 (2^{10}) bytes, but the figure is commonly rounded off. Kilobytes are a common unit of measure for the data-holding (memory) capacity of personal computers. The original IBM PC, for example, could hold (in memory) 640 kilobytes, or about 640,000 bytes of data, and earlier home computers held only 64 K.

- **Megabyte:** A *megabyte,* abbreviated M or MB and sometimes called a "meg," is about 1 million bytes. Some software programs require 16 or more megabytes, or about 16 million bytes, of memory, to operate efficiently.

- **Gigabyte:** A *gigabyte,* G or GB, is about 1 billion bytes. Pronounced "*gig*-a-bite," this unit of measure—sometimes called a "gig"—is used not only with large computers but also with newer personal computers to represent hard-disk storage capacities.

- **Terabyte:** A *terabyte,* T or TB, is about 1 trillion bytes, or 1000 gigabytes.

1.4 System Element 4: Hardware

KEY QUESTIONS

What are the five basic operations of computing, and what are the corresponding categories of hardware devices?

Preview & Review: The basic operations of computing consist of (1) input, (2) processing, (3) output, and (4) storage. Communications (5) adds an extension capability to each operation.

Hardware devices are often categorized according to which of these five operations they perform. (1) Input hardware includes the keyboard, mouse, and scanner. (2) Processing and memory hardware consists of the CPU (the processor) and main memory. (3) Output hardware includes the display screen, printer, and sound devices. (4) Secondary-storage hardware stores data on diskette, hard disk, magnetic-tape devices, and CD-ROM. (5) Communications hardware includes modems.

Tim Drummond, a 37-year-old aerospace engineer, has more info tech equipment in his modest Menlo Park, California, house than probably 99.9% of the people on earth: nine desktop computers, four printers, three high-speed phone connections, 80 miles of special wiring.

What does he use it for? At 6:45 one morning, sitting in his home office, he stares into his PC screen and moves a mouse on his desktop. From another part of the house comes the sound of a whooping chimpanzee. He moves the mouse again, and the chatter of Woody Woodpecker is heard. Then e-mail appears on Drummond's screen: "OK DAD, I'M UP." Josh, his 10-year-old son, has received his wake-up call.[22]

Maybe there's something you can use here to help get yourself up for class—but you won't need anything like Drummond's computer system. As we said earlier (✔ p. 9), a system is a group of related components and operations that interact to perform a task. Once you know how the pieces of the system fit together, you can then make better judgments about any one of them. And you can make knowledgeable decisions about buying and operating a computer system.

The Basic Operations of Computing

How does a computer system process data into information? It usually goes through four operations: *(1) input, (2) processing, (3) output,* and *(4) storage.* (■ *See Panel 1.4.*)

1 **Input operation:** In the *input* operation, data is entered or otherwise captured electronically and is converted to a form that can be processed by the computer. The means for "capturing" data (the raw, unsorted facts) is input hardware, such as a keyboard.

2 **Processing operation:** In the *processing* operation, the data is manipulated to process or transform it into information (such as

■ PANEL 1.4

The basic operations of computing

A computer goes through four operations: (1) input of data, (2) processing of data into information, (3) output of information, and (4) storage of information. Communications (5) extends the computer system's capabilities.

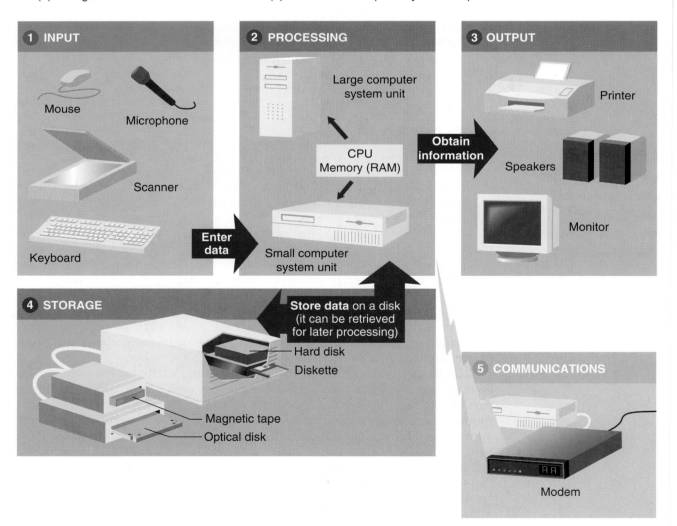

summaries or totals). For example, numbers may be added or subtracted.

3 **Output operation:** In the *output* operation, the information obtained from the data is produced in a form usable by people. Examples of output are printed text, sound, and charts and graphs displayed on a computer screen.

4 **Secondary-storage operation:** In the *storage* operation, data, information, and programs are permanently stored in computer-processable form. Diskettes are examples of materials used for storage.

Often these four operations occur so quickly that they seem to be happening simultaneously.

Where does *communications* fit in here? In the four operations of computing, communications offers an *extension* capability. Data may be input from afar, processed in a remote area, output in several different locations, and stored in yet other places. And information can be transmitted to other computers, whether 3 feet away or halfway around the world. All this is done through a wired or wireless communications connection to the computer.

Hardware Categories

Hardware is what most people think of when they picture computers. **Hardware consists of all the machinery and equipment** in a computer system. The hardware includes, among other devices, the keyboard, the screen, the printer, and the computer or processing device itself.

In general, computer hardware is categorized according to which of the five computer operations it performs:

- Input
- Processing and memory
- Output
- Secondary storage
- Communications

Regardless of the operations they perform, external devices that are connected to the main computer cabinet are referred to as "peripheral devices," or simply "peripherals." **A *peripheral device* is any piece of hardware that is connected to a computer.** Examples are the keyboard, mouse, monitor, and printer.

We describe hardware in detail elsewhere (Chapters 4–8), but the following offers a quick overview to help you gain familiarity with terms.

Input Hardware

Input hardware **consists of devices that allow people to put data into the computer in a form that the computer can use.** For example, input may be by means of a *keyboard, mouse, microphone,* or *scanner.* The keyboard is self-explanatory. The mouse is a pointing device attached to many microcomputers. An example of a scanner is the grocery-store bar-code scanner. (These and other input devices are discussed in detail in Chapter 5.)

Processing & Memory Hardware

The brains of the computer are the *processing* and *main memory* devices, housed in the computer's system unit. The *system unit,* or system cabinet, houses the electronic circuitry called the *CPU (central processing unit),* which does the actual processing, and *main memory,* which supports processing. (These are discussed in detail in Chapter 4.)

The CPU is the computing part of the computer. It controls and manipulates data to produce information. In a personal computer the CPU is usually a single, fingernail-size "chip" called a *microprocessor,* with electrical circuits printed on it. This microprocessor and other components necessary to make it work are mounted on a main circuit board called a *motherboard.*

Memory—also known as *main memory, RAM (random access memory),* or *primary storage*—is working storage. Memory is the computer's "work space," where data and programs for immediate processing are held. Computer memory is contained on memory chips mounted on the motherboard. Memory capacity is important because it determines how much data can be processed at once and how big and complex a program may be used to process the data.

Despite its name, memory does not remember. That is, once the power is turned off, all the data and programs within memory simply vanish. This is why data/information must also be stored in relatively permanent form on disks and tapes, which are called *secondary storage* to distinguish them from main memory's *primary storage.*

Output Hardware

Output hardware consists of devices that translate information processed by the computer into a form that humans can understand. We are now so exposed to products that are output by some sort of computer that we don't consider them unusual. Examples are grocery receipts, bank statements, and grade reports. More recent forms are digital recordings and even digital radio.

As a personal computer user, you will be dealing with three principal types of output hardware—*screens, printers,* and *sound output devices.* (These and other output devices are discussed in detail in Chapter 5.) The *screen* is the display area of a computer. A *printer* is a device that converts computer output into printed images. Printers are of many types, some noisy, some quiet, some able to print carbon copies, some not.

Many computers emit chirps and beeps. Some go beyond those noises and contain sound processors and speakers that can play digital music or humanlike speech. High-fidelity stereo sound is becoming more important as computer and communications technologies continue to merge.

Secondary-Storage Hardware

Main memory, or primary storage, is *temporary* storage. It works with the CPU chip on the motherboard inside the computer cabinet to hold data and programs for immediate processing. Secondary storage, by contrast, is *permanent* storage. It is not on the motherboard (although it may still be inside the system cabinet). **Secondary storage consists of devices that store data and programs permanently on disk or tape.**

You may hear people use the term "storage media." *Media* refers to the material that stores data, such as disk or magnetic tape. For microcomputers, the principal storage media are *diskette (floppy disk), hard disk, mag-*

netic tape, and *CD-ROM.* (■ *See Panel 1.5.*) (These and other secondary-storage devices are discussed in detail in Chapter 6.)

A *diskette,* or *floppy disk,* is a removable round, flexible disk that stores data as magnetized spots. The disk is contained in a plastic case to prevent the disk surface from being touched. The most common size is *3½ inches* in diameter.

To use a diskette, you need a disk drive in your computer. A *disk drive* is a device that holds and spins the diskette inside its case; it "reads" data from and "writes" data to the disk. The words *read* and *write* are used a great deal in computing. *Read* means that the data represented in magnetized spots on the disk (or tape) are converted to electronic signals and transmitted to the memory in the computer. *Write* means that the electronic information processed by the computer is recorded onto disk (or tape).

Diskettes are made out of a magnetic, plastic-type material, which is what makes them "floppy." They are also removable. By contrast, a *hard disk* is a disk platter made out of metal and covered with a magnetic recording surface. It also holds data represented by the presence (1) and absence (0) of magnetized spots. Hard-disk drives read and write data in much the same way that diskette drives do. However, there are three significant differences. First, hard-disk drives can handle thousands of times more data than diskettes do. Second, hard-disk drives are often located in the system cabinet, in which case they are not removable. (External hard-disk drives such as the Zip, Jaz, and Syquest models are also available; they use removable hard-disk cartridges that can store large amounts of data.) Third, hard disks read and write data much faster than diskettes do.

Moviemakers used to love to represent computers with banks of spinning reels of magnetic tape. Indeed, with early computers, "mag tape" was the principal method of secondary storage. The magnetic tape used for computers is made from the same material as that used for audiotape and videotape. That is, *magnetic tape* is made of flexible plastic coated on one side with a magnetic material; again, data is represented by the presence and absence of magnetized spots. Because of its drawbacks (described in Chapter 6), nowadays tape—along with hard-disk cartridges—is used mainly to provide low-cost duplicate storage. A tape that is a duplicate or copy of another form of storage is referred to as a *backup.* Because hard disks sometimes fail ("crash"), personal computer users who don't wish to do backup using a lot of diskettes may use magnetic tape or hard-disk cartridges instead.

If you have been using music CDs (compact disks), you are already familiar with optical disks. An *optical disk* is a disk that is written and read by

■ PANEL 1.5

Secondary storage for microcomputers

(Left) Examples of diskette and CD-ROM drives (the hard-disk drive has no exterior opening). *(Middle)* Inside of hard-disk drive. *(Right)* External Zip hard-disk drive with hard-disk cartridge.

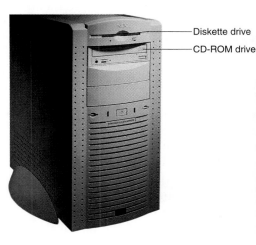

Diskette drive
CD-ROM drive

lasers. CD-ROM, which stands for compact disk—read-only memory, is only one kind of optical-disk format that is used to hold text, graphics, and sound. CD-ROMs can hold hundreds of times more data than diskettes, and can hold more data than many hard disks. A newer type of optical disk called *DVD-ROM* has more than ten times the capacity of a CD-ROM.

Communications Hardware

Computers can be "stand-alone" machines, meaning that they are not connected to anything else. Indeed, many students tote around portable personal computers on which they use word processing or other programs to help them with their work. However, the *communications* component of the computer system *vastly* extends the range of a computer—for example, via connection to the Internet, which is actually a worldwide electronic network of smaller connected networks.

The dominant communications lines developed during this century use analog transmission. Thus, for many years the principal form of direct connection was via standard copper-wire telephone lines. Hundreds of these twisted-pair copper wires are bundled together in cables and strung on telephone poles or buried underground. As mentioned, a modem is communications hardware required to translate a computer's digital signals into analog form for transmission over telephone wires. Although copper wiring still exists in most places, it is gradually being supplanted by two other kinds of direct connections: coaxial cable and fiber-optic cable. Eventually, all transmission lines will accommodate digital signals. (Communications hardware and related issues are covered in detail in Chapters 7 and 8.)

1.5 System Element 5: Software

KEY QUESTIONS

What is software, and what are the two kinds of software?

Preview & Review: Software comprises the instructions that tell the computer what to do. In general, software is divided into applications software and system software.

Applications software is software that has been developed to solve a particular problem, to perform useful work on specific tasks, or to provide entertainment. Applications software may be custom or packaged.

System software, which includes operating systems, enables the applications software to run on the computer.

Software, or *programs*, **consists of the instructions that tell the computer how to perform a task.** Software is written in special code by programmers, and the software is then copied by the manufacturer onto a storage medium, such as CD-ROM. The code on the CD-ROM is translated in the user's computer into the 0/1, off/on signals discussed earlier (✔ p. 6). In most instances, the words *software* and *program* are interchangeable. (We discuss software in detail in the next two chapters.)

For now, be aware that there are two major types of software:

- **Applications software:** This may be thought of as the kind of software that people use to perform a specific task, such as word processing software used to prepare documents and game software used to entertain.
- **System software:** This may be thought of as the underlying software that the computer uses to manage its own internal activities and run applications software. System software acts as the interpreter that allows you and your applications software to access the physical hardware devices and other resources.

Although you may not need a particular applications program, you must have system software or you will not even be able to "boot up" (start) your computer.

Applications Software

***Applications software* is software that has been developed to solve a particular problem, to perform useful work on specific tasks, or to provide entertainment.** Applications software may be either *custom* or *packaged.*

Custom software is software designed and developed for a particular customer. This is the kind of software that you would hire a computer programmer—a software creator—to develop for you. Such software would perform a task that could not be readily done with standard off-the-shelf packaged software available from a computer store or mail-order house.

Packaged software, or a *software package,* is the kind of off-the-shelf program developed for sale to the general public. This is the principal kind that will be of interest to you. Examples of packaged software that you will most likely encounter are word processing programs, spreadsheet programs, and office suites. (We discuss these in Chapter 2.)

System Software

As the user, you interact mostly with the applications software and let the applications software interact with the system software. ***System software* controls the allocation and usage of hardware resources and enables the applications software to run.**

System software consists of several programs, the most important of which is the operating system. The *operating system* acts as the master control program that runs the computer. It handles such activities as running and storing programs and storing and processing data. The purpose of the operating system is to allow applications to operate by standardizing access to shared resources such as disks and memory. Examples of operating systems are MS-DOS, Windows 95 and 98, Windows NT, Unix, and the Macintosh operating system (MacOS). (We discuss these operating systems in detail in Chapter 3.)

1.6 System Element 6: Communications

KEY QUESTIONS

How is "communications" defined, and how does digital communications present us with the possibility of having an Information Superhighway?

Preview & Review: "Communications" refers to the electronic transfer of data. The kind of data being communicated is rapidly changing from analog to digital. The Information Superhighway is a metaphor for the fusion of telephones and networked computers with television and radio programming.

Communications is defined as the electronic transfer of data from one place to another. Of all six elements in a computer-and-communications system, communications is probably experiencing the most changes at this point.

As we mentioned, until now, most data being communicated has been analog data. Recently, however, the notion of a *digital* electronic highway has roared into everyone's consciousness. Some say this so-called *Information Superhighway* promises to provide an almost endless supply of electronic interactive services. Others say it is surrounded "by more hype and inflated expectations than any technological proposal of recent memory."[23] What, in fact, is this electronic highway? Does it or will it really exist?

Parts of this idea have been raised before. Indeed, in many ways the Information Superhighway is a 1990s dusting off of earlier concepts of "the wired nation." In 1978, for example, James Martin wrote *The Wired Society*, which considered the social impacts of various telecommunications technologies. In its current form, however, **the *Information Superhighway* may be said to be a vision or a metaphor for a fusion of the two-way wired and wireless capabilities of telephones and networked computers with television and radio's capacity to transmit hundreds of programs. The resulting interactive digitized traffic would include movies, TV shows, phone calls, databases, shopping services, and online services.** This superhighway, it is hoped, would link all homes, schools, businesses, and governments.

1.7 Overview of Developments in Computer Technology

KEY QUESTIONS

What are the three developments in computing, and what are the five types of computers?

Preview & Review: Computers have developed in three directions: smaller, more powerful, and less expensive.

Today the five types of computers are microcontrollers, microcomputers, minicomputers, mainframes, and supercomputers.

Microcontrollers are embedded in machines such as cars and kitchen appliances.

Microcomputers may be personal computers (PCs) or workstations. PCs include desktop and floorstanding units, laptops, notebooks, subnotebooks, pocket PCs, and pen computers. Workstations are sophisticated desktop microcomputers used for technical purposes.

Minicomputers/midrange computers are intermediate-size machines.

Mainframes are the traditional size of computer and are used in large companies to handle millions of transactions.

The high-capacity machines called supercomputers are the fastest calculating devices and are used for large-scale projects.

Any of these last four types of computers may be used as a server, a central computer in a network.

A human generation is not a very long time, about 30 years. During the short period of one and a half generations, computers have come from nowhere to transform society in unimaginable ways. One of the first computers, the outcome of military-related research, was delivered to the U.S. Army in 1946. ENIAC—short for Electronic Numerical Integrator And Calculator—weighed 30 tons, was 80 feet long and two stories high, and required 18,000 vacuum tubes. However, it could multiply a pair of numbers in the then-remarkable time of three-thousandths of a second. This was the first general-purpose, programmable electronic computer, the grandfather of today's lightweight handheld machines.

The Three Directions of Computer Development

Since the days of ENIAC, computers have developed in three directions:

- **Smaller size:** Everything has become smaller. ENIAC's old-fashioned radio-style vacuum tubes gave way to the smaller, faster, more reliable transistor. A *transistor* is a small device used as a gateway to transfer electrical signals along predetermined paths (circuits).

   The next step was the development of tiny integrated circuits. *Integrated circuits* are entire collections of electrical circuits or pathways etched on tiny squares of silicon half the size of your thumbnail. *Silicon* is a natural element found in sand that is purified to form the base material for making computer processing devices.

■ PANEL 1.6

The principal types of computers—and the microprocessor that powers them

(Clockwise from left top) A supercomputer, a mainframe computer, a minicomputer, and two kinds of microcomputers—a personal computer (PC) and a workstation—and a microcontroller. *(Center)* A microprocessor, the miniaturized circuitry that does the processing in computers. A PC may have only one of these, a supercomputer thousands.

- **More power:** In turn, miniaturization of hardware components allowed computer makers to cram more power into their machines, providing faster processing speeds and more data storage capacity.

- **Less expense:** The miniaturized processor in a personal desktop computer performs the same sort of calculations once performed by a computer that filled an entire room. However, processor costs are only a fraction of what they were 15 years ago; the fastest processors can be had today for less than $1000, whereas 15 years ago this same processing power might have cost more than $1 million.

Five Kinds of Computers

Generally speaking, the larger the computer, the greater its processing power. Computers are often classified into five sizes: tiny, small, medium, large, and superlarge. *(■ See Panel 1.6.)*

- Microcontrollers: *Microcontrollers*, also called *embedded computers,* are the tiny, specialized microprocessors installed in "smart" appliances and automobiles. These microcontrollers enable microwave ovens, for example, to store data about how long to cook your potatoes and at what temperature.

- Microcomputers—personal computers: *Microcomputers* are small computers that can fit on or beside a desk or are portable. Microcomputers are considered to be of two types: personal computers and workstations.

    Personal computers (PCs) are desktop, tower, or portable computers that can run easy-to-use programs such as word processing or spreadsheets. PCs come in several sizes, as follows.

    Desktop and tower units: Even though many personal computers today are portable, buyers of new PCs often opt for nonportable systems, for reasons of price, power, or flexibility. For example, the television-tube-like (CRT, or cathode-ray tube) monitors that come with desktops have display screens that are easier to read than those of many portables. Moreover, you can stuff a desktop's roomy system cabinet with add-on circuit boards and other extras, which is not possible with portables.

    Desktop PCs are those in which the system cabinet sits on a desk, with keyboard in front and monitor often on top. A difficulty with this arrangement is that the system cabinet's "footprint" can deprive you of a fair amount of desk space. *Tower PCs* are those in which the system cabinet sits as a "tower" on the desk or on the floor next to the desk, giving you more usable desk space.

    Laptops: A *laptop computer* is a portable computer equipped with a flat display screen and weighing about 2–11 pounds. The top of the computer opens up like a clamshell to reveal the screen. The two principal types of laptop computers are *notebooks* and *subnotebooks,* a category sometimes called *ultralights.*

    A *notebook computer* is a portable computer that weighs 4–9 pounds and is roughly the size of a thick notebook, perhaps $8\frac{1}{2}$ by 11 inches. Notebook PCs can easily be tucked into a briefcase or backpack or simply under your arm. Notebook computers can be just as powerful as some desktop machines. Indeed, we are now at the point where a notebook may fulfill just about all the needs of a desktop.

    A *subnotebook computer* weighs 1.8–4 pounds. To save weight, subnotebooks in the past have often had external hard-disk drives, which were available as separate units.

    Pocket PCs: *Pocket personal computers*, or *handhelds*, weigh about 1 pound or so and can fit in a jacket pocket. These PCs are useful in specific situations, as when a driver of a package-delivery truck must feed hourly status reports to company headquarters. Another use allows police officers to check out suspicious car license numbers against a database in a central computer. Other pocket PCs have more general applications as electronic diaries and pocket organizers.

    In general, pocket PCs may be classified into three types: (a) *Electronic organizers* are specialized pocket computers that mainly store appointments, addresses, and "to do" lists. Recent versions feature wireless links to other computers for data transfer. (b) *Palmtop computers* are PCs that are small enough to hold in one hand and operate with the other. (c) *Pen computers* lack a keyboard or a mouse

but allow you to input data by writing directly on the screen with a stylus, or pen. Pen computers are useful for inventory control, as when a store clerk has to count merchandise; for package-delivery drivers who must get electronic signatures as proof of delivery; and for more general purposes, like those of electronic organizers and PDAs.

Personal digital assistants (PDAs), or *personal communicators,* **are small, pen-controlled, handheld computers that, in their most developed form, can do two-way wireless messaging.**

We explain more about notebooks, subnotebooks, and pocket PCs, and their usefulness, in the Experience Box at the end of Chapter 4.

- Microcomputers—workstations: Workstations look like desktop PCs but are far more powerful. Traditionally, **workstations were sophisticated machines that fit on a desk, cost many thousands of dollars, and were used mainly by engineers and scientists for technical purposes.** However, workstations have long been used for computer-aided design and manufacturing, software development, and scientific modeling. Workstations have caught the eye of the public mainly for their graphics capabilities, such as those used to breathe three-dimensional life into movies such as *Jurassic Park, Toy Story,* and *Twister.*

Two recent developments have altered the differences between workstations and PCs: (1) *Decline in workstation prices:* A workstation that not long ago cost $15,000 or more is now available starting at $3,700, which certainly puts it within range of many PC buyers. (2) *Increase in PC power:* In 1993 Intel introduced the Pentium chip; in 1994 Motorola (with IBM and Apple) introduced its PowerPC chip. Both of these very powerful microprocessors and their successors are now found in PCs. In addition, Microsoft introduced Windows NT, the first operating system designed to take advantage of more powerful microprocessors.

- Minicomputers/midrange computers: *Minicomputers* **are machines midway in cost and capability between microcomputers and mainframes. They can be used as single-user workstations. When used in a system tied by network to several hundred terminals for many users they are known as** *midrange computers.* The minicomputer overlaps with other categories of computers. A low-end minicomputer may be about as powerful as a high-end microcomputer and cost about the same. A high-end minicomputer may equal a low-end mainframe.

Traditionally, minicomputers have been used to serve the needs of medium-size companies or of departments within larger companies, often for accounting or design and manufacturing (CAD/CAM). Now many minicomputers are being replaced by groups of PCs and workstations in networks.

- Mainframes: The large computers called *mainframes* are the oldest category of computer system. The word "mainframe" probably comes from the metal frames, housed in cabinets, on which manufacturers mounted the computer's electronic circuits.

Occupying specially wired, air-conditioned rooms and capable of great processing speeds and data storage, *mainframes* **traditionally have been water- or air-cooled computers that are about the size of a Jeep and that range in price from $50,000 to $5 million.** Such machines are typically operated by professional programmers and technicians in a centrally managed department within a large company. Examples of such companies are banks, insurance companies, and airlines, which handle millions of transactions. Indeed, Federal Aviation Administration flight controllers are still using 1960s-era mainframes—*Univac* computers, which are no longer made—to keep air traffic safe.

Today, one hears, "mainframes are dead," being supplanted everywhere by small computers connected together in networks, a trend known as "downsizing." Is this true? It has been estimated that the world has $1 trillion invested in this kind of computer—perhaps 50,000 mainframes, 60% of them made and sold by IBM.[24] But what are the future prospects for people working with mainframes? Although mainframe manufacturers will probably promote new uses for their equipment, there appear to be three trends: (1) Old mainframes will be kept for some purposes. (2) Networks of smaller computers will grow. (3) Mainframes are being reinvented.

- Supercomputers: **Typically priced from $225,000 to over $30 million, *supercomputers* are high-capacity machines that require special air-conditioned rooms and are the fastest calculating devices ever invented.**

Supercomputer users are those who need to model complex phenomena. Examples are automotive engineers who simulate cars crashing into walls and airplane designers who simulate air flowing over an airplane wing. "Supers," as they are called, are also used for oil exploration and weather forecasting. In addition, they can help managers in department-store chains decide what to buy and where to stock it. Finally, they have been used to help redesign parachutes, which are surprisingly complex from the standpoint of aerodynamics. The supercomputer simulates the flow of air in and around the parachute during its descent. The most powerful computer, Janus, located at the Sandia National Laboratories in Albuquerque, New Mexico, and built by Intel, enables scientists to simulate the explosion of a nuclear bomb.

New communications lines have made possible supercomputing power that is truly awesome. In 1995 the National Science Foundation and MCI Communications, the nation's No. 2 long-distance provider, established a giant, 14,000-mile network called the Very-High-Performance Backbone Network Service (VBNS), which links the five most important concentrations of supercomputers into what they called a new Internet. (■ *See Panel 1.7.*) Each of these locations has more than one supercomputer (Cornell and Champaign-Urbana have six each). With this arrangement a scientist sitting at a terminal or workstation anywhere in the country could have access to all the power of these fast machines simultaneously. At present this new Internet is used only by an elite group of scientists for extremely complex projects (such as studies of weather turbulence), but it may one day be available to the public.

Servers

The word "server" does not describe a size of computer but rather a particular way in which a computer is used. Nevertheless, because of the principal concerns of this book—the union of computers and communications—servers deserve separate discussion here. (This topic is also included in Chapters 7 and 8.)

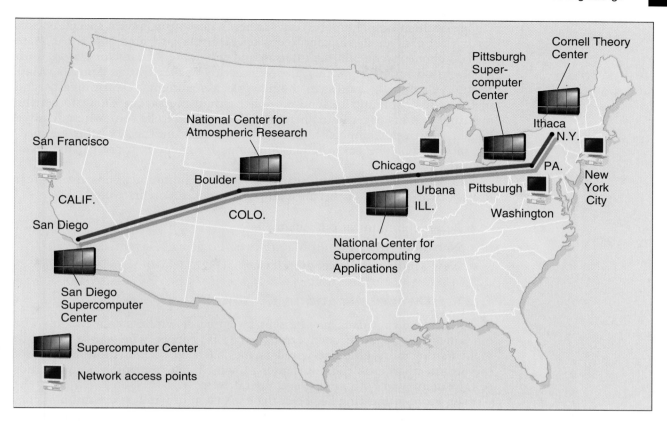

■ PANEL 1.7

Powerful supercomputer network

In the National Science Foundation–MCI arrangement, supercomputers are linked at five locations into a new Internet. This is a national network that allows scientists anywhere in the United States to have access to the computers' combined processing power.

A *server,* or *network server,* **is a central computer that holds databases and programs for many PCs, workstations, or terminals, which are called clients.** These clients are linked by a wired or wireless network. The entire network is called a *client/server network.* In small organizations, servers can store files and transmit electronic mail between departments. In large organizations, servers can house enormous libraries of financial, sales, and product information. The surge in popularity of the World Wide Web has also led to an increased demand for servers at tens of thousands of Web sites.

1.8   Overview of Developments in Communications Technology

KEY QUESTION

What are three developments in communications?

Preview & Review: Communications, or telecommunications, has had three important developments: better communications channels, the use of networks, and new sending and receiving devices.

The students who live in the Computer Science House, a drab brick dormitory at Rochester Institute of Technology in upstate New York, exemplify how far electronic communications have come.[25] The 64 CSH residents spend most of their time hunched over computer keyboards, linked electronically to people down the hall and around the world.

One of them is sophomore Tom Frazier, whose jeans have a special pocket for his cellular phone. Frazier keeps up with his girlfriend across the state via cell phone, pager, and the Internet and gets wake-up calls for classes from an alarm on his pager, which also acts as his pocket watch. Another, Matt "Soup" Campbell, hands in homework assignments electronically and instructs his computer to send reminder e-mail messages ("Don't forget the meeting!") to his pager hours before he is supposed to be somewhere. Unlike Frazier, however, Campbell refuses to date via the Internet. "I believe in people interaction," he says, "and sometimes I wonder if we're getting away from that."

Throughout the 1980s and early 1990s, telecommunications made great leaps forward. Three of the most important developments were:

- Better communications channels
- Networks
- New sending and receiving devices

Better Communications Channels

We mentioned that data may be sent by wired or wireless connections. The old kinds of telephone connections—that is, copper wire—have begun to yield to the more efficient wired forms, such as coaxial cable and, more important, fiber-optic cable (Chapter 8), which can transmit vast quantities of information in both analog and digital form.

Even more interesting has been the expansion of wireless communication. Federal regulators have permitted existing types of wireless channels to be given over to new uses, as a result of which we now have many more kinds of two-way radio, cellular telephone, and paging devices than we had previously.

Networks

When you hear the word "network," you may think of a *broadcast network*, a group of radio or television broadcasting stations that cut costs by airing the same programs. Here, however, we are concerned with **communications networks, which connect one or more telephones and computers and associated devices.** The principal difference is that *broadcast networks transmit messages in only one direction*, whereas *communications networks transmit in both directions*. Communications networks are crucial to technological convergence, for they allow information to be exchanged electronically.

A communications network may be large or small, public or private, wired or wireless or both. In addition, smaller networks may be connected to larger ones. For instance, a *local area network (LAN)* may be used to connect users located near one another, as in the same building. On some college campuses, for example, microcomputers in the rooms in residence halls are linked throughout the campus by a LAN.

New Sending & Receiving Devices

Part of the excitement about telecommunications in the last decade or so has been the development of new devices for sending and receiving information. Two examples are the *cellular phone* and the *fax machine*.

- **Cellular phones:** *Cellular telephones* use a system that divides a geographical service area into a grid of "cells." In each cell, low-

powered, portable, wireless phones can be accessed and connected to the main (wire) telephone network.

The significance of the wireless, portable phone is not just that it allows people to make calls from their cars. Most important is its effect on worldwide communications. Countries with underdeveloped wired telephone systems, for instance, can use cellular phones as a fast way to install better communications. Such technology gives these nations—Thailand, Pakistan, Hungary, and others—a better chance of joining the world economy.

Today's cellular phones are also the forerunners of something even more revolutionary—pocket phones allowing worldwide communication. Cigarette-pack-size portable phones and more fully developed satellite systems will enable people to have conversations or exchange information from anywhere on earth.

- **Fax machines:** *Fax* stands for "facsimile," which means "a copy"; more specifically, *fax* stands for "facsimile transmission." A *fax machine* scans an image and sends a copy of it in the form of electronic signals over transmission lines to a receiving fax machine. The receiving machine re-creates the image on paper. Fax messages may also be sent to and from microcomputers.

Fax machines have been commonplace in offices and even many homes for some time, and new uses have been found for them. For example, some newspapers offer facsimile editions, which are transmitted daily to subscribers' fax machines. These editions look like the papers' regular editions, using the same type and headline styles, although they have no photographs. Toronto's *Globe & Mail* offers people who will be away from Canada a four-page fax that summarizes Canadian news. The *New York Times* sends a faxed edition, transmitted by satellite, to island resorts and to cruise ships in mid-ocean. Networks, fax machines, modems and related topics are covered in detail in Chapters 7 and 8.

1.9 Computer & Communications Technology Combined: Connectivity & Interactivity

KEY QUESTION

What are connectivity and interactivity?

Preview & Review: Trends in information technology involve connectivity and interactivity.

Connectivity, or online information access, refers to connecting computers to one another by modem or network and communications lines. Connectivity provides, among other things, the benefits of voice mail, e-mail, telecommuting, teleshopping, databases, online services and networks, and electronic bulletin board systems.

Interactivity refers to the back-and-forth "dialog" between a user and a computer or communications device. Interactive devices include multimedia computers, personal digital assistants, and "smart boxes"—TV/PCs and WebTVs.

Under development are different versions of a device that combines telephone, television, and personal computer. This device would deliver digitized entertainment, communications, and information.

Alaska commercial salmon fisherman Blanton Fortson says he is such a frequent user of portable technology—on his boat, in his home, even on his airplane—that he has to wear baggy pants to carry it all around. For example, he says, "I'm often in the woods running, hiking, or biking and almost always reachable by digital pager. My home is wired, and I maintain a dedicated phone-line link between home and office-network zones. . . . With the combination of [a laptop computer] and cellular technology, I find that I very rarely need to be tied to any specific location in order to take care of business."[26]

Less well off, shipping clerk Neal Berry, 22, had enough money to buy a laptop computer, a cellular phone, and a connection to the Internet. But he couldn't afford an apartment in pricey Novato, California, so he lived in a tent near the freeway, hunkered down on a mattress rescued from a trash bin. Though homeless, he spent his evenings happily tapping on his laptop, communicating with the online world. "I made more friends in a month [electronically]," he said, "than I had all year in Novato."[27]

Fortson and Berry are beneficiaries of two trends that will no doubt intensify as information technology continues to proliferate. These trends are:

- Connectivity
- Interactivity

Connectivity: Online Information Access

As we discussed, small telecommunications networks may be connected to larger ones. This is called **connectivity, the ability to connect computers to one another by modem or network and communications lines to provide online information access.** It is this connectivity that is the foundation of the latest advances in the Digital Age.

The connectivity of telecommunications has made possible many kinds of activities. Although we cover these activities in more detail in Chapters 7 and 8, briefly they are as follows:

- **Voice mail:** *Voice mail* acts like a telephone answering machine. Incoming voice messages are digitized and stored for your retrieval later. Retrieval is accomplished by dialing into your "mailbox" number from any telephone.

  The advantage of voice mail over an answering machine is that you don't have to worry about the machine running out of message tape or not functioning properly. Also, it will take messages *while* you're on the phone. You can get your own personal voice-mail setup by paying a monthly fee to a telephone company or by purchasing special software and hardware for your personal computer.

- **E-mail:** An alternative system is e-mail. *E-mail*, or *electronic mail*, is a software-controlled system that links computers by wired or wireless connections. It allows users, through their keyboards, to post messages and to read responses on their computer screens. Whether the network is a company's small local area network or a worldwide network, e-mail allows users to send messages, documents, and images anywhere on the system.

- **Telecommuting:** In standard commuting, one takes transportation (car, bus, train) from home to work and back. In *telecommuting,* one works at home and communicates with ("commutes to") the office by computer and communications technology. Already about 7.6 million Americans—not including business owners or independent contractors—telecommute three or more days a month, according to Link Resources.[28] (■ *See Panel 1.8.*) If the definition is expanded to include self-employed contractors, part-timers, and even people who simply bring work home from the office at night, there may be 32.7 million work-at-home households.[29]

■ PANEL 1.8

Telecommuting

Telecommuters have nearly tripled in number since 1990. Growth paused in mid-decade, as layoffs mounted, but with tighter job markets employers say the practice helps attract workers.

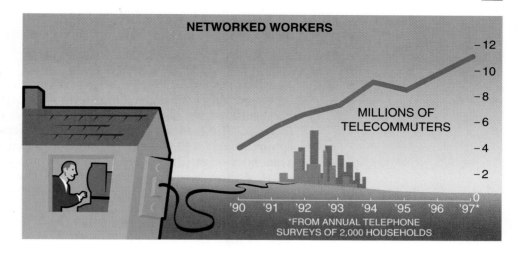

NETWORKED WORKERS

MILLIONS OF TELECOMMUTERS

'90 '91 '92 '93 '94 '95 '96 '97*

*FROM ANNUAL TELEPHONE SURVEYS OF 2,000 HOUSEHOLDS

- **Teleshopping:** Teleshopping is the computer version of cable-TV shop-at-home services. With *teleshopping,* microcomputer users dial into a telephone-linked computer-based shopping service listing prices and descriptions of products, which may be ordered through the computer. You charge the purchase to your credit card, and the teleshopping service sends the merchandise to you by mail or other delivery service. Although most people seem to prefer buying merchandise in a store or via mail catalog, online car shopping seems to have some appeal, according to one survey. As one observer put it, on the Web "you don't have some guy in your face giving you a hard sell."[30]

- **Databases:** A database may be a large collection of data located within your own unconnected personal computer. Here, however, we are concerned with databases located elsewhere. These are libraries of information at the other end of a communications connection that are available to you through your microcomputer. A *database* is a collection of electronically stored data. The data is integrated, or cross-referenced, so that different people can access it for different purposes.

  For example, suppose an unfamiliar company offered you a job. To find out about your prospective employer, you could go online to gain access to some helpful databases. Examples are Business Database Plus, Magazine Database Plus, and TRW Business Profiles. You could then study the company's products, review financial data, identify major competitors, or learn about recent sales increases or layoffs. You might even get an idea of whether you would be happy with the "corporate culture." Alternatively, you can use job-vacancy databases to search for jobs in particular areas of a country.

- **Computer online services:** Established major commercial online services include America Online, CompuServe, Microsoft Network, and Prodigy. A *computer online service* is a commercial information service that, for a fee, makes various services available to subscribers through their telephone-linked microcomputers.

  Among other things, consumers can research information in databases, go teleshopping, make airline reservations, or send messages via e-mail to others using the service. They can also dial into *electronic bulletin board systems (BBSs),* centralized information sources and message-switching systems for particular computer-linked interest groups, many accessible through the Internet. For example, there are BBSs on such varying subjects as fly-fishing, clean air, ecology, genealogy, San Diego entertainment, Cleveland city information, and adult chat.

- **The Internet and World Wide Web:** Through a computer online service or other means (such as with an Internet service provider, or ISP) you may also gain access to the greatest network of all, the Internet. **The *Internet* is an international network connecting approximately 140,000 smaller networks that link computers at academic, scientific, government, and commercial institutions.** The heart of the Internet is a backbone of high-speed communications lines that route data among thousands of other computer systems. The best-known part of the Internet is the ***World Wide Web*, which stores information in multimedia form—sounds, photos, video, as well as text.** An estimated 40 million to 55 million adults in the United States now have access to the Internet at home or work.[31] The number of microcomputers connected to the Internet worldwide is said to have topped 82 million in 1996 and is predicted to reach 268 million by 2001.[32]

Interactivity: The Examples of Multimedia Computers, Wireless Pocket PCs, & Various PC/TVs

Screens from the interactive game *Riven*. The user moves through the game's environment, and various things happen, to which the user may or may not respond.

The movie rolls on your PC/TV screen. The actors appear. Instead of passively watching the plot unfold, however, you are able to determine different plot developments by pressing keys on your keyboard. This is an example of interactivity. As we mentioned earlier, ***interactivity* means that the user is able to make an immediate response to what is going on and modify the processes. That is, there is a dialog between the user and the computer or communications device.** Video games, for example, are interactive. Interactivity allows users to be active rather than passive participants in the technological process.

Among the types of interactive devices are multimedia computers, wireless pocket PCs, and various kinds of "smart boxes" that work as a "converged" computer/TV.

- **Multimedia computers:** The word "multimedia," one of the buzzwords of the '90s, has been variously defined. Essentially, however, ***multimedia* refers to technology that presents information in more than one medium, including text, graphics, animation, video, music, and voice.**

  Multimedia personal computers are powerful microcomputers that include sound and video capability, run CD-ROM disks, and allow users to play games or perform interactive tasks.

- **Wireless pocket PCs:** In 1988, handheld electronic organizers were introduced, consisting of tiny keypads and barely readable screens. They were unable to do much more than store phone numbers and daily "to do" lists.

  Five years later, electronic organizers began to be supplanted by personal digital assistants, such as Apple's Newton. Personal digital assistants (PDAs) are simply wireless pocket-sized personal computers—small pen-controlled, handheld computers that, in their most developed form, can do two-way wireless messaging. Instead of pecking at a tiny keyboard, you can use a special pen to write out commands on the computer screen. The newer generation of wireless pocket PCs can be used not only to keep an appointment calendar and write memos but also to access the Internet and send and receive faxes and e-mail. With a wireless pocket PC, then, you can immediately get information from some remote location—such as the microcomputer on your desk at home—and, if necessary, update it.

● **"Smart boxes" & PC/TVs:** Envisioning a world of cross-breeding among televisions, telephones, and computers, enterprising manufacturers have been developing different kinds of "smart boxes," set-top control boxes or so-called PC/TVs (or TV/PCs) that merge the personal computer with the television set. With fully developed PC/TVs, consumers will be able to watch movies, view multiple cable channels, make phone calls, fax documents, exchange e-mail, do teleshopping, and browse the Internet and the Web. Set-top boxes or PC/TVs would provide two-way interactivity not only with video games but also with online entertainment, news, and educational programs.

What's interesting about interactive TV is that a few years back everyone anticipated that it would be offered mainly by cable and telephone companies, using their own communications lines. As a result, cable, phone, and mass media companies scrambled to form huge corporate alliances. Some costly experiments were tried, such as Time-Warner's centralized, proprietary interactive-TV network in Orlando, Florida (discontinued in 1997 for lack of interest). Now, however, it looks like interactive TV will be delivered via the much cheaper channel of the Internet and the World Wide Web, as indicated by the success of WebTV, a California start-up company, in finding a way to deliver the Web to ordinary television sets with an inexpensive set-top box. WebTV was acquired in 1997 by Microsoft, the software powerhouse that built its fortunes on the traditional microcomputer. The reason, says respected technology commentator Walter Mossberg, is that Microsoft "needed a new type of digital device, without the cost and complexity of a PC, that might appeal to all those families unlikely to buy a PC."[33]

WebTV is only one of several possible *TV computers* (telecomputers) that might be called an *Internet access device*. Other types of TV computers are *online game players* that not only let you play games but also cruise the Web, exchange e-mail, write letters, and do drawings on your TV. Another is the *full-blown PC-TV combination,* the joining of full-function, multimedia PCs with big-screen TVs, such as Gateway 2000's Destination, a home-entertainment setup built around a PC and a 31-inch TV, and Thomson and Compaq's PC Theatre, a high-powered microcomputer paired with a 36-inch multimedia monitor.

Another variation is the *network computer*—a cheap, stripped-down computer that connects people to networks and that is available for under $500. Instead of having all the complex memory and storage capabilities built in, the network computer (available from Sun Microsystems, IBM, Oracle, and others) is "hollowed out," designed to serve as an entry point to the online world, which is supposed to contain all the software, data, and other resources anyone would need. Many network computers use the TV as a display, but they can also be coupled with a computer monitor or have a built-in screen.

PC/TV

It's possible, of course, that a PC/TV-type device won't be a one-size-fits-all kind of box. Instead, there may be several kinds of separate, single-purpose gadgets. At least that's the vision of Diba, Inc., which has plans to offer 42 separate gizmos called Interactive Digital Electronic Appliances (IDEAs).[34,35] Diba Kitchen hangs under the counter and serves up recipes. Diba Mail is a tiny phone, e-mail, and fax machine. Diba Yellow Pages retrieves online business listings. With a single-task design, these devices could be not only inexpensive but practical and easy to use.

Whatever its final form, the future PC/TV will most likely be adapted from a machine now present everywhere—the microcomputer. Clearly, then, anyone who learns to use a microcomputer now is getting a head start on the revolution.

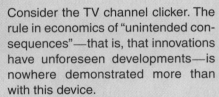

README

Practical Matters: Too Much, Too Fast—Staying Focused to Avoid Information Overload

Consider the TV channel clicker. The rule in economics of "unintended consequences"—that is, that innovations have unforeseen developments—is nowhere demonstrated more than with this device.

When these remote-control units first appeared, they seemed like mere conveniences, gadgets that would save television viewers from having to walk across the room to change channels or adjust the volume on their sets. "No one imagined the full power inherent in the remote control," writes journalist James Gleick; "no one thought in terms of hundreds or thousands of channel changes per evening."[36]

Now advertisers, television programmers, and movie producers shape content with the full knowledge that the viewer will switch away in a millisecond at the slightest twinge of boredom. "Instantaeity rules," Gleick points out "on the screen and on the [television] networks as in our daily lives: instant replay, instant coffee, instant intimacy, instant gratification." Studies show that "grazers" change channels 22 times a minute. As a result, even many talk shows have cameras constantly on the go, trying to hold our attention.

If TV and movies are getting more frenetic all the time, the Internet and the sound-and-graphics part of it known as the World Wide Web would seem perfectly suited to today's information consumer. For instance, one of the basic features of the Web is that it offers *hypertext*—highlighted words and phrases—that Web travelers can use to link up with related words and phrases. But with the Web offering *70 million* or more "home pages," the technology presents all the possibilities for being simply a stupefying waste of time. Says well-known computer journalist John Dvorak,

"the Web and the Net revolution have removed the natural barriers between us and the carloads of information we would normally never see."[37]

How, then, to keep yourself from being overwhelmed by information overload on the Web? Dvorak offers the following suggestions for creating "filters" to cut down on all the noise:

- *Stay focused on what you're trying to find:* If you're doing research on the Web, the biggest problem is *staying focused.* Dvorak gives this example of how following random links can be a time killer: "You begin your session looking for a coffee distributor, . . . and at that site, you see a link to a site on rare chocolates. You jump to that site, and before you know it, you've learned more than you ever wanted to know about the history of the chocolates. Enough already." Thus, remember why you're online in the first place.

- *Limit your time online:* Plan to go on the Web just for 45 minutes or an hour, and *stick with* the plan. If you don't, you'll find yourself up in front of the computer until 4 A.M.

- *Use your printer frequently:* When you see interesting material online, don't try to read it all then and there. Rather, use your computer's printer to print it out immediately for reading later. "That way," says Dvorak, "you can gather as much information as possible without spending all night online or chasing useless links." (We should point out that when paper costs and waste are at issue, it's best to save material you want to read later to your hard disk.)

1.10   The Ethics of Information Technology

Ethics

Preview & Review: Ethical issues pervade all aspects of the use of information technology, as will be noted with a special logo—shown below, left—throughout the book.

Every computer user will have to wrestle with ethical issues related to the use of information technology. *Ethics* is defined as a set of moral values or principles that govern the conduct of an individual or a group. Indeed, ethical questions arise so often in connection with information technology that we have decided to note them wherever they appear in this book with the symbol shown in the margin.

Here, for example, are some important ethical concerns pointed out by Tom Forester and Perry Morrison in their book *Computer Ethics.*[38]

- **Speed and scale:** Great amounts of information can be stored, retrieved, and transmitted at a speed and on a scale not possible before. Despite the benefits, this has serious implications "for data security and personal privacy [as well as employment]," they say, because information technology can never be considered totally secure against unauthorized access.

- **Unpredictability:** Computers and communications are pervasive, touching nearly every aspect of our lives. However, compared to other pervasive technologies—such as electricity, television, and automobiles—information technology is a lot less predictable and reliable.

- **Complexity:** The on/off principle underlying computer systems may be simple, but the systems themselves are often incredibly complex. Indeed, some are so complex that they are not always understood even by their creators. "This," say Forester and Morrison, "often makes them completely unmanageable," producing massive foul-ups or spectacularly out-of-control costs.

These concerns are only a few of many. You'll read about others as you work through the book.

Onward: The "All-Purpose Machine"

Computer pioneer John Von Neumann said that the computer should not be called the "computer" but rather the "all-purpose machine." After all, he pointed out, it is not just a gadget for doing calculations. The most striking thing about it is that *it can be put to any number of uses.*

But are we getting to the point where we may have *too many* info tech devices—phones, notebooks, handhelds? Will we ever get down to just having one device? When a magazine interviewer put this question to Ted Selker, an IBM scientist responsible for many innovations (such as a pointing device called the Trackpoint III), he spoke about working on "a mock-up that is based on the idea of a wallet."[39] The device would open in trifold to be used as a phone. It would have a clock, a built-in scanner to handle business cards, a display screen showing medical and other records, and other features. "I like the idea of a wallet as a metaphor," Selker says. "It is something that has all the really valuable stuff you have to have with you. . . . So I do think we are starting to get into a position where ultimately we can start eliminating the problem of having too many devices."

In the coming chapters, we will see just how right Von Neumann was: The computer *is* an "all-purpose machine" that can be put to any number of uses.

Experience Box

Better Organization & Time Management: Dealing with the Information Deluge in College—& in Life

The Experience Box appears at the end of each chapter. Each box offers you the opportunity to acquire useful experience that directly applies to the Digital Age.

One great problem most of us face now—and that will probably only increase in the future—is how to handle the information glut. A solution is to develop better organization and time-management skills. This first box illustrates skills that will benefit you in college, in this course and others. (Students reading the first two editions of our book have told us they received substantial benefit from reading this.) Elsewhere in the book we describe other techniques for surviving information overload.

At one time in human history it was possible to keep up with the fund of knowledge. In the year 1300, for instance, there were a mere 1338 volumes in the Sorbonne library in Paris. Thus, the great Italian poet Dante Alighieri, for example, could have absorbed the entire body of knowledge of the then-known world. Three hundred and seventy years later, the library at Oxford University alone held over 25,000 books—too much knowledge for any one person to absorb.[40] Today there is so much information that even people with narrow specialties are hard pressed to keep up. In 1998, the Library of Congress in Washington, D.C., for instance, held 112 million books, manuscripts, and documents.[41]

For you, the question is simple: "How on earth am I going to be able to keep up with what's required of me?" The answer is: by *learning how to learn.* By building your skills as a learner, you certainly help yourself do better in college. More than that, however, you also train yourself to be an information manager in the future.

The Art of Learning

How does one become a good learner in college? By deciding to seriously participate in the game of higher education. "Winning the game of higher education is like winning any other game," say learning experts Debbie Longman and Rhonda Atkinson. "It consists of the same basic process. First, you decide if you really want to play. If you do, then you gear your attitudes and habits to learning. Next you learn the rules. To do this, you need a playbook, a college catalog. Third, you learn about the other players—administration, faculty, and other students. Finally, you learn specific plans to improve your playing skills."[42]

Unfortunately, many students come to college with faulty study skills. They are not entirely to blame. In high school and earlier, much of the emphasis is on *what* is to be studied rather than *how* to study it.

"The secret to controlling time is to remember that there is always enough time to do what is really important," say Mervill Douglass and Donna Douglass. "The difficulty is knowing what is really important."[43]

What *is* important in college? Studying, going to classes, writing papers, and taking tests compete with family responsibilities, social life, extracurricular activities, and part-time or even full-time work. All must somehow fit into the same 24 hours available each day. Yet—unlike high school or many paying jobs—time in college is often very unstructured. For students, the clash of college demands can lead to several kinds of personal problems: sleep disturbances, alcohol and drug abuse, eating disorders, money difficulties, procrastination. Let us discuss ways to improve academic performance, which should be your top priority.

Developing Study Habits: Finding Your "Prime Study Time"

How can you use knowledge about your own body and mind to improve your academic performance? Here's one fact: in the hours you *feel best,* you'll *study best.* What is called "prime study time" is the time of day when you are at your best for learning and remembering.

Each of us has a different energy cycle. For example, two roommates may have different patterns. One (the "day person") may be an early riser who prefers to work on difficult tasks in the morning. The other (the "night person") may start slowly but be at the peak of his or her form during the evening hours.

The trick, then, is to effectively *use* your daily energy cycle. This way your hours of best performance will coincide with your heaviest academic demands. For example, if your energy level is high during the evenings, you should plan to do your studying then. This is especially true of assignments requiring heavy concentration, such as writing papers or doing math problems. Yet you need to be aware that evenings are a time when others around you like to unwind or watch TV, and you may be tempted to join them. If, by contrast, your energy level is high during the mornings, you should hit the books then. But you may have to deal with the fact that others nearby are still sleeping or that most classes are held before noon. Probably most students will find that their energy levels are higher during the first part of the day and lower later on.

These different energy patterns and distractions suggest some important actions to take:

- **Make a study schedule.** First make a master schedule that shows all your regular obligations. You'll include classes and work, of course, but you may also wish to list meals and exercise times. This schedule should be indicated for the *entire school term*—semester, quarter, or whatever.

Now insert the times during which you plan to study. As mentioned, it's best if these study periods correspond to times when you are most alert and can best concentrate. However, don't forget to schedule in hourly breaks, since your concentration will flag periodically.

Next write in major academic events. Examples are when term papers and other assignments are due, when quizzes and exams take place, any holidays and vacations.

At the beginning of every week, schedule your study sessions. Write in the specific tasks you plan to accomplish during each session. It's best to try to study something connected with every class every day. If the subject is difficult for you, try to spend an hour a day on it. This is more effective than 5 hours in one day.

In addition, don't put off major projects, such as term papers, thinking you'll do them in one concentrated period of effort. It's more efficient to break the task into smaller steps that you can handle individually. This prevents you from delaying so long that you finally have to pull an all-nighter to complete the project.

● **Find some good places to study.** Studying means first of all avoiding distractions. No doubt you know several people who study while listening to the radio or watching television. Indeed, maybe you've done this yourself. The fact is, however, that most people *are* distracted by these activities.

Avoid studying in places that are associated with other activities, particularly comfortable ones. That is, don't do your academic reading lying in bed or sitting at a kitchen table. Studying should be an intense, concentrated activity.

You may wish to designate two or three sites as regular areas for studying. Assuming they are free of distractions, two good places are at a desk in your room or at a table in the library. As these places become associated with studying, they will reinforce better studying behavior.[44] Make sure the place you study is free of clutter, which can affect your concentration and make you feel disorganized. Your desktop should contain the material you are studying and nothing else.

● **Avoid time wasters, but reward your studying.** Certainly it's much more fun to hang out with your friends or to watch television than to study. Moreover, these pleasures are real and immediate. Getting an A in a course, let alone getting a degree, seems to be in the distant future.

Clearly you need to learn to avoid distractions so that you can study. However, you must also give yourself frequent rewards so that you will indeed be *motivated* to study. You should study with the notion that after you finish, you will "pleasure yourself." The reward need not be elaborate. It could be a walk, a snack, a television show, a video game, a conversation with a friend, or some similar treat.

Improving Your Memory

Memorizing is, of course, one of the principal requirements of staying in college. And distractions are a major impediment to remembering (as they are to other forms of learning). *External distractions* are those you have no control over—noises in the hallway, people whispering in the library. If you can't banish the distraction by moving, you might try to increase your interest in the subject you are studying. *Internal* distractions are daydreams, personal worries, hunger, illness, and other physical discomforts. Small worries can be shunted aside by listing them on a page for future handling. Large worries may require talking with a friend or counselor.

Beyond getting rid of distractions, there are certain techniques you can adopt to enhance your memory.

● **Space your studying, rather than cramming.** Cramming—making a frantic, last-minute attempt to memorize massive amounts of material—is probably the least effective means of absorbing information. Indeed, it may actually tire you out and make you even more anxious prior to the test. Research shows that it is best to space out your studying of a subject on successive days. This is preferable to trying to do it all during the same number of hours on one day.[45] It is *repetition* that helps move information into your long-term memory bank.

● **Review information repeatedly—even "overlearn" it.** By repeatedly reviewing information—what is known as "rehearsing"—you can improve both your retention of it and your understanding.[46] Overlearning can improve your recall substantially. Overlearning is continuing to repeatedly review material even after you appear to have absorbed it. We said that "cramming" is not an effective way to learn. However, reviewing the material right before an examination can help counteract any forgetting that occurred since you last studied it.

● **Use memorizing tricks.** There are several ways to organize information so that you can retain it better. Longman and Atkinson mention the following methods of establishing associations between items you want to remember:[47]

—*Mental and physical imagery:* Use your visual and other senses to construct a personal image of what you want to remember. Indeed, it helps to make the image humorous, action-filled, sexual, bizarre, or outrageous in order to establish a personal connection. How, for instance, would you go about trying to remember the name of the 21st president of the United States, Chester Arthur? Perhaps you could visualize an author writing the number "21" on a wooden chest. This mental image helps you associate *chest, author* (Arthur), and *21* to recall that Chester Arthur was the 21st president.

You can also make your mental image a physical one by, for example, drawing or diagramming. Thus, to assist your recall of the parts of a computer system, you could draw a picture and label the parts.

—*Acronyms and acrostics:* An acronym is a word created from the first letters of items in a list. For instance, *Roy G. Biv* helps you remember the colors of the rainbow in order: *r*ed, *o*range, *y*ellow, *g*reen, *b*lue, *i*ndigo, *v*iolet. An acrostic is a phrase or sentence created from the first letters of items on a list. For example, *Every Good Boy Does Fine* helps you remember that the order of musical notes is *E-G-B-D-F.*

—*Location:* Location memory occurs when you associate a concept with a place or imaginary place. For example, you could learn the parts of a computer system by imagining a walk across campus. Each building you pass could be associated with a part of the computer system.

—*Word games:* Jingles and rhymes are devices frequently used by advertisers to get people to remember their products. You may recall the spelling rule "*I* before *E* except after *C* or when sounded like *A* as in *neighbor* or *weigh.*" A stalactite hangs from the top of a cave, whereas a stalagmite forms on the cave's floor. To recall the difference, you might remember that the *t* in *stalactite* signifies "top." Or you might recall that a stalactite has to hold on "tight." You can also use narrative methods, such as making up a story.

How to Benefit from Lectures

Are lectures really a good way of transmitting knowledge? Perhaps not always, but the fact remains that most colleges rely heavily on this method. Attending them makes a difference. Students with the highest grades tend to have the fewest absences.[48] (See Panel 1.9.)

Most lectures are reasonably well organized, but you will probably attend some that are not. Even so, they will indicate what the instructor thinks is important, which will be useful to you on the exams.

Regardless of the strengths of the lecturer, here are some tips for getting more out of lectures.

- **Take effective notes by listening actively.** Research shows that good test performance is related to good note taking.[49] And good note taking requires that you *listen actively*—that is, participate in the lecture process. Here are some ways to take good lecture notes:

  —*Read ahead and anticipate the lecturer:* Try to anticipate what the instructor is going to say, based on your previous reading (text or study guide). Having background knowledge makes learning more efficient.

  —*Listen for signal words:* Instructors use key phrases such as "The most important point is . . . ," "There are four reasons for . . . ," "The chief reason . . . ," "Of special importance . . . ," "Consequently . . . ." When you hear such signal phrases, mark your notes with an asterisk (*) or write *Imp* (for "Important").

  —*Take notes in your own words:* Instead of just being a stenographer, try to restate the lecturer's thoughts in your own words. This makes you pay attention to the lecture and organize it in a way that is meaningful to you. In addition, don't try to write everything down. Just get the key points.

  —*Ask questions:* By asking questions during the lecture, you necessarily participate in it and increase your understanding. Although many students are shy about asking questions, most professors welcome them.

- **Review your notes regularly.** The good news is that most students, according to one study, do take good notes. The bad news is that they don't use them effectively. That is, they wait to review their notes until just before final exams, when the notes have lost much of their meaning.[50] Make it a point to review your notes regularly, such as the afternoon after the lecture or once or twice a week. We cannot emphasize enough how important this kind of reviewing is.

■ PANEL 1.9

Class attendance and grade success

Students with grades of B or above were more apt to have better class attendance than students with grades of C- or below.

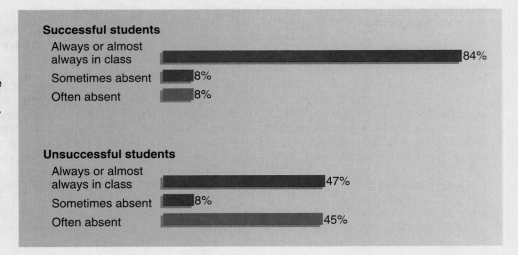

Successful students

Always or almost always in class	84%
Sometimes absent	8%
Often absent	8%

Unsuccessful students

Always or almost always in class	47%
Sometimes absent	8%
Often absent	45%

How to Improve Your Reading Ability: The SQ3R Method

We cannot teach you how to speed-read. However, perhaps we can help you make the time you do spend reading more efficient. The method we will describe here is known as the *SQ3R method,* in which "SQ3R" stands for *survey, question, read, recite,* and *review.*[51] The strategy for this method is to break down a reading assignment into small segments and mastering each before moving on.

The five steps of the SQ3R method are as follows:

1. ***Survey* the chapter before you read it.** Get an overview of the chapter or other reading assignment before you begin reading it. If you have a sense what the material is about before you begin reading it, you can predict where it is going. You can also bring your own experience to it and otherwise become involved in ways that will help you retain it.

   Many textbooks offer some "preview"-type material. Examples are a list of objectives or an outline of topic headings at the beginning of the chapter. Other books offer a summary at the end of the chapter. In this book, for instance, when you first approach a chapter, look at the Key Questions on the first page. At the beginning of each section, look at the Preview & Review beginning the section. You may also want to flip to the end of the chapter and read the summary there.

2. ***Question* the segment in the chapter before you read it.** This step is easy to do, and the point, again, is to get yourself involved in the material. After surveying the entire chapter, go to the first segment—section, subsection, or even paragraph, depending on the level of difficulty and density of information. Look at the topic heading of that segment. In your mind, restate the heading as a question. After you have formulated the question, go to steps 3 and 4 (read and recite). Then proceed to the next segment and restate the heading there as a question. For instance, consider the section heading in this chapter that reads "The Three Directions of Computer Development." You could ask yourself, "What *are* the three directions of computer development?"

3. ***Read* the segment about which you asked the question.** Now read the segment you asked the question about. Read with purpose, to answer the question you formulated. Underline or color-mark sentences you think are important, if they help you answer the question. Read this portion of the text more than once, if necessary, until you can answer the question. In addition, determine whether the segment covers any other significant questions, and formulate answers to these, too. After you have read the segment, proceed to step 4.

   Perhaps you can see where this is all leading. If you read in terms of questions and answers, you will be better prepared when you see exam questions about the material later.

4. ***Recite* the main points of the segment.** Recite means "say aloud." Thus, you should speak out loud (or softly) the answer to the principal question about the segment and any other main points. State these points in your own words, the better to enhance your understanding. If you wish, make notes on the principal ideas, so you can look them over later.

   Now that you have actively studied the first segment, move on to the second segment and do steps 2–4 for it. Continue this procedure through the rest of the segments until you have finished the chapter.

5. **Review the entire chapter by repeating questions.** After you have read the chapter, go back through it and review the main points. Then, without looking at the book, test your memory by repeating the questions.

Clearly the SQ3R method takes longer than simply reading with a rapidly moving color marker or underlining pencil. However, the technique is far more effective because it requires your *involvement and understanding.* This is the key to all effective learning.

How to Become an Effective Test Taker

The first requirement of test taking is, of course, knowledge of the subject matter. That is what our foregoing discussion has been intended to help you obtain. You should also make it a point to *ask* your instructor what kinds of questions will be asked on tests. Beyond this, however, there are certain skills one can acquire that will help during the test-taking process. Here are some suggestions offered by the authors of *Doing Well in College.*[52]

- **Reviewing: Study information that is emphasized and enumerated.** Because you won't always know whether an exam will be an objective or essay test, you need to prepare for both. Here are some general tips.

   —*Review material that is emphasized:* In the lectures, this consists of any topics your instructor pointed out as being significant or important. It also includes anything he or she spent a good deal of time discussing or specifically advised you to study.

   In the textbook, pay attention to key terms (often emphasized in *italic* or **boldface** type), their definitions, and their examples. Also, of course, material that has a good many pages given over to it should be considered important.

   —*Review material that is enumerated:* Pay attention to any numbered lists, both in your lectures and in your notes. Enumerations often provide the basis for essay and multiple-choice questions.

   —*Review other tests:* Look over past quizzes, as well as the discussion questions or review questions provided at the end of chapters in many textbooks.

- **Prepare by doing final reviews & budgeting your test time.** Learn how to make your energy and time work for you. Whether you have studied methodically or must cram for an exam, here are some tips:

—*Review your notes:* Spend the night before the test reviewing your notes. Then go to bed without interfering with the material you have absorbed (as by watching television). Get up early the next morning, and review your notes again.

—*Find a good test-taking spot:* Make sure you arrive at the exam with any pencils or other materials you need. Get to the classroom early, or at least on time, and find a quiet spot. If you don't have a watch, sit where you can see a clock. Again review your notes. Avoid talking with others, so as not to interfere with the information you have learned or increase your anxiety.

—*Read the test directions:* Many students don't do this and end up losing points because they didn't understand precisely what was required of them. Also, listen to any verbal directions or hints your instructor gives you before the test.

—*Budget your time:* Here is an important point of test strategy: Before you start, read through the entire test and figure out how much time you can spend on each section. There is a reason for budgeting your time, of course. You would hate to find you have a long essay still to be written when only a few minutes are left.

Write the number of minutes allowed for each section on the test booklet or scratch sheet and stick to the schedule. The way you budget your time should correspond to how confident you feel about answering the questions.

● **Objective tests: Answer easy questions & eliminate options.** Some suggestions for taking objective tests, such as multiple-choice, true/false, or fill-in, are as follows:

—*Answer the easy questions first:* Don't waste time stewing over difficult questions. Do the easy ones first, and come back to the hard ones later. (Put a check mark opposite those you're not sure about.) Your unconscious mind may have solved them in the meantime, or later items may provide you with the extra information you need.

—*Answer all questions:* Unless the instructor says you will be penalized for wrong answers, try to answer all questions. If you have time, review all the questions.

—*Eliminate the options:* Cross out answers you know are incorrect. Be sure to read all the possible answers, especially when the first answer is correct. (After all, other answers could also be correct, so that "All of the above" may be the right choice.) Be alert that subsequent questions may provide information pertinent to earlier questions. Pay attention to options that are long and detailed, because answers that are more detailed and specific are likely to be correct. If two answers have the opposite meaning, one of the two is probably correct.

● **Essay tests: First anticipate answers & prepare an outline.** Because time is limited, your instructor is likely to ask only a few essay questions during the exam. The key to success is to try to anticipate beforehand what the questions might be and memorize an outline for an answer. Here are the specific suggestions:

—*Anticipate ten probable essay questions:* Use the principles we discussed above of reviewing lecture and textbook material that is *emphasized* and *enumerated.* You will then be in a position to identify ten essay questions your instructor may ask. Write out these questions.

—*Prepare and memorize informal essay answers:* For each question, list the main points that need to be discussed. Put supporting information in parentheses. Circle the key words in each main point and below the question put the first letter of the key word. Make up catch phrases, using acronyms, acrostics, or word games, so that you can memorize these key words. Test yourself until you can recall the key words the letters stand for and the main points the key words represent.

Suppose, for example, the question you make up is "What is the difference between the traditional and the modern theory of adolescence?" You might put down the following answers:[53]

1. *Biologically generated.* Universal phenomenon (Hall's theory: hormonal).
2. *Sociologically generated.* Not universal phenomenon (not purely hormonal).
   BG SG   BIG GUY   SMALL GUY

When you receive the questions for the essay examination, read the entire directions carefully. Then start with the *least demanding question.* Putting down a good answer at the start will give you confidence and make it easier to proceed with the rest. Make a brief outline, similar to the one you did for your anticipated question, before you begin writing.

The Peak-Performing Student

Good students are made, not born. They have decided, as we pointed out earlier, that they really want to play the college game. They are willing to learn the rules, the players, and the playing skills. We have listed some of the studying, reading, and test-taking skills that will help you be a peak-performing student. The practice of these skills is up to you.

Suggestions for Further Reading

Lakein, Alan. *How to Get Control of Your Time and Your Life.* New York: Signet, 1978. One of the classic books on time management.

Wahlstrom, Carl, and Williams, Brian K. *Learning Success: Being Your Best at College & Life.* Belmont, CA: Wadsworth, 1999. A book on college study skills and personal and career success, co-authored by one of the authors of this book.

Summary

Note to the reader: "KQ" refers to Key Questions; see the first page of each chapter. The number ties the summary term to the appropriate section in the book.

What It Is/What It Does

Why It's Important

analog (p. 7, KQ 1.1) Refers to nondigital (noncomputer-based) forms of data transmission that can vary continuously, including voice and video. Telephone lines and radio, television, and cable-TV hookups have historically been analog transmissions media. Analog is the opposite of digital.

You need to know about analog and digital forms of communication to understand what is required for you to connect your computer to other computer systems and information services. Computers cannot communicate directly over analog lines. A modem and communications software are usually required to connect a microcomputer user to other computer systems and information services.

applications software (p. 19, KQ 1.5) Software that has been developed to solve a particular problem, perform useful work on general-purpose tasks, or provide entertainment.

Applications software such as word processing, spreadsheet, database manager, graphics, and communications packages have become commonly used tools for increasing people's productivity.

communications (p. 5, KQ 1.2, 1.6) The sixth element of a computer-and-communications system; the electronic transfer of data from one place to another.

Communications systems using electronic connections have helped to expand human communication beyond face-to-face meetings.

communications network (p. 26, KQ 1.8) System of interconnected computers, telephones, or other communications devices that can communicate with one another.

Communications networks allow users to share applications and data; without networks, information could not be electronically exchanged.

computer (p. 4, KQ 1.5) Programmable, multiuse machine that accepts data—raw facts and figures—and processes (manipulates) it into useful information, such as summaries and totals.

Computers greatly speed up problem solving and other tasks, increasing users' productivity.

computer-and-communications system (p. 9, KQ 1.2) System made up of six elements: people, procedures, data/information, hardware, software, and communications.

Users' need to understand how the six elements of a computer-and-communications system relate to one another in order to make knowledgeable decisions about buying and using a computer system.

computer professional (p. 10, KQ 1.2) Person who has had formal education in the technical aspects of using computer-and-communications systems; also called an *information technology professional.*

Computer professionals create and manage the software and systems that enable users (end-users) to accomplish many types of business, professional, and educational tasks and increase their productivity.

What It Is/What It Does	Why It's Important

connectivity (p. 28, KQ 1.9) Ability to connect devices by telecommunications lines to other devices and sources of information.

Connectivity is the foundation of the latest advances in the Digital Age. It provides online access to countless types of information and services.

data (p. 12, KQ 1.2, 1.3) Consists of the raw facts and figures that are processed into information; third element in a computer-and-communications system. For computing, data is measured in kilobytes, megabytes, gigabytes, and terabytes.

Users need data to create useful information.

digital (p. 6, KQ 1.1) Term used synonymously with *computer;* refers to communications signals or information represented in a binary, or two-state, way—1s and 0s, on and off.

Putting data into digital form allows computers to transmit voice, text, sound, graphics, color, and animation. The whole concept of an Information Superhighway is based on the existence of digital communications.

electronic organizer (p. 22, KQ 1.7) Specialized pocket computer that mainly stores appointment, addresses, and "to do" lists; recent versions feature wireless links to other computers for data transfer.

Puts in electronic form the kind of day-to-day personal information formerly kept in paper form.

end-user (p. 11, KQ 1.2) Also called a *user;* a person with moderate technical knowledge of information technology who uses computers for entertainment, education, or work-related tasks.

End-users are the people for whom most computer-and-communications systems are created (by computer professionals).

hardware (p. 15, KQ 1.2, 1.4) Fourth element in a computer-and-communications system; refers to all machinery and equipment in a computer system. Hardware is classified into five categories: input, processing and memory, output, secondary storage, and communications.

Hardware design determines the type of commands the computer system can follow. However, hardware runs under the control of software and is useless without it.

information (p. 12, KQ 1.2, 1.3) In general, refers to summarized data or otherwise manipulated data. Technically, data comprises raw facts and figures that are processed into information. However, information can also be raw data for the next person or job, so sometimes the terms are used interchangeably. Information/data is the third element in a computer-and-communications system.

The whole purpose of a computer (and communications) system is to produce (and transmit) usable information.

Information Superhighway (p. 20, KQ 1.6) Vision or metaphor for a fusion of the two-way wired and wireless capabilities of telephones and networked computers with television and radio's capacity to transmit hundreds of programs. The resulting interactive digitized traffic would include movies, TV shows, phone calls, databases, shopping services, and online services.

The Information Superhighway is envisioned as fundamentally changing the nature of communications and hence society, business, government, and personal life.

What It Is/What It Does	**Why It's Important**

information technology (p. 3, KQ 1.1) Technology that merges computing with high-speed communications links carrying data, sound, and video.

Information technology is bringing about the gradual fusion of several important industries in a phenomenon called *digital convergence* or *technological convergence*.

input hardware (p. 15, KQ 1.4) Devices that allow people to put data into the computer in a form that the computer can use; that is, they perform *input operations*. Input devices include a keyboard, mouse, pointer, scanner, or microphone.

Useful information cannot be produced without input data.

interactivity (p. 30, KQ 1.9) Situation in which the user is able to make an immediate response to what is going on and modify processes; that is, there is a dialog between the user and the computer or communications device.

Interactive devices allow the user to actively participate in a technological process instead of just reacting to it.

Internet (p. 30, KQ 1.9) International network connecting approximately 36,000 smaller networks that link computers at academic, scientific, and commercial institutions.

The Internet makes possible the sharing of all types of information and services for millions of people all around the world.

laptop computer (p. 22, KQ 1.7) Portable computer equipped with a flat display screen and weighing 2–11 pounds. The top of the computer opens up like a clamshell to reveal the screen.

Laptop and other small computers have provided users with computing capabilities in the field and on the road.

mainframe (p. 23, KQ 1.7) Second-largest computer available, after the supercomputer; occupies a specially wired, air-conditioned room, is capable of great processing speeds and data storage, and costs $50,000–$5 million.

Mainframes are used by large organizations (banks, airlines) that need to process millions of transactions.

microcomputer (p. 22, KQ 1.7) Small computer that can fit on or beside a desktop or is portable; uses a single microprocessor for its CPU. A microcomputer may be a workstation, which is more powerful and is used for specialized purposes, or a personal computer (PC), which is used for general purposes.

The microcomputer has lessened the reliance on mainframes and has enabled more ordinary users to use computers.

microcontroller (p. 22, KQ 1.7) Also called an *embedded computer;* the smallest category of computer.

Microcontrollers are built into "smart" electronic devices, as controlling devices.

minicomputer (p. 23, KQ 1.7) Also known as a *midrange computer;* computer midway in cost and capability between a microcomputer and a mainframe and costing $20,000–$250,000.

Minicomputers can be used as single units or in a system tied by network to as many as several hundred terminals for many users. Many minicomputers are being replaced by networked microcomputers.

What It Is/What It Does	Why It's Important

multimedia (p. 30, KQ 1.9) Refers to technology that presents information in more than one medium, including text, graphics, animation, video, music, and voice.

Use of multimedia is becoming more common in business, the professions, and education as a means of improving the way information is communicated.

notebook computer (p. 22, KQ 1.7) Type of portable computer weighing 4–9 pounds and measuring about 8½ x 11 inches.

Notebooks have more features than many subnotebooks yet are lighter and more portable than laptops.

output hardware (p. 16, KQ 1.4) Consists of devices that translate information processed by the computer into a form that humans can understand; that is, the devices perform *output operations.* Common output devices are monitors and printers. Sound is also a form of computer output.

Without output devices, computer users would not be able to view or use their work.

palmtop computer (p. 22, KQ 1.7) Type of pocket personal computer, weighing less than 1 pound, that is small enough to hold in one hand and operate with the other.

Unlike other pocket PCs, palmtops use the same software as IBM microcomputers and so are compatible with larger computers.

pen computer (p. 22, KQ 1.7) Type of portable computer; it lacks a keyboard or mouse but allows users to input data by writing directly on the display screen with a pen (stylus).

Pen computers are useful for specific tasks, such as for signatures to show proof of package delivery, and some general purposes, such as those fulfilled by electronic organizers and personal digital assistants.

peripheral device (p. 15, KQ 1.4) Any hardware device that is connected to a computer. Examples are keyboard, mouse, monitor, printer, and disk drives.

Most of a computer system's input and output functions are performed by peripheral devices.

personal computer (PC) (p. 22, KQ 1.7) Type of microcomputer; desktop, floor-standing (tower), or portable computer that can run easy-to-use programs, such as word processing or spreadsheets.

The PC is designed for one user at a time and so has boosted the popularity of computers.

personal digital assistant (PDA) (p. 23, KQ 1.7) Also known as *pocket communicator;* type of handheld pocket personal computer, weighing 1 pound or less, that is pen-controlled and that in its most developed form can do two-way, wireless messaging.

PDAs may supplant book-style personal organizers and calendars, as well as allow transmission of personal messages.

pocket personal computer (p. 22, KQ 1.7) Also known as a *handheld computer;* a portable computer weighing 1 pound or less. Three types of pocket PCs are electronic organizers, palmtop computers, and personal digital assistants.

Pocket PCs are useful to help workers with specific jobs, such as delivery people and parking control officers.

What It Is/What It Does	**Why It's Important**

procedures (p. 12, KQ 1.2) Descriptions of how things are done; steps for accomplishing a result or rules and guidelines for what is acceptable. Procedures are the second element in a computer-and-communications system.

In the form of documentation, procedures help users learn to use hardware and software.

secondary storage (p. 16, KQ 1.4) Refers to devices and media that store data and programs permanently—such as disks and disk drives, tape and tape drives. These devices perform *storage operations.*

Without secondary storage media, users would not be able to save their work.

server (p. 25, KQ 1.7) Computer in a network that holds databases and programs for multiple users.

The server enables many users to share equipment, programs, and data.

software (p. 18, KQ 1.2, 1.5) Also called *programs;* step-by-step instructions that tell the computer hardware how to perform a task. Software represents the fifth element of a computer-and-communications system.

Without software, hardware would be useless.

subnotebook computer (p. 22, KQ 1.7) Type of portable computer, weighing 1.8–4 pounds.

Subnotebooks are lightweight and thus extremely portable; however, they may lack features found on notebooks and other larger portable computers.

supercomputer (p. 24, KQ 1.7) High-capacity computer that is the fastest calculating device ever invented; costs $225,000–$30 million.

Used principally for research purposes, airplane design, oil exploration, weather forecasting, and other activities that cannot be handled by mainframes and other less powerful machines.

system software (p. 19, KQ 1.5) Software that controls the computer and enables it to run applications software. System software, which includes the operating system, allows the computer to manage its internal resources.

Applications software cannot run without system software.

technological convergence (p. 3, KQ 1.1) Also called *digital convergence;* refers to the technological merger of several industries through various devices that exchange information in the electronic, or digital, format used by computers. The industries are computers, communications, consumer electronics, entertainment, and mass media.

From a common electronic base, the same information may be exchanged among many organizations and people using any of a multitude of information technology devices.

World Wide Web (p. 30, KQ 1.9) The part of the Internet that stores information in multimedia form—sounds, photos, and video as well as text.

The most widely known part of the Internet, the Web stores information in multimedia form—sounds, photos, video, as well as text.

workstation (p. 23, KQ 1.7) Type of microcomputer; desktop or floor-standing (tower) machine that costs $3700 or more and is used mainly for technical purposes.

Workstations are used for scientific and engineering purposes and also for their graphics capabilities.

Exercises

Self-Test Exercises

1. The _____ refers to the part of the Internet that stores information in multimedia form.
2. Whereas most of the world is _____, computers deal with data in _____ form.
3. In _____, one works at home and communicates with the office by computer and communications technology.
4. A _____ computer is less powerful than a supercomputer, but more powerful than a minicomputer.
5. The term _____ is used to describe a programmable, multiuse machine that accepts data and manipulates it into information.

Short-Answer Questions

1. List the six main elements of a computer and communications system.
2. What is the function of the system unit in a computer system?
3. What is the difference between system software and applications software?
4. Why is it important to have a computer with more main memory rather than less?
5. Which hardware category has the most in common with a filing cabinet?

Multiple-Choice Questions

1. A kilobyte is equal to approximately:
   a. 1000 bytes
   b. 10,000 bytes
   c. 1 million bytes
   d. 1 billion bytes
   e. None of the above
2. Which of the following converts computer output into printed images?
   a. keyboard
   b. mouse
   c. scanner
   d. printer
   e. All of the above
3. Which of the following enables digital data to be transmitted over the phone lines?
   a. keyboard
   b. mouse
   c. scanner
   d. modem
   e. All of the above
4. Which of the following computer types typically costs the least and has the smallest main memory capacity?
   a. supercomputer
   b. mainframe computer
   c. workstation
   d. microcomputer
   e. microcontroller
5. What hardware category does magnetic tape fall into?
   a. input
   b. processing and memory
   c. output
   d. storage
   e. communications

True/False Questions

T F 1. Computers are continually getting larger and more expensive.

T F 2. Mainframe computers process faster than microcomputers.

T F 3. Main memory is a software component.

T F 4. System software consists of several programs, the most important of which is the operating system.

T F 5. An end-user is someone with considerable experience in the technical aspects of using a computer.

Knowledge in Action

1. Determine what types of computers are being used where you work or go to school. Microcomputers? Minicomputers? Any mainframe or supercomputers? In which departments are the different types of computers used? What are they being used for? How are they connected to other computers?

2. Identify some of the problems of information overload in one or two departments in your school or place of employment—or in a local business, such as a real estate firm, health clinic, pharmacy, or accounting firm. What types of problems are people having? How are they trying to solve them? Are they rethinking their use of computer-related technologies?

3. Imagine a business you could start or run at home. What type of business is it? What type of computer do you think you'll need? Describe the computer system in as much detail as possible, including hardware components in all five areas we discussed. Keep your notes and then refine your answers after you have completed the course.

4. Can you envision yourself using a supercomputer in your planned profession or job? If yes, how? What other type(s) of computer do you envision yourself using?

5. How do you think technological, or digital, convergence will affect you in the next five years? For example, will it affect how you currently perform your job or obtain access to education? Do you think that technological convergence is a good thing? Why? Why not?

6. Other than the topics already addressed in this chapter, do you have any ethical concerns about how computers and communications systems are being used today? What are they?

Applications Software

Tools for Thinking & Working

Think of it as a map to the buried treasures of the Information Age."

That's how one writer in 1993 described a new kind of software called a *Web browser,* designed to help computer users find their way around the Internet, particularly the sound-and-graphics part of it known as the World Wide Web.[1] The global "network of networks," the Internet is rich in information but can be baffling to navigate without assistance. The developers of the first Web browser had tried to remove that difficulty.

Indeed, they had hoped their program might become the first "killer app"—killer application—of network computing. That is, it would be a breakthrough development that would help millions of people become comfortable using electronic computer networks, a technology formerly used by only a relative few.

The computer industry puts great stock in history-making "killer apps." One of the big ones was the early 1980s development of the electronic spreadsheet program, software for manipulating numbers in financial documents. Spreadsheet software transformed the personal computer, until then used mainly by technicians and computer buffs, into an essential business tool. The application led to the widespread acquisition of desktop computers in offices all over the country.

The name of the first Web browser was Mosaic, but it was not to become the software that would make the Internet available to everyone, being overtaken in a matter of months by Netscape Navigator. Now Netscape itself is on the defensive, fighting the aggressive marketing of Microsoft and its Internet Explorer. As this is written, developers are engaged in a titanic struggle to come up with the defining tool that will simplify users' abilities to summon text, as well as sound and images, from among the Internet's many information sources.

Nevertheless, the search for highly useful applications shows how truly important software is. Without software, your computer is only about as useful as a doorstop.

2.1   How to Think About Software

KEY QUESTIONS

What is applications software, and what are the five general categories of applications software?

Preview & Review: Applications software enables users to perform work on specific tasks or to participate in different forms of entertainment.

The five types of applications software may be considered to be (1) entertainment software, (2) home/personal software, (3) education/reference software, (4) productivity software, and (5) specialty software.

Over the long course of time, luckily for us, software for personal computers has generally become easier to use. At one time, for instance, users had to learn cryptic commands such as "format a: /n:9 /t:40." Now they can use a mouse to point to words and images on a screen.

In this chapter, we'll discuss the various types of applications software. We'll also consider how software is changing and how you can deal with it.

The Most Popular Uses of Software

Let's get right to the point: What do most people use software for? The answer hasn't changed in years. If you don't count games and communication, by far the most popular applications are (1) word processing and (2) spreadsheets, according to the Software Publishers Association.[2] Small-business owners, according to one study, say the software they use most often is word processing (by 94% of those surveyed), spreadsheet (75%), database management (67%), and desktop publishing (51%).[3]

Interestingly, most people use only a few basic features of word processing and spreadsheet programs, and they use them for rather simple tasks. For example, 70% of all documents produced with word processing software are one-page letters, memos, or simple reports. And 70% of the time people use spreadsheets simply to add up numbers.[4]

This is important information. If you are this type of user, you may have no more need for fancy software and hardware than an ordinary commuter has for an expensive Italian race car. However, you may be in a profession in which you need to become a "power user," having to learn a great number of features in order to keep ahead in your career. Moreover, in the Multimedia Age, you may wish to do far more than current software and hardware allow, in which case you need to be continually learning what computing and communications can do for you.

Let's look now at the various types and uses of software.

The Two Kinds of Software: Applications & Systems

As stated in Chapter 1 (✔ p. 18), *software,* or *programs,* consists of the instructions that tell the computer how to perform a task. **Applications software is software that has been developed to solve a particular problem, to perform useful work on specific tasks, or to provide entertainment.** As the user, you interact with the applications software. In turn, *system software* (covered in the next chapter) enables the applications software to interact with the computer and helps the computer manage its internal resources. (■ *See Panel 2.1.)*

If you buy a new microcomputer in a store, you will find that some packaged software (✔ p. 19) has already been installed on it. This typically includes system software and various types of applications software compatible with it.

Versions, Releases, & Compatibility

Every year or so, software developers find ways to enhance their products and put forth new versions or new releases. Although not all software developers use the terms consistently, their accepted definitions are as follows.

■ PANEL 2.1

Applications software and system software
You interact principally with the applications software. The applications software interacts with components of the system software, which in turn interacts directly with the computer. (Sometimes, however, you do interact directly with the system software.)

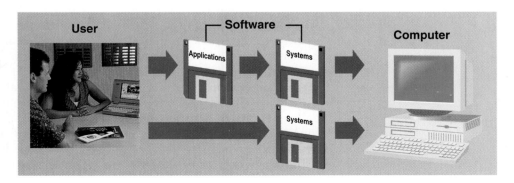

- **Version:** A *version* is a major upgrade in a software product. Traditionally versions have been indicated by numbers such as 1.0, 2.0, 3.0, and so forth. The higher the number preceding the decimal point, the more recent the version. In recent years, a number of software developers have departed from this system. Microsoft, for instance, decided to call the new operating system that it launched in 1995 "Windows 95" instead of "Windows 4.0." Windows 98 would have been Windows 5.0.

- **Release:** A *release* is a minor upgrade. Releases are usually indicated by a change in number after the decimal point—3.0, then 3.1, then perhaps 3.11, then 3.2, and so on. Some releases are now also indicated by the year they are marketed.

When you buy a new software version or release, you must make sure it is compatible with your existing system. *Compatible* means that documents created with earlier versions of the software can be processed successfully on later versions. *Compatible* also means that a new version of an applications program will run with the system software you are currently using. To avoid problems of incompatibility, be sure to read the compatibility requirements printed on the software package you are considering buying.

The Five Categories of Applications Software

Software can change the way we act, even the way we think. Some readers may intuitively understand this because they grew up playing video games. Indeed, some observers hold that video games are not quite the time wasters we have been led to believe, that these forms of entertainment can be a step to something else. That is, they say, video games are "training wheels" for using more sophisticated software that can help us be more productive.

Applications software may be classified in many ways. We use five categories. (■ *See Panel 2.2.*)

1. Entertainment software
2. Home/personal software
3. Education/reference software
4. Productivity software
5. Specialty software

Let us consider them briefly.

1. Entertainment Software: The Serious Matter of Video Games Atomic Bomberman is a real blast, say video game players. An interactive electronic game in which the player's goal is to blow up other Bomberpersons while running around a colorful playing field, Atomic Bomberman became popular on the Nintendo game console and then was adapted for use (as a CD-ROM) with personal computers. Although it is really geared toward multiplayer games on a network such as the Internet, it can also be played solo "just for a funny, fast arcade-style romp," says one review.[5]

Video games might seem frivolous, but they are more important than you might think. Important enough that they generated $3.5 billion in sales in the United States in 1996.[6] Important enough that an accredited specialized college exists in Seattle (the DigiPen Institute of Technology) that offers a four-year degree in video game and computer animation programming.[7] Important enough, in fact, that they may even help children learn better. (Children who absorb video game technology early "think differently from

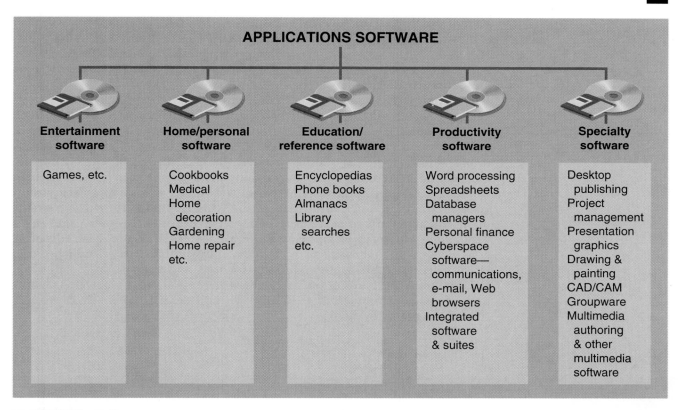

APPLICATIONS SOFTWARE

Entertainment software	Home/personal software	Education/ reference software	Productivity software	Specialty software
Games, etc.	Cookbooks Medical Home decoration Gardening Home repair etc.	Encyclopedias Phone books Almanacs Library searches etc.	Word processing Spreadsheets Database managers Personal finance Cyberspace software— communications, e-mail, Web browsers Integrated software & suites	Desktop publishing Project management Presentation graphics Drawing & painting CAD/CAM Groupware Multimedia authoring & other multimedia software

■ PANEL 2.2

The five categories of applications software

the rest of us," says William Winn, director of the Learning Center at the University of Washington's Human Interface Technology Laboratory. "It's as though their cognitive strategies were parallel, not sequential."[8] Adds Seymor Papert, of the Media Laboratory at the Massachusetts Institute of Technology, "Children who are deeply involved in games have thought more about strategies of learning than kids usually do. From the word go, they must take charge. They're on their own."[9])

Electronic spreadsheets may have put microcomputers on office desks. However, it was Pong—an electronic version of table tennis introduced by Atari in 1972—that popularized computers in the home. "Pong was the first time people saw computers as friendly and approachable," states one technology writer. "It launched a video game boom that ... prepared an entire generation for interaction with a blinking and buzzing computer screen."[10]

Pong was followed by Space Invaders and Pac-Man, and then by Super Mario, which begot Sonic the Hedgehog, which led to Mortal Kombat I and II, and so on. In 1986 Nintendo began to reshape the market when it introduced 8-bit entertainment systems. *Bit numbers* measure how much data a computer chip can process at one time. Bit (✔ p. 6) numbers are important because the higher the bit number, the greater the screen resolution (clarity), the more varied the colors, and the more complex the games.[11] Since then, video game hardware—which, after all, is just a form of computer hardware— has increased in power just as microcomputers have. In the 1990s, video game hardware manufacturers—Sega, 3DO, Atari, Sony, Nintendo—upped the ante to 16 bits, then 32 bits, until finally 64-bit machines appeared on the market.

Although cartridge-based video game machines remain the biggest sales category, the market for personal computer games is not far behind. And the Internet is transforming the development of PC games, with many games now giving the user the ability to connect to other players on the Internet or World Wide Web.[12] In addition, WorldPlay Entertainment (a subsidiary of

America Online) makes available a service in which players may pay to play games against others online.[13] Finally, despite all the games available to stay-at-homes, bigger and more technologically complex game arcades are being built, such as Stephen Spielberg's Gameworks, a 30,000-square-foot game arcade in Seattle that has as its motto: *Get Out of the House.*[14]

Ethics

So far the biggest sales punch has been in "hack and whack" games (such as Mace, which allows players to behead opponents) and sports games (such as NFL Game Day '98).[15,16] Apart from ethical questions about the effects of violence on immature personalities, such games have perhaps ignored the interests of half the population—namely, females. In recent years, however, video game makers have introduced nonviolent games such as McKenzie & Company, essentially an interactive movie aimed at preadolescent girls.[17] Focusing on emotions rather than action, this game has the viewer, as the main character, try to solve problems of the heart. In another game, Let's Talk About Me, relationships count, not big scores.[18] Despite these advances, critics complain that many such products only reinforce well-established sexist stereotypes about appropriate roles for men and women.[19]

Of course, there are other categories of entertainment software, ranging from interactive movies to gambling. We will describe these from time to time throughout the book.

README

Case Study: Women & Information Technology

Why do more boys than girls seem to like computer video games?

It's not just that more boys than girls like simulated fighting and action, which most such games seem to be about. Indeed, a 1995 Rice University study found that girls are not necessarily turned off by violence or action.

What *does* turn them off is repetition and intense competition. "They seem to be most captivated by games that offer lots of different activities, social interaction on the screen and between players, and challenges," says one writer.[20]

From video games to the executive suites of computer companies, females have in the past been underrepresented in information technology. Although grade-school girls are as competent and interested as boys in math and science, this direction has traditionally not been encouraged after middle school. As a result, nationwide only 2% of executives at the vice-presidential level and above in technology companies are women.[21] (They make up 7 to 9% of executive ranks in other Fortune 500 companies; as a whole, women make up about 45% of the total workforce.)

But the picture may be changing. Once medicine was as male as technology is now; today half of medical-school graduates are women.[22] So, too, the percentage of women earning degrees in electrical or computer engineering—one of the chief routes to a technology career—has increased from 5.9% in 1980 to 12.4% in 1995.[23] Says Kathy Wheeler, a chip designer at Hewlett-Packard in Palo Alto, California, "I think engineering is underrated as a good career opportunity. I highly recommend it to any woman thinking about it. There are some subtle barriers . . . [but] the thing about engineering is that it's precise and your results can be measured."[24]

More women are also flocking to cyberspace every day. According to NetSmart, an Internet research company, there are now 11 million women surfing the Net—nearly double the number two years ago.[25] And instead of merely looking for entertainment or information, 91% of the women surveyed said they expect the Net to be "the new time-saving household appliance" that will help them better juggle career, family, and household duties.

For women trying to find their bearings on the Net, there are a number of sites that may prove to be good compass points. One is Beatrice's Web Guide *(http://bguide.com)*, which emphasizes practical uses for the Web, with special stress on women's interests.[26] Another is the Cybergrrl Webstation *(http://www.cybergrrl.com)*, run by Aliza Sherman (known on the Web as Cybergrrl), which offers information on software and links to other female-oriented Web sites.[27]

2. Home/Personal Software The software on the CD-ROM disk called Graham Kerr's Swiftly Seasoned features recipes for 140 exotic foods, ranging from Balinese Potato Waffles to Ostrich Osso Buco, as well as video clips from Kerr's PBS television show, "The Galloping Gourmet." It also includes a function that calculates the nutritional value of recipes.[28]

Swiftly Seasoned, the Jenny Craig Cookbook, and the Completely Interactive Cookbook are examples of software cookbooks, one of several kinds of home/personal software. Other software in this category includes home repair, home decoration, gardening, genealogy, travel planning, and the like. PlanetWare North European Travel Planner, for example, is a CD-ROM offering attractions, maps, hotels, and restaurants (and currency, tours, and pictures) from France to Finland. Visual Home lets you design a floor plan, create three-dimensional rooms of various shapes, and choose the wallpaper and the furnishings. Hanes T-ShirtMaker & More allows you to print images on T-shirts, using a computer and color printer.

3. Educational/Reference Software Because of the popularity of video games, many educational software companies have been blending educational content with action and adventure—as in MathBlaster or the problem-solving game Commander Keen. They hope this marriage will help students be more receptive to learning. After all, as one writer points out, players of Nintendo's Super Mario Brothers must "become intimately acquainted with an alien landscape, with characters, artifacts, and rules completely foreign to ordinary existence. . . . Children assimilate this essentially useless information with astonishing speed."[29] Why not, then, design software that would educate as well as entertain?

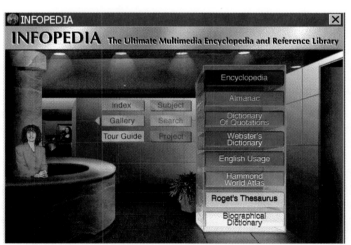

Computers alone won't boost academic performance, but they can have a positive effect on student achievement in all major subject areas, preschool through college, according to an independent consulting firm, New York's Interactive Educational Systems Design. Skills improve when students use programs that are self-paced or contain interactive video, the consultants found after analyzing 176 studies done over 5 years. This is particularly true for low-achieving students. The reason, says a representative of the firm, is that this kind of educational approach is "a different arena from the one in which they failed, and they have a sense of control."[30]

In addition to educational software, library search and reference software have become popular. For instance, there are CD-ROMs with encyclopedias, phone books, mailing lists, maps, and reproductions of famous art. With the CD-ROM encyclopedia Microsoft Encarta, for example, you can search for, say, music in 19th-century Russia, then listen to an orchestral fragment from Tchaikovsky's *1812 Overture*.

4. Productivity Software Productivity software consists of programs found in most offices, in many homes, and probably on all campuses, on personal computers and on larger computer systems. Their purpose is simply to make users more productive at performing particular tasks. Productivity software is probably the most important type of software you will learn to use.

The most popular kinds of productivity tools are:

- Word processing software
- Spreadsheet software
- Database software, including personal information managers
- Financial software, including personal finance programs
- Software for cyberspace—communications, e-mail, Web browsers
- Integrated software and suites

It may still be possible to work in an office somewhere in North America today without knowing any of these programs. However, that probably won't be the case as we move along into the 21st century.

We describe productivity software in more detail beginning in another few pages.

5. Specialty Software Whatever your occupation, you will probably find it has specialized software available to it. This is so whether your career is as an architect, building contractor, chef, dairy farmer, dance choreographer, horse breeder, lawyer, nurse, physician, police officer, tax consultant, or teacher.

Some programs help lawyers or advertising people, for instance, keep track of hours spent on particular projects for billing purposes. Other programs help construction estimators pull together the costs of materials and labor needed to estimate a job.

Later in this chapter we describe the following kinds of specialty software:

- Desktop publishing
- Presentation graphics
- Project management
- Computer-aided design
- Drawing and painting programs
- Groupware
- Multimedia authoring software

Before we discuss specific productivity and specialty software programs, however, let's first cover some of the common features of much of the software—both applications and systems—used today.

2.2 Common Features of Software

KEY QUESTION

What are some common features of the graphical software environment?

Preview & Review: In a graphical environment, software packages share some basic features. They use special-purpose keys, function keys, and a mouse to issue commands and choose options. Their interfaces include menus, windows, icons, buttons, and dialog boxes to make it easy for people to use the program.

Software packages are also accompanied by tutorials and documentation.

You may already be familiar with basic computer concepts such as the cursor, the mouse pointer, dialog boxes, and so on, as well as keyboard components such as function keys. If not, the following discussion provides a brief review of some features common to both applications software and system software.

Features of the Keyboard

We describe the keyboard as an input device in Chapter 5. Here, however, we explain some aspects of the keyboard because it and the mouse are the means for manipulating software.

Besides letter, number, and punctuation keys and often a calculator-style numeric keypad, computer keyboards have special-purpose and function keys. Sometimes keystrokes are used in combinations called *macros,* or *keyboard shortcuts.*

- **Special-purpose keys:** ***Special-purpose keys* are used to enter, delete, and edit data and to execute commands.** An example is the Esc (for "Escape") key. The Enter key, which you will use often, tells the computer to execute certain commands and to start new paragraphs in a document. Commands are instructions that cause the software to perform specific actions. For example, pressing the Esc key commands the computer, via software instructions, to cancel an operation or leave ("escape from") the current mode of operation.

  Special-purpose keys are generally used the same way regardless of the applications software package being used. Most keyboards include the following special-purpose keys: Esc, Ctrl, Alt, Del, Ins, Home, End, PgUp, PgDn, Num Lock, and a few others. (*Ctrl* means Control, *Del* means Delete, *Ins* means Insert, for example.)

- **Function keys:** ***Function keys,* labeled F1, F2, and so on, are positioned along the top or left side of the keyboard. They are used to execute commands specific to the software being used.** For example, one applications software package may use F6 to exit a file, whereas another may use F6 to underline a word.

  Many software packages come with printed templates that you can attach to the keyboard. Like the explanation of symbols on a road map, the template describes the purpose of each function key and certain combinations of keys.

- **Macros:** Sometimes you may wish to reduce the number of keystrokes required to execute a command. To do this, you use a macro. **A *macro,* also called a *keyboard shortcut,* is a single keystroke or command—or a series of keystrokes or commands—used to automatically issue a longer, predetermined series of keystrokes or commands.** Thus, you can consolidate several activities into only one or two keystrokes. The user names the macro and stores the corresponding command sequence; once this is done, the macro can be used repeatedly.

  Although many people have no need for macros, others who find themselves continually repeating complicated patterns of keystrokes say they are quite useful.

The User Interface: GUIs, Menus, Windows, Icons, Buttons, & Dialog Boxes

The first thing you look at when you call up any applications software on the screen is the user interface. **The *user interface* is the user-controllable part of the software that allows you to communicate, or interact, with it.** The type of user interface is usually determined by the system software (discussed in the next chapter). However, because this is what you see on the screen before you can begin using the applications software, we will briefly describe it here.

The kind of interface now used by most people is the graphical user interface. **With a *graphical user interface,* or *GUI* (pronounced "gooey"), you may use graphics (images) and menus as well as keystrokes to choose commands, start programs, and see lists of files and other options.**

Common features of GUIs are *menus, windows, icons, buttons,* and *dialog boxes.*

Menu bar

- **Menus: A *menu* is a list of available commands presented on the screen.** Menus may appear as menu bars, pull-down menus, or pop-up menus.

    A *menu bar* is a line of command options across the top or bottom of the screen. Examples of commands, which you activate with a mouse or with key combinations, are File, Edit, and Help.

    A *pull-down menu,* also called a *dropdown menu,* is a list of command options that "drops down" from a selected menu bar item at the top of the screen. For example, you might use the mouse to "click on" (activate) a command (for example, File) on the menu bar, which in turn would yield a pull-down menu offering further commands. These other commands might be Open, Save, Print, Copy, and Paste. Choosing one of these options may produce further menus called *cascading,* or *flyout, menus.*

    A *pop-up menu,* usually activated by a shortcut (macro) or a mouse click, is a list of command options that can "pop up" anywhere on the screen. Pop-up menus are not connected to a menu bar as are drop-down menus.

    A particularly useful option on the menu bar is the ***Help option,* which offers assistance on how to perform various tasks,** such as printing out a document. Help offers a built-in electronic instruction and reference manual. (Help can also usually be accessed by pressing a particular function key, usually F1.)

- **Windows:** A particularly interesting feature of GUIs is the use of windows. **A *window* is a rectangular area that appears on the screen and displays information from a particular part of a program.** A display screen may show more than one window—for instance, one might show information from a word processing program, another information from a spreadsheet.

    A window (small w) should not be confused with Windows (capital W)—the program known as Microsoft Windows—which is the most popular form of system software for the personal computer. However, as you might expect, Windows features extensive use of windows.

- **Icons: An *icon* is a picture used in a GUI to represent a command, a program, a file, or a task.** For example, a picture of a diskette might represent the command "Save (store) this document." Icons are activated by a mouse or other pointing device.

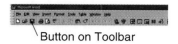

Button on Toolbar

Save changes?
yes no cancel

- **Buttons: A *button* is a simulated on-screen button (kind of icon) that is activated ("pushed") by a mouse or other pointing device to issue a command,** such as "OK" and "Cancel."

- **Dialog box: A *dialog box* is a box that appears on the screen and displays a message requiring a response from you,** such as clicking on "yes" or "no" or typing in the name of a file. For example, when you're saving changes you've written in a document, the program might display a dialog box asking if you want to replace the previous version of the document.

  A dialog box is used to collect additional information from the user before performing a command or completing a task.

Tutorials & Documentation

How are you going to learn a given software program? Most commercial packages come with tutorials and documentation.

- **Tutorials: A *tutorial* is an instruction book or program that takes you through a prescribed series of steps to help you learn how to use the product.** For instance, our publisher offers several how-to books, known as the *Irwin Advantage Series,* that enable you to learn different kinds of software. Tutorials can also be provided as part of the software package.

  Note: Not all software products come with tutorials.

- **Documentation: *Documentation* is a user guide or reference manual that is a narrative and graphical description of a program.** Documentation may be print-based, but today it is usually available on CD-ROM, as well as via the Internet. Documentation may be instructional, but features and functions are usually grouped by category for reference purposes. For example, in word processing documentation, all features having to do with printing are grouped together so you can easily look them up if you have forgotten how to perform them.

What would you like to do?

Type your question here, and then click Search.

Search

Tips   Options   Close

The Assistant

Often you can ask your software for directions on how to use the software. That is, some software makers (using a technique known as *natural language processing,* described in Chapter 11) equip their programs with features that allow you to ask "How do I . . . ?" questions in plain English. Thus, if you type "How do I add up the numbers in this column?" the software will respond by directing you to an interactive tutor or "coach" that can help you through the procedure. Lotus WordPro, for instance, has a feature called Ask the Expert, which asks you to complete a "How Do I?" question box. Microsoft Office 97 offers the same thing, except the feature is called the "Assistant."

All software packages come with documentation.

Let us now consider the various forms of applications software used as productivity tools. Then we will cover specialty programs.

2.3 Word Processing

Preview & Review: Word processing software allows you to use computers to create, edit, format, print, and store text material, among other things.

KEY QUESTION

What can you do with word processing software that you can't do with pencil and paper?

The typewriter, that long-lived machine, has gone to its reward. Indeed, if you have a manual typewriter, it is becoming as difficult to get it repaired as it is to find a blacksmith. Today, word processing software offers a much-improved way to deal with documents.

***Word processing software* allows you to use computers to format, create, edit, print, and store text material,** among other things. (■ *See Panel 2.3.*) Popular word processing programs are Microsoft Word and Corel WordPerfect for PCs, and Word and WordPerfect for the Macintosh.

Word processing software allows users to maneuver through a document and *delete, insert,* and *replace* text, the principal correction activities. It also offers such additional features as *creating, editing, formatting, printing,* and *saving.* We'll cover these features here only briefly because most readers will likely be learning about them in detail in computer lab sessions.

Creating Documents

Creating a document means entering text using the keyboard. Word processing software has three features that affect this process—the *cursor, scrolling,* and *word wrap.*

- Cursor: **The *cursor* is the movable symbol on the display screen that shows you where you may enter data or commands next.** The symbol is often a blinking rectangle or I-beam. You can move the cursor on the screen using the keyboard's directional arrow keys or a mouse. Wherever the cursor is located, that point is called the *insertion point.*

- Scrolling: ***Scrolling* is the activity of moving quickly upward or downward through the text or other screen display.** A standard

■ PANEL 2.3

Word processing screen
This Microsoft Word 97 screen shot shows a pull-down Format menu that offers, among other things, several options for styling text.

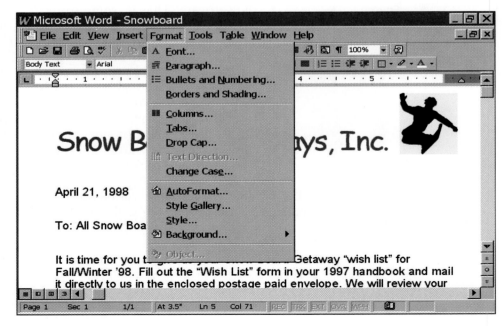

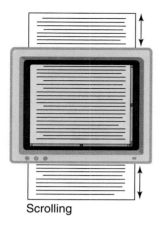

Scrolling

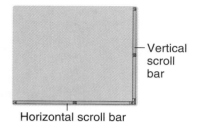

Vertical scroll bar

Horizontal scroll bar

computer screen displays only 20–22 lines of standard-size text. (Some programs use a "Zoom" function to allow you to reduce the size of the type page in order to see more or fewer lines of text on the screen.) Of course, most documents are longer than 20–22 lines. Using the directional arrow keys, or the mouse and a scroll bar located at the side of the screen, you can move ("scroll") through the display screen and into the text above and below it.

Note that when you scroll, even though you are moving to different parts of a document, your cursor insertion point remains the same. That is, the cursor stays at the same insertion point; it doesn't move as you scroll.

- Word wrap: *Word wrap* automatically continues text on the next line when you reach the right margin. That is, the text "wraps around" to the next line.

Editing Documents

Editing is the act of making alterations in the content of your document. Some features of editing are *insert and delete, undelete, find and replace, cut/copy and paste, spelling checker, grammar checker,* and *thesaurus.*

- Insert and delete: *Inserting* is the act of adding to the document. You simply place the cursor wherever you want to add text and start typing; the existing characters will be pushed along.

  Deleting is the act of removing text, usually using the Delete or Backspace keys.

  The *Undelete command* allows you to change your mind and restore text that you have deleted. Some word processing programs offer as many as 100 layers of "undo," allowing users who delete several paragraphs of text, but then change their minds, to reinstate one or more of the paragraphs.

- Find and replace: The *Find,* or *Search, command* allows you to find any word, phrase, or number that exists in your document. The *Replace command* allows you to automatically replace it with something else.

- Cut/Copy and paste: Typewriter users were accustomed to using scissors and glue to "cut and paste" to move a paragraph or block of text from one place to another in a manuscript. With word processing, you select (highlight) the portion of text you want to copy or move. Then you can use the *Copy* or *Cut command* to move it to a special area in the computer's memory called the *clipboard.* Once the material is on the clipboard, you can "paste," or transfer, it anywhere in the existing document or in a new document.

- Spelling checker, grammar checker, thesaurus: Many writers automatically run their completed documents through a *spelling checker,* which tests for incorrectly spelled words. (Some programs, such as Microsoft Word 97, have an "Auto Correct" function that automatically fixes such common mistakes as transposed letters— "teh" instead of "the.") Another feature is a *grammar checker,* which flags poor grammar, wordiness, incomplete sentences, and awkward phrases.

If you find yourself stuck for the right word while you're writing, you can call up an on-screen thesaurus, which will present you with the appropriate word or alternative words.

Formatting Documents

Formatting means determining the appearance of a document. There are many choices here.

- **Font:** You can decide what font—that is, what typeface and type size—you wish to use. You can specify what parts of it should be underlined, *italic,* or **boldface.**

- **Spacing and columns:** You can choose whether you want the lines to be *single-spaced* or *double-spaced* (or something else). You can specify whether you want text to be *one column* (like this page), *two columns* (like many magazines and books), or *several columns* (like newspapers).

- **Margins and justification:** You can indicate the dimensions of the *margins*—left, right, top, and bottom—around the text.

  You can specify whether the text should be *justified* or not. *Justify* means to align text evenly between left and right margins, as, for example, is done with most newspaper columns and this text. *Left-justify* means to not align the text evenly on the right side, as in many business letters ("ragged right").

- **Pages, headers, footers:** You can indicate *page numbers* and *headers* or *footers.* A *header* is common text (such as a date or document name) that is printed at the top of every page. A *footer* is the same thing printed at the bottom of every page.

- **Other formatting:** You can specify *borders* or other decorative lines, *shading, tables,* and *footnotes.* You can even pull in ("import") *graphics* or drawings from files in other software programs.

It's worth noting that word processing programs (and indeed most forms of applications software) come from the manufacturer with *default settings.* **Default settings are the settings automatically used by a program unless the user specifies otherwise, thereby overriding them.** Thus, for example, a word processing program may automatically prepare a document single-spaced, left-justified, with 1-inch right and left margins unless you alter these default settings.

Printing Documents

Most word processing software gives you several options for printing. For example, you can print *several copies* of a document. You can print *individual pages* or a *range of pages.* You can even preview a document before printing it out. *Previewing (print previewing)* means viewing a document on screen to see what it will look like in printed form before it's printed. Whole pages are displayed in reduced size.

Some word processors even come close to desktop-publishing programs in enabling you to prepare professional-looking documents, with different typefaces and sizes, graphics, and colors. However, as we shall see later, desktop-publishing programs do far more.

Saving Documents

Saving means to store, or preserve, the electronic files of a document permanently on diskette, hard disk, or CD-ROM, for example. Saving is a feature of nearly all applications software, but anyone accustomed to writing with a typewriter will find this activity especially valuable. Whether you want to make small changes or drastically revise your word processing document, having it stored in electronic form spares you the tiresome chore of having to retype it from scratch. You need only call it up from the storage medium and make just those changes you want, then print it out again.

2.4 Spreadsheets

Preview & Review: Spreadsheet software allows users to create tables and financial schedules by entering data into rows and columns arranged as a grid on a display screen. If one (or more) numerical value or formula is changed, the software calculates the effect of the change on the rest of the spreadsheet.

Spreadsheet software also allows users to create analytical graphics charts to present data.

KEY QUESTION

What can you do with an electronic spreadsheet that you can't do with pencil and paper and a standard calculator?

What is a spreadsheet? Traditionally, it was simply a grid of rows and columns, printed on special light-green paper, that was used by accountants and others to produce financial projections and reports. A person making up a spreadsheet often spent long days and weekends at the office penciling tiny numbers into countless tiny rectangles. When one figure changed, all the rest of the numbers on the spreadsheet had to be recomputed—and ultimately there might be wastebaskets full of jettisoned worksheets.

In the late 1970s, Daniel Bricklin was a student at the Harvard Business School. One day he was staring at columns of numbers on a blackboard when he got the idea for computerizing the spreadsheet. The result, VisiCalc, was the first of the electronic spreadsheets. **An *electronic spreadsheet*, also called simply a *spreadsheet*, allows users to create tables and financial schedules by entering data and formulas into rows and columns arranged as a grid on a display screen.**

The electronic spreadsheet quickly became the most popular small-business program. As we mentioned, it has been held directly responsible for making the microcomputer a widely used business tool. Unfortunately for Bricklin, VisiCalc was shortly surpassed by Lotus 1-2-3, a sophisticated program that combines the spreadsheet with database and graphics programs. Today the principal spreadsheets are Microsoft Excel, Lotus 1-2-3, and Quattro Pro.

Principal Features

The arrangement of a spreadsheet is as follows. (■ *See Panel 2.4.*)

- Columns, rows, and labels: In the spreadsheet's frame area (work area), lettered *column headings* appear across the top ("A" is the name of the first column, "B" the second, and so on), and numbered *row headings* appear down the left side ("1" is the name of the first row, "2" the second, and so forth). Labels are any descriptive text, such as APRIL, RENT, or GROSS SALES.

- Cells, cell addresses, values, and spreadsheet cursor: The place where a row and a column intersect is called a *cell*, and its position is called a *cell address*. For example, "A1" is the cell address for the top left

■ PANEL 2.4

Electronic spreadsheet
The Lotus 1-2-3 for Windows electronic spreadsheet *(top)* is a computerized version of the traditional paper spreadsheet *(bottom)*. The beauty of the electronic version, however, is its *recalculation* feature: When a number is changed, all related numbers on the spreadsheet are recomputed.

Column Headings —

Row headings —

Cell —

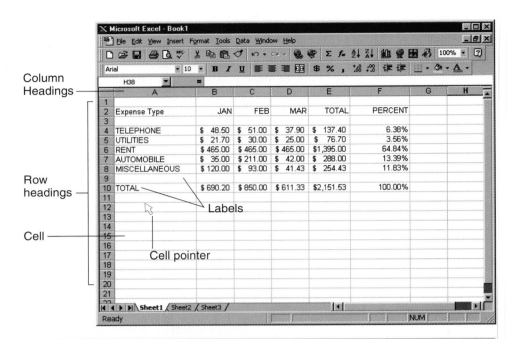

Labels

Cell pointer

EXPENSE	JAN.	FEB.	MAR.	TOTAL
TEL	48.50	51.00	37.90	137.40
UTIL	21.70	30.00	25.00	76.70
RENT	465.00	465.00	465.00	1,395.00
AUTO	35.00	211.00	42.00	288.00
MISC	120.00	93.00	41.43	254.43
TOTAL	$690.20	$850.00	$611.33	$2,151.53

cell, where column A and row 1 intersect. A selection (group) of cells is called a *range*. A number or date entered in a cell is called a *value*. The values are the actual numbers used in the spreadsheet—dollars, percentages, grade points, temperatures, or whatever. A *cell pointer*, or *spreadsheet cursor*, indicates where data is to be entered. The cell pointer can be moved around like a cursor in a word processing program.

● **Formulas, functions, and recalculation:** Now we come to the reason the electronic spreadsheet has taken offices by storm. Formulas are instructions for calculations. For example, a formula might be =SUM(A5:A15), meaning "Sum (add) all the numbers in the cells with cell addresses A5 through A15."

Functions are built-in formulas that perform common calculations. For instance, a function might average a range of numbers or round off a number to two decimal places.

After the values have been plugged into the spreadsheet, the formulas and functions can be used to calculate outcomes. What is revolutionary, however, is the way the spreadsheet can easily do recalculation. **Recalculation is the process of recomputing values,** either as an ongoing process as data is being entered or afterward, with the press of a key. With this simple feature, the hours of mind-numbing work required to manually rework paper spreadsheets became a thing of the past.

● **The "what if" world:** The recalculation feature has opened up whole new possibilities for decision making. As a user, you can create a plan, put in formulas and numbers, and then ask yourself, "What would happen if we change that detail?"—and immediately see the effect on the bottom line. You could use this if you're considering buying a car. Any number of things can be varied: total price ($15,000? $20,000?), down payment ($2,000? $3,000?), interest rate on the car loan (7%? 8%?), or number of months to pay (36? 48?). You can keep changing the "what if" possibilities until you arrive at a monthly payment figure that you're comfortable with.

Spreadsheets can be linked with other spreadsheets. The feature of *dynamic linking* allows data in one spreadsheet to be linked to and update data in another spreadsheet. Thus, the amount of data being manipulated can be enormous.

Analytical Graphics: Creating Charts

A nice feature of spreadsheet packages is the ability to create analytical graphics. *Analytical graphics,* **or business graphics, are graphical forms that make numeric data easier to analyze** than when it is in the form of rows and columns of numbers, as in electronic spreadsheets. Whether viewed on a monitor or printed out, analytical graphics help make sales figures, economic trends, and the like easier to comprehend and analyze.

The principal examples of analytical graphics are *bar charts, line graphs,* and *pie charts. (■ See Panel 2.5.)* Quite often these charts can be displayed or printed out so that they look three-dimensional. Spreadsheets can even be linked to more exciting graphics, such as digitized maps.

■ PANEL 2.5

Analytical graphics
Bar charts, line graphs, and pie charts are used to display numerical data in graphical form.

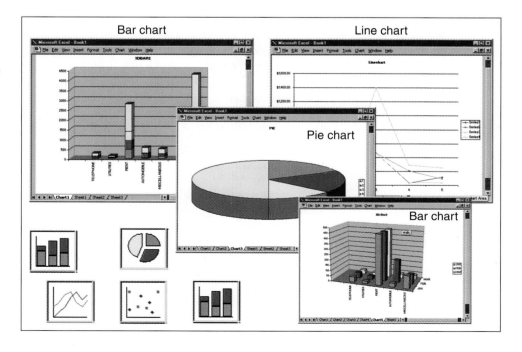

2.5 Database Software

KEY QUESTIONS

What is database software, and what is personal information management (PIM) software?

Preview & Review: A database is a computer-based collection of interrelated files. Database software is a program that controls the structure of a database and access to the data.

Personal information management (PIM) software is specialized database software that helps track and manage information used on a daily basis, such as addresses, appointments, lists, and miscellaneous notes.

In its most general sense, a database is any electronically stored collection of data in a computer system. In its more specific sense, **a *database* is a collection of interrelated files** in a computer system. These computer-based files are organized according to their common elements, so that they can be retrieved easily. (Databases are covered in detail in Chapter 9.) Sometimes called a *database manager* or *database management system (DBMS)*, **database software is a program that controls the structure of a database and access to the data.**

The Benefits of Database Software

Because it can access several files at one time, database software is much better than the old file managers (also known as flat-file management systems) that used to dominate computing. A *file manager* is a software package

that can access only one file at a time. With a file manager, you could call up a list of, say, all students at your college majoring in psychology. You could also call up a separate list of all students from Indiana. But you could not call up a list of psychology majors from Indiana, because the relevant data is kept in separate files. Most database software allows you to do that.

Databases are a lot more interesting than they used to be. Once they included only text. The Digital Age has added new kinds of information—not only documents but also pictures, sound, and animation. It's likely, for instance, that your personnel record in a future company database will include a picture of you and perhaps even a clip of your voice. If you go looking for a house to buy, you will be able to view a real estate agent's database of video clips of homes and properties without leaving the realtor's office.

Today the principal database programs are Microsoft Access, Microsoft Visual FoxPro, dBASE, Paradox, and Claris Filemaker Pro. These programs also allow users to attach multimedia—sound, motion, and graphics—to forms.

Databases have gotten easier to use, but they still can be difficult to set up. Even so, the trend is toward making such programs easier for both database creators and database users.

Principal Features of Database Software

Some features of databases are as follows:

- Organization of a database: A database is organized—from smallest to largest items—into *fields, records,* and *files.* (■ *See Panel 2.6.)*

     A *field* **is a unit of data consisting of one or more characters.** An example of a field is your name, your address, or your driver's license number.

     A *record* **is a collection of related fields.** An example of a record would be your name and address and driver's license number.

     A *file* **is a collection of related records.** An example of a file could be one in your state's Department of Motor Vehicles. The file would include everyone who received a driver's license on the same day, including their names, addresses, and driver's license numbers.

- Retrieve and display: The beauty of database software is that you can locate records in the file quickly. For example, your college may maintain several records about you—one at the registrar's, one in financial aid, one in the housing department, and so on. Any of these

■ PANEL 2.6

Illustration of the concepts of fields, records, and files

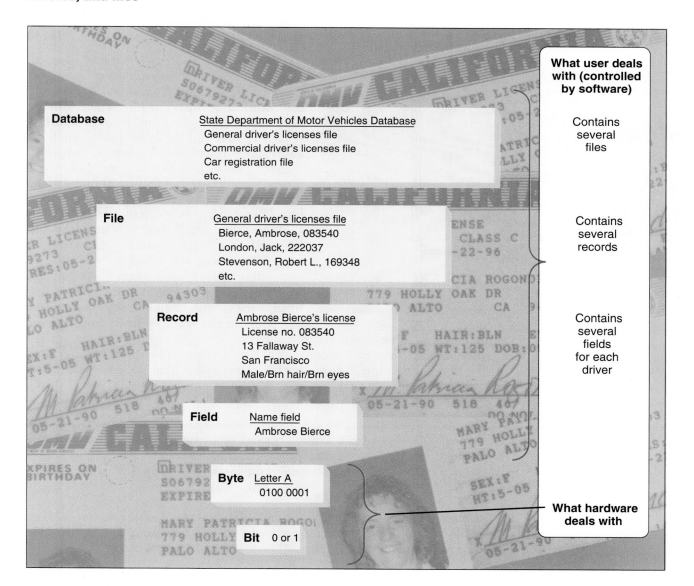

records can be called up on a computer screen for viewing and updating. Thus, if you move, your address field will need to be changed in all records. The database is quickly corrected by finding your name field. Once the record is displayed, the address field can be changed.

- **Sort:** With database software you can easily change the order of records in a file. Normally, records are entered into a database in the order they occur, such as by the date a person registered to attend college. However, all these records can be sorted in different ways. For example, they can be rearranged by state, by age, or by Social Security number.

- **Calculate and format:** Many database programs contain built-in mathematical formulas. This feature can be used, for example, to find the grade-point averages for students in different majors or in different classes. Such information can then be organized into different formats and printed out in sophisticated reports.

Personal Information Managers

Pretend you are sitting at a desk in an old-fashioned office. You have a calendar, Rolodex-type address file, and notepad. Most of these items could also be found on a student's desk. How would a computer and software improve on this arrangement?

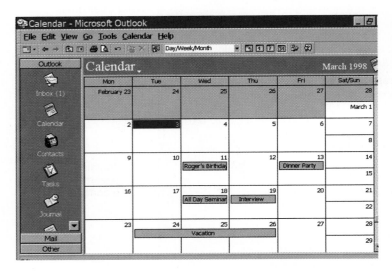

Many people find ready uses for specialized types of database software known as personal information managers. **A *personal information manager (PIM)* is software to help you keep track of and manage information you use on a daily basis, such as addresses, telephone numbers, appointments, "to do" lists, and miscellaneous notes.** Some programs feature phone dialers, outliners (for roughing out ideas in outline form), and ticklers (or reminders). With a PIM, you can key in notes in any way you like and then retrieve them later based on any of the words you typed.

Popular PIMs are Lotus Organizer, Microsoft Outlook, and Act. Lotus Organizer, for example, looks much like a paper datebook on the screen—down to simulated metal rings holding simulated paper pages. The program has screen images of section tabs labeled Calendar, To Do, Address, Notepad, Planner, and Anniversary. The Notepad section lets users enter long documents, including text and graphics, that can be called up at any time.

2.6 Financial Software

KEY QUESTION

What is the purpose of financial software?

Preview & Review: Financial software includes personal-finance managers, entry-level accounting packages, and business financial-management software. Personal-finance managers let you keep track of income and expenses, write checks, do online banking, and plan financial goals.

"Computers can automate any number of basic tasks," observes one advisor to small businesspeople, "freeing you for more interesting challenges. Like keeping your business profitable."[31]

Besides word processing, spreadsheet, and database software, the next most important program for business is financial software. ***Financial software* is a growing category that ranges from personal-finance managers to entry-level accounting programs to business financial-management packages.**

Personal-Finance Managers

Nick Ryder, an airline pilot from Marietta, Georgia, credits the best-selling personal-finance program Quicken with saving his marriage by keeping his finances afloat. ***Personal-finance managers* let you keep track of income and expenses, write checks, do online banking, and plan financial goals.**

When Ryder and his wife, Penny, were married, after many years of each being single, they found themselves deep in the credit-card hole. Despite two healthy paychecks, they never seemed to have enough money. Then they acquired Quicken and began entering everything into the program's various account categories: checking, credit cards, utility bills, all incidentals over a dollar.

After a few months of tracking expenses, some patterns began to emerge. "We were spending way too much on eating out," Ryder says. "Day to day, it doesn't look like much, but it adds up." The incidentals category also turned up a shocking number of impulse buys—magazines, snacks—that were out of line. With the knowledge acquired from Quicken, the Ryders began to cut back on expenses and even saved enough to set up investment accounts—managed by Quicken.[32]

Many personal-finance programs, such as Quicken and Microsoft Money, include a calendar and a calculator, but the principal features are the following:

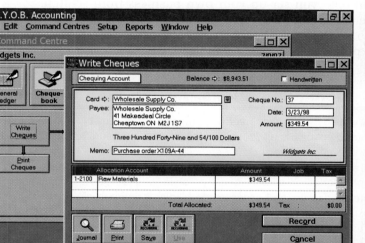

- **Tracking of income and expenses:** The programs allow you to set up various account categories for recording income and expenses, including credit card expenses.

- **Checkbook management:** All programs feature checkbook management, with an on-screen check writing form and check register that look like the ones in your checkbook. Checks can be purchased to use with your computer printer. Some programs offer a nationwide online electronic payment service that lets you pay your regular bills automatically, even depositing funds electronically into the accounts of the people owed.

- **Reporting:** All programs compare your actual expenses with your budgeted expenses. Some will compare this year's expenses to last year's.

- **Income tax:** All programs offer tax categories, for indicating types of income and expenses that are important when you're filing your tax return. Most personal-finance managers also are able to interface with a tax-preparation program.

- **Other:** Some of the more versatile personal-finance programs also offer financial-planning and portfolio-management features.

Other Financial Software

Besides personal-finance managers, financial software includes small business accounting and tax software programs, which provide virtually all the forms you need for filing income taxes. Tax programs such as TaxCut and Turbo Tax make complex calculations, check for mistakes, and even unearth deductions you didn't know existed. Tax programs can be linked to personal finance software to form an integrated tool.

A lot of financial software is of a general sort used in all kinds of enterprises, such as accounting software, which automates bookkeeping tasks, or payroll software, which keeps records of employee hours and produces reports for tax purposes.

Some programs go beyond financial management and tax and accounting management. For example, Business Plan Pro, Management Pro, and Performance Now can help you set up your own business from scratch.

Finally, there are investment software packages, such as StreetSmart from Charles Schwab and Online Xpress from Fidelity, as well as various retirement planning programs.

2.7 Software for Cyberspace: Communications, E-Mail, Web Browsers

KEY QUESTION

What do communications, e-mail, and Web-browser programs do?

Preview & Review: Communications software manages the transmission of data between computers. Electronic mail (e-mail) software enables users to exchange letters and documents between computers.

Web browsers are software programs that allow people to view information at Web sites in the form of colorful, on-screen magazine-style "pages" with text, graphics, and sound. Using a browser, users can access search tools known as directories and search engines.

In the past, many microcomputer users felt they had all the productivity they needed without ever having to hook up their machines to a telephone. However, it's clear that adding communications capabilities to your computer vastly extends your range by allowing you to access the riches of cyberspace.

The term *cyberspace* was coined by William Gibson in his novel *Neuromancer* to refer to a futuristic computer network that people use by plugging their brains into it. Today **cyberspace has come to mean the online or digital world in general and the Internet and its World Wide Web in particular.**

Three software tools for accessing cyberspace are *communications software, e-mail software,* and *Web browsers.*

Communications Software

Communications software, or *data communications software,* **manages the transmission of data between computers.** For most microcomputer users, this sending and receiving of data is by way of a modem and a telephone line. A *modem* (✔ p. 7) is an electronic device that allows computers to communicate with each other over telephone lines. The modem translates the digital signals of the computer into analog signals that can travel over telephone lines to another modem, which translates the analog signals back to digital. When you buy a modem, you often get communications software with it. Popular microcomputer communications programs are Crosstalk, QuickLink, and Procomm Plus.

Communications software gives you these capabilities:

- **Online connections:** You can connect to online services such as America Online (AOL) and Microsoft Network (MSN) and to networks, such as those used within an office (a local area network, or LAN) or the Internet.

- **Use of financial services:** You can order discount merchandise, look up airline schedules and make reservations, follow and engage in stock trading, and even do some home banking and bill paying.

- **Automatic dialing services:** You can set your software to answer for you if someone tries to call your computer, to dial certain telephone numbers automatically, and to automatically redial after a certain time if a line is busy.

- **Remote access connections:** While traveling you can use your portable computer to exchange files via modem with your computer at home.

- **File transfer:** You can obtain a file from a computer at the other end of a communications line and *download* it to your personal computer—that is, transfer it onto your computer's hard disk. You can also do the reverse—*upload* a file of data from your computer and transfer it elsewhere.

- **Fax support:** You can fax messages from your computer to others' computers or fax machines and receive their fax messages in your computer.

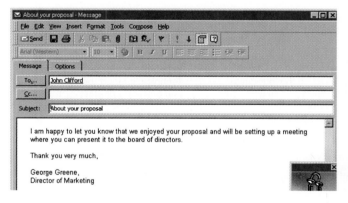

Electronic Mail Software

***Electronic mail software,* or *e-mail software,* enables users to send letters and files from one computer to another.** Many organizations have "electronic mailboxes." If you were a sales representative, for example, such a mailbox would allow you to transmit a report you created on your word processor to a sales manager in another area. Or you could route the same message to a number of users on a distribution list. Popular e-mail software packages include Eudora, Microsoft Outlook, Lotus CC:Mail, and Pegasus Mail.

Web Browsers

The Internet, that network of thousands of interconnected networks, "is just a morass of data, dribbling out of servers [computers] around the world," says one writer. "It is unfathomably chaotic, mixing items of great value with cyber-trash." This is why so-called browsers have caught people's imaginations, he states. "A browser cuts a path through the tangled growth and even creates a form of memory, so each path can be retraced."[33]

The most exciting part of the Internet is probably that fast-growing region or subset of it known as the World Wide Web. The *World Wide Web,* or simply *the Web,* consists of hundreds of thousands of intricately interlinked sites called "home pages" set up for on-screen viewing in the form of colorful magazine-style "pages" with text, images, and sound.

To be connected to the World Wide Web, you need a setup with an online service (✔ p. 29) or a commercial Internet service provider (described in the Experience Box at the end of Chapter 3 and also in Chapter 7), who will usually give you a "browser" for actually exploring the Web. (The reverse is also true: Some Web browsers you buy will help you find an Internet service provider.) **A *Web browser,* or simply *browser,* is software that enables you to "browse through" and view Web sites.** You can move from page to page by "clicking on" or using a mouse to select an icon or by typing in the address of the page. The accompanying drawing explains what the parts of a Web electronic address mean. (■ *See Panel 2.7.*)

There are several browsers available, including some relatively unsophisticated ones offered by Internet service providers and some by commercial online services such as America Online or Compuserve. However, the recent battle royal to find the "killer app" browser has been between Netscape, which produces Navigator/Communicator, and Microsoft, which produces Internet Explorer. As of mid-1998, Netscape was still leading with 54% (down from 70%) of the browser market to Microsoft's 39%.[34] However, Microsoft had increased its market share by making Internet Explorer available free as part of new releases of its Windows operating system. Because most new microcomputers come equipped with this operating system, and therefore Microsoft's Internet Explorer, buyers of new computers had less incentive to choose Netscape's browser. Netscape thereupon changed its strategy and also began giving away its browser. It also made public the "source code" that is the key to the program's inner workings, hoping that thousands of developers would tinker with the program and devise improvements that could be incorporated into the next version of the product.

Remember that online services and ISPs are *not the same* as browsers. The online services and ISPs are commercial enterprises that provide *communications* software that enables you to connect your computer with them via modem and phone lines. Because you pay the service/ISP a monthly subscription fee, you may continue to use whatever services the company offers,

■ PANEL 2.7

What's a Web browser?

This screen from Netscape Communicator illustrates some components of a home page (Web site). From here you can move to other pages.

Back
Takes you back to pages previously viewed.

Bookmarks
A list of sites can be created so the user can quickly jump to the ones used frequently (also called *favorite places*).

Home
Takes you back to the introductory screen (the one you see when you first load your browser).

Browser vendor screen
Technical support, browser copies, and browser updates.

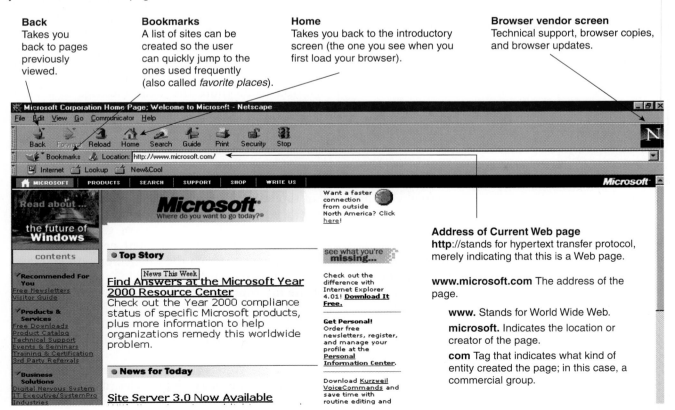

Address of Current Web page
http://stands for hypertext transfer protocol, merely indicating that this is a Web page.

www.microsoft.com The address of the page.

   www. Stands for World Wide Web.

   microsoft. Indicates the location or creator of the page.

   com Tag that indicates what kind of entity created the page; in this case, a commercial group.

including e-mail, teleshopping, investing, and the like. Browsers are software used to access the Web *after* you are online with an online service or ISP. With a browser you may access Web sites that offer services or activities that are also available from an online service—there is some overlap.

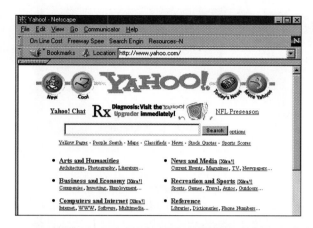

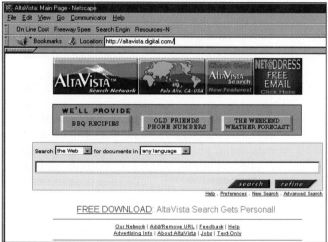

Web Search Tools: Directories & Search Engines

Once you're in your browser, you need to know how to find what you're looking for. Search tools are of two basic types—*directories* and *search engines.*

- Directories: **Web directories are indexes classified by topic.** One of the foremost examples is Yahoo! *(http://www.yahoo.com),* which provides you with an opening screen offering several general categories. Directory information is collected and ranked by people.

- Search engines: **Web** *search engines* **allow you to find specific documents through keyword searches.** An example of one useful search engine is AltaVista *(http://www.altavista.com).* Search engine information is collected and ranked by software programs (sometimes called *spiders).*

According to a 1997 Baruch College–Harris Poll, which surveyed 1000 U.S. households, 21% of adults—the equivalent of 40 million people—use the Internet and/or the Web. The most common activity, by 82% of Net users, is research, followed by education (75%), news (68%), and entertainment (61%).[35]

Thus, Web browsers, directories, and search engines may come to be among our most important software productivity tools.

2.8 Integrated Software & Suites

KEY QUESTIONS

What are integrated software packages and software suites, and how do they differ?

Preview & Review: Integrated software packages combine the basic features of several applications programs—for example, word processing, spreadsheet, database manager, graphics, and communications—into one software package. Software suites are full-fledged versions of several applications programs bundled together.

What if you want to take data from one program and use it in another—say, call up data from a database and use it in a spreadsheet? You can try using separate software packages, but one may not be designed to accept data from the other. Two alternatives are the collections of software known as *integrated software* and *software suites.*

Integrated Software: "Works" Programs

Integrated software packages **combine the most commonly used features of several applications programs—such as word processing, spreadsheet, database, graphics, and communications—into one software package.** These so-called

Practical Matters: Map Software Means "I Got Lost" Is No Longer a Valid Excuse

There you are, a stranger driving in Boston, "a city that has never seen a need to put street signs at major intersections," as one writer puts it.[36] How do you get to where you want to go?

Well, you could place your laptop computer on the seat beside you and listen as it dictates spoken directions (or pull over occasionally and glance at the directions on the screen). The computer runs a program known as "map" or "street-finding" software. Such software is either sold on CD-ROMs or available for free on the Internet. (The disadvantage of Net software is that it's much slower to obtain by downloading it to your computer compared to software you load into your computer from a CD-ROM.[37])

Map software is of two types:

- **Trip-planner database software:** With this kind of map software, you instruct the program where you want to go. With Road Trips Door-to-Door, for instance, you get a map of the United States and a blank trip planner. You enter your starting point, your destination, and any stops you want to make in between. The software then delivers a route map on the right side of your screen and detailed directions on the left. One writer says he thought he knew "every wrinkle in the 100-mile trip to my mother's home, but Door-to-Door showed me how to save a couple of minutes."[38]

  A program called Precision Mapping Streets allows you to also import satellite photos, nautical charts, and aviation maps and display them as underlays beneath the computer-generated street and highway maps.[39]

- **Satellite-directed street guides:** An example of the Boston "talking navigator" software mentioned above is Door-to-Door CoPilot. Your laptop must be equipped with a CD-ROM drive, an audio system, and a receiver for the satellite-based global position-

ing system (GPS). You type in your starting and ending sites, and CoPilot speaks instructions as you drive along, generally allowing you enough time to find and make your turns. You can also stop and click on the display to repeat the last instruction. If you miss a turn or ignore an instruction—or encounter construction chaos—the program will determine a new route and give you a fresh instruction.

Neither type of map software is without flaws. Complete details on one-way streets, dead-ends, turn restrictions, and other obstacles aren't always available.[40] A request for the quickest route may put you on congested city streets rather than nearby interstate highways. Relatively new addresses may not be in the map software's database. Satellite-directed maps may drop in accuracy for a minute or so when high buildings restrict the antenna's view of the sky.

Still, map software will no doubt get better and will become vital for ambulance drivers, delivery companies, salespeople, and others who must regularly visit new addresses.

"works" collections—the principal representatives are AppleWorks, ClarisWorks, Lotus Works, Microsoft Works, and PerfectWorks—give good value because the entire bundle often sells for $100 or less.

Integrated software packages are less powerful than separate programs used alone, such as a word processing or spreadsheet program used by itself. But that may be fine, because single-purpose programs may be more complicated and demand more computer resources than necessary. You may have no need, for instance, for a word processor that will create an index. Moreover, Microsoft Word can take up more than 20 megabytes on your hard disk, whereas Microsoft Works takes only 7 megabytes, which leaves a lot more room for other software.

Software Suites: "Office" Programs

Software suites, or simply *suites*, are applications—like spreadsheets, word processing, graphics, and communications—with a standard user interface that are bundled together and sold for a fraction of what the programs would cost if bought individually. "Bundled" and "unbundled" are jargon words frequently encountered in software and hardware merchandising. *Bundled* means that components of a system are sold together for a single price. *Unbundled* means that a system has separate prices for each component.

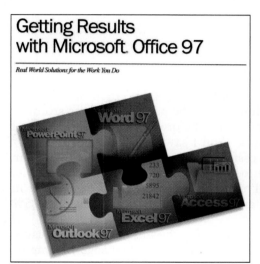

Getting Results with Microsoft Office 97

Real World Solutions for the Work You Do

Three principal suites, sometimes called "office" programs, are available. Microsoft's Office 97 is available in both "standard" and "professional" versions. IBM's Lotus SmartSuite 97 comes in one version. Corel's WordPerfect Suite 8 is the "standard" version and Office Professional is the "professional" version.

Although lower price is what makes suites attractive to many corporate customers, the software has other benefits as well. Software makers have tried to integrate the "look and feel" of the separate programs within the suites to make them easier to use. "All applications in a suite look and function similarly," says one writer. "You learn one, you learn them all. And they're integrated, passing information back and forth easily without compatibility problems—in theory, at least."[41]

A trade-off, however, is that such packages require a lot of hard-disk storage capacity. The standard edition of Office 97 hogs 120 megabytes and Corel gobbles up 157 megabytes of hard-disk space. (Compare with the "works" program Claris-Works at 13 megabytes.) In addition, it's advisable to get the version of these programs that comes on a CD-ROM rather than on diskettes. "Installing manually by diskettes," jokes one journalist about SmartSuite 97, "will take you approximately the time it takes to paint two garages."[42]

2.9 Specialty Software

KEY QUESTION

What are the principal uses of programs for desktop publishing, presentation graphics, project management, computer-aided design, drawing and painting, groupware, and multimedia authoring software?

Preview & Review: Specialty software includes the following programs. (1) Desktop publishing (DTP) combines text and graphics in a highly sophisticated manner to produce high-quality output for commercial printing. (2) Presentation graphics uses graphics and data/information from other software tools to communicate or make a presentation of data to others. (3) Project management software is used to plan, schedule, and control the people, costs, and resources required to complete a project on time. (4) Computer-aided design (CAD) programs are for designing products and structures. (5) Drawing programs allow users to design and illustrate objects and products, and painting programs allow them to simulate painting on screen. (6) Groupware is used on a network and allows users within the same building or on different continents to share ideas and update documents. (7) Multimedia authoring software enable users to integrate multimedia elements—text, images, sound, motion, animation—into a logical sequence of events.

After learning some of the productivity software just described, you may wish to extend your range by becoming familiar with more specialized programs.

For example, you might first learn word processing and then move on to desktop publishing, the technology used to prepare much of today's printed information. Or you may find yourself in an occupation that requires you to learn some very specific kinds of software. We will consider the following specialized tools, although these are but a handful of the thousands of programs available:

- Desktop publishing
- Presentation graphics
- Project management
- Computer-aided design
- Drawing and painting
- Groupware
- Multimedia authoring

Desktop Publishing

Once you've become comfortable with a word processor, could you then go on and learn to do what Margaret Trejo did? When Trejo, then 36, was laid off from her job because her boss couldn't meet the payroll, she was stunned. "Nothing like that had ever happened to me before," she said later. "But I knew it wasn't a reflection on my work. And I saw it as an opportunity."[43]

Today Trejo Production is a successful desktop-publishing company in Princeton, New Jersey, using Macintosh equipment to produce scores of books, brochures, and newsletters. "I'm making twice what I ever made in management positions," says Trejo, "and my business has increased by 25% every year."

Not everyone can set up a successful desktop-publishing business, because many complex layouts require experience, skill, and knowledge of graphic design. Indeed, use of these programs by nonprofessional users can lead to rather unprofessional-looking results. Nevertheless, the availability of microcomputers and reasonably inexpensive software has opened up a career area formerly reserved for professional typographers and printers.

***Desktop publishing*, abbreviated *DTP*, involves using a microcomputer and mouse, scanner, laser or ink-jet printer, and DTP software for mixing text and graphics to produce high-quality output for commercial printing.** Often the printer is used primarily to get an advance look before the completed job is sent to an imagesetter typesetter for even higher-quality output. (Imagesetters generate images directly onto film, which is then given to a printer for platemaking and printing.) Professional DTP programs are QuarkXPress and PageMaker. Microsoft Publisher is a "low-end," consumer-oriented DTP package. Some word processing programs, such as Word and WordPerfect, also have many DTP features, though at nowhere near the level of the packages just mentioned.

Desktop publishing has the following characteristics:

- **Mix of text with graphics:** Desktop-publishing software allows you to precisely manage and merge text with graphics. As you lay out a page on-screen, you can make the text "flow," liquid-like, around graphics such as photographs. You can resize art, silhouette it, change the colors, change the texture, flip it upside down, and make it look like a photo negative.
- **Varied type and layout styles:** As do word processing programs, DTP programs provide a variety of fonts, or typestyles, from readable

Times Roman to staid Tribune to wild Jester and Scribble. Additional fonts can be purchased on disk or downloaded online. You can also create all kinds of rules, borders, columns, and page numbering styles. A *style sheet* in the DTP program enables you to choose and record the settings that determine the appearance of the pages. This may include defining size and typestyle of text and headings, numbers of columns of type on a page, and width of lines and boxes.

- **Use of files from other programs:** It's usually not efficient to do word processing, drawing, and painting with the DTP software. Thus, text is usually composed on a word processor (such as Word or WordPerfect), artwork is created with drawing and painting software (such as Adobe Illustrator or CorelDRAW), and photographs are scanned in using a scanner and then manipulated and stored using photo-manipulation software (such as Adobe PhotoShop). Prefabricated art may also be obtained from disks containing *clip art,* or "canned" images that can be used to illustrate DTP documents. The DTP program is used to integrate all these files. You can look at your work on the display screen as one page or as two facing pages (in reduced size). Then you can see it again after it is printed out on a printer. (■ *See Panel 2.8.*)

Presentation Graphics Software

Computer graphics can be highly complicated, such as those used in special effects for movies (such as *Titanic* or *Twister*). Here we are concerned with just one kind of graphics called presentation graphics.

■ **PANEL 2.8**

How desktop publishing uses other files

Text is composed on a word processor, graphics are drawn with drawing and painting programs, and photographs and other artwork are scanned in with a scanner. Data from these files is integrated using desktop-publishing software, and the pages are printed out on a laser or ink-jet printer.

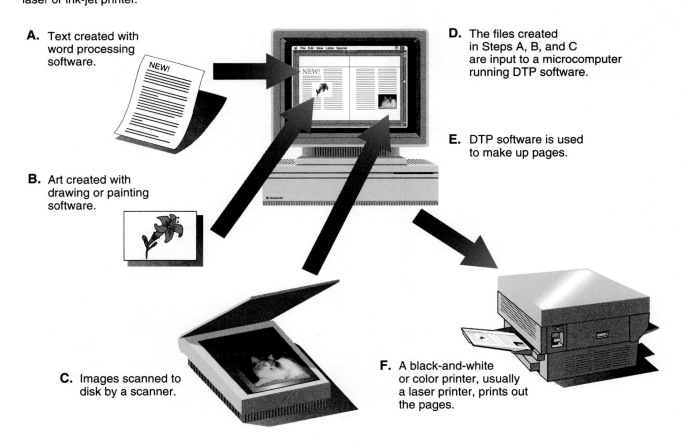

A. Text created with word processing software.

B. Art created with drawing or painting software.

C. Images scanned to disk by a scanner.

D. The files created in Steps A, B, and C are input to a microcomputer running DTP software.

E. DTP software is used to make up pages.

F. A black-and-white or color printer, usually a laser printer, prints out the pages.

Presentation graphics are part of presentation software, which uses graphics and data/information from other software tools to communicate or make a presentation of data to others, such as clients or supervisors. Presentation graphics are also used in information kiosks, multimedia training, and lectures. (A *kiosk* is a small, self-contained structure such as a newsstand or a ticket booth, designed to serve a large number of people. Unattended multimedia kiosks dispense information via computer screens. Keyboards and touch screens are used for input.)

Presentations may make use of some analytical graphics—bar, line, and pie charts—but they usually look much more sophisticated, using, for instance, different texturing patterns (speckled, solid, cross-hatched), color, and three-dimensionality. (■ *See Panel 2.9.*) Examples of well-known presentation graphics packages are Microsoft PowerPoint, WordPerfect Presentations, ASAP from Software Publishing Group, and Gold Disk's Astound.

In general, presentation graphics are displayed electronically or output as 35-millimeter slides. Presentation graphics packages often come with slide sorters, which group together a dozen or so slides in miniature. The person making the presentation can use a mouse or keyboard to bring the slides up for viewing or even start a self-running slide show.

Some presentation graphics packages provide clip art that can be electronically copied and pasted into the presentation. These programs also allow you to use electronic painting and drawing tools for creating lines, rectangles, and just about any other shape. Depending on the system's capabilities, you can add text, animated sequences, and sound. With special equipment you can do graphics presentations on transparencies and videotape. With all these options the main problem may be simply restraining yourself.

Project Management Software

The kind of database program we called a personal information manager (PIM) can help you schedule your appointments and do some planning. That is, it can help you manage your own life. But what if you need to manage the lives of others to accomplish a full-blown project, such as steering a political campaign or handling a nationwide road tour for a band? Strictly defined, a *project* is a one-time operation consisting of several tasks and multiple resources

■ PANEL 2.9
Presentation graphics

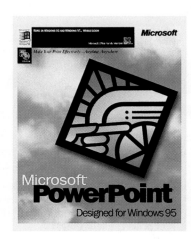

that must be organized toward completing a specific goal within a given period of time. The project can be small, such as an advertising campaign for an in-house advertising department, or large, such as construction of an office tower or a jetliner.

Project management **software is a program used to plan, schedule, and control the people, costs, and resources required to complete a project on time.** For instance, the associate producer on a feature film might use such software to keep track of the locations, cast and crew, materials, dollars, and schedules needed to complete the picture on time and within budget. The software would show the scheduled beginning and ending dates for a particular task—such as shooting all scenes on a certain set—and then the date that task was actually completed. Examples of project management software are Harvard Project Manager, Microsoft Project, Suretrack Project Manager, and ManagerPro.

Two important tools available in project management software are Gantt charts and PERT charts.

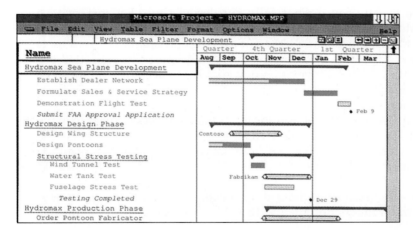

- **Gantt chart:** A *Gantt chart (left)* uses lines and bars to indicate the duration of a series of tasks. The time scale may range from minutes to years. The chart allows you to see whether tasks are being completed on schedule.

- **PERT chart:** A *PERT chart*—PERT stands for Program Evaluation Review Technique—shows *(below)* not only timing but also relationships among the tasks of a project. The relationships are represented by lines that connect boxes describing the tasks.

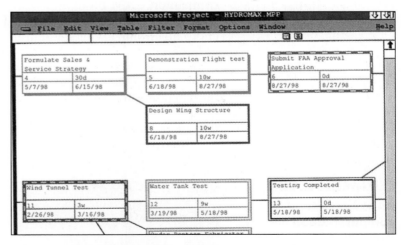

Even project management software has evolved into new forms. For example, a program called ManagePro for Windows is designed to manage not only goals and tasks but also the people charged with achieving them. "I use it to track projects, due dates, and the people who are responsible," says the manager of management information systems at a Lake Tahoe, Nevada, time-share condominium resort. "And then you can get your reports out either on project information, showing progress on all the steps, or a completely different view, showing all the steps that have to be taken by a given individual."[44] The software also offers built-in expert tips and strategies on human management for dealing with employees involved in the project.

Computer-Aided Design

Computers have long been used in engineering design. ***Computer-aided design (CAD) programs* are software programs for the design of products, structures, civil engineering drawings, and maps.** CAD programs, which are available for microcomputers, help architects design buildings and workspaces and engineers design cars, planes, electronic devices, roadways, bridges, and subdivisions. One advantage of CAD software is that the product can be drawn in three dimensions and then rotated on the screen so the designer can see all sides. (■ *See Panel 2.10.*)

Examples of CAD programs for beginners are Autosketch and CorelCAD. One CAD program, Parametric, allows engineers to do "what if" overhauls of designs, much as users of electronic spreadsheets can easily change financial data. This feature can dramatically cut design time. For instance, using Parametric, Motorola was able to design a personal cellular telephone in 9 months instead of the usual 18.[45] Yet not all CAD programs are used by technical types; one version is available that allows a relatively unskilled person to design an office. Other programs are available for designing homes. These programs include "libraries" of options such as cabinetry, furniture, fixtures, and, in the landscaping programs, trees, shrubs, and vegetables.

A variant on CAD is CADD, for *computer-aided design and drafting*, software that helps people do drafting. CADD programs include symbols (points, circles, straight lines, and arcs) that help the user put together graphic elements, such as the floor plan of a house. An example is Autodesk's Auto-CAD.

***CAD/CAM*—for *computer-aided design/computer-aided manufacturing*— software allows products designed with CAD to be input into an automated manufacturing system that makes the products.** For example, CAD and its companion CAM brought a whirlwind of enhanced creativity and efficiency to the fashion industry. Some CAD systems, says one writer, "allow designers to electronically drape digital-generated mannequins in flowing gowns or tailored suits that don't exist, or twist imaginary threads into yarns, yarns into weaves, weaves into sweaters without once touching needle to garment."[46] The designs and specifications are then input into CAM systems that enable robot pattern-cutters to automatically cut thousands of patterns from fabric with only minimal waste. Whereas previously the fashion industry worked about a year in advance of delivery, CAD/CAM has cut that time to 8 months—a competitive edge for a field that feeds on fads.

■ PANEL 2.10

CAD: example of computer-aided design
(*Left*) CAD model of wire connections. (*Right*) CAD modeling of a stadium for the 1996 Olympics in Atlanta, Georgia.

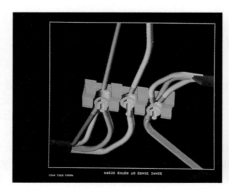

Three-dimensional CAD/CAM programs have also been used to design carpet steam cleaners (estimated time saved on production: 50%), dinnerware that looks like handmade wicker baskets (time saved: 9–18 months), and a guitar designed to the specifications of rock musician Eddie Van Halen (time saved: 3–9 months).[47]

Drawing & Painting Programs

John Ennis was trained in realistic oil painting, and for years he used his skill creating illustrations for covers and dust jackets for book publishers. Now he "paints" using a computer, software, and mouse. The greatest advantage, he says, is that if "I do a brush stroke in oil and it's not right, I have to take a rag and wipe it off. With the computer, I just hit the 'undo' command."[48]

It may be no surprise to learn that commercial artists and fine artists have begun to abandon the paintbox and pen and ink for software versions of palettes, brushes, and pens. The surprise, however, is that an artist can use mouse and pen-like stylus to create computer-generated art as good as that achievable with conventional artist's tools. More surprising, even *nonartists* can be made to look good with these programs.

There are two types of computer art programs: drawing and painting.

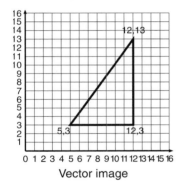

Vector image

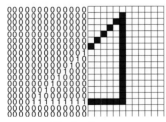

Raster image

- Drawing programs: **A *drawing program* is graphics software that allows users to design and illustrate objects and products.** CAD and drawing programs are similar. However, CAD programs provide precise dimensioning and positioning of the elements being drawn, so that they can be transferred later to CAM programs. Also, CAD programs lack the special effects for illustrations that come with drawing programs. A program named Sketcher, for example, offers 12 different "paper" textures, ranging from plain paper to canvas, and a variety of choices in pencils, chalks, pens, felt pens, markers, and charcoals. Some other drawing programs are CorelDRAW, Adobe Illustrator, and Macromedia Freehand.

  Drawing programs create *vector images (see left)*—images created from mathematical calculations.

- Painting programs: ***Painting programs* are graphics programs that allow users to simulate painting on screen.** A mouse or a tablet stylus is used to simulate a paintbrush. The program allows you to select "brush" sizes, as well as colors from a color palette. Examples of painting programs are MetaCreations' Painter, Adobe Photoshop, Corel PhotoPaint, and JASC's PaintShop Pro.

  Painting programs produce *raster images* made up of little dots *(see left)*.

Not all artists are happy about going electronic. Some think that the process makes illustration too easy—that the advantage of making image after image by hand is that it helps weed out bad ideas. Some find they are more at the mercy of art directors and clients, who, say one pair of artists, "have grown bossier and more intrusive" knowing how easily computer art can be altered.[49] And some artists find they have to work twice as hard to make the same living they did before because prices are coming down as work becomes more efficient.

Groupware

"Organizations will be under continued pressure to deliver products faster, tailored to the specific needs and schedules of customers, with reduced costs," says meetings consultant Jim Creighton. "These pressures will force organizations to be flatter, with more work done in temporary, project-driven teams."[50] The result, says Creighton, is that meetings will increasingly take the place of hierarchical systems of management.

The high-tech way to hold meetings is through the use of groupware. Most microcomputer software is written for people working alone. **Groupware, also known as *collaboration software*, is software that is used on a network and allows users in the same building or even continents away working on the same project to share ideas and update documents.** Groupware improves productivity by keeping you continually notified about what your colleagues are thinking and doing, and they about you. "Like e-mail," one writer points out, "groupware became possible when companies started linking PCs into networks. But while e-mail works fine for sending a message to a specific person or group—communicating one-to-one or one-to-many—groupware allows a new kind of communication: many-to-many."[51]

The standard for groupware was set in 1989 with the introduction of Lotus Notes, which dominates the market today with 9 million users. Other principal programs are Microsoft's Exchange, Novell's GroupWise, and Netscape's Communicator. Original versions of groupware allowed e-mail, discussion groups, customizable databases, scheduling, and network security. Newer versions enable users to get access to Web pages and even send Web pages to each other.

Ethics

Groupware has changed the kind of behavior required for success in an organization. For one thing, it requires workers to take more responsibility. Ethically, of course, when you are contributing to a group project of any kind, you should try to do your best. However, when your contribution to the project is clearly visible to all, as happens with groupware, you have to do your best. In addition, using e-mail or groupware means you need to use good manners and be sensitive to others while you're online.

Multimedia Authoring Software

Competition among cities throughout the world to host the Olympic Games is extremely intense. How did Atlanta win the bid for the 1996 summer games? By making a multimedia presentation that made members of the Olympics site-selection committee feel as if they were "inside" the stadium—which had not even been built.

As mentioned in Chapter 1 (✔ p. 30), *multimedia* refers to technology that presents information in more than one form: text, graphics, animation, video, music, and voice. **Multimedia authoring software, or simply *authoring software*, enables users to create multimedia applications that integrate text, images, sound, motion, and animation.** Basically, authoring tools allow you to sequence and time the occurrence of events, determining which graphics, sound, text, and video files shall come into the action at what point. The software also allows the creator to determine the type and level of user interaction, as in whether an event requires a user to stop and answer "yes" or "no" to a question before the program can proceed.

As we describe in Chapter 6, multimedia applications have a great many exciting uses, from entertainment to training to reference. A good deal of hardware is required to run multimedia software (CD-ROM drives, speakers, sound boards, video boards, and so on), and even more is required for multimedia development, as we discuss later. Among the various kinds of authoring software are Macromedia Authorware, Macromedia Director, Asymetrix IconAuthor, Asymetrix Toolbook, and Action! for Macintosh and Windows.

2.10 When Software Causes Problems

KEY QUESTIONS

What are two principal drawbacks of new applications software, and what can you do about them?

Preview & Review: Software can come with many more features than you'll ever need—this is sometimes called *bloatware*. Software can also come with many flaws—this is called *shovelware*. Instruction manuals, help software, telephone help lines, commercial how-to books, and knowledgeable friends can help users deal with bloatware and shovelware.

In Issaquah, Washington, reported an Associated Press news story, a man was coaxed out of his home by police officers "after he pulled a gun and shot several times at his personal computer, apparently in frustration."[52]

Experienced computer users will recognize the phrase "apparently in frustration" as an understatement. Real feelings of rage, fueling a strong desire to inflict the death penalty on your computer, are an occasional part of the learning experience.

Fortunately, newcomers will likely be spared these agonies with software because they will be operating at rather basic levels (and can seek help in computer classrooms from their instructors or staff members). Experienced users, however, find some new programs can drive them crazy. "I'm tired of reading words like 'easy' and 'intuitive' about computers," rails *San Jose Mercury News* technology columnist Phillip Robinson. "They aren't."[53]

Some Drawbacks of New Software: Bloatware & Shovelware

Not everyone feels as strongly as Robinson. Indeed, not everything is wrong with new programs or, of course, people wouldn't be using them. Nevertheless, let us consider a couple of significant drawbacks that often characterize new software and what you can do about them.

- **"Bloatware"—too many features:** Reports economist Paul Krugman, "I got my first PC in 1984; it displayed text and numbers in any color you wanted, as long as it was green. Today I work on a computer that provides me with . . . dazzling color and hundreds of nifty features, and in most respects is no more useful than my good old I.B.M."[54] Many of the features, Krugman suggests, are like the curlicues, fluted columns, ornamental arches, and encrusted scrolls that cluttered up 19th-century British machinery—and that American machinery makers had the good sense to avoid.

  Some software has become so crowded, or bloated, with features that it has led to the term "bloatware" (or "featuritis"). The reason for all these features, of course, is the software industry's strategy of planned obsolescence—to make buyers want to abandon their old programs and rush out to buy new ones. Software publishers add features to make their products stand out in a crowded marketplace, points out one computer writer. "But things have gotten so far out of hand that companies are adding features just because they can," he says. "That's how come—in my youth—cars had fins."[55]

  Many computer users don't understand how to use all the software features they already have. Yet when Microsoft's Office 97 came on the market, supplanting Office 95, it contained *4500 commands*—far more than the few dozen or even few hundred that most people would use.[56]

- **"Shovelware"—full of flaws:** "In the computer world, as in horseshoes, close is good enough," says Stephen Manes, computer columnist for the *New York Times.* "In the computer world, products are considered perfectly acceptable when they almost work the way they are supposed to."[57]

  This second drawback is even more distressing than the first. Software makers are in such a hurry to ship, or shovel, their products out the door to market that it has inspired a new word: "shovelware." And a lot of shovelware was developed so hastily that it doesn't work right.

  Sometimes an applications program doesn't perform because of incompatible standards with other parts of the computer system—for instance, it won't jibe with a newer or older version of the operating system. (We discuss operating systems in Chapter 3.) But all too often the software won't work because it was rushed out in flawed form ("close is good enough"). The software developer expects users to learn how to fix it by calling a telephone help line or by getting an upgraded, improved version from a Web site. "Now, imagine the outcry," writes Manes, "if an auto maker knowingly sold vehicles with defective door locks and offered new ones only if you demanded and installed them yourself."

Help for Software Users

Why devote so much space to a discussion of the difficulties of software and risk frightening newcomers before they've even begun? Because it's easy for users (especially newcomers) to think that the reason they can't make a program work lies with their own ineptness—when actually the culprit is the manufacturer making slipshod goods.

There are, however, a number of sources of user assistance—although even here, forewarned is forearmed.

- **Instruction manuals:** User guides or instruction manuals printed on paper have traditionally accompanied a box of applications software diskettes or CD-ROMs. Sometimes these are just what you need. Sometimes, however, you need to know what you need help with, and it may be difficult to define the problem. In addition, these guides may be a puzzle to anyone not trained as a programmer because they are often written by the very people who developed the software. And experts are not always able to anticipate the problems of the inexpert.

  More recently, software manufacturers have been giving short shrift to print-on-paper manuals. Says business-technical writer John Merchant, "as the software has become more complex, the [paper] user guides have become slimmer. . . . Apparently, the drastically reduced time allowed to get new software products to market does not allow for comprehensive manuals to be written."[58]

- **Help software:** In place of paper manuals, software publishers are now relying more on "Help" programs on a diskette or CD-ROM accompanying the software, which contain a series of Help menus. However, these, too, may feature technospeak that is baffling to newcomers. Also, like paper manuals, help programs may suffer the drawbacks of having been rushed out at the last minute.

  Help programs may also be available through the Internet. The problem with trying to use the Net or the Web, of course, is: How do you go online to solve the problem of your computer not working if your computer isn't working? Or how do you access the Internet to get advice on how to fix your software and simultaneously fix your software? (It helps to have two computers.)

- **Telephone help lines:** If you encounter a software glitch, it's possible you could use a help line to call a customer-support technician at the program's manufacturer. The tech would then try to talk you through the problem or might refer you to a Web site from which you could obtain a file that would remedy your problem.

  Both software and hardware makers have long offered telephone help lines for customers. If your computer breaks down, you may find that your hardware company still offers emergency advice for free (although you may have to pay the phone company's long-distance tolls). However, software manufacturers now invariably charge for help calls if you've had the software longer than 30–90 days. This could run to $25 or $35 per call or to around $200 or $400 a year if you buy an upfront customer-service contract allowing you, say, 10–35 calls in a 12-month period.

  An analyst for Dataquest, a market-research firm, estimates that microcomputer users placed 200 million calls in 1995 to ask software questions. He also found the average time a user had to wait on hold was about 3 minutes.[59] Actually, it's quite common to wait for 20 or 30 minutes—even at midnight—listening to bubbly music and a recorded voice urging you to "stay on the line for the next available technician."

- **Commercial how-to books:** Because of the inadequacies in user support by the software developers themselves, an entire industry has sprung up devoted to publishing how-to books. These are the kind of books found both in computer stores and in general bookstores such as Barnes & Noble or Borders. Examples are the "For Dummies" or "Complete Idiot's" books (such as *Adobe PhotoShop for Dummies* and *The Complete Idiot's Guide to Microsoft Office 97 Professional*).

  It needs to be said, however, that many of these are guides to how to *use* a particular program—that is, they are tutorials, rather than fix-it manuals telling you how to make the software *work right.*

- **Knowledgeable friends:** Believe it or not, nothing beats having a knowledgeable friend: your instructor, a student more advanced than you, or someone with a technical interest in computers. Businesspeople pay millions of dollars to have computer technicians come into their offices and deal with software and hardware problems. However, many users can't afford pricey consultants. Thus, we can't stress enough how important it is to get to know people— from your classes, from computer user groups (including online Internet groups), from family friends, or whatever—who can lend aid and expertise when your computer software gives you trouble. As it assuredly will at some point. (And don't forget to do something for them in return.)

A final word: If you're happy with a particular applications program, don't feel you have to go out and upgrade to the latest version just because everyone around you seems to be doing so. The ultimate test, after all, is: How useful is the software for *you?* You don't necessarily need to be influenced by the argument that something is "better, faster, and incorporates all the latest technology."

2.11 Ethics & Intellectual Property Rights: When Can You Copy?

KEY QUESTIONS

When is copying a violation of copyright laws, what is a software license agreement, and what types of agreements are there?

Preview & Review: Intellectual property consists of the products of the human mind. Such property can be protected by copyright, the exclusive legal right that prohibits copying it without the permission of the copyright holder.

Software piracy, network piracy, and plagiarism violate copyright laws.

Public domain software, freeware, and shareware can be legally copied, which is not the case with proprietary (commercial) software.

Ethics

Information technology has presented legislators and lawyers—and you—with some new ethical questions regarding rights to intellectual property. *Intellectual property* consists of the products, tangible or intangible, of the human mind. There are three methods of protecting intellectual property. They are *patents* (as for an invention), *trade secrets* (as for a formula or method of doing business), and *copyrights* (as for a song or a book).

What Is a Copyright?

Of principal interest to us is copyright protection. **A *copyright* is the exclusive legal right that prohibits copying of intellectual property without the permission of the copyright holder.** Copyright law protects books, articles, pamphlets, music, art, drawings, movies—and, yes, computer software. Copyright protects the *expression* of an idea but not the idea itself. Thus, others may copy your idea for, say, a new video game but not your particular variant of it. Copyright protection is automatic and lasts a minimum of 50 years; you do not have to register your idea with the government (as you do with a patent) in order to receive protection.

These matters are important because the Digital Age has made the act of copying far easier and more convenient than in the past. Copying a book on a photocopier might take hours, so people usually feel they might as well buy the book. Copying a software program onto another diskette, however, might take just seconds.

Piracy, Plagiarism, & Ownership of Images & Sounds

Three copyright-related matters deserve our attention: software and network piracy, plagiarism, and ownership of images and sounds.

Software and Network Piracy   It may be hard to think of yourself as a pirate (no sword or eyepatch) when all you've done is make a copy of some commercial software for a friend. However, from an ethical standpoint, an act of piracy is like shoplifting the product off a store shelf.

Piracy is theft or unauthorized distribution or use. **Software piracy is the unauthorized copying of copyrighted software.** One way is to copy a program from one diskette to another. Another is to download (transfer) a program from a network and make a copy of it. **Network piracy is using electronic networks for the unauthorized distribution of copyrighted materials in digitized form.** Record companies, for example, have protested the practice of computer users' sending unauthorized copies of digital recordings over the Internet. Both types of piracy are illegal.

The easy rationalization is to say that "I'm just a poor student, and making this one copy or downloading only one digital recording isn't going to hurt anyone." But it is the single act of software piracy multiplied millions of times that is causing the software publishers a billion-dollar problem. They point out that the loss of revenue cuts into their budget for offering customer support, upgrading products, and compensating their creative people. Piracy also means that software prices are less likely to come down; if anything, they are more likely to go up.

Another point to consider is that many schools and universities cannot afford the high cost of hardware and software and depend in large part on the generosity of software developers like Microsoft, Oracle, IBM, and Asymetrix to donate software. This is a multi-million-dollar benefit to some of the large schools. If the software developers found out that students, staff, and/or faculty were illegally copying (and even reselling) the free software made available to the school, the free products would disappear, causing a devastating loss to the school, because so much education is linked to productivity software.

Anti-copying technology is being developed that, when coupled with laws making the disabling of such technology a crime, will reduce the piracy problem. Some of these anti-copying programs will allow a copy to be made, but the copy will not run. Others will destroy both the original and the copy. Still others necessitate periodic entering of a password for the software to continue operating properly.

Plagiarism *Plagiarism* **is the expropriation of another writer's text, findings, or interpretations and presenting it as one's own.** Information technology puts a new face on plagiarism in two ways. On the one hand, it offers plagiarists new opportunities to go far afield for unauthorized copying. On the other hand, the technology offers new ways to catch people who steal other people's material.

Electronic online journals are not limited by the number of pages, and so they can publish papers that attract a small number of readers. In recent years, there has been an explosion in the number of such journals and of their academic and scientific papers. This proliferation may make it harder to detect when a work has been plagiarized, because few readers will know if a similar paper has been published elsewhere. In addition, the Internet has spawned many companies that market prewritten term papers to students— a practice that some states are making illegal.

Yet information technology may also be used to identify plagiarism. Scientists have used computers to search different documents for identical passages of text. In 1990, two "fraud busters" at the National Institutes of Health alleged after a computer-based analysis that a prominent historian and biographer had committed plagiarism in his books. The historian, who said the technique turned up only the repetition of stock phrases, was later exonerated in a scholarly investigation.[60]

Ownership of Images and Sounds Computers, scanners, digital cameras, and the like make it possible to alter images and sounds to be almost anything you want. What does this mean for the original copyright holders? Images can be appropriated by scanning them into a computer system, then altered or placed in a new context.

The line between artistic license and infringement of copyright is not always clearcut. In 1993, a federal appeals court in New York upheld a ruling against artist Jeff Koons for producing ceramic art of some puppies. It turned out that the puppies were identical to those that had appeared in a postcard photograph copyrighted by a California photographer.[61] But what would have been the judgment if Koons had scanned in the postcard, changed the colors, and rearranged the order of the puppies to produce a new postcard?

In any event, to avoid lawsuits for violating copyright, a growing number of artists who have recycled material have taken steps to protect themselves. This usually involves paying flat fees or a percentage of their royalties to the original copyright holders.

These are the general issues you need to consider when you're thinking about how to use someone else's intellectual property in the Digital Age. Now let's see how software fits in.

Public Domain Software, Freeware, & Shareware

No doubt most of the applications programs you will study in conjunction with this book will be commercial software packages, with brand names such as Microsoft Word or Excel. However, there are a number of software products—many available over communications lines from the Internet—that are available to you as *public domain software, freeware,* or *shareware.*

- **Public domain software:** *Public domain software* **is software that is not protected by copyright and thus may be duplicated by anyone at will.** Public domain programs—usually developed at taxpayer expense by government agencies—have been donated to the public by their creators. They are often available through sites on the Internet (or electronic bulletin boards) or through computer users groups. A users group is a club, or group, of computer users who share interests and trade information about computer systems.

  You can duplicate public domain software without fear of legal prosecution. (Beware: Downloading software through the Internet may introduce some problems in the form of bad code called viruses into your system. We discuss this problem, and how to prevent it, in Chapters 5 and 12.)

- **Freeware:** *Freeware* **is software that is available free of charge.** Freeware is distributed without charge, also usually through the Internet or computer users groups.

  Why would any software creator let the product go for free? Sometimes developers want to see how users respond, so they can make improvements in a later version. Sometimes it is to further some scholarly purpose, such as to create a standard for software on which people are apt to agree because there is no need to pay for it. An example of freeware is Mosaic, mentioned at the start of this chapter (✔ p. 48).

  Freeware developers often retain all rights to their programs, so that technically you are not supposed to duplicate and distribute them further. Still, there is no problem with your making several copies for your own use.

- **Shareware:** *Shareware* **is copyrighted software that is distributed free of charge but requires users to make a contribution in order to continue using it.** Shareware, too, is distributed primarily through communications connections such as the Internet. An example is WinZIP, a program for managing computer files, which you can obtain on the Internet.

  Sometimes "trialware" versions of shareware are available via the Internet. This type of software is limited in functionality (it's a watered-down version of the real thing), or the period of time that the software will work is self-limited. After the trial period is over, the user pays a fee to obtain the fully functional version, or the software simply becomes useless.

Is there any problem with making copies of shareware for your friends? Actually, the developer is hoping you will do just that. That's the way the program gets distributed to a lot of people—some of whom, the software creator hopes, will make a "contribution" or pay a "registration fee" for advice or upgrades.

Though copying shareware is permissible, because it is copyrighted you cannot use it as the basis for developing your own program in order to compete with the developer.

Proprietary Software & Types of Licenses

***Proprietary software* is software whose rights are owned by an individual or business,** usually a software developer. The ownership is protected by the copyright, and the owner expects you to buy a copy in order to use it. The software cannot legally be used or copied without permission.

Software manufacturers don't sell you the software so much as sell you a license to become an authorized user of it. What's the difference? In paying for a *software license,* **you sign a contract in which you agree not to make copies of the software to give away or for resale.** That is, you have bought only the company's permission to use the software and not the software itself. This legal nicety allows the company to retain its rights to the program and limits the way its customers can use it.[62] The small print in the licensing agreement allows you to make one copy (backup copy or archival copy) for your own use.

There are several types of licenses:

- **Shrink-wrap licenses:** *Shrink-wrap licenses* **are printed licenses inserted into software packages and visible through the clear plastic wrap.** The use of shrink-wrap licenses eliminates the need for a written signature, since buyers know they are entering into a binding contract by merely opening the package. Each shrink-wrap license is for a single system. (Sometimes the license allows for duplication between a desktop computer and a portable computer.)

- **Site licenses:** A *site license* **permits a customer to make multiple copies of a software product for use just within a given facility,** such as a college computer lab or a particular business.

- **Concurrent-use licenses:** A *concurrent-use license* **allows a specified number of software copies to be used within a given facility at the same time.** For example, if a concurrent-use license is given for 10 users within a company, then *any* 10 users in the firm may use the software at the same time.

MICROSOFT® Office 97 Professional Edition   **Version 8.0**

END-USER LICENSE AGREEMENT FOR MICROSOFT SOFTWARE

IMPORTANT—READ CAREFULLY: This End-User License Agreement ("EULA") is a legal agreement between you (either an individual or a single entity) and the manufacturer ("PC Manufacturer") of the computer system ("COMPUTER") with which you acquired the Microsoft software product(s) identified above ("SOFTWARE PRODUCT" or "SOFTWARE"). If the SOFTWARE PRODUCT is not accompanied by a new computer system, you may not use or copy the SOFTWARE PRODUCT. The SOFTWARE PRODUCT includes computer software, the associated media, any printed materials, and any "online" or electronic documentation. By installing, copying or otherwise using the SOFTWARE PRODUCT, you agree to be bound by the terms of this EULA. If you do not agree to the terms of this EULA, PC Manufacturer and Microsoft Corporation ("Microsoft") are unwilling to license the SOFTWARE PRODUCT to you. In such event, you may not use or copy the SOFTWARE PRODUCT, and you should promptly contact PC Manufacturer for instructions on return of the unused product(s) for a refund.

SOFTWARE PRODUCT LICENSE

The SOFTWARE PRODUCT is protected by copyright laws and international copyright treaties, as well as other intellectual property laws and treaties. The SOFTWARE PRODUCT is licensed, not sold.

1. **GRANT OF LICENSE.** This EULA grants you the following rights:
- **Software.** You may install and use one copy of the SOFTWARE PRODUCT on the COMPUTER.
- **Storage/Network Use.** You may also store or install a copy of the computer software portion of the SOFTWARE PRODUCT on the COMPUTER to allow your other computers to use the SOFTWARE PRODUCT over an internal

- **Network single-user licenses:** **A *network single-user license* limits the use of the software in a network to one user at a time.**
- **Network multiple-user licenses:** **A *network multiple-user license* allows more than one person in a network to use the software.** Each user is assigned a license, and only people so licensed may use the network software.

Onward

In this chapter we have described the most common types of applications software. Still to be discussed, however, are some of the truly exciting software developments of the Digital Age. Later in the book, we describe software technology such as virtual reality, multimedia, simulation, expert systems, and information-seeking "agents," as well as exploring further the software affecting the Internet.

README

Practical Matters: What Do Colleges Want Students to Know About Copying?

Most colleges and universities think written computer policies are important because they give notice and provide for due process if a student breaks a rule. We can't pretend to know what the rules at your institution are, but in general the prohibitions regarding copying are as follows.

Rule #1: You can't copy licensed software without permission: Of the 523 million new business software applications used globally in 1996, according to the Business Software Alliance, nearly one in two were copied—that is, were pirated.[63] Were it not for this widespread theft, software developers could cut the price of their wares in half.

Some students know that software piracy is against the law, but they do it anyway because they think they're getting back at "greedy" software companies, says Brian Rust of the division of information technology at the University of Wisconsin.[64] In addition, he points out, some students may even observe instructors or staff members copying software programs, saying they're on a tight budget. "People will seek ways to justify their behavior," Rust notes.

Regardless, unless you have permission to do so, you're not allowed to copy and distribute diskettes of licensed applications software.

Rule #2: You can't copy copyrighted music, artwork, photographs, videos, or text from the Internet without permission: Now that most music is in a digital format, a song can be posted to the World Wide Web and the quality of its sound will remain consistent no matter how many times the song is reproduced: the 100th digitized copy sounds as pristine as the first. Thus the temptation to copy.

Indeed, until now it has not been at all difficult to use a search engine to locate a Web site that will yield a free but illegal album of, say, Smashing Pumpkins, U2, or No Doubt.[65] More and more, however, music publishers, film studios, software publishers, and law-enforcement agencies are using cybersleuths to track down Web sites offering bootleg goods.

Rule #3: You can't post copyrighted materials online: Students who take, say, "Dilbert" cartoons, *Playboy's* Miss October, or the jazz of Thelonius Monk and post them online may simply not realize that they are doing wrong. But the copyright owners of these materials have contacted campuses to complain about violations of copyright law. Universal Press Syndicate contacted Peter Edstrom, a student at the University of Minnesota at Duluth, and forced him to remove "Calvin and Hobbes" images from the home page he had created on the university's network.[66] (Incidentally, students who use their university's logo on their Web pages are probably violating copyright if they didn't get permission first.)

The penalties: At most colleges and universities, penalties for breaking the rules can include expulsion for students. Companies may sue both students and their institutions for copyright infringement. Law-enforcement officials can follow with criminal prosecution.[67]

Getting Started with Computers in College & Going Online

Students who come to college with a personal computer as part of their luggage are certainly ahead of the game. If you don't have one, however, there are other options.

If You Don't Own a Personal Computer

If you don't have a PC, you can probably borrow someone else's sometimes. However, if you have a paper due the next day, you may have to defer to the owner, who may also have a deadline. When borrowing, then, you need to plan ahead and allow yourself plenty of time.

Virtually every campus now makes computers available to students, either at minimal cost or essentially for free as part of the regular student fees. This availability may take two forms:

- **Library or computer labs:** Even students who have their own computers may sometimes want to use the computers available at the library or campus computer lab. These may have special software or better printers than most students have.
- **Dormitory computer centers or dorm-room terminals:** Some campuses provide dormitory-based computer centers (for example, in the basement). Even if you have your own computer, it's nice to know about these for backup purposes.

  More and more campuses are also providing computers or terminals within students' dormitory rooms. These are usually connected by a campuswide local area network (LAN) to lab computers and administrative systems. Often, however, they also allow students to communicate over phone lines to people in other states.

  Of course, if the system cannot accommodate a large number of students, all the computers may be in high demand come term-paper time. Clearly, owning a computer offers you convenience and a competitive advantage.

If You Do Own a Personal Computer

Perhaps someone gave you a personal computer, or you acquired one, before you came to college. It will probably be one of two types: (1) a PC, such as from IBM, Gateway, Compaq, Dell, Hewlett-Packard, Packard Bell NEC; or (2) an Apple Macintosh.

If all you need to do is write term papers, nearly any microcomputer will do. Indeed, you may not even need to have a printer, if you can find other ways to print out things. The University of Michigan, for instance, offers "express stations" or "drive-up windows." These allow students to use a diskette or connect a computer to a student-use printer to print out their papers. Or, if a friend has a compatible computer, you can ask to borrow it and the printer for a short time to print your work.

You should, however, take a look around you to see *if your present system is appropriate for your campus and your major.*

- **The fit with your campus:** Most campuses are known as PC schools; however, "Macs" (Apple Macintoshes) are used at some schools. Why should choice of machine matter? The answer is that diskettes can't always be read interchangeably among the two main types of microcomputers. Thus, if you own the system that is out of step for your campus, you may find it difficult to swap files or programs with others. Nor will you be able to borrow their equipment to finish a paper if yours breaks down. Call the dean of students' office or otherwise ask around to find which system is most popular.
- **The fit with your major:** Speech communications, foreign language, physical education, political science, biology, and English majors probably don't need a fancy computer system (or even any system at all). Business, engineering, architecture, and journalism majors may have special requirements. For instance, an architecture major doing computer-aided design (CAD) projects or a journalism major doing desktop publishing will need reasonably powerful systems. A history or nursing major, who will mainly be writing papers, will not. Of course, you may be presently undeclared or undecided about your major. Even so, it's a good idea to find out what kinds of equipment and programs are being used in the majors you are contemplating.

How to Get Your Own Personal Computer

Buying a personal computer, like buying a car, often requires making a trade-off between power and expense.

Power Many computer experts try to look for a personal computer system with as much power as possible. The word *power* has different meanings when describing software and hardware:

- **Powerful software:** Applied to software, "powerful" means that the program is flexible. That is, it can do many different things. For example, a word processing program that can include graphics in documents is more powerful than one that cannot.

- **Powerful hardware:** Applied to hardware, "powerful" means that the equipment (1) is fast and (2) has *great capacity.*

  A fast computer will process data more quickly than a slow one. With an older computer, for example, it may take several seconds to save, or store on a disk, a 50-page term paper. On a newer machine, it might take less than a second.

  A computer with great capacity can run complex software and process voluminous files. *This is an especially important matter if you want to be able to run the latest releases of software.*

Will computer use make up an essential part of your major, as it might if you are going into engineering, business, or graphic arts? If so, you may want to try to acquire powerful hardware and software. People who really want (and can afford) their own desktop publishing system might buy a new Macintosh PowerPC with color ink-jet printer, scanner, and Quark page makeup software and PhotoShop photo manipulation software, among other programs. This might well cost $8000. Most students, of course, cannot afford anything close to this.

Expense If your major does not require a special computer system, a microcomputer can be acquired for relatively little.

What's the *minimum* you should get? Probably a microcomputer with 32 megabytes of memory (✔ p. 16) and one diskette drive, one hard-disk drive, a CD-ROM drive, and an external hard-disk unit such as a Zip drive with removable Zip cartridges to use for backup. (However, 64 or more megabytes of memory may be needed if you're going to run graphic-intensive programs. Because software memory requirements are increasing, check with some knowledgeable friends about recommended minimum memory before buying a computer.)

Buying a New Computer Fierce price wars among microcomputer manufacturers and retailers have made hardware more affordable. One reason PCs have become so widespread is that non-PC microcomputer manufacturers early on were legally able to copy, or "clone," IBM machines (the original PCs) and offer them at cut-rate prices. For a long time, Apple Macintoshes were considerably more expensive, although this has changed somewhat. (In part this was because other manufacturers were unable to offer inexpensive Mac clones.)

When buying hardware, make sure the system software that comes with it is at least Windows 95 for a PC and at least System 8 for the Macintosh. Also, look to see if any applications software, such as word processing or spreadsheet programs, comes "bundled" with it. In this case, *bundled* means that software is included in the selling price of the hardware. This arrangement can be a real advantage, saving you several hundred dollars. *San Jose Mercury News* computer columnist Phillip Robinson offers several other suggestions for saving money in buying a new PC and software. (■ *See Panel 2.11.)*

There are several sources for inexpensive new computers, as follows:

- **Student-discount sources:** With a college ID card, you're probably entitled to a student discount (usually 10 to 20%) through the campus bookstore or college computer resellers. In addition, during the first few weeks of the term, many campuses offer special sales on computer equipment. Campus resellers also provide on-campus service and support and can help students meet the prevailing campus standards while satisfying their personal needs.

  Note that some private educational discount companies can sell software to students that is discounted up to and above 70%. A few examples are Solomon Computer Student Discount Offers *(www.surfshop.net/users/solomon/student.htm),* Student Discount Network *(www.discount-net.com),* Software Services *(www.swservices.com/student.htm),* and Indelible Blue *(www.indelible-blue.com/press/academic.html).*

- **Computer superstores:** These are big chains such as Computer City, CompUSA, and Microage. Computers are also sold at department stores, warehouse stores such as Costco and Sam's Club, office-supply chains such as Staples and Office Depot, and electronics stores such as the Good Guys and Circuit City.

- **Mail-order houses:** Companies like Dell Computer and Gateway 2000 found they could sell computers inexpensively while offering customer support over the phone. Their success inspired IBM, Compaq, and others to plunge into the mail-order business.

  The price advantage of mail-order companies has eroded with the rise of computer superstores. Moreover, the lack of local repair and service support can be a major disadvantage. Still, if you're interested in this route, look for a copy of the phone-book-size magazine *Computer Shopper,* which carries ads from most mail-order vendors. (Make sure the mail-order house offers warranties and brand-name equipment.)

Checklist Here are some decisions you should make before acquiring a computer:

- **What software will I need?** Although it may sound backward, you should select the software before the hardware. This is because you want to choose software that will perform the kind of work you want to do. First find the kind of programs you want—word processing, spreadsheet, Web browser, PIM, graphics, or whatever. Check out the memory and other hardware requirements for those programs. Then make sure you get a system to fit them.

  The advice to start with software before hardware has always been standard for computer buyers. However, it is becoming increasingly important as programs with extensive graphics come on the market. Graphics tend to require a lot of memory, hard-disk storage, and screen display area.

Saving on computers, accessories, and software may be easier than you think: They cost us too much because we're willing to buy too often . . .

Here's my advice, from general principles to special ways to pinch your pennies.

FIRST: Slow down the cycle. Don't buy every upgrade or even every other upgrade. . . .

Go from version 2.0 to 4.0, skipping right over 2.1, 3.0, and 3.1. Such version numbers are about as reliable indicators of feature changes as dress sizes from different manufacturers and countries, but at least they give some idea. Think about buying a new computer only as often as you buy a new car.

SECOND: Don't buy the absolute best the day it comes out. If you really, really want to have it, wait six months. Then it will only be second or third best, and will cost a fair amount less—because some other, glossier new "state of the art" tool will be available. . . .

THIRD: Avoid paying full price for software. Most important programs—especially for business and office "productivity"—are sold two ways: as full packages and as upgrades. The full package might cost $500 and the upgrade just $100. . . .

Read the fine print on the "upgrade" offer. Usually, you qualify for an upgrade if you use an earlier version or a long list of competing programs, one of which may already be on your hard drive or have come free with your scanner or modem. . . .

FOURTH: Buy technical support before technology. Unless you're an expert, you're better off with a 24-hour toll-free line than the fastest processor. Make sure that support line actually has live people on the other end, not just recordings or fax-back help. If you're doing a lot of home computing, look for weekend support hours.

Ask if the tech support covers everything, because some hardware companies are now excluding bundled software. . . .

FIFTH: Look at the warranty. If it is 90 days, don't buy. If it is one year, ask the salespeople if the item is really worth the price if it will only last a year. If they assure you it will last longer, then why won't the warranty? . . .

SIXTH: Negotiate. Not all computer store or mail-order companies will bargain with you, but some will. They'll cut the price, beef up some hardware component, throw in some software or supplies, extend a warranty, etc. Always be ready to go elsewhere, or to another 800 number, and do so if you can't make a deal. . . .

SEVENTH: Beware restockings. Some companies guarantee that you can return an item in 30 days for any reason, and then charge you a "restocking fee"—sometimes hundreds of dollars—to take it back. Don't buy if there's a restocking fee.

EIGHTH: Ask if batteries or other needed parts are included. This goes for cables, too. Are they any other elements you'll need? Each might only cost $10 or $20, but that adds up, and some companies toss them into the computer, printer, or software box at no extra charge.

NINTH: Check the price of consumables. Your computer system may soon feel like the hungriest mouth to feed in the house. The printer is the worst offender, with many of the latest printers intentionally designed to send you back to the store regularly to buy $40 ink cartridges and $1 sheets of special paper. . . .

BONUS CATEGORY: Rebates—I discovered, in talking with several tech companies, that only about 20% of buyers ever send in a manufacturer's rebate coupon—even if that coupon is for $50 or more. Companies love this because they can advertise those "Only $200" prices with small print saying "after manufacturer's rebate," and yet rarely have to pay the rebate. Send that coupon in as soon as you get home.

—Phillip Robinson, "There Are Many Ways to Pinch Pennies on PCs," *San Jose Mercury News.*

■ **PANEL 2.11**

Saving on computers, accessories, and software

- **Do I want a desktop or a portable?** Look for a computer that fits your work style. For instance, you may want a portable if you spend a lot of time at the library. Some students even use portables to take notes in class. If you do most of your work in your room or at home, you may find it more comfortable to have a desktop PC. Though not portable, the monitors of desktop computers are usually easier to read.

  Actually, however, portables have come so far along that you'll probably have no trouble reading the screens on the latest models. Keep in mind, however, that the keyboards on portables are smaller. (We consider portables in more detail in the Experience Box at the end of Chapter 4.)

  Whatever type of computer you buy, make sure it is an established brand-name computer.

- **Is upgradability important?** The newest software being released is so powerful (meaning flexible) that it requires increasingly more powerful hardware. That is, the software requires hardware that is fast and has great main memory and storage capacity. Be sure to ask the salesperson how the hardware can be upgraded to accommodate increased memory and storage needs later.

- **Do I want a PC or a Macintosh?** Although the situation is changing, until recently the division between PCs on the one hand and Apple Macintoshes on the other was fundamental. Neither could run the other's software or exchange files of data without special equipment and software. We mentioned that some campuses and some academic majors tend to favor one type of microcomputer over the other. Outside of college, however, the business world tends to be dominated by PCs. In a handful of areas—education, graphic arts, and desktop publishing, for example— Macintoshes are preferred.

Getting Started Online

Computer networks have transformed life on campuses around the country, becoming a cultural and social force affecting everybody. How do you join this vast world of online information and interaction?

Hardware & Software Needed Besides a microcomputer with a hard disk, you need a modem to send messages from one computer to another via a phone line. Nowadays modems come installed on most computers. If not, you can have a store install an internal modem as an electronic circuit board on the inside of the computer, or you can buy an external modem, a box-shaped unit that is hooked up to the outside of the computer.

To go online, you'll need communications software, which may come bundled with any computer you buy or is sold on diskettes or CD-ROMs in computer stores. However, many modems come with communications software when you buy them. Or, if you sign up for an online service, it will supply the communications program you need to use its network.

Your modem will connect to a standard telephone wall jack. (When your computer is in use, it prevents you from using the phone, and callers trying to reach you will hear a busy signal. If you have call waiting, it will cut off your online session and ring the call through. Check with your telephone company about how to turn off call waiting when you are online.)

Getting Connected: Starting with an Online Service Unless you already have access to a campus network, probably the easiest first step for using this equipment is to sign up with a commercial online service, such as America Online, CompuServe, Microsoft Network, or Prodigy.

You'll need a credit card in order to join because online services charge a monthly fee, typically about $20. All online services have introductory offers that allow you a free trial period. You can get instructions and a free start-up communications disk by phoning their toll-free 800 numbers. You'll also find promotional offers at computer stores or promotional diskettes shrink-wrapped inside computer magazines on newsstands.

The chart on the next page lists the leading online services, number of users in 1998, costs, phone numbers, and other details. Rates are subject to change. All have special introductory offers.

Some Sources of Online Information About Hardware and Software

HARDWARE

Company	Telephone	Web Address
Acer America	800-733-ACER	http://www.acer.com/info
Apple Computer	800-538-9696	http://www.apple.com
AST Research	800-945-2278	http://www.ast.com
Compaq Computer	800-345-1518	http://www.compaq.com
Dell Computer	800-289-3355	http://www.dell.com
Gateway 2000	800-846-2000	http://www.gateway.com
Hewlett-Packard	800-724-6631	http://www.hpresource.com
IBM	800-426-2968	http://www.us.pc.ibm.com
Micron Electronics	800-862-6035	http://www.micronpc.com
NEC Technologies	800-NEC-INFO	http://www.nec.com
Toshiba	800-334-3445	http://www.computers.toshiba

SOFTWARE

Company	Telephone	Web Address
Adobe Systems	800-649-3875	http://www.adobe.com
Claris	508-727-8227	http://www.claris.com
Corel	800-772-6735	http://www.corel.com
Intuit	800-446-8848	http://www.intuit.com
Lotus Development	800-343-5414	http://www.lotus.com
Microsoft	800-426-9400	http:www.microsoft.com
Netscape Communications	800-638-7483	http://www.netscape.com

America Online   800-827-6364
www.aol.com
50 free hours with download of trial software; then $21.95 per month.
Call the 800 number for CD-ROM version.

CompuServe   800-848-8199
www.compuserve.com
Order online or via phone. 1 month free with trial software; then $24.95 per month for unlimited use, or $9.95 per month for 5 hours and $2.95 per hour thereafter.

Prodigy   800-776-3449
www.prodigy.com
Order online or via phone. 1 month free, then $19.95 per month.

Microsoft Network   800-free-msn
www.microsoft.com
Order online or via phone. $69.50 per year for unlimited use, or $19.95 per month for unlimited use, or $6.95 per month for 5 hours and $2.50 per hour thereafter.

Notes: Check the service's system requirements for running their software before you order it. If you download the software online, ask the service how long it will take. (If you have a slow modem, it could take hours!) All software comes in PC versions and Mac versions.

Summary

What It Is/What It Does	Why It's Important
analytical graphics (p. 63, KQ 2.4) Also called *business graphics;* graphical forms representing numeric data. The principal examples are bar charts, line graphs, and pie charts. Analytical graphics programs are a type of applications software.	Numeric data is easier to analyze in graphical form than in the form of rows and columns of numbers, as in electronic spreadsheets.
applications software (p. 49, KQ 2.1) Software that solves a particular problem, performs useful work on tasks, or provides entertainment.	Applications software such as word processing, spreadsheet, database manager, graphics, and communications packages are used to increase people's productivity, and video games have become a big business by providing new forms of entertainment with the side effect of teaching new users how to use a computer.
button (p. 57, KQ 2.2) Simulated on-screen button (kind of icon) that is activated ("pushed") by a mouse or other pointing device to issue a command.	Buttons make it easier for users to enter commands.
communications software (p. 68, KQ 2.7) Applications software that manages the transmission of data between computers. Also called *data communications software.*	Communications software is required to transmit data via modems in a communications system.
computer-aided design (CAD) (p. 78, KQ 2.9) Applications software programs for designing products and structures and making civil engineering drawings and maps.	CAD programs help architects design buildings and work spaces and engineers design cars, planes, and electronic devices. With CAD software, a product can be drawn in three dimensions and then rotated on the screen so the designer can see all sides.
computer-aided design/computer-aided manufacturing (CAD/CAM) (p. 78, KQ 2.9) Applications software that allows products designed with CAD to be input into a computer-based manufacturing system (CAM) that makes the products.	CAD/CAM systems have greatly enhanced creativity and efficiency in many industries.
concurrent-use license (p. 87, KQ 2.11) A license that allows a specified number of software copies to be used within a given facility at the same time.	*See software license.*
copyright (p. 84, KQ 2.11) Body of law that prohibits copying of intellectual property without the permission of the copyright holder.	Copyright law aims to prevent people from taking credit for and profiting unfairly from other people's work.

What It Is/What It Does	Why It's Important

cursor (p. 58, KQ 2.3) The movable symbol on the display screen that shows the user where data or commands may be entered next, that is, where the *insertion point* is. The cursor is moved around with the keyboard's directional arrow keys or an electronic mouse.

All applications software packages use cursors to show users where their current work location is on the screen.

cyberspace (p. 68, KQ 2.7) The online or digital world in general and the Internet and its World Wide Web in particular.

By adding communications capabilities to a computer to access cyberspace, the user can reach libraries, databases, and individuals all over the world.

database (p. 64, KQ 2.5) Collection of interrelated files in a computer system that is created and managed by database software. These files are organized so that those parts with a common element can be retrieved easily.

Online database services provide users with enormous research resources. Businesses and organizations use databases to keep track of transactions and increase people's efficiency.

database software (p. 64, KQ 2.5) Applications software for maintaining a database. It controls the structure of a database and access to the data.

Database manager software allows users to organize and manage huge amounts of data.

default settings (p. 60, KQ 2.3) Settings automatically used by a program unless the user specifies otherwise, thereby overriding them.

Users need to know how to change default settings in order to customize their documents.

desktop publishing (DTP) (p. 74, KQ 2.9) Producing high-quality printed output for commercial printing using DTP software and a microcomputer, mouse, and scanner to mix text and graphics, including photos. Text is usually composed first on a word processor, artwork is created with drawing and painting software, and photographs are scanned in using a scanner, or clip-art drawings or photos may be used. A laser or ink-jet printer is used to get an advance look.

Desktop publishing has reduced the number of steps, the time, and the money required to produce professional-looking printed projects.

dialog box (p. 57, KQ 2.2) With graphical user interface (GUI) software, a box that appears on the screen and displays a message requiring a response from the user—for example, Y for "Yes" or N for "No."

Dialog boxes are only one aspect of GUIs that make software easier for people to use.

documentation (p. 57, KQ 2.2) User's guide or reference manual that is a narrative and graphical description of a program. Documentation may be instructional, but usually features and functions are grouped by category.

Documentation helps users learn software commands and use of function keys, solve problems, and find information about system specifications.

drawing program (p. 79, KQ 2.9) Applications software that allows users to design and illustrate objects and products.

Drawing programs and CAD are similar. However, drawing programs provide special effects that CAD programs do not.

electronic mail (e-mail) software (p. 69, KQ 2.7) Software that enables computer users to send letters and files from one computer to another.

E-mail allows businesses and organizations to quickly and easily send messages to employees and outside people without resorting to paper messages.

What It Is/What It Does	Why It's Important

electronic spreadsheet (p. 61, KQ 2.4) Also called *spreadsheet;* applications software that allows users to create tables and financial schedules by entering data and formulas into rows and columns arranged as a grid on a display screen. If data is changed in one cell, values in other cells specified in the spreadsheet will automatically recalculate.

The electronic spreadsheet became such a popular small-business applications program that it has been held directly responsible for making the microcomputer a widely used business tool.

field (p. 65, KQ 2.5) In a database, a field is a unit of data consisting of one or more characters, such as a name or address; a field is contained in a *record.*

A field is the basic unit of information in a database. Searches can be based on any of the fields in a *record,* and records can be sorted by any field.

file (p. 65, KQ 2.5) In a database, a collection of related *records.*

An example of a database file would be records entered in the same period, such as all the driver's license records entered in one day.

financial software (p. 67, KQ 2.6) A growing category of software that ranges from *personal-finance managers* to entry-level accounting programs to business financial-management packages.

Financial software uses the power of computers to handle basic mathematical tasks, giving the user the time and information to decide on a financial course of action.

freeware (p. 86, KQ 2.11) Software that is available free of charge.

Freeware is usually distributed through the Internet. Users can make copies for their own use but are not free to make unlimited copies.

function keys (p. 55, KQ 2.2) Computer keyboard keys that are labeled F1, F2, and so on; usually positioned along the top or left side of the keyboard.

Function keys are used to issue commands. These keys are used differently depending on the software.

graphical user interface (GUI) (p. 56, KQ 2.2) User interface that uses images, menus, buttons, dialog boxes, and windows as well as keystrokes to let the user choose commands, start programs, and see file lists.

GUIs are easier to use than command-driven interfaces and menu-driven interfaces; they permit liberal use of the electronic mouse as a pointing device to move the cursor to a particular icon or place on the display screen. The function represented by the icon can be activated by pressing ("clicking") buttons on the mouse.

groupware (p. 80, KQ 2.9) Applications software that is used on a network and allows a group of users working together on the same project to share ideas and update documents. Also called *collaboration software,* it can serve users in the same building or continents apart.

Groupware improves productivity by keeping users continually notified about what colleagues are thinking and doing, and vice versa.

Help option (p. 56, KQ 2.2) Also called *Help screen;* offers on-screen instructions for using software. Help screens are accessed via a function key or by using the mouse to select Help from a menu.

Help screens provide a built-in electronic instruction manual.

icon (p. 56, KQ 2.2) In a GUI, a picture that represents a command, program, file, or task.

The icon's function can be activated by pointing at it with the mouse pointer and clicking on it. The use of icons has simplified the use of computers.

What It Is/What It Does	Why It's Important

integrated software package (p. 71, KQ 2.8) Applications software that combines in one package the commonly used features of several applications programs—usually electronic spreadsheets, word processing, database management, graphics, and communications. Often called "works" programs.

Integrated software packages offer greater flexibility than separate single-purpose programs.

macro (p. 55, KQ 2.2) Software feature that allows a single keystroke or command to be used to automatically issue a predetermined series of keystrokes or commands. Also called a *keyboard shortcut.*

Macros increase productivity by consolidating several command keystrokes into one or two.

menu (p. 56, KQ 2.2) List of available commands displayed on the screen.

Menus are used in graphical user interface programs to make software easier for people to use.

multimedia authoring software (p. 80, KQ 2.9) Applications software that enables users to create multimedia applications that integrate text, images, sound, motion, and animation. Also called *authoring software.*

Offers tools that make it possible to sequence and time the files used in a multimedia presentation, and to add user interaction screens.

network multiple-user license (p. 88, KQ 2.11) A software license that allows more than one person in a network to use the software, but each user must be licensed.

See software license.

network piracy (p. 84, KQ 2.11) The use of electronic networks for unauthorized distribution of copyrighted materials in digitized form.

If piracy is not controlled, people may not want to let their intellectual property and copyrighted material be dealt with in digital form.

network single-user license (p. 88, KQ 2.11) A software license that limits the use of software in a network to one user at a time.

See software license.

painting program (p. 79, KQ 2.9) Applications programs that simulate painting on the screen using a mouse or tablet stylus like a paintbrush.

Painting programs can render sophisticated illustrations.

personal-finance manager (p. 67, KQ 2.6) Applications software that helps users track income and expenses, write checks, do online banking, and plan financial goals.

Personal-finance software can help people manage their money more effectively.

personal information manager (PIM) (p. 66, KQ 2.5) Applications software to help the user keep track of and manage information used daily, such as addresses, telephone numbers, appointments, "to do" lists, and miscellaneous notes.

PIMs offer an electronic version of essential daily tools all in one place.

plagiarism (p. 85, KQ 2.11) Expropriation of another writer's text, findings, or interpretations and presenting them as one's own.

Information technology offers plagiarists new opportunities to go far afield for unauthorized copying, yet it also offers new ways to catch these people.

Applications Software

What It Is/What It Does	Why It's Important

presentation graphics (p. 76, KQ 2.9) Professional-looking graphics used to communicate information to others, such as in training, sales, formal reports, and kiosks. Presentation graphics are part of presentation software, which uses data/information and graphics from other software tools to create a presentation.

Presentation graphics programs may make use of analytical graphics—bar, line, and pie charts—but they look much more sophisticated, using texturing patterns, complex color, and dimensionality.

project management software (p. 77, KQ 2.9) Applications software used to plan, schedule, and control the people, costs, and resources required to complete a project on time.

Project management software increases the ease and speed of planning and managing complex projects.

proprietary software (p. 87, KQ 2.11) Software whose rights are owned by an individual or business.

Ownership of proprietary software is protected by copyright. This type of software must be purchased to be used. Copying is restricted.

public domain software (p. 86, KQ 2.11) Software that is not protected by copyright and thus may be duplicated by anyone at will.

Public domain software offers lots of software options to users who may not be able to afford a lot of commercial software. Users may make as many copies as they wish.

recalculation (p. 62, KQ 2.4) In electronic spreadsheets, the process of recomputing values automatically when data changes.

Recalculation is what makes electronic spreadsheets valuable: as information changes, the user just enters the data and the computer does the recalculating.

record (p. 65, KQ 2.5) In a database, a collection of related *fields,* such as a name, address, and driver's license number.

The record contains a set of information that can be located by searching for any of the fields in the record.

scrolling (p. 58, KQ 2.2) The activity of moving quickly upward or downward through text or other screen display, using directional arrow keys or mouse.

Normally a computer screen displays only 20–22 lines of text. Scrolling enables users to view an entire document, no matter how long.

search engines (p. 71, KQ 2.7) Software search tools that allow Web users to search for specific documents through keyword searches. Search engine information is collected by software programs.

Search engines make it easy for users to find Web sites they may be interested in.

shareware (p. 86, KQ 2.11) Copyrighted software that is distributed free of charge, usually over the Internet, but that requires users to make a contribution in order to continue using it.

Along with public domain software and freeware, shareware offers yet another inexpensive way to obtain new software.

shrink-wrap license (p. 87, KQ 2.11) Printed licenses inserted into software packages and visible through the clear plastic wrap.

The use of shrink-wrap licenses eliminates the need for a written signature, since buyers know they are entering a binding contract by opening the package.

site license (p. 87, KQ 2.11) License that permits a customer to make multiple copies of a software product for use only within a given facility.

Site licenses eliminate the need to buy many copies of one software package for use in, for example, an office or a computer lab.

What It Is/What It Does	Why It's Important

software license (p. 87, KQ 2.11) Contract by which users agree not to make copies of proprietary software to give away or to sell.

Software manufacturers don't sell people software so much as sell them licenses to become authorized users of the software.

software piracy (p. 84, KQ 2.11) Unauthorized copying of copyrighted software—for example, copying a program from one floppy disk to another or downloading a program from a network and making a copy of it.

Software piracy represents a serious loss of income to software manufacturers and is a contributor to high prices in new programs.

software suite (p. 73, KQ 2.8) Several applications software packages—like spreadsheets, word processing, graphics, communications, and groupware—bundled together and sold for a fraction of what the programs would cost if bought individually. Often called "office" programs.

Software suites can save users a lot of money.

special-purpose keys (p. 55, KQ 2.2) Computer keyboard keys used to enter, delete, and edit data and execute commands—for example, Esc, Alt, and Ctrl.

All computer keyboards have special-purpose keys. The user's software program determines how these keys are used.

tutorial (p. 57, KQ 2.2) Instruction book or program that takes users through a prescribed series of steps to help them learn the product.

Tutorials, which accompany applications software packages, enable users to practice new software in a graduated fashion, thereby saving them the time they would have used trying to teach themselves.

user interface (p. 55, KQ 2.2) Part of a software program that presents on the screen the alternative commands by which the user communicates with the system and that displays information.

Some user interfaces are easier to use than others. Most users prefer a graphical user interface.

Web browser (p. 70, KQ 2.7) Software that enables people to view Web sites on their computers.

Without browser software, users cannot use the part of the Internet called the World Wide Web.

Web directories (p. 71, KQ 2.7) Web indexes classified by topic. Directory information is collected by people rather than by software as in search engines.

Web directories make it easy for users to find Web sites they may be interested in.

window (p. 56, KQ 2.2) Feature of graphical user interfaces; rectangle that appears on the screen and displays information from a particular part of a program.

Using the windows feature, an operating system (or operating environment) can display several windows on a computer screen, each showing a different application program such as word processing, spreadsheets, and graphics.

word processing software (p. 58, KQ 2.3) Applications software that enables users to create, edit, revise, store, and print text material.

Word processing software allows a person to use a computer to easily create, edit, copy, save, and print documents such as letters, memos, reports, and manuscripts.

Exercises

Self-Test

1. A(n) _____ is a window that appears on the screen and requires a response from you.

2. _____ is the activity of moving quickly upward or downward through text or other screen elements.

3. A(n) _____ is a major upgrade in a software product. A(n) _____ is a minor upgrade.

4. The type of software most suited to creating tables and financial schedules is _____ software.

5. _____ software controls how a collection of interrelated files are organized and access to their data.

Short-Answer Questions

1. What is the difference between integrated software and a software suite?

2. What do the abbreviations CAD and CAM mean?

3. List at least three resources for software users who are in need of helpful assistance.

4. What is the difference between software piracy and network piracy?

5. List the five categories of application software.

Multiple-Choice Questions

1. Gantt and PERT charts are used by _____.
   a. desktop publishing software
   b. multimedia authoring software
   c. project management software
   d. drawing programs
   e. painting programs

2. Which of the following would you most likely use to design products, structures, civil engineering drawings, and maps?
   a. communications software
   b. electronic-mail software
   c. computer-aided design software
   d. desktop publishing software
   e. All of the above

3. Which of the following terms is used to describe software that is full of flaws?
   a. bloatware
   b. badware
   c. shovelware
   d. bugware
   e. All of the above

4. Which of the following is distributed free of charge but requires users to make a contribution later on?
   a. public domain software
   b. shareware
   c. freeware
   d. proprietary software
   e. All of the above

5. Which of the following is a type of specialty software?
   a. desktop publishing
   b. presentation graphics
   c. project management
   d. drawing and painting programs
   e. All of the above

True/False Questions

T F 1. Word processing and database management are the two most popular software applications.

T F 2. Although software documentation may be accessed online, today it is usually only available in printed form.

T F 3. A spreadsheet is composed of fields, records, and files.

T F 4. Groupware software allows users located in the same building or in different countries to collaborate on projects and update documents.

T F 5. You can access the World Wide Web using multimedia authoring software.

Knowledge in Action

1. Attend a meeting of a computer users' group in your area. What is the overall purpose of the group? Software support? Hardware support? In what ways? Does it cost money to be a member? How many members are there? How does the group get new members? If you were looking to join a user group, would you be interested in joining this group? Why/why not?

2. What is your opinion about the issue of free speech on an electronic network? Research some recent legal decisions in various countries, as well as some articles on the topic, and then give a short report about what you think. Should the contents of messages be censored? If so, under what conditions?

3. Prepare a short report about how you would use an electronic spreadsheet to organize and manage your personal finances and to project outcomes of changes. What column headings (labels) would you use? Row headings? What formula relationships would you want to establish among which cells? (For example, if your tuition increased by $2000, how would that affect the monthly amount set aside to buy a car or take a trip?)

4. Picture yourself in your future job. What types of current applications software do you see yourself using? What are you producing with this software? What kinds of new applications software would you invent to help you do your job better?

5. Research what is meant by the phrase *digital office*. What would a digital office look like? How is this different from today's offices? What companies are involved in digital-office technologies? Perform your research using the Web and current computer periodicals.

Answers

Self-Test Questions
1. dialog box 2. scrolling 3. version, release 4. electronic spreadsheet 5. database

Short-Answer Questions
1. Integrated software combines the features of several applications into one application. A software suite is several applications that have been bundled together and sold for a discounted price. 2. Computer-aided design (CAD) programs are used in the design of products, structures, civil engineering drawings, and maps. Computer-aided manufacturing (CAM) programs allow products designed with CAD to be input into an automated manufacturing system that actually makes the products. 3. Instruction manuals, help software, telephone help lines, commercial how-to books, knowledgeable friends. 4. Software piracy is the unauthorized copying of copyrighted software. Network piracy is using electronic networks for the unauthorized distribution of copyrighted materials in digitized form. 5. Entertainment software, home/personal software, education/reference software, productivity software, specialty software

Multiple-Choice Questions
1. c 2. c 3. c 4. b 5. e

True/False Questions
1. f 2. f 3. f 4. t 5. f

Extra Information: More About Fonts

Examples of fonts:

ABCDEFGHIJK   10-point Times Roman regular

ABCDEFGHIJK 14-point Ariel Bold

ABCDEFGHIJK 24-point Trump Medieval Italic

Type glossary:

Font   A character set in a particular size and type design (typeface)
Point   Unit of measurement of type size; 12 points equal 1 pica.
Pica   Unit of measurement of page elements, such as margin width; 6 picas equal 1 inch.
Leading   Unit of measurement between lines of type; for example, space (10 + 2 = 12) between lines.
            10/12 lines would be closer together than 10/14 lines.

Type basics:

Many characteristics give fonts different looks—from the ornate to the plain, text-book style. Some fonts are more readable and better for reports and documents. Some are unique or formal and may be better for a logo or invitation.

But how do font designers give their fonts different looks? One of the most common ways to change a font is to add a *serif* or leave the font *sans serif* (without a serif). This serif is a little "foot" or "hat" added to the letters. Designers also can adjust some of the type characteristics, maybe making a loop a little wider, raising an ascender, or giving a jaunty lift to the ear on a *g*. They can change the *pitch* of letters, which is how much horizontal room they get. The pitch may be *fixed* or *monospaced,* meaning each letter gets the same amount of room, or it may be *proportional*, so that the spacing depends on the width of the particular character. Finally, they can give letters or numerals a different weight, which is the thickness, or a different style, such as straight up or italics.

f i tness
Monospaced type

fitness
Proportional type

To give you an idea of what designers have to play around with, here we've assembled a chart illustrating the names of all the parts of a typeface design.

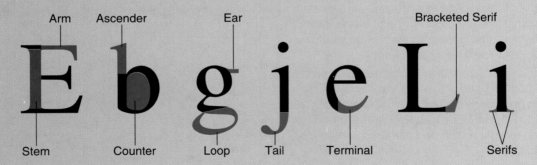

Chapter 3

System Software

The Power Behind the Power

key questions

You should be able to answer the following questions:

3.1   **Three Components of System Software** What are the three basic components of system software?

3.2   **The Operating System** What are the principal functions of the operating system?

3.3   **Common Microcomputer Operating Systems: The Changing Platforms** What are the principal operating systems and operating environments for personal computers, and what are their principal characteristics?

3.4   **Utility Programs: Software to Make Your Computer Run Smoother** How do utility programs interact with and extend an operating system's features and capabilities?

3.5   **The Network Computer: Is the Web Changing Everything?** How could the network computer make the choice of PC operating system irrelevant?

W hat we need is a science called practology, a way of thinking about machines that focuses on how things will actually be
used."

So says Alan Robbins, a professor of visual communications, on the subject of machine *interfaces*—the parts of a machine that people actually manipulate.[1] An interface is a machine's "control panel," ranging from the volume and tuner knobs on an old radio to all the switches and dials on the flight deck of a jetliner. You may have found, as Robbins thinks, that on too many of today's machines—digital watches, VCRs, even stoves—the interface is often designed to accommodate the machine or some engineering ideas rather than the people actually using them. Good interfaces are intuitive—that is, based on prior knowledge and experience—like the twin knobs on a 1940s radio, immediately usable by both novices and sophisticates. Bad interfaces, such as a software program with a bewildering array of menus and icons, force us to relearn the required behaviors every time. Of course, you can prevail over a bad interface if you repeat the procedures often enough.

How well are computer hardware and software makers doing at giving us useful, helpful interfaces? The answer is: getting better all the time, but they still have some leftovers from the past to get rid of. For instance, PC keyboards still come with a SysReq (for "System Request") key, which was once used to get the attention of the central computer but now is rarely used. (The Scroll Lock key is also seldom used.) Software still makes an estimated 60 million Americans wait a minute or two every day when they first boot, or turn on, their computers, which could add up "to as much as 1000 man years a day of lost time," one computer scientist calculates.[2] (Microsoft is working on a feature called OnNow that is intended to eliminate the boot delay.)

In time, as interfaces are refined, it's possible computers will become no more difficult to use than a car. Until then, however, for smoother computing you need to know something about how system software works. Today people communicate one way, computers another. People speak words and phrases; computers process bits and bytes (✔ p. 13). For us to communicate with these machines, we need an intermediary, an interpreter. This is the function of system software. We interact mainly with the applications software, which interacts with the system software, which controls the hardware.

3.1 Three Components of System Software

KEY QUESTION

What are the three basic components of system software?

Preview & Review: System software comprises three basic components: the operating system, utility programs, and language translators.

As we've said, *software*, or *programs*, consists of the instructions that tell the computer how to perform a task (✔ p. 18). Software is of two types—*applications software* and *system software*. Applications software (✔ p. 19) is software that can perform useful work on specific tasks, such as word processing or spreadsheets, or that is used for entertainment. (Applications software was covered in Chapter 2.) **System software enables the applications software to interact with the computer and helps the computer manage its internal and external resources.** System software is required to run applications software; however, the reverse is not true. Buyers of new computers will find the system software has already been installed by the manufacturer.

There are three basic types of system software that you need to know about—*operating systems, utility programs,* and *language translators.*

- **Operating systems:** An operating system is the principal component of system software in any computing system. We describe it at length in the next section.

- **Utility programs:** *Utility programs* are generally used to support, enhance, or expand existing programs in a computer system. Most system software bundles utility programs for performing common tasks such as merging two files into one file or performing backup. Other external, or commercial, utility programs (such as Norton Utilities) are available separately—for example, a utility to recover damaged files. We describe external utility programs later in this chapter.

- **Language translators:** **A *language translator* is software that translates a program written by a programmer in a language such as C—for example, a word processing program—into machine language (0s and 1s), which the computer can understand.** (Programming is covered in Chapter 11.)

The components of system software are diagrammed in the illustration below. (■ *See Panel 3.1.*)

■ PANEL 3.1

The three components of system software

An operating system is required for applications software to run on your computer. The user usually works with the applications software but can bypass it to work directly with the system software for certain tasks.

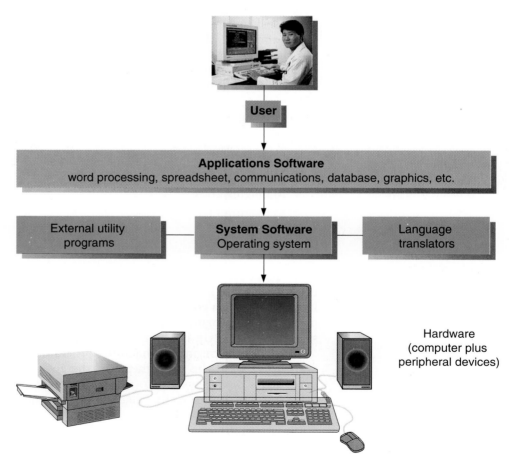

User

Applications Software
word processing, spreadsheet, communications, database, graphics, etc.

External utility programs | **System Software** Operating system | Language translators

Hardware (computer plus peripheral devices)

3.2 The Operating System

KEY QUESTION

What are the principal functions of the operating system?

Preview & Review: The operating system manages the basic operations of the computer. These operations include booting the computer and management of storage media. Another feature is the user interface, which may be command-driven, menu-driven, graphical, or network. Other operations are managing computer resources and managing files. The operating system also manages tasks, through multitasking, multiprogramming, time-sharing, or multiprocessing.

The *operating system (OS)* **consists of the master system of programs that manage the basic operations of the computer.** These programs provide resource management services of many kinds, handling such matters as the control and use of hardware resources, including disk space, memory, CPU time allocation, and peripheral devices. The operating system allows you to concentrate on your own tasks or applications rather than on the complexities of managing the computer.

Different sizes and makes of computers have their own operating systems. For example, Cray supercomputers use UNICOS and COS, IBM mainframes use MVS and VM, Data General minicomputers (midrange computers) use AOS and DG, and Compaq minicomputers use VAX/VMS. Pen-based computers have their own operating systems—PenRight, PenPoint, Pen DOS, and Windows for Pen Computing—that enable users to write scribbles and notes on the screen. *All these operating systems are not compatible with one another.* That is, in general, an operating system written for one kind of hardware will not be able to run on another kind of machine.

Microcomputer users may readily experience the aggravation of such incompatibility when they acquire a microcomputer. Do they get an Apple Macintosh with Macintosh system software, which won't run PC programs? Or do they get a PC (such as IBM, Compaq, or Dell), which won't run Macintosh programs?

Before we try to sort out these perplexities, we should see what operating systems do that deserve our attention. We consider:

- Booting
- Managing storage media
- User interface
- Managing computer resources
- Managing files
- Managing tasks

Booting

The operating system begins to operate as soon as you turn on, or "boot," the computer. The term **booting refers to the process of loading an operating system into a computer's main memory from disk.** This loading is accomplished by a program (called the *bootstrap loader* or *boot routine*) that is stored permanently in the computer's electronic circuitry.

When you turn on the machine, programs called *diagnostic routines* first start up and test the main memory, the central processing unit (✔ p. 16), and other parts of the system to make sure they are running properly. As these programs are running, the display screen may show the message "Testing RAM" (main memory).

Next, other programs (indicated on your screen as "BIOS," for basic input/output system) will be copied to main memory to help the computer interpret keyboard characters or transmit characters to the display screen or to a diskette.

Then the boot program obtains the operating system, usually from hard disk, and loads it into the computer's main memory (✔ p. 16). The operating system remains in main memory until you turn the computer off. With newer operating systems, the booting process puts you into a graphically designed starting screen, from which you choose the applications programs you want to run or the files you want to open.

Managing Storage Media

If you have not entered a command to start an applications program, what else can you do with the operating system? One important function is to perform common repetitive tasks involved with managing storage media.

An example of such a task is formatting of blank diskettes. Before you can use a blank diskette you've just bought, you may have to format it. ***Formatting***, **or** ***initializing***, **electronically prepares a diskette so it can store data or programs.** (On a PC, for example, you might insert your blank diskette into drive A and choose Format from the File menu.) Nowadays, however, it's easier to buy preformatted diskettes, which bear the label "Formatted IBM" or "Formatted Macintosh." These are sold everywhere. Note that one *never* formats a hard disk unless one wants to erase everything on it. Formatting a diskette or disk gets rid of everything that had previously been written to it.

Providing a User Interface

Many operating-system functions are never apparent on the computer's display screen. What you do see is the user interface (✔ p. 55). **The *user interface* is the user-controllable part of the operating system that allows you to communicate, or interact, with it.**

There are four types of user interfaces, for both operating systems and applications software: *command-driven; menu-driven; graphical;* and, most recently, *network.* (■ *See Panel 3.2.*)

- Command-driven interface: **A *command-driven interface* requires you to enter a command by typing in codes or words.** An example of such a command might be DIR (for "directory"). This command instructs the computer to display a directory list of all folder and file names on a disk.

  You type a command at the point on the display screen where the cursor follows the prompt, such as following "C:\>". In general, C:\ refers to the hard disk and > is a *system prompt*, asking you for a command. After you type in the command, you press the Enter key to execute the command.

  The command-driven interface is the type found on PCs with the DOS operating system (discussed shortly).

- Menu-driven interface: **A *menu-driven interface* allows you to use cursor-movement (arrow) keys to choose a command from a menu.** Like a restaurant menu, **a *menu* offers you options to choose from— in this case, commands available for manipulating data,** such as Print or Edit.

  Menus are easier to use than command-driven interfaces, especially for beginners, because users can choose from lists of options rather than having to remember the code for specific commands. The

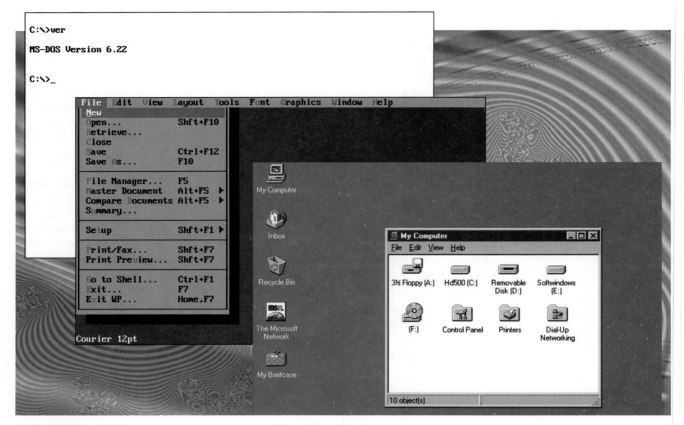

■ **PANEL 3.2**

Types of user interfaces

(Top left) A command-driven interface, as in MS-DOS, requires typing of codes or words. *(Middle)* A menu-driven interface contains menus offering displayed lists of options. *(Bottom right)* A graphical user interface, such as Windows 98, allows users to select programs, commands, files, and other items represented by pictorial figures (icons), as well as use various types of menus.

disadvantage of menus, however, is that they are slower to use. Thus, some software programs offer both features—menus for novice users and keyboard commands for experienced users.

● Graphical user interface (GUI): The easiest interface to use, **the graphical user interface allows you to use graphics (images) and menus as well as keystrokes to choose commands, start programs, and see lists of files and other options** (✔ p. 56). Some of these images take the form of icons. **Icons are small pictorial figures that represent tasks, procedures, and programs.** For example, you might select the picture of a trashcan to delete a file you no longer want.

Another feature of the GUI is the use of windows (✔ p. 56). **Windows "divide" the display screen into rectangular sections.** Each window may show a different display, such as a word processing document in one and a spreadsheet in another.

On both Windows and Macintosh computers, the windowing capabilities built into a GUI appear layered on or above a common base, or "canvas," known as a desktop. **A *desktop* presents icons on the computer screen according to principles used for organizing the top of a businessperson's desk.** Thus, a user can move on-screen items

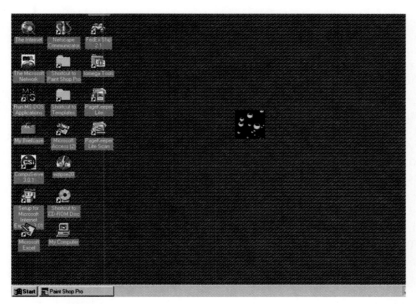

Items on a desktop

on the desktop among directories and folders.

Finally, the GUI permits liberal use of the mouse. The mouse is used as a pointing device to move the cursor to a particular place on the display screen or to point to an icon or button. The function represented by the icon can be activated by pressing ("clicking") buttons on the mouse. Or, using the mouse, you can pick up and slide ("drag") an image from one side of the screen to the other or change its size.

Microcomputer users first became generally aware of the graphical user interface from its appearance in Apple Macintosh computers (although Apple got the idea from Xerox). Later Microsoft made a graphical user interface available for PCs through its Windows program. Now most operating systems on microcomputers feature a GUI.

- **Network user interface (NUI):** The latest interface is **the *network user interface,* or *NUI* (pronounced "new-ee"), which offers a browser-like interface that helps users interact with online programs and files.** NUIs started out being designed for network computers (NCs) (✔ p. 26), but they are finding their way onto regular PCs as well. Basically, network interfaces look like GUIs.

Managing Computer Resources

Suppose you are writing a report using a word processing program and want to print out a portion of it while continuing to write. How does the computer manage both tasks?

Behind the user interface, the operating system acts like a police officer directing traffic. This activity is performed by the ***supervisor,* or *kernel,* the central component of the operating system. The supervisor, which manages the CPU, resides in (is "resident in") main memory while the computer is on and directs other "nonresident" programs to perform tasks to support applications programs.** Thus, if you enter a command to print your document, the applications software directs the operating system to select a printer (if there is more than one). Then the operating system notifies the computer to begin sending data and instructions to the appropriate program (known as a printer *driver,* because it controls, or "drives," the printer). Meanwhile, many operating systems allow you to continue writing in your applications software. Were it not for this supervisor program, you would have to stop writing and wait for your document to print out before you could resume.

The operating system also manages memory—it keeps track of the locations within main memory where the programs and data are stored. It can swap portions of data and programs between main memory and secondary storage, such as your computer's hard disk. This capability allows a computer to hold only the most immediately needed data and programs within main memory. Yet it has ready access to programs and data on the hard disk, thereby greatly expanding memory capacity.

There are several ways operating systems can manage memory:

- **Partitioning:** Some operating systems use *partitioning*. That is, they divide memory into separate areas called *partitions*, each of which can hold a program or data.
- **Foreground/background:** Some computer systems divide memory into *foreground* and *background* areas. Foreground programs have the highest priority, and background programs have the lowest priority. When you're working at your microcomputer, the foreground program is the one you are currently working with, such as word processing. The background program might be regulating the flow of print images to your printer.
- **Queues:** Programs wait on disk in *queues* (pronounced "Qs") for their turn to be executed. A queue is a temporary holding place for programs or data.

Managing Files

Files of data and programs are located in many places on your hard disk and other secondary-storage devices. The operating system allows you to find them. If you move, rename, or delete a file, the operating system manages such changes and helps you locate and gain access to it.

Some examples of file management commands are:

- **Copy:** You can *copy*, or duplicate, files and programs from one disk to another.
- **Back up:** You can *back up*, or make a duplicate copy of, the contents of a disk.
- **Erase:** You can *erase*, or remove, from a disk any files or programs that are no longer useful.
- **Rename:** You can *rename*, or give new file names to, the files on a disk.

Managing Tasks

A computer is required to perform many different tasks at once. In word processing, for example, it accepts input data, stores the data on a disk, and prints out a document—seemingly simultaneously. Some computers' operating systems can also handle more than one program at the same time—word processing, spreadsheet, database searcher—displaying them in separate windows on the screen. Others can accommodate the needs of several different users at the same time. All these examples illustrate process, or task, management—a "task" being an operation such as storing, printing, or calculating.

Among the ways operating systems manage tasks in order to run more efficiently are *multitasking, multiprogramming, time-sharing,* and *multiprocessing.* (Not all operating systems can do all these things.)

- **Multitasking: for one user—executing more than one program concurrently.** You may be writing a report on your computer with one program while another program searches an online database for research material. How does the computer handle both programs at once?

The answer is that the operating system directs the processor (CPU) to spend a predetermined amount of time executing the instructions for each program, one at a time. In essence, a small amount of each program is processed, and then the processor moves to the remaining programs, one at a time, processing small parts of each. This cycle is repeated until processing is complete. The processor speed is usually so fast that it may seem as if all the programs are being executed at the same time. However, the processor is still executing only one instruction at a time, no matter how it may appear to the user.

- **Multiprogramming: for multiple users—executing different users' programs concurrently.** As with multitasking, the CPU spends a certain amount of time executing each user's program, but it works so quickly that it seems as though all the programs are being run at the same time.

- **Time-sharing: for multiple users—executing different users' programs in round-robin fashion.** Time-sharing is used when several users are linked by a communications network to a single computer. The computer will first work on one user's task for a fraction of a second, then go on to the next user's task, and so on.

- **Multiprocessing: for single or multiple users—simultaneous processing of two or more programs by multiple computers.** With multiprocessing, two or more computers or processors linked together perform work simultaneously, meaning at precisely the same time. This can entail processing instructions from different programs or different instructions from the same program.

  One type of multiprocessing is *parallel processing,* whereby several full-fledged CPUs work together on the same tasks, sharing memory. Parallel processing is often used in large computer systems designed to keep running if one of the CPUs fails. These systems are called *fault-tolerant* systems; they have several CPUs and redundant components, such as memory and input, output, and storage devices. Fault-tolerant systems are used, for example, in airline reservation systems.

How do multitasking and time-sharing differ? With *multitasking,* the processor directs the programs to take turns accomplishing small tasks or events within the programs. These events may be making a calculation, searching for a record, printing out part of a document, and so on. Each event may take a different amount of time to accomplish. With *time-sharing,* the computer spends a fixed amount of time with each program before going on to the next one.

How do multiprocessing and multitasking differ? *Multiprocessing* goes beyond *multitasking,* which works with only one microprocessor. In both cases, the processing is so fast that, by spending a little bit of time working on each of several programs in turn, a number of programs can be run at the same time. With both multitasking and multiprocessing, the operating system keeps track of the status of each program so that it knows where it left off and where to continue processing. But the multiprocessing operating system is much more sophisticated than multitasking.

Operating system functions are shown in the chart on the following page. (*See* ■ *Panel 3.3.*)

Booting	Managing Storage Media	User Interface	Managing Computer Resources	Managing Files	Managing Tasks
Uses diagnostic routines to test system for equipment failure Stores BIOS programs in main memory Loads operating system into computer's main memory	Formats diskettes Displays information about operating system version Displays disk space available	Provides a way for user to interact with the operating system —can be command-driven, menu-driven, graphical, or network	Via the supervisor, manages the CPU and directs other programs to perform tasks to support applications programs Keeps track of locations in main memory where programs and data are stored (memory management) Moves data and programs back and forth between main memory and secondary storage (swapping).	Copies files/programs from one disk to another Backs up files/programs Erases (deletes) files/programs Renames files	May be able to perform multitasking, multiprogramming, time-sharing, or multiprocessing

▪ PANEL 3.3
Basic operating system functions

3.3 Common Microcomputer Operating Systems: The Changing Platforms

KEY QUESTIONS

What are the principal operating systems and operating environments for personal computers, and what are their principal characteristics?

Preview & Review: A computer platform is defined by its processor model and its operating system. The principal microcomputer operating system on new personal computers is Windows 95 and 98. Other microcomputer operating systems and operating environments are DOS, Macintosh operating system, Windows 3.x, OS/2, Novell's NetWare, Unix, Windows NT, and Windows CE.

More than 74% of all PCs sold in stores come with Windows 95—now upgraded to Windows 98—already installed, according to one report.[3] (Another says that 85% of all new computers come equipped with Windows 95/98, and 10% come with Windows NT.[4]) What does this mean for other platforms? **Platform refers to the particular processor model and operating system on which a computer system is based.** (We discuss processor models in Chapter 4.) Should the operating systems from Apple, IBM, and others no longer be taken seriously? And why should we even care?

Why Is Windows 95/98 the Dominant Standard?

How did Windows 95, and its successor, Windows 98, become dominant— especially when people were able to get the essential elements using other

Microcomputer operating system percentages

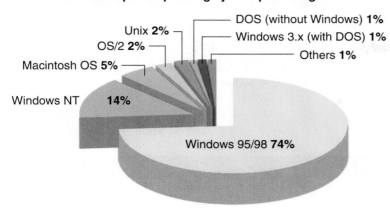

Dataquest's accounting of market share of operating systems

operating systems, such as that offered by Apple's Macintosh? The personal computer technology finally culminating in Windows 95 was actually invented more than 20 years ago by researchers at Xerox's Palo Alto Research Center (PARC) in northern California. Xerox executives asked about the profitability of marketing the PARC personal computer, which in those days, says one former researcher, "had the same ring to it as 'personal nuclear reactor' would have today."[5] When PARC scientists couldn't answer that, the PC was relegated to in-house use. At about that time, PARC gave a tour to Steve Jobs, who saw the PC. Later, as a founder of Apple Computer, Jobs used the PARC technology to create the Lisa PC, which was succeeded by the Macintosh, launched in 1984. Microsoft's Windows for DOS then followed in 1990. Now we have Windows 95, Windows 98, and Windows NT.

Principal Microcomputer Operating Systems in Use Today

Windows 95/98 may be the most popular operating system, but other operating systems are still being used on microcomputers throughout the world. Probably most of these, such as DOS, are so-called *legacy systems*. **Legacy systems are the millions of older computer systems used in offices and homes today that employ outdated yet still functional technology.** These old-timer information systems may have been overtaken by newer technology, but because of their popularity you may nonetheless find yourself having to use them at some point. (Legacy hardware cannot run the newer operating systems.)

Some people refer to legacy systems as *technologically obsolete*. This means that these systems are still useful to individuals or businesses, but that they have been superseded in the marketplace by newer versions or models. (Compare to *functionally obsolete*, which refers to a system or product that is no longer useful to its users.)

In the rest of this section, we'll discuss the following:

- DOS
- Macintosh operating system
- Windows 3.x (that is, Windows 3.0, 3.1, and 3.11)
- OS/2
- Windows 95
- Novell's NetWare
- Unix
- Windows NT
- Windows CE
- Windows 98

Historical highlights of some of these personal computer operating systems are shown in the figure on the next page. (■ *See Panel 3.4.*)

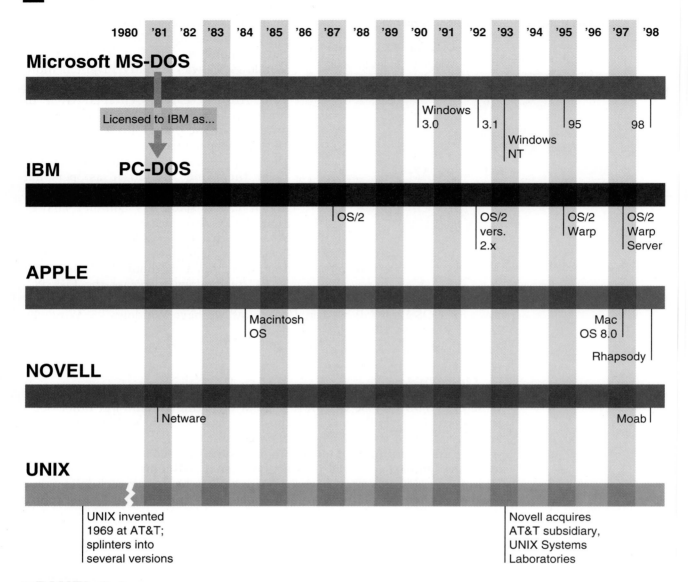

	1980	'81	'82	'83	'84	'85	'86	'87	'88	'89	'90	'91	'92	'93	'94	'95	'96	'97	'98

Microsoft MS-DOS

Licensed to IBM as...

Windows 3.0   3.1   95   98

Windows NT

IBM   **PC-DOS**

OS/2   OS/2 vers. 2.x   OS/2 Warp   OS/2 Warp Server

APPLE

Macintosh OS   Mac OS 8.0

Rhapsody

NOVELL

Netware   Moab

UNIX

UNIX invented 1969 at AT&T; splinters into several versions

Novell acquires AT&T subsidiary, UNIX Systems Laboratories

■ PANEL 3.4

Timeline of microcomputer operating systems
Key events are shown for Microsoft (red), IBM (blue), Apple (green), Novell (purple), and Unix (orange).

Operating Environments

Before we proceed, we need to define what an *operating environment* is, because some people have trouble distinguishing it from an *operating system*. **An *operating environment*—also known as a *windowing environment*, or *shell*—provides a graphical user interface or a menu-driven interface as an outer layer to an operating system.** The best-known operating environment is Microsoft Windows 3.x, which adds a graphical user interface to DOS, which is the operating system. Another is IBM's Workplace Shell, which provides a GUI for OS/2. Similar operating environments, such as Xwindows, are available for Unix.

Common features of these operating environments are use of a mouse, pull-down menus, and icons and other graphic displays. They also have the ability to run more than one application (such as word processing and spreadsheets) at the same time and the ability to exchange data between these applications.

Let's now examine the principal operating systems (and operating environments) you will probably encounter.

DOS: The Old-Timer

What a piece of luck. In 1981, when mighty IBM came calling, Bill Gates was not yet the richest man in the world but simply a Harvard dropout who, in 1975 with former childhood friend Paul Allen, had formed a small software business in Redmond, Washington. Seeing the sudden popularity of Apple's personal computer (the Apple II), International Business Machines, then the giant of the computer industry, had launched a crash program to produce a microcomputer of its own. But they needed an operating system. After failing to reach agreement with another software developer (for a then-leading operating system called CP/M), they had turned to fledgling Microsoft. Gates hurriedly acquired the rights from a third party to an operating system we now know as DOS and in turn *licensed* the rights to DOS to IBM.

Note this very important and shrewd business move, the foundation on which Microsoft was built: DOS was not *sold* but was in effect *leased* to IBM—and then to all the other hardware manufacturers who cloned IBM-style machines. In every case, Microsoft held onto the rights to DOS, which became the most widely used operating system in the world.

***DOS* (rhymes with "boss")—for *Disk Operating System*—runs primarily on PCs,** such as those made by AST, Compaq, Dell, Gateway 2000, Hewlett-Packard, IBM, NEC, and Packard Bell. There are two main operating systems calling themselves DOS:

- **Microsoft's MS-DOS:** DOS is sold under the name *MS-DOS* by Microsoft; the "MS," of course, stands for Microsoft. Microsoft launched its original version, MS-DOS 1.0, in 1981, and there have been many upgrades since then.

- **IBM's PC-DOS:** Microsoft licenses a version to IBM called *PC-DOS*. The "PC" stands for "Personal Computer." The most recent version is PC-DOS 7, released March 1995.

DOS is a command-driven operating system. For example, to format a diskette in drive A (the diskette drive designation), you would insert your diskette into the drive and then type, after the C:\>, CD\DOS, and then FORMAT A:. The command DISKCOPY copies the contents of one disk to another disk; DIR, for "directory," displays the names of files on a disk.

Two years before Windows 95 came on the scene, there were reportedly more than 100 million users of DOS, which made it the most popular software ever adopted—of any sort.[6] Today more Windows (3.x, 95, 98, NT) operating systems are being used, but DOS is still in second place, which makes it an important legacy system indeed.

Why would anyone use DOS today? First of all, if you must work with a legacy system, you may not have any choice. In addition, MIT economist Paul Krugman, for one, finds that newer software, such as Windows 95/98, that "requires fancy screens and constantly spins the [hard] disk is counterproductive when you're working on battery power." Like some other experienced business travelers who carry battery-powered laptops on airplanes, he says, "I have carefully saved an old DOS-based word processor to use when airborne."[7] In other words, if you're in circumstances where you can't easily plug in your laptop and recharge the batteries (which may last only 2 or 3 hours), DOS, with its low power requirements, may be the way to go.

Recent versions of DOS have expanded the range of the operating system. For example, Version 4.0 changed MS-DOS from a command-driven interface to a menu-driven interface. Version 5.0 added a graphical user interface.

Version 6.0 added features that optimized the use of main memory. Version 7.0 added multitasking capabilities.

No doubt DOS will be around for years, and if you find you have to use it, many DOS handbooks and tutorials are still available in bookstores. Nevertheless, as a command-driven, single-user program, DOS is a fading product. Although satisfactory for many uses, it will unquestionably be succeeded by other, more versatile operating systems. As Ken Wasch, president of the Software Publishers Association, says, "Nobody was forced to upgrade to electric lights when they still could read by kerosene lamps, but still most people found it advantageous to upgrade."[8]

Macintosh Operating System: For the Love of Mac

Apple Computer was formed in 1977 by two college dropouts, Steven Jobs and Stephen Wozniak, who had met at a club for amateur computer builders in Silicon Valley, California. The first Apple computers were built in that time-honored place of inventors, a garage, using the $1300 proceeds from the sale of an old Volkswagen. Apple computers were among the first to replace complicated switches-and-lights front panels with easy-to-use typewriter keyboards. The Apple II was hugely successful, as was the Apple IIe, and the company followed in 1983 with Lisa. Less successful, Lisa was followed in 1984 by the very easy to use Macintosh, which still enjoys a lot of popularity today in education and in desktop publishing.

The *Macintosh operating system (Mac OS)* **runs only on Apple Macintosh computers or on Mac clones,** such as those formerly made by Power Computing, Motorola, and Umax Computer Systems. The most noteworthy thing about the Mac was that it set the standard for icon-oriented graphical user interfaces. (■ *See Panel 3.5.)* Macintosh system software is especially easy to use because Apple designed its hardware and software together from the start.

The easy-to-use interface has generated a strong legion of fans. In the past, however, because Apple kept Macintosh prices high and refused to license any clones, Macs were expensive.

■ PANEL 3.5

Macintosh operating system

The icons, pull-down menus, and windows of the Apple Macintosh operating system give the Mac an ease of use that has generated many loyal fans. The screen shown here is from System 8.

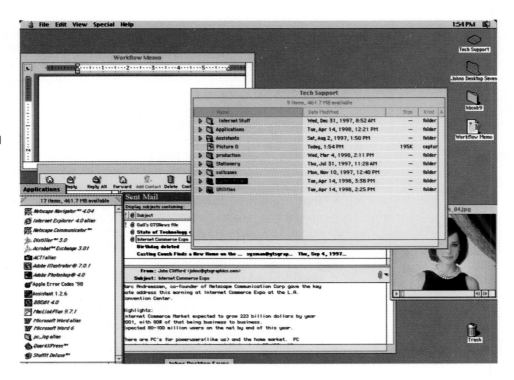

Unfortunately, PC and Macintosh microcomputers are different platforms—they are designed around different microprocessors—so it was impossible to combine the best of both. PC computers use microprocessors built by Intel (the 80286, 80386, 80486, Pentium, Pentium Pro, Pentium MMX, and Pentium II chips). Macintoshes are built around microprocessors made by Motorola (the 68000, 68020, 68030, 68040, and PowerPC chips). Intel chips could not run Macintosh programs, and Motorola chips could not run DOS or Windows programs.

Thus, because of price, and because in pre-PC times businesses were already accustomed to using IBM equipment, DOS-equipped (and later Windows-equipped) PCs have tended to rule the day in most offices. Apple also lost ground because during the 1990s it failed to upgrade and improve the Mac OS. And while Apple dallied, Microsoft came out with Windows 95 and managed to nearly match all of the Mac system's best features.

In July 1997, Apple released Mac OS 8 (that is, version 8.0). The general opinion of reviewers seemed to be that OS 8 was a well-done update, although most felt it was essentially a holding action until the next all-new operating system (then code-named *Rhapsody*) came out. Apple has said, however, it will continue releasing new versions of the Mac OS through the year 2000 at least.

Apple's share of personal computer sales in the United States has shrunk from about 14% in 1987 to 4% in 1998.[9] What are the company's chances for future success, or even survival? Among the developments affecting it are the following:

- **The alliance with Microsoft:** In 1985, Apple cofounder and chairman Steve Jobs lost an in-house battle to corporate pinstripes and was booted out of his own company, from which he remained in exile for 12 years. (Cofounder Wozniak retired voluntarily.) In 1997, after years of Apple mismanagement, Jobs was invited back. And one of his first acts was to call up Bill Gates and arrange to have Microsoft become part owner of Apple, buying $150 million in stock. In return, Microsoft agreed to settle claims that it had used some seminal Apple patents and to continue creating applications for the Mac.

- **Here come, there go, the clones:** In December 1994, Apple finally started signing deals to license its operating system to Macintosh clonemakers, belatedly hoping to gain from the same strategy that had enriched Microsoft. Two years later, however, Apple executives began to blame cloners—who often offered better performance at lower prices—for cutting into Apple hardware sales without simultaneously expanding either the hardware or software markets. And in late 1997, Apple told cloners it would no longer encourage licensing.

  The fact that Apple once again has the Macintosh all to itself could lead to higher prices. This might not deter die-hard Mac fanatics, but price-conscious consumers could abandon Apple in droves, further shrinking the company's customer base.

- **The Rhapsody OS:** It's possible even now to turn a Macintosh in effect into a PC. With Motorola's PowerPC microprocessor, the Mac can read PC-formatted diskettes, and many of the applications written for it can convert files created by Windows applications. And a program from Connectix called Virtual PC allows any recent Mac to pretend it's a Windows PC.

  Apple's *Rhapsody* was originally designed to completely change the Mac operating system so that it would run not only programs for

Macintosh PowerPC processors (made by IBM and Motorola) but also programs written for Windows running on Intel processors. However, applications software developers objected that that strategy would require them to write new applications from scratch, a task that could take one to two years. Apple thereupon changed course and released Rhapsody to run primarily on servers, the control computers on networks.[10] For the mainstream desktop machines, in lieu of Rhapsody Apple announced a new operating system to be called *Mac OS X.*

- **Mac OS X:** Scheduled for release in the fall of 1999, Mac OS X (Roman numeral 10; the system is called "OS 10") combines elements of Mac OS 8 and Rhapsody. For applications software developers, the biggest advantage is that they can revamp programs in just days. In addition, the Mac OS X prohibits runaway programs from shutting down the computer, performs faster networking tasks, and can run several functions at once (multitasking).[11] Unlike present versions of the Mac OS, which contain code developed for the early Motorola microprocessors and so work more slowly on today's new machines, OS X will also be written entirely for the Power PC chip.

Windows 3.x: Windows for DOS

Microsoft's *Windows 3.x* **is an operating *environment* (not operating system) that lays a graphical user interface shell around the DOS operating system and extends DOS capabilities.** (■ *See Panel 3.6.*) Actually, there is no Windows "3.x": this is simply shorthand for the three releases of Windows 3.0, 3.1, and 3.11. (Windows 1.0 came out in 1985 and 2.0 in 1988, but it was not until version 3.0 was released that Windows really took off.)

As mentioned earlier (✔ p. 56), a *window* is a rectangular portion of the display area with a title on top. With Windows 3.x, which supports multi-

■ PANEL 3.6

Windows 3.x screen

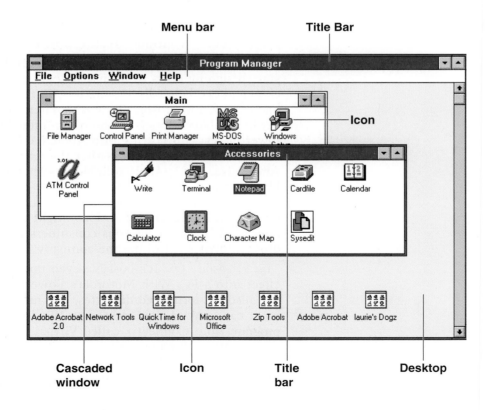

tasking, you can display several windows on the screen, each showing a different application, such as word processing or spreadsheet. You can easily switch between the applications and move data between windows. In addition, each window can be enlarged, reduced, or minimized to an icon, which temporarily removes it from view.

Microsoft released Windows 3.0 in May 1990 and promoted it as a way for frustrated DOS users not to have to switch to more user-friendly operating systems, such as that of the Macintosh. Although far easier to use than DOS, Windows 3.x is not as easy to use as the Mac. This is because Windows 3.x sits atop the old command-driven DOS operating system, which requires certain compromises on ease of use. In fact, the system has something of a split personality. In handling files, for example, after passing through the Macintosh-style display of icons, the user has to deal with the DOS file structures beneath. In addition, many users complain that installing peripherals, such as a hard-disk drive, is somewhat difficult with DOS and Windows.

Yet, even if the various Windows 3.x versions are a bit creaky, they are certainly usable by most people. And when Windows 95 was rolled out, most of the objections vanished.

Before describing Windows 95, we will introduce another of its predecessors and competitors, OS/2.

OS/2: IBM's Entry in the OS Sweepstakes

OS/2 (there is no OS/1) was developed jointly by Microsoft and IBM and was supposed to become the successor to DOS. However, the two companies parted ways and IBM kept control of OS/2, releasing it in 1987 as its contender for the next mainstream operating system. Microsoft took what was originally to be OS/2 version 3.0 and renamed it Windows NT (to be discussed), which it introduced in 1993.

OS/2—for *Operating System/2*—is designed to run on many recent IBM and IBM-compatible microcomputers. Like Windows, it has a graphical user interface, called the *Workplace Shell (WPS)*, which uses icons resembling documents, folders, printers, and the like. OS/2 can also run most DOS, Windows, and OS/2 applications simultaneously. This means that users don't have to throw out their old applications software to take advantage of new features. Lastly, this operating system was designed to connect everything from small handheld personal computers to large mainframes.

Unfortunately, because of an array of management and marketing disasters, IBM slipped far behind Microsoft. By mid-1994, for example, an estimated 50 million copies of Microsoft's Windows had been sold versus 5 million of OS/2.[12] In late 1994 IBM unveiled a souped-up version with the *Star Trek*-like name of OS/2 Warp. Despite spending $2 billion on OS/2 in its long struggle against Windows 3.x, the company failed to increase its market share.

Even though many experts rate OS/2 Warp highly, developers of applications software abandoned the IBM product in order to create new programs for the expected Microsoft blockbuster. As a result, in the summer of 1995—three weeks before Windows 95 was to be introduced—the chairman of IBM appeared to concede defeat for OS/2. Nevertheless, IBM continues to support its approximately 10 million Warp users—the latest version is OS/2 Warp Server, which can handle system management for networks. Upgrades are available online and can be downloaded from IBM's Web site. IBM also offers an OS/2-compatible version of the popular Web browser Netscape.

Windows 95: The Successor to Windows 3.x

"The original DOS was little more than a thin (and clumsy) layer of hooks that applications could use for reading and writing data to memory, screen, and disks," observes technology writer James Gleick. "Windows 95 not only provides a rich environment for controlling many programs at once; it also offers, built in, a word processor, communications software, a fax program, an assortment of games, screen savers, a telephone dialer, a paint program, back-up software, and a host of other housekeeping utilities and, of course, Internet software. By historical standards you get a remarkable bargain."[13]

Released in August 1995, **Windows 95, the successor to Windows 3.x, is not just an operating environment but rather a true PC operating system.** Unlike Windows 3.x, it does not require the separate DOS program, although DOS commands can still be used. The graphical user interface is not just a shell; it is integrated into the operating system. It will also run applications written for DOS and Windows 3.x, as well as those created for Windows 95. Finally, it includes all kinds of support for e-mail, voice mail, fax transmission, and multimedia.

Following are just some of the features of Windows 95:

- Clean "Start": Upon booting, you'll first see a clean "desktop" with program icons and, in the "tray area" at the bottom of the screen, you'll see a button labeled Start at the left and some date and volume controls at the right. (■ *See Panel 3.7.*)

- Better menus: Unlike Windows 3.x, the menus in Windows 95 let you quickly see what's stored on your disk drives and make tracking and moving files easier.

- Long file names: Whereas file names in DOS and Windows 3.x have to be limited to eight characters plus a three-character extension (for example, PSYCHRPT.NOV), file names under Windows 95 can be up to 256 characters in length (PSYCHOLOGY REPORT FOR NOVEMBER). (Macintosh OS and OS/2 have always permitted long file names.)

- The "Recycle Bin": This feature allows you to delete complete files and then get them back if you change your mind.

- 32-bit instead of 16-bit: The new software is a 32-bit program, whereas most Windows 3.x software is 16-bit. *Bit numbers* refer to how many bits of data a computer chip, and software written for it, can process at one time. Such numbers are important because they refer to the amount of information the hardware and software can use at any one time. This doesn't mean that 32-bit software will necessarily be twice as fast as 16-bit software, but it does promise that new 32-bit applications software will offer better performance efficiency and features once software developers take advantage of the design.

- Plug and play: It has always been easy to add new hardware components to Macintoshes. It used to be extremely difficult with PCs. **Plug and play refers to the ability to add a new hardware component to a computer system and have it work without needing to perform complicated technical procedures.**

   More particularly, *Plug and Play* is a standard developed for PCs by Microsoft and chip maker Intel and incorporated into Windows 95 to eliminate user frustration when one is adding new components. Now when you add a new printer or modem that is built to plug-and-play standards, your PC will recognize that a new peripheral has been added and then automatically set it up.

Start button: Click for an easy way to start using the computer.

Microsoft Network: Click here to connect to the Microsoft Network, the company's online service.

My Briefcase: Allows you to synchronize files in two computers—say, an office PC and a laptop.

Recycle Bin: Allows you to dispose of files—or retrieve them later.

Network Neighborhood: If your PC is linked to a network of PCs, click here to get a glimpse of everything available on the network.

My Computer: Gives you a quick overview of all the files and programs installed in your PC.

Document: Multitasking capabilities allow people to smoothly run more than one program at once.

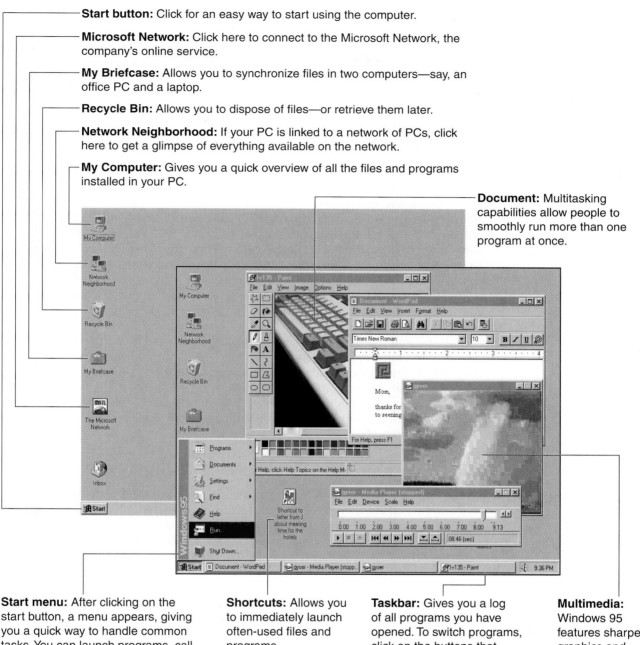

Start menu: After clicking on the start button, a menu appears, giving you a quick way to handle common tasks. You can launch programs, call up documents, change system settings, get help, and shut down your PC.

Shortcuts: Allows you to immediately launch often-used files and programs.

Taskbar: Gives you a log of all programs you have opened. To switch programs, click on the buttons that appear in the taskbar.

Multimedia: Windows 95 features sharper graphics and improved video capabilities compared to Windows 3.x.

■**PANEL 3.7**

Windows 95 screens

In its first 12 months, Windows 95 sold an unprecedented 40 million copies, most of them preloaded on (bundled with) new PCs.[14] Now there are reportedly 100 million copies in use.[15]

If Windows 95 is the name Microsoft gave to what would have been called "Windows 4.0," then Windows 98 is the name for "Windows 5.0." We discuss Windows 98 in another few pages.

Novell's NetWare: PC Networking Software

So far we have described operating systems (and operating environments) pretty much in the chronological order in which they appeared. Except for OS/2, these operating systems were principally designed to be used with stand-alone desktop machines, not large systems of networked computers. Now let us consider the three important operating systems designed to work with networks: NetWare, Unix, and Windows NT.

Novell, of Orem, Utah, is the maker of NetWare. **NetWare is a popular network operating system (NOS) for coordinating microcomputer-based local area networks (LANs) throughout a company or a campus.** LANs (✔ p. 26) allow PCs to share programs, data files, and printers and other devices.

Novell thrived as corporate data managers realized that networks of PCs could exchange information more cheaply than the previous generation of mainframes and midrange computers. Today the company still holds half the corporate network software market, with about 50 million people using its software. However, this market share is down from earlier times, owing to the rise of the Internet and competition from Microsoft. Moreover, during the early 1990s Novell went on a buying spree of consumer software titles (WordPerfect suite, Quattro Pro spreadsheet) that distracted the company from changes in the network market beginning in 1994, when the Internet began growing in popularity.[16]

Still, Novell has an important corporate client base, which so far has been only slightly eroded by Microsoft Windows NT.[17] In addition, inertia may be a powerful force: one analysis calculates it would cost about $1 million to switch a 2500-computer network from Novell to Microsoft.[18] Finally in 1998, the company was scheduled to rush out an update of NetWare—code-named Moab—that may help Novell reposition itself as a leader in network software technology.[19]

Unix: The Operating System for Multiple Users

Unix was invented more than two decades ago by American Telephone & Telegraph, making it one of the oldest operating systems. **Unix is a multi-tasking operating system for multiple users that has built-in networking capability and versions that can run on all kinds of computers.** Because it can run with relatively simple modifications on different types of computers—from micros to mainframes—Unix is called a "portable" operating system. (■ *See Panel 3.8.*)

The primary users of Unix are government agencies, universities, research institutions, large corporations, and banks, which use the software for everything from airplane-parts design to currency trading. For example, Taco Bell uses Unix to link its in-store registers to back-office servers for inventory control and labor scheduling. Red Roof Inn uses Unix servers to run the daily operations of its 280 locations. The U.S. Department of Energy, at its Sandia National Laboratory, uses a variant of Unix on a supercomputer to simulate nuclear explosions.

Unix is also used for Web-site management. Indeed, the developers of the Internet built the system around Unix because of the operating system's ability to keep large systems with hundreds of processors churning out transactions day in and day out for years without fail.

For a long time, AT&T licensed Unix to scores of companies that made midsize computers and workstations. As a result, the operating system was modified and resold by several companies, producing several versions of Unix. Then in 1993 Novell acquired an AT&T subsidiary and the Unix trade-

■ PANEL 3.8

Unix screen

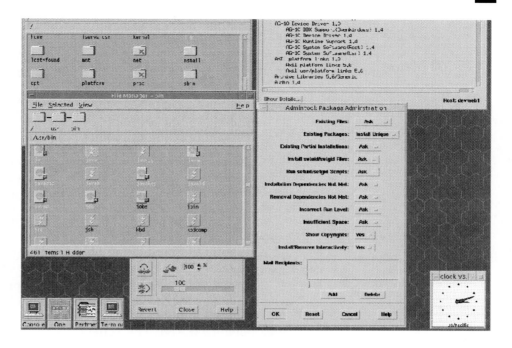

mark and gave the Unix brand name to an independent foundation. The foundation, made up of 75 software vendors, created a Unix standard in 1995.

Will Unix endure? Rapid advances by Microsoft's Windows NT have convinced some industry analysts that it will overtake Unix. Even so, there are a number of viable Unix contenders. (Examples include AT&T's descendant, The Santa Cruz Operation's SCO UnixWare; Sun Microsystem's Solaris; IBM's AIX; Digital Equipment's Unix; and Hewlett-Packard's HP-UX.) The "flavor" that seems to be becoming most popular, however, is a Unix-like operating system called *Linux,* created by Linus Torvalds, a Finnish programmer. One expert contends that "There is more Linux running Internet servers than all other OSs [operating systems] combined."[20]

What accounts for Linux's popularity is that it is shareware (✔ p. 86); that is, it is freely distributed over the Internet, and copying is encouraged. Although Windows NT would seem to have the advantage over Linux because Microsoft offers sophisticated applications software (such as Microsoft Office) and customer support to go with it, some enterprising firms are offering competitive applications to run with Linux. For instance, Red Hat Software of North Carolina sells customers 500 megabytes of software called ApplixWare for Linux that includes a word processor, a spreadsheet, a Web browser, e-mail software, presentation graphics, a database, and more. Its advantage: It costs less than $50 and works nearly as well as Microsoft Office.[21]

Windows NT: Microsoft's Software for Business Networks

Unveiled by Microsoft in 1993, **Windows NT, for *New Technology,* is a multitasking, multiuser, multiprocessing network operating system with a graphical user interface.** Multiuser systems are used to support computer *workgroups.* A *workgroup* is a group of computers connected with networking hardware and software so that users can share resources, such as files and databases.

Unlike the early Windows 3.x *operating environment,* Windows NT is a true *operating system,* which (like Windows 95) interacts directly with the hardware. It is primarily designed to run on workstations (✔ p. 23) or other more powerful computers. It runs not only NT-specific applications but also programs written for DOS, Windows 3.x, and Windows 95.

Two principal features of Windows NT are as follows:

- **Two versions—NT Workstation and NT Server:** The operating system comes in two basic versions—Windows NT Workstation and Windows NT Server.

  Windows NT Workstation, which supports one or two processors, looks exactly like Windows 95 and runs most of the programs written for Windows 95. Its power benefits graphics artists, engineers, and others who use workstations and who do intensive computing at their desks. With this version, one can do computer graphics, computer-aided design, animation, and multimedia. Because NT Workstation is relatively inexpensive and can also support group networking for up to ten users, it is becoming a force in small business and home-office networking.

  Windows NT Server supports up to 32 processors. Users of this version consist of those tied together in "client/server" networks with "file server" computers. A *client/server network* is a type of local area network (LAN). The "client" is the requesting PC or workstation (usually running a version of Windows) and the "server" is the supplying file-server or mainframe computer, which maintains databases and processes requests from the client. A *file server* is a high-speed computer in a LAN that stores the programs and files shared by the users. In addition to supporting file servers, Windows NT Server offers Internet and site management services, office support, and database management.

- **OLE and ActiveX—to join several software applications:** Pronounced "oh-*lay*," *OLE* stands for "object linking and embedding," and it is a tool for joining several software applications. *ActiveX* refers to a special technology for inserting and exchanging data and programming code. (We discuss ActiveX in Chapter 11.)

  For example, suppose you are writing a sales report in a word processing program and you embed in it a bar chart showing sales figures from a spreadsheet program. The beauty of OLE is that it remembers where that bar chart is stored and in what application it was created. Thus, if you enter new sales figures into your spreadsheet and the bar chart changes, OLE will supply the new chart to your sales report as well. OLE can merge results—graphics, data, video clips, and so on—from several different kinds of software applications.

  Microsoft developed OLE to support documents that contain or reference other documents. Now it has evolved into a basic component of Windows NT, and also Windows 95/98, and has emerged as an industry standard.

Microsoft has improved Windows NT to the point where the company is now setting its sights on heavy-duty corporate computing. This means taking on the big networks that typically run Unix or mainframe software that can exchange data among thousands of users at once—a task beyond personal computing software until now. A benefit is that by using the Windows interface for many different strengths of machines, companies don't have to train employees to use a mix of operating systems. Analysts believe that, given Microsoft's track record, it is just a matter of time before NT finds its way up the corporate ladder.[22–24]

Windows CE: Scaled-Down Windows for Handheld Computing Devices

In late 1996, Microsoft released **Windows CE, a greatly slimmed-down version of Windows 95 for handheld computing devices.** It has some of the familiar Windows look and feel and includes rudimentary word processing, spreadsheet, e-mail, Web browsing, and other software.

The devices on which Windows CE is supposed to run include pocket-sized computers, electronic Rolodex-type organizers, and digital TV set-top boxes. Many of the handheld gadgets are designed to serve as companions to desktop computers, so that users can swap information such as phone numbers, calendars, notes, and "to do" lists between the two. (We discuss some of these devices in The Experience Box at the end of Chapter 4.)

Windows 98: End of the Line for DOS, Windows 3.x, & Windows 95 Software Code?

"For more than a decade," writes Walter Mossberg, "Microsoft Corp. has built its Windows operating system by borrowing liberally from design features popularized by Apple Computer on its Macintosh machines." Now, however, "it's on its own. It is Microsoft's turn to break new ground in ease of use."[25] This is the challenge to Windows 98, the successor to Windows 95.

Released in 1998, **the Windows 98 operating system features what Microsoft calls True Web Integration—a graphical user interface, or "desktop," that not only acts like Web browser software but also allows users access to data on the Internet as easily as if it were stored on the user's hard disk.** From a visual standpoint, with True Web Integration it would make no difference to users whether information was out on the Web or stored somewhere in a user's own PC. Microsoft's strategy, Mossberg suggests, "was to make Web browsers obsolete by letting you surf the Web right from your Windows desktop." By incorporating the Web browser within the operating system, the new design seemed to take dead aim at Microsoft's principal browser competitor, the popular Netscape Navigator, which must be installed separately.

Besides the browser integration, the most significant features of Windows 98 are the following:

- Free Internet content via TV: The convergence of technologies continues. Under an agreement between Microsoft and WavePhore of Phoenix, computers equipped with television reception will be able to receive free, over an unused portion of the television broadcast spectrum, Internet information such as news, sports, weather, and entertainment. The content is to be supported by advertising.[26]

- Changes in the "desktop": The Desktop, the screen that first appears when Windows is loaded, looks much the same as in Windows 95 but with additions. It not only lists all your programs but also your favorite Web sites. In addition, Windows 98 shows little thumbnail previews of some types of files.

  A feature called Active Desktop changes the background of the desktop into a Web page that enables you to view live content from the Web, such as scrolling stock prices, sports scores, or (unfortunately) advertisements.[27]

- System software to support new hardware: Special system software ("drivers") will support not only television tuners but also new types of disk players and other leading-edge hardware devices. In addition, Windows 98 supports the ability to have the computer turn on and off automatically to perform tasks while the user is away.

Despite the hoopla over Windows 98, Microsoft is actually trying to urge consumers to move toward Windows NT. Bill Gates has stated that even home consumers are expected in time to switch to Windows NT, for which there will eventually be a home version. Says one report, "Windows 98 is likely to be the last major release of an operating system based on the line of software code that extends from its original DOS system through Windows 3.1 to Windows 95."[28]

3.4 Utility Programs: Software to Make Your Computer Run Smoother

KEY QUESTION

How do utility programs interact with and extend an operating system's features and capabilities?

Preview & Review: Utility programs either enhance existing functions or provide services not performed by other system software. They include backup, data recovery, file defragmentation, disk repair, virus protection, data compression, and memory management. Multiple-utility packages are available.

"You wouldn't take a cruise on a ship without life preservers, would you?" asks one writer. "Even though you probably wouldn't need them, the terrible *what if* is always there. Working on a computer without the help and assurance of utility software is almost as risky."[29]

The "what if" being referred to is an unlucky event, such as your hard-disk drive "crashing" (failing), risking loss of all your programs and data; or your computer system being invaded by someone or something (a virus) that disables it.

Utility programs are special programs that either enhance existing functions or provide services not provided by other system software programs. Most computers come with utilities built in for free as part of the system software (Windows 95/98 offers several of them), but they may also be bought separately as external utility programs.

Some Specific Utility Tasks

The principal services offered by utilities are the following. Some of these features are essential to preventing or rescuing you from disaster. Others merely offer convenience.

- **Backup:** If you have only one utility it should be this one.
  Suppose your hard-disk drive suddenly fails, and you have no more programs or files. Or maybe, as one article puts it, there's just "a digital hiccup that fries a text file."[30] With **a backup utility, which makes a duplicate of every file on your hard disk on diskettes or other removable storage medium** (such as a Zip cartridge), you can be back in business with your data and programs intact.
  A backup utility is integral with Windows 95 and 98. Examples of freestanding commercial backup utilities are Norton Backup from Symantec and Colorado Scheduler.

- **Data recovery:** One day in the 1970s, so the story goes, programming legend Peter Norton was working at his computer and accidentally deleted an important file. This was, and is, a common enough error. However, instead of re-entering all the information, Norton decided to write a computer program to recover the lost data. He called the

README

Case Study: What to Do If the Disk with the Only Copy of Your Novel Fails

Always make a backup copy of your files.
That's Rule No. 1.
Rule No. 2: See Rule No. 1.
While we're at it, let's mention Rule No. 3—that it's a good idea to *make a backup copy of your files.*

What this means is that you should regularly—anywhere from every 10 minutes to once a day—take the important files of data you're currently working on that you've saved to your hard-disk drive (or diskette) and *copy them* onto diskettes, disk cartridges, or a backup tape. The principle is that you should never have just *one* copy of anything.

Now suppose you're like Soo-Yin Jue and didn't follow Rules 1 through 3.[31] For nine years, Jue worked on her first novel, traveling to Asia and gathering notes, which she entered onto her faithful Macintosh. Then, after she finished writing a first draft of her novel, she went to hit "Save."

The diskette on which she had been saving all her work began to spin.

And then the Mac made an odd grinding noise.

Nothing she did would bring any of her data back to the screen.

And she had not saved any of nearly a decade's worth of original material, either on another diskette or on the hard-disk drive of some other computer.

Enter DriveSavers of Novato, California, one of a handful of companies authorized by diskmakers to do rescue work. Whether the loss results from an unrealistic faith in the invulnerability of technology or from spilled coffee, floods, or even fires afflicting storage units, DriveSavers specializes in making data-recovery miracles happen.

The small staff of engineers uses a variety of repair techniques, software, cleansers, even a sterile "clean room" free of dust particles to resurrect data from drives bound for the junkyard. Usually this can be done in about two days, for a charge of around $800.[32] The company claims a 95% success rate.

Soo-Yin Jue was one of the lucky ones. Data-recovery engineer John Christopher told her that he could rescue the file for her novel, though he couldn't save any formatting, such as paragraph marks or page breaks.

Jue was jubilant. "As long as you have the words, I don't care about anything else," she said.[33]

What about the 5% of cases in which DriveSavers is unsuccessful?

The company gets paid no matter what, charging about 10% of its normal fee when it's unable to bring data back from the dead. For those cases in which there is no hope, DriveSavers employs a "data crisis counselor," Nikki Stange.

"People express panic, guilt, anger, and fear," says Stange. To counsel them, she says, "I use techniques I developed when working on a suicide hotline."[34]

Say, what was that Rule 1 again . . .?

program The Norton Utilities. Ultimately it and other utilities made him very rich.[35]

A ***data-recovery utility*** **is used to resurrect, or "undelete," a file or information that has been accidentally deleted.** The data or program you are trying to recover may be on a hard disk or a diskette. *Undelete* means to undo the last delete operation that has taken place.

Windows 95 and 98 have a built-in data-recovery utility called Recycle Bin, from which you can undelete files you accidentally disposed of in the Bin. Another recovery utility is the Norton Unerase Wizard (part of Norton Utilities for Windows 95).

- **File defragmentation:** Over time, as you delete old files from your hard disk and add new ones, something happens: the files become *fragmented*. **Fragmentation is the scattering of portions of files about the disk in nonadjacent areas, thus greatly slowing access to the files.**

  When a hard disk is new, the operating system puts files on the disk contiguously (next to one another). However, as you update a file over time, new data for that file is distributed to unused spaces. These spaces may not be contiguous to the older data in that file. It takes the operating system longer to read these fragmented files. **A *defragmenter utility program*, commonly called a "defragger," will find all the scattered files on your hard disk and reorganize them as contiguous files.** Defragmenting the file will speed up the drive's operation.

  An example of a utility for unscrambling fragmented files is Norton Speed Disk.

- **Disk repair:** There are all kinds of small glitches that can corrupt the data and programs on your hard-disk drive. For instance, a power surge in the house electricity may cause your files to become cross-linked. **A *disk-repair utility* will check your hard-disk drive for defects and make repairs on the spot or mark the bad areas.**

  An example of a disk-repair utility is Norton's Disk Doctor.

- **Virus protection:** Few things can make your heart sink faster than the sudden failure of your hard disk. One exception is the realization that your computer system has been invaded by a virus. **A *virus* consists of hidden programming instructions that are buried within an applications or systems program. They copy themselves to other programs, causing havoc.** Sometimes the virus is merely a simple prank that pops up a message. ("Have a nice day.") Sometimes, however, it can destroy programs and data. Viruses are spread when people exchange diskettes or download (make copies of) information from computer networks.

  Fortunately, antivirus software is available. ***Antivirus software* is a utility program that scans hard disks, diskettes, and memory to detect viruses.** Some utilities destroy the virus on the spot. Others notify you of possible viral behavior. Because new viruses are constantly being created, you need the type of antivirus software that can detect unknown viruses.

  Examples of antivirus software are Norton AntiVirus, Dr. Solomon's Anti-Virus Toolkit, McAfee VirusScan, and Webscan. "New viruses are appearing every day," advises one article, "so one key feature to look for in an antivirus utility is frequent updates without additional cost."[36] (As with many other types of software, you can download virus software updates from the Web.)

- **Data compression:** As you continue to store files on your hard disk, it will eventually fill up. You then have four choices: You can delete old files to make room for the new. You can buy a new hard disk with more capacity and transfer the old files and programs to it. You can add an external hard drive with removable disk cartridges. Or you can buy a data compression utility.

  ***Data compression utilities* remove redundant elements, gaps, and unnecessary data from a computer's storage space so less space (fewer bits) is required to store or transmit data.** With a data compression utility, files can be made more compact for storage on your hard-disk drive. The files are then "stretched out" again when you need them.

Examples of data compression programs are DriveSpace 2 Stacker from Stac Electronics, Double Disk from Verisoft Systems, and SuperStor Pro from AddStor.

Compression and decompression are important matters because of the increased use of large graphic, sound, and video files in computing. Large files affect both the amount of storage space you need and the amount of time required to transmit data over a network. Increasingly, however, data compression and decompression will be taken over by built-in hardware, which will make compression/decompression utilities obsolete. (This subject is discussed in more detail in Chapter 6.)

● **Memory management:** Different microcomputers have different types of memory, and different applications programs have different memory requirements. ***Memory-management utilities* are programs that determine how to efficiently control and allocate memory resources.**

Memory-management programs may be activated by software *drivers*. **Drivers are small software programs that allow the operating system to communicate with hardware devices,** such as a mouse or printer. Electrical and mechanical requirements differ among peripheral devices. Thus, software drivers are needed so that the computer's operating system will know how to handle them. Many basic drivers come with the operating system. If, however, you buy a new peripheral device, such as a CD-ROM drive, a driver will come packaged with it, and you'll have to install it on your computer's hard-disk drive before the device will operate.

Multiple-Utility Packages

Some utilities are available singly, but others are available as "multipacks." These multiple-utility packages provide several utility disks bundled in one box, affording considerable savings. Examples are Symantec's Norton Desktop (for DOS, Windows, or Macintosh), which provides data-recovery, defragmenting, memory-management, screen-saving, and other tools. Similar combination-utility packages are 911 Utilities from Microcom, and PC Tools from Central Point Software.

3.5 The Network Computer: Is the Web Changing Everything?

KEY QUESTION

How could the network computer make the choice of PC operating system irrelevant?

Preview & Review: New computers might follow the model of network PCs, without their own operating systems, and be dominated by Web browsers to access the Web and applications software located anywhere on the network. Instead of networks having "fat client" computers loaded with software and doing most of their own processing, "fat servers" would perform most of the processing for "thin clients," network PCs that are relatively cheap and leave the upgrading to the network.

Nothing stands still. System software developers toil on the versions to come, those works in progress to which they have given fanciful code names. However, in a matter of just a couple of years, the Internet and the World Wide Web have dramatically changed the picture for system software.

Software for Online Computing: Today Versus Tomorrow

Even with the dominance of Microsoft Windows, today personal computing is complicated because of conflicting standards. Could it be different tomorrow as more and more people join the trend toward networked computers and access to the World Wide Web?

As we've seen, there are different platforms—the Macintosh platform versus the PC platform, for example, or Unix versus Windows NT. Developers of applications software, such as word processors or database managers, need to make different versions if they are to run on all the platforms.

Networking complicates things even further. "Text, photos, sound files, video, and other kinds of data come in so many different formats that it's nearly impossible to maintain the software needed to use them," points out one writer. So far, users have had to "steer their own way through the complex, upgrade-crazy world of computing."[37]

Is this now changing? Let's consider today's model versus tomorrow's proposed model.

- **Today's model—more user responsibility:** Today microcomputer users who wish to access online data sources must provide not only their own computer, modem, and communications software but also their own operating system software and applications software. (■ *See Panel 3.9, top.*)

■ PANEL 3.9

Online personal computing—today and tomorrow

(Top) Today users provide their own operating system software and their own applications software and are usually responsible for installing them on their personal computers. They are also responsible for any upgrades of hardware and software. Data can be input or downloaded from online sources. *(Bottom)* Tomorrow, according to this model, users would not have to worry about operating systems or even about having to acquire and install (and upgrade) their own applications software. Using a universal Web browser, they could download not only data but also different kinds of applications software from an online source.

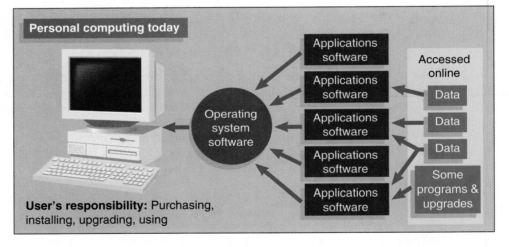

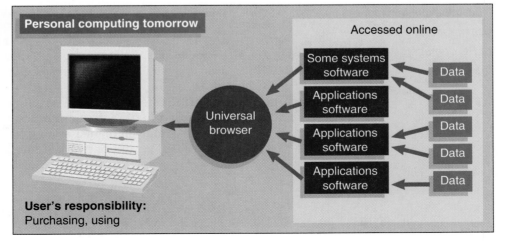

In addition, you must also take responsibility for making sure your computer system will be compatible with others you have to deal with. For instance, if a Macintosh user sends you a file to run on your PC, it's up to you to take the trouble to use special software that will translate the file so it will work on your system.

- **Tomorrow's model—more service provider responsibility:** What if the responsibility for ensuring compatibility between different systems were left to online service providers? In tomorrow's model, you would use your browser to access the World Wide Web and take advantage of applications software anywhere on the network. (■ *See Panel 3.9, bottom.)*

  In this arrangement, it will not matter what operating system you use. Applications software will become nearly disposable. You will download applications software and pay a few cents or a few dollars for each use. You could store frequently used software on your own computer. You will not need to worry about buying the right software, since it can be provided online whenever you need to accomplish a task.

Fat Clients Versus Thin Clients: Bloatware or Network Computers?

We see today's and tomorrow's models expressed in the concepts of *fat clients* versus *thin clients*.

- **Fat clients—computers with bloatware:** As we mentioned in Chapter 2 (✔ p. 81), *bloatware* is a colloquial name for software that is so crowded ("bloated") with features—it is afflicted with "featuritis"—that it requires a powerful microprocessor and enormous amounts of main memory and hard-disk storage capacity to run efficiently.

  When on a network, computers with bloatware are known as fat clients. That is, in a client/server network arrangement, a *fat client* is a client computer that performs most or all of the application processing; little or no processing is done by the network server.

- **Thin clients—slimmed-down network computers:** To staunch the continual expansion in bloatware, engineers proposed the notion of a stripped-down computer known as the network computer. Here muscular microprocessors and operating systems are replaced by a "hollowed-out" computer costing $500 or less that serves as a mere terminal or entry point to the online universe. Thus, the network computer—which might not even have a hard disk—is a peripheral of the Internet, with most software, processing, and information needs being supplied by remote servers.

  The network computer is known as a thin client. In a client/server network arrangement, a *thin client* is a client computer that performs little or no application processing. All or most processing is done by the server, which is thus known as a *fat server.*

The thin-client/fat-server strategy is to replace existing bloatware-stuffed desktop computers with a new generation of ultracheap network computers. At present there are two standards of network computers. One standard (called the NC Reference Profile) is backed by Apple, IBM, Netscape, Oracle, Sun, and others. The other standard is backed by Microsoft and chipmaker Intel.

The concept of the network computer has raised some interesting questions:[38-40]

- **Would the browser really become the OS?** Would a Web browser or some variant become the operating system? Or will existing operating systems expand, as in the past, taking over browser functions?

- **Would communications functions really take over?** Would communications functions become the entire computer, as proponents of the network computer contend? Or would they simply become part of the personal computer's existing repertoire of skills?

- **Would an NC really be easy to use?** Would a network computer really be user friendly? At present, features such as graphical user interfaces require lots of hardware and software.

- **Aren't high-speed connections required?** Even users equipped with the fastest modems would find downloading even small programs ("applets") time-consuming. Doesn't the network computer ultimately depend on faster connections than are possible with the standard telephone lines and modems now in place?

- **Doesn't the NC run counter to computing trends?** Most trends in computing have moved toward personal control and access, as from the mainframe to the microcomputer. Wouldn't a network computer that leaves most functions with online sources run counter to that trend?

- **Would users go for it?** Would computer users really prefer scaled-down generic software that must be retrieved from the Internet each time it is used? Would a pay-per-use system tied to the Internet really be cheaper in the long run? Why would anyone buy a $500 stripped-down box when he or she can get a full-fledged computer for prices that are rapidly dropping below $1000?[41]

Onward: Toward Compatibility

The push is on to make computing and communications products compatible. Customers are demanding that computer companies work together to create products that will make it easy to access and use great amounts of information. As technological capabilities increase, so will the demand for simplicity.

Whether compatibility and simplicity will be provided by a proprietary system like Windows 98 or Windows NT or by "open standards" of some sort of Web software, perhaps, finally, the best products will triumph.

Using Software to Access the World Wide Web

We cover the Internet, the Web, and other communications topics in detail in Chapters 7 and 8. However, you may want to get online and use the Web now. If so, the following discussion is designed to help.

What's the easiest way to use the Internet, that international conglomeration of thousands of smaller networks? Getting on the sound-and-graphics part of it known as the World Wide Web is no doubt the best choice. Increasingly, system software is coming out with features for accessing and exploring the Net and the Web. This Experience Box, however, describes ways to tour both the Net and the Web independent of whatever system software you have.

The Web resembles a huge encyclopedia filled with thousands of general topics or so-called Web sites that have been created by computer users and businesses around the world. The entry point to each Web site is a home page, which may offer cross-reference links to other pages. Pages may be in multimedia form—meaning they can include text, graphics, sound, animation, and video.

To get on the Net and the Web, you need a microcomputer, a modem, a telephone line, and communications software. (For details about the initial setup, review pages 89–93 in the Experience Box at the end of Chapter 2.) You then need to gain access to the Web and, finally, to get a browser. Some browsers come in kits that handle the setup for you, as we will explain.

Gaining Access to the Web

There are three principal ways of getting connected to the Internet: (1) through school or work, (2) through commercial online services, or (3) through an Internet service provider.

Connecting Through School or Work The easiest access to the Internet is available to students and employees of universities and government agencies, most colleges, and certain large businesses. If you're involved with one of these, you can simply ask another student or coworker with an Internet account how you can get one also. In the past, college students have often been able to get a free account through their institutions. However, students living off campus may not be able to use the connections of campus computers.

Connections through universities and business sites are called *dedicated*, or *direct*, connections and consist of local area networks and high-speed phone lines (called T1 or T3 carrier lines) that typically cost thousands of dollars to install and maintain every month. Their main advantage is their high speed, so that the graphic images and other content of the Web unfold more quickly. (Note that if you are connected to

the Internet via some of the special high-speed *digital* lines, you won't need a modem.)

Connecting Through Online Services The large commercial online services—such as America Online/CompuServe, Microsoft Network, or Prodigy—also offer access to the Internet. (See the Experience Box at the end of Chapter 2 for information.) Some offer their own Web browsers, but some (such as America Online, or AOL) offer Netscape Navigator and Internet Explorer. Commercial online services may also charge more than independent Internet service providers, although they are probably better organized and easier for beginners.

Web access through online services is usually called a *dial-up connection.* As long as you don't live in a rural area, there's no need to worry about long-distance telephone charges; you can generally sign on ("log on") by making a local call. When you receive membership information from the online service, it will tell you what to do.

Connecting Through Internet Service Providers Internet service providers (ISPs) (✔ p. 69) are local or national companies that provide public access to the Internet for a fee. Examples of national companies include PSI, UUNet, Netcom, and Internet MCI. Telephone companies such as AT&T and Pacific Bell have also jumped into the fray by offering Internet connections. Most ISPs offer a flat-rate monthly fee for an unlimited number of hours of use. The connections offered by ISPs may offer faster access to the Internet than those of commercial online services.

The whole industry of Internet connections is still new enough that many ISP users and online service users have had problems with uneven service (such as busy signals or severing of online "conversations"). Often ISPs signing up new subscribers aren't prepared to handle traffic jams caused by a great influx of newcomers.

In Chapter 7, we provide some tips for choosing an ISP. You can also ask someone who is already on the Web to access for you the worldwide list of ISPs at *http://www.thelist. com.* Besides giving information about each provider in your area, "thelist" provides a rating (on a scale of 1 to 10) by users of different ISPs.

Accessing the Web: Browser Software

Once you're connected to the Internet, you need a Web browser. This software program will help you to get whatever information you want on the Web by clicking your mouse pointer on words or pictures the browser displays on your screen.

The two-best known browsers are Netscape Navigator and Microsoft's Internet Explorer.

Features of Browsers What kinds of things should you consider when selecting a browser? Here are some features:

- **Price:** Some browsers are free (freeware, ✔ p. 86). Some are free to students only or come free with membership in an online service or ISP—America Online, for instance, offers browsers for both PCs and Macintoshes. Some may be acquired for a price separately from any online connection.

    You can get a kit that offers other features besides a browser. For instance, Macintosh users can buy the Apple Internet Connection Kit, which contains the browser Netscape Navigator/Communicator and several other programs. (They include Claris E-mailer Lite, News Watcher, Fetch, Alladin Stuffit Expander, NCSA Telnet, Adobe Acrobat Reader, Sparkle, Real Audio, MacTCP, MacPPP, and Apple Quicktime VR Player). The kit comes with an Apple Internet Dialer application that helps you find an Internet service provider.

- **Ease of setup:** Especially for a beginner, the browser should be easy to set up. Ease of setup favors the university/business dedicated lines or commercial online services, of course, which already have browsers. If not provided by your online service or ISP, the browser should be compatible with it. Most online services allow you to use browsers besides their own.

- **Ease of use:** If you have a multimedia PC, the browser should allow you to view and hear all of the Web's multimedia—not only text and images but also sounds and video. It should be easy to use for saving "hot lists" of frequently visited Web sites and for saving text and images to your hard disk. Finally, the browser should allow you to do "incremental" viewing of images, so that you can go on reading or browsing while a picture is slowly coming together on your screen, rather than having to wait with browser frozen until the image snaps into view.

Surfing the Web

Once you are connected to the Internet and have used your browser to access the Web, your screen will usually display the browser's home page. (■ See Panel 3.10.) (You can determine whose home page you want to see after you load your browser.)

Web Untanglers Where do you go from here? You'll find that unlike a book, there is no page 1 where everyone is supposed to start reading, and unlike an encyclopedia, the entries are not in alphabetical order. Moreover, there is no definitive listing of everything available.

There are, however, a few search tools for helping you find your way around, which can be classified as directories and search engines (✔ p. 71).

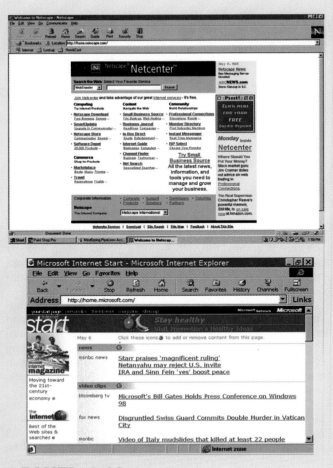

■ PANEL 3.10

Home page for a Web browser
(*Top*) Netscape Navigator's home page. (*Bottom*) Microsoft's Internet Explorer home page.

Web Addresses: URLs Getting to a Web location is easy if you know the address. Just choose, in most cases, Open location (Find location) from the browser's file menu, and then type in the address. Web addresses usually start with *http* (for Hypertext Transfer Protocol) and are followed by a colon and double slash (*://*). For example, to reach the home page of Yahoo!, you would type the address: *http://www.yahoo.com*. Your browser uses the address to connect you to the computer of the Web site; then it downloads (transfers) the Web page information to display it on your screen. (In most cases now, you can skip the *http://* and just start with *www*.)

If you get lost on the Web, you can return to your opening home page by clicking on the Home button.

Nowadays you see Web site addresses appearing everywhere, in all the mass media, and some of it's terrific and some of it's awful. For a sample of the best, try Yahoo!'s What's Cool list (*http://www.yahoo.com/Entertainment/ COOL_Links*). For less-than-useful information, try Worst of the Web (*http://turnpike.net/metro/mirsky/Worst.html*).

How to Use Directories & Search Engines

How can you best explore the Net and the Web? The key, says personal computing writer Michael Martin, is to apply two simple concepts, both of which derive from methods we are accustomed to using for finding information in other areas of life: *browsing* and *hunting.*[42]

- **Directories—for browsing:** Browsing, says Martin, "involves looking in a general area of interest, then zooming in on whatever happens to catch your attention." For example, he says, a basketball fan would head for the sports section of the newspaper, check the basketball news and scores on the front page, then skim other pages for related sports information.

   Directories—Yahoo! (✔ p. 71) is the best known— arrange resources by subject and thus are best for people who browse.

- **Search engines—for hunting:** Hunting "is what we do when we want specific information," Martin says. In his example, if you were hopelessly nearsighted and wanted to hunt up specifics on the latest advances in laser treatment, you might check with an ophthalmologist, a university library, the National Eye Institute, and so on.

   Search engines, such as AltaVista (✔ p. 71), Excite, HotBot, Infoseek, and Lycos are for those who want specifics.

If you use Netscape Navigator/Communicator, you can quickly access directories by clicking the Net Directory button and access search engines by clicking the Net Search button. We explain some of the principal directories and search engines below.

Directories: For Browsing *Directories* provide lists of Web sites covering several categories. As we mentioned earlier, directories are managed and organized by people. These are terrific tools if you want to find Web sites pertinent to a general topic you're interested in, such as bowling, heart disease, or the Vietnam War. For instance, in Yahoo! you might click your mouse on one of the general headings listed on the menu, such as Recreation or Health, then proceed to click on menus of subtopics until you find what you need.

Some general directories are the following:

- **Yahoo!** *(http://www.yahoo.com)* is the most popular of the Web directories and lists half a million Web sites. The home page lists several topics, with some subtopics listed beneath. Among other things, Yahoo! features a weekly list of "cool sites" and Cool Links connections (look for the sunglasses "cool" icon). Unlike other directories, it does not review sites, and its descriptions are rather brief.

- **Magellan** *(http://www.mckinley.com)* allows users to search 50 million Web sites. It also offers detailed overviews of more than 60,000 Web sites chosen and reviewed by Magellan's experts. Overviews include a short review and a percentage sign that rates a resource for relevance.

- **Netguide Live** *(http://www.netguide.com)* has fewer Web site listings than Yahoo! but, unlike Yahoo!, offers reviews and evaluations of the sites listed.

- **The Mining Company** *(http://www.miningco.com)* includes comprehensive Web sites for over 500 topics, run by outside experts who compile lists of sites that deal with their areas of expertise. Each site is devoted to a single topic, complete with site reviews, feature articles, and discussion areas.

- **Galaxy** *(http://www.einet.net)* calls itself "the professional's guide to a world of information" and employs professional information specialists to organize and classify Web pages. It includes resources for professionals in nine categories, including business, law, medicine, government, and science.

- **The Argus Clearinghouse** *(http://www.clearinghouse. net/index.html)* is a directory of directories, or "virtual library"; that is, it is maintained by "digital librarians" who identify, select, evaluate, and organize resources. Argus provides a list of subject-specific directories on topics ranging from arts and entertainment to social science and social issues.

- **The Internet Public Library** *(http://www.ipl.org)* is another "virtual library." It began in 1995 in the School of Information and Library Studies at the University of Michigan and has the goal of providing library services to the Internet community. Besides offering Web searches and lists of books and periodicals, it maintains its own collection of over 12,000 hand-picked and organized Internet resources.

Search Engines: For Specifics *Search engines*, such as AltaVista, are best when you're trying to find very specific information—the needle in the haystack. Search engines are Web pages containing forms into which you type keywords to suggest the subject you're searching for. The search engine then scans its database and presents you with a list of Web sites matching your search criteria. (Search engines are managed and organized by machines and software.)

The search engine's database is created by spiders (also known as crawlers or robots), software programs that scout the Web looking for new sites. When the spider finds a new page, it adds its Internet address (URL), title, and usually the headers starting each section to an index in the search engine's database. The principal search engines add index information about new pages every day.

Writer Richard Scoville points out that the bigger the database, the greater your chances for success in your search. For example, he says, he queried several engines with the keywords *recipe wheat beer.* "The massive Lycos database gave us 437 *hits* (matched pages) in return. InfoSeek and Open Text Index gave us around 200 each; others, less than 100."[43]

Important search engines include the following:

- **AltaVista** *(http://www.altavista.digital.com),* probably the largest and probably best known of the search engines, has over 100 million indexed Web pages, roughly twice as many as its competitors. It also takes care of all the searches that spill over from Yahoo! Thus, if you start your search on Yahoo! and can't find what you're looking for there, you'll automatically be switched to AltaVista. If you're worried you might miss a Web site, AltaVista is the search engine to use first. Along with HotBot, it should be one of the first Web searching tools you use.

- **Excite** *(http://www.excite.com)* is considered both a directory and a search engine and has 50 million indexed pages and 140,000 Yahoo!-style listings. Besides searching by exact words, it also searches by concept. For example, a query for "martial arts" finds sites about *kick-boxing* and *karate* even if the original search term isn't in the page.[44] After you do a search, Excite will also suggest words to use to narrow the query. In addition, it ranks the documents as to relevancy—that is, as to how well they fit your original search criteria.

- **HotBot** *(http://www.hotbot.com)* reindexes its 54 million Web pages every two weeks, which can often yield more recent material than is found using other search engines. In fact, you can find only up-to-date pages by limiting your search to those pages that have changed only in the last 3–6 months. Like Excite, HotBot offers relevancy rankings. Along with AltaVista, HotBot should be one of the first search tools you use when you're looking for something.

- **InfoSeek** *(http://www.infoseek.com)* is a blend of search engine, directory, and news service. It has 60 million Web pages and an extensive directory and ranks results according to relevance to your search criteria. InfoSeek also searches more than the Web, indexing Usenet newsgroups and several non-Internet databases. Once you complete a search, you can search with those results, by constructing a new query and clicking on the "Search These Results" button.

- **Lycos** *(http://www.lycos.com)*, which combines directory and search services, offers 30 million indexed pages. It also offers a list of interesting Web sites called A2Z, which indicates the most popular pages on the Web, as measured by the number of hypertext links, or "hits," from other Web sites pointing to them. The Lycos relevancy rankings are among the strongest of the search engines.

The best way to make a search engine useful is to be extremely specific when formulating your keywords. More on this below.

Metasearch Engines

Metasearch engines are search tools that let you use several search engines to track down information, although you are somewhat restricted compared to using single search en-

gines. Most metasearch engines also include directories, such as Yahoo! Examples of these "one-stop-shopping" sites are Savvy Search *(http://www.cs.colostate.edu/~dreiling/smartform.html)* and MetaCrawler *(http://www.cs.washington.edu/research/projects/ai/metacrawler/www/home.html).*

Tips for Searching

Here are some rules that will help improve your chances for success in operating a search engine:[45,46]

- **Read the instructions!** Every search site has an online search manual. Read it.

- **Make your keywords specific:** The more narrow or distinctive you can make your keywords, the more targeted will be your search. Say *drag racing* or *stock-car racing* rather than *auto racing*, for example. Also try to do more than one pass and try spelling variations: *drag racing, dragracing, drag-racing.* In addition, think of synonyms, and write down related key terms as they come to mind. Finally, be sure to enclose phrases within quotation marks—"drag racing" or "jet plane"— so the search tool will know that the words belong together.

- **Use AND, OR, and NOT:** Use connectors as a way of making your keyword requests even more specific. In Martin's example, if you were looking for a 1965 Mustang convertible, you could search on the three terms "1965," "Mustang," and "convertible." However, since you want all three together, try linking them with a connector: "Mustang AND convertible AND 1965." You can also sharpen the keyword request by using the word NOT for exclusion—for example, "Mustang NOT horse."

- **Don't bother with "natural language" queries:** Some search engines will let you do *natural language queries,* which means you can ask questions as you might in conversation. For example, you could ask, "Who was the Indianapolis 500 winner in 1998?" You'll probably get better results by entering "Indianapolis 500 AND race AND winner AND 1998."

- **Use more than one search engine:** "We found surprisingly little overlap in the results from a single query performed on several different search engines," writes Scoville. "So to make sure that you've got the best results, be sure to try your search with numerous sites."

All these search tools are constantly adding new features, such as easier interfaces. But whichever you end up using, you'll find that they can turn the Web from a playground or novelty into a source of real value. For additional search tips, check out Search Engine Watch: (*http://www.searchenginewatch.com*) and Internet Searching Strategies (*http://www.rice.edu/Fondren/Netguides/strategies.html*).

Summary

What It Is/What It Does

Why It's Important

antivirus software (p. 128, KQ 3.4) Software utility that scans hard disks, floppy disks, and microcomputer memory to detect viruses; some antivirus utilities also destroy viruses.

Computer users must find out what kind of antivirus software to install on their systems in order to protect them against damage or shutdown.

backup utility program (p. 126, KQ 3.4) Operating system housekeeping or utility program that duplicates the contents of a hard disk onto a removable storage medium, such as diskettes or a hard-disk cartridge.

Backing up is an essential system software function; users should back up all their work so that they don't lose it if original disks are destroyed.

booting (p. 106, KQ 3.2) Refers to the process of loading an operating system into a computer's main memory from disk.

When a computer is turned on, a program (called the *bootstrap loader* or *boot routine*) stored permanently in the computer's electronic circuitry obtains the operating system from the floppy disk or hard disk and loads it into main memory. Only after this process is completed can the user begin work.

command-driven interface (p. 107, KQ 3.2) Type of user interface that requires users to enter a command by typing in codes or words.

The command-driven interface is used on IBM and IBM-compatible computers with the DOS operating system.

data compression utility (p. 128, KQ 3.4) Software utility that removes redundant elements, gaps, and unnecessary data from computer files so less space is required to store or transmit data.

Many of today's files, with graphics, sound, and video, require too much storage space; data compression utilities allow users to reduce the space they take up.

data recovery utility (p. 127, KQ 3.4) Utility program that is used to resurrect, or "undelete," a file or information that has been accidentally deleted.

A file or any part of a file deleted by mistake can be recovered with a data recovery utility.

defragmenter utility (p. 128, KQ 3.4) A utility program that finds all the scattered files on a hard disk and reorganizes them as contiguous files; commonly called a "defragger."

Defragmenting a hard disk allows it to provide information faster because it doesn't have to constantly move to find the scattered parts (fragments) of a file.

desktop (p. 108, KQ 3.2) Graphical user interface screen of an operating system; it serves as a basic screen that presents icons on the computer screen representing items ready for use on the top of a businessperson's desk.

Items can be manipulated as on a desktop—open a folder to check information, to work on a project, or to move a file to another folder.

What It Is/What It Does	**Why It's Important**

disk-repair utility (p. 128, KQ 3.4) Utility program that checks a hard-disk drive for defects and either repairs them or marks the bad areas.

All kinds of glitches can corrupt the data and programs on a hard-disk drive, such as a power surge that causes files to become cross-linked. A disk-repair utility can find these problems and either fix the disk or mark where the damage is.

DOS (disk operating system) (p. 115, KQ 3.3) Microcomputer operating system that runs primarily on PCs. DOS is sold under the names MS-DOS by Microsoft Corporation and PC-DOS by IBM.

DOS is the second most common microcomputer operating system after the various Windows systems, making it an important *legacy system.*

driver (p. 129, KQ 3.4) Small software programs that allow the operating system to communicate with peripheral hardware, such as a mouse or printer.

Drivers are needed so that the computer's operating system will know how to handle the data and run the peripheral device. A user who buys a new piece of peripheral hardware and hooks it up to a system will also probably have to install that hardware's driver software on the hard disk.

formatting (p. 107, KQ 3.2) Also called *initializing;* a computer process that electronically prepares a diskette so it can store data or programs.

Before using a new diskette, the user has to format it unless it is labeled as formatted.

fragmentation (p. 128, KQ 3.4) The scattering of parts of files on nonadjacent areas of a hard disk, which slows down access to the files.

Fragmentation causes operating systems to run slower; to solve this problem, users can buy a file defragmentation software utility.

graphical user interface (GUI) (p. 108, KQ 3.2) User interface that uses icons and menus as well as keystrokes to allow the user to choose commands, start programs, and see lists of files and other options.

GUIs are easier to use than command-driven interfaces and menu-driven interfaces; they permit liberal use of the electronic mouse as a pointing device to move the cursor to a particular icon, button, or menu option on the display screen. The function represented by the screen item can be activated by pressing ("clicking") buttons on the mouse.

icon (p. 108, KQ 3.2) Small pictorial figure that represents a task, procedure, or program.

The function represented by the icon can be activated by pointing at it with the mouse pointer and pressing ("clicking") on the mouse. The use of icons has simplified the use of computers.

language translator (p. 105, KQ 3.1) System software that translates a program written in a computer language (such as BASIC) into the language that the computer can understand (machine language—0s and 1s).

Without language translators, software programmers would have to write all programs in machine language, which is difficult to work with.

legacy system (p. 113, KQ 3.3) An older computer system used in a home or office that employs outdated but functional technology.

Because they continue to perform their function, millions of legacy systems are still at work, and users may find they need to learn to operate them.

Macintosh operating system (Mac OS) (p. 116, KQ 3.3) Operating system used on Apple Macintosh computers or on Mac clones.

Although not used in as many offices as DOS and Windows, the Macintosh operating system is easier to use.

What It Is/What It Does	Why It's Important

memory-management utility (p. 129, KQ 3.4) Software utility that determines how to efficiently control and allocate memory resources.

Applications programs vary in their memory requirements, and microcomputers vary in types of memory; a memory-management utility allows the application to run more efficiently with the system it is on.

menu (p. 107, KQ 3.2) List on the computer screen of commands available for manipulating data.

Menus are used in graphical user interface programs to make software easier to use: the user can choose from a list instead of having to remember commands.

menu-driven interface (p. 107, KQ 3.2) User interface that allows users to choose a command from a menu.

Like a restaurant menu, a software menu offers options to choose from—in this case commands available for manipulating data. Two types of menus are available, menu bars and pull-down menus. Menu-driven interfaces are easier to use than command-driven interfaces.

NetWare (p. 122, KQ 3.3) A popular network operating system, from Novell, for orchestrating microcomputer-based local area networks (LANs) throughout a company or campus.

NetWare allows PCs to share data files, printers, and file servers.

network user interface (NUI) (p.109, KQ 3.2) A browser-like interface that helps users interact with on-line programs and files.

NUIs were originally designed for network computers where most software is located on another computer, but they are coming into use on PCs.

operating environment (p. 114, KQ 3.3) Also known as a *windowing environment* or *shell;* adds a graphical user interface as an outer layer to an operating system. Common features of these operating environments are use of an electronic mouse, pull-down menus, icons and other graphic displays, the ability to run more than one application (such as word processing and spreadsheets) at the same time, and the ability to exchange data between these applications.

Operating environments make command-driven system software easier to use; Windows 3.x is an operating environment used with DOS.

operating system (OS) (p. 106, KQ 3.1, 3.2) Principal piece of system software in any computer system; consists of the master set of programs that manage the basic operations of the computer. The operating system remains in main memory until the computer is turned off.

These programs act as an interface between the user and the computer, handling such matters as running and storing programs and storing and processing data. The operating system allows users to concentrate on their own tasks or applications rather than on the complexities of managing the computer.

OS/2 (Operating System/2) (p. 119, KQ 3.3) Micro-computer operating system designed to run on many recent IBM and IBM-compatible microcomputers.

OS/2 and its most recent version, Warp, offered a true operating system with a graphical user interface for PCs before Windows 95 was available. OS/2 and Warp can run most DOS, Windows, and OS/2 applications programs simultaneously, which means users who switch to OS/2 can keep their applications. Also, OS/2 is designed to connect everything from handheld computers to mainframes.

What It Is/What It Does	**Why It's Important**

platform (p. 112, KQ 3.3) Refers to the particular processor model and operating system on which a computer system is based—for example, IBM platform or Macintosh platform.

Users need to be aware that, without special arrangements or software, different platforms are not compatible.

plug and play (p. 120, KQ 3.3) Refers to the ability to add a new hardware component to a computer system and have it work without needing to perform complicated technical procedures.

Plug and play greatly simplifies the process of expanding and modifying systems.

supervisor (p. 109, KQ 3.2) Also called *kernel;* central component of the operating system as the manager of the CPU. It resides in main memory while the computer is on and directs other programs to perform tasks to support applications programs.

Were it not for the supervisor program, users would have to stop one task—for example, writing—and wait for another task to be completed—for example, printing out of a document.

system software (p. 104, KQ 3.1) Software that enables applications software to interact with the computer and helps the computer manage its internal resources. A computer's system software contains an operating system, utility programs, and language translators.

Applications software cannot run without system software.

Unix (p. 122, KQ 3.3) Operating system for multiple users, with built-in networking capability, the ability to run multiple tasks at one time, and versions that can run on all kinds of computers.

Because it can run with relatively simple modifications on many different kinds of computers, from micros to minis to mainframes, Unix is said to be a "portable" operating system. The main users of Unix are government agencies, universities, research institutions, large corporations, and banks that use the software for everything from designing airplane parts to currency trading.

user interface (p. 107, KQ 3.2) The part of the operating system that allows users to communicate, or interact, with it. There are four types of user interfaces: command-driven, menu-driven, graphical user, and network user.

User interfaces are necessary for users to be able to use a computer system.

utility programs (p. 126, KQ 3.1, 3.4) System software that either enhances existing functions or provides services not offered by other system software programs.

Many operating systems have utility programs built in for common purposes such as copying the contents of one disk to another. Other, external utility programs are available on separate diskettes to, for example, recover damaged or erased files.

virus (p. 128, KQ 3.4) Hidden programming instructions that are buried within an application or system program and that copy themselves to other programs, often causing damage.

Viruses can cause users to lose data or files or even shut down entire computer systems.

windows (p. 108, KQ 3.2) Feature of graphical user interfaces; causes the display screen to divide into sections. Each window is dedicated to a specific purpose.

Using the windows feature, an operating system (or operating environment) can display several windows on a computer screen, each showing a different application program, such as word processing, spreadsheets, and graphics.

What It Is/What It Does

Why It's Important

Windows 3.x (p. 118, KQ 3.3) Operating environment made by Microsoft that places a graphical user interface shell around the DOS operating system and extends DOS's capabilities.

The Windows 3.x operating environment made DOS easier to use, but only with Windows 95 did Windows become a true operating system.

Windows 95 (p. 120, KQ 3.3) Successor to Windows 3.x for DOS; this is a true operating system for PCs, rather than just an operating environment.

Windows 95 has become by far the most common system software on new microcomputers.

Windows 98 (p. 125, KQ 3.3) Latest release of the Windows operating system; uses a graphical user interface that acts like Web browser software and allows users to access data on the Internet as if it were stored on their hard disk.

Besides offering access to the Internet directly from the desktop, Windows 98 has new drivers to support television tuners and new types of disk players and other new hardware.

Windows CE (p. 125, KQ 3.3) A greatly slimmed-down version of Windows 95 for handheld computing devices. It includes basic word processing, spreadsheet, e-mail, Web browsing, and other software, and has a Windows look and feel.

A handheld computer, organizer, or TV set-top box with an operating system compatible with that of a desktop computer can readily exchange information with that computer.

Windows NT (New Technology) (p. 123, KQ 3.3) Network operating system that has multitasking, multiuser, and multiprocessing capabilities and a graphical user interface.

Multiuser systems like Windows NT are used to support computer workgroups. Windows NT comes in two versions, the Workstation version to support users who need a powerful system and no more than 10 computers in a network, and the more expensive Server version to support up to 32 processors in a LAN.

Exercises

Self-Test

1. A(n) _____ is software that translates a program written by a programmer into machine language.

2. _____ programs are special programs that either enhance existing functions or provide services not provided by other system software programs.

3. Software _____ are programs that allow the operating system to communicate with hardware devices.

4. _____ software is a program that scans hard disks, diskettes, and memory to detect viruses.

5. _____ utilities remove redundancies from a computer's storage space so less space is required to store or transmit data.

Short-Answer Questions

1. Why does a computer need system software?
2. What does the term *booting* mean?
3. What is a GUI?
4. What does the term *platform* refer to?
5. What is the difference between multitasking and time-sharing?

Multiple-Choice Questions

1. Which of the following is a multitasking, multiuser, multiprocessing network operating system?
   a. Windows 3.1
   b. Windows 95
   c. Windows NT
   d. DOS
   e. None of the above

2. Which of the following allows Microsoft Windows users to easily join and share data among applications?
   a. GUI
   b. OLE
   c. OS/2
   d. NUI
   e. NOS

3. Which of the following refers to the execution of two or more programs by multiple computers?
   a. multiprocessing
   b. multitasking
   c. time-sharing
   d. multiprogramming
   e. All of the above

4. Which of the following isn't an example of a file-management command?
   a. format
   b. copy
   c. rename
   d. erase
   e. All of the above

5. What do the terms *partitioning*, *foreground/background*, and *queues* relate to?
   a. disk management
   b. file management
   c. memory management
   d. operating system management
   e. All of the above

True/False Questions

T F 1. The operating system remains in main memory at all times when your computer is on.

T F 2. Applications software starts up the computer and functions as the principal coordinator of all hardware components.

T F 3. Command-driven interfaces allow you to use graphics and menus to start programs.

T F 4. A program that can defragment a disk is commonly referred to as a *defragger*.

T F 5. An example of a thin client is a network computer.

Knowledge in Action

1. If you have been using a particular microcomputer for two years and are planning to upgrade the version of systems software you are using, what issues must you consider before you go ahead and buy the new version?

2. If your computer runs Windows 95 or Windows 98, choose Settings, Control Panel from the Start menu to obtain information about your computer system. What are the current settings of your computer display (monitor)? Keyboard? Modem? Mouse? What other settings can you view in the Control Panel window?

3. Do you think the network computer will become a standard fixture in homes and businesses in the near future? If so, when? Research your answer on the Web and/or using current computer magazines.

4. What system software is used on the computer at your school, work, or home? Why was this software selected? Do you find this software easy to use? Would you prefer another type of system software or version upgrade? If so, why?

5. Locate someone who is using DOS and Windows 3.1. Why hasn't this person switched to Windows 95? Is the reason related to the existing hardware? Existing software? Other?

Processors

Hardware for Power & Portability

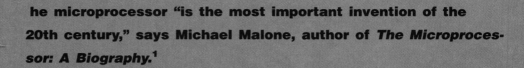

he microprocessor "is the most important invention of the 20th century," says Michael Malone, author of *The Microprocessor: A Biography.*[1]

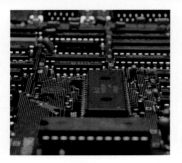

Quite a bold claim, considering the incredible products that have issued forth during the past nearly 100 years. More important than the airplane? More than television? More than atomic energy?

However, Malone argues, the case for the exalted status of this thumbnail-size information-processing device is demonstrated, first, by its pervasiveness in the important machines in our lives, from computers to transportation. (■ *See Panel 4.1.*) Second, "The microprocessor is, intrinsically, something special," he says. "Just as [the human being] is an animal, yet transcends that state, so too the microprocessor is a silicon chip, but more." Why? Because it can be programmed to recognize and respond to patterns in the environment, as humans do.

Indeed, this *is* something different. Until now, nothing that was inorganic—that was nonliving—was quite so adaptable.

■ PANEL 4.1

The pervasiveness of the microprocessor

"Everybody knows that the microprocessor is the brains of the personal computer, the video game, and the automated teller machine. But it also made possible the revolution in graphic workstations that gives us everything from 3D product design to *Jurassic Park*. That's just computation. The microprocessor also has led to a renaissance in test and measurement, bringing intelligence to a vast array of products from fetal monitors to gas chromatographs.

The microprocessor has transformed control and automation as well: all of those new industrial robots are run by microprocessors, so is the air conditioning, the gas pump and maybe even all the lights in your house. The modern airplane would have a hard time flying without hundreds of microprocessors. Nor would your automobile run safely and efficiently—many of the new models contain dozens running everything from the fuel injection to the windshield wipers."

—Michael S. Malone. *The Microprocessor: A Biography*

4.1 Microchips, Miniaturization, & Mobility

KEY QUESTION

What are the differences between transistors, integrated circuits, chips, and microprocessors?

Preview & Review: Computers used to be made from vacuum tubes. Then came the tiny switches called transistors, followed by integrated circuits made from silicon, a common mineral. Integrated circuits called microchips, or chips, are printed and cut out of "wafers" of silicon. The microcomputer microprocessors, which process data, are made from microchips. They are also used as microcontrollers in other instruments, such as phones and TVs.

The microprocessor has presented us with a gift that we may barely appreciate—that of *portability* and *mobility* in electronic devices.

In 1955, for instance, portability was exemplified by the ads showing a young woman holding a Zenith television set accompanied by the caption: IT DOESN'T TAKE A MUSCLE MAN TO MOVE THIS LIGHTWEIGHT TV. That "lightweight" TV weighed a hefty 45 pounds. Today, by contrast, there is a handheld Casio color TV weighing a mere 6.2 ounces.

Similarly, tape recorders have gone from RCA's 35-pound machine in 1953 to today's Sony microcassette recorder of 3.5 ounces. Video cameras for consumers went from two components weighing 18.8 pounds in RCA's 1979 model to JVC's 18-ounce digital video camcorder today. Portable computers began in 1982 with Osborne's advertised "24 pounds of sophisticated computing power," a "luggable" size most people would consider too unwieldy today.[2] Since then portable computers have rapidly come down in weight and size so that now we have notebooks (4–9 pounds), subnotebooks (about 1.8–4 pounds), and pocket PCs (1 pound or less).

Had the transistor not arrived, as it did in 1947, the Age of Portability and consequent mobility would never have happened. To us a "portable" telephone might have meant the 40-pound backpack radiophones carried by some American GIs through World War II, rather than the 6-ounce shirt-pocket cellular models available today.

From Vacuum Tubes to Transistors to Microchips

Old-time radios used vacuum tubes—small lightbulb-size electronic tubes with glowing filaments—to control the flow of electrons in a vacuum. The tubes acted as a type of switch. The last computer to use these tubes, the ENIAC, which was turned on in 1946, employed 18,000 of them. Unfortunately, a tube failure occurred on average once every 7 minutes. Since it took more than 15 minutes to find and replace the faulty tube, it was difficult to get any useful computing work done. Moreover, the ENIAC was enormous, occupying 1500 square feet and weighing 30 tons.

The transistor changed all that. **A *transistor* is essentially a tiny electrically operated switch that can alternate between "on" and "off" many millions of times per second.** The transistor transfers electricity across a *resistor*, made of a material somewhat resistant to the current (thus the name: *trans*fer re*sistor*). Transistors make up *logic gates*, of which there are several types designed to respond differently to particular patterns of electrical pulses. Gates make up a *circuit*, which is an electronic pathway as well as a set of electronic components that perform a particular function in an electronic system. Circuits make up a logical device, such as a CPU (✔ p. 16).

The first transistors were one-hundredth the size of a vacuum tube, needed no warm-up time, consumed less energy, and were faster and more reliable. (■ *See Panel 4.2, next page.*) Moreover, they marked the beginning of a process of miniaturization that has not ended yet. In 1960 one transistor fit into an area about a half-centimeter square. This was sufficient to permit Zenith, for instance, to market a transistor radio weighing about 1 pound (convenient,

■ PANEL 4.2

Shrinking components

The lightbulb-size 1940s vacuum tube was replaced in the 1950s by a transistor one-hundredth its size. Today's transistors are much smaller, being microscopic in size.

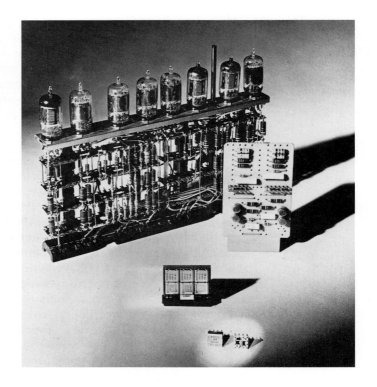

they advertised, for "pocket or purse"). Today more than 3 million transistors can be squeezed into a half centimeter, and a Sony headset radio, for example, weighs only 6.2 ounces.

In the early days of computing, transistors were made individually and then formed into an electronic circuit with the use of wires and solder. Today transistors are part of an ***integrated circuit; that is, an entire electronic circuit, including wires, is all formed together on a single "chip," or piece, of special material, usually silicon,*** as part of a single manufacturing process. An integrated circuit embodies what is called *solid-state technology.* **Solid state means that the electrons are traveling through solid material**—in this case silicon. They do not travel through a vacuum, as was the case with the old radio vacuum tubes.

What is silicon, and why use it? ***Silicon* is an element that is widely found in clay and sand.** It is used not only because its abundance makes it cheap but also because it is a *semiconductor.* **A *semiconductor* is material whose electrical properties are intermediate between a good conductor of electricity and a nonconductor of electricity.** (An example of a good conductor of electricity is copper in household wiring; an example of a nonconductor is the plastic sheath around that wiring.) Because it is only a semiconductor, silicon has partial resistance to electricity. As a result, when good-conducting metals are overlaid on the silicon, the electronic circuitry of the integrated circuit can be created.

How is such microscopic circuitry put onto the silicon? In brief, like this:

1. A large drawing of electrical circuitry is made that looks something like the map of a train yard. The drawing is photographically reduced hundreds of times so that it is of microscopic size.

2. That reduced photograph is then duplicated many times so that, like a sheet of postage stamps, there are multiple copies of the same image or circuit.

3. That sheet of multiple copies of the circuit is then printed (in a printing process called *photolithography*) and etched onto a round slice of silicon called a *wafer*. Wafers have gone from 4 inches in diameter to 6 inches to 8 inches, and now are moving toward 12 inches, which allows semiconductor manufacturers to produce more chips at lower cost.[3] (■ *See Panel 4.3.*)

4. Subsequent printings of layer after layer of additional circuits produce multilayered and interconnected electronic circuitry above and below the original silicon surface.

5. Later an automated die-cutting machine cuts the wafer into separate *chips*, which may be less than 1 centimeter square and about half a millimeter thick. **A *chip*, or *microchip*, is a tiny piece of silicon that contains millions of microminiature electronic circuit components,** mainly transistors. An 8-inch silicon wafer could have a grid of nearly 300 chips, each with as many as 5.5 million transistors.

6. After being tested, each chip is then mounted in a protective frame with protruding metallic pins that provide electrical connections through wires to a computer or other electronic device. (■ *See Panel 4.3 again.*)

    Chip manufacture requires very clean environments, which is why chip manufacturing workers appear to be dressed for a surgical operation. Such workers must also be highly skilled.

Miniaturization Miracles: Microchips, Microprocessors, & Micromachines

Microchips are the force behind the miniaturization that has revolutionized consumer electronics, computers, and communications. They are the devices that store and process data in all the electronic gadgetry we've become accustomed to. This covers a range of things from microwave ovens to video game

■ PANEL 4.3

Making of a chip
(Top left) A wafer imprinted with many microprocessors. *(Top right)* Microprocessor chip mounted in protective frame with pins that can be connected to an electronic device such as a microcomputer. *(Bottom)* Increasing size of wafers, which mean more chips for lower costs. (Each chip measures 20 mm x 20 mm.)

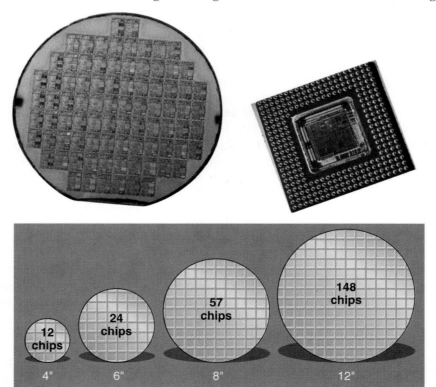

12 chips — 4"
24 chips — 6"
57 chips — 8"
148 chips — 12"

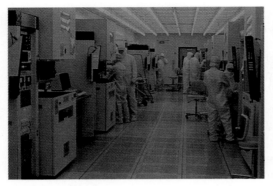

Chip-making laboratory

controllers to music synthesizers to cameras to automobile fuel-injection systems to pagers to satellites.

There are different kinds of microchips—for example, microprocessor, memory, logic, communications, graphics, and math coprocessor chips. We discuss some of these later in this chapter. Perhaps the most important is the microprocessor chip. **A *microprocessor* ("microscopic processor" or "processor on a chip") is the miniaturized circuitry of a computer processor—the part that processes, or manipulates, data into information.** When modified for use in machines other than computers, microprocessors are called *microcontrollers,* or *embedded computers* (✔ p. 22).

Mobility

Smallness—in TVs, phones, radios, camcorders, CD players, and computers—is now largely taken for granted. In the 1980s portability, or mobility, meant trading off computing power and convenience in return for smaller size and weight. Today, however, we are getting close to the point where we don't have to give up anything. As a result, experts have predicted that small, powerful, wireless personal electronic devices will transform our lives far more than the personal computer has done so far. "[T]he new generation of machines will be truly personal computers, designed for our mobile lives," wrote one reporter in 1992. "We will read office memos between strokes on the golf course and answer messages from our children in the middle of business meetings."[4] Today such activities are becoming commonplace.

4.2 The CPU & Main Memory

KEY QUESTIONS

How do the CPU and main memory work, and what are the three different kinds of processing speeds?

Preview & Review: The central processing unit (CPU)—the "brain" of the computer— consists of the control unit and the arithmetic/logic unit (ALU). Main memory holds data in storage temporarily; its capacity varies in different computers. Registers are staging areas in the CPU that store data during processing.

The operations for executing a single program instruction are called the *machine cycle,* which has an instruction cycle and an execution cycle.

Processing speeds are expressed in three ways: fractions of a second, MIPS, and flops.

How is the information in "information processing" in fact processed? As we indicated, this is the job of the circuitry known as the microprocessor. This device, the "processor-on-a-chip" found in a microcomputer, is also called the *CPU* (central processing unit). The CPU works hand in hand with other circuits known as *main memory* to carry out processing.

CPU

The *CPU*, for *central processing unit*, the "brain" of the computer, follows the instructions of the software to manipulate data into information. The CPU consists of two parts: (1) the control unit and (2) the arithmetic/logic unit. The two components are connected by a kind of electronic "roadway" called a *bus.* (■ *See Panel 4.4.*)

- **The control unit: The *control unit* tells the rest of the computer system how to carry out a program's instructions.** Rather like a

■ PANEL 4.4

The CPU and main memory

The two main CPU components (control unit and ALU), the registers, and main memory are connected by a kind of electronic "roadway" called a *bus*. (Registers are temporary data storage holding areas.)

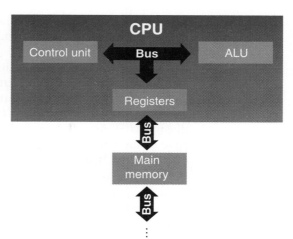

To expansion slots and input/output devices

symphony conductor or a police officer directing traffic, it directs the movement of electronic signals between main memory and the arithmetic/logic unit. It also directs these electronic signals between main memory and the input and output devices.

- **The arithmetic/logic unit:** The *arithmetic/logic unit, or ALU,* **performs arithmetic operations and logical operations and controls the speed of those operations.**

    As you might guess, *arithmetic operations* are the fundamental math operations: addition, subtraction, multiplication, and division.

    Logical operations are comparisons. That is, the ALU compares two pieces of data to see whether one is equal to $(=)$, greater than $(>)$, or less than $(<)$ the other. (The comparisons can also be combined, as in "greater than or equal to" and "less than or equal to.")

The capacities of CPUs are expressed in terms of *word size.* **A *word* is the number of bits that may be manipulated or stored at one time by the CPU.** Often the more bits in a word, the faster the computer. An 8-bit computer— that is, one with a processor that uses an 8-bit word—will transfer data within each CPU chip itself in 8-bit chunks. A 32-bit-word computer is faster, transferring data in 32-bit chunks. Other things being equal, a 32-bit computer processes 4 bytes in the same time it takes a 16-bit computer to process 2 bytes.

Main Memory

Main memory—variously known as memory, primary storage, internal memory, or RAM (for random access memory)—is working storage. It has three tasks. (1) It holds data for processing. (2) It holds instructions (the programs) for processing the data. (3) It holds processed data (that is, information) waiting to be sent to an output or secondary-storage device. Main memory is contained on special microchips called *RAM chips,* as we describe in a few pages. This memory is in effect the computer's short-term capacity. It determines the total size of the programs and data files the computer can work on at any given moment.

There are two important facts to know about main memory:

- **Its contents are temporary:** Once the power to the computer is turned off, all the data and programs within main memory simply vanish.

This is why data must also be stored on disks and tapes—called "secondary storage" to distinguish them from main memory's "primary storage."

Thus, main memory is said to be *volatile*. **Volatile storage is temporary storage; the contents are lost when the power is turned off.** Consequently, if you kick out the connecting power cord to your computer, whatever you are currently working on will immediately disappear. This impermanence is the reason you should *frequently save* your work in progress to a secondary-storage medium such as a diskette. By "frequently," we mean every 3–5 minutes.

- **Its capacity varies in different computers:** The size of main memory is important. It determines how much data can be processed at once and how big and complex the programs are that can be used to process it. This capacity varies with different computers, and older machines generally have less RAM.

  For example, the original IBM PC, introduced in 1979, held only about 64,000 bytes (characters), or 64 kilobytes, of data or instructions. By contrast, new microcomputers can have 128 million bytes (megabytes) or more of memory—a 2000-fold increase.

Registers: Staging Areas for Processing

The control unit and the ALU also use registers, or special areas that enhance the computer's performance. (■ *Refer back to Panel 4.4.*) **Registers are high-speed storage areas that temporarily store data during processing.**

It could be said that main memory (RAM) holds material that will be used "a little bit later." Registers hold material that is to be processed "immediately." The computer loads the program instructions and data from main memory into the registers just prior to processing, which helps the computer process faster. (There are several types of registers, including *instruction register, address register, storage register,* and *accumulator register.*)

Machine Cycle: How an Instruction Is Processed

How does the computer keep track of the data and instructions in main memory? Like a system of post-office mailboxes, it uses addresses. **An *address* is the location, designated by a unique number, in main memory in which a character of data or of an instruction is stored during processing.** To process each character, the control unit of the CPU retrieves that character from its address in main memory and places it into a register. This is the first step in what is called the *machine cycle.*

The *machine cycle* **is a series of operations performed to execute a single program instruction. The machine cycle consists of two parts: an instruction cycle, which fetches and decodes, and an execution cycle, which executes and stores.** (■ *See Panel 4.5.*)

- **The instruction cycle: In the *instruction cycle,* or *I-cycle,* the control unit (1) fetches (gets) an instruction from main memory and (2) decodes that instruction (determines what it means).**

- **The execution cycle: During the *execution cycle,* or *E-cycle,* the arithmetic/logic unit (3) executes the instruction (performs the operation on the data) and (4) stores the processed results in main memory or a register.**

The details of the machine cycle are actually a bit more involved than this, but our description shows the general sequence. What's important for

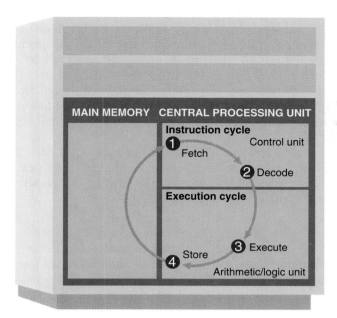 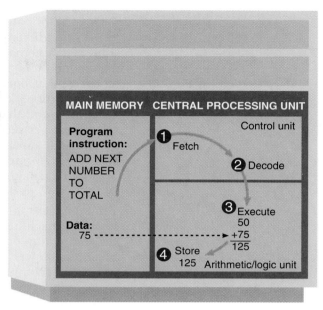

■ PANEL 4.5

The machine cycle
(Left) The machine cycle executes instructions one at a time during the instruction cycle and execution cycle. *(Right)* Example of how the addition of two numbers, 50 and 75, is processed and stored in a single cycle.

you to know is that the entire operation is synchronized by a *system clock,* as we will describe. The microprocessor clock speed, measured in megahertz, is an important factor to consider when you are buying a computer.

System Clock When people talk about a computer's "speed," they mean how fast it can do processing—turn data into information. Every microprocessor contains a system clock. **The *system clock* controls how fast all the operations within a computer take place.** The system clock uses fixed vibrations from a quartz crystal to deliver a steady stream of digital pulses to the CPU. The faster the clock, the faster the processing, assuming the computer's internal circuits can handle the increased speed.

Processing speeds may be expressed in megahertz (MHz), with 1 MHz equal to 1 million cycles per second. The original IBM PC had a clock speed of 4.77 MHz. At the time of this book's publication, Intel chips were running at speeds up to 450 MHz. The Alpha chip was running at 500 MHz.

Categories of Processing Speeds With transistors switching off and on perhaps millions of times per second, the tedious repetition of the machine cycle occurs at blinding speeds.

There are four main ways in which processing speeds are measured:

- **For microcomputers—megahertz:** Microcomputer microprocessor speeds are usually expressed in ***megahertz (MHz),* millions of machine cycles per second,** which is also the measure of a microcomputer's clock speed. For example, a 333-MHz Pentium II–based microcomputer processes 333 million machine cycles per second.

- **For workstations, minicomputers, and mainframes—MIPS:** Processing speed can also be measured according to the number of instructions processed per second that a computer can process, which today is in the millions. ***MIPS* is a measure of a computer's processing speed; MIPS stands for *millions of instructions per second* that the processor can perform.** A high-end microcomputer or workstation might perform at 100 MIPS or more, a mainframe at 200–1200 MIPS.

- **For supercomputers—flops:** The abbreviation *flops* **stands for** *floating-point operations per second,* a floating-point operation being a special kind of mathematical calculation. This measure, used mainly with supercomputers, is expressed as *megaflops* (mflops, or millions of floating-point operations per second), *gigaflops* (gflops, or billions), and *teraflops* (tflops, or trillions). Scientists are hoping to build a *petaflop* machine—one that would carry out a quadrillion (a thousand trillion) operations per second.

  The U.S. supercomputer known as Option Red cranks out 1.34 teraflops. (To put this in perspective, if a person were able to do one arithmetic calculation every second, it would take him or her about 31,000 years to do what Option Red does in a single second.)

- **Another measurement—fractions of a second:** Another way to measure machine cycle times is in fractions of a second. The speeds for completing one machine cycle are measured in milliseconds for older and slower computers. They are measured in microseconds for most microcomputers and in nanoseconds for mainframes. Picosecond measurements occur only in some experimental machines.

  A *millisecond* is one-thousandth of a second.
  A *microsecond* is one-millionth of a second.
  A *nanosecond* is one-billionth of a second.
  A *picosecond* is one-trillionth of a second.

Now that you know where data and instructions are processed, we need to review how those data and instructions are represented in the CPU, registers, buses, and RAM.

4.3   How Data & Programs Are Represented in the Computer

KEY QUESTIONS

How is data capacity represented in a computer, and how do coding schemes, parity bits, and machine language work?

Preview & Review: Computers use the two-state 0/1 binary system to represent data. A computer's capacity for data is expressed in bits, bytes, kilobytes, megabytes, gigabytes, or terabytes. Two common binary coding schemes are EBCDIC and ASCII-8; a newer one is Unicode. Accuracy checks use parity bits.

Human-language-like programming languages are processed as 0s and 1s by the computer in machine language.

As we've explained, electricity is the basis for computers and communications because electricity can be either *on* or *off.* This two-state situation allows computers to use the *binary system* (✔ p. 6) to represent data and programs.

Binary System: Using Two States

The decimal system that we are used to has 10 digits (0, 1, 2, 3, 4, 5, 6, 7, 8, 9). By contrast, **the** *binary system* **has only two digits: 0 and 1.** Thus, in the computer the 0 can be represented by the electrical current being off and the 1 by the current being on. All data and programs that go into the computer are represented in terms of these binary numbers. (■ *See Panel 4.6.*)

For example, the letter H is a translation of the electronic signal 01001000, or off-on-off-off-on-off-off-off. When you press the key for H on the computer keyboard, the character is automatically converted into the series of electronic impulses that the computer can recognize.

■ PANEL 4.6

Binary data representation
How the letters H-E-R-O are represented in one type of off/on, 0/1 binary code (ASCII-8).

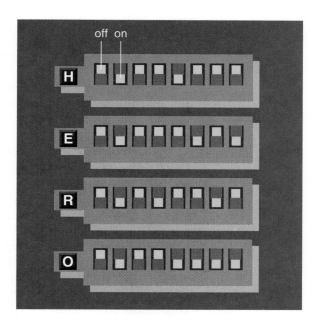

How Capacity Is Expressed

How many 0s and 1s will a computer or a storage device such as a hard disk hold? To review what we covered in Chapter 1, the following terms are used to denote capacity.

- **Bit:** In the binary system, **each 0 or 1 is called a *bit*, which is short for "binary digit."**

- **Byte:** To represent letters, numbers, or special characters (such as ! or *), bits are combined into groups. **A group of 8 bits is called a *byte*, and a byte represents one character, digit, or other value.** (As we mentioned, in one scheme, 01001000 represents the letter H.) The capacity of a computer's memory or a diskette is expressed in numbers of bytes or multiples such as kilobytes and megabytes.

- **Kilobyte:** **A *kilobyte (K, KB)* is about 1000 bytes.** (Actually, it's precisely 1024 bytes, but the figure is commonly rounded.) The kilobyte was a common unit of measure for memory or secondary-storage capacity on older computers.

- **Megabyte:** **A *megabyte (M, MB)* is about 1 million bytes** (1,048,576 bytes). Measures of microcomputer primary-storage capacity today are expressed in megabytes.

- **Gigabyte:** **A *gigabyte (G, GB)* is about 1 billion bytes** (1,073,741,824 bytes). This measure was formerly used mainly with "big iron" types of computers, but now is typical of secondary storage (hard disk) capacity of today's microcomputers.

- **Terabyte:** **A *terabyte (T, TB)* represents about 1 trillion bytes** (1,009,511,627,776 bytes).

Binary Coding Schemes

Letters, numbers, and special characters are represented within a computer system by means of *binary coding schemes.* That is, the off/on 0s and 1s are arranged in such a way that they can be made to represent characters, digits, or other values. When you type a word on the keyboard (for example,

HERO), the letters are converted into bytes—eight 0s and 1s for each letter. The bytes are represented in the computer by a combination of eight transistors, some of which are closed (representing the 0s) and some of which are open (representing the 1s).

There are many coding schemes. Two common ones are EBCDIC and ASCII-8. Both use eight bits to form each byte. (■ *See Panel 4.7.)* One newer coding scheme uses 16 bits.

- **EBCDIC:** Pronounced *"eb*-see-dick," ***EBCDIC—an acronym for Extended Binary Coded Decimal Interchange Code—is used in mainframe computers.***

- **ASCII-8:** Pronounced "askey," ***ASCII*** stands for ***American Standard Code for Information Interchange*** and is the binary code most widely used with microcomputers.***

  ASCII originally used seven bits, but a zero was added in the left position to provide an eight-bit code, which offers more possible combinations with which to form characters, such as math symbols and Greek letters. (However, although ASCII can handle the English language well, it cannot handle all the characters of some other languages, such as Chinese and Japanese.)

- **Unicode:** A superset of ASCII, ***Unicode*** uses two bytes (16 bits) for each character,*** rather than one byte (8 bits). Instead of the 256 character combinations of ASCII-8, Unicode can handle 65,536 character combinations, thus allowing almost all the written languages of the world to be represented using a single character set. (By contrast, 8-bit ASCII is not capable of representing even all the combinations of letters and other symbols that are used just with our alphabet, the Roman alphabet—used in English, Spanish, French, German, Portuguese, and so on.)

  Although each Unicode character takes up twice as much memory space and hard-disk space as each ASCII character, conversion to the Unicode standard seems likely. However, because most existing

■ PANEL 4.7

Two binary coding schemes: EBCDIC and ASCII-8

There are many more characters than those shown here. These include punctuation marks, Greek letters, math symbols, and foreign-language symbols.

Character	EBCDIC	ASCII-8	Character	EBCDIC	ASCII-8
A	1100 0001	0100 0001	N	1101 0101	0100 1110
B	1100 0010	0100 0010	O	1101 0110	0100 1111
C	1100 0011	0100 0011	P	1101 0111	0101 0000
D	1100 0100	0100 0100	Q	1101 1000	0101 0001
E	1100 0101	0100 0101	R	1101 1001	0101 0010
F	1100 0110	0100 0110	S	1110 0010	0101 0011
G	1100 0111	0100 0111	T	1110 0011	0101 0100
H	1100 1000	0100 1000	U	1110 0100	0101 0101
I	1100 1001	0100 1001	V	1110 0101	0101 0110
J	1101 0001	0100 1010	W	1110 0110	0101 0111
K	1101 0010	0100 1011	X	1110 0111	0101 1000
L	1101 0011	0100 1100	Y	1110 1000	0101 1001
M	1101 0100	0100 1101	Z	1110 1001	0101 1010
0	1111 0000	0011 0000	5	1111 0101	0011 0101
1	1111 0001	0011 0001	6	1111 0110	0011 0110
2	1111 0010	0011 0010	7	1111 0111	0011 0111
3	1111 0011	0011 0011	8	1111 1000	0011 1000
4	1111 0100	0011 0100	9	1111 1001	0011 1001
!	0101 1010	0010 0001	;	0101 1110	0011 1011

software applications and databases use the 8-bit standard, the conversion will take time.

The Parity Bit

Dust, electrical disturbance, weather conditions, and other factors can cause interference in a circuit or communications line that is transmitting a byte. How does the computer know if an error has occurred? Detection is accomplished by use of a parity bit. **A *parity bit*, also called a *check bit*, is an extra bit attached to the end of a byte for purposes of checking for accuracy.**

Parity schemes may be *even parity* or *odd parity*. In an even-parity scheme, for example, the ASCII letter H (01001000) contains two 1s. Thus, the ninth bit, the parity bit, would be 0 in order to make the sum of the bits come out even. With the letter O (01001111), which has five 1s, the ninth bit would be 1 to make the byte come out even. (■ *See Panel 4.8.*) The system software in the computer automatically and continually checks the parity scheme for accuracy. (If the message "Parity Error" appears on your screen, you need a technician to look at the computer to see what is causing the problem.)

Machine Language

Why won't word processing software that runs on an Apple Macintosh run (without special arrangements) on an IBM microcomputer? It's because each computer has its own machine language. **Machine language is a binary-type programming language that the computer can run directly.** To most people an instruction written in machine language is incomprehensible, consisting only of 0s and 1s. However, it is what the computer itself can understand, and the 0s and 1s represent precise storage locations and operations.

How do people-comprehensible program instructions become computer-comprehensible machine language? Special system programs called *language translators* rapidly convert the instructions into machine language—language that computers can understand (discussed in Chapter 11). This translating occurs virtually instantaneously, so that you are not aware it is happening.

Because the type of computer you will most likely be working with is the microcomputer, we'll now take a look at what's inside the microcomputer's system unit.

■ PANEL 4.8

Parity bit
This example uses an even-parity scheme.

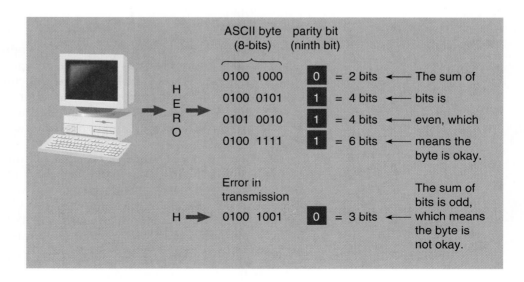

4.4 The Microcomputer System Unit

KEY QUESTION

What are the names and functions of the main parts of the system unit?

Preview & Review: The system unit, or cabinet, contains the following electrical components: the power supply, the motherboard, the CPU chip, specialized processor chips, RAM chips, ROM chips, other forms of memory (cache, VRAM, flash), expansion slots and boards, bus lines, ports, and PC (PCMCIA) slots and cards.

What is inside the gray or beige box that we call "the computer"? **The box or cabinet is the *system unit;* it contains the electrical and hardware components that make the computer work.** These components actually do the processing in information processing.

The system unit of a desktop microcomputer does not include the keyboard or printer. Usually it also does not include the monitor or display screen (although it did in early Apple Macintoshes and some Compaq Presarios). It usually does include a hard-disk drive, a diskette drive, a CD-ROM drive, and sometimes a tape drive. We describe these and other ***peripheral devices*—hardware that is outside the central processing unit**—in the chapters on input/output and secondary storage. Here we are concerned with eleven parts of the system unit, as follows:

- Power supply
- Motherboard
- CPU chip
- Specialized processor chips
- RAM chips
- ROM chips
- Other forms of memory—cache, VRAM, flash
- Expansion slots and boards
- Bus lines
- Ports
- PC (PCMCIA) slots and cards

These are terms that appear frequently in advertisements for microcomputers. After reading this section, you should be able to understand what these ads are talking about.

Power Supply

The electricity available from a standard wall outlet is alternating current (AC), but a microcomputer runs on direct current (DC). **The *power supply* is a device that converts AC to DC to run the computer.** (■ *See Panel 4.9.*) The on/off switch in your computer turns on or shuts off the electricity to the power supply. Because electricity can generate a lot of heat, a fan inside the computer keeps the power supply and other components from becoming too hot.

Electrical power drawn from a standard AC outlet can be quite uneven. For example, a sudden surge, or "spike," in AC voltage can burn out the low-voltage DC circuitry in your computer ("fry the motherboard"). Instead of plugging your computer directly into the wall electrical outlet, it's a good idea to plug it into a power protection device. The three principal types are surge protectors, voltage regulators, and UPS units.

- Surge protector: **A *surge protector,* or *surge suppressor,* is a device that helps protect a computer from being damaged by surges (spikes)**

■ PANEL 4.9

System unit and motherboard components

(Top) Motherboard. *(Bottom)* System unit.

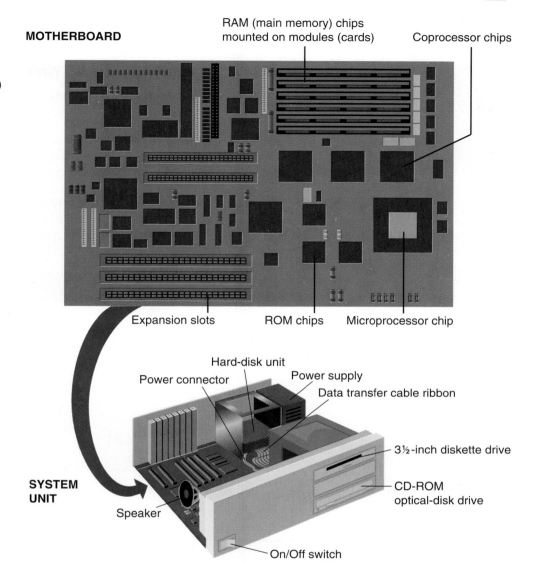

MOTHERBOARD

RAM (main memory) chips mounted on modules (cards)

Coprocessor chips

Expansion slots   ROM chips   Microprocessor chip

SYSTEM UNIT

Power connector   Hard-disk unit   Power supply   Data transfer cable ribbon

3½-inch diskette drive

CD-ROM optical-disk drive

Speaker

On/Off switch

of high voltage. The computer is plugged into the surge protector, which in turn is plugged into a standard electrical outlet.

- **Voltage regulator:** **A *voltage regulator,* or *line conditioner,* is a device that protects a computer from being damaged by insufficient power—"brownouts" or "sags" in voltage.** Brownouts can occur when a large machine such as a power tool starts up and causes the lights in your house to dim. They also may occur on very hot summer days when the power company has to lower the voltage in an area because too many people are running their air conditioners all at once.

- **UPS:** **A *UPS,* for *uninterruptible power supply,* is a battery-operated device that acts as a surge protector *and* provides a computer with electricity if there is a power failure.** The UPS will keep a computer going 5–30 minutes or more. It goes into operation as soon as the power to your computer fails.

README

Practical Matters: Preventing Problems from Too Much or Too Little Power to Your Computer

"When the power disappears, so can your data," writes *San Jose Mercury News* computer columnist Phillip Robinson. "I say this with authority, sitting here in the dark in the wake of severe storms that have hit my part of California."[5] (Deprived of use of his computer, Robinson dictated his column by phone.)

Too little electricity can be devastating to your data. Too much electricity can be devastating to your computer hardware.

Here are a few things you can do to keep both safe:[6-9]

- **Back up data regularly:** If you faithfully make backup (duplicate) copies of your data every few minutes as you're working, then if your computer has power problems you'll be able to get back in business fairly quickly once the machine is running again.
- **Use a surge protector to protect against too much electricity:** Plug all your hardware into a surge protector (suppressor), which will prevent damage to your equipment if there is a power surge. (You'll know you've experienced a power surge when the lights in the room suddenly get very bright.)

Surge protector

- **Use a voltage regulator to protect against too little electricity:** Plug your computer into a voltage regulator to adjust for power sags or brownouts. If power is too low for too long, it's as though the computer were turned off.

- **Consider using a UPS to protect against complete absence of electricity:** Consider plugging your computer into a UPS, or uninterruptible power supply. (A low-cost one, available at electronics stores, sells for about $150.) The UPS is kind of a short-term battery that will keep your computer running long enough for you to save your data before you turn off the machine after a power failure.

UPS unit

- **Turn ON highest-power-consuming hardware first:** When you turn on your computer system, you should turn on the devices that use the most power first. This will avoid causing a power drain on smaller devices. The most common advice is to turn on (1) external drives and/or scanners, (2) system unit, (3) monitor, (4) printer—in that order.
- **Turn OFF lowest-power-consuming hardware first:** When you turn off your system, follow the reverse order—first printer, then monitor, and so on. This avoids causing a power surge to the smaller devices.
- **Unplug your computer system during lightning storms:** Unplug all your system's components—including phone lines—during thunder and lightning storms. If lightning strikes your house or the power lines, it can ruin your equipment.

Motherboard

The motherboard, or system board, is the main circuit board in the system unit. (■ *Refer back to Panel 4.9.*)

The motherboard consists of a flat board that fills the bottom of the system unit. (It is accompanied by the power-supply unit and fan and probably one or more disk drives.) This board contains the "brain" of the computer, the CPU or microprocessor; electronic memory (RAM) that assists the CPU; and some sockets, called *expansion slots,* where additional circuit boards, called *expansion boards,* may be plugged in.

CPU Chip

Most personal computers today use CPU chips (microprocessors) of two kinds—those based on the model made by Intel and those based on the model made by Motorola. (■ *See Panel 4.10.*)

- **Intel-type chips:** About 90% of microcomputers use Intel-type microprocessors. Indeed, the Microsoft Windows operating system is designed to run on Intel chips. As a result, people in the computer industry tend to refer to the Windows/Intel joint powerhouse as *Wintel.*

    Intel-type chips **are made principally by Intel Corporation—but also by Advanced Micro Devices, Cyrix, DEC, and others—for PCs** such as IBM, Compaq, Dell, Gateway 2000, NEC, and Packard Bell.

    Intel used to identify its chips by numbers—8086, 8088, 80286, 80386, 80486, the latter abbreviated in common parlance to simply '286, '386, and '486. The higher the number, the faster the processing speed. (There were some variations, such as "386SX" or "486DX," with SX chips being less expensive and slower, hence appropriate for home use, and DX chips more appropriate for business use.) Since 1993, Intel has marketed its chips under the names *Pentium, Pentium Pro, Pentium MMX,* and *Pentium II.* The Pentium II, which came out in late 1997, allows users to install more demanding multimedia programs, such as three-dimensional visualization tools.[10] (Forthcoming versions are code-named *Tillamook* and *Deschutes.*)

- **Motorola-type chips:** *Motorola-type chips* **are made by Motorola for Apple Macintosh computers and its clones,** such as those formerly made by Power Computing and by Motorola itself. These chip numbers include the 68000, 68020, 68030, and 68040. Since 1993, Motorola has joined forces with IBM and Apple to produce the PowerPC family of chips. With certain hardware or software add-ons, a PowerPC can run PC as well as Mac applications software.

    Apple's 1997 decision to stop allowing other computer companies to make Macintosh clones raises some concerns about the future of the PowerPC microprocessor, made by Motorola and IBM, which was used in these machines. If Apple's market share declines further, so will the demand for the PowerPC.[11]

Two principal designs—what computer people call "architectures"—for microprocessors are CISC and RISC:

- **CISC:** *CISC chips*—**CISC** stands for *complex instruction set computing*—**are used mostly in PCs and in conventional mainframes.**

■ PANEL 4.10
Microcomputers and microprocessors
Some widely used microcomputer systems and their chips.

Manufacturer and Chip	Date Introduced	Systems Chip	Clock Speed (MHz)	Bus Width
Intel 8088	1979	IBM PC, XT	4–8	8
Motorola 68000	1979	Macintosh Plus, SE; Commodore Amiga	8–16	16
Intel 80286	1981	IBM PC/AT, PS/2 Model 50/60; Compaq Deskpro 286	8–28	16
Motorola 68020	1984	Macintosh II	16–33	32
Sun Microsystems RISC	1985	Sun Sparcstation 1, 300	20–25	32
Intel 80386DX	1985	IBM PS/2; and compatibles	16–33	32
Motorola 68030	1987	Macintosh IIx series, SE/30	16–50	32
Intel 80486DX	1989	IBM PS/2; and compatibles	25–66	32
Motorola 68040	1989	Macintosh Quadras	25–40	32
IBM RISC 6000	1990	IBM RISC/6000 workstation	20–50	32
Sun Microsystems	1992	Sun Sparcstation LX	50	32
Intel Pentium	1993	Compaq Deskpro; IBM and compatibles	60–166	64
IBM/Motorola/Apple PowerPC RISC	1994	Power Macintoshes; Power Computing PowerWave	60–150	64
Intel Pentium Pro	1995	Compaq Proliant; Data General server	150–200	64
Intel Pentium MMX	1996	Dell, Gateway 2000	166–233	64
Intel Pentium II	1997	Compaq Deskpro and compatibles	233–400	64
Apple PowerPC	1998	Power Macintoshes G3	233–266	66

CISC chips can support a large number of instructions, although this number gets in the way of processing speeds.

- **RISC:** *RISC chips—RISC* **stands for** *reduced instruction set computing*—**are used mainly in workstations,** such as those made by Sun Microsystems, Hewlett-Packard, and Digital Equipment Corporation. With RISC chips, a great many seldom-used instructions are eliminated. Thus, a RISC computer system operates with fewer instructions than those required in conventional computer systems. RISC-equipped workstations have been found to work up to 10 times faster than conventional computers. A problem, however, is that software has to be modified to work with them.

The Pentium II processor comes in a cartridge that is inserted in a slot on the motherboard.

Most new chips are compatible with older chips. For example, the word processing program and all the data files that you used for your '486 machine will continue to run if you upgrade to a Pentium machine. However, the reverse is not usually true. Thus, if you are using software written for your Pentium-powered desktop PC and buy an old '386 portable, you probably won't be able to run your software on the portable.

The development of new chips is popularly known to follow what is called *Moore's law*. Named for the legendary cofounder of Intel, Gordon Moore, Moore's law reflects his observation that, as one journalist paraphrased it,

The number of components [transistors] that can be packed on a computer chip doubles every 18 months while the price stays the same. Essentially, that means that computer power per dollar doubles every 18 months.[12]

Amazingly, this "law" has held true for more than 30 years, with power increasing and prices falling at a spectacular rate. For instance, in 1961 a chip had only 4 transistors, in 1971 it had 2300, and in 1979 it had 30,000. The 1997 Pentium II has 7.5 million transistors. The 2000 chip code-named Merced will have 40 million transistors according to one report, between 20 million and 50 million according to another—which would seem to accelerate the development rate of computing power to *more than double* every 18 months.[13,14] (■ *See Panel 4.11.*) In the 1960s, a single transistor sold for $70; in late 1997, Intel was selling its 7.5-million-transistor Pentium II for $401. Expressed another way, a transistor today can be bought for less than a millionth of a cent.[15–17]

Is there an end in sight to these miraculous jumps in computing power? The chief operating officer of Intel has predicted that PCs in the year 2011 will use a microprocessor chip that has as many as a billion transistors. And, he ventures, technology that now costs $75,000— such as high-powered workstations—will be available for just $2000.[18] We discuss these matters in another few pages.

■ PANEL 4.11

Moore's law: Intel chips' transistor explosion

How the number of transistors in Intel chips has grown. Moore's law states that the number of transistors on a chip doubles every 18 months while price stays the same.

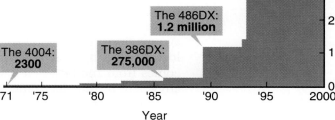

Specialized Processor Chips

A motherboard usually has slots for plugging in specialized processor chips. (■ *Refer back to Panel 4.9.)* Two in particular that you may encounter are math and graphics coprocessor chips. A *math coprocessor chip* helps programs using lots of mathematical equations to run faster. A *graphics coprocessor chip* enhances the performance of programs with lots of graphics and helps create complex screen displays. Specialized chips significantly increase the speed of a computer system by offloading work from the main processor. These chips may be plugged directly into the motherboard. However, often they are included on "daughter cards," such as sound cards and graphics cards, used to expand a computer's capabilities.

Types of Processing: Serial, Parallel, Pipelining Computers with a CISC or RISC processor execute instructions one at a time—that is, *serially,* which is called *serial processing.* However, a computer with more than one processor can execute more than one instruction at a time, which is called *parallel processing.* Although some powerful microcomputers and workstations are available with more than one main processor, the most powerful computers, such as supercomputers, often use *massively parallel processing (MPP),* which spreads processing tasks over hundreds or even thousands of standard, inexpensive microprocessors of the type used in PCs. Tasks are parceled out to a great many processors, which work simultaneously. Janus, the world's fastest supercomputer, has 9072 Pentium Pro processors and runs at 1.8 teraflops. In late 1997, several computer makers agreed on a new standard, known as Open MP, that allows workstations, using parallel processing, to take on tasks now performed by supercomputers.

In older microcomputers, the CPU had to completely finish processing one instruction before starting another; that is, each instruction had to go through all four stages of the machine cycle before the CPU could start processing the next instruction. Newer computers use CPUs built to handle *pipelining,* which allows the CPU to start a new instruction as soon as the previous instruction reaches the next stage of the machine cycle. For example, once an instruction moves from fetch to decode in the machine cycle, the CPU can fetch the next instruction. This means that processing is speeded up: by the time the CPU finishes processing the first instruction, the second, third, and fourth instructions have entered the processing "pipeline" and are at various stages of completion.

RAM Chips

As described earlier, **RAM, for *random access memory,* is memory that temporarily holds data and instructions that will be needed shortly by the CPU.** RAM is what we have been calling *main memory, internal memory,* or *primary storage;* it operates like a chalkboard that is constantly being written on, then erased, then written on again. For example, when you are working in a word processing program, the new sentences you type in are stored in RAM. If you change those sentences, then the changes are also made in RAM. However, until you tell the computer to store your work to a secondary-storage medium, such as diskette or hard disk, it remains in RAM and can be lost if power to the computer is lost or turned off. (The term *random access* comes from the fact that data can be stored and retrieved at random—from anywhere in the electronic RAM chips—in approximately equal amounts of time, no matter what the specific data locations are.)

Like the microprocessor, RAM consists of circuit-inscribed silicon chips attached to the motherboard. **RAM chips are often mounted on a small cir-**

Memory chip modules
(SIMMs)

cuit board, such as a **SIMM** or **DIMM**, which is plugged into the motherboard. A **SIMM** (for *single inline memory module*) has multiple RAM chips on one side. A **DIMM** (for *dual inline memory module*) has multiple RAM chips on both sides. (■ *Refer back to Panel 4.9.*)

The principal types of RAM chips are the following:

- DRAM chips: *DRAM* (for *dynamic random access memory*) chips are more commonly used than SRAM chips in microcomputers because they are less expensive.
- SRAM chips: *SRAM* (for *static random access memory*) chips are faster than DRAM chips, but besides being more expensive they take up more space and use more power.
- EDO RAM chips: A newer type of DRAM chip, *EDO* (for *extended data out*) *RAM* chips approach the performance of SRAM chips. The increased speed occurs because the chip keeps data available for the CPU while it is starting the next memory access.

Microcomputers come with different amounts of RAM. In many cases, additional RAM chips can be added by plugging a memory-expansion card into the motherboard, as we will explain. The more RAM you have, the faster the computer operates, and the better your software performs. If, for instance, on an older computer you type such a long document in a word processing program that it will not all fit into your computer's RAM, the computer will put part of the document onto your disk (either hard or floppy). This means you have to wait while the computer swaps data back and forth between RAM and disk.

Having enough RAM has become a critical matter! Before you buy a software package, look at the outside of the box or check the manufacturer's Web site to see how much RAM is required. Windows 95 supposedly will run with 4 MB of RAM, but a realistic minimum is 16 MB. One powerful IBM-compatible microcomputer, the Micron XKU 300, has an awesome (as of this writing) *128 megabytes* of RAM.

ROM Chips

Unlike RAM, which is constantly being written on and erased, **ROM, which stands for *read only memory* and is also known as *firmware,* cannot be written on or erased by the computer user without special equipment.** (■ *Refer back to Panel 4.9.*) ROM chips contain programs that are built in at the factory; these are special instructions for basic computer operations, such as those that start the computer or put characters on the screen.

There are variations of the ROM chip that allow programmers to vary information stored on the chip and also to erase it.

Other Forms of Memory

The performance of microcomputers can be enhanced further by adding other forms of memory, as follows.

- Cache memory: Pronounced "cash," **cache memory is a special high-speed memory area that the CPU can access quickly.** Cache memory can be located on the microprocessor chip or elsewhere on the motherboard. (■ *Refer back to Panel 4.9.*)

Cache memory is used in computers with very fast CPUs. The most frequently used instructions are kept in cache memory so the CPU can look there first. This allows the CPU to run faster because it doesn't have to take time to swap instructions in and out of main memory. Large, complex programs—such as complex spreadsheets or database management programs—benefit the most from having a cache memory available. Pentium II processors generally come with at least 512 kilobytes of cache memory.

- **Video memory:** *Video memory or video RAM (VRAM)* **chips are used to store display images for the monitor.** The amount of video memory determines how fast images appear and how many colors are available. Video memory chips are particularly desirable if you are running programs that display a lot of graphics.

- **Flash memory:** Used primarily in notebook and subnotebook computers, *flash memory,* **or** *flash RAM,* **cards consist of circuitry on credit-card-size cards that can be inserted into slots connected to the motherboard.**

  Unlike standard RAM chips, flash memory is *nonvolatile.* That is, it retains data even when the power is turned off. Flash memory can be used not only to simulate main memory but also to supplement or replace hard-disk drives for permanent storage.

  In late 1997, Intel developed a type of flash memory called StrataFlash that could hold twice as much data as normal chips—for example, 160 minutes of digital audio versus 80 minutes for conventional chips. In normal memory chips (and in processor chips still), each transistor can be only on or off, which means it can hold just one bit of data. The StrataFlash allows a transistor to have four different settings—on, two-thirds on, one-third on, or off—which means it can effectively hold twice as much data.[19,20]

Expansion Slots & Boards

Today all new microcomputer systems can be expanded. *Expandability* **refers to a computer's capacity for adding more memory or peripheral devices.** Having expandability means that when you buy a PC you can later add devices to enhance its computing power. This spares you from having to buy a completely new computer.

Expandability is made possible with expansion slots and expansion boards. *Expansion slots* **are sockets on the motherboard into which you can plug expansion cards.** *Expansion cards,* **or** *add-on boards,* **are circuit boards that provide more memory or control peripheral devices.** (■ *Refer back to Panel 4.9.)* The words *card* and *board* are used interchangeably. Some slots may be needed right away for ordinary functions, but if your system unit leaves enough slots open, you can use them for expansion later.

Among the types of expansion cards are the following. (■ *See Panel 4.12.)*

- **Expanded memory:** Memory expansion cards (SIMMs or DIMMs) allow you to add RAM chips, giving you more main memory.

- **Display adapter or graphics adapter cards:** These cards allow you to adapt different kinds of color video display monitors for your computer.

- **Controller cards:** *Controller cards* **are circuit boards that allow your CPU to work with the computer's various peripheral devices.** For example, a disk controller card allows the computer to work with different kinds of hard-disk and diskette drives.

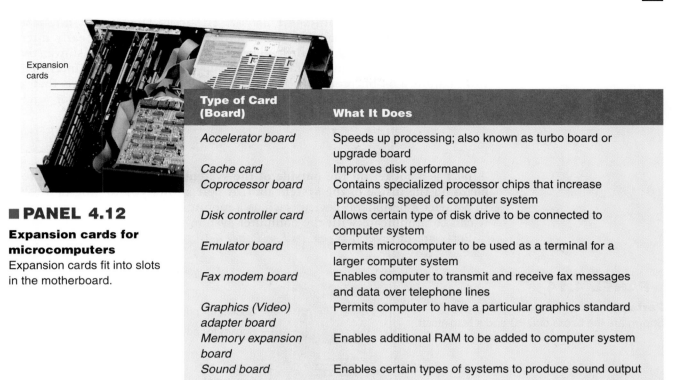

Expansion cards

Type of Card (Board)	What It Does
Accelerator board	Speeds up processing; also known as turbo board or upgrade board
Cache card	Improves disk performance
Coprocessor board	Contains specialized processor chips that increase processing speed of computer system
Disk controller card	Allows certain type of disk drive to be connected to computer system
Emulator board	Permits microcomputer to be used as a terminal for a larger computer system
Fax modem board	Enables computer to transmit and receive fax messages and data over telephone lines
Graphics (Video) adapter board	Permits computer to have a particular graphics standard
Memory expansion board	Enables additional RAM to be added to computer system
Sound board	Enables certain types of systems to produce sound output

■ PANEL 4.12

Expansion cards for microcomputers
Expansion cards fit into slots in the motherboard.

- **Other add-ons:** You can also add special circuit boards for modems, fax, sound, and networking, as well as math or graphics coprocessor chips.

Bus Lines

A *bus line,* or simply *bus,* is an electrical pathway through which bits are transmitted within the CPU and between the CPU and other devices in the system unit. There are different types of buses (address bus, control bus, data bus), but for our purposes the most important is the *expansion bus,* which carries data between RAM and the expansion slots. To obtain faster performance, some users will use a bus that avoids RAM altogether. A bus that connects expansion slots directly to the CPU is called a *local bus.* (■ *See Panel 4.13.)*

■ PANEL 4.13

Buses
Buses are the electrical pathways that carry bits within the CPU and from the CPU to peripheral devices. Expansion buses connect RAM with expansion slots. Local buses avoid RAM and connect expansion slots directly with the CPU.

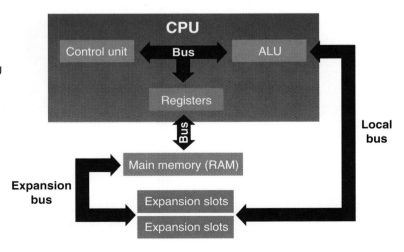

A bus resembles a multilane highway: The more lanes it has, the faster the bits can be transferred. The old-fashioned 8-bit-word bus of early microprocessors had only eight pathways. It was therefore four times slower than the 32-bit bus of later microprocessors, which had 32 pathways. Intel's Pentium chip is a 64-bit processor. Some supercomputers contain buses that are 128 bits. Today there are several principal expansion bus standards, or "architectures," for microcomputers.

Ports

A *port* is a socket on the outside of the system unit that is connected to a board on the inside of the system unit. A port allows you to plug in a cable to connect a peripheral device, such as a monitor, printer, or modem, so that it can communicate with the computer system.

Ports are of several types. (■ *See Panel 4.14.*)

■ PANEL 4.14

Ports

Shown are the backs of a PC and a Macintosh.

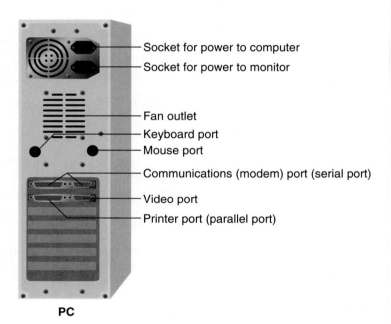

Socket for power to computer
Socket for power to monitor

Fan outlet
Keyboard port
Mouse port

Communications (modem) port (serial port)

Video port

Printer port (parallel port)

PC

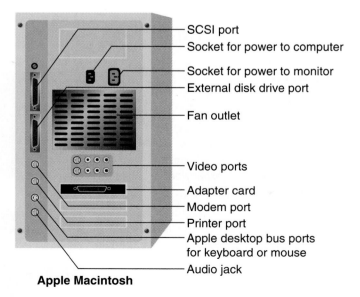

SCSI port
Socket for power to computer

Socket for power to monitor
External disk drive port

Fan outlet

Video ports

Adapter card
Modem port
Printer port
Apple desktop bus ports for keyboard or mouse
Audio jack

Apple Macintosh

- Game ports: *Game ports* **allow you to attach a joystick or similar gameplaying device to the system unit.** Not all microcomputers will have these.

- Parallel ports: **A *parallel port* allows lines to be connected that will enable 8 bits to be transmitted simultaneously,** like cars on an eight-lane highway. Parallel lines move information faster than serial lines do, but they can transmit information efficiently only up to 15 feet. Thus, parallel ports are used principally for connecting printers.

- Serial ports: **A *serial port,* or *RS-232 port,* enables a line to be connected that will send bits one after the other on a single line,** like cars on a one-lane highway. Serial ports—frequently labeled "COM" for communications—are used principally for communications lines, modems, and mice. In the Macintosh, serial ports are also used for the printer.

  On the back of newer PCs is one 9-pin connector for serial port COM1, typically used for the mouse, and one 25-pin connector serial port COM2, typically used for the modem.

- Video adapter ports: *Video adapter ports* **are used to connect the video display monitor outside the computer to the video adapter card inside the system unit.** Monitors may have either a 9-pin plug or a 15-pin plug. The plug must be compatible with the number of holes in the video adapter card.

- SCSI ports: Pronounced "scuzzy" (and **short for *small computer system interface*), a *SCSI port* provides an interface for transferring data at high speeds for up to 7 or 15 SCSI-compatible devices.** These devices include external hard-disk drives, CD-ROM drives, scanners, and magnetic-tape backup units.

  SCSI devices are linked together in what is called a *daisy chain*—a **set of devices connected in a series along an extended cable.** Sometimes the equipment on the chain is inside the computer, an internal daisy chain; sometimes it is outside the computer, an external daisy chain. (■ *See Panel 4.15, next page.*)

- Infrared ports: When you use a handheld remote unit to change channels on a TV set, you're using invisible radio waves of the type known as infrared waves. **An *infrared port* allows a computer to make a cableless connection with infrared-capable devices,** such as some printers. This type of connection requires an unobstructed line of sight between transmitting and receiving ports, and they can only be a few feet apart.

Why are so many ports needed? Why can't plugging peripherals into a computer be as easy as plugging in a lamp in your living room? The reason, says an Intel engineer, is that "The PC industry has evolved ad hoc. We were always adding one more piece of equipment."[21] As a result, connecting a new device, such as a scanner or a second printer, "is about as straightforward as triple-bypass surgery," says one writer. In many cases, it involves opening the computer and inserting a circuit board, installing or modifying pertinent software, and fiddling with little switches.[22]

Fortunately, new technology promises simplification. Several companies (Intel, Microsoft, IBM, Compaq, and others) have agreed on something called the *Universal Serial Bus (USB),* **which allows up to 127 peripherals to be connected through just one general-purpose port.** The USB connecting the port inside the computer will interpret the signals from each peripheral device and tell the computer to recognize it. With this technology, you'll be

▓ **PANEL 4.15**

Two kinds of daisy chains

SCSI-compatible devices may be linked inside the computer or outside the computer.

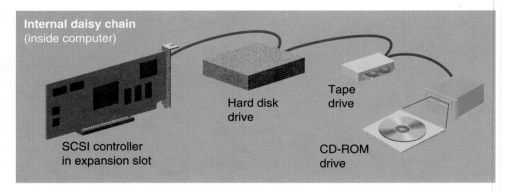

Internal daisy chain (inside computer)

SCSI controller in expansion slot

Hard disk drive

Tape drive

CD-ROM drive

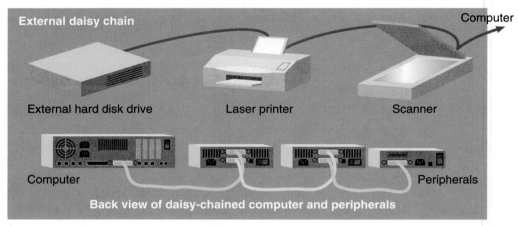

External daisy chain

Computer

External hard disk drive

Laser printer

Scanner

Computer

Peripherals

Back view of daisy-chained computer and peripherals

▓ **PANEL 4.16**

PC card

Not much bigger than a credit card, a PC card fits in a small slot on the side of a computer. PC cards are typically used to provide extra memory or a modem.

able to easily connect printers, modems, mice, keyboards, and CD-ROM drives in a daisy-chain style. If you have enough money, your biggest problem then may be to find 127 peripheral devices to connect.

Plug-In Cards: PC (PCMCIA) Slots & Cards

Although its name doesn't exactly roll off the tongue, PCMCIA has probably changed mobile computing more dramatically than any other technology today.

Short for *Personal Computer Memory Card International Association*, PCMCIA is a completely open, relatively new bus standard for portable computers. When talking about this standard, one pronounces every letter: "P-C-M-C-I-A." *PC cards*—**renamed because it's easier to say—are cards approximately 2.1 by 3.4 inches in size that contain peripherals and can be plugged into slots in microcomputers.** The PC cards may be used to hold credit-card-size modems, sound boards, hard disks, extra memory, and even pagers and cellular communicators. (▓ *See Panel 4.16.*)

At present there are four sizes for PC cards—I (thin), II (thick), III (thicker), and IV (thickest). Type I is used primarily for flash memory cards. Type II, the kind you'll find most often, is used for fax modems and adapters for local area networks (LANs). Type III is for rotating disk devices, such as hard-disk drives. Type IV is for large-capacity hard-disk drives.

README

Practical Matters: Batteries for Laptops

Why do some traveling professionals still keep old laptops with black-and-white screens?

Answer: Because today's new portables, with their power-hogging color screens, lightning processing speeds, and megabyte-gobbling software, average only 2 hours and 17 minutes of running time between battery rechargings.

By contrast, the old 1983 Tandy Model 100, with a black-and-white screen, ran for 15 hours on just four AA batteries. Thus, the Tandy "was arguably a better portable computer than anything currently on the market," says one writer.[23] Unfortunately, today it's difficult to buy a laptop with full-size keyboard that doesn't have a color screen.

Businesspeople who now despair as their notebook screens go dead in the middle of cross-country flights—and who used to try to recharge their machines in the lavatories, until that was banned by the airlines—may find some relief as more planes are outfitted with special power connectors in the seat arms. Resembling car cigarette lighters, these 15-volt connectors, located near the headphone plugs (and requiring an adapter, available in electronics stores), were pioneered by American Airlines in its first- and business-class sections.[24]

For the rest of us, it's helpful to understand the different kinds of battery technologies so that we may use them to best advantage.

Here are the choices in rechargeables:[25–27]

- **Lead-acid (lead-ion):** Old-fashioned and not generally recommended anymore. Used in portable gardening tools but adapted for use in some laptops, lead-acid batteries provide the least amount of charge per pound. Lead-acid technology is regarded as being fully mature, with research providing only slim future performance gains.

- **Nickel-cadmium ("NiCad"):** Old-fashioned and not recommended. Having a longer life than lead-acid batteries, the NiCad battery was once regarded as the standard in rechargeable batteries but now is also considered a fully mature technology and obsolete for laptops. Nickel-cadmium batteries need to be completely drained periodically to maintain the longest charge.

- **Nickel-metal-hydride (nickel-hydride, or NiMH):** Found today in many laptops and cell phones. Although offering longer life than NiCad, nickel-hydride batteries can lose much of their charge when they sit unused. A replacement battery to power a laptop sells at retail for $65–$270.

- **Lithium-ion (li-ion, or "LiOn"):** Found today in many laptops, camcorders, and other devices. LiOn batteries provide more than twice the charge per pound of nickel-hydride. In today's high-end laptops, these batteries allow users to perform word processing for as long as 8 hours by selecting a power-conserving mode that allows them to shut down unneeded, power-draining features. However, when overcharged or over-discharged, the batteries suffer permanent chemical damage. A replacement battery to power a laptop sells at retail for $200–$350.

Still in development is the *lithium-polymer* battery, which could have the best performance characteristics yet. Within the next year or two, we may see a portable computer with color screen that runs perhaps 10 hours on batteries.[28]

For now, if you need a longer-lasting power source than is available with most laptops, you might want to get the 7-hour Portable Power Pack, a 3-pound, lead-acid battery pack (from 1-800-BATTERIES; $199 with cigarette lighter adapter).[29]

Incidentally, care should be taken in disposing of used rechargeables. Don't just throw them in the trash, because incinerated batteries may end up contaminating soil. Call the Rechargeable Battery Recycling Corporation at 1-800-8BATTERY for a nearby hazardous-waste collection center.

4.5 Near & Far Horizons: Processing Power in the Future

KEY QUESTION

What are some forthcoming developments that could affect processing power?

Preview & Review: On the near horizon are multimedia superchips and other types of more powerful chips, as well as more powerful and cheaper supercomputers. On the far horizon are technologies using superconducting materials, optical processing, nano-technology, and DNA.

How far we have come. The onboard guidance computer used in 1969 by the Apollo 11 astronauts—who made the first moon landing—had 2 kilobytes of RAM and 36 kilobytes of ROM, ran at a speed of 1 megahertz, weighed 70 pounds, and required 70 watts of power. Even the Mission Control computer on the ground had only 1 megabyte of memory. "It cost $4 million and took up most of a room," says a space physicist who was there.[30] Today you can buy a personal computer with up to 128 times the memory and 400-plus times the processing speed for just a few thousand dollars.

However, while millions of people may be familiar with the "Intel inside" slogan calling attention to the principal brand of microprocessor used in microcomputers, they probably are unaware that they are more apt to go through the day using another kind of chip—microcontrollers (✔ p. 22), or *digital signal processing (DSP) chips*, like those from Texas Instruments and other manufacturers. Used in pagers, cell phones, cars, hearing aids, and even washing machines, DSP chips are designed to manipulate signals in speech, music, video, and other matters.[31] Points out Michael Malone, "There are now nearly 15 billion microprocessors and microcontrollers in use every day on Earth. That is more than all the televisions, automobiles, and telephones combined."[32]

Future Developments: Near Horizons

The old theological question of how many angels could fit on the head of a pin has a modern counterpart: the technological question of how many circuits could fit there. Computer developers are obsessed with speed and power, constantly seeking ways to promote faster processing and more main memory in a smaller area. Some of the most promising directions in the near term are the following:

- **Media processors—multimedia superchips:** The general-purpose microprocessor we've described in this chapter is now being replaced by a kind of chip called a *media processor*. As we stated in Chapter 1, *multimedia* refers to technology that presents information in more than one medium, including text, graphics, animation, video, and sound, including music and voice. A media processor, or so-called "multimedia accelerator," is a chip with a fast processing speed that can do specialized multimedia calculations and handle several multimedia functions at once, such as audio, video, and three-dimensional animation.

  For example, the Pentium MMX processor (the MMX originally stood for "multimedia extensions") has two features that make it different from its predecessors, the Pentium and Pentium Pro. First, it has 57 additional instructions to handle specific multimedia tasks. Second, it has twice the tiny internal cache of memory as regular Pentiums, which helps it perform certain operations faster. The result is improved video and sound performance in graphics-intense games and other multimedia.[33–36] Intel is now working on another advance, dubbed MMX2, that adds three-dimensional graphics capability.[37]

- **Thousand-megahertz processor chips:** Breakthroughs in manufacturing methods have brought the arrival of faster microprocessors. In late 1997, IBM announced that it had developed a way to use microscopic copper wiring in chips, which would allow electricity to run faster than on the aluminum that was customarily used. This enables IBM to produce PowerPC chips for mainframes (and Apple Macintoshes) that would run at 1000 megahertz (1 gigahertz)—twice as fast as the 500 megahertz of the previous fastest microprocessors. In addition, the new technology would allow chips to be put on a swatch of silicon as small as 0.2 micron wide. (A single strand of human hair is about 100 microns across.)[38,39]

  At about the same time, Intel and Hewlett-Packard announced a new kind of computer "architecture," or design, called *EPIC* (for *Explicitly Parallel Instruction Computing*). An alternative to the CISC and RISC designs, which process instructions sequentially, or one after the other, EPIC would process many instructions at the same time—in other words, do parallel processing. This design is the basis for Intel's Merced chip, due out in high-end computers in 1999 and desktop machines in 2003. Merced will be a 64-bit microprocessor (Pentium II is 32-bit) and is expected to deliver processing speeds of 900–1000 megahertz.[40-42]

- **Gallium arsenide versus silicon germanium:** A leading contender to silicon in chip technology is *gallium arsenide,* which chipmakers call "gas" because of its chemical notation, GaAs. The advantage of "gas" is speed: "Electrons zip through its silver-gray crystal at least five times faster than they move through silicon," points out one report.[43] Gallium arsenide chips also require less power than silicon chips and can operate at higher temperatures.

  Why hasn't gallium arsenide been more popular? For one thing, GaAs wafers on which chips are printed have been tricky to produce, costing 10 times as much as silicon. And the wafers are difficult to process, driving costs up.

  As traditional silicon chips push past 1000 megahertz, they also hit the barrier for the peak performance range for silicon. GaAs can easily handle such speeds. This and similar forthcoming telecommunications gadgetry would seem to open the door to greater use of GaAs—were it not for the recent development of a new semiconductor called *silicon germanium.*

  Developed by IBM, silicon germanium chips are three to five times as fast as current silicon chips and use power much more efficiently, reducing the amount of energy used by portable devices as much as 50%. They also have this advantage over GaAs: they don't require the construction of new chip-manufacturing plants. "This development effectively extends the use of silicon as a material for another 5 to 10 years," says the IBM researcher who led the development team.[44]

- **Billion-bit memory chips:** In 1995 two sets of companies—Hitachi and NEC on the one hand, and Motorola, Toshiba, IBM, and Siemens on the other—announced plans to build plants to make memory chips capable of storing 1 billion bits (1 gigabit) of data. This is 60 times as much information as is stored on the DRAM (dynamic random access memory) chips used in today's latest personal computers. One thumbnail-size piece of silicon could then store 10 copies of the complete works of Shakespeare, 4 hours of compact-disk quality sound, or 15 minutes of video images. Engineering samples of such chips were expected in late 1998.[45,46]

- **Petaflop and network-style supercomputers:** Installed in 1997 at Sandia National Laboratories in New Mexico, Intel's Janus is the fastest supercomputer, performing a trillion mathematical operations per second. Now scientists are hoping to multiply the speed a thousandfold by 2010, creating the first petaflop computer (which would perform an unbelievable thousand trillion, or quadrillion, operations per second).[47]

  As you might guess, the limiting factor is cost. Janus cost $55 million to make. "Today we could build a petaflop machine if we had $55 billion," says IBM researcher Gyan Bhanot. "No problem. But not even Bill Gates could afford this."[48] However, he suggests, in 10 years the cost could be down to a relatively affordable $55 million.

  Meanwhile, computer scientists are experimenting with using personal computers and networks in ways that can double as supercomputers. In a project called Technology for Education 2000, Intel is giving twelve major universities numerous PCs, workstations, servers, and other networking hardware, which will be linked in small and large clusters. When all clusters are hooked together, the system will supposedly be the equal in speed of the most powerful supercomputer.[49]

Future Developments: Far Horizons

Silicon is still king of semiconductor materials, but researchers are pushing on with other approaches. Most of the following, however, will probably take some time to realize:

- **Superconductors:** Silicon, as we stated, is a semiconductor: Electricity flows through the material with some resistance. This leads to heat buildup and the risk of circuits melting down. A *superconductor*, by contrast, is material that allows electricity to flow through it without resistance.

  Until recently superconductors were considered impractical because they have to be kept at subzero temperatures in order to carry enough current for many uses. In 1995, however, scientists at Los Alamos National Laboratory in New Mexico succeeded in fabricating a high-temperature, flexible, ribbon-like superconducting tape that could carry current at a density of more than 1 million amperes per square centimeter, considered a sort of threshold for wide practical use.[50,51]

- **Opto-electronic processing:** Today's computers are electronic; tomorrow's might be opto-electronic—using light, not electricity. With optical-electronic technology, a machine using lasers, lenses, and mirrors would represent the on-and-off codes of data with pulses of light.[52]

- **Nanotechnology:** Nanotechnology, nanoelectronics, nanostructures, nanofabrication—all start with a measurement known as a nanometer. A *nanometer* is a billionth of a meter, which means we are operating at the level of atoms and molecules. A human hair is approximately 100,000 nanometers in diameter.

  Nanotechnology is a science based on using molecules to create tiny machines to hold data or perform tasks. Experts attempt to do "nanofabrication" by building tiny "nanostructures" one atom or molecule at a time. When applied to chips and other electronic devices, the field is called "nanoelectronics."

- **Biotechnology—using DNA molecules:** Not long ago, University of Southern California computer science professor Leonard Adleman watched associates in a research lab do experiments with DNA, the chain of molecules that make up the genetic code of living things. "Adleman was amazed at the intricacy of the DNA strands," reports science writer Jane Allen. "And he was struck by how similar the laboratory cutting, splicing, and copying of these strands were to the manipulations of numbers he performed with computers."

All of a sudden the classic lightbulb went on: Perhaps, Adleman thought, DNA could be used to perform calculations just like a computer.[53] Says another writer, "It's like Stallone and Schwarzenegger teaming up in the ultimate buddy picture—biology and electronic computers, together at last. The future may never be the same."[54]

The code used in silicon-chip-based calculations is binary, having two states. DNA, however, carries information in four molecules designated A, T, C, G. To perform calculations, these four molecules can be combined to form numbers or words, which then combine to make larger words. Biological calculations, which take place by letting the molecules react in a test tube, are not very fast and a single operation may well take 30 minutes. However, because there are trillions of molecules, they can do billions of calculations at once. (A supercomputer may do millions of calculations simultaneously, but it may take hours or even days to solve a problem.)[55–57]

The development of "biochips" is already a growth industry, and scientists hope that they will be used within the next decade as cell implants to scan people for their genetic risk for scores of diseases, such as Alzheimer's.

Perhaps sometime in the future these various avenues will come together. Imagine millions of nanomachines grown from microorganisms processing information at the speed of light and sending it over far-reaching pathways. With these it's difficult to think just how different the world would be.

4.6 Computers, Obsolescence, & the Environment

Preview & Review: Information technology has had some adverse effects on the environment, including environmental pollution and energy consumption.

KEY QUESTION

What are two adverse effects of computers?

Where has miniaturization taken us today? How about making computers small enough to wear?

Indeed, PC evolution has already gone from palmtops to "smart clothes." Steve Mann, 35, a scientist and inventor who received his Ph.D. from the Massachusetts Institute of Technology, spends his waking hours wearing a

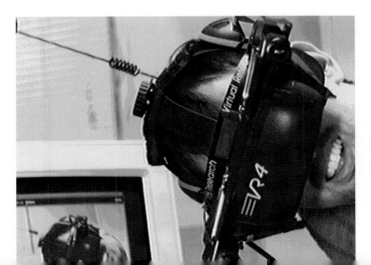

wireless wearable system. For instance, his eyeglasses contain a tiny display screen and a microcamera (which he calls *aremac*—"camera" spelled backwards) that are connected to the World Wide Web. When he got lost in downtown Boston one day, he retraced his steps by calling up images he had just sent to his Web page. "My visual memory prosthetic told me how to get back," he said.[58] Mann's cyberclothes also provide a cellular phone, pager, Walkman, dictating machine, camcorder, laptop computer, microphone, and earphones.

The first Wearable Computers Workshop was held in August 1996 in Renton, Washington, and now researchers are pursuing a number of different possibilities in computer wear.[59–63] MIT scientists have embedded a computer in a Nike sneaker. Sony is developing a monocle-style monitor. British Telecom has created Office on the Arm, with screen, touch pad, and microphone on a forearm cuff. MIT is devising a chest band so that breathing movements can generate electrical energy to run low-powered computers. Some day, Mann predicts, "We'll feel naked, confused, and lost without a computer hovering in front of our eyes to guide us."

Obsolescence

The advances in wearable computers suggest just how fast the pace of technological change is—and consequently how fast the rate of obsolescence.

To return to conventional PCs, *New York Times* technology writer Stephen Manes points out that a microcomputer that is "as fast as they come today will show its age in two or three years but should still run most software acceptably."[64] However, you have to pay top dollar for the fastest machine. If you buy an inexpensive PC, it will probably run most of today's software just fine, but "it will be unable to run some existing games, graphics, and multimedia programs that demand advanced hardware features," Manes says. Thus, consumers are continually facing a dilemma between state-of-the-art and affordability.

Because obsolescence in PC systems occurs about every three years, numerous computers, printers, monitors, fax machines, and so on have wound up as junk in landfills, although some are stripped by recyclers for valuable metals. Some computer owners, says one report, "combat the disorientation of rapid change by keeping their old stuff nearby," filling garages and office shelves.[65] However, a better idea, if you have an old-fashioned PC, is to consider donating it to an organization that can make use of it. Don't abandon it in a closet. Don't dump it in the trash. "Even if you have no further use for a machine that seems horribly antiquated," writes *San Jose Mercury News* technology editor Dan Gillmor, "someone else will be grateful for all it will do."[66]

Ethics

Energy Consumption & "Green PCs"

Besides recycling old computers, there are some environment-friendly things you can do with new computers. All the computers and communications devices discussed in this book run on electricity. Much of this is simply wasted. To reduce the amount of electricity such equipment uses, the U.S. Environmental Protection Agency launched Energy Star, a voluntary program to encourage the use of computers that consume a minimum amount of power. As a result, manufacturers now make Energy Star–compliant "green PCs," as well as energy-saving monitors and printers. In most cases, they go into a "sleep" mode after several minutes of not being used. When you press a key, they go back to full power.

If your equipment does not have the Energy Star logo, there are other things you can do to save energy. For one thing, you can turn off the PC, monitor, and printer if you're not planning to use them for an hour or more. And certainly don't leave them on 24 hours a day. People used to think a computer system would be worn out by repeatedly being turned on and off, but that's no longer true. Hard disks, for example, are built to last an average of 10,000 starts and stops. That means, says computer columnist Lawrence Magid, "that you could expect to turn the computer on and off once a day for 27 years before you would wear it out."[67]

Onward

Today people feel they are living in a time of unprecedented change. "Even [former] Intel CEO Andy Grove says he has a hard time keeping up," reports one writer. Thus, "The rest of us might as well roll over and cry uncle."[68]

Despite Pentium chips, the Internet, digital phones, and the like of our era, it may be that the early 19th century was even more tumultuous, historian Stephen Ambrose suggests. Around the 1840s, a single generation witnessed the invention of the train, the telegraph, and the steamship—colossal technological leaps over the incremental changes that had occurred in the thousands of years before.[69] In fact, perhaps the present rate of technological change pales compared to the period from the late 1850s to 1903, as the historian Robert Post suggests. The late 19th century saw the invention of the Bessemer steelmaking process, the lightbulb, the telephone, the automobile, the radio, mechanical refrigeration, and the airplane.[70] Thus, it is possible that, as futurist Paul Saffo points out, we suffer from a kind of "chronological chauvinism," believing that the present is unique simply because we happen to be living in it.[71]

There is no doubt, though, as we head into the 21st century, that science and technology will certainly be different. What we need to remember, as journalist Meg Greenfield points out, is that its "human manipulators, subjects, and beneficiaries won't."[72]

Becoming a Mobile Computer User: What to Look for in Notebooks, Subnotebooks, & Pocket PCs

How do you tell members of the class of 2000 and later from the class of 1999 at Wake Forest University? By looking at their backpacks. The first group is carrying the bulky, specially padded packs designed to protect laptop computers. Since 1995 the campus at Winston-Salem, North Carolina, has required its new students to have a laptop, which is paid for as part of the tuition.[73]

Other colleges and universities also require some or all of their students to have portable computers—for example, Bentley, Case Western, Columbia, Drew, Hartwick, Mississippi State, Nichols, UCLA, the University of Florida, the University of Minnesota's Crookston campus, and Virginia Tech. Although currently the vast majority of American students who buy computers get the larger desktop models, many are finding that mobile computing offers more convenience. They can take the computers to their dormitory rooms, to the cafeteria, to a campus bench, or on a bus.

At Hartwick College, for example, student Amy Grenier was able to work on a paper during a trip of several hours to see friends. "I did some work in the car, which I couldn't do with a large computer," she said. "You get to utilize more time to get your work done."[74]

Your First Decision: Should You Go Mobile?

Having a personal computer that you can carry around offers tremendous benefits. However, compared to the nonmobile desktop PCs, portables have some limitations:

- They have smaller screens and keyboards.
- They can be more expensive—perhaps $500–$1000 more than a desktop with similar performance.
- They are more vulnerable to theft—nearly twice as many laptops are stolen as desktops (so be sure you get replacement insurance, which may cost about $100 a year).[75]
- They are more fragile and prone to damage, although there are certain (expensive) steps that can be taken to make them more rugged.[76-78] (■ See Panel 4.17.)

You have four choices, then:

1. Forget about mobility. Just get a desktop.
2. Get a portable computer, but make sure you're comfortable with the keyboard and screen when you buy it.
3. Get both a desktop and a portable, if you can afford them. They should be compatible with each other, of course, so that you can easily swap files and programs between them.
4. As a compromise, get a portable but also get a full-size (101-key) keyboard and a desktop-size color monitor. You can then plug the more comfortable keyboard and monitor into the portable when you're at your regular desk.

Going Mobile: What to Look For

Mobile computers come in three sizes:

- Notebooks—4–9 pounds
- Subnotebooks—1.8–4 pounds
- Pocket PCs—1 pound or less

Notebooks and subnotebooks (sometimes called ultra-portables, or micro-laptops) represent the best intersection of power and convenience. The pocket PCs, though practical for many applications, may be too limited for regular student use. This is because they have limited memory, small screens, and keyboards that will definitely slow your progress in writing the next *Midnight in the Garden of Good and Evil*.

Among the factors to consider in buying a new notebook or subnotebook are: price, display screen, keyboard and mouse, portability, and availability of (or connectivity to) diskette drive, CD-ROM drive, and modem.

Price

New desktops generally cost less than portables, although prices for portables are dropping. The cheaper desktops cost approximately $1000 or less at the time of this writing; prices may not include monitors—add $200 or so—but do include some software. By contrast, the cheaper laptops, which have built-in screens, cost between $1000 and $2000.

Of course it's possible to pay $5000 or $6000 for a fully loaded notebook with record-setting speed, big screen, and 6-hour battery. But prices are rapidly coming down, and one can now buy a speedy laptop weighing less than 4 pounds, with 64 megabytes of RAM and a 3-gigabyte hard drive, for about $2400—about half what it cost a year earlier.[79] A lot depends on what microprocessor you get: a late-model Intel chip that really sizzles might account for about $500 of the cost of a $2500 PC; in a $1000 system, the microprocessor costs less than $100.[80]

Whatever you get, bear in mind that to run Windows 95 and most of today's software, 16 megabytes of RAM is really

Tender loving care or "ruggedize" your laptop?

- **Spills:** The most common hazard to laptops is soda being spilled on the keyboard. If you don't clean it promptly, the sugar crystallizes into a sludge that seeps inside the computer.
- **Screens:** Screens are very vulnerable. You shouldn't, for example, try to pick up an open laptop by holding the screen between thumb and forefinger.
- **Temperature:** Be careful of extremes of heat and cold. When one woman turned on her notebook, after leaving it in the trunk of her car during overnight lows of −10 degrees Fahrenheit, the tiny glass filaments of the screen exploded from the surge of electricity and rapid change in temperature.
- **Drops:** Watch that it doesn't get dropped. Many laptops are damaged from being knocked off the corners of desks or falling out of overhead storage bins on airplanes.

During the wintertime, says Mike Griffin, director of worldwide service and technical support for Texas Instruments, "we see a sharp rise in dropped notebooks from [people in] the Northeast and Northwest. Their hands are cold, the sidewalks are slick, or a surprising number of people leave them on top of their car and drive off."

About a dozen companies now make "ruggedized" versions of laptops. Some Panasonic portables are made with a magnesium case 20 times stronger than the standard plastic and hard drives that float in a polymer gel, so that they can withstand multiple drops on concrete. Minnesota-based FieldWorks toughens laptops by sealing all openings and moving parts and using tough materials such as magnesium frames and shock-absorbing rubber. As a result, its laptops can survive falls off ladders, being run over by trucks, and even being run through the dishwasher, all while the computer is on.

Of course, buying a ruggedized notebook carries a price—perhaps $8000–$10,000. All the improvements in durability also add weight—perhaps 4½ pounds more.

---

the *minimum,* and more and more people consider 32–64 megabytes nearly mandatory. (Check the packaging/Web sites of the software you want to buy to see what the RAM requirements are. Also check the RAM requirements for simultaneously running two or more programs you are interested in.)

Display Screen

The price of a portable is affected by the kind of display screen: color or black-and-white (monochrome). Indeed, the display alone accounts for more than half the computer's cost.[81] Color is nice but not necessary unless you plan on doing desktop publishing or lots of graphics work. Actually, most notebooks and subnotebooks now come with color. An exception is Brother's GeoBook, which is black-and-white with backlighting to make viewing easier on the eyes.

Desktop computers use monitors with a television-like CRT (cathode-ray tube). Portables use flat displays (LCD, or liquid crystal display), with two types of color screens:

- **Active-matrix color:** In this version each little dot on the screen is controlled by its own transistor. The advantage of active-matrix screens is that colors are

much brighter than those in the other version, passive matrix.

- **Passive-matrix color:** In this version a transistor controls a whole row or column of dots. The advantage of these screens is that they are less expensive—you can save $500 or so over the active-matrix screen, depending on the size. They are also less power-hungry than active-matrix screens.

  (Technical detail: Dual-scan screens rate better than single-scan screens. In dual-scan the tops and bottoms of the screens are "refreshed" independently at twice the rate of single-scan; thus, they deliver richer colors, though not usually as rich as active-matrix screens.)

The display screens of most notebooks and subnotebooks range from 10.4 to 12.1 inches diagonally, with 11.3 and 11.8 inches the two most popular in-between sizes. However, some screens (such as that on the Toshiba Libretto 50CT) are as small as 6.1 inches; on the larger side, screens of around 13 inches are on the horizon. Compare these sizes with the much more readable 14- or 15-inch (or even 17- or 20-inch) monitors that are standard with most desktop computers. Says one report, "a couple of inches in diagonal measurement may seem insignificant, but it translates into 50% or more actual viewing area."[82]

Keyboard & Mouse

You should try out the keyboard on the notebook or subnotebook to see if you can realistically touchtype on it. (This is very much a problem with the pocket PCs.) No laptop's keyboard can be better than the keyboard that comes with a desktop because of the inherent size constraints.

Notebooks and subnotebooks offer different variants on the mouse, the pointing device. The IBM ThinkPad's TrackPoint pointing controller resembles a pencil eraser stuck among the G, H, and B keys. One reporter quibbled that it "requires persnickety micromotions of the index finger that I find annoying."[83] Other portables use a touch pad—a pressure-sensitive rectangular pad located below the keyboard that allows you to control the cursor by finger pressure and direction. However, except when working in cramped spaces, as on airplane tray tables, most people find a mouse easier to use, and you'll probably want one for when you have the computer set up at a desk.

Portability

How much portable weight are you comfortable with? The Compaq Armada 4100, for instance, can be configured from a slimline 5-pound system up to a 9-pound multimedia system. Nine pounds may not seem like much, but you'll think they're tendinitis-inducing after lugging them around all day.

If you're mainly planning to leave your computer at home and only occasionally take it out (as to the library, to help with research), you'll probably want your portable to have more of the features offered by desktop machines, which means you'll probably want a notebook. Actually, studies show that 75–90% of the time portables are operated on AC power—that is, plugged into a wall plug rather than run on battery power.[84]

If, on the other hand, you're going to be toting your laptop around campus almost every day (as some students do), you'll probably want the lighter weight of a subnotebook, although it may not provide other features you want, such as built-in diskette drive or CD-ROM drive, as we discuss in the next section.

If frequent portability is important to you, then battery life may be, too. (The ReadMe box earlier in this chapter explains the different types of batteries.) On average, most laptop batteries run 2–4 hours and then need recharging, which usually takes 2–3 hours (though some require overnight). If you need lots of battery power, you might consider buying a portable that can take two batteries at once, which doubles the working time. Or you can buy a duplicate battery and swap it with the original when the charge runs low.

For both notebooks and subnotebooks, you'll need an AC adapter—the power cord with transformer—which, when plugged into a wall plug, also recharges the battery. Since the AC adapter weighs a pound or so, be sure to add that to the weight of the laptop and any other gear (such as spare battery) you'll be packing.

Incidentally, if you're traveling outside of North America, you'll need some additional special equipment in order to run your laptop.[85, 86] (■ *See Panel 4.18.*)

Internal Diskette Drive, CD-ROM Drive, Modem, or PC Cards

All laptops have built-in hard-disk drives (now usually 1 gigabyte or more), but one way manufacturers were able to make notebooks into the even lighter subnotebooks was by leaving out the internal diskette drive (floppy drive) or CD-ROM drive. It's probably more convenient if you can find a machine that has both built in. But if you have to skip the internal CD-ROM drive, at least get a laptop that has the ability to connect to an external one. (Though a CD-ROM drive might seem unnecessary, nowadays a lot of applications software is being delivered on CD-ROM disks.) Both notebooks and subnotebooks come equipped with various connectors or ports into which you can plug in various peripheral devices, such as printer, external monitor, keyboard, mouse, and CD-ROM drive. Be sure to find out before you buy what these connectors do.

In addition, both sizes of laptops come equipped with PCMCIA (PC card) slots, with one or two slots standard on just about every model. Designed for credit-card-size circuit boards, PC card slots can hold diskette drives, CD-ROM drives, hard drives, extra memory (flash memory), modems, sound cards, and even pagers, as we described earlier in this chapter. Again, you should be sure to ask what these slots are designed to do.

Modems, which you need in order to do telecommunications, are not always standard on subnotebooks. But if the portable you're considering doesn't have a built-in modem (be sure to ask), you can buy a PC card modem. *(See* ■ *Panel 4.19.)*

Pocket PCs

Pocket personal computers are variously called "electronic organizers," "palmtop PCs," "personal digital assistants (PDAs)," "handheld data communicators," and "personal communicators." Offered under different brand names, they differ in a number of ways. Some have keyboards, some a pen-like stylus, some modems, and some all of these.

Buying a pocket PC may be more difficult than you might think because the marketplace is so full of choices. Here are some of the things you need to think about:[87–91]

1. Do you want just a simple personal organizer—for appointments, phone numbers, and "to do" lists—or a more versatile handheld computer on which you can create documents? Do you want to spend closer to $300 or $1000?

2. Should the gadget fit easily into your pocket? (Most don't, in spite of the name.)

3. Do you want to enter information with a touch screen and stylus? Or through a keyboard? (You'll find most keyboards too small for anything more than hunt-and-peck.)

4. Do want your pocket PC to be compatible with your desktop or laptop PC? (Most run Windows CE, a hand-held version of Microsoft's familiar desktop operating system, but some don't. Most also bundle pocket versions of basic software like Word, Excel, and Power-Point.)

5. Is having a color screen important? (Color is easier to read, but you won't need it if you're just doing e-mail and fax or looking up addresses and appointments.)

6. Do you want it to recognize your handwriting? (Handwriting recognition is only useful for brief notes; moreover, you'll need to be able to write clearly and to train the machine to recognize your writing.)

7. Is a modem included or do you have to buy it separately?

8. Do the AC adapter and docking cradle—which enables you to exchange data with your desktop PC—cost extra?

9. What kind of expansion cards (PC cards) does the device take, and how many will it accommodate?

Common pocket PCs are the *Franklin Rolodex Rex PC Companion, 3Com PalmPilot/IBM Workpad, Sharp Zaurus, NEC MobilePro, Psion Series 5, HewlettPackard HP 360LX Palmtop PC,* and the *Philips Velo 500.* All these handheld computers differ in size, weight, functions offered, convenience, and capacity. If you're interested in a pocket PC, be sure to investigate several models thoroughly.

■ PANEL 4.18

How to make your laptop run overseas

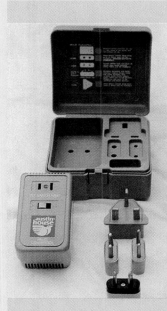

Electrical outlets

● **Plug adapters:** Different countries have different plugs. In traveling around Europe, for example, to plug in your PC designed to run in the United States and Canada, you might need five different adapters for five different countries. Electronics stores such as Radio Shack stock electrical-outlet adapters for under $10.

● **Surge protectors:** The U.S. and Canada run on 110-volt systems. Most developing countries run on 220-volt systems. If your laptop is reasonably new, the transformer on the power cord will convert current from standard outlets of whatever country you're in.

However, some countries' electrical systems aren't stable and may deliver a power surge to your laptop. For safety, therefore, you should plug your computer into a surge protector. Because 220-volt surge protectors are hard to find in the U.S., you may need to borrow one from your hotel or buy one from an electrical-supply store.

Telephone outlets

● **Phone adapters and modems:** If you're planning to try to send and retrieve e-mail, you'll need phone adapters for different countries. These can be bought in North America at electronics stores for under $20.

You can also get an old-fashioned acoustic-coupler modem that works with virtually any country's phone system by transmitting data directly over the phone itself. Price is about $150.

● **Line testers:** Most U.S. phone systems have analog lines, and your modem is designed to translate these analog signals into digital signals for your computer. However, many new phone systems overseas are *digital.* Fortunately, analog lines are usually available even in digital systems.

You need to know what kind of line you're plugging into, because attaching your analog modem to a digital line can damage the equipment. Hotel staffs may not always be able to tell you what their system is.

The best bet is to take along a line tester (available for under $50). When this is plugged into a phone jack, it flashes green if the life is safe.

■ PANEL 4.19

Going mobile: a buyer's checklist

Questions to consider when buying a notebook or subnotebook.		
	Yes	No
1. Is the device lightweight enough that you won't be tempted to leave it behind when you travel?	____	____
2. Does it work with lightweight lithium-ion or nickel-hydride batteries instead of heavier nickel-cadmium batteries?	____	____
3. Is the battery life sufficient for you to finish the jobs you need to do, and is a hibernation mode available to conserve power?	____	____
4. Can you type comfortably on the keyboard for a long stretch?	____	____
5. Is the screen crisp, sharp, and readable in different levels of light?	____	____
6. Does the system have enough storage for all your software and data?	____	____
7. Can the system's hard disk and memory be upgraded to meet your needs?	____	____
8. Does the system provide solid communications options, including a fast modem, so you can send files, retrieve data, and plug into a local area network?	____	____

Summary

What It Is/What It Does	Why It's Important
address (p. 152, KQ 4.2) The location in main memory, designated by a unique number, in which a character of data or of an instruction is stored during processing.	To process each character, the control unit of the CPU retrieves it from its address in main memory and places it into a register. This is the first step in what is called the *machine cycle.*
American Standard Code for Information Interchange (ASCII) (p. 156, KQ 4.3) Binary code used in microcomputers. ASCII originally used seven bits to form a character, but a zero was added in the left position to provide an eight-bit code, providing 256 possible combinations with which to form other characters and marks.	ASCII is the binary code most widely used with microcomputers.
arithmetic/logic unit (ALU) (p. 151, KQ 4.2) The part of the CPU that performs arithmetic operations and logical operations and that controls the speed of those operations.	Arithmetic operations are the fundamental math operations: addition, subtraction, multiplication, and division. Logical operations are comparisons, such as is equal to (=), greater than (>), or less than (<).
binary system (p. 154, KQ 4.3) A two-state system.	Computer systems use a binary system for data representation; two digits, 0 and 1, refer to the presence or absence of electrical current or a pulse of light.
bit (p. 155, KQ 4.3) Short for *binary digit,* which is either a 1 or a 0 in the binary system of data representation in computer systems.	The bit is the fundamental element of all data and information stored in a computer system.
bus (p. 167, KQ 4.4) Electrical pathway through which bits are transmitted within the CPU and between the CPU and other devices in the system unit. There are different types of buses (address bus, control bus, data bus, input/output bus).	The larger a computer's buses, the faster it operates.
byte (p. 155, KQ 4.3) A group of 8 bits.	A byte holds the equivalent of a character—such as a letter or a number—in ASCII and other leading computer data-representation coding schemes. It is also the basic unit used to measure the storage capacity of main memory and secondary-storage devices.
cache memory (p. 165, KQ 4.4) Special high-speed memory area on the CPU or another chip that the CPU can access quickly. A copy of the most frequently used instructions is kept in the cache memory so the CPU can look there first.	Cache memory allows the CPU to run faster because it doesn't have to take time to swap instructions in and out of main memory. Large, complex programs benefit the most from having a cache memory available.

What It Is/What It Does	Why It's Important

central processing unit (CPU) (p. 150, KQ 4.2) The processor; it follows the software's instructions to manipulate data to produce information. In a microcomputer the CPU is usually contained on a single integrated circuit called a *microprocessor.* This chip and other components that make it work are mounted on a circuit board called a *motherboard.* In larger computers the CPU is contained on one or several circuit boards. The CPU consists of two parts, the control unit and the arithmetic/logic unit, connected by a bus.

The CPU is the "brain" of the computer.

chip (microchip) (p. 149, KQ 4.1) Microscopic piece of silicon that contains thousands of microminiature electronic circuit components, mainly transistors.

Chips have made possible the development of small computers.

complex instruction set computing (CISC) (p. 161, KQ 4.4) Design that allows a microprocessor to support a large number of instructions, although the processing speed is slowed by that number compared to the reduced number supported by RISC chips.

CISC chips are used mostly in PCs and conventional mainframes.

control unit (p. 150, KQ 4.2) The part of the CPU that tells the rest of the computer system how to carry out a program's instructions.

The control unit directs the movement of electronic signals between main memory and the arithmetic/logic unit. It also directs these electronic signals between the main memory and input and output devices.

controller card (p. 166, KQ 4.4) Circuit board that allows the CPU to work with the computer's different peripheral devices.

For example, a disk controller card allows the computer to work with different kinds of hard-disk and diskette drives.

daisy chain (p. 169, KQ 4.4) A method of hooking up computer peripherals like external hard-disk drives and scanners in a series along an extended cable. A daisy-chain arrangement requires the port and devices to use a standard interface, either *small computer system interface (SCSI)* or *Universal Serial Bus.*

In a daisy-chain arrangement, one port can be used to connect several peripherals so that users don't have to add ports when adding devices they didn't foresee getting.

dual inline memory module (DIMM) (p. 165, KQ 4.4) Small circuit board plugged into the motherboard and carrying multiple *RAM* chips on both sides.

A DIMM's RAM chips can be used to increase a computer's working memory.

execution cycle (E-cycle) (p. 152, KQ 4.2) Part of the CPU machine cycle during which the ALU executes the instruction and stores the processed results in main memory or a register.

The completion time of the execution cycle determines how fast data is processed. The execution cycle is preceded by the instruction cycle.

expandability (p. 166, KQ 4.4) Refers to the amount of room available in a computer for adding more memory or peripheral devices. Expandability is made possible with expansion slots and expansion boards.

If a microcomputer has expandability, it means that users can later add devices to enhance its computing power, instead of having to buy a new computer.

What It Is/What It Does

Why It's Important

expansion bus (p. 167, KQ 4.4) Bus that carries data between RAM and the expansion slots.

Without buses, computing would not be possible.

expansion card (p. 166, KQ 4.4) Add-on circuit board that provides more memory or a new peripheral-device capability. (The words *card* and *board* are used interchangeably.) Expansion cards are inserted into expansion slots inside the system unit.

Users can use expansion cards to upgrade their computers instead of having to buy entire new systems.

expansion slot (p. 166, KQ 4.4) Socket on the motherboard into which users may plug an expansion card.

See expansion board.

Extended Binary Coded Decimal Interchange Code (EBCDIC) (p. 156, KQ 4.3) A binary coding scheme often used for mainframe computers.

Like ASCII-8, EBCDIC is a binary coding scheme using 8 bits to form each character.

flash memory (p. 166, KQ 4.4) Used primarily in notebook and subnotebook computers; flash memory, or a flash RAM card, consists of circuitry on credit-card-size cards that can be inserted into slots connecting to the motherboard.

Unlike standard RAM chips, flash memory is nonvolatile—it retains data even when the power is turned off. Flash memory can be used not only to simulate main memory but also to supplement or replace hard-disk drives for permanent storage.

floating-point operations per second (flops) (p. 154, KQ 4.2) A floating-point operation is a kind of mathematical calculation. This measure, usually expressed in megaflops—millions of floating-point operations per second—is mainly used with supercomputers.

Floating-point methods are used for calculating a large range of numbers quickly.

game port (p. 169, KQ 4.4) External electrical socket on the system unit that allows users to connect a joystick or similar game-playing device to an internal circuit board.

A game port allows a microcomputer to be made into a game machine.

gigabyte (G, GB) (p. 155, KQ 4.3) Approximately 1 billion bytes (1,073,741,824 bytes); a measure of storage capacity.

Gigabyte is used to express the storage capacity of large computers, such as mainframes, although it is also applied to some microcomputer secondary-storage devices.

infrared port (p. 169, KQ 4.4) Allows a computer to make a cableless connection with infrared-capable devices, such as some printers. The connection is the same as the one between a remote control and a TV, requiring an unobstructed line of sight for the infrared radio waves to travel up to several feet between devices.

Infrared connections allow hookups without a cable trailing between two devices.

instruction cycle (I-cycle) (p. 152, KQ 4.2) Part of the CPU machine cycle in which a single computer instruction is retrieved from memory, put into a register, and decoded.

"Decoding" means that the control unit alerts the circuits in the microprocessor to perform the specified operation. The instruction cycle is followed by the execution cycle.

What It Is/What It Does

integrated circuit (p. 148, KQ 4.1) An entire electrical circuit, or pathway, etched a on tiny square, or chip, of silicon half the size of a person's thumbnail. In a computer, different types of integrated circuits perform different types of operations. An integrated circuit embodies what is called *solid-state technology.*

Intel-type chips (p. 161 KQ 4.4) CPU chips designed for PCs; they are based on the model made by Intel Corporation, but other makers include Cyrix, Advanced Micro Devices, and DEC.

kilobyte (K, KB) (p. 155, KQ 4.3) Unit for measuring storage capacity; equals 1024 bytes (usually rounded off to 1000 bytes).

local bus (p. 167, KQ 4.4) Bus that connects expansion slots to the CPU, bypassing RAM.

machine cycle (p. 152, KQ 4.2) Series of operations performed by the CPU to execute a single program instruction; it consists of two parts: an *instruction cycle* and an *execution cycle.*

machine language (p. 157, KQ 4.3) Binary code (language) that the computer uses directly. The 0s and 1s represent precise storage locations and operations.

main memory (p. 151, KQ 4.2) Also known as *memory, primary storage, internal memory,* or *RAM* (for *random access memory*). Main memory is working storage that holds (1) data for processing, (2) the programs for processing the data, and (3) data after it is processed and is waiting to be sent to an output or secondary-storage device.

megabyte (M, MB) (p. 155, KQ 4.3) About 1 million bytes (1,048,576 bytes).

megahertz (MHz) (p. 153, KQ 4.2) Measurement of transmission frequency; 1 MHz equals 1 million beats (cycles) per second.

microprocessor (p. 150, KQ 4.1) A CPU (processor) consisting of miniaturized circuitry on a single chip; it controls all the processing in a computer. In other electronic devices, such as a microwave oven or a phone, the microprocessor is called a microcontroller.

Why It's Important

The integrated circuit has enabled the manufacture of the small, powerful, and relatively inexpensive computers used today.

The majority of microcomputers run on Intel-type chips, including Intel's line of 8086 through '486 chips and the Pentium and Pentium MMX chips. Apple is developing a new operating system that will allow its microcomputers to use Intel-type chips.

The sizes of stored electronic files are often measured in kilobytes.

A local bus is faster than an expansion bus.

The machine cycle is the essence of computer-based processing.

For a program to run, it must be in the machine language of the computer that is executing it.

Main memory determines the total size of the programs and data files a computer can work on at any given moment.

Most microcomputer main memory capacity is expressed in megabytes.

Generally, the higher the megahertz rate, the faster a computer can process data.

Microprocessors enabled the development of microcomputers and electronic controls for other devices.

millions of instructions per second (MIPS) (p. 153, KQ 4.2) Another measure of a computer's execution speed; for example, .5 MIPS is 500,000 instructions per second.

This measure is often used for large, relatively powerful computers and new sophisticated microcomputers.

motherboard (p. 161, KQ 4.4) Also called *system board;* the main circuit board in the system unit of a microcomputer.

It is the interconnecting assembly of essential components, including CPU, main memory, other chips, and expansion slots.

Motorola-type chips (p. 161, KQ 4.4) CPU chips made by Motorola for Apple Macintosh and its clones.

Motorola chips have provided Macintoshes and Power-PCs with a powerful CPU.

parallel port (p. 169, KQ 4.4) External electrical socket on the system unit that allows a parallel device, which transmits 8 bits simultaneously, to be connected to an internal board.

Enables microcomputer users to connect to a printer using a cable.

parity bit (p. 157, KQ 4.3) Also called a *check bit;* an extra bit attached to the end of a byte.

Enables a computer system to check for errors during transmission (the check bits are organized according to a particular coding scheme designed into the computer).

PC cards (p. 170, KQ 4.4) Small cards that contain peripherals and can be plugged into slots in portable computers. Based on the PCMCIA standard.

See Personal Computer Memory Card International Association.

peripheral device (p. 158, KQ 4.4) Hardware that is outside the central processing unit, such as input/output and secondary-storage devices.

These devices are used to get data into and out of the CPU and to store large amounts of data that cannot be held in the CPU at one time.

Personal Computer Memory Card International Association (PCMCIA) (p. 170, KQ 4.4) Completely open, nonproprietary bus standard for portable computers.

This standard enables users of notebooks and subnotebooks to insert credit-card-size peripheral devices, called *PC cards,* such as modems, memory cards, hard disks, sound boards, and even pagers and cellular communicators into their computers.

port (p. 168, KQ 4.4) Connecting socket on the outside of the computer system unit that is connected to an expansion board on the inside of the system unit. Types of ports include parallel, serial, video adapter, game, SCSI, and USB. Another type is the infrared port, using a signaling device instead of a socket.

A port enables users to connect by cable or infrared signals a peripheral device such as a monitor, printer, or modem so that it can communicate with the computer system.

power supply (p. 158, KQ 4.4) Device in the computer that converts AC current from the wall outlet to the DC current the computer uses.

The power supply enables the computer (and peripheral devices) to operate.

RAM chip (p. 164, KQ 4.4) An integrated circuit that provides *random access memory;* it is mounted on a small SIMM or DIMM circuit board that is plugged into the motherboard.

Adding RAM chips speeds up a computer by keeping more software capabilities and data immediately available for the CPU to use.

What It Is/What It Does	**Why It's Important**

random access memory (RAM) (p. 164, KQ 4.4) Also known as *main memory* or *primary storage;* type of memory that temporarily holds data and instructions needed shortly by the CPU. RAM is a volatile type of storage.

RAM is the working memory of the computer; it is the workspace into which applications programs and data are loaded and then retrieved for processing.

read-only memory (ROM) (p. 165, KQ 4.4) Also known as *firmware;* a memory chip that permanently stores instructions and data that are programmed during the chip's manufacture. ROM is a nonvolatile form of storage.

ROM chips are used to store special basic instructions for computer operations such as those that start the computer or put characters on the screen.

reduced instruction set computing (RISC) (p. 162, KQ 4.4) Type of design in which the complexity of a microprocessor is reduced by reducing the amount of superfluous or redundant instructions.

With RISC chips, a computer system gets along with fewer instructions than those required in conventional computer systems. RISC-equipped workstations work up to 10 times faster than conventional workstations.

register (p. 152, KQ 4.2) High-speed circuit that is a staging area for temporarily storing data during processing.

The computer loads the program instructions and data from the main memory into the staging areas of the registers just prior to processing.

semiconductor (p. 148, KQ 4.1) Material, such as silicon (in combination with other elements), whose electrical properties are intermediate between a good conductor and a nonconductor. When good-conducting materials are laid on the semiconducting material, an electronic circuit can be created.

Semiconductors are the materials from which integrated circuits (chips) are made.

serial port (p. 169, KQ 4.4) Also known as *RS-232 port;* external electrical socket on the system unit that allows a serial device, which transmits 1 bit at a time, to be connected to an internal board.

Serial ports are used principally for connecting communications lines, modems, and mice to microcomputers.

silicon (p. 148, KQ 4.1) Element widely found in sand and clay; it is a semiconductor.

Silicon is used to make integrated circuits (chips).

single inline memory module (SIMM) (p. 165, KQ 4.4) Small circuit board plugged into the motherboard that carries RAM chips on one side.

A SIMM's RAM chips can be used to increase a computer's main memory capacity.

small computer system interface (SCSI) port (p. 169, KQ 4.4) Pronounced "scuzzy"; an interface for transferring data at high speeds for up to 7 or 15 SCSI-compatible devices.

SCSI ports are used to connect external hard-disk drives, magnetic-tape backup units, scanners, and CD-ROM drives to the computer system.

solid-state device (p. 148, KQ 4.1) Electronic component made of solid materials with no moving parts, such as an integrated circuit.

Solid-state integrated circuits are far more reliable, smaller, and less expensive than electronic circuits made from several components.

What It Is/What It Does

surge protector (p. 158, KQ 4.4) Also called *surge suppressor;* device that protects a computer from being damaged by surges of high voltage.

system clock (p. 153, KQ 4.2) Internal timing device that uses a quartz crystal to generate a uniform electrical frequency from which digital pulses are created.

system unit (p. 158, KQ 4.4) The box or cabinet that contains the electrical components that do the computer's processing; usually includes processing components, RAM chips (main memory), ROM chips (read-only memory), power supply, expansion slots, and disk drives but not keyboard, printer, or often even the display screen.

terabyte (p. 155, KQ 4.3) Approximately 1 trillion bytes (1,009,511,627,776 bytes); a measure of capacity.

transistor (p. 147, KQ 4.1) Semiconducting device that acts as a tiny electrically operated switch, switching between "on" and "off" many millions of times per second.

Unicode (p. 156, KQ 4.3) A binary coding scheme that uses 2 bytes (16 bits) for each character, rather than 1 byte (8 bits), providing 65,536 possible characters that could represent almost all the written languages of the world.

uninterruptible power supply (UPS) (p. 159, KQ 4.4) Battery-operated device that provides a microcomputer with surge protection and with electricity if there is a power failure.

Universal Serial Bus (USB) (p. 169, KQ 4.4) Allows up to 127 peripherals to be connected through just one general-purpose port. The USB connecting the port inside the computer will interpret the signals from each peripheral device and tell the computer to recognize it.

video adapter port (p. 169, KQ 4.4) Electrical socket that connects the video display monitor outside the computer to the video adapter card inside the system unit.

video memory (p. 166, KQ 4.4) Video RAM (VRAM) chips are used to store display images for the monitor.

Why It's Important

A surge protector is an inexpensive investment compared to a new motherboard.

The system clock controls the speed of all operations within a computer. The faster the clock, the faster the processing.

The system unit protects many important processing and storage components.

Some forms of mass storage, or secondary storage for mainframes and supercomputers, are expressed in terabytes.

Transistors act as electronic switches in computers. They are more reliable and consume less energy than their predecessors, electronic vacuum tubes.

Because Unicode can provide character sets for Chinese, Japanese, and other languages that don't use the Roman alphabet, it is likely to become the standard code, though the need to convert existing software applications and databases will make the change a slow one.

A UPS can keep a microcomputer going long enough to allow the user to save data files before shutting down.

The user can easily connect printers, modems, mice, keyboards, and CD-ROM drives in a daisy-chain style without worrying about adding ports.

The video adapter port enables users to have different kinds of monitors, some having higher resolution and more colors than others.

The amount of video memory determines how fast images appear and how many colors are available on the display screen. Video memory chips are useful for programs displaying lots of graphics.

What It Is/What It Does	**Why It's Important**

volatile storage (p. 152, KQ 4.2) Temporary storage, as in main memory (RAM).

The contents of volatile storage are lost when power to the computer is turned off.

voltage regulator (p. 159, KQ 4.4) Also called a *line conditioner;* a device that protects a computer from being damaged by insufficient power.

A voltage regulator protects the computer when brownouts occur, as when a power tool starts up and causes the house lights to dim, or when the power lines provide insufficient power to an area on a very hot day.

word (p. 151, KQ 4.2) Also called *bit number;* group of bits that may be manipulated or stored at one time by the CPU.

Often the more bits in a word, the faster the computer. An 8-bit-word computer will transfer data within each CPU chip in 8-bit chunks. A 32-bit-word computer is faster, transferring data in 32-bit chunks.

Self-Test

1. A(n) _____ bit is an extra bit attached to a byte for purposes of checking for accuracy.

2. _____ is a binary programming language that the computer can run directly.

3. The _____ is often referred to as the *brain* of the computer.

4. The _____ controls how fast the operations within a computer take place. (system clock)

5. A(n) _____ is about 1,000 bytes (1,024) bytes.
    A(n) _____ is about 1 million bytes (1,048,576) bytes).
    A(n) _____ is about 1 billion bytes (1,073,741,824 bytes).

Short-Answer Questions

1. What is the function of the ALU in a microcomputer system?

2. What is the purpose of a parity scheme? How does it work?

3. What is the function of registers in a computer system?

4. List at least five electrical components commonly found in a microcomputer's system unit.

5. Why is it important that your computer be expandable?

Multiple-Choice Questions

1. Which of the following are used to hold data and instructions that will be used shortly by the CPU?
    a. RAM chips
    b. ROM chips
    c. peripheral devices
    d. cache memory
    e. All of the above

2. Which of the following coding schemes is widely used on microcomputers?
    a. EBCDIC
    b. ASCII-8
    c. Unicode
    d. Microcode
    e. All of the above

3. Which of the following is accessed when you switch on your computer?
    a. RAM chip
    b. ROM chip
    c. coprocessor chip
    d. microprocessor chip
    e. All of the above

4. Which of the following can be used in portable computers to replace the hard disk?
    a. cache memory
    b. video RAM
    c. flash memory
    d. ROM
    e. None of the above

5. Which of the following is used to measure processing speeds in supercomputers?
    a. megahertz
    b. MIPS
    c. flops
    d. picoseconds
    e. None of the above

True/False Questions

T F 1. Microcomputer processing speeds are usually measured in MIPS.

T F 2. The machine cycle is composed of the instruction cycle and execution cycle.

T F 3. Computer programmers write in programming languages that resemble machine language.

T F 4. Today's microprocessors have more transistors than those in the 1970s.

T F 5. Main memory is nonvolatile.

Knowledge in Action

1. Describe the latest microprocessor chip released by Intel. Who are the intended users of this chip? How is this chip better than its predecessor? Perform your research using current computer magazines and periodicals and/or the Internet.

2. Develop a binary system of your own (use any two states, objects, or conditions) and encode the following: I am a rocket scientist.

3. Look through some computer magazines and identify advertised microcomputer systems. Decide what microcomputer might be the best one for you to use based on your processing requirements (if necessary, pick a hypothetical job and identify some probable processing requirements). Describe the microcomputer you would choose and why. Compare this microcomputer to others you saw advertised using the following categories: (a) name and brand of computer, (b) microprocessor model, (c) RAM capacity, (d) availability of cache memory, and (f) cost.

4. Look through several computer magazines and list all the coprocessor chips and add-on boards mentioned. Next to each listed item, write down what it does and what type of computer system it's compatible with. Then, note an application (task) for which each item could be useful.

5. Using computer magazines and periodicals and/or the Internet, conduct additional research on one of the newer technologies described in the *Near and Far Horizons* section at the end of this chapter. Who will use this technology? For what? When will this technology be available? Who is developing this technology and who is providing the funding?

Input & Output

Taking Charge of Computing & Communications

've talked to people who've been separated for years but didn't get divorced because they couldn't afford an attorney," says Heather Fisher, a spokeswoman for Utah's QuickCourt. "They're really excited about this."[1]

Coca-Cola kiosk at a trade fair

What they're excited about, according to Fisher, are particular kinds of electronic kiosks that are designed to help people in filling out paperwork for no-fault divorces. With these machines, which QuickCourt makes and sells to state governments, people can get officially unmarried quickly and inexpensively.

Whatever you may think of the idea of the high-tech divorce, it's clear that the ATM, long a standby for people needing fast cash during evenings and weekends, is growing up—and is becoming something quite different from its origins. Across the United States, the new generation of kiosks is being used by governments to enable people not only to learn about tourist attractions and bus routes but also to get vehicle license plates, garage-sale permits, and information on property taxes. In New York City, kiosks (✔ p. 76) can be used by citizens to pay parking tickets and check for building code violations on their apartment buildings. In San Antonio, they allow people to see what animals at the pound are available for adoption. In Seattle, they let commuters at car-ferry terminals view live images of traffic conditions on major highways.[2,3]

These super-ATMs also have commercial uses, not only doling out $20 bills, but selling stamps, printing out checks, and making movie and plane tickets available.[4,5] Says one report, "They work 24 hours a day and never demand overtime. They're never in a surly mood. Though occasionally they run out of paper."[6]

The kiosk is an example of a device that presents the two faces of the computer that are important to humans: that is, it allows you to *input* data and to *output* information. In this chapter, we discuss what the principal input and output devices are and how you can make use of them.

5.1 Input Hardware

KEY QUESTION

What are the three general types of input hardware?

Preview & Review: Input hardware devices are classified as three types: keyboards, pointing devices, and source data-entry devices.

Input hardware consists of devices that translate data into a form the computer can process. The people-readable form may be words like the ones in these sentences, but the computer-readable form consists of 0s and 1s, or off and on electrical signals.

Input hardware devices are categorized as three types. (■ *See Panel 5.1.*) They are:

- Keyboards
- Pointing devices
- Source data-entry devices

■ PANEL 5.1

**Summary of input
devices**

CATEGORIES OF INPUT HARDWARE		
KEYBOARD HARDWARE	**POINTING DEVICES**	**SOURCE DATA–ENTRY**

Keyboards
Touch-Tone devices
Set-top boxes
	Mice, trackballs,
	  pointing sticks,
	  touchpads
	Light pens
	Digitizing tablets
	Pen computers
		Scanning devices
		Bar-code readers
		Mark- and character-
		  recognition devices
		Fax machines
		Imaging systems

		Other input devices Audio-input devices Video-input devices Digital cameras Sensors Radio-frequency identification Voice recognition systems Human-biology devices

Keyboards

In a computer, a **_keyboard_ is a device that converts letters, numbers, and other characters into electrical signals that are machine-readable by the computer's processor.** The keyboard may look like a typewriter keyboard to which some special keys have been added. Or it may look like the keys on a bank ATM or the keypad of a pocket computer used by a bread-truck driver. Or it may even be a Touch-Tone phone, network computer, or cable-TV set-top box.

Pointing Devices

Pointing devices control the position of the cursor or pointer on the screen. Pointing devices include:

- Mice, trackballs, pointing sticks, and touchpads
- Light pens
- Digitizing tablets
- Pen-based systems

Source Data-Entry Devices

Source data-entry devices refer to the many forms of data-entry devices that are not keyboards or pointing devices. **_Source data-entry devices_ create machine-readable data on magnetic media or paper or feed it directly into the computer's processor.** This is also known as _source-data automation_, the process in which data created while an event is taking place is entered directly into the system in a machine-processable form.

    Source data-entry devices include the following:

- Scanning devices
- Sensors
- Audio-input devices

- Voice-recognition systems
- Video-input devices
- Electronic cameras
- Human-biology input devices

Quite often a computer system will combine keyboard, pointing devices, and source data-entry devices. A desktop-publishing system, for example, uses a keyboard, a mouse, and a scanning device (image scanner).

In the next sections we discuss the three types of input hardware in more detail.

5.2 Keyboard Input

KEY QUESTION

What are the different types of keyboard hardware?

Preview & Review: Keyboard-type devices include computer keyboards, Touch-Tone devices, and cable-TV set-top boxes.

Even if you aren't a ten-finger touch typist, you can use a keyboard. Yale University computer scientist David Gelernter, for instance, lost the use of his right hand and right eye in a mail bombing by the Unabomber. However, he expressed not only gratitude at being alive but also recognition that he could continue to use a keyboard even with his limitations. "In the final analysis," he wrote in an online message to colleagues, "one decent typing hand and an intact head is all you really need...."[7]

Here we describe the following keyboard-type devices:

- Computer keyboards
- Touch-Tone devices
- Set-top boxes

Computer Keyboards

Conventional computer keyboards, such as those for microcomputers, have all the keys that typewriter keyboards have plus others unique to computers. The keyboard is built into laptop computers or attached to desktop computers with a cable. However, wireless keyboards (which use an infrared signal) are also available; these may be useful for those who find the keyboards that come with portable computers too small for efficient typing.

The accompanying keyboard illustration provides a review of keyboard functions. (■ *See Panel 5.2.)*

Different languages have different keyboard needs. For example, English has 26 characters in its alphabet, but traditional Chinese, used in Hong Kong and Taiwan, has thousands of characters. "For a Chinese office worker," says one account, "learning the commands needed to type a simple e-mail can mean going to class for a month."[8] The input problem is one reason the Chinese computer industry has not taken off. Besides voice input (as with IBM's ViaVoice software, discussed later), there are two keyboard solutions. One method uses the Western keyboard to write Chinese pinyin, a bridge language that adapts Chinese to the Western alphabet. The computer reads the pinyin word and turns it into a Chinese character. The other method maps basic strokes used in Chinese characters onto the keyboard. After the keyboarder types the first couple of strokes, the software offers a choice of characters that begin with those strokes.

On some computers only

Only capital letters will be displayed.

Prints what's currently displayed on the screen.

Prevents the screen from scrolling.

The Esc key allows you to exit a command or menu and return to the work screen.

Temporarily suspends the current task.

Function keys are used to issue commands specific to the software package being used.

These status lights indicate when these functions are on or off.

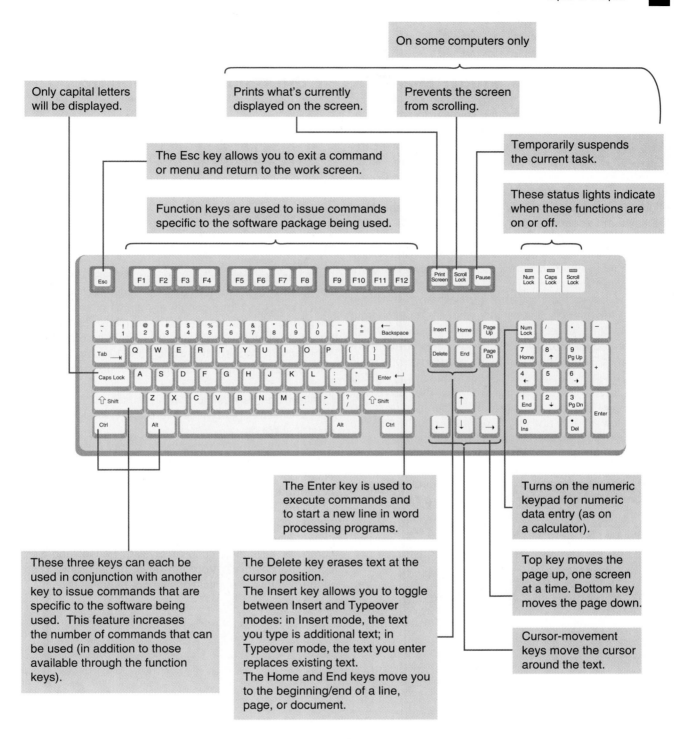

The Enter key is used to execute commands and to start a new line in word processing programs.

Turns on the numeric keypad for numeric data entry (as on a calculator).

These three keys can each be used in conjunction with another key to issue commands that are specific to the software being used. This feature increases the number of commands that can be used (in addition to those available through the function keys).

The Delete key erases text at the cursor position.
The Insert key allows you to toggle between Insert and Typeover modes: in Insert mode, the text you type is additional text; in Typeover mode, the text you enter replaces existing text.
The Home and End keys move you to the beginning/end of a line, page, or document.

Top key moves the page up, one screen at a time. Bottom key moves the page down.

Cursor-movement keys move the cursor around the text.

■ PANEL 5.2
Common keyboard layout

README

Case Study: Mysteries of the Computer Keyboard

The computer keyboard "is the most bizarre, ridiculous, nondesigned monstrosity foisted on the American public," says Don Norman, vice president of research at Apple Computer. "We've put huge amounts of effort into the design of things you see on the screen, but the keyboard seems handed down by . . . an evil god."[9] An overstatement perhaps? But just what *do* those weird key labels mean—SysRq, Scroll Lock, Pause, and so on? Couldn't the inventors have tried to make them more comprehensible?

To understand how "bizarre" (in Norman's word) the computer keyboard is, first consider that it is adapted from the typewriter keyboard, the first working model of which was patented in 1867 by the American inventor Christopher Sholes. The name given to the standard keyboard is QWERTY, which describes the beginning keys in the top row of letters *(see drawing below)*. The QWERTY arrangement has many failings, all of which add up to relatively slow speed and more errors. Actually, though, Sholes designed the keyboard this way deliberately in order to *slow down* typists, whose flying fingers were apt to jam his primitive machine.

An alternative layout exists in the Dvorak keyboard *(see drawing above),* which is supposed to be as much as

Dvorak keyboard

50% faster than QWERTY and to produce only half the number of errors.

"When I switched from a manual QWERTY to an electric Dvorak typewriter," says lawyer David Miller, "my speed doubled with a couple of months of evening practice. Three years later and just out of law school, I overheard a secretary complain that I could type faster than she could."[10] Indeed, the world speed record for typing, 186 words per minute, was set on a Dvorak computer keyboard.

Although it is easy to convert existing QWERTY keyboards with inexpensive software, the Dvorak design has never caught on.

"The real barrier is traditional thinking," Miller suggests. "[P]eople will spend millions to make computers run faster, but ignore minor changes that could speed the flow of information from brain to computer."

When the computer came along, engineers made hasty additions to Sholes's QWERTY keyboard as circumstances required. Thus, some of the keys on today's computer keyboard, such as SysRq, Pause, and Scroll Lock, are holdovers from the days of mainframes. Most modern software applications ignore these keys.

QWERTY keyboard

Touch-Tone Devices

The *Touch-Tone, or push-button, telephone* can be used like a dumb terminal to send data to a computer. For example, one way FedEx customers can request pickup service for their packages is by pushing buttons on their phones.

Another common device is the *card dialer,* or *card reader,* used by merchants to verify credit cards over phone lines with a central computer. When a credit card with a magnetized strip has become demagnetized (from frequent rubbing within a wallet, perhaps), a store clerk has to punch in the card numbers manually on the Touch-Tone phone in order to verify credit.

Interactive TV Set-Top Boxes

If you receive television programs from a cable-TV service instead of free through the air, you may have a decoder device called a set-top box (✔ p. 31). **A *set-top box* works with a keypad, such as a handheld wireless remote control, to allow cable-TV viewers to change channels or, in the case of interactive systems, to exercise other commands.** What is new, however, is that set-top boxes are heading in the direction of PC/TVs (✔ p. 31), with analog and digital signals converging to offer TV with Internet access and electronic mail.

For instance, over the next few years, Tele-Communications Inc. and other big cable-TV operators plan to put in place millions of set-top boxes that will zap all kinds of digital fare to consumers. This would include additional digital TV channels, high-resolution movies, home banking services, interactive TV, e-mail, and high-speed Internet access.[11–13]

5.3 Pointing Devices

KEY QUESTION

How do the different types of pointing devices work?

Preview & Review: Pointing devices include mice, trackballs, pointing sticks, and touchpads; light pens; digitizing tablets; and pen computers.

One of the most natural of all human gestures, the act of pointing is incorporated in several kinds of input devices. The most prominent ones are the following:

- Mice, trackballs, pointing sticks, and touchpads
- Light pens
- Digitizing tablets
- Pen-based systems

Mice, Trackballs, Pointing Sticks, & Touchpads

The principal pointing tools used with microcomputers are the mouse or its variants—the trackball, the pointing stick, and the touchpad. (◼ *See Panel 5.3, next page.*)

- Mouse: You are probably already familiar with the mouse. **A *mouse* is a device that is rolled about on a desktop and directs a pointer on the computer's display screen. The *mouse pointer* is the symbol that indicates the position of the mouse on the display screen.** It may be an arrow, a rectangle, or even a representation of a person's pointing finger. The pointer may change to the shape of an I-beam to indicate that it is a cursor and shows the place where text may be entered.

  The mouse has a cable that is connected to the microcomputer's system unit by being plugged into a special port, or socket. This tail-like cable and the rounded "head" of the instrument gave the shape that suggested the name *mouse*.

  On the bottom side of the mouse is a ball that translates the mouse movement into digital signals. On the top side are one to four buttons. The first button is used for common functions, such as clicking and dragging. The functions of the second, third, and fourth buttons are determined by whatever software you're using.

  Depending on the software, many commands that you can execute with a mouse can also be performed through the keyboard. The mouse may make it easy for you to learn the principal commands to

■ PANEL 5.3

A few pointers
Mouse, trackball, pointing
stick, and touchpad.

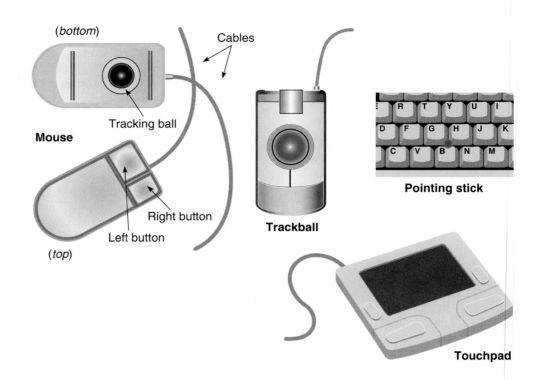

a word processing program, but later you'll probably find it faster to
execute those same commands using a combination of keystrokes on
the keyboard. Principal commands are shown below. (■ *See Panel 5.4.*)

One new development makes it easier to scroll through screens, a
need brought about by the popularity of the World Wide Web.
Microsoft's IntelliMouse features a little wheel that allows you to
move line by line or page by page through a document or to zoom in
on specific cells in a spreadsheet.[14] Another development is the
application of a technology called *force-feedback*—first seen in video
game joysticks to convey such sensations as the recoil of a gun—to

■ PANEL 5.4

**Learning mouse
language**

Term	Definition
The directions you are most likely to encounter for using a mouse or a trackball are the following:	
Point	Move the pointer to the desired spot on the screen, such as over a particular word or object.
Click	Tap—that is, press and quickly release—the left mouse button.
Double-click	Tap—press and release—the left mouse button twice, as quickly as possible.
Drag	Press and hold the left mouse button while moving the pointer to another location.
Drop	Release the mouse button after dragging.
Right-click	To make a selection using the button on the right side of the mouse, which usually brings up a pop-up menu.

heighten the sense of touch by making control devices move or vibrate along with the action on the screen. Thus, you could drag an icon and feel its simulated weight as it slides across the screen and a slight bump when it crosses the borders of software windows.[15,16]

A variant of the desktop mouse is the "air mouse," a cordless infrared mouse that works up to 40 feet away from the computer. With this you can roam a conference room during a slide-show presentation while clicking on icons or making other changes.

- **Trackball:** Another form of pointing device, **the *trackball* is a movable ball, on top of a stationary device, that is rotated with fingers or palm of the hand.** In fact, the trackball looks like the mouse turned upside down. Instead of moving the mouse around on the desktop, you move the trackball with the tips of your fingers. Like the mouse, the trackball has additional buttons whose functions vary depending on the software.

    Trackballs are especially suited to portable computers, which are often used in confined places such as on airline tray tables. Trackballs may appear on the keyboard centered below the space bar, as on the Apple PowerBook, or built into the right side of the screen. On some portables the trackball is a separate device that is clipped to the side of the keyboard.

- **Pointing stick:** **A *pointing stick* is a pointing device that looks like a pencil eraser protruding from the keyboard between the G, H, and B keys.** You move the pointing stick with your forefinger while using your thumb to press buttons located in front of the space bar.

    A forerunner of the pointing stick is the joystick. **A *joystick* is a pointing device that consists of a vertical handle like a gearshift lever mounted on a base with one or two buttons.** Named for the control mechanism that directs an airplane's back-to-front and side-to-side movement, joysticks are used principally in video games and in some computer-aided design systems.

- **Touchpad:** **The *touchpad* is a small, flat surface over which you slide your finger, using the same movements as you would with a mouse.** As you move your finger, the cursor follows the movement. You "click" by tapping your finger on the pad's surface or by pressing buttons positioned close by the pad. Touchpads are now common on portable computers, positioned just below the space bar. Because touchpads can be a bit difficult to control, many portable computer users hook up a mouse when they have the space to use one.

The box at the top of the next page presents some of the pros and cons of using a mouse, trackball, and touchpad. (■ *See Panel 5.5.*)

Light Pen

The *light pen* is a light-sensitive stylus, or pen-like device, connected by a wire to the computer terminal. The user brings the pen to a desired point on the display screen and presses the pen button, which identifies that screen location to the computer. (■ *See Panel 5.6.*) Light pens are used by engineers, graphic designers, and illustrators.

Pros		Cons
• Relatively inexpensive • Very little finger movement needed to reach buttons	**Mouse**	• When gripped too tightly can cause muscle strain • Uses more desk space than other pointing devices • Must be cleaned regularly
• Uses less desk space than mouse • Requires less arm and hand movement than mouse	**Trackball**	• Wrist is bent during use • More finger movement needed to reach buttons than with other pointing devices
• Small footprint • Least prone to dust	**Touchpad**	• Places more stress on index finger than other pointing devices do • Small active area makes precise cursor control difficult

■ PANEL 5.5

Mouse, trackball, and touchpad: pros and cons

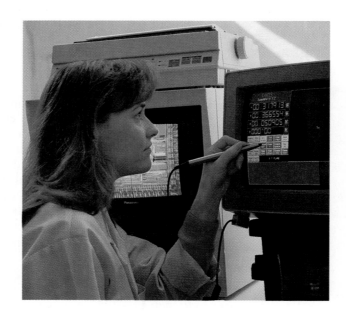

■ PANEL 5.6

Light pen
This person is using a light pen to input instructions to the computer.

Digitizing Tablet

A *digitizing tablet* consists of a tablet connected by a wire to a stylus or puck. A *stylus* is a pen-like device with which the user "sketches" an image. A *puck* is a copying device with which the user copies an image, such as an architectural drawing or a civil engineering map. (■ *See Panel 5.7.*) A puck looks a bit like a mouse but has different types of buttons and a clear plastic section extending from one end with crosshairs printed on it. The inter-

■ PANEL 5.7
Digitizing tablet

section of the crosshairs points to a location on the graphics tablet, which in turn is mapped to a specific location on the screen.

Digitizing tablets are used primarily in design and engineering. When used with drawing and painting software, a digitizing tablet and stylus allow you to do shading and many other effects similar to those artists achieve with pencil, pen, or charcoal. Alternatively, when you use a puck, you can trace a drawing laid on the tablet, and a digitized copy is stored in the computer.

Pen-Based Systems

■ PANEL 5.8

Pen-based systems
(Left) Pen-based computer with stylus and color screen. *(Right)* In some hospitals, pen-based computers are used to enter comments to patients' records.

In the next few years, students may be able to take notes in class without ink and paper, if pen-based computer systems evolve as Depauw University computer science professor David Berque hopes they will. **Pen-based computer systems use a pen-like stylus to allow people to enter handwriting and marks onto a computer screen rather than typing on a keyboard. (■** *See Panel 5.8.)* Berque has developed a prototype for a system that would connect an instructor's electronic "whiteboard" on the classroom wall with students' pen computers, so that students could receive notes directly, without having

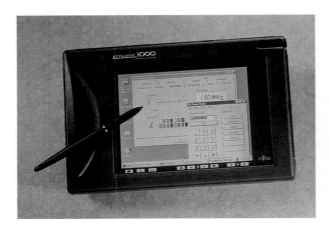

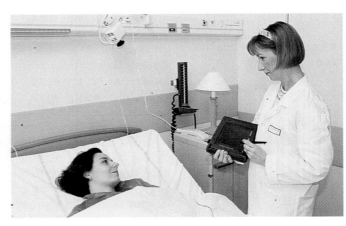

to copy information word for word. "The idea is that this might free the students up to allow them to think about what's going on," Berque says. "They wouldn't have to blindly copy things that maybe would distract them from what's going on."[17]

5.4 Common Hardware for Source Data-Entry: Scanner, Audio, Video, & Photographic Input

KEY QUESTIONS

What are the devices commonly used for source data-entry, and how do they work?

Preview & Review: Scanning devices include four types: bar-code readers, mark- and character-recognition devices, fax machines, and imaging systems. Mark-recognition and character-recognition devices include magnetic-ink character recognition (MICR), optical mark recognition (OMR), and optical character recognition (OCR). Fax machines may be dedicated machines or fax modems. Imaging systems convert text and images to digital form.

Audio-input devices may digitize audio sound by means of an audio board or a MIDI board.

Video-input devices may use frame-grabber or full-motion video cards.

Digital cameras use light-sensitive silicon chips to capture photographic images.

In many places, human meter readers fight through snowdrifts and brave vicious dogs to read gas and electric meters. They then type their readings into specialized portable computers—a time-consuming and potentially error-producing step. Over the next decade, however, it is estimated that probably 25% of the 160 million gas and electric meters in the United States will be automated—the keyboarding step will be eliminated. Instead, transmitters plugged into meters in customers' homes will send data directly to computers at the utility companies—a quicker and more accurate process.[18,19] (Because the ranks of human meter readers will shrink, the transmitters are being installed gradually, and many workers are being retrained or reassigned.)

The new way of reading meters is an example of *source data-entry.* Source data-input devices do not require keystrokes in order to input data to the computer. Rather data is entered directly from the *source;* people do not need to act as typing intermediaries.

In this section, we discuss common source data-entry devices. First, we cover *scanning devices—bar-code readers, mark- and character-recognition devices, fax machines,* and *imaging systems.* We then describe *audio-input devices.* Finally, we discuss *video* and *photographic input.*

Scanning Devices: Bar-Code Readers

Scanners **use laser beams and reflected light to translate images of text, drawings, photos, and the like into digital form.** The images can then be processed by a computer, displayed on a monitor, stored on a storage device, or communicated to another computer. Scanning devices include readers for *bar codes,* **the vertical zebra-striped marks you see on most manufactured retail products**—everything from candy to cosmetics to comic books. (■ *See Panel 5.9.)* In North America, supermarkets, food manufacturers, and others have agreed to use a bar-code system called the *Universal Product Code.* Other kinds of bar-code systems are used on everything from FedEx packages, to railroad cars, to the jerseys of long-distance relay runners.

Bar codes are read by *bar-code readers,* **photoelectric scanners that translate the bar-code symbols into digital code.** The price of a particular item is set within the store's computer and appears on the salesclerk's point-of-sale terminal and on your receipt. Records of sales are input to the store's com-

Universal Product Code bar code

■ PANEL 5.9

Bar-code readers
(Left) This NCR scanner is popular with specialty and general merchandise store retailers. *(Right)* This self-checkout system includes an ATM and an NCR scanner scale. Shoppers can scan, bag, and pay for purchases without cashier assistance.

puter and used for accounting, restocking store inventory, and weeding out products that don't sell well.

Many people still wonder if these devices are "scanners" or "scammers." That is, are the prices set by stores in their electronic checkout systems honest—the same ones that appear on store shelves or in newspaper sale ads? The technology is capable of 100% accuracy, but some stores average as low as 85%, according to a weights and measures coordinator at the National Institute of Standards and Technology.[20] In California, a study by one consumer group found that 4.1% of scanner transactions resulted in overcharges while 1.6% resulted in undercharges (for a net overcharge of 2.5%, or hundreds of millions of dollars).[21]

Inaccuracies probably result less from fraud than from human error, as when store managers forget to enter new data when programming scanner computers. Errors occur mainly with products that have frequent price changes, such as sale items. Retailers point out that generally scanners are more accurate than clerks, who punch in wrong prices about 10% of the time. However, when making transactions involving bar-code scanners, you should be aware of steps you can take to avoid overcharges, such as writing down prices of items as you shop, watching the prices being displayed as items are scanned, and checking your receipt.

Scanning Devices: Mark-Recognition & Character-Recognition Devices

There are three types of scanning devices that sense marks or characters. They are usually referred to by their abbreviations MICR, OMR, and OCR.

● **Magnetic-ink character recognition:** *Magnetic-ink character recognition (MICR)* **reads the strange-looking numbers printed at the bottom of checks.** (■ *See Panel 5.10.)* MICR characters, which are printed with magnetized ink, are read by MICR equipment, producing a digitized signal. This signal is used by a bank's reader/sorter machine to sort checks.

Bank
number

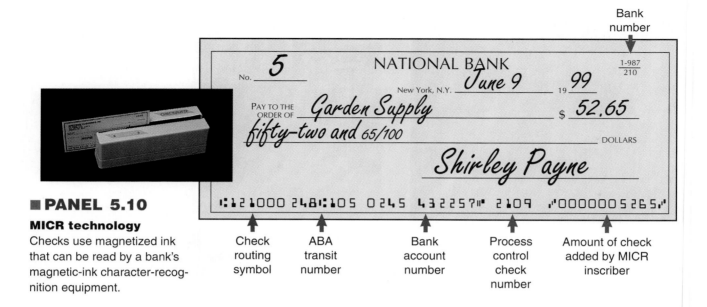

■ PANEL 5.10

MICR technology
Checks use magnetized ink
that can be read by a bank's
magnetic-ink character-recognition equipment.

Check | ABA | Bank | Process | Amount of check
routing | transit | account | control | added by MICR
symbol | number | number | check | inscriber
 | | | number |

- Optical mark recognition: **Optical mark recognition (OMR) uses a device that reads pencil marks and converts them into computer-usable form.** The most well-known example is the OMR technology used to read the College Board Scholastic Aptitude Test (SAT) and the Graduate Record Examination (GRE).

- Optical character recognition: **Optical character recognition (OCR) uses a device that reads preprinted characters in a particular font (typeface design) and converts them to digital code.** Examples of the use of OCR characters are utility bills and price tags on department-store merchandise. The wand reader is a common OCR scanning device. (■ *See Panel 5.11.*)

Scanning Devices: Fax Machines

A *fax machine—or facsimile transmission machine—*scans an image and sends it as electronic signals over telephone lines to a receiving fax machine,

■ PANEL 5.11

Optical character recognition
Special typefaces can be read by a scanning device called a
wand reader.

OCR-A
NUMERIC   0123456789
ALPHA   ABCDEFGHIJKLMNOPQRSTUVWXYZ
SYMBOLS   >$/-+-#"

OCR-B
NUMERIC   00123456789
ALPHA   ACENPSTVX
SYMBOLS   <+>-¥

which re-creates the image on paper (✔ p. 27). To review, there are two types of fax machines—*dedicated fax machines* and *fax modems.*

- Dedicated fax machines: **Dedicated fax machines are specialized devices that do nothing except send and receive fax documents.** These are what we usually mean when we say "fax machine." They are found not only in offices and homes but also alongside regular phones in public places such as airports.

  For the status-conscious or those trying to work from their cars during a long commute, fax machines can be installed in an automobile. The Robert Altman movie *The Player,* for example, contains a scene in which the stalker of a movie-studio executive, played by Tim Robbins, faxes a threatening note. It arrives through the fax machine housed beneath the dashboard in the executive's Range Rover.

- Fax modems: **A *fax modem* is installed as a circuit board inside the computer's system cabinet. It is a modem with fax capability that enables you to send signals directly from your computer to someone else's fax machine or computer fax modem.** With this device, you don't have to print out the material from your printer and then turn around and run it through the scanner on a fax machine. The fax modem allows you to send information much more quickly than if you had to feed it page by page into a machine.

  The fax modem is another feature of mobile computing, although it's more powerful as a receiving device. Fax modems are installed inside portable computers, including pocket PCs and PDAs. (■ *See Panel 5.12.*) You can also link up a cellular phone to a fax modem in your portable computer and thereby send and receive wireless fax messages.

  The main disadvantage of a fax modem is that you cannot scan in outside documents. Thus, if you have a photo or a drawing that you want to fax to someone, you need an image scanner, as we describe next.

Scanning Devices: Imaging Systems

Anthony J. Scalise, 80, of Utica, New York, found a 1922 picture of his father and other immigrants from the Italian city of Scandale. "It was wonderful, all those people with walrus mustaches," he said. He immediately had prints

■ PANEL 5.12

PDA

This personal digital assistant not only is a notepad and address book but also can send and receive fax messages.

made for friends and relatives. This is easy to do with the self-service Kodak imaging systems now found in many photo stores.[22]

An *imaging system*—or *image scanner* or *graphics scanner*—**converts text, drawings, and photographs into digital form that can be stored in a computer system and then manipulated, output, or sent via modem to another computer.** (■ *See Panel 5.13.*) The system scans each image—color or black and white—with light and breaks the image into light and dark dots or color dots, which are then converted to digital code. This is called *raster graphics*, which refers to the technique of representing a graphic image as a matrix of dots.

An example of an imaging system is the type used in desktop publishing. This device scans in artwork or photos that can then be positioned within a page of text, using desktop publishing software. Other systems are available for turning paper documents into electronic files so that people can reduce their paperwork as well as the space required to store it. Users of the World Wide Web scan photos into their computer systems to send to online friends or to post on Web pages.

Image scanners are generally of three types:

- Flatbed image scanners: *Flatbed image scanners* can scan all sorts of documents, even books. They are used for scanning high-quality graphics. Color flatbed scanners, for instance, scan at high resolution—up to 2400 dots per inch (dpi), compared to 400 dpi for sheetfed scanners. The higher the resolution, the crisper the image, though scanning time takes longer.

- Sheetfed image scanners: *Sheetfed image scanners* can scan only single sheets. Smaller and less expensive than flatbed scanners, they are often used by business and home-office users who need a convenient way to convert paperwork to electronic files.

- Handheld image scanners: *Handheld image scanners* are rolled by hand over documents to be scanned. Generally, their resolution is not very high. These scanners are mainly used to scan in small images or parts of images.

  Some small scanners are built into portable computers. Other small scanners are available just for snapshots. There are also image scanners that make three-dimensional images of small objects.

■ PANEL 5.13

Image scanner

Two workers at the U.S. National Research Center for the Identification of Missing Children scanning a missing child's photo into a computer.

Imaging-system technology has led to a whole new art or industry called *electronic imaging*. **Electronic imaging is the software-controlled combining of separate images, using scanners, digital cameras, and advanced graphic computers.** This technology has become an important part of multimedia.

Ethics

It has also led to some serious counterfeiting problems. (■ *See Panel 5.14.*) With scanners, crime rings have been able to fabricate logos and trademarks (such as those of Guess, Ray-Ban, Nike, and Adidas) that can be affixed to clothing or other products as illegal labels.[23] More importantly, electronic imaging has been used to make counterfeit money. As one article points out, "Big changes in technology over the last decade have made it easier to reproduce currency through the use of advanced copiers, printers, electronic digital scanners, color workstations, and computer software."[24] In late 1997, a computer science student in upstate New York was arrested by Secret Service agents after he allegedly used two computers and two scanners to print phony $20 bills.[25]

Such high-tech counterfeiting has impelled the U.S. Treasury to redesign the $100 bill—the one with Benjamin Franklin on it, and the most widely used U.S. paper currency throughout the world—and other bills for the first time in nearly 70 years. Besides using special inks and polyester fibers that glow when exposed to ultraviolet light, the new bills use finer printing techniques to thwart accurate copying.[26,27]

Audio-Input Devices

An *audio-input device* records analog sound and translates it for digital storage and processing. As we mentioned in Chapter 1, an analog sound signal represents a continuously variable wave within a certain frequency range (✔ p. 7). Such continuous fluctuations are usually represented with an analog device such as a cassette player. For the computer to process them, these variable waves must be converted to digital 0s and 1s, the language of the computer. The principal use of audio-input devices is to provide digital input for multimedia computers, which incorporate text, graphics, sound, video, and animation in a single digital presentation.

■ PANEL 5.14

Electronic imaging and counterfeiting

The U.S. $100 bill was overhauled for the first time in nearly 70 years to thwart increasingly sophisticated counterfeiters using imaging systems and other computer technology. On the new bill, Ben Franklin is bigger and slightly left of center.

There are two ways by which audio is digitized:

- **Audio board:** Analog sound from, for instance, a cassette player or a microphone goes through a special circuit board called an audio board (or card). **An *audio board* is an add-on circuit board in a computer that converts analog sound to digital sound and stores it for further processing and/or plays it back,** providing output directly to speakers or an external amplifier. Audio boards enable sounds to be heard from CD-ROMs and other storage media, as well as over the Internet.

  The three major sound standards are SoundBlaster, Ad Lib, and Windows. Some audio cards support all three standards. Many sound cards also have MIDI capability.

- **MIDI board:** **A *MIDI board*—MIDI, pronounced "middie," stands for *Musical Instrument Digital Interface*—provides a standard for the interchange of musical information between musical instruments, synthesizers, and computers.** MIDI consists of a set of computer instructions (not sample sounds) that is widely used for multimedia applications. For example, MIDI keyboards (also called controllers) and synthesizers can be plugged into a MIDI board to input music, which can then be stored, manipulated, and output.

Video-Input Cards

As with sound, most film and videotape is in analog form, with the signal a continuously variable wave. To be used by a computer, the signals that come from a VCR or a camcorder must be converted to digital form through a special digitizing card—a video-capture card—that is installed in the computer.

Two types of video cards are frame-grabber video and full-motion video:

- **Frame-grabber video card:** Some video cards, called *frame grabbers,* can capture and digitize only a single frame at a time.
- **Full-motion video card:** Other video cards, called *full-motion video cards* or *adapters,* can convert analog to digital signals at the rate of up to 30 frames per second, giving the effect of a continuously flowing motion picture. (■ *See Panel 5.15.*) The main limitation in capturing full video is not input but storage. It takes a huge amount of secondary storage space—perhaps 15 megabytes—to store just 1 second of video.

The rise of the World Wide Web has extended the possibilities for using video cameras. (There is, for example, a Web site that provides a stream of images of the site designer's pet iguana.) One practical innovation is the use of video cameras as "nannycams" to monitor children in daycare centers. For instance, when Amy Manning, 30, a human resource manager for a Connecticut engineering company, becomes concerned about her 3-year-old son's welfare, she can log onto the Internet and get a quick look at him at the Children's Corner daycare center. The center's "virtual parenting" system offers silent snapshots that are updated every few seconds.[28]

Digital Cameras

There are a tremendous number of options in cameras these days—almost too many of them—ranging from the $15 plastic-and-cardboard throwaways to the $30,000 sophisticated monsters that wire-service photographers use to shoot the Superbowl.[29]

■ PANEL 5.15

How analog video input is changed to digital form

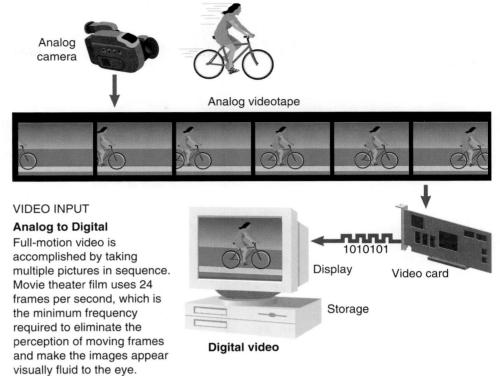

Analog camera

Analog videotape

VIDEO INPUT

Analog to Digital
Full-motion video is accomplished by taking multiple pictures in sequence. Movie theater film uses 24 frames per second, which is the minimum frequency required to eliminate the perception of moving frames and make the images appear visually fluid to the eye.

1010101

Display

Video card

Storage

Digital video

    TV video generates 30 interlaced frames per second, which is actually transmitted as 60 half frames ("fields" in TV lingo) per second.

    Video that has been digitized and stored in the computer can be displayed at varying frame rates, depending on the speed of the computer. The slower the computer, the jerkier the movement.

    Among the new products, digital cameras are particularly interesting because they foreshadow major change for the entire industry of photography. Instead of using traditional (chemical) film, a *digital camera* **uses a light-sensitive processor chip to capture photographic images in digital form on a small diskette inserted in the camera or on flash-memory chips (✔ p. 166).** (■ *See Panel 5.16, next page.*) The bits of digital information can then be copied right into a computer's hard disk for manipulation and printing out. (One camera, Sony's Mavica, puts images on a standard 3½-inch flexible diskette, which makes transferring images from camera to computer very easy.)

    Until recently, getting an instant photograph meant waiting two minutes for the print to come into focus. "Today's digital cameras allow you to take photos and immediately see them on your PC, place them inside a newsletter, e-mail them to your friends, and even publish them on your [Web] home page."[30] Except with the most expensive digital cameras, however, the photos produced with digital cameras are not as crisp and detailed as those taken on traditional chemical-based film. (If you're looking for the best image quality, stick to film and use a scanner to input photos to your computer.) Affordable filmless cameras can sometimes produce somewhat grainy images, and so their initial uses have been for jobs requiring the quick recording or relaying of information. Thus, they have been well received by insurance investigators, real-estate brokers, advertising agencies, and designers of World Wide Web pages. The more expensive the camera, the better the image quality. Still, as more and more imaging chips are produced by standard chipmaking methods, quality should improve and camera prices fall.

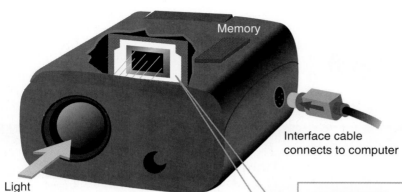

Memory

Light

3. The digital information is stored in the camera's electronic memory, either built-in or removable.

4. Using an interface cable, the digital photo can be downloaded onto a computer, where it can be manipulated, printed, placed on a Web page, or e-mailed.

Interface cable connects to computer

1. Light enters the camera through the lens.

2. The light is focused on the charge-coupled device (CCD), a solid-state chip made up of tiny, light-sensitive photosites. When light hits the CCD, it records the image electronically, just like film records images in a standard camera. The photosites convert light into electrons, which are then converted into digital information.

A look at CCDs
The smallest CCDs are 1/8 the size of a frame of 35mm film. The largest are the same size as a 35mm frame.

Smallest CCD

- Lower-end cameras start with 180,000 photosites.
- Professional cameras can have up to 6 million photosites.

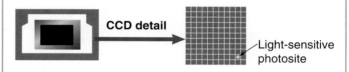

CCD detail

Light-sensitive photosite

■ PANEL 5.16

Digital cameras and how they work

Eastman Kodak is spending $500 million a year to develop digital products and services. (The alternative, in the words of one observer, "is to ride a business that will slowly fade, like train travel in the jet age."[31]) Thus, it is aggressively marketing digital cameras, photo scanners, and photo-store kiosks for electronic retouching. It also launched the Kodak Picture Network, an attempt to induce amateur photographers to store their pictures on Kodak's Web site, allowing you to view scanned images of your photos, copy them to your hard disk, or send them by e-mail to your online friends.[32]

5.5 More Hardware for Source Data-Entry: Sensor, Radio-Frequency Identification, Voice, & Human-Biology Input Devices

KEY QUESTION

What are some of the more advanced devices for source data-entry?

Preview & Review: Other source data-entry devices are sensors, radio-frequency identification, voice-recognition, and human-biology input devices. Sensors collect specific kinds of data directly from the environment. Radio-frequency identification devices enable tracking. Voice-recognition systems convert human speech into digital code. Human-biology input devices include biometric systems and line-of-sight systems.

Who are those impatient, aggressive drivers who barrel through intersections after the traffic lights have changed to red (sometimes by as much as 50 seconds after the amber)? Mostly they're 35-and-under males with previous moving violations and bad driving records. And they cause a lot of damage and pain. In the U.S., fatal crashes at intersections with traffic lights increased 18% from 1991 to 1995.

In Australia, however, injuries associated with red-light crashes have been reduced 32% since 1981. The difference: Australia uses automatic cameras permanently mounted on 15-foot poles near intersections to photograph the cars and license plates and the date and time of violations by red-light runners, who are then mailed notices of stiff fines. Now these cameras, which are wired to the traffic signals and to sensors buried in the pavement, are coming to many cities in the United States.[33]

As this example suggests, there are some even more intriguing source data-entry devices beyond those we have described. In this section, we describe *sensors, radio-frequency identification devices, voice-recognition devices,* and *human-biology input devices.*

Sensors

■ PANEL 5.17

Earthquake sensor
Sensor instruments of a telemetered weak motion seismic station (earthquake motion detection)

A *sensor* is a type of input device that collects specific kinds of data directly from the environment and transmits it to a computer. Although you are unlikely to see such input devices connected to a PC in an office, they exist all around us, often in nearly invisible form. Sensors can be used for detecting all kinds of things: speed, movement, weight, pressure, temperature, humidity, wind, current, fog, gas, smoke, light, shapes, images, and so on.

Besides being used to detect the speed and volume of traffic and adjust traffic lights, sensors are used on mountain highways in wintertime in the Sierra Nevada as weather-sensing devices to tell workers when to roll out snowplows.[34] In California, sensors have been planted along major earthquake fault lines in an experiment to see whether scientists can predict major earth movements. (**■** *See Panel 5.17.*) In aviation, sensors are used to detect ice buildup on airplane wings or to alert pilots to sudden changes in wind direction. Dairy farmers are beginning to use sensors to record how often and how much animals eat.[35] Sophisticated sensors are being used as "electronic sniffers" to test beverage quality at Coors beer and Starbucks coffee.[36] Sensors are used by government regulators to monitor whether companies are complying with air-pollution standards.

Radio-Frequency Identification Devices

Maybe you already pay for gas by swiping your gasoline credit card through the card reader attached to the pump. Not fast enough for you? Then you're a candidate for Mobil Corporation's Speedpass, a key ring device that takes pay-at-the-pump technology a step further.[37,38]

Introduced on a test basis in nine American markets in 1997, the Speedpass is a cylinder about an inch long and a quarter inch in diameter that can be attached to a ring with your car keys. The cylinder contains a microchip carrying your identification number. When you wave the Speedpass past one of Mobil's special pumps, a radio signal from the pump triggers a response from the pass that turns on the pump and begins charging your purchase to your credit card.

***Radio-frequency identification technology,* or RF-ID tagging, consists of (1) a "tag" containing a microchip that contains code numbers that (2) can be read by the radio waves of a scanner linked to a database.** In the case of the E-Z Pass used on the world's busiest bridge, the George Washington Bridge linking New Jersey and New York City, the radio-readable "tag" is the size of a cassette tape. Drivers can breeze through the tollbooths without having to even roll down their windows, and the toll is charged to their accounts with the toll authority.[39]

In other applications, radio-readable ID "tags" are being used by the Postal Service to monitor the flow of mail through its system. They are being used in inventory control and warehousing. They are being used in the railroad industry to keep track of rail cars. They are even being injected into dogs and cats, so that veterinarians with the right scanning equipment can identify them if the pets become separated from their owners. If the cost of tags can be driven down to a fraction of a cent, they may even come to be used in grocery products. Unlike bar-code readers, RF-ID tags need not come within line of sight of the scanner; thus, in the future, an entire cart of groceries could be scanned in a single pass.[40]

Voice-Recognition Input

When you speak to a computer, can it tell whether you want it to "recognize speech" or "wreck a nice beach"?

Voice-recognition systems, whereby you dictate input via a microphone, have faced considerable hurdles: different voices, pronunciations, and accents. In Atlanta, for instance, natives say "Tick a rat" (Make a right turn). In New York City, they say "Gnome sane?" (Do you know what I'm saying?).[41] Such regionalisms pose real challenges to voice-recognition experts. With past voice-recognition systems, users had to use voice-recognition software to "train" a particular computer to recognize their voices, and they had to pause between words rather than speak normally. Recently, however, the systems have measurably improved.

A *voice-recognition system*, using a microphone (or a telephone) as an input device, converts a person's speech into digital code by comparing the electrical patterns produced by the speaker's voice with a set of prerecorded patterns stored in the computer.

Voice-recognition systems are finding many uses. Warehouse workers are able to speed inventory-taking by recording inventory counts verbally. Traders on stock exchanges can communicate their trades by speaking to computers. Radiologists can dictate their interpretations of X-rays directly into transcription machines. Nurses can fill out patient charts by talking to a computer. Speakers of Chinese can speak to machines that will print out Chinese characters. (■ *See Panel 5.18.*) Indeed, for many disabled individuals, a computer isn't so much a luxury or a simple productivity tool as it is a necessity. It provides freedom of expression, independence, and empowerment.

In addition, many large companies are offering voice input to customers via the telephone. For instance, using IBM voice-recognition technology, Alamo car rentals in Fort Lauderdale, Florida, allows customers to reserve cars and cancel reservations over the telephone. United Parcel Service is piloting a voice-recognition application that will let users track their packages. American Express uses a server-based voice-recognition system named Paris to allow users to book airplane reservations to almost any destination in the United States. Company representatives say that the new system will cut the average transaction time to 2 minutes from 7, helping the travel department boost the number of reservations by 5% without adding more agents.

Voice-recognition technology clearly has several advantages over traditional keyboard and mouse input: It protects users against repetitive stress injuries; it simplifies computing for beginning users; it frees users' hands and eyes for other tasks; it improves data-entry speed and can improve accuracy; it eliminates spelling errors.[42]

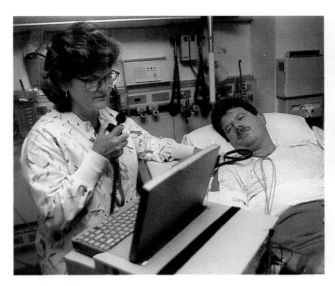

■ PANEL 5.18

Voice-recognition technology

(Left) A registered nurse tests a voice-recognition system designed to help fill out patient charts. *(Right)* Taiwanese scientist Lee Lin-shan displays a computer that can listen to continuous speech in Chinese (Mandarin) and then print out the words at the rate of three characters a second.

Besides a microphone, microcomputer voice-recognition software requires a sound board, a fair amount of hard-disk space, a fast processor, 24–48 megabytes of RAM, and 512 kilobytes of cache RAM (✔ p. 165). Two popular programs are Dragon Dictate and Dragon NaturallySpeaking from Dragon Systems. Users talk in a natural manner and pace. The spoken words immediately appear on the screen. The powerful text editing and formatting capabilities allow users to use boldface, italics, and different fonts (✔ p. 102), for example. The program comes with its own user training program and an active vocabulary of 30,000 words and a back-up dictionary of 230,000 or more words. (These programs run on Windows 95 and Windows NT; installation does require some time and effort.) Users can also purchase special vocabulary libraries for the medical and financial industries.

ViaVoice from IBM hooks directly into Microsoft Word, as does Kurzweil VoiceCommands from Lernout and Hauspie. Unlike ViaVoice, VoiceCommands does not support dictation; instead it offers extensive editing and formatting functions in Word 97. This program is handy for users who have to format rather complex documents and have trouble remembering all the commands and menu options needed. For example, one can say "Insert a table with two columns and six rows"; "Select column one"; "Make width of column 2 inches." You can also cut, copy, paste, change colors and fonts, insert symbols, spell-check, save, and more. When the program doesn't recognize certain words, you can train it to recognize any valid command.

Current accuracy rates are at about 90–95%. Although some experts insist these rates aren't high enough to replace the keyboard, that time is approaching quickly.

Human-Biology Input Devices

Characteristics and movements of the human body, when interpreted by sensors, optical scanners, voice recognition, and other technologies, can become forms of input. Some examples:

- **Biometric systems:** **Biometrics is the science of measuring individual body characteristics.** Biometric security devices identify a person through a fingerprint, voice intonation, or other biological characteristic. For example, there is a postage-stamp-sized fingerprint reader from Veridicom that is small and cheap enough that it could be built into computer keyboards, so fingerprints could replace passwords.[43] So far, however, fingerprint systems don't always work. Immigrant Pushp Grover of Everett, Washington, tried 11 times to give her prints for her citizenship application and was rejected every time; her fingers were too smooth for the tiny ridges to be read.[44]

Retinal-identification devices use a ray of light to identify the distinctive network of blood vessels at the back of one's eyeball.[45,46] Thomas J. Drury, CEO of Sensar Corporation, demonstrates his company's iris scanner by walking up to an automated teller machine set up in his office, swiping his bank card through the card reader, and staring straight ahead. The machine verifies his identity by reading his eyes.[47] The Sensar system is being tested on ATMs around the world.

- **Line-of-sight systems:** Line-of-sight systems enable a person to use his or her eyes to "point" at the screen, a technology that allows some physically disabled users to direct a computer. For example, the Eyegaze System from LC Technologies allows you to operate a computer by focusing on particular areas of a display screen. A camera mounted on the computer analyzes the point of focus of the eye to determine where you are looking. You operate the computer by looking at icons on the screen and "press a key" by looking at one spot for a specified period of time.

5.6 Output Hardware

KEY QUESTION

What is the difference between softcopy and hardcopy output?

Preview & Review: Output devices translate information processed by the computer into a form that humans can understand. The two principal kinds of output are softcopy, such as material shown on a display screen, and hardcopy, which is printed.

Output devices include display screens; printers, plotters, and multifunction devices; devices to output sound, voice, and video; virtual reality; and robots.

How important is music, really?

Apparently it provides food not only for the soul but also for the brain. Children who begin listening to and studying classical music early, like Justin Brandt of Atlanta, who started at age 3, may be getting an intellectual head start on their peers. "Not only does [music] pluck at emotional heart strings," says one writer, "but scientists say it also turns on brain circuits that aid recognition of patterns and structures critical to development of mathematics skills, logic, perception, and memory."[48]

Thus, at a time when schools are cutting budgets for music, classical pianist Jeffrey Biegel's innovations offer some important advantages. "I feel like a pioneer," Biegel says. "I guess I can be called a cyberpianist." Steinway Hall in New York City holds only 75 people, but one night in 1997 Biegel gave a performance that was able to reach thousands of people around the world on their home computers. His "cyber-recital" was made possible thanks to a camcorder and digital technology, which allowed the live musical recital to be transmitted over the World Wide Web with both audio and video.[49]

If the foregoing example is any guide, the closing days of the 20th century represent exciting times in the development of computer output. In the following pages we describe present and future types of output devices.

The principal kinds of output are *softcopy* and *hardcopy*. (■ *See Panel 5.19.*)

■ PANEL 5.19

Summary of output devices

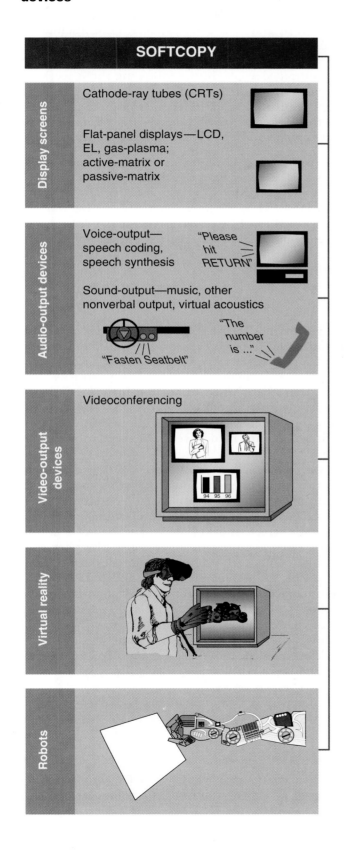

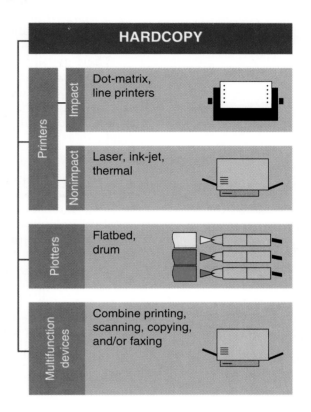

- Softcopy: **Softcopy refers to data that is shown on a display screen or is in audio or voice form.** This kind of output is not tangible; it cannot be touched.
- Hardcopy: **Hardcopy refers to printed output.** The principal examples are printouts, whether text or graphics, from printers. Film, including microfilm and microfiche, is also considered hardcopy output.

There are several types of output devices. In the following three sections, we discuss, first, *softcopy hardware—display screens;* second, *hardcopy hardware—printers, plotters, multifunction devices,* and *microfilm* and *microfiche devices;* and, third, *other output hardware—sound-output devices, voice-output devices, video-output devices, virtual-reality devices,* and *robots.*

5.7 Softcopy Output: Display Screens

KEY QUESTION

What are the different types of display screens and video display adapters?

Preview & Review: Display screens are either CRT (cathode-ray tube) or flat-panel display. CRTs use a vacuum tube like that in a TV set. Flat-panel displays are thinner, weigh less, and consume less power than CRTs, but only recently have some LCD screens become as clear as CRT screens. The principal flat-panel displays are liquid crystal display (LCD), electroluminescent (EL) display, and gas-plasma display.

Various video display adapters allow various kinds of resolution and colors.

All things great and small—we are now at the point where we can have both in computer displays. Fujitsu has developed a hang-on-the-wall monitor that measures 42 inches corner to corner but is only 4 inches thick.[50] Kopin Corporation is making a screen that is a mere ¼ inch diagonally, so tiny (less than the diameter of a dime) that a lens must be included; the device may be used to display Web pages in cellular phones.[51]

Display screens are among the principal windows of information technology. As more and more refinements are made, we can expect to see them adapted to many innovative uses.

Display screens—also variously called *monitors, CRTs,* or simply *screens*—are output devices that show programming instructions and data as they are being input and information after it is processed. Sometimes a display screen is also referred to as a **VDT, for *video display terminal,* although technically a VDT includes both screen and keyboard.** The size of a screen is measured diagonally from corner to corner in inches, just like television screens. For terminals on large computer systems and for desktop microcomputers, 14- to 17-inch screens are a standard size, although 19- and 21-inch screens are not uncommon. Notebook and subnotebook computers have screens ranging from 6.1 inches to 14.4 inches, with 11.3 and 12.1 inches the two most popular in-between sizes. Pocket-size computers may have even smaller screens. To give themselves a larger screen size, some portable-computer users buy a larger desktop monitor to which the portable can be connected. Near the display screen are control knobs that allow you to adjust brightness and contrast, as on a television set.

Display screens are of two types: *cathode-ray tubes* and *flat-panel displays.*

Cathode-Ray Tubes (CRTs)

The most common form of display screen is the CRT. **A *CRT,* for *cathode-ray tube,* is a vacuum tube used as a display screen in a computer or video**

display terminal. This same kind of technology is found not only in the screens of desktop computers but also in television sets and flight-information monitors in airports.

Images are represented on the screen by individual dots called *pixels.* **A *pixel*, for "picture element," is the smallest unit on the screen that can be turned on and off or made different shades.** (■ *See Panel 5.20.*) A stream of bits defining the image is sent from the computer (from the CPU) to the CRT's electron gun, where the bits are converted to electrons. The inside of the front of the CRT screen is coated with phosphor. When a beam of electrons from the electron gun (deflected through a yoke) hits the phosphor, it lights up selected pixels to generate an image on the screen.

Flat-Panel Displays

If CRTs were the only existing technology for computer screens, we would still be carrying around 25-pound "luggables" instead of lightweight notebooks, subnotebooks, and pocket PCs. CRTs provide bright, clear images, but they add weight and consume space and power.

Compared to CRTs, flat-panel displays are much thinner, weigh less, and consume less power. Thus, they are better for portable computers, although they are becoming available for desktop computers as well. ***Flat-panel displays* are made up of two plates of glass with a substance in between them, which is activated in different ways.**

Flat-panel displays are distinguished in two ways: (1) by the substance between the plates of glass and (2) by the arrangement of the transistors in the screens.

■ PANEL 5.20

How a CRT works

(Left) Each character on the screen is made up of small dots called *pixels,* short for *picture elements. (Right)* A stream of bits from the computer (from the CPU) is sent to the electron gun, which converts the bits into electrons. The gun then shoots a beam of electrons through the yoke, which deflects the beam in different directions. When the beam hits the phosphor coating on the inside of the CRT screen, a number of pixels light up, making the image on the screen.

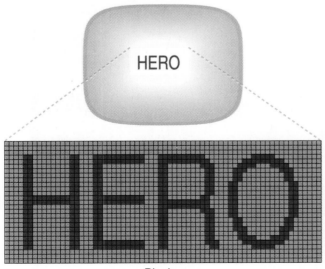

Pixels

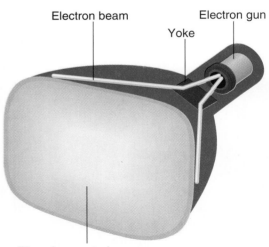

Electron beam   Electron gun

Yoke

Phosphor-coated screen

- Substances between plates—LCD, EL, and gas plasma: The types of technology used in flat-panel display are *liquid crystal display, electroluminescent display,* and *gas-plasma display.* (■ *See Panel 5.21.*)

  Liquid crystal display (LCD) **consists of a substance called** *liquid crystal,* **the molecules of which line up in a way that alters their optical properties. As a result, light—usually backlighting behind the screen—is blocked or allowed through to create an image.**

  Electroluminescent (EL) display **contains a substance that glows when it is charged by an electric current.** A pixel is formed on the screen when current is sent to the intersection of the appropriate row and column. The combined voltages from the row and column cause the screen to glow at that point.

  Gas-plasma display **is like a neon bulb, in which the display uses a gas that emits light in the presence of an electric current.** That is, the technology uses predominantly neon gas and electrodes above and below the gas. When electric current passes between the electrodes, the gas glows. Although gas-plasma technology has high resolution, it is expensive. The 42-inch Fujitsu hang-on-the-wall display mentioned above uses gas-plasma display technology, but the price is $17,500.

- Arrangements of transistors—active-matrix or passive-matrix: Flat-panel screens (as discussed in the Experience Box at the end of Chapter 4) are either active-matrix or passive-matrix displays, according to where their transistors are located.

  In an *active-matrix display,* **each pixel on the screen is controlled by its own transistor.** Active-matrix screens are much brighter and sharper than passive-matrix screens, but they are more complicated and thus more expensive.

  In a *passive-matrix display,* **a transistor controls a whole row or column of pixels.** Passive matrix provides a sharp image

■ PANEL 5.21

Flat-panel displays

(Left) Active-matrix LCD, Planar System's CleanScreen compact computer for hospital information systems. *(Top right)* Planar EL screen. *(Bottom right)* Fujitsu 42-inch QFTV gas-plasma display.

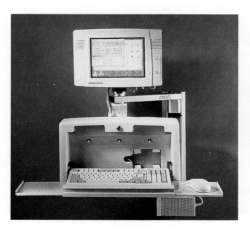

for one-color (monochrome) screens but is more subdued for color. The advantage is that passive-matrix displays are less expensive and use less power than active-matrix displays.

Until recently, flat-panel screens were not available for desktop computers, and you still had to use the bulky CRT monitor. Now, however, manufacturers such as NEC, Panasonic, and Sharp make LCDs for desktop PCs that are as good as CRTs.[52]

Screen Clarity

Whether CRT or flat-panel display, screen clarity depends on three qualities: *resolution, dot pitch,* and *refresh rate.*

- Resolution: **The image sharpness of a display screen is called its *resolution;* the more pixels there are per square inch, the finer the level of detail attained.**

  Resolution is expressed in terms of the formula *horizontal pixels ✕ vertical pixels.* Each pixel can be assigned a color or a particular shade of gray. Thus, a screen with 640 ✕ 480 pixels multiplied together equals 307,200 pixels. This screen will be less clear and sharp than a screen with 800 ✕ 600 (equals 480,000) or 1024 ✕ 768 (equals 786,432) pixels.

  Some displays can only go as high as 1024 ✕ 768. Others can go to 1280 ✕ 1024 or even to 1600 ✕ 1200.

- Dot pitch: ***Dot pitch* is the amount of space between the centers of adjacent pixels; the closer the dots, the crisper the image.** A .28 dot pitch, for instance, means dots are 28/100ths of a millimeter apart. Generally, a dot pitch of less than .31 will provide clear images. Multimedia and desktop-publishing users typically use .25-millimeter dot-pitch monitors.

- Refresh rate: ***Refresh rate* is the number of times per second that the pixels are recharged so that their glow remains bright.** Refresh is necessary because the phosphors hold their glow for just a fraction of a second. The higher the refresh rate, the more solid the image looks on the screen—that is, doesn't flicker.

  In dual-scan screens, the tops and bottoms of the screens are refreshed independently at twice the rate of single-scan screens, producing more clarity and richer colors. In general, displays are refreshed 45–100 times per second.

  Some computers use *interlaced* screens, whereby the electron beam refreshes all odd-numbered scan lines in one vertical sweep of the screen and all even-numbered scan lines in the next sweep. Because of the screen phosphor's ability to maintain an image for a short time before fading and the tendency of the human eye to average or blend subtle differences in light intensity, the human viewer sees a complete display, but the amount of information carried by the display signal and the number of lines that must be displayed per sweep are halved.[53]

Monochrome Versus Color Screens

Display screens can be either monochrome or color. Monochrome display screens display only one color on a background. Although they are dying out, you may encounter them in text-based mainframe systems.

Color display screens, also called *RGB monitors* (for red, green, blue), can display between 16 colors and 16.7 million colors, depending on the type. The number of colors is referred to as the *color depth*, or *bit depth*. Most software today is developed for color, and—except for some pocket PCs— most microcomputers today are sold with color display screens.

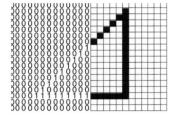

Bitmapped Displays: Both Text & Graphics Capability

At one time, computers could display only text. Today's computers can display both text and graphics—that is, not only letters and numbers but also charts, graphs, drawings, and icons. This is because PCs now use bitmapping as a standard.

The computer uses bits (0s and 1s) to describe each pixel's color and position. On monochrome displays, one bit represents one pixel on the screen. On color displays, several bits represent one pixel. **Bitmapped display screens permit the computer to manipulate pixels on the screen individually,** enabling software to create a greater variety of images.

Video Display Adapters

To display graphics, a display screen must have a video display adapter. **A *video display adapter*, also called a *video graphics card*, is a circuit board that determines the resolution, number of colors, and speed with which images appear on the display screen.** Video display adapters come with their own memory chips (RAM, or *VRAM*, for *video RAM*), which determine how fast the card processes images, the resolution of the images displayed, and how many colors it can display. At a resolution of 640 × 480, a video display adapter with 256 kilobytes of memory will provide 16 colors; one with 1 megabyte will support 16.7 million colors.

In notebook computers, the video display adapter is built into the motherboard (✔ p. 161); in desktop computers, it is an expansion card that plugs into an expansion slot (✔ p. 166). Video display adapters embody certain standards. Today's microcomputer monitors commonly use VGA and SVGA standards.

- VGA: ***VGA*, for *Video Graphics Array*, will support 16–256 colors, depending on resolution.** At 320 × 200 pixels it will support 256 colors; at the sharper resolution of 640 × 480 pixels it will support 16 colors. VGA is called *4-bit color.*

- SVGA: ***SVGA*, for *Super Video Graphics Array*, will support 256 colors at higher resolution than VGA.** SVGA has two graphics modes: 800 × 600 pixels and 1024 × 768. SVGA is called *8-bit color.*

- XGA: Also referred to as *high-resolution display*, **XGA, for *Extended Graphics Array*, supports up to 16.7 million colors at a resolution of 1024 × 768 pixels.** Depending on the video display adapter memory chip, XGA will support 256, 65,536, or 16,777,216 colors. At its highest quality, XGA is called *24-bit color* or *true color.*

  For any of these displays to work, video display adapters and monitors must be compatible. That is, the monitor must be capable of displaying the screen resolution, number of colors, and refresh rate fed to it by the video display adapter. Your computer's software and the video display adapter must also be compatible. Thus, if you are changing your monitor or your video display adapter, be sure the new one will still work with the old equipment.

Ethics

From the standpoints of protecting the environment and the health of one-self or other users, anyone buying a monitor should be aware of two factors: energy consumption and electromagnetic emissions (described later in this chapter). If a monitor manufacturer says that the display complies with Energy Star standards, it means it meets a voluntary federal standard for reduced power consumption in computer equipment. If the monitor also meets so-called MPR-2 standards, it means it meets a Swedish government rule, also adopted by many American manufacturers, for reduced electro-magnetic emissions, which may be advantageous to health.

5.8 Hardcopy Output: Printers, Plotters, Multifunction Devices, & Microfilm & Microfiche Devices

KEY QUESTION

What are the different types of hardcopy output devices?

Preview & Review: Printers, plotters, and multifunction devices produce printed text or images on paper.

Printers may be desktop or portable, impact or nonimpact. The most common impact printers are dot-matrix printers. Nonimpact printers include laser, ink-jet, and thermal printers.

Plotters are flatbed or drum.

Multifunction devices combine capabilities such as printing, scanning, copying, and faxing.

Computer output microfilm and microfiche can solve storage problems.

It's known as "the hotel facsimile trick," a strategy of desperation for the computer-mobile.

Many travelers carry a portable computer but no printer, to avoid toting along a second device that may weigh as much as their PC. What do they do, then, when on reaching their hotel they find they need to print out a document, but there's no nearby copy shop where they can rent time on a printer? In the hotel facsimile trick, they connect the built-in fax modem in their PC to the phone jack in their hotel room. Then they go down to the hotel's front desk and pick up the paper copy of the document they have just sent to themselves at the hotel's fax telephone number.

Unfortunately, hotel fax machines don't always use high-quality paper of the sort you would want, say, for preparing a presentation to a client. More-over, the hotel may charge $2–$5 a page for the fax. Fortunately, however, some new innovations in printers are helping travelers.

To get to definitions, **a *printer* is an output device that prints characters, symbols, and perhaps graphics on paper or another hardcopy medium.** Print-ers are categorized according to whether or not the image produced is formed by physical contact of the print mechanism with the paper. *Impact printers* do have contact; *nonimpact printers* do not.

Desktop Versus Portable Printers

Technologies used for printing range from those that resemble typewriters to those that resemble photocopying machines. A question you might want to ask yourself in this era of mobility is: Will a desktop printer be sufficient, or will you also need a printer that is portable?

- **Desktop printers:** Many people, probably including most students, find portable computers useful but have no need for a portable printer. You can do your writing or computing wherever you can tote your laptop, then print out documents back at your regular desk. The advantage of desktop printers is the wide range in quality and price available.

type styles and sizes. The more expensive models can print in different colors.

Laser printers have built-in RAM chips to store documents output from the computer, which allows you to continue working without any slowdown. If you are printing out complicated documents with color and graphics, as in desktop publishing, you will need a printer with a lot of RAM. Laser printers also have their own ROM chips to store fonts and their own small dedicated processor chip.

To be able to manage graphics and complex page design, a laser printer works with a page description language. **A *page description language* is software that describes the shape and position of characters and graphics to the printer.** PostScript (from Adobe Systems) is one common type of page description language; Hewlett-Packard Graphic Language (HPGL) is another.

- **Ink-jet printer:** Like laser and dot-matrix printers, ink-jet printers also form images with little dots. ***Ink-jet printers* spray small, electrically charged droplets of ink from four nozzles through holes in a matrix at high speed onto paper.** (■ *See Panel 5.24.*)

The advantages of ink-jet printers are that they can print in color, are quieter, and are much less expensive than color laser printers. The disadvantages are that they print in a somewhat lower resolution than laser printers (300–720 dpi) and they are slower (about 1–4 text-only pages per minute). A further disadvantage is that printing of high-resolution images requires the use of special coated paper, which

■ PANEL 5.24

Ink-jet printer operations

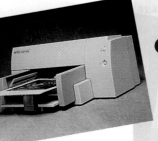

1 Four removable ink cartridges are attached to print heads with 64 firing chambers and nozzles apiece.

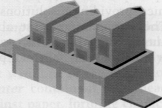

2 As the print heads move back and forth across the page, software instructs them where to apply dots of ink, what colors to use, and in what quantity.

3 To follow those instructions, the printer sends electrical pulses to thin resistors at the base of the firing chambers behind each nozzle.

Resistor
Vapor bubble
Ink

5 A matrix of dots forms characters and pictures. Colors are created by layering multiple color dots in varying densities.

4 The resistor heats a thin layer of ink, which in turn forms a vapor bubble. That expansion forces ink through the nozzle and onto the paper at a rate of about 6,000 dots per second.

costs more than regular paper. And printing a document with high-resolution color graphics can take as long as 10 minutes or more for a single page.

A variation on ink-jet technology is the *bubble-jet printer*, which uses miniature heating elements to force specially formulated inks through print heads with 128 tiny nozzles. The multiple nozzles print fine images at high speeds. This technology is commonly used in portable printers.

- Thermal printer: For people who want the highest-quality color printing available with a desktop printer, thermal printers are the answer. **Thermal printers use colored waxes and heat to produce images by burning dots onto special paper.** The colored wax sheets are not required for black-and-white output.

    However, thermal printers are expensive, and they require expensive paper. Thus, they are not generally used for jobs requiring a high volume of output.

The table compares printer technologies. (■ *Panel 5.25.)*

Black-&-White Versus Color Printers

Today prices have plummeted for laser and ink-jet printers, so that the cheap but noisy dot-matrix printer may well be going the way of the black-and-white TV set. Your choice in printers may come down to how much you print and whether you need color.

■ PANEL 5.25

Printer comparisons

Type	Technology	Advantages	Disadvantages	Typical Speed	Approximate Cost
Dot-matrix	Print head with small pins strikes an inked ribbon against paper	Inexpensive; produces draft quality and near-letter-quality; can output some graphics; can print multi-part forms; low cost per page	Noisy; cannot produce high-quality output of text and graphics; limited fonts	30 to 500+ cps*	$100–$2000
Laser	Laser beam directed onto a drum, "etching" spots that attract toner, which is then transferred to paper	Quiet; excellent quality; output of text and graphics; very high speed	High cost, especially for color	8–200 ppm*	$500–$20,000
Ink-jet	Electrostatically charged drops hit paper	Quiet; prints color, text, and graphics; less expensive; fast	Relatively slow; clogged jets; fewer dots per inch	35–400+ cps	$150–$2000
Thermal	Temperature-sensitive paper changes color when treated; characters are formed by selectively heating print head	Quiet; high-quality color output of text and graphics; can also produce transparencies	Special paper required; expensive; slow	11–80 cps	$2000–$22,000

*cps = characters per second; ppm = pages per minute.

Lasers, which print a page at a time, can handle thousands of black-and-white pages a month. Moreover, compared to ink-jets, laser printers are faster and crisper (though not by much) at printing black-and-white copies and a cent or two cheaper per page. Finally, a freshly printed page from a laser won't smear, as one from an ink-jet might. Low-end black-and-white laser printers start at under $400; color laser printers start at about $3000.

Ink-jets, which spray ink onto the page a line at a time, can give you both high-quality black-and-white text and high-quality color graphics. The rock-bottom price for ink-jets is only about $150, although if you print a lot of color, you'll find color ink-jets much slower and more expensive to operate than color laser printers. Still, the initial cost of a color ink-jet printer is considerably less than a color laser printer.

Some questions to consider when choosing a printer for a microcomputer are given in the accompanying box. (■ *See Panel 5.26.*)

Plotters

A *plotter* is a specialized output device designed to produce high-quality graphics in a variety of colors. (■ *See Panel 5.27.*) Plotters are especially useful for creating maps and architectural drawings, although they may also produce less complicated charts and graphs.

The two principal kinds of plotters are *flatbed* and *drum.*

- Flatbed plotter: **A *flatbed plotter* is designed so that paper lies flat on a table-like surface.** The size of the bed determines the maximum

■ PANEL 5.26

Questions to consider when choosing a printer for a microcomputer

Do I want a desktop or portable printer—or both? You'll probably find a desktop printer satisfactory (and less expensive than a portable). If you're on the road enough to warrant using a portable, see whether a *transportable* or an *ultraportable* would best suit you. (See text p. 224.)

Do I need color, or will black-only do? Are you mainly printing text or will you need to produce color charts and illustrations (and, if so, how often)? If you print lots of black text, consider getting a laser printer. If you might occasionally print color, get an ink-jet that will accept cartridges for both black and color.

Do I have other special output requirements? Do you need to print envelopes or labels? special fonts (type styles)? multiple copies? transparencies or on heavy stock? Find out if the printer comes with envelope feeders, sheet feeders holding at least 100 sheets, or whatever will meet your requirements.

Is the printer easy to set up? Can you easily put the unit together, plug in the hardware, and adjust the software (the "driver" programs) to make the printer work with your computer?

Is the printer easy to operate? Can you add paper, replace ink/toner cartridges or ribbons, and otherwise operate the printer without much difficulty?

Does the printer provide the speed and quality I want? Will the machine print at least three pages a minute of black text and two pages a minute of color? Are the blacks dark enough and the colors vivid enough?

Will I get a reasonable cost per page? Special paper, ink or toner cartridges (especially color), and ribbons are all ongoing costs. Ink-jet color cartridges, for example, may last 100–500 pages and cost $25–$30 new. Laser toner cartridges are cheaper. Ribbons for dot-matrix printers are cheaper still. Ask the seller what the cost per page works out to.

Does the manufacturer offer a good warranty and good telephone technical support? Find out if the warranty lasts at least 2 years. See if the printer's manufacturer offers telephone support in case you have technical problems. The best support systems offer toll-free numbers and operate evenings and weekends as well as weekdays.

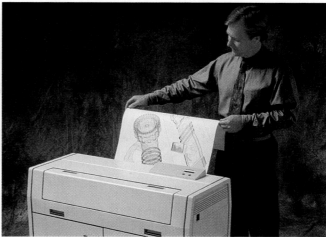

■ PANEL 5.27

Plotters

size of the sheet of paper. Under computer control, between one and four color pens move across the paper, and the paper moves beneath the pens.

- Drum plotter: **A *drum plotter* works like a flatbed plotter except that the paper is output over a drum,** enabling continuous output that is useful, for example, to track earthquake activity.

Multifunction Printer Technology: Printers That Do More Than Print

Everything is becoming something else, and even printers are becoming devices that do more than print. For instance, plain-paper fax machines are available that can also function as answering machines and laser or ink-jet printers. Since 1990, Xerox has sold an expensive printer-copier-scanner that can be hooked into corporate computer networks.

Some recent hardware can do even more. ***Multifunction devices* combine several capabilities, such as printing, scanning, copying, and faxing, all in one device.** (■ *See Panel 5.28.*) Both Xerox and Hewlett-Packard make machines that combine four pieces of office equipment in one—photocopier, fax machine, scanner, and laser printer. By doing the work of four separate office machines at a price below the combined cost of buying these devices separately, the multifunction machine offers budgetary and space advantages.

■ PANEL 5.28

The multifunction device
This machine combines four functions in one—printer, copier, fax machine, and scanner.

These devices can also do top-quality color printing. In addition, Hewlett-Packard makes what it calls a "mopier" (for "multiple original prints"), a machine that can be an office copier as well as rapidly print out copies of a large document transmitted from one office to another over a corporate network.

Computer Output Microfilm & Microfiche

If you take your time getting rid of old newspapers, you know it doesn't take long for them to pile up—and to take up space. No wonder, then, that libraries try to put newspaper back issues on microfilm or microfiche. One ounce of microfilm can store the equivalent of 10 pounds of paper.

Computer output microfilm is computer output produced as tiny images on rolls of microfilm; microfiche uses sheets instead of rolls. The images are up to 48 times smaller than those produced on a printer. Moreover, they can be recorded far faster and cheaper than the same thing on paper-printed output.

The principal disadvantage, however, is that a microfilm or microfiche reader is needed to read this type of output. It's possible that this technology could be made obsolete by developments in secondary-storage techniques, such as the use of removable, high-capacity hard disks. Currently, though, computer-assisted retrieval that uses readers with automatic indexing and data-lookup capabilities makes computer output microfilm and microfiche the preferred technology.

5.9 More Output Devices: Audio, Video & Digital TV, Virtual Reality, & Robots

KEY QUESTION

How do audio, video, virtual reality, and robots operate?

Preview & Review: Other output hardware includes devices for sound output, voice output, and video output; virtual-reality and simulation devices; and robots.

Sound output includes music and other nonverbal sounds. Voice-output technology uses speech-synthesis devices that convert digital data into speech-like sounds.

Video output includes videoconferencing. Digital TV presents sharper images or more channels.

Virtual reality is a kind of computer-generated artificial reality.

Robots perform functions ordinarily ascribed to human beings.

How long until speech robots call you during the dinner hour and use charming voices to get you to part with a charitable donation through your credit card? Perhaps sooner than you think. Already University of Iowa scientists have created a computer program and audio output that simulates 90% of the acoustic properties made by one man's voice. "What we have now are extremely human-like speech sounds," says Brad H. Story, the lead researcher and also the volunteer whose vocal tract was analyzed and simulated. "We can produce vowels and consonants in isolation." A harder task, still to be accomplished, is to recreate the transitions between key linguistic sounds—what researchers call "running speech."

In the meantime, Story and a colleague have performed around the country with an electronic device they call "Pavarobotti." The operatic "singing" (such as Puccini arias) is not borrowed from recordings of humans but is created electronically. Singing, however, is easier to simulate than speech. Even singing that would bring an audience to its feet, such as holding a high pitch for a long time, is easy to perform with microchips and speakers.[56]

Voice output is only one of the many other forms of output technology that remain to be discussed. In this section, we describe *sound-output devices, voice-output devices, video-output devices, virtual-reality devices,* and *robots.*

Sound-Output Devices

Sound-output devices **produce digitized sounds, ranging from beeps and chirps to music.** All these sounds are nonverbal. PC owners can customize their machines to greet each new program with the sound of breaking glass or to moo like a cow every hour. Or they can make their computers issue the distinctive sounds available (from the book-disk combination *Cool Mac Sounds*) under the titles "Arrgh!!!" or "B-Movie Scream." Or they can download sound files from the Internet and save them to disk, insert them in documents, and play them.

To exercise these possibilities, you need both the necessary software and sound card, or digital audio circuit board (such as Sound Blaster). The sound card plugs into an expansion slot in your computer; on newer computers, it is integrated with the motherboard.

A sound card is also required in making computerized music.

Most microcomputers come with sound speakers, although these speakers often have a rather tinny quality. For good sound, you need to connect external amplified speakers.

Voice-Output Devices

How do blind and visually impaired people read e-mail? They use voice synthesizers that read aloud the words on the screen. But what do they do on the heavily graphics-oriented World Wide Web? Many use PC Webspeak, a nonvisual browser, which reads the HTML—hypertext markup language, the programming code of the Web page—and interprets it directly. Or they rely on Web page designers to provide a text description of graphics or photographs used, which is translated into aural communication.[57]

Voice-output, **or** *speech-synthesis, devices* **convert digital data into speech-like sounds.** These devices are no longer very unusual. You hear such forms of voice output on telephones ("Please hang up and dial your call again"), in soft-drink machines, in cars, in toys and games, and recently in mapping software for vehicle-navigation devices.

Some uses of speech output are simply frivolous or amusing. You can replace your computer start-up beep with the sound of James Brown screaming "I feel goooooood!" Or you can attach a voice annotation to a spreadsheet that says "I know this looks high, Bob, but trust me."

But some uses are quite serious. For people with physical disabilities, computers help to level the playing field. A 39-year-old woman with cerebral palsy had little physical dexterity and was unable to talk. By pressing keys on the laptop computer attached to her wheelchair, she was able to construct the following voice-synthesized message: "I can do checkbooks for the first time in my life. I cannot live without my computer."[58]

Video-Output Devices

As we explained, CRTs are used both for TV sets and for computer monitors. One area in which we are seeing innovations in video technology is videoconferencing. We describe this topic in detail elsewhere (Chapter 7), but it's of interest here because of the kind of technology used.

Want to have a meeting with someone across the country and go over some documents—without having to go there? **Videoconferencing is a method whereby people in different geographical locations can have a meeting—and see and hear one another—using computers and communications.**

Videoconferencing systems range from small videophones to group conference rooms with cameras and multimedia equipment.[59] You can also do it yourself using a desktop PC system. Say you're on the West Coast and want to go over a draft of a client proposal with your boss on the East Coast. The first thing you need for such a meeting is a high-capacity telephone line or a high-quality Internet connection. To this you link your PC running Windows to which you have added a hardware and software package that consists of a small video camera that sits atop your display monitor, a circuit board, and software that turns your microcomputer into a personal conferencing system.

Your boss's image appears in one window on your computer's display screen, and the document you're working on together is in another window. (An optional window shows your own image.) Although the display screen images are choppier than those on a standard TV set, they're clear enough to enable both of you to observe facial expressions and most body language. The software includes drawing tools and text tools for adding comments. Thus, you can go through and edit paragraphs and draw crude sketches on the proposal draft.

Digital Television

Here are just a few consequences of digital televison: TV news reporters will have to be more careful about how they put on their makeup because the picture will be much sharper and clearer. Studio sets may have to be rebuilt to look more realistic. And you will have to buy a new TV set (starting at $2000) to view high-definition programs. Or else you'll have to rent a set–top converter box to see shows in standard definition on your old set.[60]

What we have today, of course, is analog TV sent in waves. However, *digital television (DTV)—also known as high–definition television (HDTV) and advanced television (ATV)—is sent in bits and can show pictures much sharper than those on today's sets.* These pictures are capable of having 1080 scan lines (compared to 525 of an average TV screen today), thus offering nearly twice the resolution, or sharpness. Moreover, images can appear, movie screen style, within a frame that is about one–third wider than today's screens. In addition, DTV offers better color and better audio, with sound being of compact-disk quality.

Forty-two stations began broadcasting in the United States in the fall of 1998.[61]

Virtual-Reality & Simulation Devices

Virtual reality (VR) is a kind of computer-generated artificial reality that projects a person into a sensation of three-dimensional space. (■ *See Panel 5.29.*) To put yourself into virtual reality, you need the following interactive sensory equipment:

- Headgear: The headgear—which is called *head-mounted display (HMD)*—has two small video display screens, one for each eye, that create the sense of three-dimensionality. Headphones pipe in stereophonic sound or even "3-D" sound. Three-dimensional sound makes you think you are hearing sounds not only near each ear but also in various places all around you.

- Glove: The glove has sensors that collect data about your hand movements.

- Software: Software gives the wearer of this special headgear and glove the interactive sensory experience that feels like an alternative to real-world experiences.

■ PANEL 5.29

Virtual reality

(Top left) Man wearing interactive sensory headset and glove. When the man moves his head, the 3-D stereoscopic views change. *(Top right)* When the man moves his glove, sensors collect data about his hand movements. The view then changes so that the man feels he is "moving" over to the bookshelf and "grasping" a book. *(Middle right)* What the man is looking at—a simulation of an office. *(Bottom left)* During a virtual reality experiment in Chapel Hill, North Carolina, a man uses a treadmill to walk through a virtual reality environment. *(Bottom right)* Two medical students study the leg bones of a virtual cadaver.

You may have seen virtual reality used in arcade-type games, such as Atlantis, a computer simulation of The Lost Continent. You may even have tried to tee off on a virtual golf driving range or driven a virtual racing car. There are also a few virtual-reality home video games, such as The 7th Sense. However, there are far more important uses, one of them being in simulators for training.

***Simulators* are devices that represent the behavior of physical or abstract systems.** Virtual-reality simulation technologies are applied a great deal in training. For instance, they have been used to create lifelike bus control panels and various scenarios such as icy road conditions to train bus drivers. They are used to train pilots on various aircraft and to prepare air-traffic controllers for equipment failures. They also help children who prefer hands-on learning to explore subjects such as chemistry.

Of particular value are the uses of virtual reality in health and medicine. For instance, surgeons-in-training can rehearse their craft through simulation on "digital patients." Virtual-reality therapy has been used for autistic children and in the treatment of phobias, such as extreme fear of public speaking or of being in public places or high places. It has also been used to rally the spirits of quadriplegics and paraplegics by engaging them in plays and song-and-dance routines. As one patient said, "When you spend a lot of time in bed, you can go crazy."[62]

Ethics

Interestingly, one ethical—and potential litigation—problem for makers of virtual-reality equipment is how to keep users from getting sick. Known as *cybersickness* or *simulator sickness*, symptoms include eyestrain, queasiness, nausea and confusion, and even visual and audio "flashbacks" among some VR users. The disorder sometimes afflicts military pilots training on flight simulators, for example, who are then prohibited from flying. Not all users of virtual-reality equipment will experience all symptoms. Nevertheless, in preparation for the expected wave of VR products, researchers are taking a long look at what kinds of measures can be taken to head off these complaints.

Robots

We discuss robots—which might be considered complete computer systems with both input and output aspects—in detail in Chapter 12. Here, however, they are of interest to us because they output motion rather than information. They can perform computer-driven electromechanical functions that the other devices so far described cannot.

Basically, **a *robot* is an automatic device that performs functions ordinarily ascribed to human beings or that operates with what appears to be almost human intelligence.** Actually, robots are of several kinds—industrial robots, perception systems, and mobile robots, for example, as we discuss in Chapter 12.

More than 40 years ago, in *Forbidden Planet*, Robby the Robot could sew, distill bourbon, and speak 187 languages. We haven't caught up with science-fiction movies, but maybe we'll get there yet. ScrubMate—a robot equipped with computerized controls, ultrasonic "eyes," sensors, batteries, three different cleaning and scrubbing tools, and a self-squeezing mop—can clean bathrooms. Rosie the HelpMate delivers special-order meals from the kitchen to nursing stations in hospitals. Robodoc—notice how all these robots have names—is used in surgery to bore the thighbone so that a hip implant can be attached. Remote Mobile Investigator 9 is used by Maryland police to flush out barricaded gunmen and negotiate with terrorists. A driverless harvester, guided by satellite signals and artificial vision system, is used to harvest alfalfa and other crops.

Robots are also used for more exotic purposes such as fighting oil-well fires, doing nuclear inspections and cleanups, and checking for mines and booby traps. An eight-legged, satellite-linked robot called Dante II was used to explore the inside of Mount Spurr, an active Alaskan volcano, sometimes without human guidance. A file-drawer-size, six-wheeled robot vehicle named Sojourner was used in NASA's 1997 Pathfinder exploration of Mars to sample the planet's atmosphere and soil and to radio data and photos to Earth.

5.10 In & Out: Devices That Do Both

KEY QUESTION

What three devices are both input and output hardware?

Preview & Review: Some hardware devices perform both input and output. Three common ones are terminals, smart cards and optical cards, and touch screens.

Terminals

People working on large computer systems are usually connected to the main, or host, computer via terminals. **A *terminal* is an input/output device that uses a keyboard for input and a monitor for output.** They come in two varieties: dumb or intelligent.

Airline terminals

- Dumb: The most common type of terminal is dumb. A *dumb terminal* can be used only to input data to and receive information from a computer system. That is, it cannot do any processing on its own. An example of a dumb terminal is the type used by airline clerks at airport ticket and check-in counters.

- Intelligent: An *intelligent terminal* has built-in processing capability and RAM but does not have its own storage capacity. Intelligent terminals are not as powerful as microcomputers and are not designed to operate as stand-alone machines. Intelligent terminals are often found in local area networks in offices. Users share applications software and data stored on a server.

The network computer (✔ p. 31), which has no secondary-storage capacity, is a type of intelligent terminal. The network computer is hooked up to the Internet, where the applications software and data you would use are stored on a server.

Two examples of intelligent terminals are automated teller machines and point-of-sale terminals. An *automated teller machine (ATM)* is used to retrieve information on bank balances, make deposits, transfer sums between accounts, and withdraw cash. Usually the cash is disbursed in $20 bills. Some Nevada gambling casinos have machines that dispense only $100 bills.

ETM

ATMs have become popular throughout the world. You can now use your ATM card, for example, to get cash from machines on cruise ships (or even some Cathay Pacific airplanes) or local currency in machines in foreign airports.[63] Residents of Singapore now use ATMs to buy shares of stock. In airports, variations on ATMs called *electronic ticketing machines (ETMs)* help travelers buy their own tickets, helping them avoid lines. As mentioned, some teller machines, called *electronic kiosks,* act like vending machines, selling theater tickets, traveler's checks, stamps, phone cards, and other documents.

In a great leap beyond paper, various companies are working on devices, such as one called a Personal ATM, that you can install in your home. With this equipment, you can use your phone or Internet connection to download funds from your bank account onto a "smart card," which, as we shall describe later, is a credit-card-like piece of plastic embedded with a computer chip and usable as a cash substitute.[64-66]

A *point-of-sale (POS)* terminal is a smart terminal used much like a cash register. It records customer transactions at the point of sale

POS terminal

but also stores data for billing and inventory purposes. POS terminals are found in most department stores. POS terminals can also store data for marketing purposes.

There seem to be two conflicting trends. On the one hand, microcomputers increasingly are being used in business as terminals—and in place of dumb and intelligent terminals. This trend is occurring not only because their prices have come down but also because they reduce the processing and storage load on the main computer system. On the other hand, network computers and cable-TV set-top boxes—which really are forms of dumb terminals—are also emerging, which could throw the load back onto the main computer system.

Smart Cards & Optical Cards

It has already come to this: Just as many people collect stamps or baseball cards, there is now a major worldwide collecting mania for used wallet-size telephone debit cards. These are the cards, called "stored-value" cards, by which telephone time is sold and consumed in many countries. Generally the cards are collected for their designs, which bear likenesses of anything from Elvis Presley to Felix the Cat to Martin Luther King, Jr. The cards have been in use in Europe for nearly 20 years, and now many U.S. phone companies sell them.

Telephone debit cards

Most of these telephone cards are examples of new "smart cards." Today in the United States most ATM cards and credit cards are the old-fashioned magnetic-stripe cards. A third kind of card, the optical card, holds by far more information. Let's consider all three.

- Magnetic-stripe cards: A magnetic-stripe card has a stripe of magnetically encoded data on its back. The encoded data might include your name, account number, and PIN (personal identification number).

  New uses are still being found for such cards. For example, in recent years, the California Department of Motor Vehicles has been issuing driver's licenses and ID cards with magnetic stripes. Working with the DMV, the 7-Eleven convenience-store chain in 1997 began installing "age-verification" equipment in its California stores to help spot underage drinkers and smokers. Now clerks faced with a youthful customer wanting to buy beer or cigarettes can press one of three buttons (alcohol, tobacco, or both) on the specialized card reader, swipe the customer's driver's license through the machine, and instantly learn the birth date encoded in the card.[67] (They also, of course, need to compare the customer's face with the photograph on the license.)

- Smart cards: **A smart card looks like a credit card but contains a microprocessor and memory chip.** When inserted into a reader, it transfers data to and from a central computer, and it can store some basic financial records. It is more secure than a magnetic-stripe card and can be programmed to self-destruct if the wrong password is entered too many times.

  In France, where the smart card was invented, you can buy telephone debit cards at most cafés and newsstands. You insert the card into a slot in the phone, wait for a tone, and dial the number. The time of your call is automatically calculated on the chip inside the card and deducted from the balance. The French also use smart cards as bank cards, and some people carry their medical histories on them.

The United States has been slow to embrace smart cards because of the prevalence of conventional magnetic-stripe credit cards. Moreover, the United States has a large installed base of credit-card readers and phone networks with which merchants can check on cards. However, in some other countries phone lines are scarcer, and merchants cannot as easily check over the phone with a centralized credit database. In these situations, stored-value smart cards, sometimes called "electronic purses," make sense because they carry their own spending limits. Thus, while the Americans have been in the pilot-project stage, the Europeans, in transcending their own antiquated phone systems, have gone all out with smart cards—which, as one observer suggests, "says a lot about the wacky ways in which technology spreads these days."[68]

Recently, various credit-card companies and banks in North America have begun to promote smart cards, combining ATM, credit, and debit cards in one instrument. One of the first such experiments began in late 1997 with 50,000 New Yorkers on Manhattan's Upper West Side.[69,70] When fully implemented, this system will follow the Mondex card system used in England. "Mondex money" looks and works something like a debit card. The difference is that it disburses electronic "cash" that has previously been loaded from the customer's bank account onto the card's microchip by means of an ATM or special smart phone (such as the Personal ATM, described above) that conveys financial data. Customers can use the Mondex cards, which can pay automatically in five currencies, at retail stores, toll booths, pay phones, and even taxis equipped with card readers.

- **Optical cards:** The conventional magnetic-stripe credit card holds the equivalent of a half page of data. The smart card with a microprocessor and memory chip holds the equivalent of 250 pages. The optical card presently holds about 2000 pages of data. Optical cards use the same type of technology as music compact disks but look like silvery credit cards. ***Optical cards* are plastic, laser-recordable, wallet-type cards used with an optical-card reader.**

Because they can cram so much data (6.6 megabytes) into so little space, they may become popular in the future. With an optical card, for instance, there's enough room for a person's health card to hold not only his or her medical history and health-insurance information but also digital images. Examples are electrocardiograms, low-resolution chest X-rays, and ultrasound pictures of a fetus. A book containing 1000 pages of text plus 150 detailed drawings could be mailed on an optical card in a 1-ounce first-class letter.

Ethics

Smart cards and optical cards may be used to solve some present problems, but they may also generate some ethical problems of their own. On the positive side, for instance, by substituting plastic debit cards for paper coupons, Texas and Maryland have successfully reduced the fraudulent exchange of food stamps for cash or drugs. Electronic records make it easier to detect misuse.[71] On the negative side, smart cards may evolve into some sort of universal identity card containing a lot of personal information that you wouldn't want to share but can't control. "We think everything will migrate toward a single card," says Visa chief executive officer Carl Pascarella. "I might want to have my driver's license, frequent-flier miles, medical information, and HMO [health-maintenance organization] data on my card."[72] Privacy experts fear the possible result.

Touch Screens

A _touch screen_ is a video display screen that has been sensitized to receive input from the touch of a finger. (■ _See Panel 5.30.)_ The screen is covered with a plastic layer, behind which are invisible beams of infrared light. You can input requests for information by pressing on buttons or menus displayed. The answers to your requests are displayed as output in words or pictures on the screen. (There may also be sound.)

Because touch screens are easy to use, they can convey information quickly. You find touch screens in kiosks, ATMs, airport tourist directories, hotel TV screens (for guest checkout), and campus information kiosks making available everything from lists of coming events to (with proper ID and personal code) student financial-aid records and grades.

5.11 The Future of Input & Output

KEY QUESTION

What are some examples of the future of input and output technology?

Preview & Review: Increasingly, input will be performed in remote locations and will rely on source-data automation. As for output, it too is being performed in remote locations.

Input technology seems headed in two directions: (1) toward more input devices in remote locations and (2) toward more refinements in source-data automation. Output is distinguished by (1) more output in remote locations and (2) more and more realistic, even life-like, forms.

Toward More Input from Remote Locations

When management consultant Steve Kaye of Santa Ana, California, wants to change a brochure or company letterhead, he doesn't have to drop everything and drive over to a printer. He simply enters his requests through the phone line to an electronic bulletin board at the Sir Speedy print shop that he deals with. "What this does is free me up to focus on my business," Kaye says.[73]

■ PANEL 5.30

Touch screen
The staff in many restaurants and bars use touch screens for placing orders.

The linkage of computers and telecommunications means that input may be done from nearly anywhere. For instance, X-ray machines are now going digital, which means that a medical technician in the jungles of South America can take an X-ray of a patient and then transmit a perfect copy of it by satellite uplink to a hospital in Boston.[74,75] Visa and MasterCard are moving closer to using "smart cards," or stored-value cards, for Internet transactions.[76]

Toward More Source-Data Automation

Keyboards and pointing devices will no doubt always be with us, but more and more often input technology is being designed to capture data at its source. This will reduce the costs and mistakes that come with copying or otherwise preparing data in a form suitable for processing.

Some reports from the input-technology front:

3-D jeans scanner

- **High-capacity bar codes:** Traditional bar codes read only horizontally. A new generation of bar codes has appeared that reads vertically as well, which enables them to store more than 100 times the data of traditional bar codes. With the ability to pack so much more information in such a small space, bar codes can now be used to include digitized photos, along with a person's date of birth, eye color, blood type, and other personal data.

- **3-D scanners:** Have difficulty getting the perfect fit in blue jeans? Modern clothes are designed to fit mannequins, which is why it's difficult to get that "sprayed-on" look for people. However, clothing makers (including Levi Strauss, the world's largest jeans maker) have inaugurated a body scanner that enables people to buy clothes that fit precisely.

  The device, which doesn't use lasers (to alleviate possible customer health concerns), allows you to enter a store, put on a body suit, and be measured three-dimensionally all over. You can then select the clothes you're interested in, view them imposed on your body-scanned image on a screen, and order them custom-manufactured.

- **More sophisticated touch devices:** Touch screens are becoming commonplace. Sometime soon, futurists have suggested, you may be able to use a dashboard touch screen in your car. The screen would be linked to mobile electronic "yellow pages" that would enable you to reserve a motel room for the night or find the nearest Chinese restaurant.[77]

- **Smarter smart cards:** Over the next few years, stored-value smart cards with microchips, acting as "electronic purses," will no doubt begin to displace cash in many transactions. Targets for smart cards are not only convenience stores and toll booths but also battery-powered card readers in newspaper racks and similar devices.

  We've already mentioned that microchips with identification numbers are being injected into dogs and cats so that, with the help of a scanner, lost pets can be identified (✔ p. 214). Although it's doubtful chip implantation for identification purposes would be extended to people (though it could), smart cards and optical cards could evolve into all-purpose cards including biotechnological identifiers. These could contain medical records, driver's license data, insurance information, security codes for the office, and frequent flier program information. Already, some health insurance companies are taking steps to give enrollees smart cards that contain their coverage information; when doctor's offices swipe the cards through card readers, patients can see what their insurance will pay for before leaving.[78]

● **More sensors:** Sensors are beginning to find all kinds of new uses. Luxury cars, for instance, will begin employing sensors in seats that will automatically inflate or deflate air cells in the seat to provide the best support.[79] A special "smart needle" has been devised with embedded sensors that can be inserted into a tumor so that doctors can collect more information about the cancer, such as its response to chemotherapy.[80] Wheelchair-bound quadriplegics whose injuries are low enough on the spinal cord to preserve some shoulder and arm movement are undergoing experiments with "shoulder-position sensors"; these devices translate small movements in the shoulder into electrical impulses that can direct the muscles of the hand and arm.[81]

● **Smaller electronic cameras:** Digital still cameras and video cameras are fast becoming commonplace. The next development may be the camera-on-a-chip, which will contain all the components necessary to take a photograph or make a movie. Such a device, called an *active pixel sensor*, based on NASA space technology, is now being made by a company called Photobit. Because it can be made on standard semiconductor production lines, the camera-on-a-chip can be made incredibly cheaply, perhaps for $20 apiece. Such micro-cameras could be put anywhere as security devices or for other purposes.

● **Biometric devices:** Would you believe a computer that can read people's emotions from changes in their facial patterns, like surprise and sadness? Such devices are being worked on at Georgia Institute of Technology and other places.[82] (■ *See Panel 5.31.*)

● **Brain-wave devices:** Perhaps the ultimate input device analyzes the electrical signals of the brain and translates them into computer commands. Experiments have been successful in getting users to move a cursor on the screen through sheer power of thought. Other experiments have shown users able to type a letter by slowly spelling out the words in their heads.[83–87]

Although there is a very long way to go before brain-wave input technology becomes practical, the consequences could be tremendous, not only for disabled people but for all of us.

■ PANEL 5.31

Computers tuned to emotions
At Georgia Institute of Technology, researchers have devised a computer system that can read people's emotions from changes in their faces. Expressions of happiness, surprise, and other movements of facial muscles are converted by a special camera into digitized renderings of energy patterns.

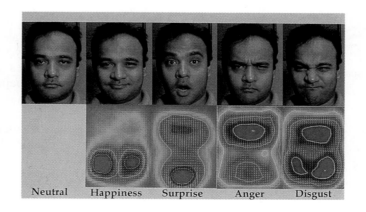

Neutral   Happiness   Surprise   Anger   Disgust

READ**ME**

Practical Matters: Getting Real About Credit Cards

The Digital Age forces some discipline on us that hasn't been required before. Consider: What's probably the easiest input device you've learned to use that perhaps you wish you hadn't? Maybe the credit card?

"We all probably spent more than we earned," said Jennifer Benjamin, 26, commenting on the credit-card habits of her University of Massachusetts friends and her own ruined credit rating. "It wasn't real money. It was play money."[88]

Benjamin wrecked her credit by being unable to pay the huge amounts she'd charged to plastic. This is a not-uncommon problem. American college students are drowning in debt. Indeed, of the debtors seeking professional help at the National Consumer Counseling Service, more than half are between 18 and 32.[89]

Banks and credit-card companies are bombarding college students with credit-card offers. As a result, 64% of college students have a credit card, according to one survey.[90] Another study, by a professor of family finance, found that 76% of students surveyed held three or more credit cards, and 40%—twice the rate of the population at large—had six or more.[91]

Nationwide, for students and everyone else, the average credit-card holder has four cards and is about $4000 in debt.[92] A typical card balance for students is about $1900. A survey of students at three Michigan universities found that a handful had bills as high as $5000 or $6000.[93,94] In addition, it's easy for student debt to snowball because many students are taking on larger student loans just to get by.[95] The median student loan debt is about $15,000.[96]

Cards can be ranked in order from most risky to least risky in terms of costing you money:

- **Credit cards—most risky:** Credit cards allow you to build up charges to be paid off in installments plus interest, provided you make a minimum payment every month. Visa, MasterCard, Discover, Sears, and Macys are credit cards.

 The worst problem with credit cards is that you can seemingly get what you want without the pain of feeling that you're paying for it.

 The second worst problem is that once you build up debt, it may take you forever to get out of it. If your credit-card balance is $1900 and if the interest rate is 18%, then if you make the minimum payment of 2% ($38) on the unpaid balance every month,

you'll be paying off that $1900 for—the next 23 years! (The cost includes $4790 in interest, and it assumes you never again charge on that card.)[97]

- **Charge cards—somewhat risky:** Charge cards require that the bill be paid off every month. American Express and oil-company cards (Shell, Texaco, Mobil) are charge cards.

 The risk here is that you'll be socked with late charges and interest charges if you don't pay on time. Moreover, if you're frequently late, your card will be canceled, triggering a bad credit report that will make it difficult for you to get credit cards and loans in the future.

- **Debit cards—somewhat risky:** Debit cards make instant electronic withdrawals from your checking account when you use them at an automated teller machine (ATM) or to make purchases at stores. Banks have been quietly trying to replace ATM cards—debit cards that are free—with debit cards called check cards, for which they often charge users $12—$15 a year (and they can charge merchants for every transaction).

 The risk with debit cards is that you can easily run up an overdraft in your checking account—and incur the bank's penalty charge—if you forget to record debit purchases in your checkbook.

 Moreover, if you lose a check card (as opposed to an ATM card), a crook who gets hold of it can quickly clean out your bank account (often check cards do not require the use of a PIN, or personal identification number).[98] And it doesn't have the theft protection of credit cards, for which you can be charged only $50. (If the loss or theft of your debit card is reported within two days, you're liable for up to $50; after two days, up to $500.)[99]

- **Prepaid cards—least risky:** Cards that you pay for in advance, such as "stored-value cards," are those that enable you to make long-distance telephone calls, for instance. The biggest risk is that if you lose the card, anyone can use it.

There are ways to arrest credit-card spending. You can, for instance, do what Yashika Gomes did. A student at the University of Texas–Austin, Gomes immersed her three credit cards in a mug of water and put them in the freezer, a trick she learned from *Oprah*. You can thaw the cards if there's a real emergency.[100]

Toward More Output in Remote Locations

What will happen to paper-and-ink newspapers in the Digital Age? Roger Fidler, former head of a research lab for the Knight-Ridder newspaper chain and now at Kent State University, has conceptualized a device called the *electronic news tablet*, which could be given or rented to newspaper readers.[101,102] This gadget would be about the same size as this textbook and perhaps half the thickness, and most of its surface area would be a screen. Readers would stick the tablet in a slot in an electronic rack and download a digital newspaper into its memory. The racks, of course, would be connected by some sort of electronic network over which information would be transmitted from a central source.

Clearly, output in remote locations is the wave of the future. As TV and the personal computer further converge, we can expect more scenarios like this: Your PC continually receives any Web sites covering topics of interest to you—CNN's home page for news on Bosnia, for example—which are stored on your hard disk for later viewing. The information, on up to 5000 Web sites a day, is transmitted from a television broadcast satellite to an 18-inch satellite dish on your roof. Interested? The technology is available now with direct broadcast services such as DirecTV. It requires that you install a $400 circuit board (made by Adaptec) that permits Web surfing at speeds up to 30 megabits per second—more than 100 times as fast as the current PC modems.[103]

Toward More Realistic Output

Once upon a time, having a "home theater" probably meant you were a wealthy film-industry figure with a large room containing a wall-size screen and movie-house seats. Now, says technology writer Phillip Robinson, "home theater has become a middle-income commodity with new audio and video gear promising to improve your television viewing so much you'll practically think you're in a theater instead of your living room or den."[104]

The enhanced qualities of home theater won't be limited to television viewing, of course. As analog and digital technologies merge, we will no doubt see this type of increased realism appearing in all forms of output. Let's consider what's coming into view.

- **Display screens—better and cheaper:** Computer screens are becoming crisper, brighter, bigger, and cheaper.

- **Audio—higher fidelity:** Some researchers think that recent advances will put PC sound on the same par as home-theater audio systems.

- **Video—movie quality for PCs:** Today the movement of most video images displayed on a microcomputer comes across as fuzzy and jerky, with a person's lip movements out of sync with his or her voice. This is because currently available equipment is capable of running only about eight frames a second.

   New technology based on digital wavelet theory has led to software that can compress digitized pictures into fewer bytes and do it more quickly than current standards, giving images the look and feel of a movie. Although this advance has more to do with software than hardware, it will clearly affect future video output.

- **Three-dimensional display:** With 3-D technology, flat, cartoon-like images give way to rounded objects with shadows and textures. Artists can even add "radiosity," so that the image of a dog standing next to a red car, for instance, will pick up a red glow.

- **Virtual worlds—3-D in cyberspace:** Virtual reality, as we explained, involves head-mounted displays and data gloves. By contrast, a *virtual world* requires only a microcomputer, mouse, and special software to display and navigate three-dimensional scenes. The 3-D scenes are presented via interlinked, or networked, computers on the World Wide Web part of the Internet. For example, users can meet in three-dimensional fantasy landscapes (using on-screen stand-ins for themselves called *avatars*) and move around while "talking" through the keyboard with others.

5.12 Input & Output Technology & Quality of Life: Health & Ergonomics

KEY QUESTION

What are the principal health and ergonomics issues related to using computers?

Preview & Review: The use of computers and communications technology can have important effects on our health. Some of these are repetitive strain injuries such as carpal tunnel syndrome, eyestrain and headaches, and backstrain.

Negative health effects have increased interest in the field of ergonomics, the study of the relationship of people to a work environment.

Susan Harrigan, a financial reporter for *Newsday,* a daily newspaper based on New York's Long Island, had to learn to write her stories using a voice-activated computer. She did not do so by choice, nor was she as efficient as she used to be. She did it because she was too disabled to type at all.

After more than two decades of writing articles with deadline-driven fingers at the keyboard, she had developed a crippling hand disorder. At first the pain was so severe she couldn't even hold a subway token. "Also, I couldn't open doors," she said, "so I'd have to stand in front of doors and ask someone to open them for me."[105]

Health Matters

Harrigan suffers from one of the computer-induced disorders classified as repetitive stress injuries (RSIs). The computer is supposed to make us efficient. Unfortunately, it has made some users—journalists, postal workers, data-entry clerks—anything but. The reasons are repetitive strain injuries, eyestrain and headache, and back and neck pains. In this section we consider these health matters, along with the effects of electromagnetic fields and noise.

- Repetitive stress injuries: ***Repetitive stress (or strain) injuries (RSIs) are several wrist, hand, arm, and neck injuries resulting when muscle groups are forced through fast, repetitive motions,*** such as when typing on a computer, playing certain musical instruments, or moving grocery items over a scanner during checkout. The Bureau of Labor Statistics says 25% of all injuries that result in lost work time are due to RSI problems.[106] Most victims of RSI are in meat-packing, automobile manufacturing, poultry slaughtering, and clothing manufacturing. Musicians, too, are often troubled by RSI (because of long hours of practice).

  People who use computer keyboards—some of whom make as many as 21,600 keystrokes *an hour*—account for about 12% of RSI cases that result in lost work time.[107] Before computers came along, typists would stop to make corrections or change paper. These motions had the effect of providing many small rest breaks. Today keyboard users must devise their own mini-breaks to prevent excessive use of hands and wrists.

RSIs cover a number of disorders. Some, such as muscle strain and tendinitis, are painful but usually not crippling. These injuries, often caused by hitting the keys too hard, may be cured by rest, anti-inflammatory medication, and change in typing technique. However, carpal tunnel syndrome is disabling and often requires surgery. ***Carpal tunnel syndrome (CTS)* consists of a debilitating condition caused by pressure on the median nerve in the wrist, producing damage and pain to nerves and tendons in the hands.** (■ *See Panel 5.32.*)

It's important to point out, however, that scientists still don't know what causes RSIs. They don't know why some people operating keyboards develop upper body and wrist pains and others don't. The working list of possible explanations for RSI includes "wrist size, stress level, relationship with supervisors, job pace, posture, length of workday, exercise routine, workplace furniture, [and] job security." Other possible contributors are diabetes, weight, and menopause.[108]

● **Eyestrain and headaches:** Vision problems are actually more common than RSI problems among computer users.[109,110] Computers compel people to use their eyes at close range for a long time. However, our eyes were made to see most efficiently at a distance. It's not surprising, then, that people develop what's called computer vision syndrome.

***Computer vision syndrome (CVS)* consists of eyestrain, headaches, double vision, and other problems caused by improper use of computer display screens.** By "improper use," we mean not only staring at the screen for too long but also failing to employ the technology as it should be employed. This includes allowing faulty lighting and screen glare, and using screens with poor resolution.

■ PANEL 5.32

Carpal tunnel syndrome
The carpal ligament creates a tunnel across the bones of the wrist. When the tendons passing through the carpal tunnel become swollen, they press against the median nerve, which runs to the thumb and the first three fingers.

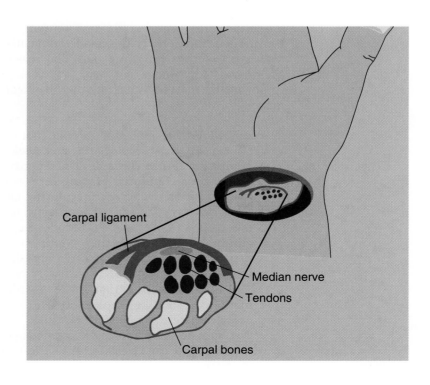

Carpal ligament

Median nerve

Tendons

Carpal bones

- **Back and neck pains:** Many people use improper chairs or position keyboards and display screens in improper ways, leading to back and neck pains. All kinds of adjustable, special-purpose furniture and equipment is available to avoid or diminish such maladies.

- **Electromagnetic fields:** Like kitchen appliances, hairdryers, and television sets, many devices related to computers and communications generate low-level electromagnetic field emissions. ***Electromagnetic fields (EMFs)* are waves of electrical energy and magnetic energy.**

    In recent years, stories have appeared in the mass media reflecting concerns that high-voltage power lines, cellular phones, and CRT-type computer monitors might be harmful. There have been worries that monitors might be linked to miscarriages and birth defects, and that cellular phones and power lines might lead to some types of cancers. There have been suggestions that people with high occupational exposure to EMFs (such as sewing-machine operators) have higher rates of Alzheimer's disease.[111]

    Is there anything to this? The answer is: so far no one is sure. The evidence seems scant that weak electromagnetic fields, such as those used for cellular phones and found near high-voltage lines, cause cancer. Still, unlike car phones or older, luggable cellular phones, the handheld cellular phones do put the radio transmitter next to the user's head. This causes some health professionals concern about the effects of radio waves entering the brain as they seek out the nearest cellular transmitter. A Finnish study partly funded by the telecommunications industry has found mobile phones pose no health threat to users, although they do transmit heat to people's brains.[112] On the other hand, an Australian study found that cellular phone radiation can promote the growth of lymphoma, a form of cancer, at twice the rate in laboratory mice.[113]

    As for CRT monitors, those made since the early 1980s produce very low emissions. Even so, users are advised to not work closer than arm's length to a CRT monitor. The strongest fields are emitted from the sides and backs of terminals. Alternatively, you can use laptop computers, because their liquid crystal display (LCD) screens emit negligible radiation.

    The current advice from the Environmental Protection Agency is to exercise *prudent avoidance.* That is, we should take precautions when they are easy steps. However, we should not feel compelled to change our whole lives or to spend a fortune minimizing exposure to electromagnetic fields. Thus, we can take steps to put some distance between ourselves and a CRT monitor.

- **Noise:** The chatter of impact printers or hum of fans in computer power units can be psychologically stressful to many people. Sound-muffling covers are available for impact printers. Some system units may be placed on the floor under the desk to minimize noise from fans.

Ergonomics: Design with People in Mind

Previously workers had to fit themselves to the job environment. However, health and productivity issues have spurred the development of a relatively new field, called *ergonomics,* that is concerned with fitting the job environment to the worker.

***Ergonomics* is the study of the physical relationships between people and their work environment.** It is concerned with designing hardware and software, as well as office furniture, that is less stressful and more comfortable to use, that blends more smoothly with a person's body or actions. Examples of ergonomic hardware are tilting display screens, detachable keyboards, and keyboards hinged in the middle so that the user's wrists are presumably in a more natural position. (■ *See Panel 5.33.)*

We address some further ergonomic issues in the Experience Box at the end of this chapter.

Onward

Herb Permillion of Berkeley, California, decided to get out of the car-repair business 30 years ago because he wanted secure work. So he switched to repairing typewriters.

Now his California Typewriter is one of only three shops in Berkeley that sell and repair typewriters, where once there were a dozen. Most cities have none. In 1995, Smith Corona Company, the nation's largest typewriter maker, closed its doors after 113 years in business. Even Permillion's daughter, who keeps the books of his business, inputs service orders into an electronic database. Ironically, as the typewriter has devolved from being a useful tool to a charming antique, many Web sites have sprung up selling mail-order typewriter parts and antique typewriters.

Clearly, the world is changing, and changing fast. We look forward to input devices that are more intuitive—computers that respond to voice commands, facial expressions, and thoughts, computers with "digital personalities" that understand what we are trying to do and offer assistance when we need it. And we look forward to output devices that will produce materials in more than one form. For instance, you could be reading this chapter printed in a traditional bound book. Or you might be reading it in a "course pack," printed on paper through some sort of electronic delivery system. Or it might appear on a computer display screen. Or it could be in multimedia form, adding sounds, pictures, and video to text. Thus, information technology changes the nature of how ideas are communicated.

If materials can be input and output in new ways, where will they be stored? That is the subject of the next chapter.

■ PANEL 5.32

Ergonomic keyboards
(Left) Keyboard by Kinesis.
(Right) Keyboard by Microsoft.

Good Habits: Protecting Your Computer System, Your Data, & Your Health

Whether you set up a desktop computer and never move it or tote a portable PC from place to place, you need to be concerned about protection. You don't want your computer to get stolen or zapped by a power surge. You don't want to lose your data. And you certainly don't want to lose your health for computer-related reasons. Here are some tips for taking care of these vital areas.

Protecting Your Computer System

Computers are easily stolen, particularly portables. They also don't take kindly to fire, flood, or being dropped. Finally, a power surge through the power line can wreck the insides.

Guarding Against Hardware Theft & Loss Portable computers—laptops, notebooks, and subnotebooks—are easy targets for thieves. Obviously, anything conveniently small enough to be slipped into your briefcase or backpack can be slipped into someone else's. Never leave a portable computer unattended in a public place.

It's also possible to simply lose a portable, as in forgetting it's in the overhead-luggage bin in an airplane. To help in its return, use a wide piece of clear tape to tape a card with your name and address to the outside of the machine. You should tape a similar card to the inside also. In addition, scatter a few such cards in the pockets of the carrying case.

Desktop computers are also easily stolen. However, for under $25, you can buy a cable and lock, like those used for bicycles, that secure the computer, monitor, and printer to a work area. For instance, you can drill a quarter-inch hole in your equipment and desk, then use a product called LEASH-IT (from Z-Lock, Redondo Beach, California) to connect them together. LEASH-IT consists of two tubular locks and a quarter-inch aircraft-grade stainless steel cable.

If your hardware does get stolen, its recovery may be helped if you have inscribed your driver's license number, Social Security number, or home address on each piece. Some campus and city police departments lend inscribing tools for such purposes. (And the tools can be used to mark some of your other possessions.)

Finally, insurance to cover computer theft or damage is surprisingly cheap. Look for advertisements in computer magazines. (If you have standard tenants' or homeowners' insurance, it may not cover your computer. Ask your insurance agent.)

Guarding Against Heat, Cold, Spills, & Drops "We fried them. We froze them. We hurled them. We even tried to drown them," proclaimed *PC Computing,* in a story about its sixth annual "torture test" of notebook computers. "And only about half survived."[114]

The magazine put 16 notebook computers through durability trials. One approximated putting these machines in a car trunk in the desert heat (2 hours at 180 degrees), another with leaving them outdoors in a Chicago winter (2 hours at 0 degrees). A third test simulated sloshing coffee on a keyboard, and a fourth dropped them 29 inches to the floor. All passed the bake and freeze tests. Some keyboards failed the coffee-spill test, although most revived after drying out and/or cleaning. Seven failed one of the two drop tests (flat drop and edge drop). Of the 16, nine ultimately survived (although some required reformatting after the drop tests).

This gives you an idea of how durable computers are. Designed for portability, notebooks may be hardier than desktop machines. Even so, you really don't want to tempt fate by dropping your computer, which could cause your hard-disk drive to fail. And you really shouldn't drink around your notebook. Or, if you do, take your coffee black, since sugar and cream do the most damage.

Guarding Against Power Fluctuations Electricity is supposed to flow to an outlet at a steady voltage level. No doubt, however, you've noticed instances when the lights in your house suddenly brighten or, because a household appliance kicks in, dim momentarily. Such power fluctuations can cause havoc with your computer system, although most computers have some built-in protection. An increase in voltage may be a spike, lasting only a fraction of a second, or a surge, lasting longer. A surge can burn out the power supply circuitry in the system unit. A decrease may be a momentary voltage sag, a longer brownout, or a complete failure or blackout. Sags and brownouts can produce a slowdown of the hard-disk drive or a system shutdown.

Power problems can be handled by plugging your computer, monitor, and other devices into a surge protector or surge suppressor (✔ p. 160) or a UPS (✔ p. 160). If you're concerned about a lightning storm sending a surge to your system, simply unplug all your hardware until the storm passes.

Guarding Against Damage to Software System software and applications software generally come on CD-ROM disks or flexible diskettes. The unbreakable rule is simply this: Copy the original disk, either onto your hard-disk drive or onto another diskette. Then store the original disk in a safe place. If your computer gets stolen or your software destroyed, you can retrieve the original and make another copy.

Protecting Your Data

Computer hardware and commercial software are nearly always replaceable, although perhaps with some expense and difficulty. Data, however, may be major trouble to replace or even be irreplaceable. (A report of an eyewitness account, say, or a complex spreadsheet project might not come out the same way when you try to reconstruct it.) The following are some precautions to take to protect your data.

Backup Backup Backup Almost every microcomputer user sooner or later has the experience of accidentally wiping out or losing material and having no backup copy. This is what makes people true believers in backing up their data. If you're working on a research paper, for example, it's fairly easy to copy your work onto a diskette at the end of your work session. You can then store that disk in another location. If your computer is destroyed by fire, at least you'll still have the data (unless you stored your disk right next to the computer).

If you do lose data because your disk has been physically damaged, you may still be able to recover it by using special software. (See the discussion of utility programs in Chapter 3, ✔ p. 126.)

Treating Diskettes with Care Diskettes can be harmed by any number of enemies. These include spills, dirt, heat, moisture, weights, and magnetic fields and magnetized objects. Here are some diskette-maintenance tips:

- Insert the diskette *carefully* into the disk drive.
- Do not manipulate the metal "shutter" on the diskette; it protects the surface of the magnetic material inside.
- Do not place heavy objects on the diskette.
- Do not expose the diskette to excessive heat or light.
- Do not use or place diskettes near a magnetic field, such as a telephone or paper clips stored in magnetic holders. Data can be lost if exposed.
- Do not clean diskettes.
- Instead of leaving disks scattered on your desk, where they can be harmed by dust or beverage spills, it's best to store them in their boxes.
- From time to time it's best to clean the diskette drive, because dirt can get into the drive and cause data loss. You can buy an inexpensive drive-cleaning kit, which includes a disk that looks like a diskette and cleans the drive's read/write heads.

Note: No disk lasts forever. Experts suggest that a diskette that is used properly might last 10 years. However, if you're storing data for the long term, you should copy the data onto new disks every 2 years (or use tape for backup). Note also that over the long run, software compatibility may become an issue if you are trying to use diskettes written in an old program.

Guarding Against Viruses Computer viruses are programs—"deviant" programs—that can cause destruction to computers that contract them. They are spread from computer to computer in two ways: (1) They may be passed by way of an "infected" diskette, such as one a friend gives you containing a copy of a game you want. (2) They may be passed over a network or an online service. They may then attach themselves to software on your hard disk, adding garbage to or erasing your files or wreaking havoc with the system software. They may display messages on your screen (such as "Jason Lives") or they may evade detection and spread their influence elsewhere.

Each day, viruses are getting more sophisticated and harder to detect. There are several types of viruses (covered in Chapter 12). The best protection is to install antivirus software. Some programs prevent viruses from infecting your system, others detect the viruses that have slipped through, and still others remove viruses or institute damage control. (Some of the major antivirus programs are Norton AntiVirus, Dr. Solomon's Anti-Virus Toolkit, McAfee Associates' Virus Scan, and Webscan.)

Coping with Airport Security You're standing in front of the metal detector and hand-luggage X-ray machine at airport security. Suddenly it occurs to you, "Could these machines mess up my computer files?"

If you're carrying loose diskettes, it's *possible* they will be damaged, although it's more likely if you're traveling outside North America. The harm could come not from the metal detector or the X-rays but from the magnetic fields of the powerful AC transformer inside the X-ray machine, particularly on poorly adjusted machines, as might happen in some developing countries. However, millions of travelers put their disks through without incident. If you're really worried, you can give your diskettes to the airport security personnel to hand-check.

The same precautions apply with a portable PC with a hard disk, particularly when you're traveling. It's *possible* the magnetic fields from the X-ray machine's transformer will partly erase the data on your hard disk (in already-weak places on the disk). Again, if you're worried, ask for a hand inspection of your machine. This means you'll have to take it out, set it up, and turn it on. The point of the inspection, of course, is to show the airport security people that your portable is not a bomb in disguise.

Protecting Your Health

More important than any computer system and (probably) any data is your health. What adverse effects might computers cause? As we discussed earlier in the chapter, the most serious are painful hand and wrist injuries, eyestrain and headache, and back and neck pains. Some experts also worry about the long-range effects of exposure to electromagnetic fields and noise. All these matters can be addressed by ergonomics, the study of the physical relationships between people and their work environment. Let's see what you can do to avoid these problems.

Protecting Your Hands & Wrists To avoid difficulties, consider employing the following:

- **Hand exercises:** You should warm up for the keyboard just as athletes warm up before doing a sport in order to prevent injury. There are several types of warm-up exercises. You can gently massage the hands, press the palm down to stretch the underside of the forearms, or press the fist down to stretch the top side of the forearm. (■ *See Panel 5.34.*) Experts advise taking frequent breaks, during which time you should rotate and massage your hands.

- **Work-area setup:** Many people set up their computers in the same way as they would a typewriter. However, the two machines are for various reasons ergonomically different. With a computer, it's important to set up your work area so that you sit with both feet on the floor, thighs at right angles to your body. The chair should be adjustable and support your lower back. Your forearms should be parallel to the floor. You should look down slightly at the screen. (■ *See Panel 5.35.*) This setup is particularly important if you are going to be sitting at a computer for hours.

- **Wrist position:** To avoid wrist and forearm injuries, you should keep your wrists straight and hands relaxed as you type. Instead of putting the keyboard on top of a desk, therefore, you should put it on a low table or in a keyboard drawer under the desk. Otherwise the

nerves in your wrists will rub against the sheaths surrounding them, possibly leading to RSI pains. Some experts also suggest using a padded, adjustable wrist rest, which attaches to the keyboard.

Various kinds of ergonomic keyboards are also available, such as those that are hinged in the middle.

Guarding Against Eyestrain, Headaches, & Back & Neck Pains Eyestrain and headaches usually arise because of improper lighting, screen glare, and long shifts staring at the screen. Make sure your windows and lights don't throw a glare on the screen, and that your computer is not framed by an uncovered window. Headaches may also result from too much noise, such as listening for hours to an impact printer printing out.

Back and neck pains occur because furniture is not adjusted correctly or because of heavy computer use. Adjustable furniture and frequent breaks should provide relief here.

Some people worry about emissions of electromagnetic waves and whether they could cause problems in pregnancy or even cause cancer. The best approach is to simply work at an arm's length from computers with CRT-type monitors.

■ PANEL 5.34

Hand exercises

Positions should be held for 10 seconds or more. Warm-up exercises can prevent injuries.

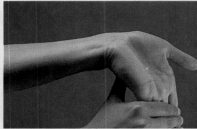

HEAD Directly over shoulders, without straining forward or backward, about an arm's length from screen.

NECK Elongated and relaxed.

SHOULDERS Kept down, with the chest open and wide.

BACK Upright or inclined slightly forward from the hips. Maintain the slight natural curve of the lower back.

ELBOWS Relaxed, at about a right angle.

WRISTS Relaxed, and in a neutral position, without flexing up or down.

KNEES Slightly lower than the hips.

CHAIR Sloped slightly forward to facilitate proper knee position.

LIGHT SOURCE Should come from behind the head.

SCREEN At eye level or slightly lower. Use an anti-glare screen.

FINGERS Gently curved.

KEYBOARD Best when kept flat (for proper wrist positioning) and at or just below elbow level. Computer keys that are far away should be reached by moving the entire arm, starting from the shoulders, rather than by twisting the wrists or straining the fingers. Take frequent rest breaks.

FEET Firmly planted on the floor. Shorter people may need a footrest.

Summary

What It Is/What It Does	Why It's Important
active-matrix display (p. 220, KQ 5.7) Type of flat-panel display in which each pixel on the screen is controlled by its own transistor.	Active-matrix screens are much brighter and sharper than passive-matrix screens, but they are more complicated and thus more expensive.
audio board (p. 210, KQ 5.4) Type of audio-input device; add-on circuit board in a computer that converts analog sound to digital sound and stores it for further processing or plays it back.	Audio boards are used to provide digital sound input for multimedia personal computers to output sound signals to speakers or an amplifier.
audio-input device (p. 209, KQ 5.4) Device that records or plays analog sound and translates it for digital storage and processing.	Audio-input devices, such as audio boards and MIDI boards, are important for multimedia computing.
bar code (p. 204, KQ 5.4) Vertical striped marks of varying widths that are imprinted on retail products and other items; when scanned by a bar-code reader, the code is converted into computer-acceptable digital input.	Bar codes may be used to input data from many items, from food products to overnight packages to railroad cars, for tracking and data manipulation.
bar-code reader (p. 204, KQ 5.4) Photoelectric scanner, found in many supermarkets, that translates bar code symbols on products into digital code.	With bar-code readers and the appropriate computer system, retail clerks can total purchases and produce invoices with increased speed and accuracy; and stores can monitor inventory with greater efficiency.
biometrics (p. 215, KQ 5.5) Science of measuring individual body characteristics.	Biometric systems are used in lieu of typed passwords to identify people authorized to use a computer system.
bitmapped display (p. 222, KQ 5.7) Display screen that permits the computer to manipulate pixels on the screen individually, enabling software to create a greater variety of images.	Bitmapped display screens allow the software to show text and graphics built pixel by pixel rather than in character blocks, so that a greater variety of images is possible. The more bits represented by a pixel, the higher the resolution.
carpal tunnel syndrome (CTS) (p. 244, KQ 5.12) Type of repetitive stress injury; condition caused by pressure on the median nerve in the wrist, producing damage and pain to nerves and tendons in the hands.	CTS is a debilitating, possibly disabling, condition brought about by overuse of computer keyboards; it may require surgery.
cathode-ray tube (CRT) (p. 218, KQ 5.7) Vacuum tube used as a display screen in a computer or video display terminal. Images are represented on the screen by individual dots or "picture elements" called *pixels.*	This technology is found not only in the screens of desktop computers but also in television sets and flight-information monitors in airports.

What It Is/What It Does	**Why It's Important**

computer output microfilm/microfiche (p. 230, KQ 5.8) Computer output produced as tiny images on rolls or sheets of microfilm.

This fast and inexpensive process can store a lot of data in a small amount of space.

computer vision syndrome (CVS) (p. 244, KQ 5.12) Computer-related disability; consists of eyestrain, headaches, double vision, and other problems caused by improper use of computer display screens.

Contributors to CVS include faulty lighting, screen glare, and screens with poor resolution.

dedicated fax machine (p. 207, KQ 5.4) Specialized machine for scanning images on paper documents and sending them as electronic signals over telephone lines to receiving fax machines or fax-equipped computers; a dedicated fax machine will also received faxed documents.

Unlike fax modems installed inside computers, dedicated fax machines can scan paper documents.

digital camera (p. 211, KQ 5.4) Type of electronic camera that uses a light-sensitive silicon chip to capture photographic images in digital form on a small diskette or flash-memory chip.

Digital cameras can produce images in digital form that can be transmitted directly to a computer's hard disk for manipulation, storage, and printing out.

digitizing tablet (p. 202, KQ 5.3) Tablet connected by a wire to a pen-like stylus with which the user sketches an image, or to a puck with which the user copies an image.

A digitizing tablet can be used to achieve shading and other artistic effects or to "trace" a drawing, which can be stored in digitized form.

display screen (p. 218, KQ 5.7) Also variously called *monitor, CRT,* or simply *screen;* softcopy output device that shows programming instructions and data as they are being input and information after it is processed. Sometimes a display screen is also referred to as a VDT, for video display terminal, although technically a VDT includes both screen and keyboard. The size of a screen is measured diagonally from corner to corner in inches, just like television screens.

Display screens enable users to immediately view the results of input and processing.

dot-matrix printer (p. 224, KQ 5.8) Printer that contains a print head of small pins that strike an inked ribbon, forming characters or images. Print heads are available with 9, 18, or 24 pins, with the 24-pin head offering the best quality.

Dot-matrix printers can print draft quality, a coarser-looking 72 dots per inch vertically, or near-letter-quality (NLQ), a crisper-looking 144 dots per inch vertically. They can also print graphics.

dot pitch (p. 221, KQ 5.7) Amount of space between pixels (dots); the closer the dots, the crisper the image.

Dot pitch is one of the measures of display screen clarity.

drum plotter (p. 229, KQ 5.8) Output device that places paper on a moving drum below one to four color pens that move across the paper

Drum plotters allow continuous recording, useful for recording data like earthquake activity.

What It Is/What It Does

Why It's Important

electroluminescent (EL) display (p. 220, KQ 5.7) Flat-panel display that contains a substance that glows when it is charged by an electric current. A pixel is formed on the screen when current is sent to the intersection of the appropriate row and column.

Flat-panel display technologies including EL provide thinner, lighter display screens that use less power than CRTs, so they are well suited to portable computers.

electromagnetic fields (EMFs) (p. 245, KQ 5.12) Waves of electrical energy and magnetic energy, including the low-level EMFs generated by computers and communications devices as well as many appliances.

Some users of cellular phones and CRT monitors have expressed concerns over possible health (cancer) effects of EMFs, but the evidence is weak.

electronic imaging (p. 209, KQ 5.4) The combining of separate images, using scanners, digital cameras, and advanced graphic computers; software controls the process.

Electronic imaging has become an important part of multimedia.

ergonomics (p. 246, KQ 5.12) Study of the physical and psychological relationships between people and their work environment.

Ergonomic principles are used in designing computers, software, and office furniture to further productivity while avoiding stress, illness, and injuries.

Extended Graphics Array (XGA) (p. 222, KQ 5.7) Graphics board display standard, also referred to as *high resolution, 24-bit color,* or *true color;* supports up to 16.7 million colors at a resolution of 1024 × 768 pixels. Depending on the video display adapter memory chip, XGA will support 256, 65,536, or 16,777,216 colors.

Extended Graphics Array offers the most sophisticated standard for color and resolution.

fax machine (p. 206, KQ 5.4) Short for *facsimile transmission machine;* input device for scanning an image and sending it as electronic signals over telephone lines to a receiving fax machine, which re-creates the image on paper. Fax machines may be dedicated fax machines or fax modems.

Fax machines enable the transmission of text and graphic data over telephone lines quickly and inexpensively.

fax modem (p. 207, KQ 5.4) Modem with fax capability installed as a circuit board inside a computer; it can send and receive electronic signals via telephone lines directly to/from a computer similarly equipped or to/from a dedicated fax machine.

With a fax modem, users can send information much more quickly than they would if they had to feed it page by page through a dedicated fax machine. However, fax modems cannot scan paper documents for faxing.

flatbed plotter (p.228, KQ 5.8) An output device designed so that paper lies flat on a table-like surface while one to four color pens move across the paper, and the paper moves beneath the pens.

Flatbed plotters can produce high-quality graphics such as maps and architectural drawings.

What It Is/What It Does	Why It's Important

flat-panel display (p. 219, KQ 5.7) Refers to display screens that are much thinner, weigh less, and consume less power than CRTs. Flat-panel displays are made up of two plates of glass with a substance between them that is activated in different ways. Three types of substance used are liquid crystal, electroluminescent, and gas plasma. Flat-panel screens are either active-matrix or passive-matrix displays. Images are represented on the screen by individual dots, or picture elements called *pixels.*

Flat-panel displays are used in portable computers and recently became available for desktop PCs.

font (p. 225, KQ 5.8) Set of type characters in a particular type style and size.

Desktop publishing programs, along with laser printers, have enabled users to dress up their printed projects with a choice of many different fonts.

gas-plasma display (p. 220, KQ 5.8) Type of flat-panel display in which the display uses a gas that emits light in the presence of an electric current, like a neon bulb. The technology uses predominantly neon gas and electrodes above and below the gas. When electric current passes between the electrodes, the gas glows.

Gas-plasma displays offer better resolution than LCD displays, but they are more expensive.

hardcopy (p. 218, KQ 5.6) Refers to printed output (as opposed to softcopy). The principal examples are printouts, whether text or graphics, from printers. Film, including microfilm and microfiche, is also considered hardcopy output.

Hardcopy is convenient for people to use and distribute; it can be easily handled or stored.

imaging system (p. 208, KQ 5.4) Also known as *image scanner,* or *graphics scanner;* input device that converts text, drawings, and photographs into digital form that can be stored in a computer system.

Image scanners have enabled users with desktop-publishing software to readily input images into computer systems for manipulation, storage, and output.

impact printer (p. 224, KQ 5.8) Type of printer that forms characters or images by striking a mechanism such as a print hammer or wheel against an inked ribbon, leaving an image on paper.

Dot-matrix printers are the most common impact printers for microcomputer users. For large computers, line printers provide high-speed output.

ink-jet printer (p. 226, KQ 5.8) Nonimpact printer that forms images with little dots. Ink-jet printers spray small, electrically charged droplets of ink from four nozzles through holes in a matrix at high speed onto paper.

Ink-jet printers can print in color and cost much less than color laser printers and are quieter. However, they are slower than laser printers and may have lower resolution (300–720 dots per inch compared to 300–1200 for laser).

input hardware (p. 194, KQ 5.1) Devices that take data and programs that people can read or comprehend and convert them to a form the computer can process. Devices are of three types: keyboards, pointing devices, and source data-entry devices.

Input hardware enables data to be put into computer-processable form.

What It Is/What It Does

joystick (p. 201, KQ 5.3) Pointing device that consists of a vertical handle like a gearshift lever mounted on a base with one or two buttons; it directs a cursor or pointer on the display screen.

keyboard (p. 195, KQ 5.1, 5.2) Input device that converts letters, numbers, and other characters into electrical signals that the computer's processor can "read."

laser printer (p. 225, KQ 5.8) Nonimpact printer similar to a photocopying machine; images are created as dots on a drum, treated with a magnetically charged ink-like toner (powder), and then transferred from drum to paper.

light pen (p. 201, KQ 5.3) Light-sensitive pen-like device connected by a wire to a computer terminal; the user brings the pen to a desired point on the display screen and presses the pen button, which identifies that screen location to the computer.

liquid crystal display (LCD) (p. 220, KQ 5.7) Flat-panel display that consists of a substance called *liquid crystal,* the molecules of which line up in a way that alters their optical properties. As a result, light—usually backlighting behind the screen—is blocked or allowed through to create an image.

magnetic-ink character recognition (MICR) (p. 205, KQ 5.4) Type of scanning technology that reads magnetized-ink characters printed at the bottom of checks and converts them to computer-acceptable digital form.

mouse (p. 199, KQ 5.3) Input device that is rolled about on a desktop to position a cursor or pointer on the computer's display screen, which indicates the area where data may be entered or a command executed.

mouse pointer (p. 199, KQ 5.3) Symbol on the display screen whose movement is directed by moving a mouse on a flat surface, such as a table top.

multifunction device (p. 229, KQ 5.8) Single hardware device that combines several capabilities, such as printing, scanning, copying, and faxing.

Why It's Important

Joysticks are used principally in video games and in some computer-aided design systems.

Keyboards are the most popular kind of input device.

Laser printers produce much better image quality than dot-matrix printers do and can print in many more colors; they are also quieter. Laser printers, along with page description languages, enabled the development of desktop publishing.

Light pens are used by engineers, graphic designers, and illustrators for making drawings.

LCD is useful not only for portable computers and new flat-panel display screens for desktop PCs but also as a display for various electronic devices, such as watches and radios.

MICR technology is used by banks to sort checks.

For many purposes, a mouse is easier to use than a keyboard for communicating commands to a computer. With microcomputers, a mouse is needed to use most graphical user interface programs and to draw illustrations.

The position of the mouse pointer indicates where information may be entered or a command (such as clicking, dragging, or dropping) may be executed. Also, the shape of the pointer may change, indicating a particular function that may be performed at that point.

A multifunction machine can do the work of several separate office machines at a price below the combined cost of buying these devices separately.

What It Is/What It Does	Why It's Important

Musical Instrument Digital Interface (MIDI) board (p. 210, KQ 5.4) Type of audio-input device; add-on circuit board in a computer that provides a standard for the interchange of musical information between musical instruments, synthesizers, and computers, allowing the information to be stored, manipulated, and output.

MIDI boards are used to provide digital input of music for multimedia personal computers.

nonimpact printer (p. 225, KQ 5.8) Printer that forms characters and images without making direct physical contact between printing mechanism and paper. Two types of nonimpact printers often used with microcomputers are laser printers and ink-jet printers. A third kind, the thermal printer, is seen less frequently.

Nonimpact printers are faster and quieter than impact printers because they have fewer moving parts. They can print text, graphics, and color, but they cannot be used to print on multipage forms.

optical card (p. 237, KQ 5.10) Plastic, wallet-type card using laser technology like music compact disks, which can be used to store data.

Because they hold so much data, optical cards have considerable uses, as for a health card holding a person's medical history, including digital images such as X-rays.

optical character recognition (OCR) (p. 206, KQ 5.4) Type of scanning technology that reads preprinted characters in a particular font and converts them to computer-usable form. A common OCR scanning device is the wand reader.

OCR technology is frequently used with utility bills and price tags on department-store merchandise.

optical mark recognition (OMR) (p. 206, KQ 5.4) Type of scanning technology that reads pencil marks and converts them into computer-usable form.

OMR technology is frequently used for grading multiple-choice and true/false tests, such as parts of the College Board Scholastic Aptitude Test.

page description language (p. 226, KQ 5.8) Software used in desktop publishing that describes the shape and position of characters and graphics to the printer.

Page description languages, used along with laser printers, gave birth to desktop publishing. They allow users to combine different types of graphics with text in different fonts, all on the same page.

passive-matrix display (p. 220, KQ 5.7) Type of flat-panel display in which each transistor controls a whole row or column of pixels.

Although passive-matrix displays are less bright and less sharp than active-matrix displays, they are less expensive and use less power.

pen-based computer system (p. 203, KQ 5.3) Input system that uses a pen-like stylus to enter handwriting and marks onto a computer screen.

Pen-based computer systems benefit people who don't know how to or who don't want to type or need to make routinized kinds of inputs such as checkmarks.

pixel (p. 219, KQ 5.7) Short for *picture element;* smallest unit on the screen that can be turned on and off or made different shades. A stream of bits defining the image is sent from the computer (from the CPU) to the CRT's electron gun, where the bits are converted to electrons.

Pixels are the building blocks that allow graphical images to be presented on a display screen.

plotter (p. 228, KQ 5.8) Specialized hardcopy output device designed to produce high-quality graphics in a variety of colors. The two principal kinds of plotters are flatbed and drum.

Plotters are especially useful for creating maps and architectural drawings, although they may also produce less complicated charts and graphs.

pointing device (p. 195, KQ 5.1, 5.3) Input device that controls the position of the cursor or pointer on the screen. Includes mice, trackballs, pointing sticks, touchpads, light pens, digitizing tablets, and pen-based systems.

Pointing devices allow the user to quickly choose from selections rather than having to remember commands, and some pointers allow the user to input drawing or handwriting to the computer.

pointing stick (p. 201, KQ 5.3) Input device used instead of a mouse; looks like a pencil eraser protruding from the keyboard between the G, H, and B keys. The user moves the stick with a forefinger while using the thumb to press buttons located in front of the space bar.

A pointing stick and the related buttons work like a mouse and don't require desktop space to move around, so they are well suited for laptop computers.

point-of-sale (POS) terminal (p. 235, KQ 5.10) Smart terminal used much like a cash register.

POS terminals record customer transactions at the point of sale but also store data for billing and inventory purposes.

printer (p. 223, KQ 5.8) Output device that prints characters, symbols, and perhaps graphics on paper or another medium. Printers are categorized according to whether the image produced is formed by physical contact of the print mechanism with the paper. Impact printers have contact; nonimpact printers do not.

Printers provide one of the principal forms of computer output.

puck (p. 202, KQ 5.3) Copying device with which the user of a digitizing table may copy an image.

With a puck, users may "trace" (copy) a drawing and store it in digitized form.

radio-frequency identification technology (p. 213, KQ 5.5) Also called *RF-ID tagging;* a source data-entry technology that uses a tag containing a microchip with code numbers that can be read by the radio waves of a scanner linked to a database.

The radio waves that read RF-ID tags don't need line-of-sight contact with the tag. Thus the tags are being used, for example, as tollbooth passes that can be read without stopping traffic.

refresh rate (p. 221, KQ 5.7) Number of times per second that screen pixels are recharged so that their glow remains bright.

In dual-scan screens, the tops and bottoms of the screens are refreshed independently at twice the rate of single-scan screens, producing more clarity and richer colors.

repetitive stress (strain) injuries (RSI) (p. 243, KQ 5.12) Several kinds of wrist, hand, arm, and neck injuries that can result when muscle groups are forced through fast, repetitive motions.

Computer users may suffer RSIs such as muscle strain and tendinitis, which are not disabling, or carpal tunnel syndrome, which is.

What It Is/What It Does	Why It's Important

resolution (p. 221, KQ 5.7) Clarity or sharpness of a display screen; the more pixels there are per square inch, the finer the level of detail. Resolution is expressed in terms of the formula *horizontal pixels × vertical pixels*. A screen with 640 × 480 pixels multiplied together equals 307,200 pixels. This screen will be less clear and sharp than a screen with 800 × 600 (equals 480,000) or 1024 × 768 (equals 786,432) pixels.

Users need to know what screen resolution is appropriate for their purposes.

robot (p. 234, KQ 5.9) Automatic device that performs functions ordinarily ascribed to human beings or that operate with what appears to be almost human intelligence.

Robots are of several kinds—industrial robots, perception systems, and mobile robots. They are performing more and more functions in business and the professions.

scanners (p. 204, KQ 5.4) Input devices that use laser beams and reflected light to translate images such as optical marks, text, drawings, and photos into digital form.

Scanning devices—bar-code readers, mark- and character-recognition devices, fax machines, imaging systems—simplify the input of complex data.

sensor (p. 213, KQ 5.4) Type of input device that collects specific kinds of data directly from the environment and transmits it to a computer.

Sensors can be used for detecting speed, movement, weight, pressure, temperature, humidity, wind, current, fog, gas, smoke, light, shapes, images, and so on.

set-top box (p. 199, KQ 5.2) Input device that works with a keypad, such as a handheld wireless remote control, to allow cable-TV viewers to change channels, or, in the case of interactive systems, to exercise other commands.

If used with PC/TVs and incorporating digital signals, such a device could offer Internet access, electronic mail, digital and interactive TV, and home banking.

simulator (p. 233, KQ 5.9) Device that represents the behavior of a physical or abstract system. Virtual-reality simulators are now being used.

Simulators allow people to train for tasks they need to perform that would be dangerous or costly to actually do.

smart card (p. 236, KQ 5.10) Wallet-type card containing a microprocessor and memory chip that can be used to store data and transfer data to and from a computer.

In many countries, telephone users may buy a smart card that lets them make telephone calls until the total cost limit programmed into the card has been reached.

softcopy (p. 218, KQ 5.6) Refers to data that is shown on a display screen or is in audio or voice form. This kind of output is not tangible; it cannot be touched. Virtual reality and robots might also be considered softcopy devices.

This term is used to distinguish nonprinted output from printed output.

sound-output device (p. 231, KQ 5.9) Output device that produces digitized, nonverbal sounds, ranging from beeps and chirps to music. It includes software and a sound card or digital audio circuit board.

PC owners can customize their machines to greet each new program with particular sounds. Sound output is also used in multimedia presentations.

What It Is/What It Does	**Why It's Important**

source data-entry device (p. 195, KQ 5.1, 5.4, 5.5) Also called *source-data automation;* device other than keyboard or pointer that creates machine-readable data on magnetic media or paper or feeds it directly into the computer's processor. The category includes scanning devices; sensors; voice-recognition devices; audio-input devices; video-input devices; electronic cameras; and human-biology input devices.

Source data-entry devices lessen reliance on keyboards for data entry and can make data entry more accurate. Some also enable users to draw graphics on screen and create other effects not possible with a keyboard.

stylus (p. 202, KQ 5.3) Pen-like device with which the user of a digitizing tablet "sketches" an image.

With a stylus, users can achieve artistic effects similar to those achieved with pen or pencil.

Super Video Graphics Array (SVGA) (p. 222, KQ 5.7) Graphics board display standard that supports 256 colors at higher resolution than VGA. SVGA has two graphics modes: 800 × 600 pixels and 1024 × 768. Also called *8-bit color.*

Super VGA is a higher-resolution version of Video Graphics Array (VGA), introduced in 1987.

terminal (p. 235, KQ 5.10) Input and output device that uses a keyboard for input, a monitor for output, and a communications line to a main computer system.

A terminal is generally used to input data to, and receive visual data from, a mainframe computer system. An intelligent terminal also has processing capability and RAM.

thermal printer (p. 227, KQ 5.8) Nonimpact printer that uses colored waxes and heat to produce images by burning dots onto special paper.

The colored wax sheets are not required for black-and-white output because the thermal print head will register the dots on the paper.

touchpad (p. 201, KQ 5.3) Input device used instead of a mouse; the user slides a finger over the small, flat surface to move the cursor on the display screen, and taps a finger on the pad's surface or presses buttons positioned nearby to perform the same tasks done with the mouse buttons.

Touchpads fit readily on a laptop computer and give the user a pointing device that is convenient to use in a confined space, although they can be more difficult to control than a mouse.

touch screen (p. 238, KQ 5.10) Video display screen that has been sensitized to receive input from the touch of a finger. It is often used in automated teller machines and in directories conveying tourist information.

Because touch screens are easy to use, they can convey information quickly and can be used by people with no computer training; however, the amount of information offered is usually limited.

trackball (p. 201, KQ 5.3) Input device used instead of a mouse; a movable ball on top of a stationary device is rotated with the fingers or palm of the hand to move the cursor or pointer on the display screen, and buttons are positioned nearby.

Unlike a mouse, a trackball is especially suited to portable computers, which are often used in confined places.

videoconferencing (p. 232, KQ 5.9) A method of communicating whereby people in different geographical locations can have a meeting—and see and hear one another—using computers and communications technologies.

Videoconferencing technology enables people to conduct business meetings without having to travel.

What It Is/What It Does

video display adapter (p. 222, KQ 5.7) Also called a *video graphics card;* circuit board that contains its own video RAM and determines the resolution, number of colors, and how fast images appear on the display screen.

video display terminal (VDT) (p. 218, KQ 5.7) Computer keyboard and display screen.

Video Graphics Array (VGA) (p. 222, KQ 5.7) Graphics board display standard that supports 16 to 256 colors, depending on resolution. At 320 × 200 pixels it will support 256 colors; at the sharper resolution of 640 × 480 pixels it will support 16 colors. Also called *16-bit color.*

virtual reality (p. 232, KQ 5.9) Computer-generated artificial reality that projects user into sensation of three-dimensional space. Interactive sensory equipment consists of headgear, glove, and software. The headgear has small video display screens, one for each eye, to create a three-dimensional sense, and headphones to pipe in stereo or 3-D sound. The glove has sensors that collect data about hand movements. The software gives the wearer the interactive sensory experience.

voice-output device (p. 231, KQ 5.9) Output device that converts digital data into speech-like sounds.

voice-recognition system (p. 214, KQ 5.5) Input system that converts a person's speech into digital code; the system compares the electrical patterns produced by the speaker's voice with a set of prerecorded patterns stored in the computer.

Why It's Important

Video display adapters determine how fast the card processes images and how many colors it can display.

Video display terminals are the principal input/output devices for accessing large computer systems such as mainframes.

VGA and SVGA are the most common video standards used today.

Virtual reality is used most in entertainment, as in arcade-type games, but has applications in architectural design and training simulators.

Voice-output devices are a common technology, found in telephone systems, soft-drink machines, and toys and games.

Voice-recognition technology is useful for inputting data in situations in which people are unable to use their hands or need their hands free for other purposes.

Exercises

Self-Test

1. A(n) _____ terminal is entirely dependent for all of its processing activities on the computer system to which it is hooked up.

2. _____ is the study of the physical relationships between people and their work environment.

3. A(n) _____ is an input device that is rolled about on a desktop and directs a pointer on the computer's display screen.

4. A device that translates images of text, drawings, photos, and the like into digital form is called a(n) _____.

5. The smallest unit on the screen that can be turned on and off is called a(n) _____.

Short-Answer Questions

1. What does a voice-recognition system do?
2. What three characteristics determine the clarity of a computer screen?
3. What is the main difference between a dedicated fax machine and a fax modem?
4. What is an imaging system?
5. What is a terminal?

Multiple-Choice Questions

1. Which of the following characteristics affect how bright images appear on a computer screen?
   a. resolution
   b. dot pitch
   c. refresh rate
   d. screen size
   e. All of the above

2. Which of the following should you consider using if you have to work in a confined space?
   a. mouse
   b. digitizing tablet
   c. trackball
   d. keyboard
   e. None of the above

3. Which of the following should you purchase if you have a limited amount of space in your office but need to be able to print, make photocopies, scan images, and fax documents?
   a. printer
   b. fax machine
   c. scanner
   d. multifunction device
   e. All of the above

4. A _____ looks like a credit card but contains a microprocessor and memory chip.
   a. sound card
   b. smart card
   c. touch card
   d. point-of-sale card
   e. None of the above

5. Which of the following gathers information from the environment and then transmits it to a computer?
   a. human-biology input device
   b. digital camera
   c. display screen
   d. sensor
   e. None of the above

True/False Questions

T F 1. Photos taken with a digital camera can be downloaded to a computer's hard drive.

T F 2. On a computer screen, the more pixels that appear per square inch, the higher the resolution.

T F 3. In your future, you will likely see better and cheaper display screens.

T F 4. A Touch-Tone telephone can be used like a dumb terminal to send data to a computer.

T F 5. Terminals are either dumb or intelligent.

Knowledge in Action

1. If you could buy any printer you want, what type (make, model, etc.) would you choose? Does the printer need to fit into a small space? Does it need to print across the width of wide paper (11 × 14 inches)? In color? On multicarbon forms? How much printer RAM would you need? Review some of the current computer publications for articles or advertisements relating to printers. How much does the printer cost? Your needs should be able to justify the cost of the printer (if necessary, make up what your needs might be).

2. What uses can you imagine for voice output and/or sound output in your planned job or profession?

3. *Paperless office* is a term that has been appearing in computer-related journals and books for over 5 years. However, the paperless office has not yet been achieved. Do you think the paperless office is a good idea? Do you think it's possible? Why do you think it has not yet been achieved?

4. Using computer magazines and periodicals and/or the Internet, conduct additional research on one of the newer technologies described in the section at the end of this chapter describing the future of input and output. Who will use this technology? For what? When will this technology be available? Who is developing this technology and who is providing the funding?

5. Research the current uses of smart card technology and how companies hope to implement smart card technology in the future. Will smart cards have an effect on shopping over the Internet? The transportation industry? What other industries might be affected? Conduct your research using current periodicals and/or the Internet.

Answers

Self-Test Questions
1. dumb   2. ergonomics   3. mouse   4. scanner   5. pixel

Short-Answer Questions
1. A voice recognition system converts a person's speech into digital code by comparing the electrical patterns produced by the speaker's voice with a set of prerecorded patterns stored in the computer.   2. resolution, dot pitch, refresh rate   3. A dedicated fax machine is a standalone device (outside the computer), whereas a fax modem is typically installed inside a computer's system unit.   4. An imaging system uses an image or graphics scanner to convert text and images into a digital form that can then be stored and further edited on the computer.   5. A terminal is an input/output device that uses a keyboard for input and a monitor for output.

Multiple-Choice Questions
1. c   2. c   3. d   4. b   5. d

True/False Questions
1. T   2. T   3. T   4. T   5. T

Storage

Foundations for Interactivity, Multimedia, & Knowledge

key questions

You should be able to answer the following questions:

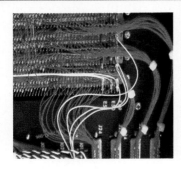

Entertainment, suggests Keith McCurdy, the director for North America of the video game company Electronic Arts.[1] Once it was word processing and spreadsheets that were the "killer apps"—the application programs that helped fuel the personal computer market's growth and technological advancement. Now it is advancements in video games—better performance, creativity, and animated realism—that are spinning off business applications, such as animated graphics and other multimedia for corporate presentations.

"The CD-ROM market was first embraced and, as a result, ushered in by video game software," says McCurdy. "While business applications normally [had] little need to store software on a huge 650-megabyte CD-ROM, it is just such a CD-ROM that is essential for video games and their vast array of full-screen animation, live-video footage, digital music, and multiple levels of game play."

The engine of entertainment has made "interactivity" and "multimedia" among the most overused marketing words of recent times. *Interactivity* refers to the user's back-and-forth interaction with a computer program. In other words, the user's actions and choices affect what the software does. *Multimedia* refers to the use of a variety of media—text, sound, video—to deliver information, whether on a computer disk or via the anticipated union of computers, telecommunications, and broadcasting technologies.

Interactive programs and multimedia programs require machines that can handle and store enormous amounts of data. For this reason, the capacities of storage hardware, called *secondary storage* or just *storage*, seem to be increasing almost daily. This chapter covers the most common types of secondary storage, as well as their uses. First, however, we need to go over the fundamentals that relate to the whole topic of secondary storage.

6.1 Storage Fundamentals

KEY QUESTION

What are the units of storage measurement and data access methods?

Preview & Review: Storage capacity is measured in multiples of bytes. In addition, there are two main types of data access.

As you learned previously (✔ pp. 16, 164), when you are, for example, word processing a document, the data you are working on is stored in RAM (primary storage) during processing. Because that data in RAM is an electrical state, when you turn off the power to your computer, the data there disappears. Therefore, before you turn your computer off, you must save your work onto a storage device that stores data permanently (until it is changed or erased), such as a diskette or a hard disk. When saved to a secondary-storage device, your data will remain intact even when the computer is turned off.

In addition to data, computer software programs must be stored in a computer-usable form. A copy of software instructions must be retrieved from a secondary-storage device and placed into RAM before processing can begin. The computer's operating system determines where and how programs are stored on the secondary-storage devices.

In very general terms, a secondary-storage device can be thought of as a file cabinet. You store data there until you need it. Then you open the drawer, take out the appropriate folder (file), and place it on the top of your desk (in primary storage, or RAM), where you work on it—perhaps writing a few things in it or throwing away a few pages. In the case of electronic documents on a computer, however, you are actually taking out a *copy* of the desired file and putting it on the "desktop." An old version of the file remains in the file cabinet (secondary storage) while the copy of the file is being edited or updated on the desktop (in RAM). When you are finished with the file, you take it off the desktop (out of primary storage) and return it to the cabinet (secondary storage). Thus, the updated file replaces the old file.

Units of Measurement for Storage

Nearly every recent development in special effects in movie making—from the stampeding dinosaurs of *Jurassic Park* to the digital water of *Titanic* and *Deep Impact* to the 20-story high reptile in *Godzilla*—has been made possible because of great leaps in secondary-storage technology. *Jurassic Park*, for example, contains about 100 billion bytes, or 100 gigabytes. To get a sense of how big a gigabyte is, suppose that, starting at age 10, you rigorously kept a journal consisting of two double-spaced typewritten pages a day. By the time you were close to 96 years old, you would have a gigabyte's worth of data—a manuscript of 62,500 pages, which would be 21 feet tall if stacked.[2]

We explained the meaning of gigabytes and other measurements in the discussion of processing hardware (✔ p. 155). The same terms are also used to measure the data capacity of storage devices. To repeat:

- **Kilobyte:** A *kilobyte* (abbreviated K or KB) is equivalent to 1024 bytes.
- **Megabyte:** A *megabyte* (abbreviated M or MB) is about 1 million bytes.
- **Gigabyte:** A *gigabyte* (G or GB) is about 1 billion bytes.
- **Terabyte:** A *terabyte* (T or TB) is about 1 trillion bytes.

The amount of data being held in a file in your personal computer might be expressed in kilobytes or megabytes. The amount of data being stored in a remote database accessible to you over a communications line might be expressed in gigabytes or terabytes.

Data Access Methods

Before we move on to discuss individual secondary-storage devices, we need to mention one more aspect of storage fundamentals—data access storage methods. The way that a secondary-storage device allows access to the data stored on it affects its speed and its usefulness for certain applications. The two main types of data access are sequential and direct.

- Sequential Storage: *Sequential storage* **means that data is stored in sequence,** such as alphabetically. Tape storage falls in the category of sequential storage. Thus, you would have to search a tape past all the information from A to J, for example, before you got to K. Or, if you are looking for employee number 8888, the computer will have to start with 0001, then go past 0002, 0003, and so on, until it finally comes to 8888. This data access method is less expensive than other methods because it uses magnetic tape, which is cheaper than disks. The disadvantage of sequential file organization is that searching for data is slow.

- Direct Access Storage: ***Direct access storage* means that the computer can go directly to the information you want.** The data is retrieved (accessed) according to a unique data identifier called a *key field* (covered in Chapter 9). This method of file organization is used with hard disks and other types of disks. It is ideal for applications such as airline reservation systems or computer-based directory-assistance operations. In these cases, there is no fixed pattern to the requests for data.

    Direct file access is much faster than sequential access for finding specific data. However, because the method requires hard disk or other type of disk storage, it is more expensive to use than sequential access (magnetic tape).

Now that we have covered some secondary-storage fundamentals, we turn to the topic of rating secondary-storage devices and discussing the ones you are likely to encounter. Then, at the end of the chapter, we discuss the important topic of data compression, as well as the future of storage devices and some ethical concerns.

6.2 Criteria for Rating Secondary-Storage Devices

KEY QUESTION

What six factors can be used to distinguish secondary-storage devices?

Preview & Review: Storage capacity, access speed, transfer rate, size, removability, and cost are all factors in rating secondary-storage devices.

To evaluate the various devices, ranging from disk to flash-memory cards to tape to online storage, it's helpful to have the following information:

- Storage capacity: As we mentioned earlier, high-capacity storage devices are desirable or required for many sophisticated programs and large databases. However, as capacity increases, so does price. Some users find compression software to be an economical solution to storage-capacity problems. Hard disks can store more data than diskettes, and optical disks can store more than hard disks.

- Access speed: *Access speed* refers to the average time needed to locate data on a secondary-storage device. Access speed is measured in milliseconds (thousandths of a second). Hard disks are faster than optical disks, which are faster than diskettes. Disks are faster than magnetic tape. However, the slower media are more economical.

- Transfer rate: *Transfer rate* refers to the speed at which data is transferred from secondary storage to main memory (primary storage). It is measured in megabytes per second.

- Size: Some situations require compact storage devices (for portability); others don't. Users need to know what their options are.

- Removability: Storage devices that are sealed within the computer may suffice for some users; other users may also need removable storage media.

- Cost: As we have indicated, the cost of a storage device is directly related to the previous five factors.

Now let's take an in-depth look at the following types of secondary-storage devices:

- Diskettes
- Hard disks
- Optical disks
- Flash-memory cards
- Magnetic tape
- Online storage

6.3 Diskettes

KEY QUESTION

What are diskettes, their principal features, and rules for taking care of them?

Preview & Review: Diskettes are round pieces of flat plastic that store data and programs as magnetized spots. A disk drive copies, or reads, data from the disk and writes, or records, data to the disk.

Components of a diskette include tracks and sectors; diskettes come in various densities. All have write-protect features.

Care must be taken to avoid destroying data on disks, and users are advised to back up, or duplicate, the data on their disks.

A *diskette,* or *floppy disk,* **is a removable round, flat piece of mylar plastic 3½ inches across that stores data and programs as magnetized spots.** More specifically, data is stored as electromagnetic charges on a metal oxide film that coats the mylar plastic. Data is represented by the presence or absence of these electromagnetic charges, following standard patterns of data representation (such as ASCII, ✔ p. 156). The disk is contained in a plastic case to protect it from being touched by human hands. It is called "floppy" because the disk within the case is flexible, not rigid.

The Disk Drive

To use a diskette, you need a disk drive. **A *disk drive* is a device that holds, spins, and reads data from and writes data to a diskette.**

The words *read* and *write* (✔ p. 17) have exact meanings. To review:

- Read: ***Read* means that the data represented on the secondary-storage medium is converted to electronic signals and transmitted to primary storage (RAM).** That is, *read* means that data is *copied from* the diskette, disk, or other type of storage.
- Write: ***Write* means that the electronic information processed by the computer is transferred to a secondary-storage medium.**

Disk drive

Disk drives are usually located inside the computer's system cabinet.

How a Disk Drive Works

A diskette is inserted into a slot, called the *drive gate* or *drive door,* in the front of the disk drive. This clamps the diskette in place over the spindle of the drive mechanism so the drive can operate. An access light goes on when the disk is in use. After using the disk, you can retrieve it by pressing an eject button beside the drive. (Note: *Do not remove the disk when the access light is on.*)

The device by which the data on a disk is transferred to the computer, and from the computer to the disk, is the disk drive's *read/write head.* The

How it works:

1. When a diskette is inserted into the drive, it presses against a system of levers. One lever opens the metal plate, or shutter, to expose the data access area.

2. Other levers and gears move two read/write heads until they almost touch the diskette on both sides.

3. The drive's circuit board receives signals, including data and instructions for reading/ writing that data from/to disk, from the drive's controller board. The circuit board translates the instructions into signals that control the movement of the disk and the read/write heads.

4. A motor located beneath the disk spins a shaft that engages a notch on the hub of the disk, causing the disk to spin.

5. When the heads are in the correct position, electrical impulses create a magnetic field in one of the heads to write data to either the top or bottom surface of the disk. When the heads are reading data, they react to magnetic fields generated by the metallic particles on the disk.

■ PANEL 6.1
Cutaway view of a disk drive

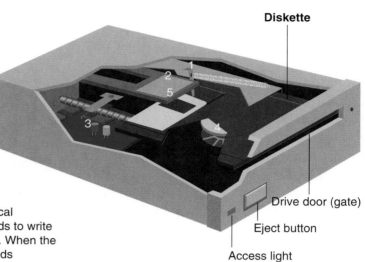

Diskette

Drive door (gate)

Eject button

Access light

diskette spins inside its case, and the read/write head moves back and forth over the *data access area,* which is under the diskette's metal protective plate. Disk drive operations are illustrated above. *(■ See Panel 6.1.)*

Characteristics of Diskettes

Diskettes have the following characteristics *(■ see Panel 6.2):*

● **Tracks and sectors:** On a diskette, **data is recorded in rings called tracks.** Unlike on a phonograph record, these tracks are neither visible grooves nor a single spiral. Rather, they are closed concentric rings. *(■ Refer to Panel 6.2 again.)*

Each track is divided into sectors. *Sectors* **are invisible wedge-shaped sections used for storage reference purposes.** The number of sectors on the diskette varies according to the recording density—the number of bits per inch (as we describe below under "Data capacity"). Each sector typically holds 512 bytes of data.

When you save data from your computer to a diskette, the data is distributed by tracks and sectors on the disk. That is, the system software uses the point at which a sector intersects a track to reference the data location in order to spin the disk and position the read/write head.

● **Unformatted versus formatted disks:** When you buy a new box of diskettes to use for storing data, the box may state that it is "unformatted" (or say nothing at all). This means you have a task to perform before you can use the disks with your computer and disk drive. **Unformatted disks are manufactured without tracks and sectors in place.** *Formatting*—or *initializing,* **as it is called on the Macintosh—means that you must prepare the disk for use so that the**

3¹/₂-inch diskette

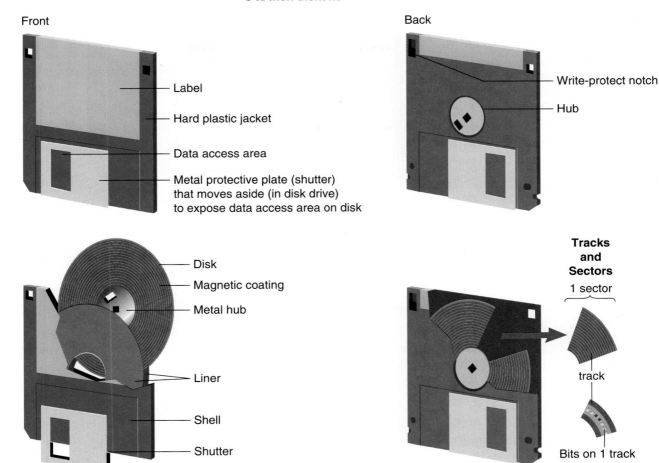

Front

— Label
— Hard plastic jacket
— Data access area
— Metal protective plate (shutter) that moves aside (in disk drive) to expose data access area on disk

Back

— Write-protect notch
— Hub

— Disk
— Magnetic coating
— Metal hub
— Liner
— Shell
— Shutter

Tracks and Sectors

1 sector

track

Bits on 1 track

■ PANEL 6.2

Diskette anatomy

operating system can write information on it. This includes defining the tracks and sectors on it. Formatting is done quickly by using a few simple software commands (described in the user manual that comes with your computer).

Alternatively, when you buy a new box of diskettes, the box may state that they are "formatted IBM" or "formatted Macintosh." This means that you can simply insert a disk into the drive gate of your PC or Macintosh and use it immediately.

- Data capacity—sides and densities: Not all disks hold the same amount of data. Diskettes are ***double-sided, called "DS" or "2," capable of storing data on both sides.*** However, a disk's capacity depends on its recording density. ***Recording density refers to the number of bits per inch that can be written onto the surface of the disk.*** Thus, diskettes are either *high-density (HD)* or *extended density (ED).* A high-density 3½-inch diskette can store 1.44 megabytes. An extended density diskette can store 2.8 megabytes. (Note: You need an ED drive to use ED diskettes; an ED drive will also accept HD disks.)

- Write-protect features: The ***write-protect feature allows you to protect a diskette from being written to,*** which would replace data on the disk. To write-protect a diskette, you press a slide lever toward

■ PANEL 6.3

Write-protect features

For data to be written to this diskette, a small piece of plastic must be closed over the tiny window on one side of the disk. To protect the disk from being written to, you must open the window. (Using the tip of a pen helps.)

Writable

Write-protect
window closed

Write-protected

Write-protect
window open

the edge of the disk, uncovering a hole (which appears on the lower right side, viewed from the back). (■ *See Panel 6.3.*)

Taking Care of Diskettes

Diskettes need at least the same amount of care that you would give to an audiotape or music CD. In fact, they need more care than that if you are dealing with difficult-to-replace data or programs.

There are a number of rules for taking care of diskettes:

- **Don't touch disk surfaces:** Don't touch anything visible through the protective case, such as the data access area. Don't manipulate the metal shutter.

- **Handle disks gently:** Don't try to bend diskettes or put heavy weights on them.

- **Avoid risky physical environments:** Disks don't do well in sun or heat (such as in glove compartments or on top of steam radiators). They should not be placed near magnetic fields (including those created by nearby telephones or electric motors). They also should not be exposed to chemicals (such as cleaning solvents) or spilled coffee or alcohol.

- **Don't leave the diskette in the drive:** Take the diskette out of the drive when you're done. If you leave the diskette in the drive, the read/write head remains resting on the diskette surface.

More suggestions for taking care of diskettes are given at the end of Chapter 5. (See the Experience Box, "Good Habits: Protecting Your Computer System, Your Data, & Your Health.")

6.4 Hard Disks

KEY QUESTION

What are hard disks and their various features and options?

Preview & Review: Hard disks are rigid metal platters that hold data as magnetized spots.

Usually a microcomputer hard-disk drive is located inside the system unit, but external hard-disk drives are available, as are removable hard-disk cartridges.

Large computers use removable-pack hard disk systems, fixed-disk drives, or RAID storage systems.

Hard disks **are thin but rigid metal platters covered with a substance that allows data to be held in the form of magnetized spots.** Hard disks are tightly sealed within an enclosed hard-disk-drive unit to prevent any foreign matter from getting inside. Data may be recorded on both sides of the disk platters.

When users with the early microcomputers that had only diskette drives switched to machines with hard disks, they found the difference was like moving a household all at once with an enormous moving van instead of doing it in several trips in a small sportscar. Whereas a high-density 3½-inch diskette holds 1.44 megabytes, for a couple of hundred dollars you can easily get a hard-disk drive that holds 2–10 (or more) gigabytes.

At first you might think you would never be able to find enough programs and data to fill up your microcomputer's 2-gigabyte hard disk. Ultimately, though, you'll probably find it's not capacious enough—especially if you're using space-gobbling graphics-oriented or multimedia programs. "Hard disks on new PCs are like closets in new homes," says computer columnist Lawrence Magid. "When you first move in, there's a ton of empty space. But after you've lived there awhile, you start to feel cramped."[3]

But hard-disk capacities for microcomputers have been making incredible jumps in very short periods of time. Examples:

1993—Average microcomputer hard disk is 200 megabytes.

1996—Average is 1.2 gigabytes.

July 1997—IBM introduces an 8.4-gigabyte drive.

October 1997—Quantum introduces a 12-gigabyte drive.

November 1997—IBM introduces a 16.8-gigabyte drive.

A company called TeraStor recently announced it had the technology to create a disk that would hold 20 gigabytes of data on each side. The cost of data storage on disk drives has also plummeted, dropping from $5.23 a megabyte in 1991 to about 10 cents a megabyte in 1997.[4]

We'll now describe the following aspects of hard-disk technology:

- Microcomputer hard-disk drives
- Hard-disk connections—SCSI and EIDE
- Defragmentation to speed up hard disks
- Virtual memory
- Microcomputer hard-disk variations
- Hard-disk technology for large computer systems

Microcomputer Hard-Disk Drives

In microcomputers, *hard disks* are one or more platters sealed inside a hard-disk drive that is located inside the system unit and cannot be removed. The drive is installed in a drive bay, a shelf or opening in the computer cabinet. The disks may be 5¼ inches in diameter, although today they are more often 3½ inches, with some even smaller.

From the outside of a microcomputer, a hard-disk drive is not visible; it is hidden behind a front panel on the system cabinet. Inside, however, is a disk mechanism the size of a small sandwich containing disk platters on a drive spindle, read/write heads mounted on an actuator (access) arm that moves back and forth, and power connections and circuitry. (■ *See Panel 6.4.*) The operation is much the same as for a diskette drive, with the read/write heads locating specific pieces of data according to track and sector.

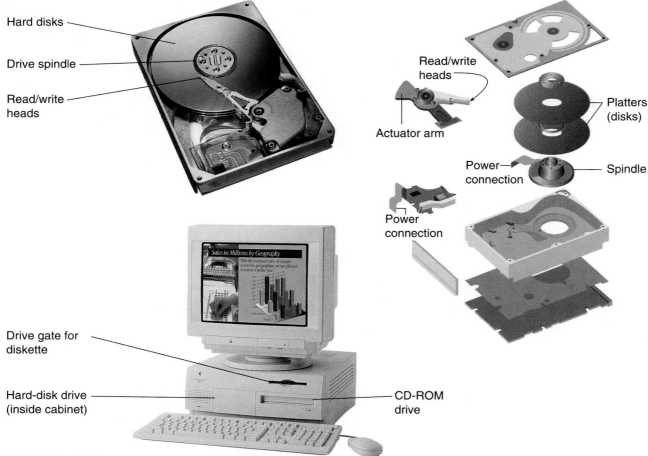

Hard disks

Drive spindle

Read/write heads

Read/write heads

Actuator arm

Platters (disks)

Spindle

Power connection

Power connection

Drive gate for diskette

Hard-disk drive (inside cabinet)

CD-ROM drive

■ PANEL 6.4

Microcomputer internal hard-disk drive

(Top left) A hard-disk drive that has been removed from the system cabinet. *(Middle right)* Anatomy of a hard-disk drive. *(Bottom left)* The hard-disk drive is sealed inside the system cabinet and is not accessible. The drive gate is for inserting a diskette.

Hard disks have a couple of real advantages over diskettes—and at least one significant disadvantage.

- **Advantages—capacity and speed:** We mentioned that hard disks have a data storage capacity that is significantly greater than that of diskettes. Microcomputer hard-disk drives now typically hold 2–4 gigabytes, and newer ones are 6–9 gigabytes.

  As for speed, hard disks allow faster access to data than do diskettes because a hard disk spins several times faster than a diskette. For instance, a 2.1-gigabyte hard disk will typically spin at 5600–7800 revolutions per minute (rpm), compared to 360 rpm for a diskette drive.

- **Disadvantage—possible "head crash":** In principle a hard disk is quite a sensitive device. The read/write head does not actually touch the disk but rather rides on a cushion of air about 0.000001 inch thick. The disk is sealed from impurities within a container, and the whole apparatus is manufactured under sterile conditions. Otherwise, all it would take is a smoke particle, a human hair, or a fingerprint to cause what is called a *head crash*.

  A *head crash* **happens when the surface of the read/write head or particles on its surface come into contact with the disk surface, causing the loss of some or all of the data on the disk.** This can also happen when you bump a computer too hard or drop something heavy on the system cabinet. An incident of this sort could, of course, be a disaster if the data has not been backed up. There are firms that specialize in trying to retrieve (for a hefty price) data from crashed hard disks, though this cannot always be done. (See the ReadMe box, page 127, in Chapter 3, "What to Do If the Disk with the Only Copy of Your Novel Fails.")

In recent years, computer magazines have evaluated the durability of portable computers containing hard disks by submitting them to drop tests. Most of the newer machines are surprisingly hardy. However, with hard disks—whether in portable or in desktop computers—the possibility of disk failure always exists.

Hard-Disk Connections: SCSI & EIDE

"I'm buying a second hard disk this weekend, I think," the letter writer wrote computer columnist Gina Smith. "What's better? A SCSI drive or an EIDE drive? And just what is the difference? I'd like to know before I go to the store."[5]

Replied Smith: "You'd *better* know before you go to the store. Though it isn't true of all salespeople, some hold the mistaken impression that a SCSI type drive is faster than an EIDE type drive. And that simply isn't so."

SCSI and EIDE are simply terms that describe two kinds of technological connections by which a hard disk is attached to a microcomputer.

- **SCSI:** ***SCSI* (for *small computer system interface*), pronounced "scuzzy," allows you to plug a number of peripheral devices into a single expansion board in your computer by linking these devices end to end in a sort of daisy chain** (✔ p. 170). Thus, you could link between 7 and 15 peripheral devices, such as a hard disk plus a tape drive plus a scanner.

  SCSI is the drive-interface connection used on the Macintosh computer and higher-end PCs, including multimedia workstations and network servers.

- **EIDE:** ***EIDE* (for *enhanced integrated drive electronics*) connects hard drives to a microcomputer by using a flat ribbon cable attached to an expansion board—called a host adapter—that plugs into an expansion slot on the motherboard.** EIDE is popular because of its low cost, and it is increasingly being used to connect CD-ROM drives and tape drives. An inexpensive EIDE host adapter can control two to four hard drives (one of which may be a CD-ROM or optical disk drive), two diskette drives, two serial ports, a parallel port, and a game port.

Fragmentation & Defragmentation: Speeding Up Slow-Running Hard Disks

Like diskettes, for addressing purposes hard disks are divided into a number of tracks and typically nine invisible pie-shaped sectors. Data is stored within the tracks and sectors in groups of clusters. **A *cluster* is the smallest storage unit the computer can access, and it always refers to a number of sectors,** usually two to eight. (Among other things, cluster size depends on the operating system.)

With a brand-new hard disk, the computer will try to place the data in clusters that are contiguous—that is, that are adjacent (next to one another). Thus, data would be stored on track 1 in sectors 1, 2, 3, 4, and so on. However, as data files are updated and the disk fills up, the operating system stores data in whatever free space is available. Thus, files become fragmented.

***Fragmentation* means that a data file becomes spread out across the hard disk in many noncontiguous clusters.**

Fragmented files cause the read/write head to go through extra movements to find data, thus slowing access to the data. This means that the computer runs more slowly than it would if all the data elements in each file were stored in contiguous locations. To speed up the disk access, you must defragment the disk. *Defragmentation* means that data on the hard disk is reorganized so that data in each file is stored in contiguous clusters. Programs for defragmenting are available on some operating systems or as separate (external) software utilities (✔ p. 126).

Virtual Memory

Sometimes the computer uses hard-disk space called *virtual memory* to expand RAM. When RAM space is limited, the use of virtual memory can let users run more software at once, if the computer's CPU and operating system are equipped to use it. The system does this by using some free hard-disk space as an extension of RAM—that is, the computer *swaps* parts of the program and/or data between the hard disk and RAM as needed.

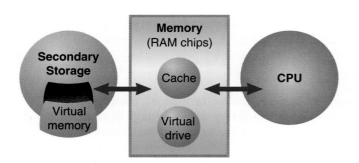

For example, when doing this book's page makeup, we used the Mac PowerPC 9500's virtual memory function to be able to run Quark Xpress desktop-publishing software at the same time as Adobe PhotoShop, which was used to manipulate photos inserted on the pages. Note that virtual memory can make the computer run more slowly, as compared to cache memory (✔ p. 165), which makes the computer run faster.

Microcomputer Hard-Disk Variations: Power & Portability

If you have an older microcomputer or one with limited hard-disk capacity, some variations are available that can provide additional power or portability:

- **Miniaturization:** Newer hard-disk drives are less than half the height of older drives (1½ inches versus 3½ inches high) and so are called *half-height drives.* Thus, you could fit two disk drives into the bay in the system cabinet formerly occupied by one.

  In addition, the diameter of the disks has been getting smaller. Instead of 3½ inches, some platters are as small as 1 inch in diameter.

- **External hard-disk drives:** If you don't have room in the system unit for another internal hard disk but need additional storage, consider adding an external hard-disk drive. Some detached external hard-disk drives, which have their own power supply and are not built into the system cabinet, can store gigabytes of data.

- **Removable hard disks or cartridge systems:** "I LOVE MY ZIP DRIVE!!!!" exclaims Craig Clarke of Elma, New York, in an online message.[6] Why the excitement? Because the first law of computing, says one technology journalist, is "You can never have enough disk space."[7] Today's software takes up so much space (some programs can take up to 100 megabytes of hard-disk space) that it reduces the amount of room on which you can store data files. This is particularly the case for owners of portable computers.

Instead of deleting useful files from your hard disk or adding a second hard-disk drive, you can use removable hard disks or cartridge systems that hold far more data than conventional diskettes. One of these, the kind enthusiast Clarke likes, is Iomega's Zip drive, whose cartridges each store 100 megabytes. Other popular hard-disk cartridge systems are the Avatar Shark drive (250-megabyte cartridges), the SyQuest SyJet drive (1-gigabyte cartridges), and the Iomega Jaz drive (2-gigabyte cartridges).

Hard-disk cartridges, or removable hard disks, consist of one or two platters enclosed along with read/write heads in a hard plastic case. The case is inserted into an internal or detached external cartridge drive connected to a microcomputer. (■ *See Panel 6.5.*) The drives are available in several configurations: EIDE, SCSI, or for parallel ports. A cartridge, which is removable and easily transported in a briefcase, may hold as much as 2 gigabytes of data. These cartridges are often used to transport huge files, such as desktop-publishing files with color and graphics and large spreadsheets. They are also frequently used for backing up data because, although they are relatively expensive, they hold much more data than diskettes and are much faster than tape.

Hard-Disk Technology for Large Computer Systems

As a microcomputer user, you may regard secondary-storage technology for large computer systems as being of only casual interest. However, this technology forms the backbone of the revolution in making information available to you over communications lines. The large databases offered by such organizations as America Online and through the Internet and World Wide Web depend to a great degree on secondary-storage technology.

Secondary-storage devices for large computers consist of the following:

- **Removable packs: A *removable-pack hard disk system* contains 6–20 hard disks, of 10½- or 14-inch diameter, aligned one above the other in a sealed unit.** Capacity varies, with some packs ranging into the terabytes.

  These removable hard-disk packs resemble a stack of phonograph records, except that there is space between disks to allow access arms

■ PANEL 6.5

Removable hard-disk cartridges and drives
Each cartridge has self-contained disks and read/write heads. The entire cartridge, which may store from 100 megabytes to 2 gigabytes of data, may be removed for transporting or may be replaced by another cartridge. *(Left)* Iomega Jaz cartridges and drive, and Zip cartridges and drive. *(Right)* Syquest drive and cartridges.

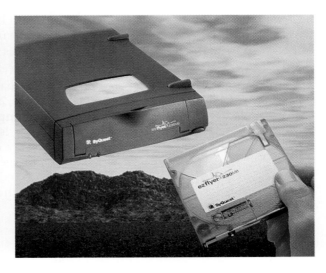

■ PANEL 6.6

Multiple disks and cylinders

In a stack of disks, access arms slide in and out to specific tracks. They use the cylinder method to locate data—the same track numbers lined up vertically one above the other form a "cylinder."

Multiple disks and cylinders

In a stack of disks, access arms slide in and out to specific tracks. They use the cylinder method to locate data—the same track numbers lined up vertically one above the other form a "cylinder."

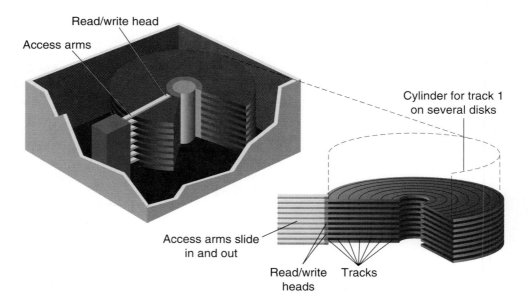

Read/write head

Access arms

Cylinder for track 1 on several disks

Access arms slide in and out

Read/write heads   Tracks

RAID unit

to move in and out. Each access arm has two read/write heads—one reading the disk surface below, the other the disk surface above. However, only *one* of the read/write heads is activated at any given moment.

Secondary-storage systems that use several hard disks don't use the sector method to locate data. Rather they use what is known as the cylinder method. Because the access arms holding the read/write heads all move together, the read/write heads are always over the same track on each disk at the same time. **All tracks with the same track number, lined up one above the other, thus form a** *cylinder.* (■ *See Panel 6.6.)*

- Fixed-disk drives: *Fixed-disk drives* **are high-speed, high-capacity disk drives that are housed (sealed) in their own cabinets.** Although not removable or portable, they generally have greater storage capacity and are more reliable than removable packs, and so they are used more often. A single mainframe computer might have 20–100 such fixed-disk drives attached to it.

- RAID storage system: A fixed-disk drive sends data to the computer along a single path. A *RAID storage system,* **which consists of anywhere from 2 to 100 disk drives within a single cabinet, sends data to the computer along several parallel paths simultaneously.** Response time is thereby significantly improved. RAID stands for redundant array of independent disks.

  The advantage of a RAID system is that it not only holds more data than a fixed-disk drive within the same amount of space, but it also is more reliable because if one drive fails, others can take over.

6.5 Optical Disks

KEY QUESTION

What are the differences among the four principal types of optical disks?

Preview & Review: Optical disks are removable disks on which data is written and read using laser technology. Four types of optical disks are CD-ROM, CD-R, erasable, and DVD.

By now optical-disk technology is well known to most people. **An *optical disk* is a removable disk on which data is written and read through the use of laser beams.** There is no mechanical arm, as with diskettes and hard disks.

The most familiar type of optical disk is the one used in the music industry. An audio CD uses digital code and looks like a miniature phonograph record. Such a CD holds up to 74 minutes (3 billion bits' worth) of high-fidelity stereo sound.

A single optical disk of the type used on computers, called CD-ROM, can hold up to about 650 megabytes of data. This works out to about 269,000 pages of text, or approximately 7500 photos or graphics, or 20 hours of speech, or 77 minutes of video. Although some disks are used strictly for digital data storage, many are used to distribute multimedia programs that combine text, visuals, and sound.

In the principal types of optical-disk technology, a high-power laser beam is used to represent data by burning tiny pits into the surface of a hard plastic disk. To read the data, a low-power laser light scans the disk surface: Pitted areas are not reflected and are interpreted as 0 bits; smooth areas are reflected and are interpreted as 1 bits. (■ *See Panel 6.7.*) Because the pits are

■ PANEL 6.7

Optical disks

(Top) Writing data—a high-powered laser beam records data by burning tiny pits in an encoded pattern onto the surface of a disk. *(Bottom)* Reading data—a low-powered laser beam is used to read data because it reflects off smooth areas, which are interpreted as 1 bits, but does not reflect off pitted areas, which are interpreted as 0 bits.

Recording data

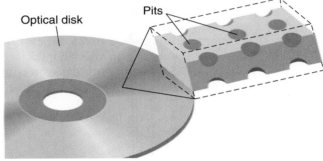

Optical disk · Pits

Reading data

Reading "1":
The laser beam reflects off the smooth surface, which is interpreted as a 1 bit.

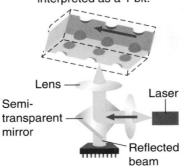

Lens · Laser · Semi-transparent mirror · Reflected beam

Reading "0":
The laser beam enters a pit and is not reflected, which is interpreted as a 0 bit.

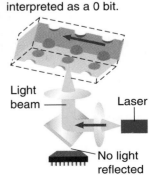

Light beam · Laser · No light reflected

so tiny, a great deal more data can be represented than is possible in the same amount of space on a diskette and many hard disks.

The optical-disk technology used with computers consists of four types:

- CD-ROM disks
- CD-R disks
- Erasable disks (CD-E)
- DVD/DVD-ROM disks

CD-ROM Disks

For microcomputer users, the best-known type of optical disk is the CD-ROM. **CD-ROM, which stands for *compact disk–read-only memory*, is an optical-disk format that is used to hold prerecorded text, graphics, and sound.** Like music CDs, a CD-ROM is a read-only disk. **Read-only means the disk's content is recorded at the time of manufacture and cannot be written on or erased by the user.** You as the user have access only to the data imprinted by the disk's manufacturer.

Current microcomputers are being sold with built-in CD-ROM drives. (■ *See Panel 6.8.*) The drives come either in a SCSI configuration or an EIDE configuration. In the former case, you also need a SCSI adapter card; in the latter case, an EIDE-controller adapter card.

At one time a CD-ROM drive was only a single-speed drive. Now there are four-, eight-, ten-, twelve-, sixteen-, twenty-four-, and thirty-two-speed drives (known as 4x, 8x, 10x, 12x, 16x, 24x, and 32x, respectively). A single-speed drive will access data at 150 kilobytes per second, a 16x drive at 2400 kbps, and a 32x drive at 4800 kbps. The faster the drive spins, the more quickly it can deliver data to the processor. CD-ROM drives used to handle only one disk at a time. Now, however, there are multidisk drives that can handle up to 100 disks. (Such drives are called *jukeboxes,* or *CD changers,* discussed later in this chapter.)

CD-ROMs for both desktop computers and notebook computers are a standard 120 millimeters in diameter.

■ PANEL 6.8

CD-ROM drives in notebook computers

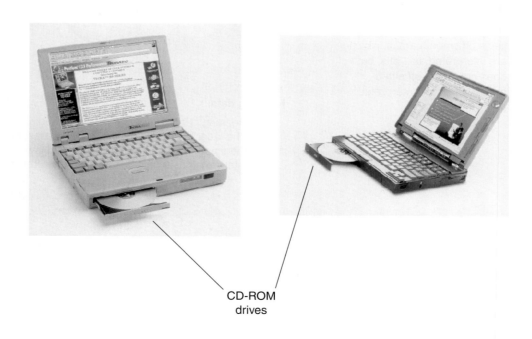

CD-ROM
drives

Originally, computer makers thought that CD-ROMs "would be good for storing databases, documents, directories, and other archival information that would not need to be altered," says one report. "Customers would be libraries and businesses."[8] Clearly, though, there are many uses:

- **Data storage:** Among the top-selling titles are road maps, typeface and illustration libraries for graphics professionals, and video and audio clips. Publishers are also mailing CD-ROMs on such subjects as medical literature, patents, and law.

  Want to have access to every issue in the first 108 years of *National Geographic* magazine? They're now available on 30 CD-ROMs "that include every article, photograph, and even advertisement that has appeared in the magazine during its . . . history," according to one report.[9]

- **Encyclopedias, atlases, and reference works:** The principal CD-ROM encyclopedias are *Britannica CD, Collier's Encyclopedia, Compton's*

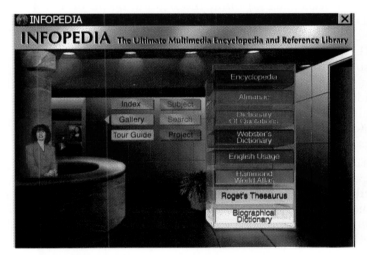

Interactive Encyclopedia, Encarta Encyclopedia from Microsoft, *Grolier Multimedia Encyclopedia, Infopedia,* and *World Book Multimedia Encyclopedia.* Each packs the entire text of a traditional multivolume encyclopedia onto a single disk, accompanied by pictures, maps, animation, and snippets of audio and video. All have pull-down menus and buttons to trigger search functions. Some are available in whole or in part on America Online or on the Web.

CD-ROMs are also turning atlases into multimedia extravaganzas. Mindscape's *World Atlas & Almanac,* for instance, combines the maps, color photos, geographical information, and demographic statistics of a traditional book-style atlas with video, sounds, and the ability to immediately find what you want.

Maps, for example, may include political charts showing countries and cities, three-dimensional maps showing mountain ranges, and even satellite maps. An audio feature lets you hear the pronunciation of place names. There are also street atlases (*Street Atlas U.S.A., StreetFinder*), which give detailed maps that can pinpoint addresses and show every block in a city or town, and trip planners (*TripMaker, Map 'n' Go*), which suggest routes, attractions, and places to eat and sleep.

- **Catalogs:** Publishers have also discovered that CD-ROMs can be used as electronic catalogs, or even "megalogs." One, for instance, combines the catalogs of several companies. "A single disk now holds the equivalent of 7000 pages of [text and graphical] information on almost 50,000 different products from salad-bar sneeze-guards to deep-fat fryers," noted one report.[10] Cinemania offers a multimedia catalog of movies available on videotape.

- **Entertainment and games:** As you might expect, CD-ROM has been a hugely successful medium for entertainment and games. Examples are 25 years of Garry Trudeau's comic strip, *Doonesbury* (on one disk), as well as games such as *Myst, Doom II, Dark Forces,* and *Sherlock Holmes, Consulting Detective.*

- **Music, culture, and films:** In *Xplora 1: Peter Gabriel's Secret World*, you can not only hear rock star Gabriel play his songs but also create "jam sessions" in which you can match up musicians from around the world and hear the result. Other examples of such CD-ROMs are *Bob Dylan: Highway 61 Interactive; Multimedia Beethoven, Mozart, Schubert; Art Gallery; American Interactive; A Passion for Art;* and *Robert Mapplethorpe: An Overview.*

  Developers have also released several films on CD-ROM, such as the 1964 Beatles movie *A Hard Day's Night, This Is Spinal Tap,* and *The Day After Trinity.*

- **Education and training:** Want to learn photography? You could buy a pair of CD-ROMs by Bryan Peterson called *Learning to See Creatively* (about composition) and *Understanding Exposure* (discussing the science of exposure). When you pop these disks in your computer, you can practice on-screen with lenses, camera settings, film speeds, and the like. Or you could learn history from such CD-ROMs as *Critical Mass: America's Race to Build the Atomic Bomb* or *The War in Vietnam.* CD-ROMs are also available to help students raise their scores on the Scholastic Aptitude Test (*Score Builder for the SAT, Inside the SAT*).

  You may also explore the inner workings of the human body in *Body Voyage,* a 15-gigabyte CD-ROM package based on Joseph Paul Jernigan, a condemned prisoner executed in 1993. Jernigan gave scientists permission to freeze his body and then cut it into 1878 slices, 1 millimeter each, which were then photographed and scanned into a computer.

- **Edutainment:** *Edutainment software* consists of programs that look like games but actually teach, in a way that feels like fun. An example for children ages 3–6 is *Yearn 2 Learn Peanuts,* which teaches math, geography, and reading. *Multimedia Beethoven: The Ninth Symphony,* an edutainment program for adults, plays the four movements of the symphony while the on-screen text provides a running commentary and allows you to stop and interact with the program.

- **Books and magazines:** Book publishers have hundreds of CD-ROM titles, ranging from *Discovering Shakespeare* and *The Official Super Bowl Commemorative Edition* to business directories such as *ProPhone Select* and *11 Million Businesses Phone Book.* At Marquette University, the annual yearbook, *The Hilltop,* has been replaced by a CD-ROM version.

  Examples of CD-ROM magazines are *Blender,* on the subject of music; *Medio,* a general-interest monthly; *Go Digital;* and *Launch,* which aims to offer a look at new music, movies, and computer games.

- **Applications and systems software:** Finally, we need to mention that today a great deal of software, such as Windows 98 and Office 97, comes on CD-ROMs rather than diskettes. Or, you may simply find installation of such software more convenient with CD-ROMs, rather than working with numerous diskettes.

Clearly CD-ROMs are not just a mildly interesting technological improvement. They have evolved into a full-fledged mass medium of their own, on the way to becoming as important as books or films.

README

Case Study: How Well Does Multimedia Learning Work?

One of the newest business buzzwords is the phrase "presentation technology," meaning the kind of technology used to make multimedia presentations to accompany training, for instance. It is also used with sales promotions, speeches, and similar forms of business communication. Presentations "are no longer just linear stories with a supportive collection of charts and pretty pictures," says a special *Business Week* advertising section devoted to the topic. The concept "involves a dazzling array of attention-grabbing, information-delivering sight and sound equipment designed to get the message across."[11]

Multimedia may indeed represent the wave of the future. Companies ranging from Ford Motor Company to Target Stores are already using CD-ROMs to satisfy a variety of training needs. Many college instructors are also using multimedia tools to accompany their lectures or in other ways. However, before we embrace multimedia learning uncritically, we need to ask: How well does it work?

Stanford University economics and education professor Henry Levin agrees that multimedia learning is "splashy and attractive." However, he points out that there has been "no truly rigorous study that compares the effectiveness of CD-ROM training to more traditional classroom training."[12]

Corporations in particular may like CD-ROMs better than live instructors because they need only make an initial investment (though an expensive one), whereas instructors and classroom materials have to be continually paid for. Of course, it's still not cost-effective if employees or students aren't learning, so a combination of both teachers and technology may be best. "There is no way a CD-ROM can answer all the questions [adult students] are going to come up with," says a business educator.[13] Multimedia programs may be sufficient for teaching "do this, do that" basic information, but they aren't good "at picking up subtle behavior cues for evaluating thought processes," Levin states.

What about using a book instead of a CD-ROM? Books have endured for a number of reasons, point out intellectual-property lawyer Richard Hsu and former multimedia producer William Mitchell. Compared to computers, books are easier to read, are more portable and durable, cost less, last longer, and often (because of limited capacity) are more carefully edited and more succinctly written.[14]

"Multimedia may be the flavor of the month," says Richard Clark, professor of educational psychology at the University of Southern California, "but there's no evidence of it—even if very well designed—having a performance benefit over other types of training."[15]

But the times they are a'changing. "Students love working with [multimedia] computer technology: computer ranks with lunch as their favorite part of the day," says a Connecticut middle-school teacher. "The shape and process of knowledge have changed, and today's students know it."[16]

CD-ROMs & Multimedia

CD-ROMs have enabled the development of the multimedia business. However, as the use of CD-ROMs has burgeoned, so has the vocabulary, creating difficulty for consumers. Much of this confusion arises in conjunction with the words *interactive* and *multimedia.*

As we mentioned earlier, ***interactive* means that the user controls the direction of a program or presentation on the storage medium.** That is, there is back-and-forth interaction, as between a player and a video game. You could create an interactive production of your own on a diskette. However, because of its limited capacity, you would not be able to present full-motion video and high-quality audio. The best interactive storage media are those with high capacity, such as CD-ROMs and DVD-ROMs.

***Multimedia* refers to technology that presents information in more than one medium, including text, graphics, animation, video, sound, and voice.** As used by telephone, cable, broadcasting, and entertainment companies, *multimedia* also refers to the so-called Information Superhighway. On this avenue various kinds of information and entertainment will presumably be delivered to your home through wired and wireless communication lines.

There are many different CD-ROM formats, some of which work on computers and some of which work only on TVs or special monitors. The majority of nongame CD-ROM disks are available for Macintosh or for Windows-based microcomputers. However, there are a host of other formats (CD-I, CDTV, Video CD, Sega CD, 3DO, CD1G, CD1MIDI, CD-V) that are not mutually compatible. Most will probably be on their way out if a standard DVD format (discussed below) begins to take over.

CD-R Disks: Recording Your Own CDs

CD-R, **which stands for** *compact disk—recordable,* **is a CD format that allows users to use a peripheral CD recorder to write data (only once) onto a specially manufactured disk that can then be read by any compatible CD-ROM drive.**

CD-R is often used by companies for archiving. However, one of the most interesting examples of CD-R technology is the Photo CD system. Developed by Eastman Kodak, ***Photo CD*** **is a technology that allows photographs taken with an ordinary 35-millimeter camera to be stored digitally on an optical disk.** (■ *See Panel 6.9.*) You can shoot up to 100 color photographs and take them to a local photo shop for processing. Then, an hour or so later, your photos will be available online for you to download, or, a bit later than that, you can pick up a CD-ROM with your images. You can then view the

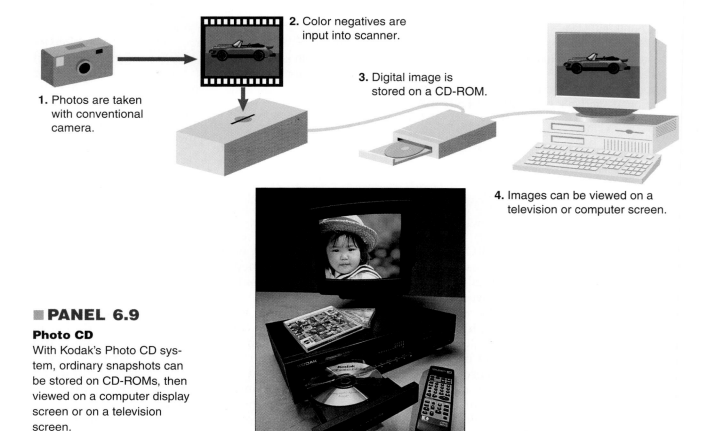

2. Color negatives are input into scanner.

3. Digital image is stored on a CD-ROM.

1. Photos are taken with conventional camera.

4. Images can be viewed on a television or computer screen.

■ PANEL 6.9

Photo CD

With Kodak's Photo CD system, ordinary snapshots can be stored on CD-ROMs, then viewed on a computer display screen or on a television screen.

disk using any compatible CD-ROM drive—PC, Macintosh, or one of Kodak's own Photo CD players, which attaches directly to a television set.

Erasable Optical Disks

An *erasable optical disk (CDE),* or *rewritable optical disk (CD-RW),* **allows users to record and erase data so that the disk can be used over and over again.** The most common type of erasable or rewritable optical disk is probably the *magneto-optical (MO) disk,* which uses aspects of both magnetic-disk and optical-disk technologies. Such disks are useful to people who need to save successive versions of large documents, handle databases, back up large amounts of data and information, or work in multimedia production or desktop publishing.

MO disks come in cartridges that are inserted into compatible MO drives hooked up to the computer. Because these drives are expensive, they are not yet widely used, even though they are much more durable than magnetic drives.

DVD: The "Digital Convergence" Disk

A DVD is a silvery, 5-inch optically readable digital disk that looks like an audio compact disk. But, as one writer points out, "DVDs encompass much more: multiple dialogue tracks and screen formats, and best of all, smashing sound and video."[17] **The DVD represents a new generation of high-density CD-ROM disks, which are read by laser and which have both write-once and rewritable capabilities.** According to the various industries sponsoring it, DVD stands for either "digital video disk" or "digital versatile disk."

DVDs can store much more data than standard CDs. The single-sided, single-layer DVD has a capacity of 4.7 gigabytes per side. Single-sided dual-layer DVDs hold 8.5 gigabytes per side. Double-sided DVDs hold 9.4 gigabytes if they are single-layer, and 17 gigabytes if they are dual-layer.

How does a DVD work? Like a CD or CD-ROM, the surface of a DVD contains microscopic pits, which represent the 0s and 1s of digital code that can be read by a laser. (■ *See Panel 6.10, next page.*) The pits on the DVD, however, are much smaller and closer together than those on a CD, allowing far more information to be represented there. Also, the technology uses a new generation of lasers that allows a laser beam to focus on pits roughly half the size of those on current audio CDs. Another important development is that the DVD format allows for two layers of data-defining pits, not just one. (■ *See Panel 6.11, next page.*) Finally, engineers have succeeded in squeezing more data into fewer pits, principally through data compression.[18]

DVDs have enormous potential to replace CDs for archival storage, mass distribution of software, and entertainment.[19] Indeed, many new computer systems now come with a DVD drive as standard equipment; these drives can also take standard CD-ROM disks.

Several problems still exist, however, that must be eliminated before DVD really takes off. First, DVD drives are expensive. Second, although some movies are available now on DVD, most people don't want to watch them on their computer; they want to use their TV. Third, a confusing battle of standards is going on that involves competing equipment manufacturers, content providers, and merchants. Some standards are read-only; others include various methods for rewritability. Other standards purport to offer RAM and ROM functions.

■ PANEL 6.10

DVD versus CD-ROM

Why the DVD can hold more data

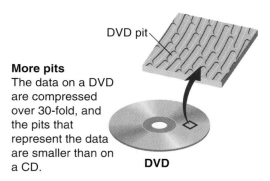

More pits
The data on a DVD are compressed over 30-fold, and the pits that represent the data are smaller than on a CD.

DVD pit

DVD

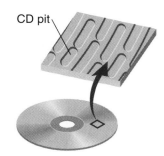

CD limits
CDs have larger pits—and record data only on one side in one layer. DVD disks can put data on both sides in two layers.

CD pit

CD

■ PANEL 6.11

Multilayered DVDs

DVD disks are multilayered. A dual-focus lens allows the laser to read two different levels simultaneously. DVDs that have a dual layer of data or that have data on both sides can hold more information.

DVD disks are multilayered. A dual-focus lens enables the laser to read two different levels simultaneously.

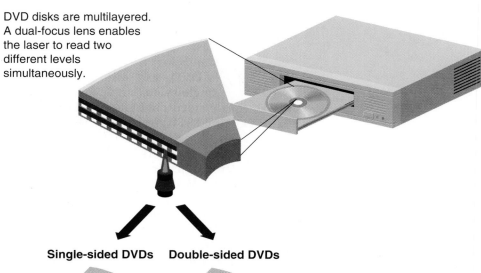

Single-sided DVDs

Double-sided DVDs

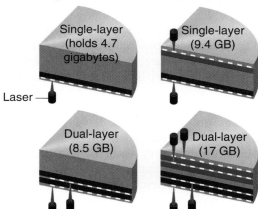

Single-layer (holds 4.7 gigabytes)

Single-layer (9.4 GB)

Laser

Dual-layer (8.5 GB)

Dual-layer (17 GB)

With a dual-layer disk, the laser reads the second layer of information by changing focus and shining through the first layer.

In addition, many copyright issues have not been resolved by the movie industry. For example, important film studios (particularly Paramount and Universal) had worried all along that pirates would break encryption codes and make perfect copies of DVD movies (supposedly already being done in China).[20,21] Thus, they were reluctant to support a rewritable DVD drive.

To confuse things further, we now have a format known as *Divx*, named for Digital Video Express, a partnership of electronics retailer Circuit City Stores and a powerful Hollywood law firm. The DVD format is now marketed largely for movie consumers intending to buy (at $20–$25 a disk), just as they now buy audio CDs. In contrast, *Divx* is a competing movie-on-a-disk format that is intended to be an alternative to renting VCR videotapes; the movie stored on a Divx disk will be locked and unavailable 48 hours after it's first played.[22] (A modem would be required to obtain the authorization to play the movies at home.) This means that viewers could pay a rental-like price of under $5 for a disk that is disposable—it doesn't need to be returned to any store after viewing. It is a "pay-per-view" type disk.

To add to the confusion, Divx players will play DVD disks, but DVD players won't play Divx disks. "It seems unlikely both formats could prosper," says one writer. "Remember the abandonment and frustration that Beta users felt when the VHS format won the videotape wars?"[23]

The conclusion is—if you are interested in purchasing a computer with a DVD drive, find out what features it has, what it's compatible with, and what's available for it.

6.6 Flash-Memory Cards

KEY QUESTION

What is a flash-memory card?

Preview & Review: Flash-memory cards consist of circuitry on PC cards that can be inserted into slots in a microcomputer.

Disk drives, whether for diskettes, hard disks, CD-ROMs, or DVDs, all involve moving parts—and moving parts can break. Flash-memory cards (✔ p. 166), by contrast, are variations on conventional computer-memory chips, which have no moving parts. As mentioned in Chapter 4, **flash-memory cards consist of circuitry on PC cards that can be inserted into slots connecting to the motherboard.** Flash RAM is one of the options available with PC (PCMCIA) cards. Used as a supplement to or replacement for a hard disk in a portable computer, each flash memory card can hold up to 100 megabytes of data.

A videotape produced for Intel, which makes flash-memory cards, demonstrates their advantage, as one report makes clear:

> In it, engineers strap a memory card onto one electric paint shaker and a disk drive onto another. Each storage device is linked to a personal computer, running identical graphics programs. Then the engineers switch on the paint shakers. Immediately, the disk drive fails, its delicate recording heads smashed against its spinning metal platters. The flash-memory card takes the shaking and keeps on going.[24]

6.7 Magnetic Tape

KEY QUESTION

What are the main uses of magnetic tape?

Preview & Review: Magnetic tape is thin plastic tape on which data can be represented with magnetized spots. Tape is used mainly for archiving and backup.

The Omaha computer center for the credit-card processor First Data Corporation is described as being bigger than a football field.

"Most of it," says a writer, "is filled with data storage silos that contain more than 200,000 computer tapes. The silos contain information on 92 million credit cards processed [that year]. . . . You can walk blocks before bumping into a human being."[25]

Magnetic tape is thin plastic tape that has been coated with a substance that can be magnetized. Data is represented by magnetized spots (representing 1s) or nonmagnetized spots (representing 0s). On large computers, tapes are used on magnetic-tape units or reels, and in cartridges. On microcomputers, tapes are used only in cartridges.

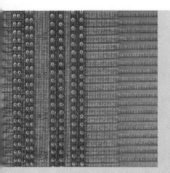

README

Case Study: How Long Will Digitized Data Last?

In 1982, software pioneer Jaron Lanier created a video game called *Moondust* for the then-popular Commodore 64 personal computer. Fifteen years later, when asked by a museum to display the game, he couldn't find a way to do it—until he had tracked down an old microcomputer of exactly that brand, type, and age, along with a joystick and video interface that would work with it.[26]

Would this have been a problem if Lanier had originally published a game in a *book*? Probably not. The first printed book dates back to about 1453, when Johannes Gutenberg developed the printing press and used it to print 150 copies of the Bible in Latin. Some of these Gutenberg Bibles still exist—and are still readable (if you can read Latin).[27]

Digital storage has a serious problem: It isn't as long-lived as older forms of data storage. Today's books printed on "permanent" (low-acid, buffered) paper may last up to 500 years. Even books printed on cheap paper that crumbles will still be readable.

By contrast, data stored on diskettes, magnetic tape, and optical disks is subject to two hazards:[28]

● Short life span of storage media: The storage media themselves have a short life expectancy, and often the degradation is not apparent until it's too late. The maximum time seems to be 50 years, the longevity of a high-quality CD-ROM. Some average-quality CD-ROMs won't last 5–10 years, according to tests run at the National Media Laboratory—about as long as newsprint.[29]

The magnetic tapes holding government records, which are stored in the National Archives in Washington, D.C., need to be "refreshed"—copied onto more advanced tapes—every 10 years.

● Hardware and software obsolescence: As Jaron Lanier found out, even when tapes and disks remain intact, the hardware and software needed to read them may no longer be available. Without the programs and computers used to encode data, digital information may no longer be readable.

"Eight-inch floppy disks and drives, popular as recently as a dozen years ago, are now virtually extinct," says one article, "and their 5¼-inch successors are rapidly disappearing. Optical and magnetic disks recorded under nonstandard storage schemes will be increasingly useless because of the lack of working equipment to read them."[30]

What about the personal records you would store on your own PC, such as financial records, inventories, genealogies, and photographs? *New York Times* technology writer Stephen Manes has a number of suggestions:[31] (1) Choose your storage media carefully. CD-R disks are probably best for archiving—especially if you also keep a paper record. (2) Keep it simple. Store files in a standard format, such as text files and uncompressed bitmapped files. (3) Store data along with the software that created it. (4) Keep two copies, stored in separate places, preferably cool, dry environments. (5) Use high-quality media, not off brands. (6) When you upgrade to a new hardware or software product, have a strategy for transferring the old data.

Spinning reels of magnetic tape used to be a feature of 1960s television shows, such as *The Man from U.N.C.L.E.* Today, however, "mag tape" is used mainly for backup and archiving—that is, for maintaining historical records—where there is no need for quick access. Large organizations like First Data Corporation house magnetic-tape units, which provide sequential data access, in tape libraries or special rooms, and there are strict security procedures governing their use. If users want quick access to information, they are more apt to store it on direct access storage devices, such as hard disks.

Cartridge tape units, also called *tape streamers,* are one method used to back up data from a microcomputer hard disk onto a tape cartridge. A cartridge tape unit using ¼-inch cassettes *(QIC, or quarter-inch cassettes)* fits into a standard slot in the microcomputer system cabinet and uses minicartridges that can store up to 17 gigabytes of data on a single tape. A more advanced form of cassette, adapted from technology used in the music industry, is the digital audiotape (DAT), which uses 2- or 3-inch cassettes and stores 2–4 gigabytes. Redesigned DATs called *Traven* technology are expected to hold as much as 8 gigabytes.

Some microcomputer users don't use tape for their backup because, although it is relatively inexpensive, it makes backing up a slow process—sometimes as much as several hours for a large-capacity hard disk. They prefer instead to use removable hard-disk cartridges or writable CD-ROMs as storage resources. Or they may use online secondary storage, as we discuss next.

6.8 Online Secondary Storage

KEY QUESTIONS

How does online storage work, and what are its advantages and disadvantages?

Preview & Review: Online storage can be used as a secondary-storage method, particularly for stripped-down network computers but also for additional security backup for standard microcomputers. There are advantages and disadvantages to online storage.

Suppose that the network computer actually becomes as popular as its promoters hope it will. This device, you'll recall, consists of a small computer with just a keyboard, display screen, processor, and connecting ports for network or phone cables. The gadget doesn't have any secondary storage. Rather, the Internet itself becomes, in effect, your hard disk. This is the notion being pushed by Oracle, Sun, and other companies backing such machines.

Is the network computer (NC) with online storage just wishful thinking? Computer editor Dan Gillmor, for one, believes the concept makes plenty of sense.[32] Companies could issue the network computer to on-the-road employees to pull information from corporate databases. Schools and libraries would find NCs ideal information-hunting devices. Families could keep one in the kitchen to input and retrieve recipes and "to do" lists. If you think of the network computer as a clever information device to help navigate life's routines, says Gillmor, the idea has a lot of merit.

With or without a network computer, however, you can use the Internet today as a storage vehicle. This may be particularly useful for backup purposes. Online backup services include Connected Online Backup *(http://www.connected.com),* Network Associates (formerly McAfee) Quick Backup to Personal Vault *(http://www.mcafee.com)* and SafeGuard Interactive *(http://www.sgii.com).* Monthly prices are generally in the $10–$15 range. When you sign up with the service, you usually download free software from a Web site that lets you upload whatever files you wish to the company's server. For security, you are given a password, and the files are supposedly encrypted to guard against anyone giving them an unwanted look.

The Advantages of Online Storage

There are many advantages to online backup storage.[33]

- **Convenience:** If you do only word processing, you may be able to back up your files on floppy disks. However, this would not be practical for backing up both software programs *and* data, which is advisable if you want to be able to recreate *everything* that was on a disk that crashed. (If you have a full 3-gigabyte hard drive, it would take 2000 diskettes to make a duplicate copy of everything.)
- **Price:** The backup alternatives of removable hard-disk cartridges (Zip, Jaz, SyQuest) can be expensive. Although tape is inexpensive, it is exceedingly slow.
- **Safety:** If burglars steal your computer system or fire or flood destroys your work setup, your backup may disappear, too. Online storage puts your backup files in a remote location. (Of course, you could send your backed-up diskettes or tapes to a relative, but they probably won't give them the care that a backup service would.)
- **Portability:** People who travel a lot with portable computers can particularly benefit because they can easily store and access their work simply by activating their modem.

The Disadvantages of Online Storage

There are two potential drawbacks:

- **Risk to privacy:** "Do you want some stranger's computer holding on to your business accounting records or your personal digital photos?" asks computer columnist Phillip Robinson. "The company may promise not to look at your stuff, but maybe its underpaid employees will peek just for fun or, worse, for profit."[34]
  Besides getting a password from the backup service, you should be sure that the backup software includes encryption, which will scramble your data so that unauthorized people can't decipher it.
- **Slow speed:** Robinson also points out that the slow speed of downloading your data through a typical (28.8 kbps or 33.6 kbps) modem may take 5 minutes to transfer a single megabyte of data. "Backing up the hundreds of megabytes of programs on a typical hard drive," he says, "would take you all day."

From a practical standpoint, therefore, online backup should be used only for vital files. Removable hard-disk cartridges are the best medium for backing up entire hard disks, including files and programs.

6.9 Compression & Decompression

KEY QUESTIONS

What is compression, and why is it important?

Preview & Review: Compression is a method of removing redundant elements from a computer file so that it requires less storage space and less time to transmit across the Internet. The two principal compression techniques are "lossless" and "lossy." The principal lossy compression schemes are JPEG for still images and MPEG-1, MPEG-2, and MPEG-4 for moving images.

"Like Gargantua, the computer industry's appetite grows as it feeds," says one writer. "So the smartest software engineers [have been] looking for ways to shrink the data-meals computers consume, without reducing their nutritional value."[35]

What this writer is referring to is the "digital obesity" brought on by the requirements of the multimedia revolution for putting pictures, sound, and video onto disk or sending them over a communications line. For example, a 2-hour movie contains so much sound and visual information that if stored without modification on a standard CD-ROM, it would require 360 disk changes during a single showing. A broadcast of *Oprah* that presently fits into one conventional, or analog, television channel would require 45 channels if sent in digital language.[36]

The solution for putting more data into less space comes from the mathematical process called compression. **Compression, or *digital-data compression*, is a method of removing redundant (repetitive) elements from a file so that the file requires less storage space and less time to transmit.** After the data is stored or transmitted and is to be used again, it is decompressed. The techniques of compression and decompression are sometimes referred to as *codec* (for *compression/decompression*) techniques.

"Lossless" Versus "Lossy" Compression

There are two principal methods of compressing data—*lossless* and *lossy*. The trade-off between these two techniques is basically data quality versus storage space.

- **"Lossless" techniques:** *Lossless compression* uses mathematical techniques to replace repetitive patterns of bits with a kind of coded summary. During decompression, the "summaries" are replaced with the original patterns of bits. That is, the data that comes out is every bit the same as what went in; it has merely been repackaged for purposes of storage or transmission. Lossless techniques are used for cases in which it's important that *nothing* be lost—as for computer data, database records, spreadsheets, and word processing files.

  Microcomputer users, for example, can increase the amount of data they store on their hard disks by using lossless compression utility programs that typically reduce a file to 40% of its original size. Individual files can be compressed before they are transmitted over the Internet.

- **"Lossy" techniques:** It is much easier to compress text than to compress sounds, pictures, and videos. "In this case," commented one article, "some information has to be thrown away forever. The trick is to work out what will not be missed."[37] This is the problem for "lossy" techniques.

  "Lossy" compression techniques permanently discard some data during compression. Lossy data compression involves a certain loss of accuracy in exchange for a high degree of compression (shrinking

material down to as little as 5% of the original file size). This type of compression is often used for graphics files and sound files. Thus, a lossy codec might discard shades of color that a viewer would not notice or soft sounds that are masked by louder ones. In general, most viewers or listeners would not notice the absence of these details.

Compression Standards

Several standards exist for compression, particularly of visual data. If you record and compress in one standard, you cannot play it back in another. The main reason for the lack of agreement is that different industries have different priorities. What will satisfy the users of still photographs, for instance, will not work for the users of movies.

Lossless compression schemes are used for text and numeric data files. Some popular compression programs are PKZip and WinZip for PCs and Stuffit for the Macintosh. Many lossless compression utilities can be downloaded for free from the Internet.

Lossy compression schemes are used with graphics and video files. The principal lossy compression schemes are *JPEG* and *MPEG.*

- **Still images—JPEG:** Techniques for storing and transmitting still photographs require that the data remain of high quality. The leading standard for still images is *JPEG* (pronounced "jay-peg"), for the Joint Photographic Experts Group of the International Standards Organization. The JPEG codec looks for a way to squeeze a single image, mainly by eliminating repetitive pixels, or picture-element dots, within the image.

  Unfortunately, there are more than 30 kinds of JPEG programs. "Unless the decoder in your computer recognizes the version that was used to compress a particular image," noted one reporter, "the result on your computer screen will be multimedia applesauce."[38]

- **Moving images—MPEG:** People who work with videos are less concerned with the niceties of preserving details than are those who deal with still images. They are interested mainly in storing or transmitting an enormous amount of visual information in economical form. A group called *MPEG* ("em-peg"), for Motion Picture Experts Group, was formed to set standards for weeding out redundancies between neighboring images in a stream of video.

  MPEG keeps a complete, detailed image for the first frame (or key frame) in a video segment. For subsequent frames, only the information that changes is stored from frame to frame. Key frames with complete information (called *intra-coded frames,* or *I-frames*) are placed in regular intervals to maintain picture quality.

  Three MPEG standards have been developed for compressing visual information—MPEG-1, MPEG-2, and MPEG-4. (■ *See Panel 6.12.*)

The vast streams of bits and bytes of text, audio, and visual information threaten to overwhelm us. Compression/decompression has become a vital technology for rescuing us from the swamp of digital data and enabling us to efficiently transmit data electronically.

■ PANEL 6.12

How video images are compressed, and three MPEG standards

(Below) MPEG is a method of computerized compression/decompression that can reduce the size of a video signal by 95%. As a result, the signal can be stored or transmitted more efficiently and economically. *(Right)* Three MPEG standards (MPEG-3 has been incorporated into MPEG-2).

Standard	Mission
MPEG-1	For microcomputers and consumer gadgets. Provides full-screen video (VHS-like quality) of images similar to those found on videocassette.
MPEG-2	For broadcast and cable television. Provides digital-TV-quality video for use with cable networks, satellite dishes, and new types of CD-ROMs. (MPEG-3 was incorporated into MPEG-4.)
MPEG-4	For wireless videoconferencing.

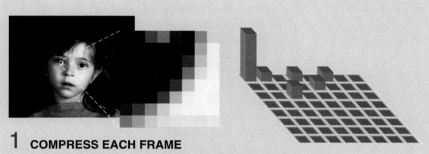

1 COMPRESS EACH FRAME

MPEG divides a frame of video into many tiny blocks, each containing 64 picture elements (pixels). The patterns in each block are transformed into a set of numbers. A few of these numbers (bars) contain most of the important picture information and everything else is discarded.

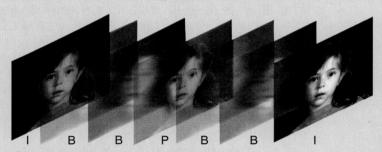

2 COMPRESS BETWEEN FRAMES

MPEG divides the video signal into three types of frames. Every 1/3 of a second, an intraframe picture (I) captures all the information in the compressed signal. A predicted frame (P) based on the previous I frame, and bidirectional frames (B) interpolated between the two, contain less data but preserve video quality.

3 REVERSE THE PROCESS

On playback, the restored frames lack some information but the eye is fooled into seeing detail that doesn't exist.

6.10 The Future of Secondary Storage

KEY QUESTION

What are some possible developments to come in secondary storage?

Preview & Review: Other developments in secondary storage include higher-density disks, higher-capacity CD-ROM jukeboxes, advanced compression schemes, digital VCRs, and advanced storage technology.

What follows are some noteworthy developments to which we should pay close attention—*higher-density disks, CD-ROM jukeboxes, advanced compression schemes, digital VCRs,* and *advanced storage technology.*

Higher-Density Disks

When IBM introduced the world's first disk drive in 1956, it was capable of storing 2000 bits per square inch. Forty years later, in April 1996, the company announced it had broken the magnetic disk barrier of 1 billion bits per square inch. Then in December 1997, it said it had passed the level of *11.6 billion bits* per square inch (1.8 billion bits per square centimeter).[39] The new technology will first be used in 2½-inch, nonremovable disk drives in portable computers, which means that a single-platter disk drive would hold about 6.5 gigabytes of data. A 3½-inch platter would hold 12–13 gigabytes. As one writer points out, at 11.6 billion bits per square inch, "every square inch of disk space could hold 1450 average-sized novels or more than 725,000 pages of double-spaced typewritten pages, which would make a stack taller than an 18-story building."[40]

In 10 years, density may well go up by another factor of 10. With such densities, says one writer, "Movie buffs could download entire movies from online services, storing them on diskettes the size of a quarter. Hospitals and doctors could put a patient's records and X-rays on small disks to be carried with the patient."[41]

CD-ROM Jukeboxes

How does a chemical company with a long history going back to 1861 go about searching for patent information that by 1990 had swelled to 30 million documents? BASF of Germany decided to scan its entire archives and store the contents on a library of 1700 CD-ROM disks. The company then installed what are known as CD-ROM jukeboxes—21 of them, each with 100 CD-ROM disks and all interconnected over a network.[42]

Like the coin-operated music machine that entertains patrons in bars, a CD-ROM jukebox uses a robot arm to exchange CD-ROMs between one or more drives. Also called libraries or changers, such jukeboxes can hold as few as six CD-ROM disks and a single drive and as many as 1400 disks and 32 drives. Optical disks may be 3½, 5¼, and 12 inches.

You can get a CD-ROM jukebox for use at home that will hold 25–200 disks. Some jukeboxes "daisy-chain" to other jukeboxes, so that you could play 300–600 disks from a single controller.

Now imagine what a jukebox with 17-gigabyte DVD disks will be like!

Advanced Compression Schemes

A fully digitized photograph requires about 80 megabytes of storage space. In 1995 Eastman Kodak demonstrated a technique for hypercompressing photographs that reproduced human faces using only about 50 bytes of data. Such images, when stored on the magnetic strip of credit cards or bar codes on bank checks, could be used to counteract fraud. Tellers or retail clerks could pass the card or check through a device that shows the customer's likeness on a display monitor.[43]

The Kodak technology uses a technique called *wavelet compression;* however, other compression methods are also being exploited:

- **Wavelet compression:** In *wavelet compression,* an image is compressed as a whole, rather than block by block as with JPEG and MPEG codecs. This technology is useful for files that must be stored or transmitted with high fidelity, such as medical images or music.

- **Fractal compression:** A *fractal image* is one in which the features look similar at different scales of magnification. For example, the features of a coastline are more or less the same no matter how high the altitude from which they are photographed. With *fractal compression,* just one image may be stored and then used over and over to re-create the whole picture. The image is based on information about size and position that is stored somewhere else.

- **Compression by object-oriented programming:** *Object-oriented programming* (discussed in Chapter 11) treats each segment in a software program as an individual unit, called an "object," that can be used repeatedly in different applications and by different programmers. The reusable "objects" might be blocks of data, mathematical procedures, or—for present purposes—video images. With the use of object-oriented programming, an image in a movie might be described once and then reused endlessly. Thus, an image of a plane flying through the air might be defined with as few as two objects—the plane and the sky.

- **Compression using neural networks:** *Neural networks* (discussed in Chapter 12) use physical electronic devices or software to mimic the neurological structure of the human brain. Neural networks, which learn from experience, can be trained to squeeze data down to the irreducible minimum. "The neural net can be fed billions of bits of data—all the pixels in all the frames of a movie or all the text in an online library—and it will generate a cloud-like map of points," says one account. "Because these points exist in a world of higher mathematics with an almost infinite number of dimensions, each point contains a wealth of data, which the net can use to reconstitute every detail of the original information."[44]

Digital VCRs

In 1994 several companies, including some of the world's leading names such as IBM, Apple, Sony, and Matsushita, met in Tokyo to agree on standards for a digital videocassette recorder for the home.[45] Machines coming onto the market—called DVCR or DVHS machines, depending on the maker—are backward compatible, able to play regular VCR tapes.

Digital VCRs could provide a gigantic leap forward in unifying the separate sectors of computers and telecommunications. VCRs in digital format provide better picture quality and smaller and/or longer-playing cassettes. They also have the capability to make perfect copies, which will make it easier for camcorder owners to edit their videotapes. Equally important, digital VCRs will make a better match for future television systems, such as advanced TV, that will process signals in digital form. Indeed, they are specifically designed to record programming signals beamed down directly from the new class of Digital Satellite System. Finally, such VCRs could store not only video and television pictures but also large amounts of computer data. Thus, computer users could use this device to make backup copies of the data stored on their hard disks.

Video Servers

One of the most talked-about features of the coming Information Superhighway has been the potential ability to deliver interactive video and movies-on-demand through cable or other connections to people's homes. Although the arrival of such offerings seems to recede further into the future, companies such as Oracle are actively working to develop video servers, large-scale systems for storing thousands of movies and videos.

This is not an easy challenge. "Broadcasting a digitized film is fairly easy," explains one account. "Serving up 100 different films to 100 houses at 100 different times is a programming nightmare."[46] The solution that Oracle engineers came up with was to use a supercomputer, divide a movie into 1000 different segments, and store each segment on a different disk drive, each with its own microprocessor. One "master" microprocessor controls all the others so that a film flows seamlessly in the correct order.

Molecular Electronics: Storage at the Subatomic Level

An emerging field, molecular electronics, may push secondary storage into another dimension entirely. Here are some possibilities:

- **Holograms as storage:** A *hologram* is a three-dimensional picture that is created by two lasers, which move electrons physically through a transparent crystal to turn different areas from light to dark. You sometimes see holograms used as logos on bank credit cards.

  In the future, holograms could replace not only hard-disk drives but also memory chips. In a 1997 development, SRI International discovered that it could store different colors of light on the same crystal, thereby improving storage density up to 100 times and speeding the process of writing data into the crystal a thousandfold.[47]

- **Molecule-size magnets as storage:** In early 1997, researchers from Xerox, the University of Barcelona in Spain, and the City College of New York announced the creation of a microscopic magnet, one molecule in size, derived from a special combination of materials (manganese, oxygen, carbon, and hydrogen). The magnet could be used to create a data storage system that could pack data thousands or millions of times more densely than is possible in today's computer systems. As one account puts it, "Using magnets the size of . . . molecules, it might some day be possible to store hundreds of gigabytes of data in an area no larger than the head of a pin."[48]

- **Subatomic lines as storage:** In what has been called "the world's smallest Etch-a-Sketch," physicists at NEC in Tokyo used a sophisticated probe—a tool called a scanning tunneling microscope (STM)—to paint and erase tiny lines roughly 20 atoms thick. This development could lead to ultra-high-capacity storage devices for computer data.

- **Bacteria as storage:** Scientists have reported research involving use of bacteria to store data in three dimensions. Said Robert Birge, who fashioned a 1-centimeter cube made of protein molecules that could store data in three dimensions, "Six of these cubes can store the entire Library of Congress."[49]

6.11   The Manipulation of Truth in Art & Journalism

KEY QUESTION

What are the issues in the manipulation of truth?

Preview & Review: Users of information technology must weigh the effects of the digital manipulation of sound, photos, and video in art and journalism.

Ethics

The giant capacities of today's storage devices has given photographers, graphics professionals, and others a new tool—the ability to manipulate images at the level of pixels (✔ p. 219). For example, photographers can easily do *morphing*—transforming one image into another. **In *morphing*, a film or video image is displayed on a computer screen and altered pixel by pixel, or dot by dot. The result is that the image metamorphoses into something else**—a pair of lips into the front of a Toyota, for example. "Because the image is digital," says one writer, "it can be taken apart pixel by pixel and put back together in many ways."[50] This helps photo professionals further their range, although at the same time it presents a danger that photographs will be compromised in their credibility.

The ability to manipulate digitized output—images and sounds—has brought a wonderful new tool to art. However, it has created some big new problems in the area of credibility, especially for journalism. How can we now know that what we're seeing or hearing is the truth? Consider the following.

Manipulation of Sound

Frank Sinatra's 1994 album *Duets* paired him through technological tricks with singers like Barbra Streisand, Liza Minnelli, and Bono of U2. Sinatra recorded solos in a recording studio. His singing partners, while listening to his taped performance on earphones, dubbed in their own voices. This was done not only at different times but often, through distortion-free phone lines, from different places. The illusion in the final recording is that the two singers are standing shoulder to shoulder.

Newspaper columnist William Safire loves the way "digitally remastered" recordings recapture great singing he enjoyed in the past. However, he called *Duets* "a series of artistic frauds." Said Safire, "The question raised is this: When a performer's voice and image can not only be edited, echoed, refined, spliced, corrected, and enhanced—but can be transported and combined with others not physically present—what is performance? . . . Enough of additives, plasticity, virtual venality; give me organic entertainment."[51] Another critic said that to call the disk *Duets* seemed a misnomer. "Sonic collage would be a more truthful description."[52]

Some listeners feel that the technology changes the character of a performance for the better—that the sour notes and clinkers can be edited out. Others, however, think the practice of assembling bits and pieces in a studio drains the music of its essential flow and unity.

Whatever the problems of misrepresentation in art, however, they pale beside those in journalism. Could not a radio station edit a stream of digitized sound to achieve an entirely different effect from what actually happened?

Manipulation of Photos

When O. J. Simpson was arrested in 1994 on suspicion of murder, the two principal American newsmagazines both ran pictures of him on their covers.[53, 54] *Newsweek* ran the mug shot unmodified, as taken by the Los Angeles Police Department. *Time*, however, had the shot redone with special effects as a "photo-illustration" by an artist working with a computer. Simpson's image was darkened so that it still looked like a photo but, some critics said, with a more sinister cast to it.

Should a magazine that reports the news be taking such artistic license? Should *National Geographic* in 1982 have photographically moved two Egyptian pyramids closer together so that they would fit on a vertical cover? Was it even right for *TV Guide* in 1989 to run a cover showing Oprah

Winfrey's head placed on Ann-Margret's body? In another case, to show what can be done, a photographer digitally manipulated the famous 1945 photo showing the meeting of the leaders of the wartime Allied powers at Yalta. Joining Stalin, Churchill, and Roosevelt are some startling newcomers: Sylvester Stallone and Groucho Marx. The additions are done so seamlessly it is impossible to tell the photo has been altered. (■ *See Panel 6.13.)*

The potential for abuse is clear. "For 150 years, the photographic image has been viewed as more persuasive than written accounts as a form of 'evidence,'" says one writer. "Now this authenticity is breaking down under the assault of technology."[55] Asks a former photo editor of the *New York Times Magazine*, "What would happen if the photograph appeared to be a straightforward recording of physical reality, but could no longer be relied upon to depict actual people and events?"[56]

Actually, this problem has already raised great concern in nature photography, which has led *National Geographic* editor Bill Allen to say, "Technology is not taking us closer to reality but further from it."[57] Taking the perfect shot of an animal in the wild is a costly, time-consuming business, and *National Geographic* now takes painstaking steps to make sure its photographs are real: It took one photojournalist seven months, $300,000, and four trips to India to produce high-quality photos of an Indian tiger. The *Geographic* editors were horrified, therefore, when an advertisement (promoting the magazine's own Web site) ran in one issue which featured a polar bear lying on an ice floe in Antartica. There are no polar bears in Antartica. The photographer had taken a picture of a polar bear in the Cincinnati Zoo and digitized it onto a scene he had shot in Antarctica.

Many editors try to distinguish between photos used for commercialism (advertising) versus for journalism, or for feature stories versus for news stories. However, this distinction implies that the integrity of photos applies only to some narrow definition of news. In the end, it can be argued, altered photographs pollute the credibility of all of journalism.

■ **PANEL 6.13**

Photo manipulation
In this 1945 photo, World War II Allied leaders Joseph Stalin, Winston Churchill, and Franklin D. Roosevelt are shown from left to right. Digital manipulation has added Sylvester Stallone standing behind Roosevelt and Groucho Marx seated at right.

Manipulation of Video

The technique of morphing, used in still photos, takes a quantum jump when used in movies, videos, and television commercials. Morphing and other techniques of digital image manipulation have had a tremendous impact on filmmaking. Director and digital pioneer Robert Zemeckis *(Death Becomes Her)* compares the new technology to the advent of sound in Hollywood.[58] It can be used to erase jet contrails from the sky in a western and to make digital planes do impossible stunts. It can even be used to add and erase actors. In *Forrest Gump,* many scenes involved old film and TV footage that had been altered so that the Tom Hanks character was interacting with historical figures.

Films and videotapes are widely thought to accurately represent real scenes (as evidenced by the reaction to the amateur videotape of the Rodney King beating by police in Los Angeles). Thus, the possibility of digital alterations raises some real problems. One is the possibility of doctoring videotapes supposed to represent actual events. Another concern is for film archives: Because digital videotapes suffer no loss in resolution when copied, there are no "generations." Thus, it will be impossible for historians and archivists to tell whether the videotape they're viewing is the real thing or not.[59]

Information technology increasingly is blurring humans' ability to distinguish between natural and artificial experience, say Stanford University communications professors Byron Reeves and Clifford Nass.[60] For instance, they have found that showing a political candidate on a large screen (30 or 60 inches) makes a great difference in people's reactions. In fact, you will actually like him or her more than if you watch on a 13-inch screen. "We've found in the laboratory that big pictures automatically take more of a viewer's attention," said Reeves. "You will like someone more on the large screen and pay more attention to what he or she says but remember less." (This is why compelling TV or computer technology may not aid education.) Our visual perception system, they find, is unable to discount information—to say that "this is artificial"—just because it is symbolic rather than real.

If our minds have this inclination anyway, how can we be expected to exercise our critical faculties when the "reality" is not merely artificial but actively doctored?

Onward: Toward a More Interactive Life

"This is the first generation that has never watched television without a remote control," says *Newsweek* writer Michael Rogers. "Neither conventional print nor passive television is really attractive to them anymore."[61]

Rogers was referring to people under age 25. Mass-media experts think this restless, demanding generation in the next century will want information and entertainment that they can control—in a word, that is interactive.

Also, if people suspect that a piece of information is being slanted, they will want to electronically turn to the sources themselves. In short, electronic multimedia offers a wider base of knowledge.

Photo Opportunities: Working with Digitized Photographs

How does a newly minted real-estate agent compete with more experienced agents in persuading home sellers to let him or her represent them? That was the situation that Drew Armstrong found himself in.

Armstrong, 26, of Provo, Utah, decided he had to do something different. So he purchased a digital camera, a color laser printer, and a pair of Macintosh computers. When visiting prospective clients, he uses the camera to take pictures of their houses, and the sellers then pick the ones they like best. Using his portable Macintosh, he then e-mails the photos to his office. An assistant puts them into software in the other Mac computer and prints out colorful, "magazine-quality" house-for-sale flyers—far more impressive than the black-and-white, hand-lettered flyers of Armstrong's competitors. The flyers can be in the clients' hands within minutes. Customers have been so impressed with the service, says Armstrong, that "I've tripled my business."[62]

Working with photos in digital form, once the sphere of professional graphic artists, is within the realm of the rankest amateur, assuming you have the right equipment. Let us describe the possibilities.

Obtaining Photos in Digital Form

How do you get a digital photo in the first place, and how do you get it into your computer? There are three ways:[63–67]

1. You can ask your film developer to develop your standard (chemical-based) film not only into prints or slides but also onto a diskette or CD-ROM that you can then put in your computer's disk drive.

2. You can use an image scanner attached to your computer to scan in a traditional photographic print (or its negative).

3. You can take a picture with a digital camera, which will store the image on a diskette or removable memory card. You can then use a cable that comes with the camera to transfer the stored image to your computer.

Once the digitized images are on the hard drive in your computer, they are ready for viewing, printing, editing, or sending via the World Wide Web.

Viewing

One of the major advantages of digital cameras is that you can view your shots immediately, on your computer's monitor. But however you get the image into the computer, on the screen it will probably look as good as or better than any print.

Once you've viewed the image, you can decide whether to delete it, store it, have it professionally printed by a photo shop, print it out on your own printer, edit it, or send it via the Internet to someone else.

Printing

Ink-jet printers are now available that can produce output that is almost indistinguishable from a photograph. Special paper is available that gives the result the same look and feel as a photographic print.

Editing

Before printing the image, you might want to manipulate it, or as the professionals say, "edit" it. Examples of image-editing software are PhotoDeluxe (from Adobe, maker of Photo-Shop, which serves professionals), Picture It, PictureWorks, Photo Soap, PhotoImpact, and Live Picture.

With this software, you can make changes from the minor to the outrageous: Erase scratches on the photo, the wrinkles from your face, or the teenager's nose ring in the special photo for Grandma. Put yourself in a picture of Hong Kong, which you've never visited. Even change Santa's clothes from red to blue and give him two heads.

You can stretch, shrink, and distort images; remove the flash-photo "red eye"; sharpen the focus of fuzzy pictures; crop awkwardly shot photos; turn pictures around; cut and paste segments of photos; apply special-effects filters; retouch documents; and merge and otherwise alter images.

Says one report, "Not only can you place pictures against different backgrounds—there's you grinning sheepishly on the cover of a major newsweekly, your smitten visage smack in the middle of a big, red valentine heart!—but you can morph, swirl, stretch, and squish photos any which way you please."[68]

Transmitting via the Web

You can use the World Wide Web to transmit pictures to people wherever they are, as long as they are connected to an online computer. For instance, a program called Pictra Album allows you to assemble photos in a digital album that you can then send to Pictra's Website for others to look at.[69] (The pictures can be viewed publicly or privately through use of a password you select.) Kodak also offers the Kodak Picture Network, which allows photographers to store their photos on the Web, which they can then make available by e-mail to friends and family.[70]

Coming soon: a technology that will allow you to print high-quality photos straight from the Web. All the pieces are coming together.

Summary

What It Is/What It Does	Why It's Important
cluster (p. 273, KQ 6.4) Smallest storage unit the computer can access on a hard disk; refers to a number of sectors.	If a file's data is stored in nonadjacent clusters, the disk becomes fragmented, thus slowing access time.
compact disk–read-only memory (CD-ROM) (p. 278, KQ 6.5) Optical-disk form of secondary storage that holds more data than diskettes; used to hold prerecorded text, graphics, sound, and video. Like music CDs, a CD-ROM is a read-only disk.	CD-ROM disks are being used to sell off-the-shelf microcomputer software, books and magazines, and multimedia games, encyclopedias, atlases, movie guides, and training and education programs.
compact disk—recordable (CD-R) (p. 282, KQ 6.5) CD format that allows users to write data onto a specially manufactured disk inserted in a peripheral CD recorder; the disk can then be read by a compatible CD-ROM drive.	CD-R is often used by business for archiving, but it is also the basis for Kodak's Photo CD technology, and it allows home users to do their own recordings in CD format.
compression (digital-data compression) (p. 289, KQ 6.9) Process of removing redundant and unnecessary elements from a file so that it requires less storage space or can be easily transmitted. After the data is stored or transmitted it can be decompressed and used again.	Storage and transmission of digital data—particularly of graphics—requires a huge amount of electronic storage capacity, and transmission is difficult to accomplish over copper wire. Thus compression programs are necessary to reduce the size of these files.
cylinder (p. 276, KQ 6.4) All the tracks in a disk pack with the same track number, lined up one above the other.	Secondary storage systems that use several hard disks don't use the sector method to locate data; they use the cylinder method. Because the access arms holding the read/write heads all move together, the read/write heads are always over the same track on each disk, that is, in the same cylinder, at the same time. Data access is faster because all read/write heads move simultaneously.
direct access storage (p. 266, KQ 6.1) Method of storage that allows the computer to go directly to the information sought. Any record can be found quickly by entering its key field, such as a Social Security number, so direct access saves time compared to sequential storage when there is no pattern to searches.	Although the storage media—hard disks and other kinds of disks—is more expensive than tape, the cost is justified by the much quicker random searches.
disk drive (p. 267, KQ 6.3) Computer hardware device that holds, spins, reads from, and writes to magnetic or optical disks.	Users need disk drives in order to use their disks. Disk drives can be internal (built into the computer system cabinet) or external (connected to the computer by a cable).

What It Is/What It Does	**Why It's Important**
diskette (p. 267, KQ 6.3) Also called *floppy disk;* secondary-storage medium; removable round, flexible mylar disk that stores data as electromagnetic charges on a metal oxide film that coats the mylar plastic. Data is represented by the presence or absence of these electromagnetic charges, following standard patterns of data representation (such as ASCII). The 3½-inch-wide disk is contained in a square plastic case to protect it from being touched by human hands. It is called "floppy" because the disk within the envelope or case is flexible, not rigid.	Diskettes are used on all microcomputers.
double-sided diskette (DS or 2) (p. 269, KQ 6.3) Diskette that stores data on both sides; only older diskettes are single-sided. For double-sided diskettes to work, the disk drive must have read/write heads that will read both sides simultaneously.	Double-sided diskettes hold twice as much data as single-sided diskettes did.
DVD (digital video, or versatile, disk) (p. 283, KQ 6.5) A new generation of high-density compact disk that is read by laser and has both write-once and rewritable capabilities. The five-inch optical disk looks like a regular audio CD, but with dual layers, more tightly spaced pits, and the more precise laser that reads it, a DVD can store up to 17 gigabytes.	DVDs provide great storage capacity, studio-quality images, and theater-like surround sound. However, competing DVD standards and a new challenger, Divx, have so far kept DVDs from taking over the place of CD-ROMs.
enhanced integrated drive electronics (EIDE) (p. 273, KQ 6.4) An interface that connects hard drives to a microcomputer by using a flat ribbon cable attached to an expansion board (the host adapter) that plugs into an expansion slot on the motherboard. Also used increasingly to connect CD-ROM and tape drives.	EIDE is popular because of its low cost; one inexpensive host adapter can control two to four hard drives, two diskette drives, two serial ports, a parallel port, and a game port.
erasable optical disk (CDE) (p. 283, KQ 6.5) Also called a *rewritable optical disk (CD-RW);* optical disk that allows users to erase data so that the disk can be used over and over again (as opposed to CD-ROMs, which can be read only).	The most common type of erasable and rewritable optical disk is probably the magneto-optical disk, which uses aspects of both magnetic-disk and optical-disk technologies. Such disks are useful to people who need to save successive versions of large documents, handle enormous databases, or work in multimedia production or desktop publishing.
fixed-disk drive (p. 276, KQ 6.4) High-speed, high-capacity disk drive housed in its own cabinet.	Although fixed disks are not removable or portable, these units generally have greater storage capacity and are more reliable than removable disk packs. A single mainframe computer might have 20–100 such fixed disk drives attached to it.

What It Is/What It Does

Why It's Important

flash-memory card (p. 285, KQ 6.6) Circuitry on credit-card-size cards (PC cards) that can be inserted into slots in the computer that connect to the motherboard.

Flash-memory cards are variations on conventional computer-memory chips; however, unlike standard RAM chips, flash memory is nonvolatile—it retains data even when the power is turned off. Flash memory can be used not only to simulate main memory but also to supplement or replace hard-disk drives for permanent storage.

formatting (initializing) (p. 268, KQ 6.3) Process by which users prepare diskettes so that the operating system can write information on them. This includes defining the tracks and sectors (the storage layout). Formatting is carried out by one or two simple computer commands.

Diskettes cannot be used until they have been formatted; they may be purchased with or without formatting.

fragmentation (p. 274, KQ 6.4) Refers to the situation when a hard disk's file data is stored in nonadjacent clusters.

Fragmentation slows access time.

hard disk (p. 271, KQ 6.4) Secondary-storage medium; thin but rigid metal platter covered with a substance that allows data to be held in the form of magnetized spots. Hard disks are tightly sealed within an enclosed hard-disk-drive unit to prevent any foreign matter from getting inside. Data may be recorded on both sides of the disk platters. In a microcomputer, the drive is inside the system unit and not removable.

Hard disks hold much more data than diskettes do. Nearly all microcomputers now use hard disks as their principal secondary-storage medium.

hard-disk cartridge (p. 275, KQ 6.4) One or two hard-disk platters enclosed along with read/write heads in a hard plastic case. The case is inserted into an external or detached cartridge system connected to a microcomputer.

A hard-disk cartridge, which is removable and easily transported in a briefcase, may hold gigabytes of data. Hard-disk cartridges are often used for transporting large graphics files and for backing up data.

head crash (p. 272, KQ 6.4) Disk disturbance that occurs when the surface of a read/write head or particles on its surface come into contact with the disk surface, causing the loss of some or all of the data on the disk.

Head crashes can spell disaster if the data on the disk has not been backed up.

interactive (p. 281, KQ 6.5) Refers to a situation in which the user controls the direction of a program or presentation on the storage medium; that is, there is back-and-forth interaction between the user and the computer or communications device. The best interactive storage media are those with high capacity, such as CD-ROMs and DVDs.

Interactive devices allow the user to be an active participant in what is going on instead of just reading to it.

magnetic tape (p. 286, KQ 6.7) Thin plastic tape coated with a substance that can be magnetized; data is represented by the magnetized or nonmagnetized spots. Tape can store files only sequentially.

Tapes are used in reels, cartridges, and cassettes. Today "mag tape" is used mainly to provide backup, or duplicate storage.

What It Is/What It Does	Why It's Important

morphing (p. 294, KQ 6.11) Process in which a film or video image is displayed on a computer screen and altered pixel by pixel (dot by dot) so that it metamorphoses into something else, such as a toddler into a grandfather.

Morphing is frequently used in advertising and helps photo professionals further their range, but it also threatens the credibility of photographs.

multimedia (p. 281, KQ 6.5) Refers to technology that presents information in more than one medium, including text, graphics, animation, video, sound effects, music, and voice.

Use of multimedia is becoming more common in business, the professions, and education as a means of adding depth and variety to presentations, such as those in entertainment and education. *Multimedia* also refers to the so-called Information Superhighway, the delivery of information and entertainment to users' homes through wired and wireless communication lines.

optical disk (p. 277, KQ 6.5) Removable disk on which data is written and read through the use of laser beams. The most familiar form of optical disk is the CD used in the music industry.

Optical disks hold much more data than magnetic disks. Optical disk storage is expected to dramatically affect the storage capacity of microcomputers.

Photo CD (p. 282, KQ 6.5) Technology developed by Eastman Kodak that allows photographs taken with an ordinary 35-millimeter camera to be stored digitally on an optical disk.

Users can shoot up to 100 color photographs and take them to a local photo shop for processing. About an hour later the photos are available online for the user to download, or a little later they can be picked up on a CD-ROM and viewed using any compatible CD-ROM drive on a PC, Macintosh, or a Photo CD player attached to a TV.

RAID (redundant array of independent disks) (p. 276, KQ 6.4) Storage system that consists of 2–100 disk drives within a single cabinet. It sends data to the computer along several parallel paths simultaneously, which significantly improves response time.

The advantage of a RAID system is that it not only holds more data than a fixed disk drive within the same amount of space, but it also is more reliable because if one drive fails, others can take over.

read (p. 267, KQ 6.3) Computer activity in which data represented in the magnetized spots on the disk (or tape) are converted to electronic signals and transmitted to the primary storage (RAM) in the computer.

Read means the disk drive copies data—stored as magnetic spots—from the disk. Whereas reading simply makes a copy of the original data, without altering the original, writing actually replaces the data on the disk.

read-only (p. 278, KQ 6.5) Means the storage medium cannot be written on or erased by the user.

With read-only storage media, the user has access only to the data imprinted by the manufacturer.

recording density (p. 269, KQ 6.3) Refers to the number of bytes per inch of data that can be written onto the surface of the disk. New diskettes have one of two densities: high-density (HD), holding 1.44 MB of data, or extended density (ED), holding 2.8 MB. An ED drive is required for ED diskettes; it will also accept HD disks.

Users need to know what diskettes their system can use and consider the size of files they need to store on a diskette.

What It Is/What It Does	Why It's Important

removable-pack hard disk system (p. 275, KQ 6.4) Secondary storage with 6–20 hard disks, of 10½- or 14-inch diameter, aligned one above the other in a sealed unit. These removable hard-disk packs resemble a stack of phonograph records, except that there is space between disks to allow access arms to move in and out. Each access arm has two read/write heads—one reading the disk surface below, the other the disk surface above. However, only one of the read/write heads is activated at any given moment.

Such secondary storage systems enable a large computer system to store massive amounts of data.

sectors (p. 268, KQ 6.3) On a diskette, invisible wedge-shaped sections used by the computer for storage reference purposes.

When users save data from computer to diskette, it is distributed by tracks and sectors on the disk. That is, the system software uses the point at which a sector intersects a track to reference the data location in order to spin the disk and position the read/write head.

sequential storage (p. 265, KQ 6.1) Method of data storage by which data is stored in sequence, such as alphabetically.

Sequential storage is the least expensive form of storage because it uses magnetic tape, which is cheaper than disks. But finding data is time-consuming because it requires searching through all records that precede the record sought, such as searching the A through M records to find an N record.

small computer system interface (SCSI) (p. 273, KQ 6.4) An interface that allows the user to plug a number of peripheral devices linked end to end in a daisy chain into a single expansion board in the computer cabinet.

A SCSI interface allows one board to serve a number of peripherals, such as a tape drive, scanner, and external hard disk. It is used on Macintoshes and higher-end PCs.

tracks (p. 268, KQ 6.3) The rings on a diskette along which data is recorded. Unlike on a phonograph record, these tracks are neither visible grooves nor a single spiral. Rather, they are closed concentric rings. Each track is divided into sectors.

See sectors.

unformatted disk (p. 268, KQ 6.3) Diskette manufactured without tracks and sectors in place.

Unformatted diskettes must be formatted (initialized) by users before the disks can be used to store data.

write (p. 267, KQ 6.3) Computer activity in which data processed by the computer is transferred to a secondary-storage medium.

Write means the disk drive transfers data—represented as electronic signals within the computer's memory—onto the disk. Whereas *reading* simply makes a copy of the original data, without altering the original, *writing* replaces the data underneath it.

write-protect feature (p. 269, KQ 6.3) Feature of diskettes that prevents the disk from being written to, which would replace its data. By pressing a slide lever to uncover a small hole in the lower right corner of the disk, write-protection is activated; closing the slide allows the disk to be written to again.

This feature allows users to protect data on diskettes from accidental change or erasure.

Exercises

Self-Test

1. A(n) _____ is about 1 trillion bytes.

2. _____ is the data access method used by hard disks.

3. _____, a secondary storage device, is most commonly used for backup and archiving information.

4. A removable hard disk is called a _____ _____.

5. A(n) _____ is a removable disk on which data is written and read through the use of laser beams.

Short-Answer Questions

1. What is the significance of the terms *track* and *sector*?

2. What are some of the uses for CD-ROM technology?

3. What kinds of secondary-storage devices do large computer systems use?

4. What are the advantages of a hard disk over a diskette?

Multiple-Choice Questions

1. All diskettes must be _____ before they can store data.
   a. named
   b. saved
   c. retrieved
   d. formatted
   e. All of the above

2. Which of the following optical technologies can be used with the full range of electronic, television, and computer hardware?
   a. CD-ROM
   b. CD-R
   c. CDE
   d. DVD
   e. All of the above

3. Which of the following is considered a disadvantage of online storage?
   a. price
   b. safety
   c. portability
   d. slow speed
   e. None of the above

True/False Questions

T F   1. EIDE and SCSI describe two kinds of connections by which a hard disk is attached to a microcomputer.

T F   2. Hard disks may be affected by a head crash when particles on the read/write heads come into contact with the disk's surface.

T F   3. To use a diskette, you usually need a disk drive, but not always.

T F   4. Magnetic tape can handle only sequential data storage and retrieval.

T F   5. One advantage to using diskettes is that they aren't susceptible to extreme temperatures.

Knowledge in Action

1. You want to purchase a hard disk for use with your microcomputer. Because you don't want to have to upgrade your secondary-storage capacity in the near future, you are going to buy one with the highest storage capacity you can find. Use computer magazines or visit computer stores and find a hard disk you would like to buy. What is its capacity? How much does it cost? Who is the manufacturer? What are the system requirements? Is it an internal or an external drive? Can you install/connect this unit yourself? Why have you chosen this unit and this storage capacity?

2. What types of storage hardware are currently being used in the computer you use at school or at work? What are the storage capacities of these hardware devices? Would you recommend that alternate storage hardware be used? Why or why not?

3. Do you think books published on CD-ROMs will ever replace printed books? Why or why not? Look up some recent articles on this topic and prepare a short report.

4. Sometimes users forget to back up their work, or their backup tape/diskettes are lost or destroyed. What can you do, then, if your hard disk crashes and you have forgotten to back up your work? Look in the Yellow Pages of your phone book under Computer Disaster Recovery or Data Recovery. Call up the services listed and find out what they do, how they do it, and what they charge. Give a short report on the topic.

5. What optical technologies do you think you will use in your planned career or profession? How do you expect to use this technology? Which optical technologies have you already used? For what?

Tele-communications

The Uses of Online Resources & the Internet

key questions

You should be able to answer the following questions:

he white line dividing computers and telephones, voice and data, is blurring at last."

Why? Because "of a confluence of technology and demand—driven, to a huge degree, by the Internet phenomenon," says this *Business Week* analysis.[1] Companies that in the 1980s spent millions constructing their own private communications systems are now shifting over to the Internet and the graphics-oriented section of it called the World Wide Web. Because of its standard interfaces and low rates, "the Internet has been the great leveler for communications—the way the PC was for computing," says Boston analyst Virginia Brooks.[2]

Who's driving the Internet phenomenon?

Among millions of other users, here's Richard Shuster, 24, a recent college graduate who runs a site on the World Wide Web devoted to "Save the West Virginia River Chicken" *(http://pilot.msu.edu/user/shusterr/)*. His Web site attempts to enlist supporters in a fight to save this rare fowl, only about 200 of which exist today. Shuster did most of the work of developing the Web site himself, though some friends helped type in information.[3]

Here's businessman Neal Mazer, a resident of Santa Barbara, California, who has a brief stopover at San Francisco International Airport. Mazer is a frequent Internet user. But, he says, "It's always chaos when you're traveling. Unless you have a laptop and a ready phone jack, you don't have access" to the Internet or e-mail. So Mazer steps up to a small, nondescript kiosk (✔ pp. 76, 194) near a United Airlines passenger lounge and, after typing in his name, e-mail address, and password, checks out the weather for his destination: London. Now found in airline terminals around the world, the Internet kiosk allows users to exchange e-mail, send faxes, or explore the Web while waiting for flights. "It's the '90s version of the pay phone," says another passenger who tried it.[4]

And here's Martine Kempf, 39, a French inventor, telling why she was quickly drawn to the giant new Toscana apartment complex in Sunnyvale, in California's Silicon Valley. "[W]hen I saw this apartment," she says, "I realized I could have a nice place to live, plenty of sunlight, and a faster connection to the Internet." Each one of Toscana's 709 units comes equipped with a business-quality data line for high-speed Internet access. Moreover, the development has enough servers and communications gear to function as a full-fledged Internet service provider for its residents. Kempf's job is to build devices that permit voice control of wheelchairs and cars for disabled drivers. Through her rapid Internet access, she can call up technical information in seconds, faster than going to her bookshelf for the right manual. With the standard online access that most people have, the long wait would make this resource useless.[5]

This chapter describes the Internet, of course, but it is also concerned with the much wider world of communications—the uses of everything from fax machines to extranets. Just as you can learn to operate a car without knowing much about how it works, so you can learn to use telecommunications. And millions of people do. But to stay in the forefront of your career, it's imperative that you *do* know how things work, so that you can understand what's happening with the changes in the field. Accordingly, in this book we first explain *what* you can do with telecommunications; that's covered in this chapter. Then we go behind the scenes and describe *how* it all works;

that's covered in the next chapter. (If you wish, the chapters can be read in reverse order.)

7.1 The Practical Uses of Communications & Connectivity

KEY QUESTION

What are communications and connectivity?

Preview & Review: Communications, also called telecommunications, refers to the electronic collection and transfer of information from one location to another. The ability to connect devices by communications lines to other devices and sources of information is known as connectivity.

Users can use communications technology for a number of purposes, ranging from relatively simpler to more complex activities. They can use telephone-related services, such as fax and voice mail. They can do video/voice communication. They can use online information services, the Internet, and the Web for research, e-mail, games, travel services, teleshopping, and discussion and news groups, among other things. They can share resources through workgroup computing, Electronic Data Interchange (EDI), intranets, and extranets.

Clearly, communications is extending into every nook and cranny of civilization—the "plumbing of cyberspace," as it has been called. The term *cyberspace* was coined by William Gibson in his novel *Neuromancer.* In that book it refers to a futuristic computer network that people use by plugging their brains into it.

Today many people equate cyberspace with the Internet. But it is much more than that, says David Whittle in his book *Cyberspace: The Human Dimension.*[6] Cyberspace "includes not only the World Wide Web, chat rooms, online bulletin boards, and member-based services like America Online and CompuServe"—all features we explain in this chapter—"but also such things as conference calls and automatic teller machines."[7] To get to formal definitions, then, **cyberspace encompasses not only the computer online world and the Internet in particular but also the whole wired and wireless world of communications in general.**

Communications & Connectivity

***Communications,* also called *telecommunications,* refers to the electronic collection and transfer of information from one location to another.** The data being communicated may consist of voice, sound, text, video, graphics, or all of these. The electromagnetic instruments sending the data may be telegraph, telephone, cable, microwave, radio, or television. The distance may be as close as the next room or as far away as the outer edge of the solar system.

The television set is an instrument of communications, but it is a low-skill tool. That is, the many people of a mass audience receive one-way communications from a few communicators. This is why television (like AM/FM radio, newspapers, and music CDs) is called one of the *mass media.* Telephone systems are not mass media, since they involve two-way communications of many to many. But they, too, are low-skill communications tools. By contrast, linkages of microcomputers have allowed a few people with a fairly high level of skill to achieve two-way communication with a few others. **The ability to connect devices by communications lines to other devices and sources of information is known as *connectivity.*** Traditionally, computers have offered greater varieties of connectivity than have other communications devices.

Tools of Communications & Connectivity

What kinds of options do communications and connectivity give you? Let us consider the possibilities. We will take them in order, more or less, from relatively simpler to more complex activities. (■ *See Panel 7.1.*) They include:

- Telephone-related communications services: fax messages and voice mail
- Video/voice communication: videoconferencing and picture phones
- Online information services
- The Internet
- The World Wide Web part of the Internet
- Shared resources: workgroup computing, Electronic Data Interchange (EDI), intranets, and extranets
- New Internet technologies

■ PANEL 7.1

The world of connectivity

Wired or wireless communications links offer several options for information and communications.

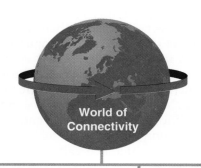

World of Connectivity

Telephone-related services

Fax messages

Voice mail

Video/voice communication

Videoconferencing

Picture phone

Online information services

Research and news
E-mail, BBSs, chat
Games, entertainment
Travel services
Shopping
Other

The Internet

E-mail
Usenet news
groups
Mailing lists
FTP
Telenet
Gopherspace
World Wide Web
Services offered
by online information
service

7.2 Telephone-Related Communications Services

KEY QUESTION

How do fax and voice mail work?

Preview & Review: Telephone-related communications include fax messages, transmitted by dedicated fax machines and fax boards, and voice mail, voice messages stored in digitized form.

Phone systems and computer systems have begun to fuse together. Services available through telephone connections, whether the conventional wired kind or the wireless cellular-phone type, include fax messages and voice mail.

Fax Messages

Asking "What is your fax number?" is about as common a question in the work world today as asking for someone's telephone number. Indeed, the majority of business cards include both a telephone number and a fax number. Recall from Chapter 1 that *fax* **stands for "facsimile transmission," or reproduction.**

A fax may be sent by dedicated fax machine or by fax board (✔ p. 207).

- Dedicated fax machines: *Dedicated fax machines* **are specialized devices that do nothing except send and receive documents over transmission lines to and from other fax machines.** These are the stand-alone machines nowadays found everywhere, from offices to airports to instant-printing shops.

Shared resources

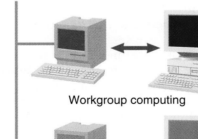

Workgroup computing

Electronic Data Interchange

Intranets
Extranets

● Fax boards: **A *fax board*, which is installed as a circuit board inside a computer's system cabinet, is a modem with fax capability. It enables you to send signals directly from your computer to someone else's fax machine or fax board.** Computer faxing can help eliminate paper—you don't have to buy fax paper or make sure your fax machine is full. However, you have to keep your computer turned on, and faxing can tie up your modem. Also, with a fax board you can't transmit hardcopy pages unless you also have a scanner for inputting the document.

Both fax machines and fax boards now automatically allow you to delay sending faxes until the phone rates drop at night, and they can "broadcast" a single fax to many different fax numbers.

New technology, such as the Panafax, can send faxes as e-mail over the Internet, as well as conventional faxes over phone lines.[8] Another device, called FaxPal, will store incoming faxes even while a computer is shut off.[9]

Voice Mail

You don't even need to have a fixed address in order to use voice mail. Carl Hygrant, a homeless person in New York, found that the technology helped him get work. Earlier, when he put down on his résumé the phone number of the Bronx shelter that was his temporary home, prospective employers would lose interest when they called. After a telephone company launched an experimental program giving homeless people their individual voice mail, Hygrant landed a job.[10]

Like a sophisticated telephone answering machine, ***voice mail* digitizes incoming voice messages and stores them in the recipient's "voice mailbox" in digitized form. It then converts the digitized versions back to voice messages when they are retrieved.**

Voice-mail systems also allow callers to deliver the same message to many people within an organization by pressing a single key. They can forward calls to the recipient's home or hotel. They allow the person checking messages to speed through them or to slow them down. He or she can save some messages and erase others and can dictate replies that the system will send out.

The main benefit of voice mail is that it helps eliminate "telephone tag." With regular phone calls, more than 80% don't reach the intended party. With voice mail, two callers can continue to exchange messages even when they can't reach each other directly. Replying to a message can also be efficient, because it eliminates the small talk that often goes with phone conversations.

7.3 Video/Voice Communication: Videoconferencing & Picture Phones

KEY QUESTION

What are the characteristics of videoconferencing and picture phones?

Preview & Review: Videoconferencing is the use of television, sound, and computer technology to enable people in different locations to see, hear, and talk with one another. Videoconferencing could lead to V-mail, or video mail, which allows video messages to be sent, stored, and retrieved like e-mail. The picture phone is a telephone with a TV-like screen and a built-in camera that allows you to see the person you're calling, and vice versa.

Want to have a meeting with people on the other side of the country or the world but don't want the hassle of travel? You may have heard of or participated in a *conference call*, also known as *audio teleconferencing*, a meeting in which people in different geographical locations talk on the telephone. Now we have video plus voice communication, specifically *videoconferencing* and *picture phones*.

Videoconferencing

"I was a little nervous about going in front of the camera," says Mark Dillard, "but I calmed down pretty quickly after we got going, and it went well."[11]

Interviewing for a job can be an uncomfortable event for a lot of people. However, Dillard had just undergone the high-tech version of it: sitting in front of a video camera in a booth at a local Kinko's store in Atlanta and talking to a job recruiter in New York.

***Videoconferencing*, also called *teleconferencing*, is the use of television video and sound technology as well as computers to enable people in different locations to see, hear, and talk with one another.** Videoconferencing can still consist of people meeting in separate conference rooms or booths with specially equipped television cameras. However, now videoconferencing equipment, such as Intel's Proshare hardware and software, can be set up on people's desks, with a camera and microphone to capture the person speaking and a monitor and speakers for the person being spoken to. (Modems or network interface cards and sound and video-capture cards are also needed.) (■ *See Panel 7.2.*)

A relatively new development is an initiative to deliver *V-mail*, or *video mail*, video messages that are sent, stored, and retrieved like e-mail. One version would use the Proshare Windows-based videoconferencing product and Oracle's Media Server, a computer storage system developed for movies-on-demand technologies.

Videoconferencing is still problematic for most people. The main difficulty is that the standard copper wire in what the telecommunications industry calls POTS—for "plain old telephone service"—has been unable to communicate images very rapidly. Thus, unless you are with an organization that can afford expensive high-speed communications lines, present-day videoconferencing screens will convey a series of jerky, freeze-frame or stop-action still images of the faces of the communicating parties.

■ PANEL 7.2

Videoconferencing
Face-to-face communication via desktop computer

Picture Phones

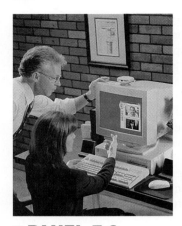

The *picture phone* is a telephone with a TV-like screen and a built-in camera that allows you to see the person you're calling, and vice versa. An example is AT&T's VideoPhone. (■ *See Panel 7.3.*)

The idea of the picture phone has been around since 1964, when AT&T showed its Picturephone at the New York World's Fair. However, as with videoconferencing, the main difficulty is that the old POTS lines don't communicate images rapidly enough. However, as we discuss in Chapter 8, faster connections are being installed in many places that can transmit more visual information. Moreover, software is available that can compress images quickly, delivering video as well as audio images in real time even over old-fashioned copper wires. Finally, the traditional telephone companies are starting to get some competition in phone service from other quarters, such as cable companies and even electric utilities, which may result in more and cheaper high-speed lines becoming available.

■ PANEL 7.3

A picture phone

7.4 Online Information Services

KEY QUESTIONS

What are online services, and what do they offer?

Preview & Review: For a monthly fee, commercial companies called online information services, which predate popular use of the Internet, provide software that connects users' computers and modems over phone lines with facilities that provide e-mail, research, shopping, financial services, and Internet connections, among a host of other services.

Before the use of the Internet became common and before the World Wide Web was developed, computer users used online information services to access many sources of information and services. **An *online information service* provides access, for a fee, to all kinds of databases and electronic meeting places to subscribers equipped with telephone-linked microcomputers.** There are several online services, but four are considered the most mainstream. They are:[12]

- America Online (AOL)—with nearly 10 million subscribers
- CompuServe—about 7.5 million
- Microsoft Network (MSN)—about 3 million
- Prodigy—about 1 million

There are also about a million subscribers enrolled in small online services, which measure their membership in the hundreds rather than the millions. Examples are The WELL and Women's Wire in the San Francisco Bay Area and ECHO (East Coast Hang Out) in New York. Still others—Dialog, Dow Jones News/Retrieval, Nexis, Lexis—may principally be considered huge collections of databases rather than department-store-like online services.

Getting Access

To gain access to online services, you need a computer and printer so you can download (transfer) and print out online materials. You also need a modem. Finally, you need communications software so your computer can communicate via modem and interact with the online service's computers. (We described the fundamentals of going online with an online service in the

Experience Box at the end of Chapter 2.) America Online, CompuServe, and other mainstream services provide subscribers with their own software for going online, but you can also buy communications programs separately, such as ProComm Plus.

Opening an account with an online service requires a credit card, and billing policies resemble those used by cable-TV and telephone companies. As with cable TV, you may be charged a fee for basic service, with additional fees for specialized services. In addition, the online service may charge you for the time spent while on the line. (The most popular online services charge $10–20 per month for unlimited use of general services.)

Finally, you will also be charged by your telephone company for your time on the line, just as when making a regular phone call. However, most information services offer local access numbers. Thus, unless you live in a rural area, you will not be paying long-distance phone charges. All told, the typical user may pay $20—$40 a month to use an online service. Obsessive users, however, can run up much larger bills, as can users of nongeneral services—for example, research on privately owned databases such as Knowledge Index and Data Quest.

The Offerings of Online Services

Although the Internet offers the same information as online services, many users still prefer online services because the information is packaged—organized, filtered, and put in user-friendly form.

What kinds of things could you use an online service for? Here are some of the options:

- **People connections—e-mail, message boards, chat rooms:** Online services can provide a community through which you can connect with people with kindred interests (without identifying yourself, if you prefer). The primary means for making people connections are via e-mail, message boards, and "chat rooms."

  E-mail will be discussed in detail shortly. *Message boards* allow you to post and read messages on any of thousands of special topics. *Chat rooms* are discussion areas in which you may join with others in a real-time "conversation," typed in through your keyboard. The topic may be general or specific, and the collective chat-room conversation scrolls on the screen.

- **Research and news:** The only restriction on the amount of research you can do online is the limit on whatever credit card you are charging your time to, if you are not using a free database. Depending on the online service, you can avail yourself of several encyclopedias, such as *Compton's Interactive Encyclopedia* and *Grolier Academic American Encyclopedia.*

  Many online services also offer access, for a fee, to databases of unabridged text from newspapers, journals, and magazines. Indeed, the information resources available online are mind-boggling, impossible to describe in this short space.

- **Games, entertainment, and clubs:** Online computer games are extremely popular. In single-player games, you play against the computer. In multiplayer games, you play against others, whether someone in your household or someone overseas.

  Other entertainments include cartoons, sound clips, pictures of show-business celebrities, and reviews of movies and CDs. You can also join online clubs with others who share your interests, whether science fiction, popular music, or cooking.

- **Free software:** Many users download freeware, shareware (✔ p. 86), and commercial demonstration programs from online sources. They can also download software updates (called patches).

- **Travel services:** Online services use Eaasy Sabre or Travelshopper, streamlined versions of the reservations systems travel agents use. You can search for flights and book reservations through the computer and have tickets sent to you by Federal Express. You can also refer to weather maps, which show regions of interest. In addition, you can review hotel directories, such as the ABC Worldwide Hotel Guide, and restaurant guides, such as the Zagat Restaurant Directory.

- **Shopping:** If you can't stand parking hassles, limited store hours, and checkout lines, online services may provide a shopping alternative. CompuServe, for instance, offers 24-hour shopping with its Electronic Mall. This feature lists products from over 100 retail stores, discount wholesalers, specialty shops, and catalog companies. You can scan through listings of merchandise, order something on a credit card with a few keystrokes, and have the goods delivered by a shipping service like UPS or the postal service.

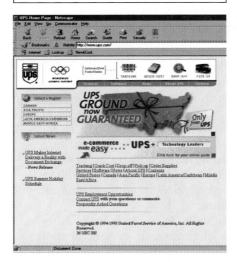

  In some cities it's even possible to order groceries through online services. Peapod Inc. is an online grocery service serving 10,000 households in the Chicago and San Francisco areas. Peapod offers more than 18,000 items, from laundry detergent to lettuce, available in Jewel/Osco in Chicago or Safeway in San Francisco. Users can shop by brand name, category, or store aisle, and they can use coupons. Specially trained Peapod shoppers handle each order, even selecting the best produce available. The orders are then delivered in temperature-controlled containers.

- **Financial management:** Online services also offer access to investment brokerages so that you can invest money and keep tabs on your portfolio and on the stock market.

Will Online Services Survive the Internet?

The online services have a lot to offer. One survey two years ago, for instance, found that *half* of the people on the Net got there through commercial services, which suggests they may be among the easiest ways to get to the Web.[13] In addition, the online services package information so that you can more quickly and easily find what you're looking for. It's also easier to conduct a live "chat" session on an online service than it is on the Web, and it is easier for parents to exert control over the kinds of materials their children may view.

  It's been suggested that the war of online services is over and that America Online has won it, signing up several million more subscribers than its rivals. But many observers believe the Internet and particularly the World Wide Web threaten to swamp the online services. As the Net and the Web have become easier to navigate, online services have begun to lose customers and content providers—even as they have added their own arrangements for accessing the Internet. Indeed, many of the same kinds of things the online services offer are now directly accessible through the Internet.

7.5 The Internet

KEY QUESTION

What are the Internet, its connections, its addresses, and its features?

Preview & Review: The Internet, the world's biggest network, uses a protocol called TCP/IP to allow computers to communicate. Users can connect to the Internet via direct connections, online information services, and Internet service providers. Its features include e-mail, Usenet newsgroups, mailing lists, FTP, Gopherspace, Telnet, and the World Wide Web.

The war among the online services was Round 1. Round 2 has already started. INTERNET SURFING IS A SNAP, reads the newspaper headline. WEB SITE APES ONLINE SERVICE.[14]

Snap Online is a Web site, but it would rather be considered an online service, competing against giant AOL. How can it do this? "As the Internet expands, more and more content providers are making their material broadly available on the Web," says the story following the headline. "As a result, information once available only through commercial services such as AOL is now all over the place. You can get timely sports scores, weather reports, and stock quotes just as easily on Snap . . . as you can on AOL." Indeed, Snap is what is known as one of the new Internet "portals," or gateways, as we shall describe.

How widespread is use of the Internet? Consider these facts:

- The Internet is growing faster than all other technologies that have preceded it. Radio existed for 38 years before it had 50 million listeners, and television took 13 years to reach that number of viewers.[15]

- In 1994, only 3 million people were connected to the Internet. By the end of 1997, more than 100 million were using it.[16]

- One in four adults in the USA and Canada now use the Internet, of which the World Wide Web is just one part.[17]

- The number of personal computers hooked up to the Internet is expected to triple to 268 million by 2001, according to Dataquest Inc., a San Jose, California, market research firm.[18]

Called "the mother of all networks," the ***Internet*, or simply "the Net," is an international network connecting approximately 140,000 smaller networks in more than 200 countries.** These networks are formed by educational, commercial, nonprofit, government, and military entities. Each of the small autonomous networks on the Internet makes its own decision about what resources to make available on the Internet. There is no single authority that controls the Net overall.

Try as you may, you cannot imagine how much data is available on the Internet. Besides e-mail, chat rooms, message boards, games, and free software, there are thousands of databases containing information of all sorts. Here is a sampling:

The Library of Congress card catalog. The daily White House press releases. Weather maps and forecasts. Schedules of professional sports teams. Weekly Nielsen television ratings. Recipe archives. The Central Intelligence Agency world map. A ZIP Code guide. The National Family Database. Project Gutenberg (offering the complete text of many works of literature). The Alcoholism Research Data Base. Guitar chords. U.S. government addresses and phone (and fax) numbers. *The Simpsons* archive.[19]

And those are just a few droplets from what is a Niagara Falls of information.

To connect with the Internet, you need pretty much the same things you need to connect with online information services: a computer, modem and telephone line (or other network connection), and appropriate communications software.

Created by the U.S. Department of Defense in 1969 (under the name ARPAnet—ARPA was the department's Advanced Research Project Agency), the Internet was built to serve two purposes. The first was to share research among military, industry, and university sources. The second (a rationale actually developed later) was to provide a diversified system for sustaining communication among military units in the event of nuclear attack. Thus, the system was designed to allow many routes among many computers, so that a message could arrive at its destination by many possible ways, not just a single path. This original network system was largely based on the Unix operating system (✔ p. 122).

With the many different kinds of computers being connected, engineers had to find a way for the computers to speak the same language. The solution developed was *TCP/IP*, the standard since 1983 and the heart of the Internet. **TCP/IP, for *Transmission Control Protocol/Internet Protocol*, is the standardized set of computer guidelines (protocols) that allow different computers on different networks to communicate with each other efficiently.** (We discuss the concept of protocols in Chapter 8.) The effect is to make the Internet appear to the user to operate as a single network.

Connecting to the Internet

There are three ways to connect your microcomputer with the Internet:

- **Through school or work:** Universities, colleges, and most large businesses have high-speed phone lines that provide a direct connection to the Internet. This type of connection is known as dedicated access. *Dedicated access* means a communication line is used that is designed for one purpose.

  If you're a student, this may be the best deal because the connection is free or low cost. However, if you live off-campus and want to get this Internet connection from home, you probably won't be able to do so. To use a direct connection, your microcomputer must have TCP/IP software and be connected to the local network that has the direct-line connection to the Net.

- **Through online information services:** As mentioned, subscribing to a commercial online information service, which provides you with its own communications software, may not be the cheapest way to connect to the Internet, but in the past, at least, it was one of the easiest. In this case, the online service acts as an electronic "gateway" to the Internet.

- **Through Internet service providers (ISPs):** *Internet service providers (ISPs)* **are local or national companies that will provide public access to the Internet (and World Wide Web) for a flat monthly fee.** Essentially an ISP is a small network connected to the high-speed communications links that make up the Internet's backbone—the major supercomputer sites and educational and research foundations within the United States and throughout the world. There are more than 5000 ISPs in the United States, and the field is growing rapidly.[20]

  Once you have contacted an ISP and paid the required fee, the ISP will provide you with information about phone numbers for a local

ISP company	Toll-free number
NETCOM	800-538-2551
AT&T WorldNet	800-967-5363
network MCI	800-550-0927
SPRYNET	800-777-9638
EarthLink Network	800-395-8425

connection, called a *point of presence (POP)*—a server owned by the ISP or leased from a common carrier, such as AT&T. The ISP also provides software for setting up your computer and modem to dial into their network of servers, which involves acquiring a user name ("user ID") and a password. After this, you can use a browser, such as Netscape Navigator or Microsoft Explorer, to find your way around the World Wide Web, the graphical part of the Internet. (Or if you're an experienced Internet user, you can type in Unix-based commands.)

So far, most ISPs have been small and limited in geographic coverage. Among the largest national companies are NETCOM, AT&T WorldNet, network MCI, SPRYNET, and EarthLink Network. Other competitors are MindSpring Enterprises, UUNet Technologies, BBN Corporation, Pacific Bell Internet, and Tele-Communication Inc.'s @Home (pronounced "At Home"). Clearly, this is an area of fierce competition, but the presence of the phone and cable-TV companies in particular could help expand the mass market for Internet services.

You can ask someone who is already on the Web to access the worldwide list of ISPs at *http://www.thelist.com.* This site presents pricing data and describes the features supported by each ISP.

Internet Addresses

To send and receive e-mail on the Internet and interact with other networks, you need an Internet address. In the *Domain Name System*, the Internet's addressing scheme, an Internet address usually has two sections. (■ *See Panel 7.4, next page.*)

Consider the following address.

president@whitehouse.gov.us

The first section, the *userID*, tells "who" is at the address—in this case, *president* is the recipient. (Sometimes an underscore, or _, is used between a recipient's first name or initials and last name: bill_clinton, for example.)

The first and second sections are separated by an @ (called "at") symbol.

The second section—in this case, *whitehouse.gov.us*—tells "where" the address is. Components are separated by periods, called "dots." The second section includes the *location* (such as *whitehouse*; the location may have more than one part), the *top-level domain* (such as *.gov, .edu,* and *.com,* as we shall explain), and country if required (such as *.us* for United States and *.se* for Sweden).

Among the top-level domains are the following:

- *.com* for commercial organizations
- *.edu* for educational and research organizations
- *.firm* for certain types of businesses
- *.gov* for governmental organizations
- *.info* for distributors of information
- *.int* for international
- *.mil* for military organizations
- *.net* for gateway or host networks
- *.nom* for individual users

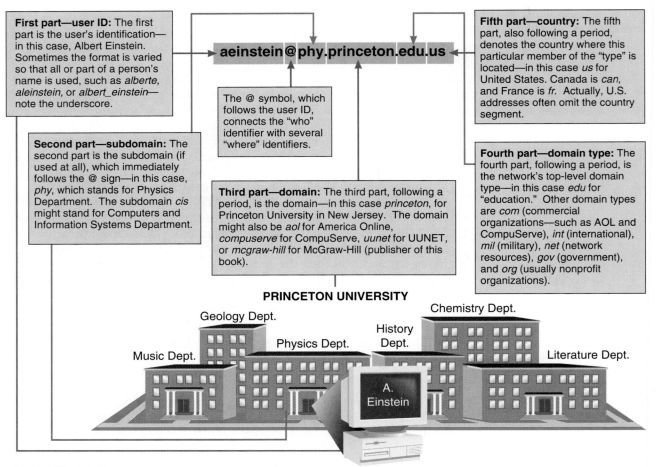

First part—user ID: The first part is the user's identification—in this case, Albert Einstein. Sometimes the format is varied so that all or part of a person's name is used, such as *alberte*, *aleinstein*, or *albert_einstein*—note the underscore.

Second part—subdomain: The second part is the subdomain (if used at all), which immediately follows the @ sign—in this case, *phy*, which stands for Physics Department. The subdomain *cis* might stand for Computers and Information Systems Department.

The @ symbol, which follows the user ID, connects the "who" identifier with several "where" identifiers.

aeinstein @ phy.princeton.edu.us

Fifth part—country: The fifth part, also following a period, denotes the country where this particular member of the "type" is located—in this case *us* for United States. Canada is *can*, and France is *fr*. Actually, U.S. addresses often omit the country segment.

Third part—domain: The third part, following a period, is the domain—in this case *princeton*, for Princeton University in New Jersey. The domain might also be *aol* for America Online, *compuserve* for CompuServe, *uunet* for UUNET, or *mcgraw-hill* for McGraw-Hill (publisher of this book).

Fourth part—domain type: The fourth part, following a period, is the network's top-level domain type—in this case *edu* for "education." Other domain types are *com* (commercial organizations—such as AOL and CompuServe), *int* (international), *mil* (military), *net* (network resources), *gov* (government), and *org* (usually nonprofit organizations).

PRINCETON UNIVERSITY

Geology Dept.

Chemistry Dept.

Music Dept.

Physics Dept.

History Dept.

Literature Dept.

A. Einstein

■ PANEL 7.4

What an Internet address means

How an e-mail message might find its way to a hypothetical address for Albert Einstein in the Physics Department of Princeton University.

- *.org* for nonprofit or miscellaneous organizations
- *.rec* for groups involved in recreational activities
- *.store* for retailers
- *.web* for businesses related to the Web

Domain names are registered with an international agency called Network Solutions, Inc. (NSI), a Virginia-based company contracted by the U.S. National Science Foundation. This agency regulates registration services for the Internet and issues new domain names for a fee.

Features of the Internet

As we have mentioned, the Internet offers access to the same information and services that have been available through online information services—plus much more. "For many people, the Internet has subsumed the functions of libraries, telephones, televisions, catalogs—even support groups and singles bars," says writer Jared Sandberg. "And that's just a sample of its capabilities."[21] (■ *See Panel 7.5.*)

Let us consider the Internet tools at your disposal:

- **E-mail:** "The World Wide Web is getting all the headlines, but for many people the main attraction of the Internet is electronic mail,"

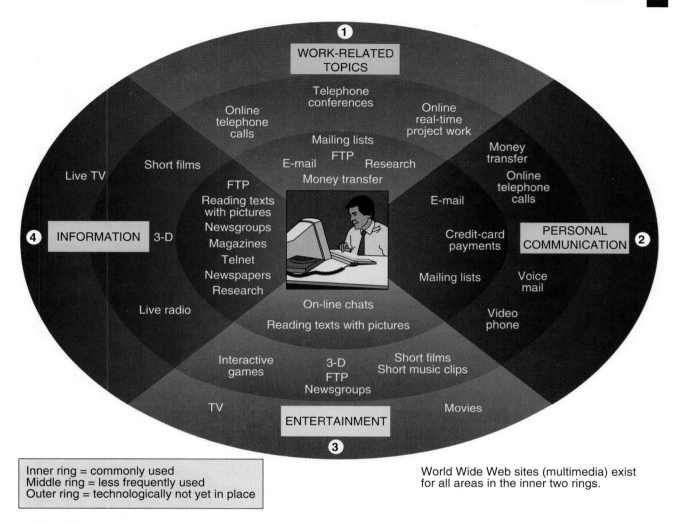

① WORK-RELATED TOPICS

Telephone conferences

Online telephone calls

Online real-time project work

Mailing lists

FTP

E-mail   Research

Money transfer

Money transfer

Live TV   Short films

FTP
Reading texts with pictures
Newsgroups
Magazines
Telnet
Newspapers
Research

E-mail

④ INFORMATION   3-D

PERSONAL COMMUNICATION ②

Credit-card payments

Online telephone calls

Mailing lists   Voice mail

Live radio

On-line chats

Reading texts with pictures

Video phone

Interactive games

3-D
FTP
Newsgroups

Short films
Short music clips

TV

Movies

ENTERTAINMENT

③

Inner ring = commonly used
Middle ring = less frequently used
Outer ring = technologically not yet in place

World Wide Web sites (multimedia) exist for all areas in the inner two rings.

■ PANEL 7.5

The Internet
What's available through the network of all networks.

says technology writer David Einstein.[22] There are millions of users of e-mail in the world, and although half of them are on private corporate networks, a great many of the rest are on the Internet.

"E-mail is so clearly superior to paper mail for so many purposes," writes *New York Times* computer writer Peter Lewis, "that most people who try it cannot imagine going back to working without it."[23] Says another writer, e-mail "occupies a psychological space all its own: It's almost as immediate as a phone call, but if you need to, you can think about what you're going to say for days and reply when it's convenient."[24]

***E-mail,* or *electronic mail*, links computers by wired or wireless connections and allows users, via electronic mailboxes, to send and receive messages.** With e-mail, you enter the recipient's e-mail address, type in the subject and the message, then click on the Send icon. The message will be stored in the recipient's mailbox—that is, on a server accessible by the recipient—until it is accessed by that person. To gain access to your mailbox, you connect to your e-mail system and type in your *password*, a secret word or numbers that limit access. You may download (transfer) to your hard disk your mail and read the list of senders. You can discard messages without reading (opening) them, and you can read messages and then discard, save, and/or copy them to other people.

With most e-mail programs, you can permanently store your password so that you can use the Get mail menu option/icon without having to type the password in each time you retrieve mail. You can also use the Create mail/New message option to automatically go to a blank message form.

Foremost among the Internet e-mail programs is Qualcomm's Eudora software, which is used by the majority of the educational institutions on the Net. Eudora Light is free and comes bundled with other Internet products. (If you're connected to an online information service, such as AOL, you would use the e-mail system provided by the service.) Windows 98 and Office 98's e-mail features may turn out to be a major competitor to Eudora.

If you're part of a company, university, or other large organization, you may get e-mail for free. Otherwise you can sign up with a commercial online service (America Online, CompuServe, Microsoft Network), e-mail service (such as MCI Mail), or Internet access provider (such as EarthLink or MindSpring).

E-mail has both advantages and disadvantages:[25–28] Like voice mail, it helps people avoid playing phone tag or coping with paper and stamps. A message can be as simple as a birthday greeting or as complex and lengthy as a report with supporting documents (including attached text, graphic, video and/or sound files). It can be quicker than a fax message and more organized than a voice-mail message. By reading the list of senders and topics displayed on the screen you can quickly decide which messages are important. Also, e-mail software automatically creates an archive of all sent and received messages. Sending an e-mail message usually costs as little as a local phone call or less but it can go across many time zones and be read at any time. Some e-mail messages are now received as voice mail and read aloud to the recipient (so that, for example, you could catch up on your e-mail messages on your car phone en route to work).

Nevertheless there are some problems: You might have to sort through a blizzard of messages every day, a form of junk mail brought about by the ease with which anyone can send duplicate copies of a message to many people. "I know people who routinely get 50 to 100 e-mails per day," says David De Long, of the consulting firm Ernst & Young, "and some receive as many as 500."[29] Your messages are far from private and may be read by e-mail system operators and others (such as employers); thus, experts recommend you think of e-mail as a postcard rather than a private letter.

Mail that travels via the Internet often takes a circuitous route, bouncing around various computers in the country in an effort to find the fastest and most efficient route. Although a lot of messages may go through in a minute's time, others may be hung up because of system overload, taking hours and even days. Finally, if users let their e-mail messages pile up, they may ultimately fill up their allocated space on the server that is storing them (though some systems automatically delete stored messages after a period of time).

The e-mail boom is only just beginning. The U.S. Postal Service has begun to offer e-mail with features of first-class mail, including "postmarks" and return receipts. Telephone companies are offering phones with small screens for displaying e-mail sent through their e-mail centers.

● **Usenet newsgroups—electronic discussion groups:** One of the Internet's most interesting features goes under the name *usenet*, short for "user network," which is essentially a giant, dispersed bulletin board. **Usenet newsgroups are electronic discussion groups that focus on a specific topic,** the equivalent of AOL's or CompuServe's "forums." They are one of the most lively and heavily trafficked areas of the Net.

Usenet users exchange e-mail and messages ("news"). "Users post questions, answers, general information, and FAQ files on Usenet," says one online specialist. "The flow of messages, or 'articles,' is phenomenal, and you can get easily hooked."[30] Pronounced "fack," a **FAQ, for** *frequently asked questions,* **is a file that lays out the basics for a newsgroup's discussion.** It's always best to read a newsgroup's FAQ before joining the discussion or posting (asking) questions.

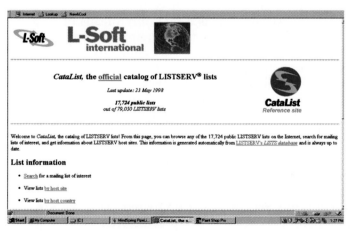

There are more than 15,000 Usenet newsgroup forums and they cover hundreds of topics. Examples are *rec.arts.startrek.info, soc.culture.african.american,* and *misc.jobs.offered.* The first part is the group— *rec* for recreation, *soc* for social issues, *comp* for computers, *biz* for business, *sci* for science, *misc* for miscellaneous. The next part is the subject—for example, *rec.food.cooking.* The category called *alt* news groups offers more free-form topics, such as *alt.rock-n-roll.metal* or *alt.internet.services.*

● **Mailing lists—e-mail-based discussion groups:** Combining e-mail and newsgroups, mailing lists—called *listservs*—allow anyone to subscribe (generally free) to an e-mail mailing list on a particular subject or subjects and post messages. The mailing-list sponsor then sends those messages to everyone else on that list. Thus, newsgroup listserv messages appear automatically in your mailbox; you do not have to make the effort of accessing the newsgroup. (As a result, it's necessary to download and delete mail almost every day, or your mailbox will quickly become full.) There are more than 3000 electronic mailing-list discussion groups.

● **FTP—for copying all the free files you want:** Many Net users enjoy "FTPing"—cruising the system and checking into some of the tens of thousands of FTP sites, which predate the Web, offering interesting free files to copy (download). **FTP, for** *File Transfer Protocol,* **is a method whereby you can connect to a remote computer called an FTP site and transfer publicly available files to your own microcomputer's hard disk.** The free files offered cover nearly anything that can be stored on a computer: software, games, photos, maps, art, music, books, statistics. You should be sure to scan all downloaded files and programs for viruses (✔ p. 128) before opening them.

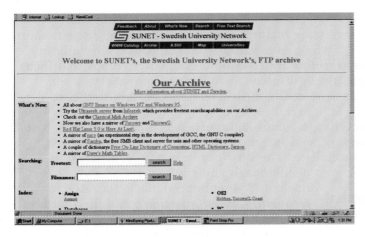

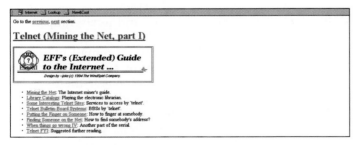

Some 2000-plus FTP sites (so-called *anonymous FTP sites*) are open to anyone; others can be accessed only by knowing a password. You can also use FTP to upload (transfer) your files to an FTP site.

● **Gopherspace—the hierarchical, text-based menu system:** Several tools exist to help sift through the staggering amount of information on the Internet, but one of the most important has been Gopher, which, like FTP, predates the Web. *Gopher* is a uniform system of menus, or series of lists, that allows users to easily browse and retrieve files stored on different computers by making successive menu selections. Why is it called "Gopher"? Because the first gopher was developed at the home of the Golden Gophers, the University of Minnesota, and it helps you "go fer" the files you seek. There are thousands of Gopher servers hooked up to the Internet— "Gopherspace."

● **Telnet—to connect to remote computers:** *Telnet* is a terminal emulation protocol that allows you to connect (log on) to remote computers. This feature, which allows microcomputers to communicate successfully with mainframes, enables you to tap into Internet computers and access public files as though you were connected directly instead of, for example, through your ISP site.

Although it is a text-only means of communication, the Telnet feature is especially useful for perusing large databases or library card catalogs. There are perhaps 1000 library catalogs accessible through the Internet, and a few thousand more Internet sites around the world have Telnet interfaces. Telnet programs are also usually provided by ISPs and information services, as well as some operating systems. (FTP sites, Gopher resources, and Telnet sites can also be accessed using a Web browser.)

One last feature of the Internet remains to be discussed—perhaps, for most general users, the most important one: the World Wide Web.

7.6 The World Wide Web

KEY QUESTIONS

What are the attributes of the Web, and how can you find information on it?

Preview & Review: The Web is the graphics-based component of the Internet, which makes it easier to use. Two distinctive features are that it provides information in multimedia form and uses hypertext to link various resources. Information is found using browsers and directories and search engines.

Police inspector José Berrios of San Juan, Puerto Rico, was "surfing"—searching through—the World Wide Web when he came upon a Web site with photographs of missing children. Two hours and 500 photos later, he found what international agencies had been unable to resolve for three months—the name of an 8-year-old girl who did not know who she really was.

The Web photo showed her at a much earlier age, 14 months, when she was alleged to have been kidnapped from her parents' home in California. Several years later in Puerto Rico, she had come to the attention of police by being involved in a child abuse case. Said a law-enforcement administrator about the discovery of the girl's name, "Our agents were jumping around they were so excited."[31]

The Web is surely one of the most exciting phenomena of our time. The fastest-growing part of the Internet (growing at perhaps 4% per month in number of users), the World Wide Web is the most graphically inviting and easily navigable section of it. **The *World Wide Web*, or simply "the Web," consists of an interconnected system of sites, or servers, all over the world that can store information in multimedia form—sounds, photos, and video, as well as text. The sites share a form consisting of a hypertext series of links that connect similar words and phrases.** Note two distinctive features:

1. Multimedia form: Whereas Gopher and Telnet deal with text, the Web provides information in *multimedia* form—graphics, video, and audio as well as text. You can still access Gopher, FTP, and the like through the Web, but the Web offers capabilities not offered by these more restricted forerunner Internet tools.

README

Practical Matters: Managing Your E-Mail

1. Don't use your electronic mailbox as a things-to-do list. Instead, create a second mailbox or folder for e-mail messages that still need to be answered or acted on. This will prevent important new messages from getting lost.

2. Read all of your e-mail as soon as it arrives and file it away immediately.

3. Don't create too many folders—otherwise, you'll lose messages that you file. Instead, adopt a simple message filing system and stick to it.

4. Don't save long messages with the thought that you will get around to them later. By the time later arrives, you'll have received even more e-mail.

5. Create a new set of folders every year; copy the previous year's correspondence onto a floppy disk. That way, if you change mail providers (or jobs), you won't lose all of your personal letters.

6. You don't have to reply to every e-mail message that you get. If you do, and your correspondents do as well, then the number of messages you get every day will increase geometrically.

7. Do not send chain-letters. If you get a chain-letter, just delete it. They may seem funny, but they clog mail systems and have shut down networks.

8. If somebody sends a request for help to a mailing list that you are on, send your response directly to that person, rather than to the entire list.

9. If you are on a mailing list that has too much traffic for you, don't make things worse by sending mail to the list asking people to send less mail to the list. Just have yourself taken off.

10. If somebody sends you a flame [an insulting message], don't make things worse by broadening the scope of the disaster. If you feel compelled to send mail back to the flamer, send it just to him or her.

—Simson L. Garfinkel, "Managing Your Mail," *San Jose Mercury News*

2. **Use of hypertext:** Whereas Gopher is a menu-based approach to accessing Net resources, the Web uses a hypertext format. *Hypertext* **is a system in which documents scattered across many Internet sites are directly linked, so that a word or phrase in one document becomes a connection to a document in a different place.**

The Web: A Working Vocabulary

If a Rip Van Winkle fell asleep as recently as 1989 (the year computer scientist Tim Berners-Lee developed the Web software) and awoke today, he would be completely baffled by the new vocabulary that we now encounter on an almost daily basis: *Web site, home page, http://.* Let's see how we would explain to him what these and similar Web terms mean.

- **HTML—instructions for document links:** The format, or language, used on the Web is called hypertext markup language. *Hypertext markup language (HTML)* **is the set of special instructions, called tags or markups, that are used to specify document structure, formatting, and links to other documents.** Anyone can create HTML documents by using Web-page design software (discussed below).

- **http://—communications standard for the Web:** HTML documents travel back and forth using hypertext transfer protocol. *Hypertext transfer protocol*—which is expressed as *http://*—is the **communications standard (protocol) used to transfer information on the Web.** This appears as a prefix on Web addresses, such as *http://www.mcgraw_hill.com.* When you use your mouse to point-and-click on a hypertext link—a highlighted word or phrase—it may become a doorway to another place within the same document or to another computer thousands of miles away.

- **Web sites—locations of hyperlinked documents:** The places you visit on the Web are called Web sites, and the estimated number of such sites throughout the world ranges up to 1,250,000. More specifically, a *Web site* **is the Internet location of a computer or server on which a hyperlinked document is stored.**

  For example, the Parents Place Web site *(http://www.parentsplace. com)* is a resource run by mothers and fathers that includes links to related sites, such as the Computer Museum Guide to the Best Software for Kids and the National Parenting Center.

- **Web pages—hypertext documents:** Information on a Web site is stored on "pages." **A** *Web page* **is actually a document, consisting of an HTML file. The** *home page,* **or** *welcome page,* **is the main page or first screen you see when you access a Web site,** but there are often other pages or screens. "Web site" and "home page" tend to be used interchangeably, although a site may have many pages.

  How fast a Web page will emerge on your screen depends on the speed of your modem or other connection and how many graphics are on the Web page. A text-only Web page may take only a few seconds. A page heavy on graphics may take several minutes. Most Web browsers allow you to turn off the graphics part in order to accelerate the display of pages.

- **Web browsers:** To access a Web site, you use Web browser software (✔ p. 70) and the site's address, called a URL. **A** *Web browser,* **or simply** *browser,* **is graphical user interface software that translates HTML documents and allows you to view Web pages on your**

computer screen. The main Web browsers are Netscape Navigator and Microsoft Internet Explorer (✔ p. 70). Others include HotJava and Opera.

With a browser, you can surf, browse, or search through the Web. When you connect with a particular Web site, the screen full of information (the home page) is sent to you. You can easily do **Web surfing—move from one page to another**—by using your mouse to click on the hypertext/hypermedia links.

- **URL—the address:** To locate a particular Web site, you type in its address, or *URL* (✔ p. 134). **The *URL*, for *Uniform Resource Locator*, is an address that points to a specific resource on the Web.** All Web page URLs begin with *http://*. Often a URL looks something like this:
  http://www.blah.com/blah.html.
  Here *http* stands for "hypertext transfer protocol."
  www stands for "World Wide Web."
  html stands for "hypertext markup language."
  The browsers Internet Explorer and Netscape Navigator automatically put the *http://* before any address beginning with *www*. The meaning of the parts of a URL address are explained in the box below. (■ *See Panel 7.6.*)
  A URL, we need to point out, is *not* the same thing as an e-mail address. Some people might type in *president@whitehouse.gov.us* and expect to get a Web site, but it won't happen. The Web site for the White House (which includes presidential information, history, a tour, and guide to federal services) is *http://www.whitehouse.gov*
  Note that URLs, as well as domain-type Internet addresses, are case-sensitive—that is, lowercase and capital letters should be typed

■ PANEL 7.6

Meaning of a URL address

An example of a URL is that for "Deb&Jen's Land O' Useless Facts," which consists of bizarre trivia submitted by readers.

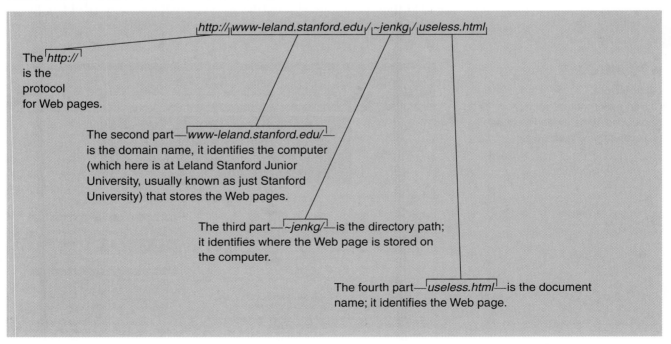

http://
The *http://* is the protocol for Web pages.

The second part—*www-leland.stanford.edu/*—is the domain name, it identifies the computer (which here is at Leland Stanford Junior University, usually known as just Stanford University) that stores the Web pages.

The third part—*~jenkg/*—is the directory path; it identifies where the Web page is stored on the computer.

The fourth part—*useless.html*—is the document name; it identifies the Web page.

as such. This requirement relates to the underlying Unix structure of the Internet.

URLs often change. If you get a "cannot locate server" message, try using a search engine (described below) to locate the site at its new address. Incidentally, many sites are simply abandoned because their creators have not updated or deleted them—the online equivalent of space-age debris orbiting the earth.

● **Hyperlinks, history lists, and bookmarks:** Whatever page you are currently viewing will show the hyperlinks to other pages by displaying them in color or with an underline. *(■ See Panel 7.7.)* When you move your mouse pointer over a hyperlink, the pointer will change to a hand. You can then move to that link by clicking on the hyperlink.

Suppose you want to go back to some Web pages you have viewed. You can use either a history list or a bookmark. **With a *history list*, the browser records the Web pages you have viewed during a particular connection session.** During that session, if you want to return to a site you visited earlier, you can click on that item in the history list. When you exit the browser, the history list is usually canceled. (Check to see if your browser or online service saves your history of sites visited *after* you disconnect. If so, anyone with access to your computer can view this list.) **Bookmarks, also called *favorite places*, consist of titles and URLs of Web pages that you choose to add to your bookmark list because you think you will visit them frequently in the future.** With these bookmarks stored in your browser, you can easily return to those pages in a future session by clicking on the listings.

Some common examples of Web page components are shown in the figure opposite. *(■ See Panel 7.8.)*

What Can You Find on the Web, & How Can You Find It?

There's a Web site for every interest: America's Job Bank, CIA World Factbook, Four11 (phone numbers), Internet Movie Database, Library of Congress,

■ PANEL 7.7

Hyperlinks

Clicking your mouse pointer on the underlined items (hyperlinks) will take you to new Web pages.

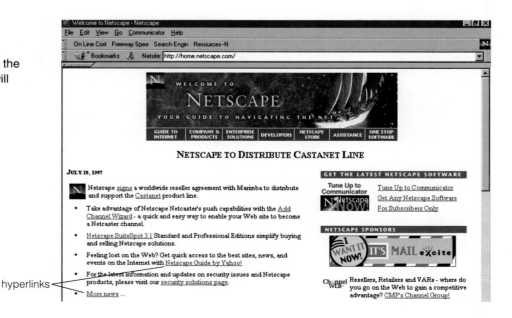

■ PANEL 7.8

Common examples of Web page components

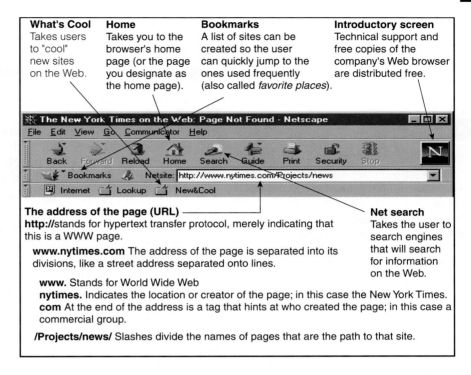

What's Cool Takes users to "cool" new sites on the Web.

Home Takes you to the browser's home page (or the page you designate as the home page).

Bookmarks A list of sites can be created so the user can quickly jump to the ones used frequently (also called *favorite places*).

Introductory screen Technical support and free copies of the company's Web browser are distributed free.

The address of the page (URL)
http:// stands for hypertext transfer protocol, merely indicating that this is a WWW page.

www.nytimes.com The address of the page is separated into its divisions, like a street address separated onto lines.

www. Stands for World Wide Web
nytimes. Indicates the location or creator of the page; in this case the New York Times.
com At the end of the address is a tag that hints at who created the page; in this case a commercial group.

/Projects/news/ Slashes divide the names of pages that are the path to that site.

Net search Takes the user to search engines that will search for information on the Web.

NASA Spacelink, New York Times, Recipe Archives, Rock and Roll Hall of Fame, TV Guide Online, U.S. Census Information, Woody Allen Quotes, and on and on. (■ *See Panel 7.9.*)

■ PANEL 7.9

Most heavily visited Web sites

These counts are based on the number of unique visitors.

	No. of visitors
1 Yahoo!	14,584,000
2 Netscape	12,589,000
3 Microsoft	11,630,000
4 Excite	9,357,000
5 America Online	7,831,000
6 Infoseek	7,335,000
7 GeoCities	6,381,000
8 Microsoft Network	5,929,000
9 Lycos	5,459,000
10 Alta Vista	4,977,000
11 CNet	4,346,000
12 ZD Net	3,974,000
13 CNN	3,669,000
14 HotMail	2,875,000
15 Four11	2,843,000
16 Pathfinder	2,436,000
17 USA TODAY	2,275,000
18 ESPN SportsZone	2,187,000
19 WhoWhere	2,018,000
20 MSNBC	2,016,000
21 Real Audio	1,940,000
22 Wired	1,916,000
23 Commonwealth Network	1,877,000
24 Tripod	1,785,000
25 AT&T	1,770,000

Unfortunately, there's no central registry of cyberspace keeping track of the comings and goings and categories of Web sites. However, there are two ways to find information. First, you can buy books (such as *1001 Really Cool Web Sites*), updated every year, that catalog hundreds of popular Web sites. Second, you can use *search tools—search engines and directories—to locate the URLs of sites on topics that interest you* (✔ p. 135):

Search tool	URL address
Yahoo!	www.yahoo.com
Infoseek	www.infoseek.com
AltaVista	www.altavista.com
Lycos	www.lycos.com
Excite	www.excite.com
WebCrawler	www.webcrawler.com
HotBot	www.hotbot.com
Magellan	www.mckinley.com
Galaxy	galaxy.tradewave.com

● Directories: **Directories are lists of Web sites classified by topic.** Directories are created by people submitting Web pages to a group of other people who classify and index them. Yahoo! is an example of a directory.

● Search engines: **Search engines allow you to find specific documents through keyword searches or menu choices.** Search engines find Web pages on their own. Search engines use software indexers (called *spiders*) to "crawl" around the Web and build indexes based on what they find. AltaVista, Excite, HotBot, Infoseek, and Lycos are examples of search engines.

Web browsers allow you to quickly use directories and search engines by clicking on the NET SEARCH button and then clicking on the icon of the directory or search engine you want to use. You can also use the browser to type in directories' URLs, for example, under the "Open location" option on the "File" menu.

We discussed searching tools and tips for searching in the Experience Box at the end of Chapter 3, page 133.

Browser Add-Ons for Multimedia: Plug-Ins

As we've said, what really distinguishes the World Wide Web from the rest of the Internet—and what accounts for a great part of its popularity—is the fact that it provides information in *multimedia* form—graphics, animation, video, and audio as well as text.

At the moment, the technology is in a stage of evolution that requires some users to do some extra work in order to activate Web multimedia on their own computers. Thus, you may have to use so-called plug-in programs to run certain multimedia components of a Web page within the browser window. **Plug-ins are programs that can be attached to your Web browser, giving it additional capabilities.** Many plug-ins, which are downloaded from a vendor's Web and enhance your browser, are free, or at least free for a trial period. Often plug-ins are used to improve animation, video, and audio.

Popular plug-ins are Real Audio (for audio: *www.realaudio.com*), Shockwave (for animation: *www.macromedia.com*), Live 3D (for virtual reality: *www.netscape.com*), VivoActive (for video: *www.vivo.com*), and CU-SeeMe (for videoconferencing: *www.wpine.com*).

Designing Web Pages

When Charlotte Buchanan and her two colleagues took their boutique Glam-Orama, a Seattle-based apparel store, online by designing their own Web site, they broke even within 6 months. Using a credit card, customers can buy some of the "kewlest toys and gifts around" from GlamOrama's catalog of more than 150 items.[32]

To put a business online, you need to design a Web page and perhaps online order forms, determine any hyperlinks, and hire 24-hour-a-day space on a Web server or buy one of your own. Professional Web page designers can produce a page for you, or you can do it yourself using a menu-driven program included with your Web browser or a Web-page design software package such as Microsoft FrontPage or Adobe PageMill.

Once your home page is designed, complete with links, you can let it reside on your hard disk. If you do, however, you will have to leave your computer and modem turned on all the time, which will tie up your equipment and allow only one user access at a time. Alternatively, after you have designed your Web page, you can rent space on your ISP's server. The ISP, which will charge you according to how many megabytes of space your file takes up on the server, will give you directions on how to send your Web page file to them via modem. (We discuss how to build your own Web site in the Readme box in Chapter 11.)

Push Technology: Web Sites Come Looking for You

Whereas it used to be that people went out searching the World Wide Web, now the Web is looking for us. The driving force behind this is *push technology,* or "webcasting," defined as software that enables information to find consumers rather than consumers having to retrieve it from the Web using browser programs.[33]

"Pull" is basic surfing: You go to a Web site and pull down the information to your desktop. "Push," however, consists of using special software to deliver information from various sites to your PC. Here's how it works.[34–37] Several services offer personalized news and information, based on a profile that you define when you register with them and download their software. You select the categories, or channels, of interest: sports news from the *Miami Herald*, stock updates on Eli Lily, weather reports from Iceland—and the provider sends what you want as soon as it happens or at times scheduled by you. The push software PointCast, for example, displays headlines as a screen saver; when you click on the headlines, you're transported to news summaries. Click again and you're zipped to the point of origin. (Note that information can be pushed to you as soon as it is available only if your computer is constantly connected.)

Among the push-media programs available are BackWeb, Ifusion, InCommon, Intermind, Marimba, PointCast, and Wayfarer. Netscape has a push-media product called Netcaster, part of its Communicator product. Microsoft has Active Channels as part of its Internet Explorer browser suite. Each company has lined up many media and other sources, such as Disney and CNN, to supply "channels" of pushed information.

Push technologies have come in for their share of criticism. "Push is valuable if you know what you want, if you don't have a lot of time, and if you want to receive something regularly," says Vin Crosbie, a new media consultant. "Push won't replace the morning newspaper. . . ."[38] Average home users, points out another observer, may find push services slow and distracting. Downloading is slow on an average PC, and many of the channels mainly offer teasers, which means you have to go online and pull up the publisher's normal Web site to find what you want, as well as advertisements. Worse, many channels don't really allow you to customize the news you want pushed at you.[39] Corporate technology managers have also become concerned that a constant stream of pushed information will gum up their internal networks. As a result, push software developers are trying to find ways to reshape the kind of content delivered to make it more useful.[40]

Portal Sites: The Fight to Be Your Gateway

A recent development has been the attempt by some high-traffic Internet connections and Web sites to redefine themselves as "portals." Representing the first view that a user sees upon directly dialing an Internet service or the first destination once connected, **a *portal* is an Internet gateway that offers search tools plus free features such as e-mail, customized news, and chat rooms; revenue comes from online advertising.** The movement toward portals attempts to lure more customers—and therefore advertisers—by consolidating content in one location.

Among the top contenders to become your portal are:

- Direct-dial services: Examples are America Online (*aol.com*), with 12 million paying subscribers, and Yahoo!/MCI (*www.yahoo.com*), which has 32 million visitors a month.
- Browser stops: An example is Netscape's Netcenter (*www.netscape.com*); Netscape has 21 million visitors a month.
- Web search sites: Examples are Excite (*www.excite.com*; 17 million visitors a month); Infoseek (*www.infoseek.com*; 14 million); and Lycos (*www.lycos.com*; 11 million).
- Online malls: The biggest example is Microsoft's Internet Start (*www.start.com*), with 19 million visitors. Snap! (*www.snap.com*) has 2 million visitors.

Industry analysts believe there will be a shakeout and that only about five of the portal players will survive.

7.7 Shared Resources: Workgroup Computing, EDI, & Intranets & Extranets

KEY QUESTION

What are the features of workgroup computing, EDI, intranets and firewalls, and extranets?

Preview & Review: Workgroup computing enables teams of co-workers to use networked microcomputers to share information and cooperate on projects.

Electronic data interchange (EDI) is the direct electronic exchange of standard business documents between organizations' computer systems.

Intranets and extranets are special-purpose spin-offs of Internet and Web technologies.

When they were first brought into the workplace, microcomputers were used simply as another personal-productivity tool, like typewriters or calculators. Gradually, however, companies began to link a handful of microcomputers together on a network, usually to share an expensive piece of hardware, such as a laser printer. Then employees found that networks allowed them to share files and databases as well. Networking using common software also allowed users to buy equipment from different manufacturers—a mix of workstations from both Sun Microsystems and Hewlett-Packard, for example. Sharing resources has led to workgroup computing.

Workgroup Computing & Groupware

***Workgroup computing,* also called *collaborative computing,* enables teams of co-workers to use networks of microcomputers to share information and to cooperate on projects.** Workgroup computing is made possible not only by networks and microcomputers but also by *groupware* (✔ p. 80). As mentioned earlier, groupware is software that allows two or more people on a network to work on the same information at the same time. Recently groupware has become the glue that ties organizations together.

In general, groupware permits office workers to collaborate with colleagues and to tap into company information through computer networks. It also enables them to link up with crucial contacts outside their organization—a customer in Nashville, a supplier in Hong Kong, for example.

One of the best-known groupware programs is Lotus Notes. Notes can run on a variety of operating systems and allows users to send e-mail via several online services. It also lets users create and store all kinds of data—text, audio, video, pictures—on common databases. In recent versions, users can create documents that can be displayed on the Web and use a built-in browser to surf the Web. In addition, Notes has the advantages of offering better security and the ability to synchronize multiple kinds of databases.

Electronic Data Interchange

Paper handling is the bane of organizations. Paper must be transmitted, filed, and stored. It takes up much of people's time and requires the felling of considerable numbers of trees. Is there a way to accomplish the same business tasks without using paper?

One answer lies in business-to-business transactions conducted via a computer network. ***Electronic data interchange (EDI) is the direct electronic exchange between organizations' computer systems of standard business documents,*** such as purchase orders, invoices, and shipping documents. For example, Wal-Mart has electronic ties to major suppliers like Procter & Gamble, allowing both companies to track the progress of an order or other document.

To use EDI, organizations wishing to exchange transaction documents must have compatible computer systems, or else go through an intermediary. For example, more than 500 colleges are now testing or using EDI to send transcripts and other educational records to do away with standard paper handling and its costs.

Intranets & Firewalls

It had to happen: First, businesses found that they could use the World Wide Web to get information to customers, suppliers, or investors. Federal Express, for example, saved millions by putting up a server in 1994 that enabled customers to click through Web pages to trace their parcels, instead of having FedEx customer-service agents do it. It was a short step from that to companies starting to use the same technology inside—in internal Internet networks called *intranets.*

Intranets **are internal corporate networks that use the infrastructure and standards of the Internet and the World Wide Web.** "The Web, it turns out, is an inexpensive yet powerful alternative to other forms of internal communications, including conventional computer setups," says one expert. "Because Web browsers run on any type of computer, the same electronic information can be viewed by any employee."[41] Thus, intranets connect all the types of computers, be they PCs, Macs, or workstations.

One of the greatest considerations of an intranet is security—making sure that sensitive company data accessible on intranets is protected from the outside world. The means for doing this is a security system called a *firewall.* **A *firewall* is a system of hardware and software that connects the intranet**

to external networks, such as the Internet. It blocks unauthorized traffic from entering the intranet and can also prevent unauthorized employees from accessing the intranet.

Two components of firewalls are a *proxy server* and *caching:*

- Proxy server: **A *proxy server* is a server, or remote computer, that may exist outside of the organization's network, and all communications to the organization are routed through it.** The proxy server decides which messages or files are safe to pass through to the organization's network. A proxy server can also provide document caching, as described next.

- Caching: **Big networks use *caching* (pronounced "cashing") to store copies of Web pages for quick access; the purpose is to speed up the Web for their users.** For example, you could type in a Web address for a particular Web page, but it might be located on a computer halfway around the world, and it would take some time to access it. On the other hand, if the Web page is used frequently enough, your organization might store a copy of it in a cache on the proxy server. Then the next time a user asks to see that page, the cached copy is already on hand.

  The only difficulty with this practice is that the original Web page may have been updated, but this is not reflected on the copy in the cache in the proxy computer. As one writer points out, if a proxy computer "saves its cached copy at noon and then the Web site modifies the page at 12:15—adding, say, a news story or a sports score—a user who punches in the page's address at 12:30 would see the noontime version that had been cached. The updates wouldn't appear. . . ."[42]

Extranets

Taking intranet technology a few steps further, extranets may change forever the way business is conducted. Like intranets, extranets offer security and controlled access. However, intranets are internal systems, designed to connect the members of a specific group or a single company. By contrast, **extranets are extended intranets connecting not only internal personnel but also selected customers, suppliers, and other strategic offices.** Large companies using extranets can, for example, save millions in telephone charges for fax documents. Federal Express, for example, used to give its clients special software to let them dial into the FedEx database to track their packages. Now anyone can track a delivery on the FedEx Web extranet; the company saves money by not having to send out software to its customers.[43]

Ford Motor Company has an extranet that connects more than 15,000 Ford dealers worldwide. Called FocalPt, the extranet supports sales and servicing of cars, with the aim of providing support to Ford customers during the entire life of their cars.

7.8 New Internet Technologies: Phone, Radio, TV, & 3-D

KEY QUESTION

What are the characteristics of telephone, radio, TV, and 3-D Internet technologies?

Preview & Review: The Internet has resulted in new technologies that combine PC capabilities with telephone, radio, television, and three-dimensionality.

Where are we headed with the seemingly all-purpose Internet? Let's take a look.

Before 1989, people communicated on the Internet by sending text messages and some graphics. Then came the World Wide Web, and now all kinds of other media—multimedia—are possible. First came colorful graphics and images, which now enliven so many home pages. Then came animation, which helped to make graphics even more interesting, as when a moving ribbon of sports scores or stock prices is made to scroll across the screen or the backgrounds and characters change in a video game.

Audio on the Web is also now available, in two forms—as sound files that can be downloaded, and as ongoing "streaming audio." Want to hear what's on a band's new CD? Music companies often promote their new releases by allowing interested consumers to download samples as sound files, which can be played on your multimedia computer's sound system. *Streaming audio* allows you to listen to sound while you are downloading it from the Web to your computer. This has opened up some interesting possibilities—telephone and radio applications—as we shall describe.

Video on the Web also comes in two forms—as video files that can be downloaded, and as "streaming video." Movie companies and film archives make short segments from movies available as files for you to download (though it may take some time) and then play back for viewing on your computer. *Streaming video* allows you to view video as it is being downloaded to your computer. Web video also led to some interesting new uses, which we'll discuss.

Let us see how these technologies are evolving. We describe the Internet and the Web used as phone line, radio network, television network, and 3-D theater.

- **Telephones on the Net:** "Hey, Pops, you there?" The voice of his son coming through the father's PC speakers, from a college dorm 150 miles away, was remarkably clear. The father switched on his computer microphone. "Pops here," he replied.[44]

  With Internet *telephony*—**using the Net to make phone calls, either one-to-one or for audioconferencing**—it's possible to make long-distance phone calls that are surprisingly inexpensive. Sending overseas telephone calls via the Net, in fact, costs only a small fraction of international phone charges. (Calls to London from San Francisco, for instance, cost only 16 cents a minute versus $1.09 under a basic AT&T plan.[45] Qwest Communications and a few other carriers already allow people to make calls over the Net for 5 cents to 7.5 cents a minute.[46]) The nation's long-distance telephone companies have seen the future, and have reason to be scared.[47]

  Telephony can be performed using a PC with a sound card and a microphone, a modem linked to a standard Internet service provider, and the right software: Netscape Conference (part of Netscape Communicator) or Microsoft NetMeeting (part of Microsoft Internet Explorer). With NetMeeting, you can also view applications over the Net, so your telephone partner and you can work on a report at the same time, while talking about it on the Internet phone.

  In addition, you can make Net phone calls from your PC to ordinary phones. Or you can make your call on an ordinary phone, placing a call to a phone company, which in turn connects to a gateway to the Internet—converting your voice to digital information to send over the Net. That digital information is then converted by another gateway company back to a voice signal, which then sends it to a telephone company, which routes it to the receiving party. (■ *See Panel 7.10.*)

● **Radio on the Net:** "I read an article recently that talked about special software for picking up Internet radio," declared the letter to science journalist David Einstein. "What I am looking for is something that would be able to receive uninterrupted Grateful Dead music."[48]

What he needed, Einstein replied in his column, was software known as RealPlayer (downloadable for free at *www.real.com*). With this, Einstein added, "you can listen to continuous Grateful Dead music by going to *www.deadradio.com*."

Desktop radio broadcasting is here—both music and spoken programming—and has been since the 1995 unveiling of RealAudio software, which can compress sound so it can be played in real time, even though sent over telephone lines. You can, for instance, listen to 24-hour-a-day net.radio, which features "vintage rock," or English-language services of 19 shortwave outlets from World Radio Network in London. "The implications of the new technology are enormous," says one computer writer. "It could provide a global soapbox for political parties, religious movements, and other groups that lack access to broadcast services."[49]

● **Television on the Net:** You can already visit the Web on television, as with WebTV. But can you get television programs over the Internet? RealPlayer (*http://www.real.com*) offers live, television-style

■ PANEL 7.10

Internet telephony: how it works

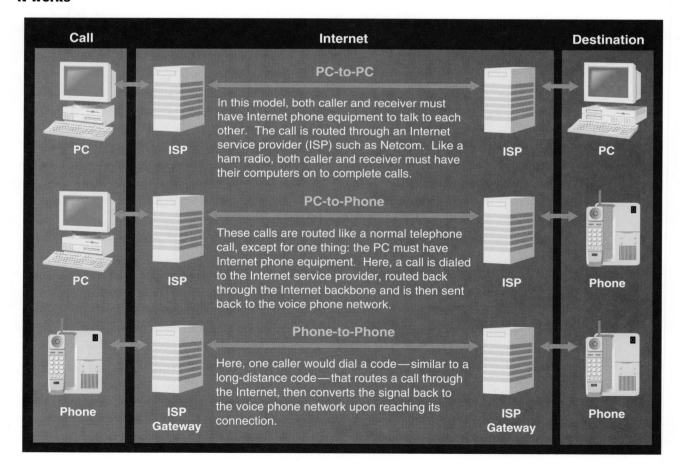

Call — **Internet** — **Destination**

PC-to-PC

PC · ISP

In this model, both caller and receiver must have Internet phone equipment to talk to each other. The call is routed through an Internet service provider (ISP) such as Netcom. Like a ham radio, both caller and receiver must have their computers on to complete calls.

ISP · PC

PC-to-Phone

PC · ISP

These calls are routed like a normal telephone call, except for one thing: the PC must have Internet phone equipment. Here, a call is dialed to the Internet service provider, routed back through the Internet backbone and is then sent back to the voice phone network.

ISP · Phone

Phone-to-Phone

Phone · ISP Gateway

Here, one caller would dial a code—similar to a long-distance code—that routes a call through the Internet, then converts the signal back to the voice phone network upon reaching its connection.

ISP Gateway · Phone

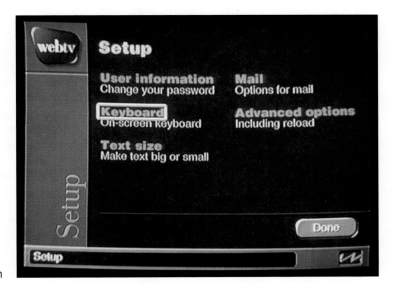

Set-top box that controls Web TV

Web TV screen

broadcasts over the Internet for viewing on your PC screen. You download RealPlayer, install it, then point your browser to a site featuring RealVideo. That will produce a streaming-video television image in a window a few inches wide.

"Considering that you're seeing and hearing the broadcast on your computer, it seems nothing short of incredible," says reviewer Thomas Weber. "But compared with even the cheapest television set, the picture looks pretty bad."[50]

Weber tuned into a politician delivering a speech on the Web site of the C-SPAN cable television network and found the image "ghostly and jerky." When he switched over to MSNBC, "a car commercial went haywire, creating a mishmash of colors."

The problem is the speed of most people's Internet modems. RealPlayer, Weber found, actually does a pretty good job of adapting to the available speed, skipping parts of the picture when necessary to keep from interrupting the flow. But until high-speed Net connections are commonplace (as we discuss in Chapter 8), watching Internet television on a PC will be a headache.

For the past couple of years, several television broadcasters (such as NBC, CNN, and MTV) have used a technology called *Intercast*, which simultaneously delivers TV broadcasts and selected Web pages to your PC. (You'll need a TV tuner/decoder card or a new microcomputer with such circuitry built in.) The main appeal is that you can access Web home pages related to the television program. For example, as MSNBC anchor Brian Williams talks about Iraq, you could get background information about the region from the Web.

- **3-D on the Net:** Three-dimensionality may be one of the tougher challenges because so far the computer can't update images as fast as the human eye, so that, says one writer, "walking through 3-D cyberstores usually feels like staggering, which literally makes some people sick."[51]

Still, some companies are trying to bring 3-D to the Web, using software technology called VRML (for Virtual Reality Modeling Language, discussed in Chapter 11). Silicon Graphics offers a 3-D viewer called VRML2 to supplement browsers so that, for instance, users can zoom in on a realistic image of a car they want to buy. Microsoft and Netscape agreed to popularize the standard, and the capability is available on new versions of Netscape's Navigator/ Communicator and Microsoft's Internet Explorer.[52]

What about bringing the sense of touch to the Internet? That too seems to be (literally) within reach. A device called the Phantom from SensAble Technologies that uses what's known as force-feedback technology gives you the feeling you are actually touching and manipulating objects on your PC screen. Thus, you could seem to grasp the handle of a screwdriver on screen and tell whether it was hard plastic or rubber.[53]

As we discuss in the next chapter, most consumers now don't have the very high-speed digital lines or other hardware needed to receive many of these Internet features in real time. That will assuredly change.

7.9 Cyberethics: Netiquette, Controversial Material & Censorship, & Privacy Issues

KEY QUESTION

What are considerations to be aware of regarding cyberethics?

Preview & Review: Users of communications technology must weigh standards of behavior and conduct in three areas: netiquette, controversial material and censorship, and privacy.

Ethics

Communications technology gives us more choices of nearly every sort. Not only does it provide us with different ways of working, thinking, and playing; it also presents us with some different moral choices—determining right actions in the digital and online universe. Here let us consider three important aspects of "cyberethics"—netiquette, controversial material and censorship, and matters of privacy.

Netiquette

One morning *New Yorker* magazine writer John Seabrook, who had recently published an article about Microsoft chairman Bill Gates, checked into his computer to find the following e-mail reaction to his story from a reader:

> Listen, you toadying *[deleted]* scumbag... remove your head *[three words deleted]* long enough to look around and notice that real reporters don't fawn over their subjects, pretend that their subjects are making some sort of special contact with them, or, worse, curry favor by TELLING their subjects how great the *[deleted]* profile is going to turn out and then brag in print about doing it. . . .

On finishing the message, Seabrook rocked back in his chair. "Whoa," he said aloud to himself. "I got flamed."[54]

A form of speech unique to online communication, *flaming* **is writing an online message that uses derogatory, obscene, or inappropriate language.** The attack on Seabrook is probably unusual; most flaming happens when someone violates online manners or "netiquette."

As mentioned, many online news groups have a set of "FAQs"—frequently asked questions—that newcomers, or "newbies," are expected to become familiar with before joining in any chat forums. Most FAQs offer *netiquette,* **or "net etiquette," guides to appropriate behavior while online.** The commercial online services also have special online sites where the uninitiated can go to learn how to avoid an embarrassing breach of manners. Examples of netiquette blunders are typing with the CAPS LOCK key on—the Net equivalent of yelling—discussing subjects not appropriate to the forum, repetition of points made earlier, and improper use of the software. **Spamming, or sending unsolicited mail,** is especially irksome; a spam includes chain letters, advertising, or similar junk mail. Something that also helps in the online culture is use of **emoticons, keyboard-produced pictorial representations of expressions.** (■ *See Panel 7.11.*)

One of the great things about the Internet is that it has got Americans to start writing again. However, it shouldn't be sloppy. "Most people who see poor spelling and poor grammar think the writer is a dummy," says Virginia Shea, author of a book called *Netiquette,* who has become a sort of Miss Manners of cyberspace. "You'd think some people never opened a dictionary in their lives from the spelling in their e-mail. They just dash off a note and figure that it doesn't matter how it reads. That is a big mistake."[55]

Controversial Material & Censorship

In Saudi Arabia, officials try to police taboo subjects (sex, religion, politics) on the Internet. In China, Net users must register with the government. In the United States, however, free speech is protected by the First Amendment to the Constitution.

If a U.S. court decides *after* you have spoken that you have defamed or maliciously damaged someone, you may be sued for slander (spoken speech) or libel (written speech) or charged with harassment, but you cannot be stopped beforehand. However, "obscene" material is not constitutionally protected free speech. Obscenity is defined as sexually explicit material that is offensive as measured by "contemporary community standards"—a definition with considerable leeway, depending on localities.

In 1996 Congress passed legislation, as part of its broad telecommunications law overhaul called the Communications Decency Act, that imposed heavy fines and prison sentences on people making available "patently offensive" sexually explicit material over the Internet in a manner available to children. This part of the law was blocked by a federal judge as being constitutionally vague, and in 1997 the U.S. Supreme Court ruled that this part was indeed unconstitutional.

■ PANEL 7.11

Emoticons

Emoticons enable online users to get in the kind of facial expressions and inflections that are used in normal conversation.

"How do emoticons work? Tilt your head to the left and take a look at this one :=)

Do you see a smiley face with a long nose? A comment or joke followed by a smiley is often a good way to ensure that it was taken in good humor. Have sad news? Show it while you tell it. :-(Feeling teary-eyed? :'(Feeling sarcastic? Show that as well. :-/ Stick your tongue out at someone :=P Or pucker up for a kiss :-*

Emoticons are good for more than just facial expressions, though. You can send hugs. (()) or {{}} And you can send roses. @---^--- Placing <w> before your words signifies a whisper, and <g> a grin."

—Tosca Moon Lee, "Smiling Online," *PC Novice*

:-)	Happy face	<g>	Grin
:-(	Sorrow or frown	BTW	By the way
:-O	Shock	IMHO	In my humble opinion
:-/	Sarcasm	FYI	For your information
;-)	Wink		

Since computers are simply another way of communicating, there should be no surprise that a lot of people use them to communicate about sex. Yahoo!, the Internet directory company, says that the word "sex" is the most popular search word on the Net.[56] All kinds of online X-rated bulletin boards, chat rooms, and Usenet newsgroups exist.

A special problem is with children having access to sexual conversations, downloading hard-core pictures, or encountering criminals tempting them into meeting them. "Parents should never use [a computer] as an electronic baby sitter," says computer columnist Lawrence Magid. People online are not always what they seem to be, he points out, and a message seemingly from a 12-year-old girl could really be from a 30-year-old man. "Children should be warned never to give out personal information," says Magid, "and to tell their parents if they encounter mail or messages that make them uncomfortable."[57]

Not only are parents concerned about pornography in electronic form, so are employers. Many companies are concerned about the loss of productivity—and the risk of being sued for sexual harassment—as workers spend time online looking at sexually explicit material. What can be done about all this? Some possibilities:

- **Blocking software:** Some software developers have discovered a golden opportunity in making programs like SurfWatch, Net Nanny, and CYBERsitter. These "blocking" programs screen out objectionable matter typically by identifying certain unapproved keywords in a user's request or comparing the user's request for information against a list of prohibited sites. (The screening is sometimes imperfect: The White House Web site was once accidentally put off-limits by SurfWatch because it used the supposedly indecent term "couples" in conjunction with the vice president and his wife.)

- **Browsers with ratings:** Another proposal in the works is browser software that contains built-in ratings for Internet, Usenet, and World Wide Web files. Parents could, for example, choose a browser that has been endorsed by the local school board or the online service provider.

- **The V-chip:** The 1996 Telecommunications Law officially launched the era of the V-chip, a device that will be required equipment in most new television sets. The *V-chip* allows parents to automatically block out programs that have been labeled as high in violence, sex, or other objectionable material. Who will do the ratings (of 600,000 hours of programming currently broadcast per year) and whether the system is really workable remains to be seen. However, as conventional television and the Internet converge, the V-chip could become a concern to Net users as well as TV watchers.

The difficulty with any attempts at restricting the flow of information, perhaps, is the basic Cold War design of the Internet itself, with its strategy of offering different roads to the same place. "If access to information on a computer is blocked by one route," writes the *New York Times*'s Peter Lewis, "a moderately skilled computer user can simply tap into another computer by an alternative route." Lewis points out an Internet axiom attributed to an engineer named John Gilmore: "The Internet interprets censorship as damage and routes around it."[58]

Privacy

Privacy **is the right of people not to reveal information about themselves.** Technology, however, puts constant pressure on this right.

Consider Web cookies, little pieces of data left in your computer by some sites you visit.[59–61] A *cookie* is a file that a Web server stores on a user's hard-disk drive when the user visits a Web site. Thus, unknown to you, a Web site operator or companies advertising on the site can log your movements within a site. These records provide information that marketers can use to target customers for their products. In addition, however, other Web sites can get access to the cookies and acquire information about you. Don't want your employer to know you visited an X-rated site during your lunch hour? A cookie may be left on your hard drive to show you did.

Fortunately, you do have some safeguards against cookies. Your browser can be set to alert you whenever a Web site wants to deposit a cookie in your computer. There are also some shareware programs available to manage cookies (Cookie Crusher, Cookie Pal, and CookieMaster can be obtained from *http://www.hotfiles.com*).

However, none of these programs will help prevent other intrusions on your privacy. For instance, David Post, an anthropologist and law school professor at Temple University, says that he has sent casual messages (about his legal views on cyberspace) to obscure online discussion groups, then had those views quoted back to him by a reporter. "It made me realize that what I was sending out could be found 50 years from now," he says.[62]

Think your medical records are inviolable? Actually, private medical information is bought and sold freely by various companies since there is no federal law prohibiting it. (And they simply ignore the patchwork of varying state laws.)

Think the boss can't snoop on your e-mail at work? Various courts have ruled that the U.S. Constitution does not provide privacy protection in the workplace, and only a few state laws specifically protect employees' e-mail messages. "In fact," writes sociologist Amitai Etzioni, "many corporate policies warn employees that e-mail is about as private as making a personal call over a speaker phone in a crowded office."[63] The law allows employers to "intercept" employee communications if one of the parties involved agrees to the "interception." The party "involved" is the employer. Indeed, employer snooping seems to be widespread.

Think you're anonymous online when you don't sign your real name? America Online, in violation of the company's subscriber confidentiality policy, released to Navy investigators the name of a chief petty officer who had sent an e-mail message under the screen name "boysrch" to a military colleague. Word got back to the Navy administrators, who looked up the sailor's profile on AOL's public screen-profile menu and found that the person behind "boysrch" had entered "gay" under his marital status.[64] (The sailor won his case against the Navy for violating the Electronic Communications Privacy Act and was later promoted.)

A great many people are concerned about the loss of their right to privacy. Indeed, one survey found that 80% of the people contacted worried that they had lost "all control" of the personal information being collected and tracked by computers.[65] Although the government is constrained by several laws on acquiring and disseminating information and listening in on private conversations, there are reasons to be alarmed. We discuss privacy further in Chapter 9 and security in Chapter 12.

Onward

"On the 21st-century Internet, as in the 19th-century American West," says former *Time* journalist Kurt Andersen, "the mavericks and cranks drawn to the frontier will not be wiped out, and their romantic sensibility will inform the spirit of the place, but the rude settlements and wild behavior are being overshadowed by more traditional, trustworthy modes."[66]

In our time the Internet is a "hyper-democratic media world," in which speculation, rumor, and pseudoinformation seem to have equal footing with reliable news and scientific facts. As in the 19th century before the arrival of mass-circulation magazines and wire services, and then later with centralized radio and television networks, the Internet represents a collection of "media" that are scattered and local even as they are linked internationally.

But that is changing. "The once-untamed territory is becoming rapidly civilized . . . ," says Andersen. "Each year on the Net is equal to about a decade in 19th-century-frontier terms: the rise of Internet news groups and bulletin boards is analogous to the half century after Lewis and Clark; the beginning of the World Wide Web, in the early 1990s, can be compared to the discovery of gold in California; and the present cybermoment is equivalent to, say, 1880." Users of the Internet are now demanding reliable filters that can help them separate the bogus and the bunk from the mostly accurate. In the next century, these will no doubt fall into place.

README

Practical Matters: Netiquette—From Flaming to Flirting

Netiquette is the code of manners that is supposed to govern online communication. Much of it is based on common sense, but some of the rules are particular to cyberspace. While people will forgive a gaffe or two, continued violations are likely to leave the user with a mailbox that's loaded with angry mail—or empty.

1. Chat Rudeness: Cursing, arguing for sport, sexual harassment.

   Continued violation: Temporary or permanent ban from the chat group. An "ignore" command can be issued against you so that you can type your comments to the screen, but no one except you will see them.

   Good netiquette: Pay attention to the thread of conversation. Address people respectfully. . . .

2. Too-Easy Familiarity: Writing to strangers and addressing them like you already know them well.

   Continued violation: Your letters go unread and unanswered.

   Good netiquette: Using "Dear so-and-so" still works pretty well.

3. Send-Reply-Send-Reply: It's a problem of being too respectful: Neither party can end a series of e-mail communications because they don't want to hurt the other's feelings.

   Continued violation: Dead silence on the other end. . . .

   Good netiquette: Don't allow someone to continue meaningless or boring correspondence. Write something like "over and out"—and mean it.

4. Spamming: Unsolicited mail including advertisements, chain letters, and inclusion on mailing lists without permission.

   Continued violation: Counter-spamming. Hope you've got a lot of unused hard drive space for the flame mail you're going to get.

   Good netiquette: Don't put anyone's name on a mailing list without their permission. Don't forward chain letters.

5. Flame Bait: An angry note, usually written in haste.

   Continued violation: A flame war after the other party responds in kind.

   Good netiquette: Write an angry note—then junk it instead of sending it. Think carefully about what you put in print. Remember there's a person on the other end.

6. Mash Notes: Never send out a love letter unless you're sure that no one but the intended person wants it or will see it.

   Continued violation: A lot of unexplained giggling when the sender walks by the water cooler. In an extreme case, the sender is fired for sexual harassment.

   Good netiquette: If you aren't sure about the security of e-mail on either end, send a Shakespearean sonnet instead of something more steamy. And never, ever, send a mash note to someone who doesn't want it.

—Ramon G. McLeod, "Netiquette—Cyberspace's Cryptic Social Code," *San Francisco Chronicle*

Experience Box

Web Research, Term Papers, & Plagiarism

No matter how much students may be able to rationalize cheating in college—for example, trying to pass off someone else's term paper as their own (plagiarism)—ignorance of the consequences is not an excuse. Most instructors announce the penalties for cheating at the beginning of their course—usually a failing grade in the course and possible suspension or expulsion from school.

Even so, probably every student becomes aware before long that the World Wide Web contains sites that offer term papers, either for free or for a price. Some dishonest students may download (transfer or copy) papers and just change the author's name to their own. Others are more likely to use the papers just for ideas. Perhaps, suggests one article, "the fear of getting caught makes the online papers more a diversion than an invitation to wide-scale plagiarism."[67]

How the Web Can Lead to Plagiarism

Two types of term-paper Web sites are:

- **Sites offering papers for free:** Such a site requires users to fill out a membership form, then provides at least one free student term paper.

- **Sites offering papers for sale:** Commercial sites may charge $6–$10 a page, which users may charge to their credit card.

"These paper-writing scams present little threat to serious students or savvy educators," says one writer. "No self-respecting student will hand in purchased papers; no savvy educator will fail to spot them and flunk the student outright."[68] English professor Bruce Leland, director of writing at Western Illinois University, points out that many Web papers are so bad that no one is likely to benefit from them in any case.[69]

How do instructors detect and defend against student plagiarism? Leland says professors are unlikely to be fooled if they tailor term-paper assignments to work done in class, monitor students' progress—from outline to completion—and are alert to papers that seem radically different from a student's past work.

Eugene Dwyer, a professor of art history at Kenyon College, requires that papers in his classes be submitted electronically, along with a list of World Wide Web site references. "This way I can click along as I read the paper. This format is more efficient than running around the college library, checking each footnote."[70]

The World Wide Web, points out Mitchell Zimmerman, also provides the means to detect cheaters. "Faster computer search programs will make it possible for teachers to locate texts containing identified strings of words from amid the mil-lions of pages found on the Web."[71] Thus, a professor could input passages from a student's paper into a search program that would scan the Web for identical blocks of text.

How the Web Can Lead to Low-Quality Papers

Besides tempting some students to plagiarism, the Web also creates another problem in term-paper writing, says William Rukeyser, coordinator for Learning in the Real World, a non-profit information clearinghouse. This is that it enables students "to cut and paste together reports or presentations that appear to have taken hours or days to write but have really been assembled in minutes with no actual mastery or understanding by the student."[72]

This comes about because the Web features *hypertext*—highlighted words and phrases that can be linked online to related words and phrases. Students use hypertext to skip to isolated phrases or paragraphs in source material, which they then pull together, complains Rukeyser, without making the effort to gain any real knowledge of the material.

Philosophy professor David Rothenberg, of the New Jersey Institute of Technology, reports that as a result of students doing more of their research on the Web he has seen "a disturbing decline in both the quality of the writing and the originality of the thoughts expressed."[73] The Web, he says, makes research look too easy:

You toss a query to the machine, wait a few minutes, and suddenly a lot of possible sources of information appear on your screen. Instead of books that you have to check out of the library, read carefully, understand, synthesize, and then tactfully excerpt, these sources are quips, blips, pictures, and short summaries that may be downloaded magically to the dorm-room computer screen. Fabulous! How simple! The only problem is that a paper consisting of summaries of summaries is bound to be fragmented and superficial, and to demonstrate more of a random montage than an ability to sustain an article through 10 to 15 double-spaced pages.

How does an instructor spot a term paper based primarily on Web research? Rothenberg offers four clues:

- **No books cited:** The student's bibliography cites no books, just articles or references to Web sites. Sadly, says Rothenberg, "one finds few references to careful, in-depth commentaries on the subject of the paper, the kind of analysis that requires a book, rather than an article, for its full development."

- **Outdated material:** A lot of the material in the bibliography is strangely out of date, says Rothenberg. "A lot of stuff on the Web that is advertised as timely is actually at least a few years old."

- **Unrelated pictures and graphs:** Students may intersperse the text with a lot of impressive-looking pictures and graphs that may look as though they were the result of careful work and analysis but actually bear little relation to the precise subject of the paper. "Cut and pasted from the vast realm of what's out there for the taking, they masquerade as original work."

- **Superficial references:** "Too much of what passes for information [online] these days is simply *advertising* for information," points out Rothenberg. "Screen after screen shows you where you can find out more, how you can connect to this place or that." And a lot of other kinds of information is detailed but often superficial: "pages and pages of federal documents, corporate propaganda, snippets of commentary by people whose credibility is difficult to assess."

Once, points out Brian Hecht, the Internet was a text-only medium for disseminating no-frills information. The World Wide Web added a means of delivering graphics, sound, and video, but as access has widened, says Hecht, "legitimate information has been subsumed by a deluge of vanity 'home pages,' corporate marketing gimmicks, and trashy infomercials. . . . It is impossible to know where [a given piece of] information comes from, who has paid for it, whether it is reliable, and whether you will ever be able to find it again."[74]

In sum: Having access to the Internet and the Web doesn't provide miracles in research and term-paper writing.

Summary

Summary

What It Is/What It Does

bookmark (p. 328, KQ 7.6) Also called a *favorite place;* consists of the title and URL of a Web page the user has added to a personal bookmark list to make it easy to visit the site frequently in the future.

caching (p. 334, KQ 7.7) In an organization's computer network, caching is a system of storing copies of Web pages for quick access.

communications (p. 309, KQ 7.1) Also called *telecommunications;* the electronic transfer of information from one location to another. Also refers to electromagnetic devices and systems for communicating data.

connectivity (p. 309, KQ 7.1) The state of being able to connect devices by communications technology to other devices and sources of information.

cyberspace (p. 309, KQ 7.1) Refers to the computer online world and the Internet in particular, as well as the whole wired and wireless world of communications.

dedicated fax machine (p. 311, KQ 7.2) Specialized device that does nothing except scan in, send, and receive documents over telephone lines to and from other fax machines.

directory (p. 330, KQ 7.6) On the Web, a directory is a list of Web pages classified by topic; it is created by a group that indexes Web pages that people submit.

electronic data interchange (EDI) (p. 333, KQ 7.7) System of direct electronic exchange between organizations' computer systems of standard business documents, such as purchase orders, invoices, and shipping documents.

electronic mail (e-mail) (p. 321, KQ 7.5) System in which computer users, linked by wired or wireless communications lines, may use their keyboards to post messages and their display screens to read responses.

Why It's Important

The connection to a site of interest can be made by clicking on the bookmark with a mouse instead of trying to find the address from other sources again.

A Web page cached on an organization's computer system can be accessed nearly instantly; however, the stored version may not be the latest version from the original Web site.

Communications systems have helped to expand human communication beyond face-to-face meetings to electronic connections called the *global village.*

Computers offer greater varieties of connectivity than other communications devices such as telephones or radio systems.

Suggests the vast amount of connections and interactivity now available to users of computer and communications systems.

Fax machines have enabled people to instantly transmit graphics and documents for the price of a phone call.

Directories are useful for browsing—looking at Web pages in a general category and finding items of interest. *Search engines* may be more useful for hunting specific information.

EDI allows the companies involved to do away with standard paper handling and its costs.

E-mail allows users to send messages to a single recipient's "mailbox"—a file stored on the computer system—or to multiple users. It is a much faster way of transmitting written messages than traditional mail services.

What It Is/What It Does

emoticons (p. 339, KQ 7.9) Keyboard-produced pictorial representations of expressions that can be used in online communications to stand for a facial expression or inflection; for example, :) for a smile.

extranet (p. 334, KQ 7.7) An extension of an internal network (intranet) to connect not only internal personnel but also selected customers, suppliers, and other strategic offices.

FAQ (frequently asked questions) (p. 323, KQ 7.5) Refers to the file that contains basic information about a Usenet newsgroup on the Internet.

fax (p. 311, KQ 7.2) Stands for *facsimile transmission* or reproduction; a message sent by dedicated fax machine or by fax modem.

fax board (p. 312, KQ 7.2) Type of modem installed as a circuit board inside a computer; it exchanges fax messages with another fax machine or fax board.

File Transfer Protocol (FTP) (p. 323, KQ 7.5) Feature of the Internet whereby users can connect their PCs to a remote computer that is an FTP site and transfer (download) the publicly available files to their own microcomputer's hard disk.

firewall (p. 333, KQ 7.7) System of hardware and software used to connect internal networks (intranets) to external networks, such as the Internet. It prevents unauthorized people, whether outside or inside the company, from accessing the network.

flaming (p. 338, KQ 7.9) In online communication, writing a message that uses derogatory, obscene, or inappropriate language; a form of attack usually directed at someone who violates online manners *(netiquette)*.

history list (p. 328, KQ 7.6) A record made by Web browser software of the Web pages a user has viewed during a particular connection session.

home page (p. 326, KQ 7.6) The first page (main page)—that is, the first screen—seen upon accessing a Web site.

Why It's Important

Words typed online lack the meanings added by body language and voice inflection; emoticons can add a little of that content and also serve as shorthand to express a feeling.

Extranets provide a direct line of communication that makes it easier, for example, to access databases and to send faxes without incurring long-distance phone charges.

FAQ files provide users with information they need to decide if a particular newsgroup is right for them.

A fax message may transmit a copy of text or graphics for the price of a telephone call.

The benefit of fax boards is that messages can be transmitted directly from a microcomputer; no paper or scanner is required.

FTP enables users to copy free files of software, games, photos, music, and so on.

Firewalls are necessary to protect an organization's internal network against theft and corruption.

Flaming is punishment for offending another person online; it's often in response to violating the code of online manners.

The history list makes it easy to revisit a site after surfing on to related sites, but it is usually not saved after the user ends that Web session; instead a bookmark can be used to revisit the site in the future.

The home page provides a menu or explanation of the topics available on that Web site.

What It Is/What It Does

Why It's Important

hypertext (p. 326, KQ 7.6) System in which documents scattered across many Internet sites are directly linked so that a word or phrase in one document becomes a connection to a document in a different place.

Clicking on a hypertext link calls up a related topic that may be in the same document or a document on a computer far away from the first site.

hypertext markup language (HTML) (p. 326, KQ 7.6) Set of instructions, called tags or markups, that are used for documents on the Web to specify document structure, formatting, and links to other documents.

HTML makes it relatively simple for computer users to create Web pages and link them to other sites.

hypertext transfer protocol (p. 326, KQ 7.6) Expressed as *http://*; the communications standard used to transfer information on the Web. The abbreviation appears as a prefix on Web addresses.

Hypertext transfer protocol provides a standard for the transfer of multimedia files with hypertext links.

Internet (p. 317, KQ 7.5) Also called "the Net"; international network composed of approximately 140,000 smaller networks in more than 200 countries. Created as ARPAnet in 1969 by the U.S. Department of Defense, the Internet was designed to share research among military, industry, and university sources and to sustain communication in the event of nuclear attack.

Today the Internet is essentially a self-governing and noncommercial community offering both scholars and the public such features as information gathering, electronic mail, and discussion and newsgroups.

Internet service provider (ISP) (p. 318, KQ 7.5) Local or national company that provides unlimited public access to the Internet and the Web for a flat fee.

Unless they are connected to the Internet through an online information service or a direct network connection, microcomputer users need an ISP to connect to the Internet.

intranet (p. 333, KQ 7.7) Internal corporate network that uses the infrastructure and standards of the Internet and the World Wide Web.

Intranets can connect all types of computers.

netiquette (p. 338, KQ 7.9) "Net etiquette"; guides to appropriate behavior while online.

Netiquette rules help users to avoid offending other users.

online information service (p. 314, KQ 7.4) Company that provides access to databases and electronic meeting places to subscribers equipped with telephone-linked microcomputers—for example Prodigy, America Online, and Microsoft Network.

Online information services offer a wealth of services, from electronic mail to home shopping to video games to enormous research facilities to discussion groups.

picture phone (p. 314, KQ 7.3) Telephone with a TV-like screen and a built-in camera that allows the people talking by telephone to see each other.

With high-speed transmission lines and data compression software making it possible to send video images in real time, picture phones may gain wider acceptance.

plug-in (p. 330, KQ 7.6) In reference to Web software, a plug-in is a program that can be attached to a Web browser for added capability, such as improving animation, video, or audio or providing *telephony*.

Plug-in programs can activate multimedia components of Web pages if the receiving computer doesn't already have that capability.

What It Is/What It Does

portal (p. 332, KQ 7.6) An Internet gateway that offers search tools plus free features such as e-mail, customized news, and chat rooms; revenue comes from online advertising.

privacy (p. 341, KQ 7.9) Right of people not to reveal information about themselves.

proxy server (p. 334, KQ 7.7) Part of an internal network's *firewall;* a computer that exists outside an organization's network (intranet) to receive all communications to the organization and decide which messages or files are safe to pass through to the intranet. May also provide *caching* for the intranet.

push technology (p. 331, KQ 7.6) Also called *webcasting;* software that enables information on the World Wide Web to find consumers. The consumer selects a "channel" or category of interest and the provider sends information as it becomes available or at scheduled times.

search engine (p. 330, KQ 7.6) Type of search tool that allows the user to find specific documents through keyword searches or menu choices. It uses software indexers ("spiders") to "crawl" around the Web and build indexes based on what they find in available Web pages.

search tool (p. 330, KQ 7.6) Refers to either a *directory* or *search engine* that can locate the URLs (addresses) of sites on topics of interest to the user.

spamming (p. 339, KQ 7.9) On the Internet, refers to sending unsolicited mail such as chain letters and advertising.

TCP/IP (Transmission Control Protocol/Internet Protocol) (p. 318, KQ 7.5) Standardized set of guidelines (protocols) that allows computers on different networks to communicate with one another efficiently.

telephony (p. 335, KQ 7.8) On the Internet, making phone calls to one or more other users; all callers need the same telephonic software, a microcomputer with a microphone, sound card, modem, and a link to an Internet service provider.

Uniform Resource Locator (URL) (p. 327, KQ 7.6) Address that points to a specific resource on the Web.

Why It's Important

Portals attempt to lure more customers—and therefore more advertisers—by consolidating content in one location.

Computer technology and electronic databases have made it more difficult for people to protect their privacy.

See firewall.

Users can see the latest news on a topic; usually they receive a short notification like a sports score or headline and can then go to a Web site to see more complete information.

Search engines allow the user to build an index of Web pages that mention any given topic, so they may be more useful for hunting specific information, whereas a *directory* can provide an index of Web pages on a general area of interest for browsing.

Search tools are an online way of finding Web pages of interest; catalogs and recommendations from other people are another way.

Internet users greatly resent junk mail and may respond by *flaming* the sender.

TCP/IP is the standard language of the Internet.

Telephony on the Internet is less expensive than standard phone calls, especially for international calls. It can also support audioconferences in which participants can view and work on documents together.

Addresses are necessary to distinguish among Web sites.

What It Is/What It Does	**Why It's Important**

Usenet newsgroup (p. 323, KQ 7.5) Electronic discussion groups that focus on a specific topic.

Usenet newsgroups enable people with similar interests to readily find each other, no matter where they live.

videoconferencing (p. 313, KQ 7.3) Also called *teleconferencing;* form of conferencing using television video cameras, monitors, microphones, and speakers as well as computers to allow people at different locations to see, hear, and talk with one another.

Videoconferencing may be done from a special video-conference room or handled with equipment set up on a desk.

voice mail (p. 312, KQ 7.2) System in which incoming voice messages are stored in a recipient's "voice mail-box" in digitized form. The system converts the digitized versions back to voice messages when they are retrieved. With voice mail, callers can direct calls within an office using buttons on their Touch-Tone phone.

Voice mail enables callers to deliver the same message to many people, to forward calls, to save or erase messages, and to dictate replies. The main benefit is that voice mail helps eliminate "telephone tag."

Web browser (p. 326, KQ 7.6) Software that translates HTML documents and allows a user to view a remote Web page; has a graphical user interface.

Browser software lets a user view Web documents and do *Web surfing.*

Web page (p. 326, KQ 7.6) Document in hypertext markup language (HTML) that is on a computer connected to the Internet. The first screen of a Web page is the home page.

Each Web page focuses on a particular topic. The information on a site is stored on "pages." The starting page is called the *home page.*

Web site (p. 326, KQ 7.6) Internet location of a computer or server on which a hyperlinked document (Web page) is stored; also used to refer to the document.

A Web site needs to use a computer that remains turned on, and it's best to have the site on a server with multiple connections so that more than one user at a time can visit.

Web surfing (p. 327, KQ 7.6) A user's action of moving from one Web page to another by using the computer mouse to click on the hypertext links.

See World Wide Web.

workgroup computing (p. 332, KQ 7.7) Also called *collaborative computing;* technology that enables teams of co-workers to use networks of microcomputers to share information and cooperate on projects. Workgroup computing is made possible not only by networks and microcomputers but also by groupware.

Workgroup computing permits office workers to collaborate with colleagues, suppliers, and customers and to tap into company information through computer networks.

World Wide Web (p. 325, KQ 7.6) Interconnected system of sites, or servers, of the Internet that store information in multimedia form and share a hypertext form that links similar words or phrases between sites.

Web software allows users to view information that includes not just text but graphics, animation, video, and sound, and to move between related sites via hypertext links.

Exercises

Self-Test

1. The _____ is the most extensive network in the world.

2. _____ is an Internet feature that lets you connect to a remote computer and download files to your computer's hard disk.

3. _____
_____ enables teams of coworkers to collaborate on projects via a network of microcomputers. The software component is referred to as *groupware*.

4. _____ is writing an online message that uses derogatory, obscene, or inappropriate language.

5. Corporate networks that use the infrastructure and standards of the Internet and the World Wide Web are called _____.

Short-Answer Questions

1. List three ways you can connect your microcomputer to the Internet.
2. What is an Internet service provider?
3. What are the principal features of the Internet?
4. What is a firewall?
5. What is meant by the term *connectivity*?

Multiple-Choice Questions

1. Which of the following enables different computers on different networks to communicate with each other?
   a. modems
   b. Internet service provider
   c. TCP/IP software
   d. FTP
   e. All of the above

2. Which of the following would you use to search for and retrieve files stored on different computers?
   a. FTP
   b. Gopher
   c. HTTP
   d. HTML
   e. All of the above

3. Which of the following would you use to write files for the Web?
   a. FTP
   b. Gopher
   c. HTTP
   d. HTML
   e. All of the above

4. Which of the following has the most potential to reduce paper handling?
   a. EDI
   b. Intranets
   c. HTML
   d. Veronica
   e. All of the above

5. Which of the following addresses might you use to connect to a nonprofit organization?
   a. *clifford@mindspring.com*
   b. *71222.1111@compuserve.com*
   c. *susanh@universe.org*
   d. *help@volunteer.mil*
   e. All of the above

True/False Questions

T F 1. Because of technological limitations, the Web will never support 3-D.

T F 2. The term *cyberspace* refers to the World Wide Web, but not to most of the other features of the Internet.

T F 3. A picture phone has already been developed by AT&T.

T F 4. With a direct connection to the Internet, you don't need to use a modem.

T F 5. For a fee, you can use an online information service to send e-mail, access databases, and download shareware.

Knowledge in Action

1. You need to purchase a computer to use at home to perform business-related (school-related) tasks. You want to be able to communicate with the network at work (school) and the Internet. Include the following in a report:
   - Description of the hardware and software used at work (school).
   - Description of the types of tasks you will want to perform at home.
   - Name of the computer system you would buy. (Include a detailed description of the computer system, such as the RAM capacity, secondary storage capacity, and modem speed.)
   - The communications software you would need to purchase or obtain.
   - The cost estimate for the system and for the online and telephone charges

2. In an effort to reduce new construction costs, some rapidly expanding companies are allowing more and more employees to telecommute several days a week. Offices are shared by several telecommuting employees. One employee will use the office two or three days a week, and another employee will use the same office other days of the week. What advantages does the company gain from this type of arrangement? What advantages do these employees have over the traditional work environment? What are some of the disadvantages to both the company and the employees? Do you think employees' productivity will decline from telecommuting and/or sharing office space with other telecommuting employees? If so, why?

3. What do you think the future holds for online information services? Research your response using current magazines and periodicals and/or on the Internet.

4. Explore the state-of-the-art of v-mail (video mail). What software and hardware is required? What companies currently have v-mail products and when were they released? What are the current limitations of v-mail?

5. "Distance learning," or "distance education," uses electronic links to extend college campuses to people who otherwise would not be able to take college courses. Is your school or someone you know involved in distance learning? If so, research the system's components and uses. What hardware and software do students need in order to communicate with the instructor and classmates?

Chapter 8

Communications Technology

Hardware, Channels, & Networks

key questions

You should be able to answer the following questions:

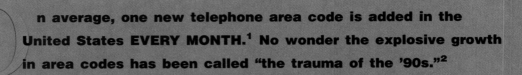

n average, one new telephone area code is added in the United States **EVERY MONTH.**[1] No wonder the explosive growth in area codes has been called "the trauma of the '90s."[2]

Between 1995 and 1997, 62 new geographic area codes were introduced around the country. Over the next decade, 120 new codes will be implemented.[3] Worse, the United States, Canada, and the Caribbean will run out of three-digit combinations early in the 21st century, which means that one or more digits will have to be added to every phone number or area code. This in turn will disrupt some phone switches and computer databases that store phone numbers in fixed 10-digit formats. "If you expand the phone number," says one telephone company executive, "every kind of database that stores a phone number—every store account, pagers, dial-up modems, everything you can think of—will have to be fixed."[4]

No doubt you already suspect the reasons for the spectacular growth in area codes. Explains economics editor Jonathan Marshall:

> Every time you add an extra phone line for a fax machine, Internet connection, home office, or talkative teenager, you need another number.
> Every time you add a pager or a cellular phone to your stockpile of equipment, you add more numbers.
> And every time you swipe your credit card through a supermarket or department store register, or put your ATM card into a cash machine, you are making use of a separate "point of sale" phone line and number.
> Across the nation, a new customer signs up for cellular phone service every 2.8 seconds. Each one needs a new number.[5]

What is the effect of all these numbers and area codes? Let us take a look at one aspect: the new ways in which work is distributed.

8.1 Portable Work: Telecommuting, Virtual Offices, & Mobile Workplaces

KEY QUESTION

What are telecommuting and virtual offices?

Preview & Review: Working at home with computer and communications connections between office and home is called telecommuting.

The virtual office is a nonpermanent and mobile office run with computer and communications technology.

"In a country that has been moaning about low productivity and searching for new ways to increase it," observed futurist Alvin Toffler, "the single most anti-productive thing we do is ship millions of workers back and forth across the landscape every morning and evening."[6]

Toffler was referring, of course, to the great American phenomenon of physically commuting to and from work. About 96 million Americans commute to work by car, 39 million not to a city's downtown but rather to jobs scattered about in the suburbs.[7] Information technology has responded to the cry of "Move the work instead of the workers!" Computers and communications tools have led to telecommuting and telework centers, the virtual office and "hoteling," and the mobile workplace.

Telecommuting & Telework Centers

Working at home with telecommunications between office and home is called *telecommuting.* The number of telecommuters—those who work at home at least one day a week—has nearly tripled to 11 million in the United States from 4 million in 1990, according to a survey by a market research firm.[8] The study projects about 14 million home-based workers by 2000.

Telecommuting has a lot of benefits.[9] The advantages to society are reduced traffic congestion, energy consumption, and air pollution. The advantages to employers include increased productivity—20% more, according to New York–based Link Resources—because telecommuters experience less distraction at home and can work flexible hours. Absenteeism is decreased because telecommuters will work with the sniffles and other minor ailments that might keep them away from the office. There are further advantages of improved teamwork and an expanded labor pool because hard-to-get employees don't have to uproot themselves from where they want to live. And, of course, many employees are happier. Medical transcriber Carolyn McCann found her productivity increased 20% since she stopped going into the office, yet she was more satisfied with her life. "There's no end to the money you can save by not having to go [drive] to work," she said.[10]

Another term for telecommuting is telework. However, *telework* includes not only those who work at least part time from home but also those who work at remote or satellite offices, removed from organizations' main offices. Such satellite offices are sometimes called *telework centers.* An example of a telework center is the Riverside Telecommuting Center, in Riverside, California, supported by several companies and local governments. The center provides office space that helps employees who live in the area avoid lengthy commutes to downtown Los Angeles. However, these days an office can be just about anywhere.

The Virtual Office

The *virtual office* is an often nonpermanent and mobile office run with computer and communications technology. Employees work not in a central office but from their homes, cars, and other new work sites. They use pocket pagers, portable computers, fax machines, and various phone and network services to conduct business.

Could you stand not having a permanent office at all? Here's how one variant, called *"hoteling"* or *"alternative officing,"* works. You call ahead to book a room, and speak to the concierge. However, your "hotel" isn't a Hilton, and the "concierge" isn't a hotel employee who handles reservations, luggage, and local tours. Rather, the organization is accounting and management consulting firm Ernst & Young, advertising agency Chiat/Day, or computer maker Tandem Computers, to name three examples. And the concierge is an administrator who handles scheduling of available office cubicles, of which there is perhaps only one for every three workers.[11–13]

Hoteling works for Ernst & Young, for example, because its auditors and management consultants spend 50% to 90% of their time in the field, in the offices of clients. When they need to return to their local E&Y office, they call a few hours in advance. The concierge consults a computerized scheduling program and determines which cubicles are available on the days requested. He or she chooses one and puts the proper nameplate on the office wall. The concierge then punches a few codes into the phone to program its number and voice mail. When employees come in, they pick up personal effects and files from lockers. They then take them to the cubicles they will use for a few days or weeks.

What makes hoteling possible, of course, is computer and communications technology. Computers handle the cubicle scheduling and reprogramming of phones. They also allow employees to carry their work around with them stored on the hard drives of their laptops. Cellular phones, fax machines, and e-mail permit employees to stay in touch with supervisors and co-workers.

Mobile Workplaces

"These days, some truckers are more inclined to sport white collars than tank tops," says one report. "Once (and still) lumped as rednecks and high-

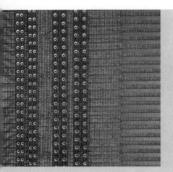

README

Practical Matters: The Telecommuter Test— Will Working at Home Work for You?

Here are the rules that Elaine Carey has insisted that her children, Alexander, 12, and Sofia, 13, follow when they're out of school during the summer so they don't disrupt her work as a media representative, which she does out of her home: (1) No rowdy play inside or slamming of doors. (2) Knock on her office door and wait quietly before entering (in case she's on the phone). (3) Let her know the day before if they need a ride somewhere.[14]

Of course, one reason some parents work from home is that they want to be able to concentrate on their children. Still, there is the constant need to keep the professional and the personal separated.

Here are some factors to consider if you're entertaining the idea of working at home:

Do You Have the Right Kind of Job? First, you have to have the right kind of job to work outside an office. "Positions that require a lot of independent work, such as sales, and some kinds of consulting, writing, and research analysis, are a natural fit," says one report. "People whose jobs require numerous unscheduled meetings with co-workers, such as those held by many managers, present more problems."[15]

Do You Have the Right Temperament? Some home-workers have trouble with the quiet and isolation. Others find the collaborative times around the office water cooler "are not supplanted by e-mail discussions from home."[16] Still others are stressed by the fact that the workload is always there, 24 hours a day. Catherine Rossbach says she hated working at home. "I'd do my grocery shopping in the middle of the day when there were no crowds," she says, "and then I'd end up working until 2 A.M. I had no structure to my workday and felt totally isolated."[17]

Do You Have the Right Work Space & Equipment? The main thing is to be sure the day-to-day activities can be done comfortably, says productivity consultant Odette Pollar. "Being next to the washing machine or having to cross a field of distractions will make getting started each morning harder," she says. "So will having to unplug the toaster every time you wish to use the calculator."[18]

Having good furniture and equipment is crucial, though you shouldn't let yourself be talked into getting more than you need (such as a color printer when a black-and-white one will do). If no one sees your office, quality of furniture doesn't matter. One business start-up in San Francisco, an online company named NetNoir that focuses on African Americans, adopted the following standard for every desk: Each end is three standard concrete blocks stacked up, with three concrete paving stones piled on top, and a standard door is laid atop all that. This produces a desk 29 inches tall, considered the optimal height for a computer desk.[19]

Does Your Office Project the Right Image for Clients? Maybe you'll never see clients in your home office, but you still don't want to project an amateurish image. "Voice mail, different lines for your phone and fax, and e-mail capability can make at-home setups seem professional," says a Portland, Oregon, freelance business writer.[20]

In addition, you may want to have a business-mail address separate from your home address, a live answering service, and a place to meet clients outside your home. Though he does most of his work from his home, the foregoing writer pays $250 a month to rent an executive suite downtown operated by HQ Business Centers, a national franchise. This gives him a better address and a space to meet clients, as well as a place to work "when the kids are at home banging a basketball against the garage."

school dropouts, they are now fluent with computers, satellites, and fax machines—all of which can be found in the cabs of their 18-wheelers."[21]

Truckers may now be required to carry laptops with which they keep in touch via satellite with headquarters. They may also have to take on tasks previously never dreamed of. These include faxing sales invoices, hounding late-paying customers, and training people to whom they deliver high-tech office equipment.

Other workers—field service representatives, salespeople, and roving executives—also find that they need mobile communications in order to stay competitive. For instance, Bob Spoer of San Francisco, who started a telecommunications firm that needs constant tending, takes a cell phone to baseball games and on ski lifts ("I actually get some good reception up there").[22]

Not everyone is as enthusiastic about being so accessible, however. Said one observer back in 1991: "The movable office—a godsend for workaholics, a nightmare for those who live with them—has only just begun."[23] Many people find that technology creates an electronic leash. "I get 20 beeps on a weekend," said Peter Hart, then a supervisor at a California chip manufacturer—before he changed jobs because pagers, cell phones, and e-mail were taking over his life.[24]

Information technology is blurring time and space, eroding the barriers between work and private life. Some people hate it, but others thrive on it. Let us now take a look at the technology that has brought us to this point.

8.2 Using Computers to Communicate: Analog & Digital Signals, Modems, & Other Technological Basics

KEY QUESTIONS

How do analog and digital signals and modems work, and what are alternatives to modems?

Preview & Review: To communicate online through a microcomputer, users need a modem to send and receive computer-generated messages over telephone lines. The modem translates the computer's digital signal of discrete bursts into an analog signal of continuous waves, and vice versa. A modem may have various transmission speeds. Communications, or datacomm, software is also required.

ISDN lines, DSL lines, T1 lines, cable modems, and satellite dishes are faster than conventional PC modems.

Communications, or *telecommunications,* refers to the transfer of data from a transmitter (sender or source) to a receiver across a distance. The data transferred can be voice, sound, images, video, text, or a combination thereof (multimedia). Some form of electromagnetic energy—electricity, radio waves, or light—is used to represent the data, which is transmitted through a wire, the air, or other physical medium. To set up a path for the data transfer, intermediate devices are put in place, such as microwave towers.

Data is transmitted by two types of signals, each requiring different kinds of communications technology. The two types of signals are *analog* and *digital.* (■ *See Panel 8.1, next page.*) In a way they resemble analog and digital watches. An analog watch shows time as a continuum. A digital watch shows time as discrete numeric values.

Analog Signals: Continuous Waves

Telephones, radios, and televisions—the older forms of communications technology—were designed to work with an analog signal (✔ p. 7). **An *analog signal* is a continuous electrical signal in the form of a wave.** The wave is called a *carrier wave.*

Two characteristics of analog carrier waves that can be altered are frequency and amplitude.

■ PANEL 8.1

Analog and digital signals

An analog signal represents a continuous electrical signal in the form of a wave. A digital signal is discontinuous, expressed as discrete bursts in on/off electrical pulses.

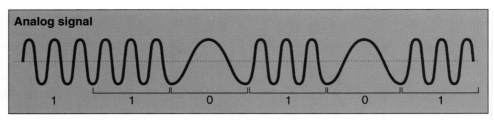

Analog signal

1   1   0   1   0   1

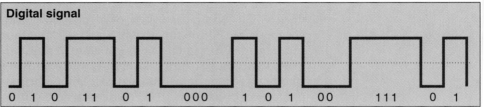

Digital signal

0 1 0 11 0 1 000 1 0 1 00 111 0 1

- Frequency: **Frequency is the number of times a wave repeats during a specific time interval**—that is, how many times it completes a *cycle* in a second.

- Amplitude: **Amplitude is the height of a wave within a given period of time.** Amplitude is actually the strength or volume—the loudness—of a signal.

Both frequency and amplitude can be modified by making adjustments to the wave. Indeed, it is by such adjustments that an analog signal can be made to express a digital signal, as we shall explain.

Digital Signals: Discrete Bursts

A *digital signal* **uses on/off or present/absent electrical pulses in discontinuous, or discrete, bursts, rather than a continuous wave.** This two-state kind of signal works perfectly in representing the two-state binary language of 0s and 1s that computers use. That is, the presence of an electrical pulse can represent a 1 bit, its absence a 0 bit.

The Modem: Today's Compromise

Digital signals are better—that is, faster and more accurate—at transmitting computer data. However, many of our present communications lines, such as telephone and microwave, are still analog. To get around this problem, we need a *modem* (✔ p. 8). **A *modem*—short for *modulate/demodulate*—converts digital signals into analog form (a process known as *modulation*) to send over phone lines. A receiving modem at the other end of the phone line then converts the analog signal back to a digital signal (a process known as *demodulation*).** (■ *See Panel 8.2.*)

■ PANEL 8.2

How modems work

A sending modem translates digital signals into analog waves for transmission over phone lines. A receiving modem translates the analog signals back into digital signals.

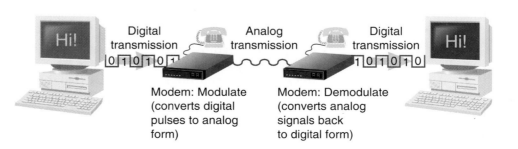

Digital transmission   Analog transmission   Digital transmission

0 1 0 1 0 1   1 0 1 0 1 0

Modem: Modulate (converts digital pulses to analog form)   Modem: Demodulate (converts analog signals back to digital form)

Modifying an analog signal

A modem may modify an analog signal to carry the on/off digital signals of a computer in two ways. *(Top)* The frequency of wave cycles is altered so that a normal wave represents a 1 and a more frequent wave within a given period represents a 0. *(Bottom)* The amplitude (height) of a wave is altered so that a wave of normal height represents a 1 and a wave of lesser height represents a 0.

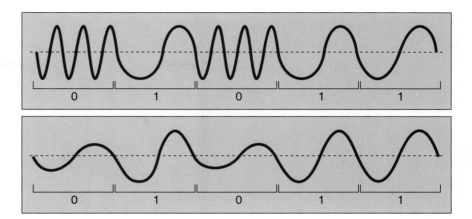

Modulation/demodulation does not actually make an analog signal into the on/off form of the digital signal. Rather, it changes the shape of the wave to convey digital information. For instance, the frequency might be changed. A normal wave cycle within a given period of time might represent a 1, but more frequent wave cycles with a given period might represent a 0. Or, the amplitude might be changed. A loud sound might represent a 1 bit, a soft sound might represent a 0 bit. That is, a wave with normal height (amplitude) might signify a 1, a wave with smaller height a 0. (■ *See Panel 8.3.*) The wave itself does not assume the boxy on/off shape represented by the true digital signal.

From this we can see that modems are a compromise. They cannot transmit digital signals in a way that delivers their full benefits. As a consequence, communications companies have been developing alternatives, such as the Integrated Services Digital Network (ISDN) and cable modems, discussed shortly.

Choosing a Modem

Two criteria for choosing a modem are whether you want an internal or external one, and what transmission speed you wish:

- **External versus internal:** Most modems these days are internal, but some are external. (■ *See Panel 8.4.*)

  An *external modem* is a box that is separate from the computer. The box may be large or it may be portable, pocket size. A cable connects the modem to a port in the back of the computer. A second line connects the modem to a standard telephone jack. There is also a power cord that plugs into a standard AC wall socket.

  The advantage of the external modem is that it can be used with different computers. Thus, if you buy a new microcomputer, you will probably be able to use your old external modem. Also, external modems help isolate the computer's internal circuitry from phone line–conducted lightning surges.

  An *internal modem* is a circuit board that plugs into a slot inside the system cabinet. Currently most new microcomputers come with an internal modem already installed (some are built right into the motherboard). Advantages of the internal modem are that it doesn't take up extra space on your desk, it is less expensive than an external modem, and it doesn't have a separate power cord.

External modem

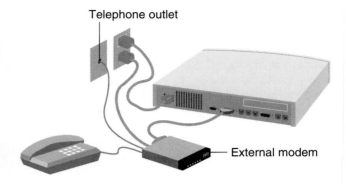

Telephone outlet

External modem

Internal modem

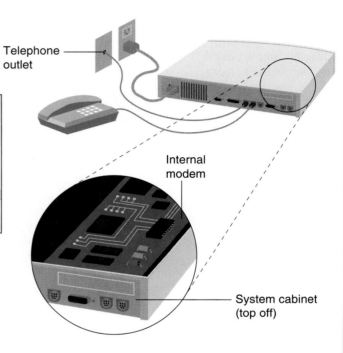

Telephone outlet

Internal modem

System cabinet (top off)

■ PANEL 8.4

External versus internal modems

An external modem is a box that is outside the computer. An internal modem is a circuit board installed in an expansion slot inside the system cabinet. Modems may also take the form of removable PC cards that fit into a PCMCIA slot in a portable computer.

For laptop computers, there are also easily removable internal modems. That is, modems are available as PC cards (✔ p. 170), which can be slipped in and out of a PCMCIA slot.

● **Transmission speed:** Because most modems use standard telephone lines, users are charged the usual rates by phone companies, whether local or long-distance. Users are also often charged by online services for time spent online. Accordingly, *transmission speed*—the speed at which modems transmit data—becomes an important consideration. The faster the modem, the less time you need to spend on the telephone line.

Today users refer to **bits per second (bps) or, more likely, *kilobits per second (kbps)* to express data transmission speeds.** A 28,800-bps modem, for example, is a 28.8-kbps modem.

Modems transmit at three ranges of speed—slow, moderately fast, and high-speed.

(1) *Slow: 1200, 2400, and 4800 bps.* A 10-page single-spaced letter can be transmitted by a 2400-bps modem in 2½ minutes. Slow is not worth using anymore.

(2) *Moderately fast: 9600 and 14,400 bps.* The letter can be transmitted by a 9600-bps modem in 38 seconds. Faxes are typically delivered at these speeds.

(3) *High-speed: 28,800, 33,600, and 56,000 bps* (or 28.8, 33.6, and 56 kbps). The letter can be sent by a 28,800-bps modem in about 10 seconds. It can be downloaded (transferred) by a 56,000-bps modem in about 5 seconds.

The fastest modem, known as the "56K" modem, doesn't usually operate at 56 kbps. One writer reports that in 60 log-ons to three different access providers he never got a connection faster than 46.6 kbps, and the average was 42.6 kbps.[25] In addition, you won't even get this speed increase unless you're linked to an online service or Internet service provider whose own modems are compatible with your modem's 56K technology. (A lack of compatibility between two standards of 56K modems was ended by the International Telecommunications Union in February 1998.)

Static-filled phone lines may reduce a modem's efficiency. The next generation will probably consist of modem-type hardware using ISDN, DSL, or cable-TV circuits (all discussed shortly).

Communications Software

To communicate via a modem, your microcomputer requires communications software. *Communications software,* or "datacomm software," manages the transmission of data between computers, as we mentioned in Chapter 2 (✔ p. 68). Macintosh users have Smartcom. Windows users have Smartcom, Crosstalk, CommWorks, Crosstalk, ProComm Plus, and Hyper-Terminal. OS/2 Warp users have HyperAccess. Often the software comes on diskettes bundled with (sold along with) the modem. Also, communications software now often comes as part of the system software.

One of the principal uses to which you will put communications software is to download and upload files. **Download means that you retrieve files from another computer and transfer them to the main memory (RAM) on the hard disk in your computer.** *Upload* **means that you send files from your computer to another computer.**

Besides establishing connections between computers, communications software may perform other functions:

- **Error correction:** Static on telephone lines can introduce errors into data transmission, or "noise." Noise is an extraneous signal that causes distortion in the data signal when it is received. Noise can be caused by power line voltage spikes, poorly fitting electrical contacts, or strong electrical or magnetic signals coming from nearby power lines or equipment such as air conditioners. Although such "noise" may not affect voice transmission very much, it can garble high-speed data transmission.

  When acquiring a modem and its accompanying software, you should inquire whether it incorporates error-correction features.

- **Data compression:** As we discussed in Chapter 6 (✔ p. 289), data *compression* reduces the volume of data in a message, thereby reducing the amount of time required to send data from one modem to another. When the compressed message reaches the receiver, the full message is restored. With text and graphics, a message may be compressed to as little as one-tenth of its original size.

- Remote control: *Remote-control software* **allows you to control a microcomputer from another microcomputer in a different location,** perhaps even thousands of miles away. One part of the program is in the machine in front of you, the other in the remote machine. Such software is useful for travelers who want to use their home machines from a distance. It's also helpful for technicians trying to assist users with support problems. Examples of remote-control software for microcomputers are Carbon Copy, Commute, PCAnywhere, and Timbuktu/Remote. There are also inexpensive shareware (✔ p. 86) versions.

- Terminal emulation: Mainframes and minicomputers are designed to be accessed by terminals, not by microcomputers. Thus, their operating systems are different. *Terminal emulation software* **allows you to use your microcomputer to simulate a mainframe's terminal.** That is, the software "tricks" the large computer into acting as if it is communicating with a terminal. Your PC needs terminal emulation capability to access computers holding databases of research materials. Telnet (✔ p. 324) is a terminal emulation protocol. (Some system software, such as Windows 95, includes terminal emulation software.)

ISDN Lines, DSL Lines, T1 Lines, Cable Modems, & Satellite Dishes

If you're wondering why people are concerned with online speed, consider this: Among online warriors playing shoot-'em-up games, those with fast Internet connections can get the drop on slower ones. "I roam among the slow people and pick 'em off," boasts online Quake player Erik Nachbar, 23, of Towson, Maryland, who is able to fire on opponents before they know they've even been targeted.[26]

Of course, there are lots of more important reasons why speedy access is important (such as those involving online commerce). In any case, users who have found themselves banging the table in frustration as their 28.8 modem takes 25 minutes to transmit a 1-minute low-quality video from a Web site are getting some relief. Probably the principal contenders to standard phone modems are *ISDN lines, DSL (ADSL), T1 lines, cable modems,* and *satellite dishes.* (■ *See Panel 8.5.*)

■ **PANEL 8.5**

Connection competitors

Approximate time to transfer a 40-megabyte file.

Technology	Speed	Time
Telephone modem	33.6 kbps	2 hours, 38 minutes
ISDN phone line	128 kbps	41 minutes, 40 seconds
DSL phone line	1.088 Mbps upload 8.192 Mbps download	4 minutes, 54 seconds 39 seconds
T1 phone line	1.544 Mbps	3 minutes, 27 seconds
Cable modem	500 kbps upload 10 Mbps download	10 minutes, 40 seconds 32 seconds
Satellite system	33.6 kbps upload 400 kbps download	2 hours, 38 minutes 13 minutes, 20 seconds

kbps = kilobytes per second; Mbps = megabytes per second (a megabyte is 1,000,000 kilobytes).

- **ISDN lines:** *ISDN* **stands for** *Integrated Services Digital Network.* **It consists of hardware and software that allow voice, video, and data to be communicated as digital signals over traditional copper-wire telephone lines.**

  ISDN comes with two main channels that can be used for regular phone calls, Internet access, or other data transmissions. When combined, they are capable of transmitting up to 128 kbps—more than four times faster than conventional 28.8 modems, or more than double the speed of 56K modems. A third channel—called the "D" channel or the "always on" channel—became available in 1997 to carry e-mail, stock quotes, or other data. "Using the D channel," says one report, "ISDN customers could speak on one voice channel, send files into a corporate office on another, and get e-mail on a third."[27]

  Provided by many telephone companies ("telcos," in communications jargon), ISDN is not cheap, costing perhaps two or three times as much per month as regular phone service. Installation could also cost $200 or more if you need a phone technician to wire your house and install the software in your PC. You also need to buy a special ISDN connector box or adapter card, to which you connect your microcomputer, fax machine, modem, and telephone.

  Nevertheless, with the number of people now working at home and/or surfing the Internet, demand has pushed ISDN orders off the charts. Waiting time may be as long as five months before the phone company can get around to installing one for you. Forecasts are for 7 million U.S. installations by 2000.[28]

  Even so, ISDN's time may have come and gone. The reason: DSL, cable modem, and other technologies threaten to render it obsolete.

- **DSL (ADSL):** If you were trying to download an approximately 6-minute-long music video from the World Wide Web, it would take you about 4 hours and 45 minutes using a 28.8 modem. An ISDN connection would reduce this to an hour. With DSL (ADSL), however, you would really notice the difference—11 minutes.[29]

  Most modems transmit data in kilobits per second (kbps). **Short for** *Digital Subscriber Line, DSL* **uses regular phone lines to transmit data in megabits per second (Mbps)—specifically 1.5–8 Mbps.** (ADSL stands for Asymmetric Digital Subscriber Line.) Deploying DSL requires an adapter, or special modem, for the user and an identical device at the phone company's central office. The two are connected by the standard twisted-pair copper wiring that makes up most telephone company networks. DSL modems use more of the voice channel on the telephone line by splitting the line into three channels—normal voice, outgoing data, and incoming data.

  DSL downloads data faster than it uploads data (a 40-megabyte file takes 39 seconds to download but almost 5 minutes to upload). Thus, at present it is not suited to videoconferencing, where conversations need to take place in real time. It is, however, well suited to Web browsing, where the amount of information coming down to your browser is much greater than the amounts you send up, such as e-mail.[30,31]

  Andrew Kessler, a partner in a Palo Alto, California, technology investment company, likes the idea of DSL, which is offered by his local telco, Pacific Bell: "With this I will get 12 times the speed of my ISDN service at half the price."[32] Moreover, because DSL works over regular phone lines without interrupting voice calls, users need not order a second phone line for their computers. The biggest initial charge for Kessler was for the hardware—$450 for a DSL modem, plus a so-called splitter (which aggregates the voice and data lines), plus a network interface card for his PC.

- **T1 lines:** **A *T1 line* is essentially a traditional trunk line that carries 24 normal telephone circuits and has a speed of 1500 kbps.** Generally, T1 lines are high-capacity communications links found at corporate, government, or academic sites. To get similar speeds, consumers usually use DSL lines. Whereas a T1 line might cost $400 a month after installation (plus Internet service provider charges as steep as $800 a month), a DSL line might cost $165 a month.

  Another high-speed digital line is the T3 line, which may cost as much as $40,000 a month and requires a huge investment in equipment. **A *T3 line* transmits at a speed of 4500 kbps.**

- **Cable modems:** If DSL's 11 minutes to move a nearly 6-minute video sounds good, 2 minutes sounds even better. That's the rate of transmission for cable modems. Or, to make another comparison, if you wanted to transmit all 857 pages of Herman Melville's novel *Moby Dick*, a cable modem could do it in about 2 seconds. In the same time, an ISDN line could move only 10 pages, and a 28.8 modem wouldn't get past page 3.[33]

  Cable companies have found that when cable modem service is introduced, customer use jumps dramatically. "Some nights I can't get off the thing," said biology professor Grant Balkema, after cable modems were installed at Boston College. "I've started some nights at around 10 and stayed up until 2 A.M. It's—dare I say?—addictive."[34]

  A *cable modem* is a modem that connects a personal computer to a cable-TV system that offers online services. The gadgets are still not in common use, and it will probably be a while before internationally standardized cable modems go on sale. The reason: So far probably 90% of U.S. cable subscribers are served by networks that don't permit much in the way of two-way data communications. "The vast majority of today's . . . cable systems can deliver a river of data downstream," says one writer, "but only a cocktail straw's worth back the other way."[35] (Downloads may take only 32 seconds for a 40-megabyte file but uploads nearly 11 minutes.) Nevertheless, Forrester Research Inc. predicts about 6.8 million American homes will have cable modems by 2000.[36]

- **Satellite dishes:** Hughes Network Systems makes direct-broadcast satellite dishes for TV reception. It also offers an Internet service called DirecDuo that, besides providing regular satellite-TV programming, allows Internet downloads at speeds of supposedly up to 400 kbps—about 12 times as fast as a typical modem. (The DirecDuo grew out of another Hughes Internet-only satellite service called DirecPC, which failed to catch on because it required another type of dish than that used to receive TV.)

  Respected technology writer Walter Mossberg found the TV portion of DirecDuo worked just fine. As for Internet connections, the system worked faster than a phone modem but, for technical reasons having to do with the nature of the Internet, the average speed—about 45 kbps, and occasionally as fast as 266 kbps—was nowhere near the 400 kbps maximum.[37]

  No doubt satellite transmission will improve as more satellites are booted into space. We discuss satellite systems in another few pages.

8.3   Communications Channels: The Conduits of Communications

KEY QUESTION

What are types of wired and wireless channels and some types of wireless communications?

Preview & Review: A channel is the path, either wired or wireless, over which information travels. Various channels occupy various radio-wave bands on the electromagnetic spectrum. Types of wired channels include twisted-pair wire, coaxial cable, and fiber-optic cable. Two principal types of wireless channels are microwave and satellite systems. Some types of wireless communications are pagers, analog cellular phones, packet radio, and Cellular Digit Packet Data (CDPD).

The next generation of wireless communications will include digital cellular phones, personal communications services (PCS), specialized mobile radio (SMR), and satellite-based systems.

If you are of a certain age, you may recall when two-way individual communications were accomplished mainly in two ways. They were carried by (1) a telephone wire or (2) a wireless method such as shortwave radio. Today there are many kinds of communications channels, although they are still wired or wireless. **A *communications channel* is the path—the physical medium—over which information travels in a telecommunications system from its source to its destination.** (Channels are also called *links, lines,* or *media.*) The basis for all telecommunications channels, both wired and wireless, is the electromagnetic spectrum.

The Electromagnetic Spectrum

Telephone signals, radar waves, and the invisible commands from a garage-door opener all represent different waves on what is called the electromagnetic spectrum. **The *electromagnetic spectrum* consists of fields of electrical energy and magnetic energy, which travel in waves.** (■ *See Panel 8.6, next page.*)

All radio signals, light rays, X-rays, and radioactivity radiate an energy that behaves like rippling waves. The waves vary according to two characteristics, frequency and wavelength:

- Frequency: As we've seen, *frequency* is the number of times a wave repeats (makes a cycle) in a second. **Frequency is measured in *hertz (Hz)*, with 1 Hz equal to 1 cycle per second.** One thousand hertz is called a *kilohertz (KHz),* 1 million hertz is called a *megahertz (MHz),* and 1 billion hertz is called a *gigahertz (GHz).*

    Ranges of frequencies are called *bands* or *bandwidths.* The bandwidth is the difference between the lowest and highest frequencies transmitted. Thus, for example, cellular phones are on the 800–900 megahertz bandwidth—that is, their bandwidth is 100 megahertz. The wider the bandwidth, the faster data can be transmitted.

    Why is it important to know this? "Low-frequency waves can travel far and curve with the Earth but can't carry much information," points out technology writer Kevin Maney. "High-frequency waves can travel only a short distance before breaking up and won't curve over the horizon, but they can carry much more information."[38] Thus, different technologies (cell phones versus PCS phones, for instance, as we describe) are best suited to different purposes, depending on the frequency range—bandwidth—they are in.

- Wavelength: Waves also vary according to their length—their *wavelength.* We see references to "wave" length in the terms "shortwave radio" and "microwave oven."

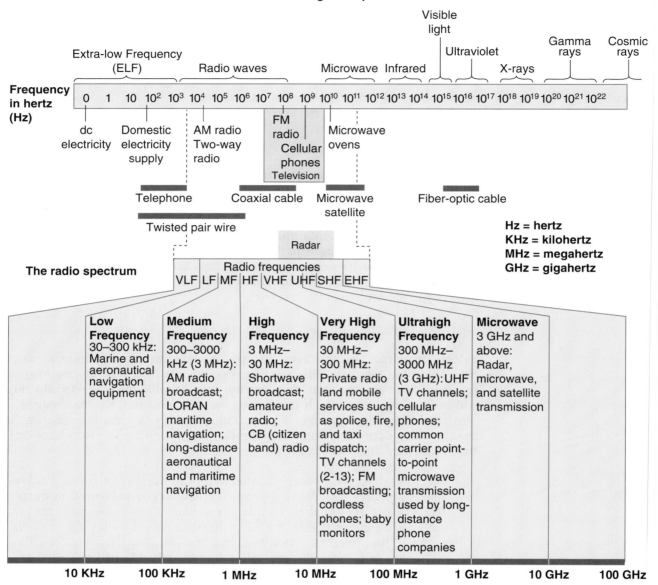

The electromagnetic spectrum

The electromagnetic spectrum

At the low end of the spectrum, the waves are of low frequency and of long wavelength (such as domestic electricity). At the high end, the waves are of high frequency and short wavelength (such as cosmic rays).

The electromagnetic spectrum can be represented in terms of the appliances and machines that emit or detect particular wavelengths:

- We could start on the left, at the low-frequency end, with video display terminals and hair dryers.
- We would then range rightward through AM and FM radios, shortwave radios, UHF and VHF television, and cellular phones. This is the area we're particularly concerned with in this book.
- Next we would proceed through radar, microwave ovens, infrared "nightscope" binoculars, and ultraviolet-light tanning machines. We're interested in some of this area, too.
- Finally, we would go through X-ray machines and end up on the far right, high-frequency end of the spectrum with gamma-ray machines for food irradiation.

The part of the spectrum of principal concern to us is that area in the middle—between 3 million and 300 billion hertz (3 megahertz to 300 gigahertz). This is the portion that is regulated by the government for communications purposes.

In the United States, certain bands are assigned by the Federal Communications Commission (FCC) for certain purposes—that is, to be controlled by different types of media equipment or types of users. Some frequencies traditionally used by railroads, electric utilities, and police and fire departments have in recent times been opened up for new uses. These new applications include personal telephones, mobile data services, and satellite message services. We explain these further in the next few pages.

Let us now look more closely at the various types of channels:

- Twisted-pair wire
- Coaxial cable
- Fiber-optic cable
- Microwave and satellite systems
- Other wireless communications
- The next generation of wireless communications

Twisted-Pair Wire

The telephone line that runs from your house to the pole outside, or underground, is probably twisted-pair wire. **Twisted-pair wire consists of two strands of insulated copper wire, twisted around each other in pairs.** They are then covered in another layer of plastic insulation. (■ *See Panel 8.7, next page.*)

Because so much of the world is already served by twisted-pair wire, it will no doubt continue to be used for years, both for voice messages and for modem-transmitted computer data. However, compared to other forms of wiring or cabling, it is relatively slow and does not protect well against electrical interference. As a result, it will certainly be superseded by better communications channels, wired or wireless.

twisted wire

coaxial cable

■ PANEL 8.7

Three types of wired communications channels
(Top left) Twisted-pair wire. This type does not protect well against electrical interference. *(Middle left)* Coaxial cable. This type is shielded against electrical interference. It also can carry more data than most kinds of twisted-pair wire. *(Right)* When coaxial cable is bundled together, as here, it can carry more than 40,000 conversations at once. *(Bottom left)* Fiber-optic cable. Thin glass strands transmit pulsating light instead of electricity. These strands can carry computer and voice data over long distances.

Coaxial Cable

Coaxial cable, **commonly called "co-ax," consists of insulated copper wire wrapped in a solid or braided metal shield, then in an external cover.** Co-ax is widely used for cable television. The extra insulation makes coaxial cable much better at resisting noise (static) than twisted-pair wiring. Moreover, it can carry voice and data at a faster rate (up to 200 megabits per second, compared to only 16–100 megabits per second for twisted-pair wire). Often many coaxial cables will be bundled together.

Fiber-Optic Cable

The population of North Dakota has remained virtually the same (638,000) since the 1920s, as many educated young people have left in search of jobs. But when 4600 miles of fiber-optic lines were laid in the state, bringing faster and higher-quality communications, it started a telecommunications boom that has changed the employment picture completely. Today North Dakota is home to several telemarketing, data processing, and reservations centers, such as the reservations headquarters for Quality Inns.

Farm wife Susan Horner now works the phones for a travel company in the town of Linton, booking vacations for clients far away while her husband milks their 50 cows on his own. "When this office opened, every fourth business on Main Street was closed, Linton was headed to becoming a ghost town," she says. "My paycheck makes our house payment and our farm payment."[39]

A *fiber-optic cable* **consists of hundreds or thousands of thin strands of glass that transmit not electricity but rather pulsating beams of light.** These strands, each as thin as a human hair, can transmit billions of pulses per second, each "on" pulse representing one bit. When bundled together, fiber-optic strands in a cable 0.12 inch thick can support a quarter- to a half-million voice conversations at the same time. Moreover, unlike electrical signals, light pulses are not affected by random electromagnetic interference in the environment. Thus, they have much lower error rates than normal telephone wire and cable. In addition, fiber-optic cable is lighter and more durable than twisted-pair and coaxial cable. A final advantage is that it cannot be easily wiretapped or listened into, so transmissions are more secure.

The main drawbacks until recently have been cost and the material's inability to bend around tight corners. In mid-1995, however, new material was announced—called *graded-index plastic optical fiber*—that was cheaper, lighter, and more flexible than glass fibers. The plastic flexible fiber is able to handle loops and curves with ease and thus will be better than glass for curb-to-home wiring.

Microwave & Satellite Systems

Wired forms of communications, which require physical connection between sender and receiver, will not disappear any time soon, if ever. For one thing, fiber-optic cables can transmit data communications 10,000 times faster than microwave and satellite systems can. Moreover, they are resistant to data theft.

Still, some of the most exciting developments are in wireless communications. After all, there are many situations in which it is difficult to run physical wires. Here let us consider microwave and satellite systems.

- **Microwave systems:** *Microwave systems* **transmit voice and data through the atmosphere as super-high-frequency radio waves.** Microwave systems transmit microwaves, of course. *Microwaves* are the electromagnetic waves that vibrate at 1 gigahertz (1 billion hertz) per second or higher. These frequencies are used not only to operate microwave ovens but also to transmit messages between ground-based earth stations and satellite communications systems.

  Today you see dish- or horn-shaped microwave antennas nearly everywhere—on towers, buildings, and hilltops. (■ *See Panel 8.8, next page.*) Why, you might wonder, do people have to intrude on nature by putting a microwave dish on top of a mountain? The reason: microwaves cannot bend around corners or around the earth's curvature; they are "line-of-sight." *Line-of-sight* means there must be an unobstructed view between transmitter and receiver.

  Thus, microwave stations need to be placed within 25–30 miles of each other, with no obstructions in between. The size of the dish varies with the distance (perhaps 2–4 feet in diameter for short distances, 10 feet or more for long distances). A string of microwave relay stations will each receive incoming messages, boost the signal strength, and relay the signal to the next station.

  More than half of today's telephone system uses dish microwave transmission. However, the airwaves are becoming so saturated with microwave signals that future needs will have to be satisfied by other channels, such as satellite systems.

■ **PANEL 8.8**

Microwave systems

Microwaves cannot bend around corners or around the curvature of the earth. Therefore, microwave antennas must be in "line of sight" of each other—that is, unobstructed. Microwave dishes and relay towers may be on the ground (shown here on Midway Island, 1150 miles from Hawaii). Usually, however, they are situated atop high places, such as mountains or tall buildings, so that signals can be beamed over uneven terrain.

Microwave
relay station

Line-of-
sight signal

- **Satellite systems:** To avoid some of the limitations of microwave earth stations, communications companies have added microwave "sky stations"—communications satellites. ***Communications satellites* are microwave relay stations in orbit around the earth.** (■ *See Panel 8.9.*)

■ **PANEL 8.9**

Communications satellite

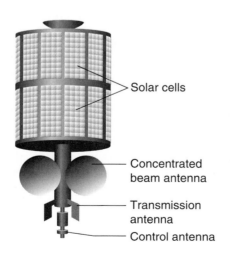

Solar cells

Concentrated
beam antenna

Transmission
antenna

Control antenna

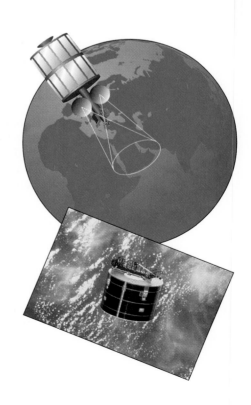

Traditionally, the orbit has been 22,300 miles above the earth (although newer systems will be much lower, as we discuss). Because they travel at the same speed as the earth, they appear to an observer on the ground to be stationary in space—that is, they are *geostationary*. Consequently, microwave earth stations are always able to beam signals to a fixed location above. The orbiting satellite has solar-powered receivers and transmitters (transponders) that receive the signals, amplify them, and retransmit them to another earth station. The satellite contains many communications channels and receives both analog and digital signals from earth stations.

Transmitting a signal from a ground station to a satellite is called *uplinking*; the reverse is called *downlinking*. Since more than one satellite may be required to get a message delivered, this can slow down the delivery process.

Other Wireless Communications

His friends thought Hank Kahrs, an insurance auditor, was just being silly in taking along a cell phone on their hike up California's Mount Whitney, the highest mountain in the continental United States. One reason for getting outdoors, they chided him, was to get away from civilization and its gadgets; anyway, they said, the phone probably wouldn't work at 14,494 feet. Kahrs, however, didn't want to break his long-standing custom of calling his wife every day.

At the top of the peak, he got a pleasant surprise. "It took 30 seconds before the phone started ringing" at his wife's number, he said. After he made his call, his hiking companions' attitudes changed. "When they saw the signal was fine, my friends all wanted to use the phone," said Kahrs.[40]

Call this a glimpse of the future's "unwired planet." Very soon, it will be nearly impossible to find a place where you *can't* make a phone call or at least page someone from anywhere on earth (except from a cave perhaps). Already this has produced headaches for forest rangers, who have had to rescue too many hikers who embarked on wilderness treks without being properly equipped—except for a cellular phone to use to call for help.

The Detroit Police Department started using two-way car radios in 1921. Mobile telephones were introduced in 1946. Clearly, mobile wireless communications have been around for some time. Today, however, we are witnessing an explosion in mobile wireless use that is making worldwide changes.

There are essentially four ways to move information through the air long-distance on radio frequencies: (1) via *one-way communications*, as typified by the satellite navigation system known as the Global Positioning System (GPS) and by pagers, and via *two-way communications*, which are classified as (2) analog cellular phones, (3) packet radio, and (4) Cellular Digital Packet Data (CDPD). (Other wireless methods operate at short distances.)

- One-way communications—the Global Positioning System (GPS): A $10 billion infrastructure developed by the military in the mid-1980s, **GPS, for *Global Positioning System*, consists of a series of 24 earth-orbiting satellites 10,600 miles above the earth that continuously transmit timed radio signals that can be used to identify earth locations.** A GPS receiver—handheld or mounted in a vehicle, plane, or boat—can pick up transmissions from any four satellites, interpret the information from each, and calculate to within a few hundred feet or less the receiver's longitude, latitude, and altitude.[41] (■ *See Panel 8.10, next page.*)

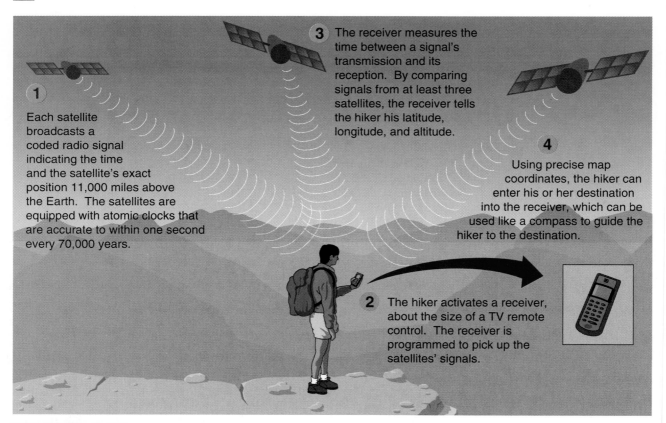

1 Each satellite broadcasts a coded radio signal indicating the time and the satellite's exact position 11,000 miles above the Earth. The satellites are equipped with atomic clocks that are accurate to within one second every 70,000 years.

3 The receiver measures the time between a signal's transmission and its reception. By comparing signals from at least three satellites, the receiver tells the hiker his latitude, longitude, and altitude.

4 Using precise map coordinates, the hiker can enter his or her destination into the receiver, which can be used like a compass to guide the hiker to the destination.

2 The hiker activates a receiver, about the size of a TV remote control. The receiver is programmed to pick up the satellites' signals.

■ PANEL 8.10

GPS receiver: handheld compass

The Global Positioning System uses 24 satellites, developed for military use, to pinpoint a location on the earth's surface. It has a theoretical accuracy of 3–100 feet for military uses, but for commercial purposes the military deliberately distorts the signals, reducing accuracy to about 300 feet.

The system, accurate within 3–100 feet, is used to tell military units carrying special receivers where they are. To allow GPS to be used for civilian purposes (and to avoid giving enemies any military advantage), the Pentagon deliberately distorts these signals, so that they are accurate to only about 300 feet. (Nevertheless, reports one writer, "during the Persian Gulf war—the one conflict in which the military has actually used GPS navigation—the Pentagon . . . gave troops 3000 ordinary civilian receivers, enabling them to execute a perfectly coordinated surprise hook maneuver through a featureless desert."[42])

GPS is used for such civilian activities as tracking trucks and taxis, locating stolen cars, orienting hikers, and aiding in surveying. The Chicago Transit Authority is installing GPS receivers on 1600 of its buses and service vehicles. "The devices have begun to tell Chicago bus drivers when they fall behind schedule," says one report, "and eventually will trigger traffic lights to turn green as buses approach [certain] intersections. . . ."[43] The makers of the film *Forrest Gump* used a GPS device to track the sun and time their sunrise and sunset shots so they didn't get shadows.[44] GPS has also been used by scientists to keep a satellite watch over a Hawaiian volcano, Mauna Loa, and to capture infinitesimal movements that were used to predict eruptions.[45] Some GPS receivers include map software for finding your way around, as with the Guidestar system available with some rental cars in cities such as Miami, Los Angeles, and New York.

● **One-way communications—pagers:** In a Clearwater, Florida, child-care center, if one child bites another or otherwise misbehaves, the head of the center can instantly alert the parents. Mom or Dad need only dial the number of the pager that parents are given as part of the child-care service.[46]

Once stereotyped as devices for doctors and drug dealers, pagers are now consumer items. Commonly known as beepers for the sound they make when activated, *pagers are simple radio receivers that receive data (but not voice messages) sent from a special radio transmitter.* Often the pager has its own telephone number. When the number is dialed from a phone, the call goes by way of the transmitter straight to the designated pager. Paging services include SkyTel (MTel), PageNet, and EMBARC (Motorola).

Pagers also do more than beep, transmitting full-blown alphanumeric text (such as four-line, 80-character messages) and other data. Newer ones are mini-answering machines, capable of relaying digitized voice messages.

Pagers are very efficient for transmitting one-way information—emergency messages, news, prices, stock quotations, delivery-route assignments, even sports news and scores—at low cost to single or multiple receivers. Recently advances have given us *two-way paging* or *enhanced paging.* In one version of this technology, such as Motorola's Jazz, customers can send a preprogrammed message ("Will be late—stuck in traffic") or acknowledgment that they have received a message. Another kind of two-way pager, such as Skytel's SkyWriter, allows consumers to compose and exchange e-mail with anyone on the Internet and with other pagers. So far, however, two-way paging has largely been a flop because of garbled messages, software problems, and the limited availability of national two-way networks.[47]

- **Two-way communications—analog cellular:** *Analog cellular phones are designed primarily for communicating by voice through a system of ground-area cells. Each cell is hexagonal in shape, usually 8 miles or less in diameter, and is served by a transmitter-receiving tower.* Communications are handled in the bandwidth of 824–894 megahertz. Calls are directed between cells by a mobile telephone switching office (MTSO). Movement between cells requires that calls be "handed off" by the MTSO. (■ *See Panel 8.11, next page.)*

   Handing off voice calls between cells poses only minimal problems. However, handing off data transmission (where every bit counts), with the inevitable gaps and pauses as one moves from one cell to another, is much more difficult. In the long run, data transmissions will probably have to be handled by the technology we discuss next, packet radio.

- **Two-way communications—packet radio:** *Packet-radio-based communications use a nationwide system of radio towers that send data to handheld computers.* Packet radio is the basis for services such as RAM Mobile Data and Ardis. The advantage of packet-radio transmission is that the wireless computer identifies itself to the local base station, which can transmit over as many as 16 separate radio channels. Packet switching encapsulates the data in "envelopes," which ensures that the information arrives intact.

   Packet-radio data networks are useful for mobile workers who need to communicate frequently with a corporate database. For example, National Car Rental System sends workers with handheld terminals to prowl parking lots, recording the location of rental cars and noting the latest scratches and dents. They can thereby easily check a

Calling from a cellular phone:
When you dial a call on a cellular phone, whether on the street or in a car, the call moves as radio waves to the transmitting-receiving tower that serves that particular cell. The call then moves by wire or microwaves to the mobile telephone switching office (MTSO), which directs the call from there on—generally to a regular local phone exchange, after which it becomes a conventional phone call.

Receiving a call on a cellular phone:
The MTSO transmits the number dialed to all the cells it services. Once it finds the phone, it directs the call to it through the nearest transmitting-receiving tower.

On the move:
When you make calls to or from phones while on the move, as in a moving car, the MTSO's computers sense when a phone's signal is becoming weaker. The computers then figure out which adjacent cell to "hand off" the call to and find an open frequency in that new cell to switch to.

1 A call originates from a mobile cellular phone.

2 The call wirelessly finds the nearest cellular tower using its FM tuner to make a connection.

3 The tower sends the signal to a Mobile Telephone Switching Office (MTSO) using traditional telephone network land lines.

4 The MTSO routes the call over the telephone network to a land-based phone or initiates a search for the recipient on the cellular network.

5 The MTSO sends the recipient's phone number to all its towers, which broadcast the number via radio frequency.

6 The recipient's phone "hears" the broadcast and establishes a connection with the nearest tower. A voice line is established via the tower by the MTSO.

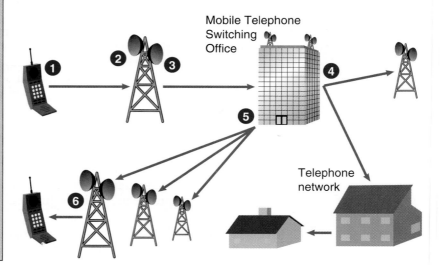

■ PANEL 8.11

Cellular connections

customer's claim that a car was already damaged or find out quickly when one is stolen.

- **Two-way communications—CDPD:** Short for *Cellular Digital Packet Data, CDPD* **places messages in packets, or digital electronic "envelopes," and sends them through underused radio channels or between pauses in cellular phone conversations.** CDPD is thus an enhancement to today's analog cellular phone systems, allowing packets of data to "hop" between temporarily free voice channels. As a result, a user carrying a CDPD device could have access to both voice and data. One problem with CDPD so far, however, is that it has limited coverage.

The Next Generation of Wireless Communications

Other kinds of wireless data services are rapidly being put in place, promising to offer us lots of choices. The following are a few such developments.

- **Digital cellular phones:** Cellular telephone companies are trying to rectify the problem of faulty data transmission by switching from analog to digital. *Digital cellular phone* **networks turn your voice message into digital bits, which are sent through the airwaves, then decoded back into your voice by the cellular handset.**

Unlike analog cellular phones, digital phones can handle short e-mail messages, paging, and some headline news items in addition to voice transmission. Currently, however, these extra features won't work if the user is traveling outside the digital network service area. Moreover, to make and receive analog phone calls, your handset has to be able to work in both digital and analog modes.

A digital cell phone costs more than an analog one, but the monthly bill may be less, especially for heavy users. Digital phone networks promise clearer sound, although some consumers don't agree. They also offer more privacy.

Despite advances in wireless technology, American cell phones are useless outside North America. If you're an American traveling abroad, you'll need to rent a temporary unit. All this promises to change in the next six years or so, however. Sixty-nine nations under the umbrella of the World Trade Organization have agreed to certain standards in worldwide telecommunications systems.

- **Personal communications services:** Fast becoming one of the most popular of the wireless services are special portable phones known as PCS phones. The lure, points out one article: "service that's noticeably better-sounding and less expensive than the standard analog cellular phones."[48]

  Like digital cellular but lower-powered, ***personal communications services (PCS), or personal communications networks (PCN), are digital wireless services that use a new band of frequencies (1850–1990 megahertz) and transmitter-receivers in thousands of microcells.*** PCS systems operate at super-high frequencies, where the spectrum isn't crowded. The microcells are smaller than the cells of today's cellular phone systems.

  At the moment, PCS phones have spotty availability. For instance, says one report, "a PCS subscriber driving east out of downtown Dallas on a surface street loses the signal but can hold onto it by driving on the highway."[49]

- **Specialized mobile radio:** ***Specialized mobile radio (SMR) is a two-way radio voice-dispatching service used by taxis and trucks; it is being converted to a digital system.*** Nextel Communications has been building a nationwide SMR network, called Direct Connect, putting itself in direct competition with cellular phone services. It has been doing so by buying frequency from local dispatch services catering to truckers and cab drivers, then persuading them to turn in their clunky two-way radios.[50]

- **Satellite-based systems:** More than half the people in the world, mostly in underdeveloped countries, live more than 2 hours from the nearest telephone. (China has only four telephone lines for every 100 people.) These people, as well as business travelers and corporations needing speedy data transmission, will probably demand more than wire-line or cellular service can deliver.

  The first communications satellite, AT&T's Telstar, went up in 1962. Now all of a sudden it looks like we will have a traffic jam in space. "A staggering 1700 or so birds [satellites] will be launched in the next decade," says one report, "more than 10 times the 150 commercial satellites now in orbit."[51]

  In the next few years, four kinds of satellite systems will dot the skies to provide a variety of consumer services.[52]

The first is the TV direct-broadcast system, and the second is the GPS system, both of which we described above. The third type is designed to handle cellular-phone and paging services, using satellite transmissions in place of tower-to-tower microwave transmissions. The fourth consists, in one description, of "global high-speed satellite networks that will let users exchange a much broader range of data, including Internet pages and videophone calls, anywhere in the world."[53]

Since most of these satellite systems are still in the future, we will take them up in the last section of this chapter.

README

Practical Matters: Comparing Mobile Phones

"Cell phones have taken off faster than fax machines, faster than cable TV, and just as fast as the . . . VCR," says a *Consumer Reports* article.[54]

But it's a jungle out there, and the new wireless technologies make the mobile phone business trickier than ever for consumers. The first thing to decide is: Do you really need a portable phone? You might be able to get one to carry around for emergency purposes for a monthly rate of as little as $10 plus a charge per call. However, a cell-phone call can cost 10 times more per minute than a conventional long-distance call—and, says *Consumer Reports*, 60 times as much as some ads imply. "The mind-boggling number of service plans the cellular companies promote may appear to offer a rich abundance of consumer choice," points out the magazine. "In fact, the various plans seem designed more to confuse, often in costly ways."

Let's compare the three major types of cell phone service: analog cellular, digital cellular, and digital PCS.[55-59] In general, service is more important than hardware.

Analog Cellular   With analog cellular covering more than 95% of the United States (using a transmission technology known as AMPS, for Advanced Mobile Phone Service), these phones are relatively cheap, typically ranging from free to $100, and you'll have half a dozen models to choose from. Monthly service costs (which may include up to 30 minutes of calls) typically go for about $29 and up. These phones are best for anyone who travels long distances and needs extensive coverage or anyone who wants a basic phone for occasional or emergency use. They also offer voice mail, call forwarding, and three-way calling, but most don't offer Internet access.

Digital Cellular   There are three types of digital wireless technologies: CDMA, TDMA, and GSM digital with AMPS analog, which are not compatible with each other. (CDMA stands for Code Division Multiple Access, TDMA for Time Division Multiple Access, and GSM for Global System for Mobile.) Equipment that works on one technology won't work on others. Most coverage of digital cellular is local or regional (with fallback analog coverage nationally, if your phone is "dual mode"—that is, works in both analog and digital modes). Phone prices typically range from free to $100 and monthly rates are $19 and up. Unlike analog, digital service offers Internet access, among other options. Consumers who use the phone heavily and need a less costly local or roaming plan will benefit from having digital rather than analog.

Digital PCS   Digital PCS is still being built in many markets. Thus, although PCS units can deliver short e-mail messages, paging service, and even headline news items, these features won't work if you're traveling outside a digital network service area. Moreover, the technology can be CDMA, TDMA, or GSM digital, and PCS units that operate on one network (such as AT&T's) won't operate on another (such as Sprint's). Finally, PCS may not work with analog cellular (that is, not all PCS service is dual mode).

Incidentally, whatever phone you get, you should know that a review of automobile accident reports from several states has found trends that "cellular phone use is a growing factor in crashes." Most of the time, drivers were talking on their phones rather than dialing at the time of the crash.[60]

Clearly, we are very near the time when voices, images, and information can be transmitted to any place on earth. Says C. Michael Armstrong, now president and CEO of AT&T, these advances already are revolutionizing "how we talk to each other and how we relate to each other."[61]

8.4 Communications Networks

KEY QUESTION

What are types, features, and advantages of networks?

Preview & Review: Communications channels and hardware may be used in different layouts or networks, varying in size from large to small: wide area networks (WANs), metropolitan area networks (MANs), and local networks. A network requires a network operating system.

Features of networks are hosts, nodes, servers, and clients. Networks allow users to share peripheral devices, programs, and data; to have better communications; to have more secure information; and to have access to databases.

Whether wired, wireless, or both, all the channels we've described can be used singly or in mix-and-match fashion to form networks. **A *network*, or *communications network*, is a system of interconnected computers, telephones, or other communications devices that can communicate with one another and share applications and data.** It is the tying together of so many communications devices in so many ways that is changing the world we live in.

Here let us consider the following:

- Types of networks—wide area, metropolitan area, and local
- Network operating systems
- Some network features
- Advantages of networks

Types of Networks: Wide Area, Metropolitan Area, & Local

Networks are categorized principally in the following three sizes:

- **Wide area network:** **A *wide area network (WAN)* is a communications network that covers a wide geographical area, such as a state or a country.** Some examples of computer WANs are Tymnet, Telenet, Uninet, and Accunet. The Internet links together hundreds of computer WANs. Of course, most telephone systems—long-distance, regional Bells, and local—are WANs.

- **Metropolitan area network:** **A *metropolitan area network (MAN)* is a communications network covering a geographic area the size of a city or suburb.** The purpose of a MAN is often to bypass local telephone companies when accessing long-distance services. Cellular phone systems are often MANs.

- **Local network:** **A *local network* is a privately owned communications network that serves users within a confined geographical area.** The range is usually within a mile—perhaps one office, one building, or a group of buildings close together, as a college campus. Local networks are of two types: private branch exchanges (PBXs) and local area networks (LANs), as we discuss shortly.

All these networks may consist of various combinations of computers, storage devices, and communications devices.

Network Operating Systems

A network requires a network operating system to support access by multiple users and provide for recognition of users based on passwords and terminal identifications. It may be a completely self-contained operating system, such as NetWare (✔ p. 122). Or it may require an existing operating system in order to function; for example, LAN Manager requires OS/2, and LANtastic requires DOS or Windows.

Hosts & Nodes, Servers & Clients

Many computer networks, particularly large ones, are served by a host computer. **A *host computer*, or simply a *host*, is the main computer—the central computer that controls the network. A *node* is simply a device that is attached to a network.**

On a local area network, some of the functions of the host may be performed by a server. As discussed in Chapter 1 (✔ p. 24), a *server*, or *network server*, is a central computer that holds databases and programs for many PCs, workstations, or terminals, which are called *clients*. These clients are nodes linked by a wired or wireless network, and the entire network is called a *client/server network*. Applications programs and files on the server are loaded into the main memories of the client machines.

Network servers appear in a variety of sizes. As one writer describes them, they may consist of "everything from souped-up PCs selling for $10,000 to mainframes and supercomputer-class systems costing millions."[62] For more than a decade, he points out, the computer industry was driven by a rush to put stand-alone microcomputers in offices and homes. Now that we are far along in putting a PC on every desktop, the spotlight is shifting to computers that can do work for many different people at once.

On the one hand, this puts "big iron"—minis, mainframes, supers—back in the picture. Recognizing this, IBM combined formerly separate personal, midrange, and mainframe computer units into an umbrella organization called the Server Group. On the other hand, the demand for servers based on microcomputers has made souped-up PCs and Macintoshes a growth industry—and is bringing these machines "close to the power of more expensive minicomputers and mainframes," according to some PC makers.[63]

Advantages of Networks

The following advantages are particularly true for LANs, although they apply to MANs and WANs as well.

- **Sharing of peripheral devices:** Laser printers, disk drives, and scanners are examples of peripheral devices—that is, hardware that is connected to a computer. Any newly introduced piece of hardware is often quite expensive, as was the case with laser or color printers. To justify their purchase, companies want them to be shared by many users. Usually the best way to do this is to connect the peripheral device to a network serving several computer users.

- **Sharing of programs and data:** In most organizations, people use the same software and need access to the same information. It could be expensive for a company to buy a copy of, say, a word processing program for each employee. Rather, the company will usually buy a network version of that program that will serve many employees.

    Organizations also save a great deal of money by letting all employees have access to the same data on a shared storage device.

This way the organization avoids such problems as some employees updating customer addresses on their own separate machines while other employees remain ignorant of such changes. It is much easier to update (maintain) software on the server than it is to update it on each user's individual system.

Finally, network-linked employees can more easily work together online on shared projects, using a type of software known as groupware (✔ p. 80).

- **Better communications:** One of the greatest features of networks is electronic mail. With e-mail everyone on a network can easily keep others posted about important information. Thus, the company eliminates the delays encountered with standard interoffice mail delivery or telephone tag.

- **Security of information:** Before networks became commonplace, an individual employee might be the only one with a particular piece of information, stored in his or her desktop computer. If the employee was dismissed—or if a fire or flood demolished the office—no one else in the company might have any knowledge of that information. Today such data would be backed up or duplicated on a networked storage device shared by others.

- **Access to databases:** Networks also enable users to tap into numerous databases, whether the private databases of a company or public databases available online through the Internet.

8.5 Local Networks

KEY QUESTION

What are local networks and the types, components, topologies, and impact of LANs?

Preview & Review: Local networks may be private branch exchanges (PBXs) or local area networks (LANs).

LANs may be client/server or peer-to-peer and include components such as network cabling, network interface cards, an operating system, other shared devices, and bridges and gateways. The topology, or shape, of a network may take five forms: star, ring, bus, hybrid, or FDDI.

Although large networks are useful, many organizations need to have a local network—an in-house network—to tie together their own equipment. Here let's consider the following aspects of local networks:

- Types of local networks—PBXs and LANs
- Types of LANs—client/server and peer-to-peer
- Components of a LAN
- Topology of LANs—star, ring, bus, hybrid, and FDDI
- Impact of LANs

Types of Local Networks: PBXs & LANs

The most common types of local networks are PBXs and LANs.

- **Private branch exchange (PBX):** **A *private branch exchange (PBX)* is a private or leased telephone switching system that connects telephone extensions in-house.** It also connects them to the outside phone system.

A public telephone system consists of "public branch exchanges"—thousands of switching stations that direct calls to different "branches" of the network. A private branch exchange is essentially the old-fashioned company switchboard. You call in from the outside, a switchboard operator says "How may I direct your call?" (or an automated voice gives you options for directing your call), and you are connected to the extension of the person you wish to talk to.

- **Local area network (LAN):** PBXs may share existing phone lines with the telephone system. Local area networks usually require installation of their own communication channels, whether wired or wireless. *Local area networks (LANs)* **are local networks consisting of a communications link, network operating system, microcomputers or workstations, servers, and other shared hardware.** Such shared hardware might include printers, scanners, and storage devices.

Types of LANs: Client/Server & Peer-to-Peer

Local area networks are of two principal types: client/server and peer-to-peer. (■ *See Panel 8.12.*)

- **Client/server LANs:** **A** *client/server LAN* **consists of requesting microcomputers, called** *clients,* **and supplying devices that provide a service, called** *servers.* The server is a computer that manages shared devices, such as laser printers, or shared files. The server microcomputer is usually a powerful one, running on a high-speed chip such as a Pentium II. Client/server networks, such as those run under Novell's NetWare or Windows NT operating systems, are the most common types of LANs. One piece of the network operating system resides in each client machine, and another resides in each server. The operating system allows the remote hard-disk drives on the servers to be accessed as if they were local drives on the client machine.

  Different servers can be used to manage different tasks—files and programs, databases, printers. The one you may hear about most often is the file server. **A** *file server* **is a computer that stores the programs and data files shared by users on a LAN.** It acts like a disk drive but is in a remote location.

  A *database server* is a computer in a LAN that stores data. Unlike a file server, it does not store programs. A *print server* is a computer in a LAN that controls one or more printers. It stores the print-image output from all the microcomputers on the system. It then feeds the output to the printer or printers one document at a time. *Fax servers* are dedicated to managing fax transmissions, and *mail servers* manage e-mail.

- **Peer-to-peer:** The word *peer* denotes one who is equal in standing with another (as in the phrases "peer pressure" or "jury of one's peers"). **A** *peer-to-peer LAN* **is one in which all microcomputers on the network communicate directly with one another without relying on a server.** Peer-to-peer networks are less expensive than client/server networks and work effectively for up to 25 computers. Beyond that they slow down under heavy use. They are thus appropriate for networking in small groups.

Many LANs mix elements from both client/server and peer-to-peer models.

■ PANEL 8.12

Two types of LANs: client/server and peer-to-peer

(Top) In a client/server LAN, individual microcomputer users, or "clients," share the services of a centralized computer called a "server." In this case, the server is a file server, allowing users to share files of data and some programs. *(Bottom)* In a peer-to-peer LAN, computers share equally with one another without having to rely on a central server.

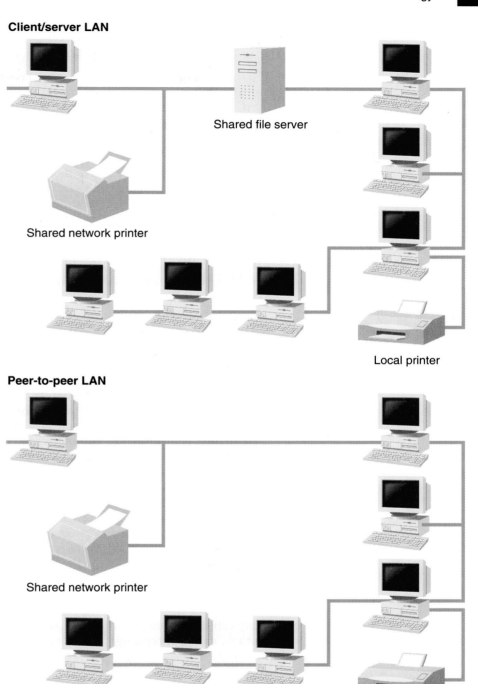

Client/server LAN

Shared file server

Shared network printer

Local printer

Peer-to-peer LAN

Shared network printer

Local printer

Components of a LAN

Local area networks are made up of several standard components.

- **Connection or cabling system:** LANs do not use the telephone network. Instead, they use some other cabling or connection system, either wired or wireless. Wired connections may be twisted-pair wiring, coaxial cable, or fiber-optic cable. Wireless connections may be infrared or radio-wave transmission. Wireless networks are especially useful if computers are portable and are moved often. However, they are subject to interference.

- **Microcomputers with interface cards:** Two or more microcomputers are required, along with network interface cards. **A *network interface card*, which is inserted into an expansion slot in a microcomputer, enables the computer to send and receive messages on the LAN.** The interface card can also exist in a separate box, which can serve a number of devices.

- **Network operating system:** As mentioned, the network operating system software manages the activity of the network. Depending on the type of network, the operating system software may be stored on the file server, on each microcomputer on the network, or a combination of both.

  Examples of network operating systems are Novell's NetWare, Microsoft's Windows NT, and IBM's LAN. Peer-to-peer networking can also be accomplished with AppleTalk, Windows 95 or 98, Windows NT Workstation, and Microsoft Windows for Workgroups.

- **Other shared devices:** Printers, fax machines, scanners, storage devices, and other peripherals may be added to the network as necessary and shared by all users.

- **Bridges, routers, and gateways:** A LAN may stand alone, but it may also connect to other networks, either similar or different in technology. Network designers determine the types of hardware and software devices necessary—bridges, routers, gateways—to use as interfaces to make these connections. (■ *See Panel 8.13.*)

  A *bridge* is a hardware and software combination used to connect the same types of networks.

  A *router* is a special computer that directs communicating messages when several networks are connected together. High-speed routers can serve as part of the Internet backbone, or transmission path, handling the major data traffic.

  A *gateway* is an interface that enables dissimilar networks to communicate, such as a LAN with a WAN or two LANs based on different topologies or network operating systems.

Topology of LANs

Networks can be laid out in different ways. **The logical layout, or shape, of a network is called a *topology*.** The five basic topologies are *star, ring, bus, hybrid,* and *FDDI.*

- **Star network:** **A *star network* is one in which all microcomputers and other communications devices are connected to a central server.** (■ *See Panel 8.14.*) Electronic messages are routed through the central hub to their destinations, so the central hub monitors the flow of traffic. A PBX system is an example of a star network.

  The advantage of a star network is that the hub prevents collisions between messages. Moreover, if a connection is broken between any communications device and the hub, the rest of the devices on the network will continue operating. However, if the hub goes down, the entire network will stop.

- **Ring network:** **A *ring network* is one in which all microcomputers and other communications devices are connected in a continuous**

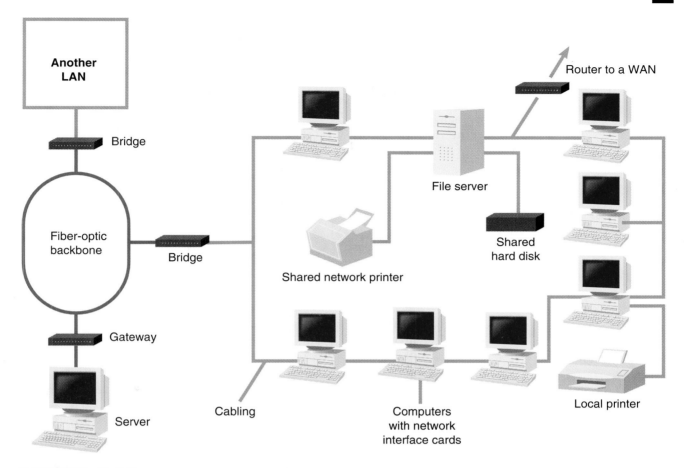

■ PANEL 8.13

Components of a typical LAN

loop. (■ *See Panel 8.15, next page.*) Electronic messages are passed around the ring until they reach the right destination. There is no central server. An example of a ring network is IBM's Token Ring Network, in which a bit pattern (called a "token") determines which user on the network can send information.

■ PANEL 8.14

Star network
This arrangement connects all the network's devices to a central host computer, through which all communications must pass.

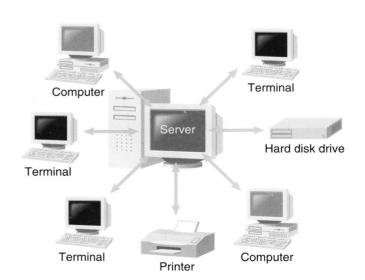

■ PANEL 8.15

Ring network

This arrangement connects the network's devices in a closed loop.

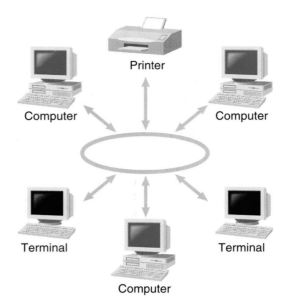

The advantage of a ring network is that messages flow in only one direction. Thus, there is no danger of collisions. The disadvantage is that if a connection is broken, the entire network stops working.

- **Bus network:** The bus network works like a bus system at rush hour, with various buses pausing in different bus zones to pick up passengers. **In a *bus network*, all communications devices are connected to a common channel.** (*■ See Panel 8.16.*) There is no central server. Each communications device transmits electronic messages to other devices. If some of those messages collide, the device waits and tries to retransmit. An example of a bus network is Xerox's Ethernet (which can also be configured in a star topology).

  The advantage of a bus network is that it may be organized as a client/server or peer-to-peer network. The disadvantage is that extra

■ PANEL 8.16

Bus network

A single channel connects all communications devices.

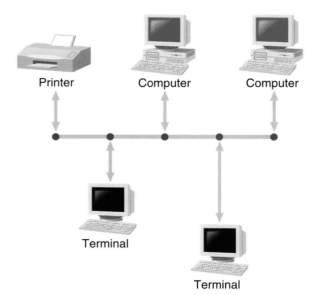

circuitry and software are needed to avoid collisions between data. Also, if a connection in the bus is broken—as when someone moves a desk and knocks the connection out—the entire network may stop working.

- **Hybrid network:** *Hybrid networks* **are combinations of star, ring, and bus networks.** For example, a small college campus might use a bus network to connect buildings and star and ring networks within certain buildings.

- **FDDI network:** A newer and higher-speed network is the FDDI, short for Fiber Distributed Data Interface. Capable of transmitting 100–200 megabits per second, **an** *FDDI network* **uses fiber-optic cable with an adaptation of ring topology using not one but two "token rings."** (■ *See Panel 8.17.*) The FDDI network is being used for such high-tech purposes as electronic imaging, high-resolution graphics, and digital video.

  One advantage of the FDDI network is clearly its speed. Another is that because two rings are used, if one should fail, the network can continue operating with the second ring. The disadvantages of FDDI are its cost and its fragility because it uses fiber-optic cable.

The Impact of LANs

Sales of mainframes and minicomputers have been falling for some time. This is largely because companies have discovered that LANs can take their place for many (though certainly not all) functions, and at considerably less expense. This situation reflects a trend known as *downsizing.* Still, a LAN, like a mainframe, requires a skilled support staff. Moreover, LANs have neither the great storage capacity nor the security that mainframes have, which makes them not useful for some applications.

■ PANEL 8.17

FDDI network

Fiber-optic cable is used in an adaptation of ring topology, using two rings. If one ring fails, the other will keep the network operating.

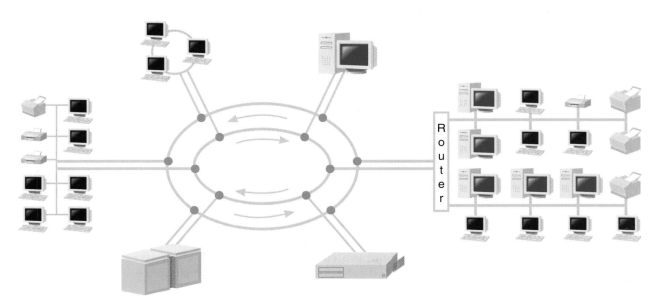

8.6 Factors Affecting Communications Among Devices

KEY QUESTION

What are the factors affecting data transmission?

Preview & Review: Factors affecting how data is transmitted include the transmission rate (frequency and bandwidth), the line configuration (point-to-point or multipoint), serial versus parallel transmission, the direction of transmission flow (simplex, half-duplex, or full-duplex), transmission mode (asynchronous or synchronous), packet switching, multiplexing, and protocols.

Things are changing, and changing fast. It's not enough to know about the types of communications channels and network configurations available. As the technology moves forward, you'll also want to know what's happening behind the scenes. This section describes the essentials of data communications.

Several factors affect how data is transmitted. They include the following:

- Transmission rate—frequency and bandwidth
- Line configurations—point-to-point versus multipoint
- Serial versus parallel transmission
- Direction of transmission—simplex, half-duplex, and full-duplex
- Transmission mode—asynchronous versus synchronous
- Packet switching
- Multiplexing
- Protocols

Transmission Rate: Higher Frequency, Wider Bandwidth, More Data

Transmission rate is a function of two variables: frequency and bandwidth.

The amount of data that can be transmitted on a channel depends on the wave *frequency*—the cycles of waves per second (expressed in hertz). The more cycles per second, the more data that can be sent through that channel.

The greater a channel's *bandwidth*—the difference (range) between the highest and lowest frequencies—the more frequencies it has available and hence the more data that can be sent through that channel (expressed in bits per second, or bps).

A twisted-pair telephone wire of 4000 hertz might send only 1 kilobyte per second of data. A coaxial cable of 100 megahertz might send 10 megabytes per second. And a fiber-optic cable of 200 trillion hertz might send 1 gigabyte per second.

Line Configurations: Point-to-Point & Multipoint

There are two principal line configurations, or ways of connecting communications lines: point-to-point and multipoint.

- Point-to-point: **A *point-to-point line* directly connects the sending and receiving devices,** such as a terminal with a central computer. This arrangement is appropriate for a private line whose sole purpose is to keep data secure by transmitting it from one device to another.
- Multipoint: **A *multipoint line* is a single line that interconnects several communications devices to one computer.** Often on a multipoint line only one communications device, such as a terminal, can transmit at any given time.

■ PANEL 8.18

Serial data transmission
Data resembles cars moving
down a one-lane road.

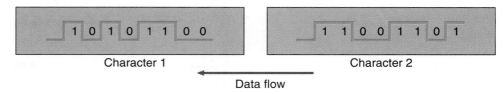

Character 1 | Character 2

← Data flow

Serial & Parallel Transmission

Data is transmitted in two ways: serially and in parallel.

- Serial data transmission: **In *serial data transmission*, bits are transmitted sequentially, one after the other.** This arrangement resembles cars proceeding down a one-lane road. (■ *See Panel 8.18.*)

  Serial transmission is the way most data flows over a twisted-pair telephone line. Serial transmission is found in communications lines and modems. When you send a command through your mouse, it will probably be conveyed by serial transmission. The plug-in board for a microcomputer modem usually has a serial port (✔ p. 169).

- Parallel data transmission: **In *parallel data transmission*, bits are transmitted through separate lines simultaneously.** The arrangement resembles cars moving in separate lanes at the same speed on a multilane freeway. (■ *See Panel 8.19.*)

  Parallel lines move information faster than serial lines do, but they are efficient for up to only 15 feet. Thus, parallel lines are used, for example, to transmit data from a computer's CPU to a printer.

  Parallel transmission may also be used within a company's facility for transmitting data between terminals and the main computer.

Direction of Transmission Flow: Simplex, Half-Duplex, & Full-Duplex

When two computers are in communication, data can flow in three ways: simplex, half-duplex, or full-duplex. These are fancy terms for easily understood processes.

■ PANEL 8.19

**Parallel data
transmission**
Data resembles cars moving
in separate lanes at the same
speed on a multilane freeway.

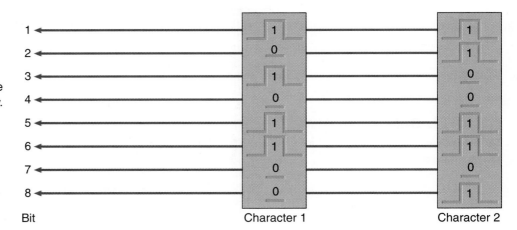

■ **PANEL 8.20**

Simplex transmission

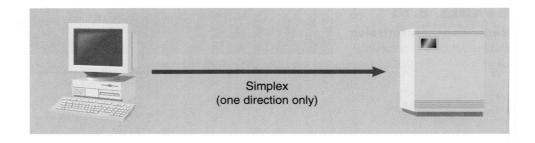

- Simplex transmission: **In *simplex transmission*, data can travel in only one direction.** (■ *See Panel 8.20.*)

  An example is a traditional television broadcast, in which the signal is sent from the transmitter to your TV antenna. There is no return signal. Some computerized data collection devices also work this way, such as seismograph sensors that measure earthquakes.

- Half-duplex transmission: **In *half-duplex transmission*, data travels in both directions but only in one direction at a time.** This arrangement resembles traffic on a one-lane bridge; the separate streams of cars heading in both directions must take turns. (■ *See Panel 8.21.*)

  Half-duplex transmission is seen with CB or marine radios, in which both parties must take turns talking. This is the most common mode of data transmission used today.

- Full-duplex transmission: **In *full-duplex transmission*, data is transmitted back and forth at the same time.** This arrangement resembles automobile traffic on a two-way street. (■ *See Panel 8.22.*)

  An example is two people on the telephone talking and listening simultaneously. Full-duplex is sometimes used in large computer systems. It is also available in newer microcomputer modems to support truly interactive workgroup computing.

■ **PANEL 8.21**

Half-duplex transmission

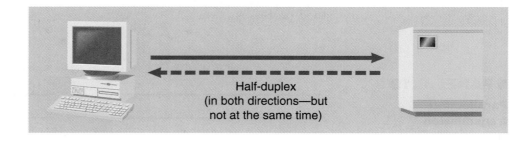

■ **PANEL 8.22**

Full-duplex transmission

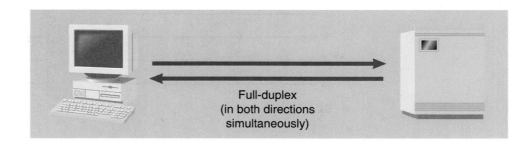

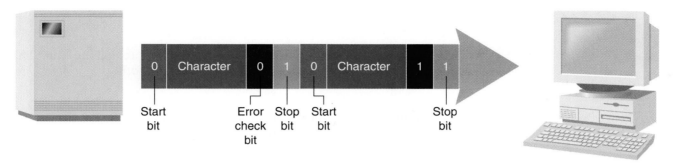

| 0 | Character | 0 | 1 | 0 | Character | 1 | 1 |

Start bit   Error check bit   Stop bit   Start bit   Stop bit

■ PANEL 8.23

Asynchronous transmission
Each character is preceded by a "start" bit and followed by a "stop" bit.

Transmission Mode: Asynchronous Versus Synchronous

Suppose your computer sends the word CONGRATULATIONS! to someone as bits and bytes over a communications line. How does the receiving equipment know where one byte (or character) ends and another begins? This matter is resolved through either *asynchronous transmission* or *synchronous transmission*.

- **Asynchronous transmission:** This method, used with most microcomputers, is also called *start-stop transmission.* **In asynchronous transmission, data is sent one byte (or character) at a time. Each string of bits making up the byte is bracketed, or marked off, with special control bits.** That is, a "start" bit represents the beginning of a character, and a "stop" bit represents its end. (■ *See Panel 8.23.*)

  Transmitting only one byte at a time makes this a relatively slow method. As a result, asynchronous transmission is not used when great amounts of data must be sent rapidly. Its advantage is that the data can be transmitted whenever it is convenient for the sender.

- **Synchronous transmission:** Instead of using start and stop bits, *synchronous transmission* **sends data in blocks. Start and stop bit patterns, called synch bytes, are transmitted at the beginning and end of the blocks.** These start and end bit patterns synchronize internal clocks in the sending and receiving devices so that they are in time with each other. (■ *See Panel 8.24.*)

  This method is rarely used with microcomputers because it is more complicated and more expensive than asynchronous transmission. It also requires careful timing between sending and receiving equipment. It is appropriate for computer systems that need to transmit great quantities of data quickly.

■ PANEL 8.24

Synchronous transmission
Messages are sent in blocks with start and stop patterns of bits, called synch bytes, before and after the blocks. The synch bytes synchronize the timing of the internal clocks between sending and receiving devices.

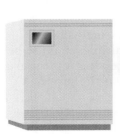
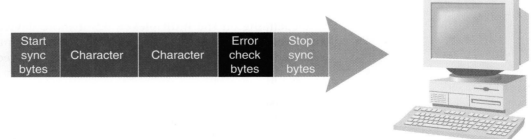

| Start sync bytes | Character | Character | Error check bytes | Stop sync bytes |

Circuit Switching, Packet Switching, & Asynchronous Transfer Mode: For Voice, Data, & Both

What is the most efficient way to send messages over a telephone line? That depends on whether the messages are *voice, data,* or *both.*

- Circuit switching—best for voice: Circuit switching is used by the telephone company for its voice networks to guarantee steady, consistent service for telephone conversations. **In *circuit switching,* the transmitter has full use of the circuit until all the data has been transmitted and the circuit is terminated.**

- Packet switching—best for data: **A *packet* is a fixed-length block of data for transmission.** The packet also contains instructions about the destination of the packet. ***Packet switching* is a technique for dividing electronic messages into packets for transmission over a wide area network to their destination through the most expedient route.**

  Here's how packet switching works: A sending computer breaks an electronic message apart into packets. The various packets are sent through a communications network—often by different routes, at different speeds, and sandwiched in between packets from other messages. Once the packets arrive at their destination, the receiving computer reassembles them into proper sequence to complete the message.

  The benefit of packet switching is that it can handle high-volume traffic in a network. It also allows more users to share a network, thereby offering cost savings. The method is particularly appropriate for sending data long distances, such as across the country. Accordingly, it is used in large data networks such as Telenet, Tymnet, and AT&T's Accunet.

- Asynchronous transfer mode (ATM)—best for both: **A newer technology, called *asynchronous transfer mode (ATM),* combines the efficiency of packet switching with some aspects of circuit switching,** thus enabling it to handle both data and real-time voice and video. ATM is designed to run on high-bandwidth fiber-optic cables.

Multiplexing: Enhancing Communications Efficiencies

Communications lines nearly always have far greater capacity than a single microcomputer or terminal can use. Because operating such lines is expensive, it's more efficient if several communications devices can share a line at the same time. This is the rationale for multiplexing. ***Multiplexing* is the transmission of multiple signals over a single communications channel.**

Three types of devices are used to achieve multiplexing—*multiplexers, concentrators,* and *front-end processors:*

- Multiplexers: **A *multiplexer* is a device that merges several low-speed transmissions into one high-speed transmission.** (■ *See Panel 8.25.*)

  Depending on the multiplexer, 32 or more devices may share a single communications line. Messages sent by a multiplexer must be received by a multiplexer of the same type. The receiving multiplexer sorts out the individual messages and directs them to the proper recipient.

  High-speed multiplexers called *T1 multiplexers,* which use high-speed digital lines, can carry as many messages, both voice and data, as 24 analog telephone lines.

■ PANEL 8.25

How multiplexing works

With sending and receiving multiplexers, several low-speed transmissions may share a high-speed line.

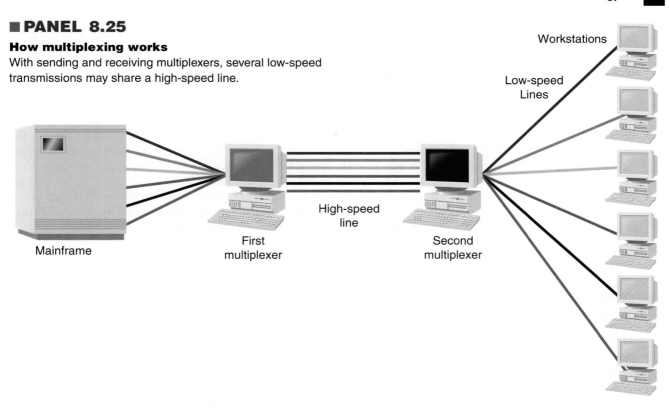

- **Concentrators:** Like a multiplexer, a concentrator is a piece of hardware that enables several devices to share a single communications line. However, unlike a multiplexer, **a *concentrator* collects data in a temporary storage area.** Whereas a multiplexer spreads the signals back out again on the receiving end, the concentrator has a receiving computer perform that function.

- **Front-end processors:** The most sophisticated of these communications-management devices is the front-end processor, a computer that handles communications for mainframes. **A *front-end processor* is a smaller computer that is connected to a larger computer and assists with communications functions.** It transmits and receives messages over the communications channels, corrects errors, and relieves the larger computer of routine computational tasks.

Protocols: The Rules of Data Transmission

Does the foregoing information in this section seem unduly technical for an ordinary computer user? Although you should understand these details, fortunately you won't have to think about them much. Experts will already have taken care of them for you in sets of rules called protocols.

The word *protocol* is used in the military and in diplomacy to express rules of precedence, rank, manners, and other matters of correctness. (An example would be the protocol for who will precede whom into a formal reception.) Here, however, **a *protocol*, or *communications protocol*, is a set of conventions governing the exchange of data between hardware and/or software components in a communications network.**

Protocols are built into the hardware or software you are using. The protocol in your communications software, for example, will specify how receiver devices will acknowledge sending devices, a matter called *handshaking.* Handshaking establishes the fact that the circuit is available and operational. It also establishes the level of device compatibility and the speed of transmission. Protocols will also specify the type of electrical connections used, the timing of message exchanges, error-detection techniques, and so on.

In the past, not all hardware and software developers subscribed to the same protocols. As a result, many kinds of equipment and programs have not been able to work with one another. In recent years, more developers have agreed to subscribe to a standard of protocols called *OSI* that is backed by the International Standards Organization. **OSI, short for *Open Systems Interconnection,* is an international standard that defines seven layers of protocols for worldwide computer communications. (■** *See Panel 8.26.)*

8.7 The Future of Communications

KEY QUESTION

What are the characteristics of satellite systems, gigabit ethernets, photonics, and power lines?

Preview & Review: New technologies to help speed up communications and make them more accessible include new commercial satellite systems (GEO, MEO, and LEO), gigabit ethernets, photonics, and the use of electric power lines.

The ongoing joke is that the "www" in most World Wide Web addresses stands for the "World Wide Wait." The lack of truly speedy access, according to Bill Gates, is the greatest obstacle to the Net's widespread use.[64]

On college campuses, Web sites that become popular and attract thousands of visitors can slow campus networks to a crawl.[65] An organization calling itself BugNet, which sent out 500 e-mails to customers and 500 letters to current postal addresses, found that e-mail's failure rate "was more than three times higher than first-class mail."[66] Another study found that 86% of e-mail messages sent over the Net were delivered within 5 minutes and only 1.6% took more than an hour, but that didn't reflect delays that occur inside online services or private computer networks, where slowness is commonplace.[67]

Says one maker of Internet products, Bob Quillin, "The angina pains Internet users are feeling today—'server busy' messages, jerky stop-and-go transmissions, stuck connections—may signal a future heart attack, forcing the Internet to live out its life as an invalid unable to meet our expectations."[68] And Quillin makes an important point: *The computer industry has shifted to a bandwidth economy.* That is, instead of depending on semiconductor companies to keep delivering faster and faster chips, the industry "depends as much or more on the willingness of the telecom industry, Internet service providers, and cable companies to keep spending millions of dollars on hardware and wiring to build out the Internet infrastructure"—that is, to provide more bandwidth, or high-speed data transmission capacity.

So serious is the problem of Internet congestion that the nation's main telecommunications regulator, Federal Communications Commission chairman Bill Kennard, has made easing it a top priority. "One issue that I'm particularly interested in is finding ways that we can foster more investment in high-capacity bandwidth," he said. "I believe that our nation will have an ever-increasing appetite for bandwidth."[69]

So, the course is clear: Strike up the bandwidth! What kinds of things can we look forward to in this regard? We discuss: new commercial satellite systems, gigabit ethernets, the photonic revolution, and the use of electric power lines for phone and data transmission.

■ PANEL 8.26

OSI

The seven layers of the OSI standard for worldwide communications that defines a framework for implementing protocols.

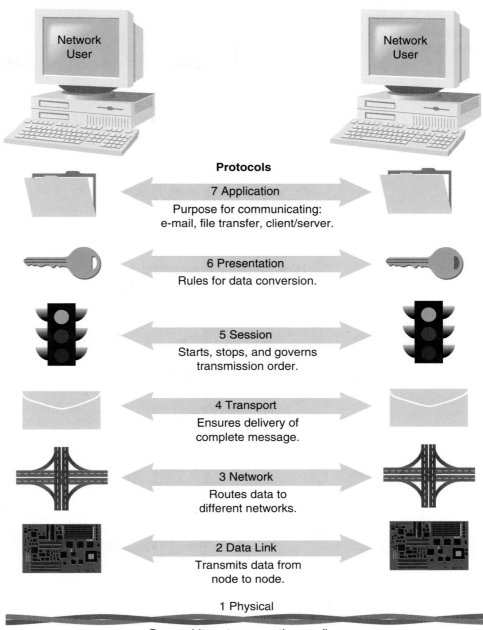

Protocols

7 Application
Purpose for communicating:
e-mail, file transfer, client/server.

6 Presentation
Rules for data conversion.

5 Session
Starts, stops, and governs
transmission order.

4 Transport
Ensures delivery of
complete message.

3 Network
Routes data to
different networks.

2 Data Link
Transmits data from
node to node.

1 Physical

Passes bits onto connecting medium.

The New Commercial Satellite Systems

At a World Trade Organization meeting in 1997, 67 nations agreed to open their communications markets to foreign satellite systems.[70] Even before this, however, plans had been laid for a new race. Huge rockets from Russia, Ukraine, and China now compete with American and French boosters in elevating into orbit satellites ranging in size from volleyballs to Volkswagens.

"These are exciting times," says Marc Newman, Washington manager for Globalstar, one new satellite network. "It's a revolution in global telecommunications services. You'll be able to call a person wherever they may be on the planet."[71] By 2000, there could be 500 million wireless subscribers worldwide, according to Robert Kinzie, chairman of Iridium, a competing satellite system.[72]

■ PANEL 8.27

Geostationary earth orbit (GEO)

Orbit:
22,300 miles
at the equator

Satellite systems may occupy one of three zones in space: *GEO, MEO,* and *LEO.* The highest level, 22,300 miles up at the equator, is known as *geostationary earth orbit (GEO).* The next zone, *medium-earth orbit (MEO),* is 5000–10,000 miles up (the average satellite is at around 6000 miles). The lowest level, *low-earth orbit (LEO),* is 400–1000 miles up.

Readers of this book will probably have already come away with an appreciation for the importance of companies such as Microsoft, IBM, and Apple. However, the commercial satellite companies staking out their claims in space may come to be equally important. We will introduce them as we describe the three satellite zones.[73–79]

- **Geostationary earth orbit (GEO)—22,300 miles up at the equator:** This orbit is directly above the equator. (■ *See Panel 8.27.*) Consequently, room is limited, and GEO can't hold many more than the 150 satellites now there. Satellites remain in a fixed position above the earth's surface. This position makes them good for TV transmission (as in DirecTV), but their quarter-second delay makes two-way telephone conversations difficult. This high an orbit requires fewer satellites for global coverage.

  Spaceway, an eight-satellite system, is due by 2000 from Hughes Electronics of Los Angeles. At 22,300 miles, three satellites are enough to cover the earth. (Hughes already has 15 big satellites in position over the equator.) The purpose is to provide ordinary fixed telephone service in developing countries, along with fax, data, videoconferencing, and Internet services. The first global region will be operational in 1999, the rest in 2000.

- **Medium-earth orbit (MEO)—5000–10,000 miles up:** A medium-earth orbit requires fewer satellites for global coverage than LEO but more than GEO. (■ *See Panel 8.28.*) In the next few years, two dozen or

■ PANEL 8.28

Medium-earth orbit (MEO)

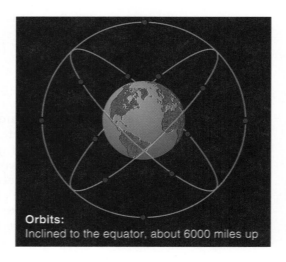

Orbits:
Inclined to the equator, about 6000 miles up

■ **PANEL 8.29**
Low-earth orbit (LEO)

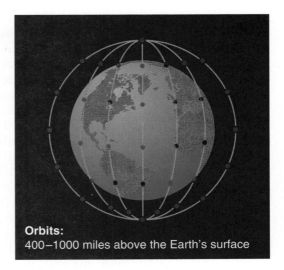

Orbits:
400–1000 miles above the Earth's surface

more large satellites could be put into MEO to handle new mobile phones and data systems.

London-based *ICO Global Communications* plans to have ten satellites 6434 miles up and operational by 2000. Its goal is to provide phone service of all kinds—for cars, ships, aircraft, and fixed phones in developing areas.

Odyssey (co-owned by TRW and Teleglobe) has the goal of putting a 12-satellite network 6434 miles up by 2001. Odyssey wants to provide worldwide mobile phone services and other communications services mainly to the developing world.

• **Low-earth orbit (LEO)—400–1000 miles up:** This is the zone in which the real dogfight for business is taking place. (■ *See Panel 8.29.*) An advantage of LEO is that there is no signal delay. Moreover, satellites are smaller and so are cheaper to launch. Down on the ground, portable phones need less power; hence, batteries can be smaller. The biggest, untested drawback is that data and phone calls will have to be able to move efficiently among many satellites.

Within the next few years, five giant companies or consortiums of companies will be competing: GlobalStar, Iridium, Orbcomm, Teledesic, and Motorola. Here more satellites are needed to cover the earth than is true in the other orbits.

Orbcomm (Orbital Sciences and Teleglobe are the partners) had 28 minisatellites weighing 95 pounds each orbiting at the 480-mile level by the spring of 1998. The purpose of the system is to transmit messages from individuals as well as to relay signal beams from transmitters attached to oil and gas pipelines, monitor water levels in Death Valley, and track elephant movements in Indonesia, among other uses.

GlobalStar (the partners are Loral Space & Communications and Qualcomm) plans to be fully operational in 1999, with 48 satellites orbiting at an altitude of 736 miles up. Their function will be to handle worldwide mobile and fixed phones, paging, and low-speed data. Revenues are supposed to come from rural areas in places such as China and India that are underserved by land lines.

Iridium (a $5 billion international consortium led by Motorola and including Raytheon, Lockheed Martin, and Sprint) was expected to have 66 satellites 421.5 miles up and operational by late 1998. The purpose is to enable business travelers to call anywhere from anywhere, as well as to move faxes, pager calls, and low-speed data. (Motorola also has another venture, Celestri, discussed below.)

Teledesic, known as the "Internet in the Sky" project (costing $9 billion and backed by Microsoft's Bill Gates, cellular phone magnate Craig McCaw, and airplane maker Boeing), is perhaps the most ambitious scheme. By 2002 it aims to have no less than 220 satellites (scaled back from 840) orbiting at an altitude of 435 miles. Its purpose is to make broadband multimedia connections anywhere, just like fiber-optic cables, that will be 1000 times faster than most Internet access today. It will be used for broadband multimedia for the Internet, corporate intranets, and videoconferencing.

Celestri from Motorola is designed as competition to Teledesic. Costing perhaps $12.9 billion, it would consist of 63 orbiting satellites plus a few fixed satellites. By 2002, according to the plan, it would transmit text, multimedia images, and video around the globe at speeds 100–1000 times faster than is possible now.

All these companies are taking huge risks, because the consensus of experts is that not all these ventures can survive. "The spread of cellular phones, especially in developing countries, could make phone systems like Iridium and TRW's Odyssey network unnecessary," says one commentator. "And technologies such as [DSL] for phones, cable modems, and a host of so-called gigabit ethernet technologies would make consumer and corporate networks faster—and satellite systems less attractive."[80]

Gigabit Ethernet: Sending Data at Supercomputer Speeds

Most corporate networks use Ethernet, a form of bus network, devised in the 1970s. It moves data at the rate of 10 million bits (10 megabits) per second. In 1995, Fast Ethernet was invented, which can transmit data at 100 million bits (100 megabits) per second.

Now there is a hot new technology for speeding up office networks 100 times faster than standard. "It will help end data roadblocks at major financial companies, Internet service providers, heavy-duty engineering firms, and other big enterprises," says one report.[81] Developed at government and defense laboratories to help supercomputers communicate, Gigabit Ethernet can move data at 1000 million bits (1 gigabit) per second. Recently, commercial firms have jumped in to make the technology more widely available.

Photonics: Optical Technologies at Warp Speed

"Moore's Law said that chip power would double every 18 months," writes journalist Howard Banks. "That's plodding. The new law of the photon says that bandwidth triples every year."[82]

Photonics is the science of sending data bits down pulses of light carried on hair-thin glass fibers. For 15 years, the glass fibers of fiber-optic lines have been used to carry light pulses representing voice and data in long-distance telephone lines. Photonics has achieved breakthroughs that enable glass fibers to carry more light signals than ever before.

Older fiber-optic technologies have been limited to only a few dozen miles. Then the light beams had to be converted to electrical signals, amplified, and converted back into light signals. This made the technique slow, unreliable, and expensive.

Then in 1988, researcher David Payne at the University of Southampton in England developed an *optical amplifier*. This device boosts light signals without converting them first to electrical signals. The amplifier, in one description, "gives a huge push to the incoming signal, letting it carry on for dozens of more miles to the next amplifier."[83] Engineers also devised another technology (called wave-division multiplexing or dense wavelength division multiplexing) that allows laser pulses of different hues to be sent down the same tiny fiber.[84] This allows carriers to transmit at least 16 channels per fiber, which may grow to 100 channels per fiber in a few years. The upshot is that in the laboratory researchers have been able to produce glass fiber systems that carry 2 trillion bits (2 terabits) per second. That is *six times the volume of all phone calls in the United States on an average day.*[85]

No doubt David Payne is right in saying that "We are really only in the Stone Age of optical communication."[86] Adds William Gartner, vice president for optical networking products at Lucent Technologies' Bell Labs, "For business and consumers, applications will emerge that today we don't even dream about." As an example, he poses this scenario:

> People are exploring things like remote surgery today.... Optics will allow networking of huge bandwidths from anywhere to anywhere, so it's maybe the Mayo Clinic tied in with [New York University], tied in with the University of Houston, all collaborating on this patient who's being operated on in Argentina. Doctors don't even fathom that today.[87]

When Utilities Become Communications Companies

Can you plug your phone and toaster into the same wall outlet?

Not yet. But it's a measure of how things have changed that electric companies now want to bring you phone, Internet, and cable-TV service.

Freed by changes in federal law brought about by the 1996 Telecommunications Act, electric and gas utilities are now plunging into the communications business. Utilities in the United States have 600,000 miles of high-capacity, fiber-optic networks that they use to monitor transmission lines and generator plants. They also have long-established rights-of-way they could use to lay more connections.

The utilities could probably learn from the cable-TV companies, whose ventures into phone service have so far been mostly unsuccessful. Moreover, points out one writer, the old-monopoly electric and gas companies have had little experience in the bruising marketing wars that characterize the telecommunications business.[88] Nevertheless, companies like Colorado's ICG Communications, for instance, are aggressively offering their extra fiber-optic capacity to be used for phone and Internet service.[89]

In England, engineers have developed a technology that would let consumers make phone calls and access the Internet at high speeds—10 times faster than ISDN—through electrical outlets in the walls.[90] Internet usage has lagged in Europe behind the U.S. because of the steep costs of accessing the Net via the national phone monopolies. This development could provide an inexpensive way of creating telecommunications services over existing power lines.

Still, the commercial feasibility of the system hasn't been proven. "We have to understand the engineering of this technology on a much grander scale," says an executive for a power provider in England and Wales.[91]

Telecommunications companies have also dreamed of turning phone lines into television lines. Now a device called a VidModem has been developed that would bring high-quality audio and video signals over existing copper-wire phone lines—and analog lines at that. When connected to every TV or

PC in a building, VidModem will enable users to watch TV broadcasts on their PC and conduct two-way videoconferencing, as well as make phone calls and simultaneously share data files.[92]

Onward

There are roughly 50 million wireless device users in the United States, but wireless is booming even faster in other parts of the world, such as Latin America.[93,94] In Mexico, for instance, pole-and-wire networks reach only half the nation's homes (compared with over 90% in the U.S.). There cell phones are helping to fill in the gaps in the communications infrastructure. The same is happening in countries such as Thailand, Pakistan, and Hungary.

Finland has one Internet server link for every 25 citizens and the United States one for every 50, making them the world's two most high-tech countries.[95] Asia seems to be the great communications frontier: China has 561,000 people per Net server and India 1.2 million per server.

Clearly, more roads will be added in cyberspace. We will consider the implications of these developments in the final chapter of this book.

Online Résumés & Other Career Strategies for the Digital Age

"If you have 8000 résumés and you're looking for a COBOL programmer . . . ," commented vice president of recruitment Saundra Banks Loggins of Wells Fargo bank, "this system can save you a lot of time."[96]

The system referred to is a high-tech résumé-scanning system called Resumix. This technology uses an optical scanner to input 900 pages of résumés a day, storing the data in a computerized database. The system can search for up to 60 key factors, such as job titles, technical expertise, education, geographic location, and employment history. Resumix can also track race, religion, gender, and other factors to help companies diversify their workforce. These descriptors can then be matched with available openings.

Résumé scanners can save companies thousands of dollars. They allow organizations to more efficiently search their existing pool of applicants before turning to advertising or executive-search ("head-hunter") firms to recruit employees. For applicants, however, résumé banks and other electronic systems have turned job hunting into a whole new ball game.

Writing a Computer-Friendly & Recruiter-Friendly Résumé

Some of the old rules for presenting yourself in a résumé might not benefit you at all now. The latest advice is as follows.

Use the Right Paper & Print In the past, job seekers have used tricks such as colored paper and fancy typefaces in their résumés to try to catch a bored personnel officer's eye. However, optical scanners have trouble reading type on colored or gray paper and are confused by unusual typefaces. They even have difficulty reading underlining and poor-quality dot-matrix printing.[97]

Resumix Inc. suggests observing the following rules of format for résumé writing:[98]

- Exotic typefaces, underlining, and decorative borders and graphics don't scan well.

- It's best to send originals, not copies, and not to use a dot-matrix printer.

- Too-small print may confuse the scanner; don't go below 12-point type.

- Use standard 8½ × 11-inch paper and do not fold. Words in a crease can't be read easily.

- Use white or light-beige paper. Blues and grays minimize the contrast between the letters and the background.

- Avoid double columns. The scanner reads from left to right.

Another old rule of résumé writing that has vanished: You need no longer confine it to one page. Multiple sheets are fine. (But print on one side only.)

Use Keywords for Skills or Attributes Just as important as the format of a résumé today are the *words* used in it. In the past, résumé writers tried to clearly present their skills. Now it's necessary to use as many of the buzzwords or keywords of your profession or industry as you can.

Action words ("accelerated," "launched," "built") should still be used, but they are less important than nouns. Nouns include job titles, capabilities, languages spoken, type of degree, and the like ("vice president," "systems analyst," "Spanish," "Unix"). The reason, of course, is that a computer will scan for keywords applicable to the job that is to be filled.[99]

Because résumé-screening programs sort and rank the number of keywords found, those with the most rise to the top of the electronic pile. Thus, careers columnist Joyce Lain Kennedy suggests you pack your résumé with every keyword that applies to you. You should especially use keywords of the sort that appear in help-wanted ads.[100]

If you are looking for a job in desktop publishing, for instance, there are specific keywords, such as names of applications programs with which you are familiar, that will make you stand out. Examples: *Adobe Illustrator, Pagemaker, PhotoShop, Quark*. A compensation analyst might use keywords such as *base pay plan design* and *incentive plan design*.

Make the Résumé Impress People, Too Your résumé shouldn't just be pages of keywords. It has to impress a human recruiter, too, who may still have some fairly traditional ideas about résumés.

Some tips for organizing résumés, offered by reporter Kathleen Pender, who interviewed numerous professional résumé writers, are as follows.[101]

- **The beginning:** Start with your name, address, and phone number. (Add your e-mail and Web page addresses, too, if you have them.)

  Follow with a clear objective stating what it is you want to do. (Example: "Sales representative in computer furniture industry.")

  Under the heading "Summary" give three compelling reasons why you are the ideal person for the job. (Example of one line: "Experienced sales representative to corporations and small businesses.")

  After the beginning, your résumé can follow either a *chronological* format or a *functional* format.

- **The chronological résumé:** The chronological résumé is best for people who have stayed in the same line of work and have moved steadily upward in their careers, with no gaps in work history. Start with your latest job and work backward, and say more about your recent jobs than your earlier ones.

The format is to list years you worked at each place, followed by your job title, employer name, and a few of your accomplishments. Omit accomplishments that have nothing to do with the job you're applying for.

- **The functional résumé:** The functional résumé works best for people who are changing careers or re-entering the job market. It also is for people who need to emphasize skills from earlier in their careers or who want to emphasize their volunteer experience. It's recommended, too, for people who have had responsibility but never an important job title.

  The format is to emphasize the skills, then follow with a brief chronological work history giving dates, job titles, and employer names.

- **The conclusion:** Both types of résumés should have a concluding section showing college, degree, and graduation date; professional credentials or licenses; and professional affiliations and awards if they are relevant to the job you're seeking.

- **The biggest mistakes on résumés:** The biggest mistake you can make on a résumé is to *lie*. Sooner or later a lie will probably catch up with you and may get you fired, maybe even sued.

  The second mistake is to have spelling errors—*any* spelling errors. Spelling mistakes communicate to prospective employers a basic carelessness.

Other dos and don'ts appear in the box opposite. (■ *See Panel 8.30.*) It's a good idea to have a friend or instructor read over your résumé before you send it out.

Write a Good Cover Letter Write a targeted cover letter to accompany your résumé. This advice especially should be followed if you're responding to an ad.

Most people don't bother to write a cover letter focusing on the particular job being advertised. Moreover, if they do, say San Francisco employment experts Howard Bennett and Chuck McFadden, "they tend to talk about what *they* are looking for in a job. This is a major turn-off for employers."[102] Employers don't care very much about your dreams and aspirations, only about finding the best candidate for the job.

Bennett and McFadden suggest the following strategy for a cover letter:

- **Emphasize how you will meet the employer's needs:** Employers advertise because they have needs to be met. "You will get much more attention," say Bennett and McFadden, "if you demonstrate your ability to fill those needs."

  How do you find out what those needs are? *You read the ad.* By closely reading the ad (whether in a newspaper or on an employer's Web site), you can find out how the company talks about itself. You can also discover what attributes it is looking for in employees and what the needs are for the particular position.

- **Use the language of the ad:** In your cover letter, use as much of the ad's language as you can. "Use the same words as much as possible," advise Bennett and McFadden. "Feed the company's language back to them." The effect of this will be to produce "an almost subliminal realization in the company that you are the person they've been looking for."

- **Take care with the format of the letter:** Keep the letter to one page and use bullets or dashes to emphasize the areas where you meet the needs described in the ad. Make sure the sentences read well and—very important—that no word or name is misspelled.

The intent of both cover letter and résumé is to get you an interview, which means you are in the top 10–15% of candidates. Once you're into an interview, a different set of skills is needed. We urge you to research these on your own. Richard Bolles, author of the best-selling job-hunting book *What Color Is Your Parachute?*, suggests that aside from looking clean and well-groomed, you need to tell the employer what distinguishes you from the 20 other people who are interviewing. "If you say you are a very thorough person, don't just say it," suggests Bolles. "Demonstrate it by telling them what you know about their company, which you learned beforehand by doing your homework."[103]

Résumé Database Services

By putting your résumé in an online database, you give employers the opportunity to find you. Among the kinds of resources for employers are the following.

- **Databases for college students:** Colleges sometimes have or are members of online database services on which their students or alumni may place electronic résumés. For example, several universities have formed University ProNet, which provides online résumés to interested employers; students and alumni pay a one-time lifetime fee of $35.

  The Career Placement Registry allows college students or recent graduates to post résumés for $15.

- **Databases for people with experience:** Résumés for experienced people may be collected by private databases. An example is Connexions, which charges experienced professionals $40 a year to post their résumé.

  Some databases serve employers in particular geographical areas or those looking for people with particular kinds of experience. HispanData in Santa Barbara, California, specializes in marketing Hispanic professionals to employers seeking to diversify their workforce.

We discuss the subject of job hunting further in the Chapter 12 Experience Box, "Job Searching on the Internet & World Wide Web."

Suggested Resources

Online Résumé Help

Career Mosaic: *http://www.interbiznet.com/hunt/tools.html*

JobSmart: *http://jobsmart.org/tools/resume/index/html*

Joyce Lain Kennedy's How to Write an Electronic Résumé, at Online Career Center. *http://www.occ.com/occ/JLK/HowTo EResume.html*

Resumix Creating Your Résumé. *http://www.resumix.com/resume/resumeindex.html*

Yahoo!'s Résumé Area. *http://www.yahoo.com/business_and_economy/employment/resumes/resume_writing_tips/*

Online Résumé Databases

Career Placement Registry. $15 for college students or recent graduates. Telephone: 800-368-3093.

HispanData. $15 one-time listing fee for Hispanic college-educated individuals. Telephone: 805-682-5843, ext. 800. e-mail: *hdata@hbinc.com* Web: *http://www.greenearth.com/hispandata/jobsearch.html*

Job Bank USA. $30 for 12 months. Telephone: 800-296-1USA.

kiNexus. Free to many students through their colleges' career centers; everybody else $30. Telephone: 800-828-0422.

National Résumé Bank. $25 for 3 months; $40 for 6 months. Telephone: 813-896-3694.

Peterson's Connexion. Free to many students through their colleges' career centers; everybody else $40 for 12-month listing. 800-338-3282.

SkillSearch. $65 one-time fee for college alumni. Telephone: 800-252-5665. Web: *http://www.internet-is.com/skillsearch/*

University ProNet. Private company operated by 11 participating universities, including M.I.T., Ohio State, University of Michigan, Stanford, and UCLA. Alumni pay $35 lifetime fee to register and are allowed to update their résumés annually. Telephone: 800-726-0280. Web: *http://www.univprontet.com/*

❝There are no hard and fast rules to résumé writing, but these are a few points on which the majority of experts would agree.

Do
- Start with a clear objective.
- Have different résumés for different types of jobs.
- List as many relevant skills as you legitimately possess.
- Use jargon or buzzwords that are understood in the industry.
- Use superlatives: biggest, best, most, first.
- Start sentences with action verbs (organized, reduced, increased, negotiated, analyzed).
- List relevant credentials and affiliations.
- Use standard-size, white or off-white heavy paper.
- Use a standard typeface and a letter-quality or laser-jet printer.
- Spell check and proofread, several times.

Don't
- Lie
- Sound over pompous.
- Use pronouns such as I, we.
- Send a photo of yourself
- List personal information such as height, weight, marital status or age, unless you're applying for a job as an actor or model.
- List hobbies, unless they're directly related to your objective.
- Provide references unless requested. ("References on request" is optional.)
- Include salary information.
- Start a sentence with "responsibilities included:"
- Overuse a mix type styles such as bold, underline, italic, and uppercase.**❞**

■ PANEL 8.30

Résumé dos and don'ts

—Kathleen Pender, "Résumé Dos and Don'ts," *San Francisco Chronicle*

Summary

What It Is/What It Does	Why It's Important

amplitude (p. 358, KQ 8.2) In analog transmission, the height of a wave within a given period of time.

Amplitude refers to the strength or volume—the loudness of a signal.

analog cellular phone (p. 373, KQ 8.3) Mobile telephone designed primarily for communicating by voice through a system of ground-area *cells*. Calls are directed to cells by a mobile telephone switching office (MTSO). Moving between cells requires that calls be "handed off" by the MTSO.

Cellular phone systems allow callers mobility.

analog signal (p. 357, KQ 8.2) Continuous electrical signal in the form of a wave. The wave is called a *carrier wave.* Two characteristics of analog carrier waves that can be altered are frequency and amplitude. Computers cannot process analog signals.

Analog signals are used to convey voices and sounds over wire telephone lines, as well as in radio and TV broadcasting. Computers, however, use digital signals, which must be converted to analog signals in order to be transmitted over telephone wires.

asynchronous transfer mode (ATM) (p. 390, KQ 8.6) Method of communications transmission that combines the efficiency of packet switching with some aspects of circuit switching.

ATM transmission is a recent development using fiber-optic cable to handle both data and real-time voice and video transmissions.

asynchronous transmission (p. 389, KQ 8.6) Also called *start-stop transmission;* data is sent one byte (character) at a time. Each string of bits making up the byte is bracketed with special control bits; a "start" bit represents the beginning of a character, and a "stop" bit represents its end.

This method of communications is used with most microcomputers. Its advantage is that data can be transmitted whenever convenient for the sender. Its drawback is that transmitting only one byte at a time makes it a relatively slow method that cannot be used when great amounts of data must be sent rapidly.

bands (bandwidths) (p. 365, KQ 8.3) Ranges of frequencies. The bandwidth is the difference between the lowest and highest frequencies transmitted.

Different telecommunications systems use different bandwidths for different purposes, whether cellular phones or network television.

bits per second (bps) (p. 360, KQ 8.2) Measurement of data transmission speeds. Modems transmit at 1200, 2400, and 4800 bps (slow), 9600 and 14,400 bps (moderately fast), and 28,800, 33,600, and 56,000 bps (high-speed).

A 10-page single-spaced letter can be transmitted by a 2400-bps modem in 2½ minutes. It can be transmitted by a 9600-bps modem in 38 seconds and by a 56,000-bps modem in about 5 seconds. The faster the modem, the less time online and therefore less expense.

bridge (p. 382, KQ 8.5) Hardware and software combination used to connect the same types of networks.

Smaller networks (local area networks) can be joined together to create larger networks.

What It Is/What It Does

Why It's Important

bus network (p. 384, KQ 8.5) Type of network in which all communications devices are connected to a common channel, with no central server. Each communications device transmits electronic messages to other devices. If some of those messages collide, the device waits and tries to retransmit.

The advantage of a bus network is that it may be organized as a client/server or peer-to-peer network. The disadvantage is that extra circuitry and software are needed to avoid collisions between data. Also, if a connection is broken, the entire network may stop working.

cable modem (p. 364, KQ 8.2) Modem that connects a PC to a cable-TV system that offers online services as well as TV.

Cable modems transmit data faster than standard modems, but users can't send data nearly as fast as they receive it unless their cable service has upgraded for better two-way communications.

cell (p. 373, KQ 8.3) Geographical component of a cellular telephone system; a cell is hexagonal in shape, usually 8 miles or less in diameter, and is served by a transmitter-receiving tower. Calls are directed between cells by a mobile telephone switching office (MTSO). Movement between cells requires that calls be "handed off" by the MTSO.

Handing off voice calls between cells poses only minimal problems. However, handing off data transmission (where every bit counts), with the inevitable gaps and pauses as one moves from one cell to another, is much more difficult.

Cellular Digital Packet Data (CDPD) (p. 374, KQ 8.3) Wireless two-way communications system that places messages in packets—digital electronic "envelopes"—and sends them through underused radio channels or between pauses in cellular phone conversations.

CDPD device allows its user to send and receive voice and data messages, and the packet technique keeps data intact.

circuit switching (p. 390, KQ 8.6) Method of telephone transmission in which the transmitter has full use of the circuit until all the data has been transmitted and the circuit is terminated.

Used for voice conversations to provide steady service throughout.

client/server LAN (p. 380, KQ 8.5) Type of local area network (LAN); it consists of requesting microcomputers, called *clients,* and supplying devices that provide a service, called *servers.* The server is a computer that manages shared devices, such as laser printers, or shared files.

Client/server networks are the most common type of LAN. Compare with *peer-to-peer LAN.*

coaxial cable (p. 368, KQ 8.3) Type of communications channel; commonly called *co-ax,* it consists of insulated copper wire wrapped in a solid or braided metal shield, then in an external cover.

Coaxial cable is much better at resisting noise than twisted-pair wiring. Moreover, it can carry voice and data at a faster rate.

communications channel (p. 365, KQ 8.3) Also called *links, lines,* or *media;* the physical path over which information travels in a telecommunications system from its source to its destination.

There are many different telecommunications channels, both wired and wireless, some more efficient than others for different purposes.

What It Is/What It Does	Why It's Important

communications satellites (p. 370, KQ 8.3) Microwave relay stations in orbit above the earth. Microwave earth stations beam signals to the satellite. The satellite has solar-powered receivers and transmitters (transponders) that receive the signals, amplify them, and retransmit them to another earth station.

An orbiting satellite contains many communications channels and receives both analog and digital signals from ground microwave stations.

concentrator (p. 391, KQ 8.6) Communications device such as a minicomputer that collects data in a temporary storage area, then forwards the data when enough has been accumulated.

Concentrators enable data to be sent more economically.

digital cellular phone (p. 374, KQ 8.3) Mobile phone system that uses cells like an analog cellular phone system but transmits digital signals. Voice messages are decoded by the cellular handset, and short data items can also be transmitted.

Like an analog cellular phone, it offers the user mobility, but digital service also provides lower costs for heavy users, greater privacy, and perhaps clearer sound.

digital signal (p. 358, KQ 8.2) Type of electrical signal that uses on/off or present/absent electrical pulses in discontinuous, or discrete, bursts, rather than a continuous wave.

This two-state kind of signal works perfectly in representing the two-state binary language of 0s and 1s that computers use.

Digital Subscriber Line (DSL) (p. 363, KQ 8.2) Data transmission service that uses regular phone lines to transmit data at speeds of 1.5–8 megabits per second (Mbps), and splits the line into three channels for normal voice, outgoing data, and incoming data. Users need a special modem.

DSL service is much faster than ISDN's 128 kbps. Because it's faster for downloading than uploading, it's well suited to receiving Web or other files but not for video-conferencing, where large amounts of data need to travel fast enough to keep up with real time in both directions.

download (p. 361, KQ 8.2) To retrieve files online from another computer and store them in the main memory of one's own microcomputer. Compare with *upload.*

Downloading enables users of online systems to quickly scan file names and then save the files for later reading; this reduces the time and charges of being online.

electromagnetic spectrum (p. 365, KQ 8.3) All the fields of electrical energy and magnetic energy, which travel in waves. This includes all radio signals, light rays, X-rays, and radioactivity.

The part of the electromagnetic spectrum of particular interest is the area in the middle, which is used for communications purposes. Various frequencies are assigned by the federal government for different purposes.

FDDI network (p. 385, KQ 8.5) Short for Fiber Distributed Data Interface; a type of local area network that uses fiber-optic cable with an adaptation of ring topology using two "token rings."

The FDDI network is being used for such high-tech purposes as electronic imaging, high-resolution graphics, and digital video.

What It Is/What It Does

fiber-optic cable (p. 369, KQ 8.3) Type of communications channel consisting of hundreds or thousands of thin strands of glass that transmit pulsating beams of light. These strands, each as thin as a human hair, can transmit billions of pulses per second, each "on" pulse representing one bit.

file server (p. 380, KQ 8.5) Type of computer used on a local area network (LAN) that acts like a disk drive and stores the programs and data files shared by users of the LAN.

frequency (p. 358, KQ 8.2) Number of times a wave repeats during a specific time interval—that is, how many times it completes a cycle in a second. 1 hertz (Hz) = 1 cycle per second.

front-end processor (p. 391, KQ 8.6) Smaller computer that is connected to a larger computer to assist it with communications functions.

full-duplex transmission (p. 388, KQ 8.6) Type of data transmission in which data is transmitted back and forth at the same time, unlike simplex and half-duplex.

gateway (p. 382, KQ 8.5) Interface that enables dissimilar networks to communicate with one another.

Global Positioning System (GPS) (p. 371, KQ 8.3) System of 24 earth-orbiting satellites 10,600 miles above the earth that continuously transmit timed radio signals used to identify earth locations.

half-duplex transmission (p. 388, KQ 8.6) Type of data transmission in which data travels in both directions but only in one direction at a time, as with CB or marine radios; the two parties must take turns talking.

hertz (Hz) (p. 365, KQ 8.3) Provides a measure of the frequency of electrical vibrations (cycles) per second; 1 Hz = 1 cycle per second.

host computer (p. 378, KQ 8.4) The central computer that controls a network. On a local area network, the host's functions may be performed by a computer called a *server*.

Why It's Important

When bundled together, fiber-optic strands in a cable 0.12 inch thick can support a quarter- to a half-million simultaneous voice conversations. Moreover, unlike electrical signals, light pulses are not affected by random electromagnetic interference in the environment and thus have much lower error rates than telephone wire and cable.

A file server enables users of a LAN to all have access to the same programs and data.

The higher the frequency—that is, the more cycles per second—the more data can be sent through a channel.

The front-end processor transmits and receives messages over the communications channels, corrects errors, and relieves the larger computer of routine tasks.

Full-duplex is available in some large computer systems and in newer microcomputer modems to support workgroup computing.

With a gateway, a local area network may be connected to a larger network, such as a wide area network.

GPS was designed for military use but has civilian uses such as tracking trucks and taxis, and combined with map software it can be used as a guide for motorists.

Half-duplex is the most common method of data transmission, as when logging onto a bulletin board system.

One million hertz equals 1 megahertz. Bandwidths are defined according to megahertz and gigahertz ranges.

The host is responsible for managing the entire network.

What It Is/What It Does	Why It's Important

hybrid network (p. 385, KQ 8.5) Type of local area network (LAN) that combines star, ring, and bus networks.

A hybrid network can link different types of LANs. For example, a small college campus might use a bus network to connect buildings and star and ring networks within certain buildings.

Integrated Services Digital Network (ISDN) (p. 363, KQ 8.2) Hardware and software that allow voice, video, and data to be communicated as digital signals over traditional copper-wire telephone lines.

The main benefit of ISDN is speed. It allows people to send digital data at 128 kbps, more than double the speed the fastest modems can now deliver on the analog voice network. However, other, faster technologies will probably take over ISDN's role of high-speed data transmission for consumers and small business.

kilobits per second (kbps) (p. 360, KQ 8.2) 1000 bits per second; an expression of data transmission speeds. A 56,000-bps modem is a 56-kbps, or 56K, modem.

See bits per second.

local area network (LAN) (p. 380, KQ 8.5) A network consisting of a communications link, network operating system, microcomputers or workstations, servers, and other shared hardware such as printers or storage devices. LANs are of two principal types: client/server and peer-to-peer.

LANs have replaced mainframes and minicomputers for many functions and are considerably less expensive. However, LANs have neither the great storage capacity nor the security of mainframes.

local network (p. 377, KQ 8.4) Privately owned communications network that serves users within a confined geographical area. The range is usually within a mile.

Local networks are of two types: private branch exchanges (PBXs) and local area networks (LANs).

metropolitan area network (MAN) (p. 377, KQ 8.4) Communications network covering a geographic area the size of a city or suburb. Cellular phone systems are often MANs.

The purpose of a MAN is often to bypass telephone companies when accessing long-distance services.

microwave systems (p. 369, KQ 8.3) Communications systems that transmit voice and data through the atmosphere as super-high-frequency radio waves. Microwaves are the electromagnetic waves that vibrate at 1 billion hertz per second or higher.

Microwave frequencies are used to transmit messages between ground-based earth stations and satellite communications systems. More than half of today's telephone system uses microwave transmission.

modem (p. 358, KQ 8.2) Short for *mo*dulater/*dem*odulater. A device that converts digital signals into a representation of analog form (modulation) to send over phone lines; a receiving modem then converts the analog signal back to a digital signal (demodulation).

A modem enables users to transmit data from one computer to another by using standard telephone lines instead of special communications lines such as fiber optic or cable.

multiplexer (p. 390, KQ 8.6) Device that merges several low-speed transmissions into one high-speed transmission. Depending on the model, 32 or more devices may share a single communications line.

High-speed multiplexers using high-speed digital lines can carry as many messages, both voice and data, as 24 analog telephone lines.

What It Is/What It Does

Why It's Important

multiplexing (p. 390, KQ 8.6) Transmission of multiple signals over a single communications channel; the device used may be a multiplexer, concentrator, or front-end processor.

Multiplexing allows several communications devices to share a line, taking advantage of the otherwise unused capacity of expensive communications lines.

multipoint line (p. 386, KQ 8.6) Single line that interconnects several communications devices to one computer.

Often on a multipoint line only one communications device, such as a terminal, can transmit at any given time.

network (communications network) (p. 377, KQ 8.4) System of interconnected computers, telephones, or other communications devices that can communicate with one another.

Networks allow users to share applications and data.

network interface card (p. 382, KQ 8.5) Circuit board inserted into an expansion slot in a microcomputer that enables it to send and receive messages on a local area network.

Without a network interface card, a microcomputer cannot be used to communicate on a LAN.

node (p. 378, KQ 8.4) Any device that is attached to a network.

A node may be a microcomputer, terminal, storage device, or some peripheral device, any of which enhance the usefulness of the network.

Open Systems Interconnection (OSI) (p. 392, KQ 8.6) International standard that defines seven layers of protocols for worldwide computer communications.

Creates a set of standards that enables communications hardware manufacturers to build devices that can "talk" to each other.

packet (p. 390, KQ 8.6) Fixed-length block of data for transmission. The packet also contains instructions about the destination of the packet.

By creating data in the form of packets, a transmission system can deliver the data more efficiently and economically, as in packet switching.

packet-radio-based communications (p. 373, KQ 8.3) Wireless two-way communications system that uses a nationwide system of radio towers to send data to hand-held computers.

Uses packet switching to encapsulate the data in envelopes so that it arrives intact, which may not be the case with data transmission over analog cellular systems. Used by mobile workers who need to communicate frequently with a corporate database.

packet switching (p. 390, KQ 8.6) Technique for dividing electronic messages into packets—fixed-length blocks of data—for transmission over a wide area network to their destination through the most expedient route. A sending computer breaks an electronic message apart into packets, which are sent through a communications network—via different routes and speeds—to a receiving computer, which reassembles them into proper sequence to complete the message.

The benefit of packet switching is that it can handle high-volume traffic in a network. It also allows more users to share a network, thereby offering cost savings.

What It Is/What It Does	**Why It's Important**

pager (p. 373, KQ 8.3) Simple radio receiver that receives data, but not voice messages, sent from a special radio transmitter. The pager number is dialed from a phone and travels via the transmitter to the pager.

Pagers have become a common way of receiving notification of phone calls so the user can return the calls immediately; some pagers can also display messages of up to 80 characters and send preprogrammed messages.

parallel data transmission (p. 387, KQ 8.6) Method of transmitting data in which bits are sent through separate lines simultaneously.

Unlike serial lines, parallel lines move information fast, but they are efficient for only up to 15 feet. Thus, parallel lines are used, for example, to transmit data from a computer's CPU to a printer.

peer-to-peer LAN (p. 380, KQ 8.5) Type of local area network (LAN); all microcomputers on the network communicate directly with one another without relying on a server.

Peer-to-peer networks are less expensive than client/server networks and work effectively for up to 25 computers. Thus, they are appropriate for networking in small groups.

personal communications services (PCS) (p. 375, KQ 8.3) Also called *personal communications networks (PCN);* a digital wireless phone service that uses a new band of frequencies (1850–1990 megahertz) and transmitter-receivers in thousands of microcells. A lower-powered version of digital cellular phone service.

PCS takes advantage of the uncrowded super high frequencies, but with its small cells users may find themselves out of range of the service in many areas.

photonics (p. 396, KQ 8.7) The science of sending data bits down pulses of light carried on hair-thin glass fibers.

Photonics continues to improve fiber-optic technology to speed transmission and greatly increase the number of channels per fiber; results could include real-time live video communications.

point-to-point line (p. 386, KQ 8.6) Communications line that directly connects the sending and receiving devices, such as a terminal with a central computer.

This arrangement is appropriate for a private line whose sole purpose is to keep data secure by transmitting it from one device to another.

private branch exchange (PBX) (p. 379, KQ 8.5) Private or leased telephone switching system that connects telephone extensions in-house as well as to the outside telephone system.

Newer PBXs can handle not only analog telephones but also digital equipment, including computers.

protocol (communications protocol) (p. 391, KQ 8.6) Set of conventions governing the exchange of data between hardware and/or software components in a communications network.

Protocols are built into hardware and software to allow different devices to work together, and with OSI standards, protocols have become much more universal.

remote-control software (p. 362, KQ 8.2) Software that allows a user to control a microcomputer from another microcomputer in a different location.

Such software is useful for travelers who want to use their home computers from afar and for technicians trying to assist computer users with support problems.

What It Is/What It Does

Why It's Important

ring network (p. 382, KQ 8.5) Type of local area network (LAN) in which all communications devices are connected in a continuous loop and messages are passed around the ring until they reach the right destination. There is no central server.

The advantage of a ring network is that messages flow in only one direction and so there is no danger of collisions. The disadvantage is that if a connection is broken, the entire network stops working.

router (p. 382, KQ 8.5) Special computer that directs communication messages between several networks.

High-speed routers can serve as part of the Internet backbone.

serial data transmission (p. 387, KQ 8.6) Method of data transmission in which bits are sent sequentially, one after the other, through one line.

Serial transmission is found in communications lines, modems, and mice.

simplex transmission (p. 388, KQ 8.6) Type of transmission in which data can travel in only one direction; there is no return signal.

Some computerized data collection devices, such as seismograph sensors that measure earthquakes, use simplex transmission.

specialized mobile radio (SMR) (p. 375, KQ 8.3) Two-way radio voice-dispatching service used by taxis and trucks; the frequencies used by SMR are being bought and converted to a digital system.

Digital SMR is one of the new forms of wireless communication being offered.

star network (p. 382, KQ 8.5) Type of local area network (LAN) in which all microcomputers and other communications devices are connected to a central hub, such as a file server. Electronic messages are routed through the central hub to their destinations. The central hub monitors the flow of traffic.

The advantage of a star network is that the hub prevents collisions between messages. Moreover, if a connection is broken between any communications device and the hub, the rest of the devices on the network will continue operating.

synchronous transmission (p. 389, KQ 8.6) Type of transmission in which data is sent in blocks. Start and stop bit patterns, called synch bytes, are transmitted at the beginning and end of the blocks. These start and end bit patterns synchronize internal clocks in the sending and receiving devices so that they are in time with each other.

Synchronous transmission is rarely used with microcomputers because it is more complicated and more expensive than asynchronous transmission. It is appropriate for computer systems that need to transmit great quantities of data quickly.

telecommuting (p. 355, KQ 8.1) Way of working at home with telecommunications—phone, fax, and computer—between office and home.

Telecommuting can help ease traffic and the stress of commuting by car, increase productivity and job satisfaction, and let a company hire employees who don't want to move.

terminal emulation (p. 362, KQ 8.2) Communications software that allows users to use their microcomputers to access a mainframe or minicomputer; the software "tricks" the large computer into acting as if it is communicating with a terminal.

Terminal emulation software is necessary for microcomputer users to make full use of the networks and the resources available with mainframes.

What It Is/What It Does	Why It's Important

topology (p. 382, KQ 8.5) The logical layout, or shape, of a local area network. The five basic topologies are star, ring, bus, hybrid, and FDDI.

Different topologies can be used to suit different office and equipment configurations.

twisted-pair wire (p. 367, KQ 8.3) Type of communications channel consisting of two strands of insulated copper wire, twisted around each other in pairs.

Twisted-pair wire has been the most common channel or medium used for telephone systems. It is relatively slow and does not protect well against electrical interference.

T1 line (p. 364, KQ 8.2) A traditional telephone trunk line that carries 24 normal telephone circuits and has a speed of 1500 kbps.

Used as high-capacity communications links at corporate, government, or academic sites.

T3 line (p. 364, KQ 8.2) A digital communications trunk line that transmits at a speed of 4500 kbps.

The digital equivalent of a T1 line, offering three times the speed but at a much higher cost for equipment and service charges.

upload (p. 361, KQ 8.2) To send files from a user's microcomputer to another computer. Compare with *download.*

Uploading allows microcomputer users to easily exchange files with each other over networks.

virtual office (p. 355, KQ 8.1) An often nonpermanent and mobile office run with computer and communications technology.

Employees work not in a central office but from their homes, cars, and customers' offices. They use pocket pagers, portable computers, fax machines, and various phone and network services to conduct business.

wide area network (WAN) (p. 377, KQ 8.4) Type of communications network that covers a wide geographical area, such as a state or a country.

Wide area networks provide worldwide communications systems.

Exercises

Self-Test Exercises

1. A(n) _____ converts digital signals into analog signals for transmission over phone lines.

2. Before a microcomputer in a LAN can send and receive messages, a(n) _____ _____ must be inserted into an expansion slot in a computer.

3. _____ transmission sends data in both directions simultaneously, similar to two trains passing in opposite directions on side-by-side tracks.

4. A(n) _____ _____ network is a communications network that covers a wide geographical area, such as a state or a country.

5. _____ cable transmits data as pulses of light rather than as electricity.

Short-Answer Questions

1. Why is speed an important consideration when selecting a modem?

2. What is meant by the term *protocol* as it relates to communicating between two computers?

3. What is a multiplexer?

4. What is a hybrid network?

5. When talking about communications, what is the significance of the electromagnetic spectrum?

Multiple-Choice Questions

1. Which of the following functions does communications software perform?
   a. error correction
   b. data compression
   c. remote control
   d. terminal emulation
   e. All of the above

2. Which of the following is a standard LAN component?
   a. network operating system
   b. cabling system
   c. network interface cards
   d. shared devices
   e. All of the above

3. Which of the following best describes the telephone line that is used in most homes today?
   a. twisted-pair wire
   b. coaxial cable
   c. fiber-optic cable
   d. modem cable
   e. None of the above

4. Which of the following network configurations always uses a central server?
   a. bus
   b. star
   c. ring
   d. hybrid
   e. All of the above

5. Which of the following do local area networks enable?
   a. sharing of peripheral devices
   b. sharing of programs and data
   c. better communications
   d. access to databases
   e. All of the above

True/False Questions

T F   1. In a LAN, a bridge is used to connect the same types of networks, whereas a gateway is used to enable dissimilar networks to communicate.

T F   2. The current limitation of cable modems is that they don't provide much online interactivity.

T F   3. Transmission rate is a function of two variables: frequency and bandwidth.

T F 4. All communications channels are either wired or wireless.

T F 5. Parallel transmission is faster than serial transmission.

Knowledge in Action

1. Are the computers at your school or work connected to a network? If so, what are the characteristics of the network? What advantages does the network provide in terms of hardware and software support? What types of computers are connected to the network (microcomputers, minicomputers, and/or mainframes)? Specifically, what software/hardware is allowing the network to function?

2. Using current articles, publications, and/or the Web, research the history of cable modems, how they are being used today, and what you think the future holds for them. Do you think you will use a cable modem in the future? Present your findings in a paper or a 15-minute discussion.

3. Describe in more detail the FCC's role in regulating the communications industry. What happens when frequencies are opened up for new communications services? Who gets to use these frequencies?

4. Research the technology behind Gigabit Ethernet. What makes this technology possible? How likely is it that Gigabit Ethernet will become standard in homes? What are the limitations of this technology?

5. Of the different technologies discussed in this chapter, which do you think will have the biggest impact on you? Why?

Files & Databases

From Data Organizing to Data Mining

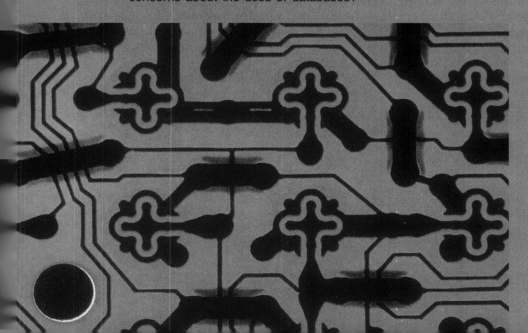

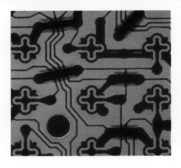

If information exists in one place, it exists in more than one place."

So says Carole A. Lane, a database expert.[1] This is what she calls "Lane's First Law of Information." Perhaps it could stand as a summary of one of the most important consequences of the Digital Age.

How does information in one place get to be in more than one place? This question suggests one reason why databases are important—and why they deserve to be studied in a chapter on their own. A database is not just the computerization of what used to go into manila folders and a filing cabinet. A database is an organized collection of related files, a method of pulling together facts, that allows the slicing and dicing and mixing and matching of information in all kinds of ways.

Gambling casinos in Las Vegas, Atlantic City, and elsewhere, for instance, have long kept track of the "high rollers," their best customers. Recently, however, casinos have also been creating databases of details about less extravagant gamblers, in order to target them with marketing appeals and get them to gamble more.[2] How are such databases constructed? By recording existing customers' gaming habits and other details. By buying direct-mail lists (as of respondents to a vacation-sweepstakes pitch). By purchasing lists from credit-card companies, which are then matched with census data to calculate people's incomes. If you've been to one of the Harrah's clubs, says a report, the company's database "is likely to know about the T-bone steak you gobbled and the giftshop sweatshirt you bought—not to mention the amount of your mortgage and how much you can be expected to gamble during your next visit."[3] Some critics worry that the casinos' aggressive collection of data will be used to concentrate their marketing on compulsive gamblers, exploiting their addiction.

Ethics

But databases also offer a great deal of constructive social use. The Library of Congress, for instance, is in the middle of a five-year plan to establish a National Digital Library, which will make 5 million items of American history and culture available on the World Wide Web.[4,5] "There's been so much talk about all the junk on the Internet," says Librarian of Congress James Billington. "What we're trying to do is get some quality content out there."[6] Materials to be included are drafts of the Declaration of Independence, the notebooks of poet Walt Whitman (previously missing for 50 years), 19th century Native American documents and artifacts, Civil War photographs, early motion pictures, 1500 pieces of pre-1920 sheet music, and more than 8000 photographs about the history of south Texas and the northeastern Mexico border.

Are your name and facts about you in a database or databases? Undoubtedly. How is that data being used? That is the interesting question. Let us begin to examine what databases are and how they work.

9.1 All Databases Great & Small

KEY QUESTIONS

What is the definition of database, and what does a database manager do?

Preview & Review: A database is an organized collection of integrated files. Examples of databases are personal, public, and private. Organizations usually appoint a database administrator to manage the database and related activities.

As we stated, a *database* is an organized collection of integrated files. This makes them usable in more ways than traditional filing systems (computerized or not), as we shall see.

A database may be small, contained entirely within your own personal computer. Or it may be massive, available online through computer and telephone connections. (■ *See Panel 9.1.*) Such online databases are of special interest to us in this book because they offer us phenomenal resources that until recently were unavailable to most ordinary computer users.

Microcomputer users can set up their own databases using popular database management software like that we discussed earlier (in Chapter 2, ✔ p. 64). Examples are Paradox, Access, dBASE 5, and FoxPro. Such programs are used, for example, by graduate students to conduct research, by salespeople to keep track of clients, by purchasing agents to monitor orders, and by coaches to keep watch on other teams and players.

Some databases are so large that they cannot possibly be stored in a microcomputer. Some of these can be accessed by going online. Such databases, sometimes called *information utilities*, represent enormous compilations of data, any part of which is available, for a fee, to the public.

Examples of well-known information utilities—more commonly known as *online services*—are America Online, CompuServe, and Microsoft Network. As we described in Chapter 7, these offer access to news, weather, travel information, home shopping services, reference works, and a great deal more. Some public databases are specialized, such as Lexis, which gives lawyers access to local, state, and federal laws.

Other public-access databases are online archives like those of "virtual art museums," such as the Smithsonian's National Museum of American Art, which has put images from its collections online since 1993, and the National Gallery of Art in Washington, D.C.

Other types of large databases are collections of records shared or distributed throughout a company or other organization. Generally, the records are available only to employees or selected individuals and not to outsiders. For example, many university libraries have been transforming drawers of catalog cards into electronic databases for use by their students and faculty. Libraries at Yale, Johns Hopkins, and other universities have contracted with

■ PANEL 9.1

Small (personal), medium-size, and large databases: examples

Type	Example	Typical number of users	Typical size of database
Personal	Mary Richards House Painting	1	< 10 million bytes (10 MB)
Medium-size (workgroup)	Seaview Yacht Sales	25 or fewer	< 2 billion bytes (2 GB)
Large organizational	Automobile Licensing and Registration	100s	> 1 trillion bytes (1 TB)

Chapter 9

■ PANEL 9.2

Building a library database
Father Patrick Creeden enters data into a computer
at the Monastery of the Holy Cross in Chicago.

"CHICAGO–Father Thomas Baxter stands
in his habit, holding a computer printout. Later
today, he'll review the Scriptures in his cell
here at the Monastery of the Holy Cross of
Jerusalem, but this morning he's carefully
proofreading a library's computerized records.

He and two other monks in this community
of five are modern-day scribes, using computers to participate in an age-old monastic tradition: preserving knowledge. In this case, they're
helping university and public libraries transform
drawers of catalogue cards into electronic databases.

Father Baxter is pointing at a line of text highlighted in yellow, indicating a discrepancy between two data fields. One of his jobs as prior, or leader, of the monastery is to look for typographical errors, call them to the attention of the brother responsible for them, and remind
him to concentrate more closely on his work.

This monastery is part of an effort to bring religious communities back into the information
business. A company calling itself The Electronic Scriptorium—referring to the room where
monks would use quills and ink to copy intricate manuscripts long ago—is matching up
monastic communities with libraries and others in need of complex data-entry work. The
partnerships benefit the monasteries and convents, which need flexible jobs to support
themselves, and the libraries, which need their records entered accurately.**"**

–Jeffrey R. Young, "Modern-Day Monastery, " *Chronicle of Higher Education*

a Virginia company called The Electronic Scriptorium, which employs monks
and nuns at six monasteries to convert card catalogs to an electronic system.[7,8] *(■ See Panel 9.2.)*

Shared Versus Distributed Databases

A database may be *shared* or *distributed.*

- **Shared database:** **A *shared database* is shared by users in one
  company or organization in one location.** Shared databases can be
  found in local area networks (✔ p. 380). The company owns the
  database, which is often stored on a minicomputer or mainframe.
  Users are linked to the database through terminals or microcomputer
  workstations.

- **Distributed database:** **A *distributed database* is one that is stored on
  different computers in different locations connected by a client/server
  network** (✔ p. 380). For example, sales figures for a chain of discount
  stores might be located in computers at the various stores, but they
  would also be available to executives in regional offices or at
  corporate headquarters. An employee using the database would not
  know where the data is coming from. However, all employees would
  still use the same commands to access and use the database.

The Database Administrator

The information in a large database—such as a corporation's patents, formulas, advertising strategies, and sales information—is the organization's lifeblood. Someone needs to manage all activities related to the database. This person is the ***database administrator (DBA),* a person who coordinates all related activities and needs for a corporation's database.**

The responsibilities include the following:

- **Database design, implementation, and operation:** At the beginning, the DBA helps determine the design of the database. Later he or she determines how space will be used on secondary-storage devices, how files and records may be added and deleted, and how changes are documented.

- **Coordination with users:** The DBA determines user access privileges; sets standards, guidelines, and control procedures; assists in establishing priorities for requests; prioritizes conflicting user needs; and develops user documentation and input procedures.

README

Case Study: SAP's R/3 Software Is Not for Saps—and You'll Probably Have to Learn to Deal with It

Wouldn't you think Microsoft, the world's largest software company, and IBM, the world's largest computer company, would run their own software for their own internal purposes?

They don't.

They run something called R/3, made by SAP AG, a German company. And it's not easy to put in place.

Installing the software "is the corporate equivalent of a root canal," says one article. "The software is fiendishly complex and expensive to configure. Companies must play host to armies of consultants who sometimes charge as much as five times what the software costs and can stay on the job for years."[9]

Still, it's becoming wildly popular in big organizations. What does R/3 do? First, some background.

About a decade ago, when many companies "re-engineered" their procedures, they began to change from having all data crunched on the corporate mainframe to client/server computing—having data processed in smaller computers (servers) linked to desktop computers (clients).

The problem, says one account, "was that corporations wanted software that was as bullet-proof [reliable] as what they had on their old mainframes. But the client-server software of that era—mainly written by little software houses—wasn't robust enough to handle critical data."[10]

In the late 1980s, SAP, which was already well established in Germany, began developing client/server software designed to accommodate many users at the same time, just like a mainframe.

To the average user, R/3 looks like any other database entry form, with blank cells labeled "quantity," "price," or "product description." (The user enters information into the blank cells.) The purpose of the software is to tie together and automate the basic processes of a business: take orders, check credit, verify payments, and balance the books.

Thus, a salesperson's order will automatically have effects throughout the company, adjusting production schedules, parts supplies, inventory lists, accounting entries, and balance sheets. "Multiple currencies, national laws, and business practices are embedded in the software," says one description, "so that it can translate transactions seamlessly from a subsidiary in Portugal to another company unit in, say, the United States."[11]

SAP's business software is now used in over 7000 companies around the world, three-quarters of them outside Germany. They include such big-corporation names as Chevron, Coca-Cola, Colgate-Palmolive, and Owens-Corning. The next most promising market is small companies.

Thus, the chances are pretty good that sooner or later you will encounter R/3.

- **System security:** The DBA sets up and monitors a system for preventing unauthorized access to the database.
- **Backup and recovery:** Because loss of data or a crash in the database could vitally affect the organization, the DBA needs to make sure the system is regularly backed up. He or she also needs to develop plans for recovering data or operations should a failure or disaster occur. (Most corporations have established disaster recovery plans for databases.)
- **Performance monitoring:** The DBA monitors the system to make sure it is serving users appropriately. A standard complaint is that the system is too slow, usually because too many users are trying to access it.

9.2 The Data Storage Hierarchy & the Concept of the Key Field

KEY QUESTIONS

What is the data storage hierarchy, and why is the key field important?

Preview & Review: Data in storage is organized as a hierarchy: bits, bytes, fields, records, and files, which are the elements of a database. A key field uniquely identifies each record, making its role in data organization very important.

How does a database actually work? To understand this, first we need to consider how stored data is structured—the *data storage hierarchy* and the concept of *key field*. We then need to discuss *file management systems*, then *database management systems*.

The Data Storage Hierarchy

Data can be grouped into a hierarchy of categories, each increasingly more complex. **The *data storage hierarchy* consists of the levels of data stored in a computer file: bits, bytes (characters), fields, records, files, and databases.** (■ *See Panel 9.3.*)

Computers, we have said, are based on the principle that electricity may be "on" or "off." Thus, individual items of data are represented by the bits 0 for off and 1 for on (✔ p. 6). Bits and bytes are the building blocks for representing data, whether it is being processed, stored, or telecommunicated. Bits and bytes are what the computer hardware deals with, and you need not be concerned with them. You will, however, be dealing with characters, fields, records, files, and databases.

- **Field:** A *field* **is a unit of data consisting of one or more characters (bytes).** An example of a field is your name, your address, or your Social Security number.

  Note: One reason the Social Security number is often used in computing—for good or for ill—is that, perhaps unlike your name, it is a *distinctive* (unique) field. Thus, it can be used to easily locate information about you. Such a field is called a *key field*. More on this below.

- **Record:** A *record* **is a collection of related fields.** An example of a record would be your name *and* address *and* Social Security number.

- **File:** A *file* **is a collection of related records.** An example of a file is data collected on everyone employed in the same department of a company, including all names, addresses, and Social Security numbers.

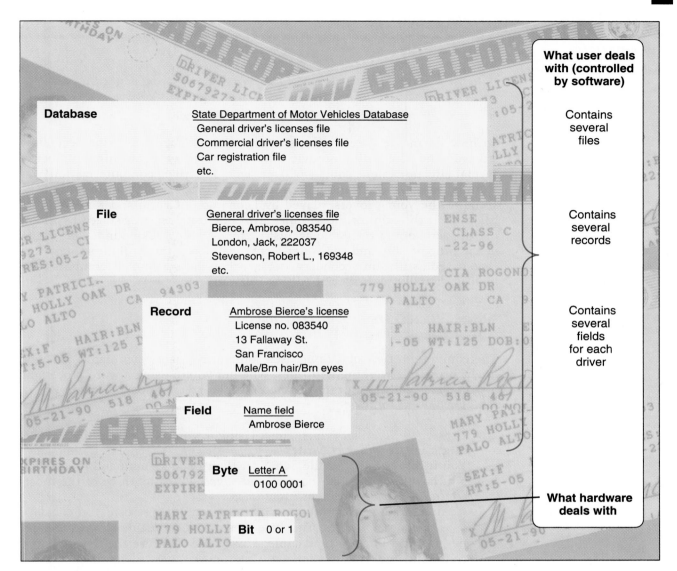

Database   State Department of Motor Vehicles Database
General driver's licenses file
Commercial driver's licenses file
Car registration file
etc.

What user deals with (controlled by software)

Contains several files

File   General driver's licenses file
Bierce, Ambrose, 083540
London, Jack, 222037
Stevenson, Robert L., 169348
etc.

Contains several records

Record   Ambrose Bierce's license
License no. 083540
13 Fallaway St.
San Francisco
Male/Brn hair/Brn eyes

Contains several fields for each driver

Field   Name field
Ambrose Bierce

Byte   Letter A
0100 0001

What hardware deals with

Bit   0 or 1

▌ PANEL 9.3

Review of data storage hierarchy: how data is organized

As described in Chapter 2, bits are organized into bytes, bytes into fields, fields into records, records into files. Related files may be organized into a database.

• Database: **A *database* is an organized collection of integrated files.** A company database might include files on all past and current employees in all departments. There would be various files for each employee: payroll, retirement benefits, sales quotas and achievements (if in sales), and so on.

The Key Field

An important concept in data organization is that of the *key field*. **A *key field* is a field that is chosen to uniquely identify a record so that it can be easily retrieved and processed.** The key field is often an identification number, Social Security number, customer account number, or the like.

As mentioned, the primary characteristic of the key field is that it is *unique*. Thus, numbers are clearly preferable to names as key fields because there are many people with common names like James Johnson, Susan Williams, Ann Wong, or Roberto Sanchez, whose records might be confused.

9.3 File Management: Basic Concepts

KEY QUESTIONS

What are the basic types of files, batch and real-time processing, a master file, and a transaction file?

Preview & Review: Common types of files are program, data, ASCII, image, audio, and video files.

Transaction files are used to update master files. Files may be processed by batch processing or by real-time processing. Batch processing tends to favor off-line storage. Online, or real-time, processing requires online storage.

So how do movie special-effects artists manipulate data, using secondary storage? The $200 million movie *Titanic* required an astonishing number of special-effects camera shots—600, in fact. But then computerization was required to make them mesh perfectly with other elements of the film. The scene of the young lovers, played by Leonardo DiCaprio and Kate Winslet, balanced at the front of the *Titanic*, arms outstretched, was staged by shooting the actors on a turntable against a green-screen background. The scenes of the moving ship and the water were created separately. Both scenes were then saved as files in secondary storage, then pulled up on a computer and melded digitally with special editing software.

What You Can Do with Files: File Management Functions

The computer artists of *Titanic* used some of the same commands that you might use while working with files, whether you're using a word processing or a graphics editing program. Some examples of common file management functions are as follows:

- **Create, name, save, delete, copy:** Using a word processing file, you might create—that is, write—a scene for your screenplay, *Titanic II*. You might name your file *SCENE1*, then issue a command to save it. You have thus *created, named,* and *saved* your file, three of the most common file management functions.

  After reading over the scene on your computer screen, you decide to make some changes. You type these in, then decide to save the new version under a new name. You would then use the *Save as* command and save the version 2 file under a new filename such as *SCENE1.V2*.

  You could then *delete* the old *SCENE1* file. Or you could first *copy* that file from your hard disk to a diskette before deleting it.

- **Retrieve, update, and print:** The next day you might turn on the computer and *retrieve* the second file from secondary storage and display its contents on the monitor. You might type in additional changes and save them, thereby *updating* the file. You would then *print* the file on your computer's printer.

- **Upload, download, compression:** Since a producer has asked to read the scene, you could send it via the Internet by *uploading* the file from your computer to an online network. A week later, you might *download* the producer's edited version of the file from the online network and store it on your computer's hard disk. Before putting the file online, the producer might *compress* it so it would take less time to send. Or you might compress it so it would take up less space on your hard disk. (We discuss compression later in the chapter.)

- **Import and export:** Before you send the scene's script out to the producer again, you might decide to dress it up with an illustration or two. You could therefore *import* a file containing photos of a ship from another program and insert the illustration in your word processing file. If you wished, you could *export* your word processing file into another program (such as a desktop publishing program).

How is data stored and accessed? To understand this, first we consider *types of files* and how the computer *keeps track of files.*

Types of Files: Program Files, Data Files, & Others

The *file* is the collection of data or information that is treated as a unit by the computer. Files are given names—*filenames.* For instance, you might give your word processing file containing a paper you're writing for a psychology course the filename PSYCHREP or PSYCHREPORT.

Filenames used for the PC—but not for the Macintosh—also have *extension names,* or three-letter tags, which are usually inserted automatically by the applications software. These extensions are added on after a period following the filename—for example, .DOC in PSYCHREPORT.DOC, where the .DOC stands for "document." When you look in the directory of the files stored on your hard disk, you will notice a number of extensions, such as .DOC and .EXE.

There are many kinds of files, but perhaps the two principal ones are program files and data files.

Some common file extensions

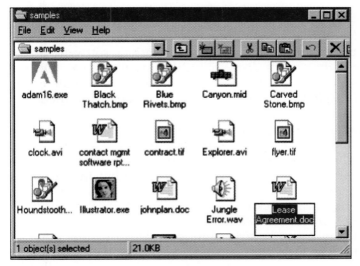

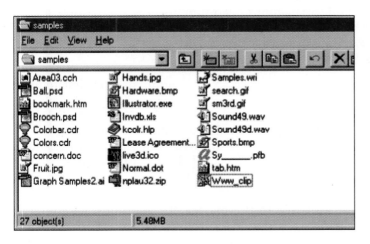

- **Program files:** ***Program files* are files containing software instructions.** Examples are word processing or spreadsheet programs, which are made up of several different program files. The two most important are source program files and executable files.

  Source program files contain high-level computer instructions in the original form written by the programmer. Source program filenames may have the extension .COM.

  For the processor to use source program instructions, they must be translated into machine language (✔ p. 157). The files that contain the machine-language instructions are called *executable files.* The filenames of executable files may have the extension .EXE. (Certain system support files containing machine-language instructions may use .DLL and .DRV.)

- **Data files:** ***Data files* are files that contain data**—content such as a report you've created using word processing applications software. On PC systems, data files, too, take extensions after the filenames.

  The extension .DOC appears in filenames used for data files created with word processing or desktop publishing programs. Data files created with database software take the extensions .DAT, .DBF, and .MDB. Common spreadsheet extensions are .XLS and .WKS.

Other common types of files are ASCII files, image files, audio files, and video files; the last three are used in multimedia.

- **ASCII files:** ASCII (✔ p. 156) is a common binary coding scheme used to represent data in a computer. *ASCII ("as-key") files* are text-only files that contain no graphics and no formatting, such as boldface or italics. This format is used to transfer documents between incompatible computers, such as PC and Macintosh. (Such files may use the .TXT extension.)

- **Image files:** If ASCII files are for text, *image files* are for digitized graphics, such as art or photographs. (They are indicated by such extensions as .TIF, .EPS, .JPG, .GIF, and .BMP.)

- **Audio files:** *Audio files* contain digitized sound and are used for conveying sound in CD-ROM multimedia and over the Internet. (They have extensions such as .WAV and .MID.)

- **Video files:** *Video files* contain digitized video images and are used for such purposes as to convey moving images over the Internet. (Common extensions are .AVI and .MPG.)

There are other types of files, but the preceding are those you are most apt to encounter.

Keeping Track of Files

How do you keep track of files on your computer? If you use a PC, the operating system allows the systems software, the applications software, or the user to create different *directories* displaying different filenames. (On the Macintosh, the directories are called *folders*.) For example, you could create a directory named *TITANIC II*. The directory might contain the filenames *SCENE1, SCENE2*, and so on, listed in order.

If you acquire a brand-new hard disk storage device, the operating system on your computer will store the files in one place on the hard disk. After you've been using the hard disk for a while, the operating system will begin to store the data in different places—not all in one place but in *clusters* of data spread out in different locations on the disk.

How does the computer keep track of the file locations? On PCs, the operating system uses the *file allocation table (FAT)*, a directory that stores such information as the file name, file size, time and date the file was last changed, and the hard disk locations (clusters) where the data in a file is stored. In Macintosh computers, the tracking of file locations is handled by the part of the operating system called the *Finder*.

Two Types of Data Files: Master File & Transaction File

Among the several types of data files, two are commonly used to update data: a master file and a transaction file.

- **Master file:** The *master file* **is a data file containing relatively permanent records that are generally updated periodically.** An example of a master file would be the address-label file for all students currently enrolled at your college.

- **Transaction file:** The *transaction file* **is a temporary holding file that holds all changes to be made to the master file: additions, deletions,**

revisions. For example, in the case of the address labels for your college, a transaction file would hold new names and addresses to be added (because over time new students enroll) and names and addresses to be deleted (because students leave). It would also hold revised names and addresses (because students change their names or move). Each month or so, the master file would be *updated* with the changes called for in the transaction file.

Batch Versus Online Processing

Updating can be done in two ways: (1) "later," via *batch processing*, or (2) "right now," via *online (real-time) processing*.

- Batch processing: **In *batch processing*, data is collected over several days or weeks and then processed all at one time, as a "batch," against a master file.** Thus, if users need to make some request of the system, they may have to wait until the batch has been processed. Batch processing is less expensive than online processing and is suitable for work in which immediate answers to queries are not needed.

  An example of batch processing is that done by banks for balancing checking accounts. When you deposit a check in the morning, the bank will make a record of it. However, it will not compute your account balance until the end of the day, after all checks have been processed in a batch.

- Online processing: ***Online processing*, also called *real-time processing*, means entering transactions into a computer system as they take place and updating the master files as the transactions occur.** For example, when you use your ATM card to withdraw cash from an automated teller machine, the system automatically computes your account balance then and there. Airline reservation systems also use online processing.

Offline Versus Online Storage

Whether it's on magnetic tape or on some form of disk, data may be stored either offline or online.

- Offline: ***Offline storage* means that data is not directly accessible for processing until the tape or disk it's on has been loaded onto an input device.** That is, the storage is not under the direct, immediate control of the central processing unit.

- Online: ***Online storage* means that stored data is directly accessible for processing.** That is, storage is under the direct, immediate control of the central processing unit. You need not wait for a tape or disk to be loaded onto an input device.

  For processing to be online, the storage must be online and *fast*. This nearly always means storage on disk rather than magnetic tape. With magnetic tape, it is not possible to go directly to the required record; instead, the read/write head has to search through all the records that precede it, which takes time. This is known as *sequential storage*, as we discussed in Chapter 6 (✔ p. 265). With disk, however, the system can go directly and quickly to the record—just as a CD player can go directly to a particular spot on a music CD. This is known as *direct access storage*, as we discussed earlier (✔ p. 266).

9.4 File Management Systems

KEY QUESTIONS

What is a file management system, and what are its disadvantages?

Preview & Review: Files may be retrieved through a file management system, one file at a time. Disadvantages of a file management system are data redundancy, lack of data integrity, and lack of program independence.

In the 1950s, when commercial use of computers was just beginning, magnetic tape was the storage medium and records and files were stored sequentially. To work with these files, a user needed a file management system.

A *file management system,* or *file manager,* **is software for creating, retrieving, and manipulating files, one file at a time.** Traditionally, a large organization such as a university would have different files for different purposes. For you as a student, for example, there might be one file on you for course grades, another for student records, and a third for tuition billing. Each file would be used independently to produce its own separate reports. If you changed your address, someone had to make the change separately in each file. (■ *See Panel 9.4.*)

Disadvantages of File Management Systems

File management systems worked well enough for the time, but they had several disadvantages:

- **Data redundancy:** *Data redundancy* means that the same data fields appear in many different files and often in different formats. Thus, separate files tend to repeat some of the same data over and over. A student's course grades file and tuition billing file would both contain similar data (name, address, telephone number). When data fields are repeated in different files, they waste storage space.

- **Lack of data integrity:** *Data integrity* means that data is accurate, consistent, and up to date. However, when the same data fields (a student's address and phone number, for example) must be changed in different files, some files may be missed or mistakes in some files may go unnoticed. The result is that some reports will be produced with erroneous information.

- **Lack of program independence:** With file management systems, different files were often written by different programmers using different file formats. Thus, the files were not *program-independent.*

■ PANEL 9.4

File management system

In the traditional file management system, some of the same data elements, such as addresses, were repeated in different files. Information was not shared among files.

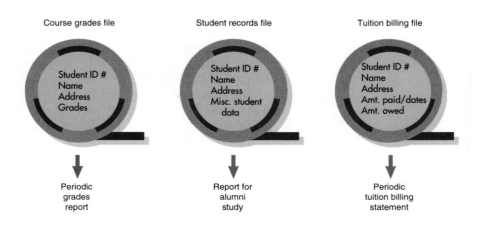

Course grades file	Student records file	Tuition billing file
Student ID # Name Address Grades	Student ID # Name Address Misc. student data	Student ID # Name Address Amt. paid/dates Amt. owed
↓	↓	↓
Periodic grades report	Report for alumni study	Periodic tuition billing statement

The arrangement meant more time was required to maintain files. It also prevented a programmer from writing a single program that would access all the data in multiple files.

As computers became more and more important in daily life, the frustrations of working with separate, redundant files lacking data integrity and program independence began to be overwhelming. Fortunately, magnetic disk began to supplant magnetic tape as the most popular medium of secondary storage, leading to new possibilities for managing data.

9.5 Database Management Systems

KEY QUESTIONS

What is a database management system, and what are its advantages and disadvantages?

Preview & Review: Database management systems are an improvement over file management systems. They use database management system (DBMS) software, which controls the structure of a database and access to the data.

The advantages of databases are reduced data redundancy, improved data integrity, more program independence, increased user productivity, and increased security. However, installing and maintaining a database management system can be expensive.

When magnetic tape began to be replaced by magnetic disk, sequential access storage then began to be replaced by direct access storage. The result was a new technology and new software: the database management system.

As mentioned, a *database* is a collection of integrated files, meaning that the file records are logically related, or cross-referenced, to one another. Thus, even though all the pieces of data on a topic are kept in records in different files, they can easily be organized and retrieved with simple requests.

The software for manipulating databases is ***database management system (DBMS) software,*** or a ***database manager,*** **a program that controls the structure of a database and access to the data.** With a DBMS, then, a large organization such as a university might still have different files for different purposes. As a student, you might have had the same files as you would have had in a file management system (one for course grades, another for student records, and a third for tuition billing). However, in the database management system, data elements are integrated (cross-referenced) and shared among different files. (■ *See Panel 9.5.*) Thus, your address data would need to be in only one file because it can be automatically accessed by the other files.

■ **PANEL 9.5**

Database management system

In the database management system, data elements are integrated and shared among different files. Information updated in one file will automatically be updated in other files.

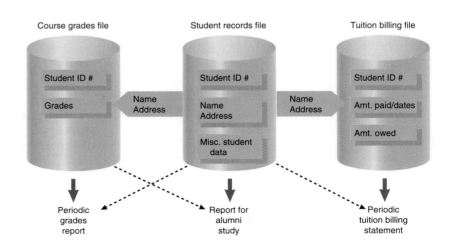

Advantages of a DBMS

The advantages of databases and DBMS software are as follows:

- **Reduced data redundancy:** Instead of the same data fields being repeated in different files, in a database the information appears just once. The single biggest advantage of a database is that the *same* information is available to *different* users. Moreover, reduced redundancy lowers the expense of storage media and hardware because more data can be stored on the media.

- **Improved data integrity:** Reduced redundancy increases the chances of data integrity—that the data is accurate, consistent, and up to date—because each updating change is made in only one place.

- **More program independence:** With a database management system, the program and the file formats are the same, so that one programmer or even several programmers can spend less time maintaining files.

- **Increased user productivity:** Database management systems are fairly easy to use, so that users can get their requests for information answered without having to resort to technical manipulations. In addition, users don't have to wait for a computer professional to provide what they need.

- **Increased security:** Although various departments may share data in common, access to specific information can be limited to selected users. Thus, through the use of passwords, a student's financial, medical, and grade information in a university database is made available only to those who have a legitimate need to know.

Disadvantages of a DBMS

Although there are clear advantages to having databases, there are still some disadvantages:

- **Cost issues:** Installing and maintaining a database is expensive, particularly in a large organization. In addition, there are costs associated with training people to use it correctly.

- **Security issues:** Although databases can be structured to restrict access, it's always possible unauthorized users will get past the safeguards. And when they do, they may have access to *all* the files, not just a few. In addition, if a database is destroyed by fire, earthquake, theft, or hardware or software problems, it could be fatal to an organization's business activities—unless steps have been taken to regularly make backup copies of the files and store them elsewhere. (Indeed, if *one* file in a database is destroyed, the entire database could be rendered useless.)

- **Privacy issues:** Databases may hold information they should not and be used for unintended purposes, perhaps intruding on people's privacy. Medical data, for instance, may be used inappropriately in evaluating an employee for a job promotion. Privacy and other ethical issues are discussed later in this chapter.

9.6   Types of Database Organization

Preview & Review: Types of database organization are hierarchical, network, relational, and object-oriented.

KEY QUESTIONS

What are four types of database organization, and how do they work?

Just as files can be organized in different ways (sequentially or directly, for example), so can databases. The four most common arrangements for database management systems are *hierarchical, network, relational,* and *object-oriented.* For installation and maintenance, each of the four types of database requires a database administrator trained in its structure.

Hierarchical Database

In a *hierarchical database,* fields or records are arranged in related groups resembling a family tree, with lower-level records subordinate to higher-level records. (■ *See Panel 9.6.*) A lower-level record is called a *child,* and a higher-level record is called a *parent.* The parent record at the top of the database is called the *root record.*

Unlike families in real life, a parent in a hierarchical database may have more than one child, but a child always has only one parent. This is called a one-to-many relationship. To find a particular record, you have to start at the top with a parent and trace down the chart to the child.

Used principally on mainframes, hierarchical DBMSs are the oldest of the four forms of database organization, but they are still used in some types of passenger reservation systems. Also, accessing or updating data is very fast because the relationships have been predefined. However, because the structure must be defined in advance, it is quite rigid. There may be only one parent per child and no relationships among the child records. Moreover, adding new fields to database records requires that the entire database be redefined.

■ PANEL 9.6

Hierarchical database: example of a cruise ship reservation system
Records are arranged in related groups resembling a family tree, with "child" records subordinate to "parent" records. Cabin numbers (A-1, A-2, A-3) are children of the parent July 15. Sailing dates (April 15, May 30, July 15) are children of the parent The Love Boat. The parent at the top, Miami, is called the "root parent."

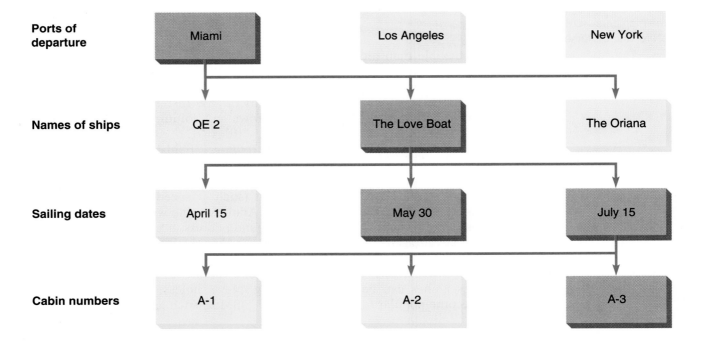

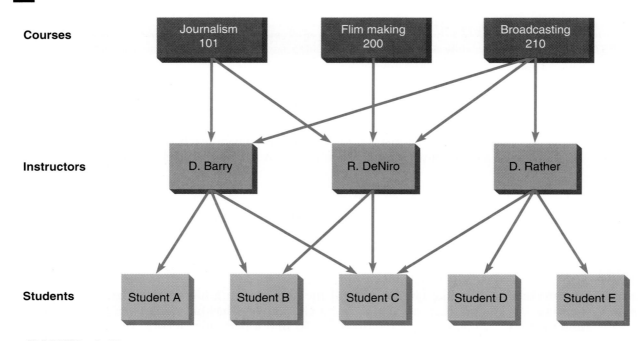

Courses — Journalism 101, Flim making 200, Broadcasting 210

Instructors — D. Barry, R. DeNiro, D. Rather

Students — Student A, Student B, Student C, Student D, Student E

■ PANEL 9.7

Network database: example of a college class scheduling system

This is similar to a hierarchical database, but each child, or "member," record can have more than one parent, or "owner." For example, Student B's owners are instructors D. Barry and R. DeNiro. The owner Broadcasting 210 has three members—D. Barry, R. DeNiro, and D. Rather.

Network Database

A *network database* is similar to a hierarchical DBMS, but each child record can have more than one parent record. (■ *See Panel 9.7.*) Thus, a child record, which in network database terminology is called a *member,* may be reached through more than one parent, which is called an *owner.*

Also used principally with mainframes, the network database is more flexible than the hierarchical one because different relationships may be established between different branches of data. However, it still requires that the structure be defined in advance. Moreover, there are limits to the number of links that can be made among records.

Relational Database

More flexible than hierarchical and network database models, **the *relational database* relates, or connects, data in different files through the use of a key field, or common data element.** (■ *See Panel 9.8.*) In this arrangement there are no access paths down through a hierarchy. Instead, data elements are stored in different tables made up of rows and columns. In database terminology, the tables are called *relations* (files), the rows are called *tuples* (records), and the columns are called *attributes* (fields).

All related tables must have a key field that uniquely identifies each row. Thus, one table might have a row consisting of a driver's license number as the key field and any traffic violations (such as speeding) attributed to the license holder (see row 3 in Panel 9.8). Another table would have the driver's license number and the bearer's name and address (see row 1).

The advantage of relational databases is that the user does not have to be aware of any "structure." Thus, they can be used with little training. Moreover, entries can easily be added, deleted, or modified. A disadvantage is that some searches can be time-consuming. Nevertheless, the relational model has become popular for microcomputer DBMSs, such as Paradox and Access.

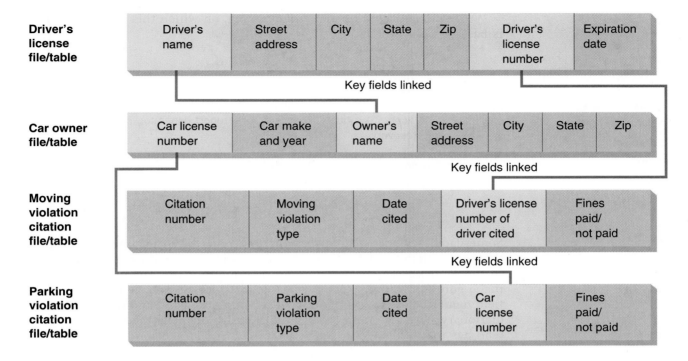

Driver's license file/table

| Driver's name | Street address | City | State | Zip | Driver's license number | Expiration date |

Key fields linked

Car owner file/table

| Car license number | Car make and year | Owner's name | Street address | City | State | Zip |

Key fields linked

Moving violation citation file/table

| Citation number | Moving violation type | Date cited | Driver's license number of driver cited | Fines paid/ not paid |

Key fields linked

Parking violation citation file/table

| Citation number | Parking violation type | Date cited | Car license number | Fines paid/ not paid |

■ PANEL 9.8

Relational database: example of a state department of motor vehicles database

This kind of database relates, or connects, data in different files through the use of a key field, or common data element. The relational database does not require predefined relationships.

Object-Oriented Database

The previous three types of databases deal with *structured data*—that is, data that can be neatly classified into fields, rows, and columns. Object-oriented databases can handle new data types, including graphics, audio, and video, which can be combined with text into a multimedia format. Object-oriented databases, then, are important in businesses related to technological convergence and storage of data in multimedia form.

An *object-oriented database system (OODBMS)* uses "objects," software written in small, reusable chunks, as elements within database files. An *object* consists of (1) data in the form of text, sound, video, and pictures and (2) instructions on the action to be taken on the data. A hierarchical or network database would contain only numeric and text data about a student—identification number, name, address, and so on. By contrast, an object-oriented database might also contain the student's photograph, a "sound bite" of his or her voice, and even a short piece of video. Moreover, the object would store operations, called *methods*, programs that objects use to process themselves—for example, how to calculate the student's grade-point average and how to display or print the student's record. Objects interact by sending messages to each other. (Some relational databases can handle sound, graphics, and video as well as text, but the components do not include processing instructions.)

Traditional businesses often don't use OODBMS because they are expensive to develop, and most organizations are unwilling to convert the millions or billions of bytes they already have in existing relational databases.[12] However, because OODBMS allows the reusable chunks of software to seamlessly mix and match images, sound, video, text, and numbers, users can gain great benefits from OODBMS multimedia capabilities.[13] In medicine, for example, such databases could be used to hold images as well as text: X-rays, CAT scans, MRI scans, and electrocardiograms. In engineering, they could contain blueprints, sketches, diagrams, photos, and illustrations. In education, they could include video clips demonstrating how things work. In geographic applications, they could hold maps and aerial and satellite photos.

We describe object-oriented programming, on which this form of database is based, in Chapter 11.

9.7 Features of a DBMS

KEY QUESTION

What are the features of a database management system?

Preview & Review: Features of a database management system include (1) a data dictionary, (2) utilities, (3) a query language, (4) a report generator, (5) access security, and (6) system recovery.

A database management system may have a number of components, including the following. (■ *See Panel 9.9.*)

Data Dictionary

Some databases have a ***data dictionary*, a procedures document or disk file that stores the data definitions or a description of the structure of data used in the database.** The data dictionary may monitor the data being entered to make sure it conforms to the rules defined during data definition, such as field name, field size, type of data (text, numeric, date, and so on). The data dictionary may also help protect the security of the database by indicating who has the right to gain access to it.

Utilities

The *DBMS utilities* are programs that allow you to maintain the database by creating, editing, and deleting data, records, and files. The utilities allow people to monitor the types of data being input and to adjust display screens for data input, for example.

Query Language

Also known as a *data manipulation language,* a ***query language* is an easy-to-use computer language for making queries to a database and for retrieving selected records,** based on the particular criteria and format indicated. Typically, the query is in the form of a sentence or near-English command, using such basic words as SELECT, DELETE, or MODIFY. There are several different query languages, each with its own vocabulary and procedures. One

■ PANEL 9.9

Some important features of a database management system

Component	Description
Data dictionary	Describes files and fields of data
Utilities	Help maintain the database by creating, editing, and monitoring data input
Query language	Enables users to make queries to a database and retrieve selected records
Report generator	Enables nonexperts to create readable, attractive on-screen or hardcopy reports of records
Access security	Specifies user access privileges
Data recovery	Enables contents of database to be recovered after system failure

of the most popular is *Structured Query Language,* or *SQL.* An example of an SQL query is as follows:

```
SELECT PRODUCT-NUMBER, PRODUCT-NAME
FROM PRODUCT
WHERE PRICE < 100.00
```

This query selects all records in the product file for products that cost less than $100.00 and displays the selected records according to product number and name—for example:

```
A-34    Mirror
C-50    Chair
D-168   Table
```

One feature of most query languages is *query by example.* Often a user will seek information in a database by describing a procedure for finding it. However, in **query by example (QBE), the user asks for information in a database by using a sample record to define the qualifications he or she wants for selected records.**

For example, a university's database of student-loan records of its students all over the United States and the amounts they owe might have the column headings (field names) NAME, ADDRESS, CITY, STATE, ZIP, AMOUNT OWED. When you use the QBE method, the database would display an empty record with these column headings. You would then type in the search conditions that you want in the appropriate columns.

Thus, if you wanted to find all Beverly Hills, California, students with a loan balance due of $3000 or more, you would type *BEVERLY HILLS* under the CITY column, *CA* under the STATE column, and *>=3000* ("greater than or equal to $3000") in the AMOUNT OWED column.

Some DBMSs, such as Symantec's Q&A, use natural language interfaces, which allow users to make queries in any spoken language, such as English. With this software, you could ask your questions—either typing or speaking (if the system has voice recognition)—in a natural way, such as "How many sales reps sold more than 1 million dollars worth of books in the Western Region in January?"

Report Generator

A *report generator* is a program users may employ to produce an on-screen or printed document from all or part of a database. You can specify the format of the report in advance—row headings, column headings, page headers, and so on. With a report generator, even nonexperts can create attractive, readable reports on short notice.

Access Security

At one point in *Disclosure,* the Michael Douglas/Demi Moore movie, Douglas's character, the beleaguered division head suddenly at odds with his company, types SHOW PRIVILEGES into his desktop computer, which is tied to the corporate network. To his consternation, the system responds by showing him downgraded from PRIOR USER LEVEL: 5 to CURRENT USER LEVEL: 0, shutting him out of files to which he formerly had access.

This is an example of the use of *access security,* a feature allowing database administrators to specify different access privileges for different users of a DBMS. For instance, one kind of user might be allowed only to retrieve (view) data whereas another might have the right to update data and delete records. The purpose of this security feature, of course, is to protect the database from unauthorized access and sabotage.

Physical security is also important, and one of the most effective strategies is to simply *isolate* a database system to protect it from threats. For example, backup copies of databases on removable magnetic disks could be stored in a guarded vault, with authorized employees admitted only by producing a badge with their encoded personal voice prints.[14]

System Recovery

Database management systems should have *system recovery* features that enable the database administrator to recover contents of the database in the event of a hardware or software failure. For instance, the feature may recover transactions that appear to have been lost since the last time the system was backed up.

Performing a recovery may be difficult because it is often impossible to just fix the problem and resume processing where it was interrupted. Even if no data was lost during the failure—an unrealistic assumption, usually, because some computer memories will be volatile (✔ p. 152)—the timing and scheduling of computer processing are too complex to be accurately re-created.

Four approaches are possible for system recovery—*mirroring, reprocessing, rollforward,* and *rollback*:[15]

- **Mirroring—two copies in different locations:** In database *mirroring,* frequent simultaneous copying of the database is done to maintain two or more complete copies of the database online but in different locations. Mirroring is an expensive strategy, but it is necessary when recovery is needed quickly—say, in seconds or minutes—as with an airline reservation system's database.

- **Reprocessing—redoing the processing from a known past point:** In *reprocessing,* the database administrator goes back to a known point of database activity before the failure and reprocesses the workload from there.

  To make reprocessing an available option, periodic database copies (called *database saves*) must be made. Also, records must be kept of all the transactions made since each save. When there is a failure, the database can be restored from the save. Then all the transactions made since that save are re-entered and reprocessed.

  This type of recovery can be time-consuming, and the processing of new transactions must be delayed until the database recovery is completed.

- **Rollforward—a variant on reprocessing:** *Rollforward,* also called *forward recovery,* is somewhat similar to reprocessing. Here, too, the current database is re-created, using a previous database state. However, in this case, transactions made since the last save are not re-entered and then reprocessed all over again. Rather, the lost data is recovered using a more sophisticated version of a transaction log that contains what are called *after-image records* and that includes some processing information.

- **Rollback—undoing unwanted changes:** *Rollback,* or *backward recovery,* is used to *undo* unwanted changes to the database. This is done, for example, when some failure interrupts a half-completed transaction.

9.8 More on Database Management: Data Mining, Data Warehouses, & Data "Siftware"

KEY QUESTIONS

What is data mining, and how does it work?

Preview & Review: Data mining is the computer-assisted process of sifting through and analyzing vast amounts of data in order to extract meaning and discover new knowledge.

Data taken from many sources is "scrubbed" or cleaned of errors and checked for consistency of formats. The cleaned-up data and a variation called meta-data are then sent to a special database called a data warehouse.

Three kinds of software, or "siftware," tools are used to perform data mining: query-and-reporting tools, multidimensional-analysis tools, and intelligent agents.

A personal database is usually small-scale. But some efforts are going on with databases that are almost unimaginably large-scale, involving records of millions of households and thousands of terabytes of data. Some of these activities require the use of so-called *massively parallel database computers* (✔ p. 164) costing $1 million or more. "These machines gang together scores or even hundreds of the fastest microprocessors around," says one description, "giving them the oomph to respond in minutes to complex database queries."[16]

The efforts of which we speak go under the name *data mining.* Let us see what this is.

Data Mining: What It Is, What It's Used For

Data mining (DM), also called "knowledge discovery," is the computer-assisted process of sifting through and analyzing vast amounts of data in order to extract meaning and discover new knowledge. The purpose of DM is to describe past trends and predict future trends.[17] Thus, data-mining tools might sift through a company's immense collections of customer, marketing, production, and financial data and identify what's worth noting and what's not.[18]

Although the concept seems simple enough, its effect has been to overwhelm traditional query-and-report methods of organizing and analyzing data, such as those previously described in this chapter. The result has been the need for "data warehouses" and for new software tools, as we shall discuss.

Data mining has come about because companies find that in today's fierce competitive business environment, they need to turn the gazillions of bytes of raw data at their disposal to new uses for further profitability. However, nonprofit institutions have also found DM methods useful, as in the pursuit of scientific and medical discoveries. For example:[19]

- **Marketing:** Marketers use DM tools (such as one called Spotlight) to mine point-of-sale databases of retail stores, which contain facts (such as prices, quantities sold, dates of sale) for thousands of products in hundreds of geographic areas. By understanding customer preferences and buying patterns, marketers hope to target consumers' individual needs.

- **Health:** A coach in the U.S. Gymnastics Federation is using a DM system (called IDIS) to discover what long-term factors contribute to an athlete's performance, so as to know what problems to treat early on. A Los Angeles hospital is using the same tool to see what subtle factors affect success and failure in back surgery. Another system helps health-care organizations pinpoint groups whose costs are likely to increase in the near future, so that medical interventions can be taken.

- Science: DM techniques are being employed to find new patterns in genetic data, molecular structures, global climate changes, and more. For instance, one DM tool (called SKICAT) is being used to catalog more than 50 million galaxies, which will be reduced to a 3-terabye galaxy catalog.

Clearly, short-term payoffs can be dramatic. One telephone company, for instance, mined its existing billing data to identify 10,000 supposedly "residential" customers who spent more than $1000 a month on their phone bills. When it looked more closely, the company found these customers were really small businesses trying to avoid paying the more expensive business rates for their telephone service.[20]

However, the payoffs in the long term could be truly astonishing. Sifting medical-research data or subatomic-particle information may reveal new treatments for diseases or new insights into the nature of the universe.[21]

Preparing Data for the Data Warehouse

Data mining begins with acquiring data and preparing it for what is known as the "data warehouse." (■ *See Panel 9.10.*) This takes the following steps.[22]

1. Data sources: Data may come from a number of sources: (1) point-of-sale transactions in files (flat files) managed by file management systems on mainframes, (2) databases of all kinds, and (3) other—for example, news articles transmitted over newswires or online sources such as the Internet. To the mix may also be added (4) data from data warehouses, as we describe.

2. Data fusion and cleansing: Data from diverse sources, whether from inside the company (internal data) or purchased from outside the company (external data), must be fused together, then put through a process known as *data cleansing*, or *scrubbing*. Even if the data comes from just one source, such as one company's mainframe, the data may be of poor quality, full of errors and inconsistencies. Therefore, for data mining to produce accurate results, the source data has to be "scrubbed"—that is, cleaned of errors and checked for consistency of formats.

3. Data and meta-data: Out of the cleansing process come both the cleaned-up data and a variation of it called *meta-data*. Meta-data shows the origins of the data, the transformations it has undergone, and summary information about it, which makes it more useful than the cleansed but unintegrated, unsummarized data. The meta-data also describes the contents of the data warehouse.

4. The data warehouse: Both the data and the meta-data are sent to the data warehouse. **A *data warehouse* is a special database of cleaned-up data and meta-data.** It is a replica, or close reproduction, of a mainframe's data. The data warehouse is stored on disk using storage technology such as RAID (redundant array of independent disks, ✔ p. 276). Small data warehouses may hold 100 gigabytes of data or less. Once 500 gigabytes are reached, massively parallel processing computers are needed. Projections call for large data warehouses holding hundreds of terabytes within the next few years.

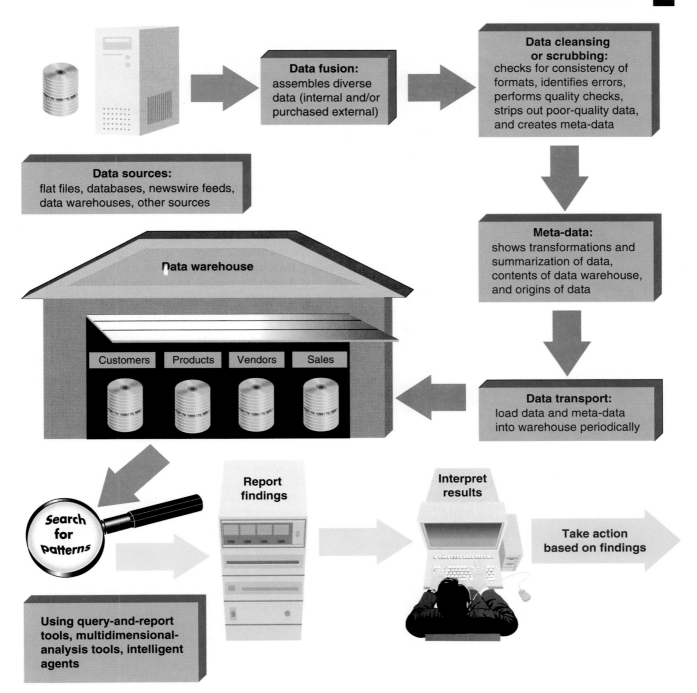

Data fusion: assembles diverse data (internal and/or purchased external)

Data cleansing or scrubbing: checks for consistency of formats, identifies errors, performs quality checks, strips out poor-quality data, and creates meta-data

Data sources: flat files, databases, newswire feeds, data warehouses, other sources

Data warehouse

Customers   Products   Vendors   Sales

Meta-data: shows transformations and summarization of data, contents of data warehouse, and origins of data

Data transport: load data and meta-data into warehouse periodically

Search for Patterns

Report findings

Interpret results

Take action based on findings

Using query-and-report tools, multidimensional-analysis tools, intelligent agents

■ PANEL 9.10

The data-mining process

"Siftware" for Finding & Analyzing

Three kinds of software, or "siftware," tools are used to perform data mining—that is, to do finding and analyzing tasks. They are *query-and-reporting tools, multidimensional-analysis (MDA) tools,* and *intelligent agents.*[23]

- Query-and-reporting tools: Query-and-reporting tools (examples are Focus Reporter and Esperant) require a database structure and work well with relational databases. They may have graphical interfaces (✔ p. 108). Their best use is for specific questions to verify hypotheses. For example, if a company decides to mine its database to find customers most likely to respond to a mail-order promotion, it might use a query-and-reporting tool and construct a query (using SQL): "How many credit-card customers who made purchases of over $100 on sporting goods in August have at least $2000 of available credit?"[24]

- **Multidimensional-analysis tools:** Multidimensional-analysis (MDA) tools (examples are Essbase and Lightship) can do "data surfing" to explore all dimensions of a particular subset of data. In one writer's example, "The idea [with MDA] is to load a multidimensional server with data that is likely to be combined. Imagine all the possible ways of analyzing clothing sales: by brand name, size, color, location, advertising, and so on."[25] Using MDA tools, you can analyze this multidimensional database from all points of view.

- **Intelligent agents:** An intelligent agent is a computer program that roams through networks performing complex work tasks for people. There are several kinds of intelligent agents (as we explore in Chapter 12), such as those used to prioritize e-mail messages for individuals. However, the kind we are concerned with here (such as DataEngine and Data/Logic) are those used as data-mining tools.

  Intelligent agents are best used for turning up unsuspected relationships and patterns. "These patterns may be so nonobvious as to appear almost nonsensical," says one writer, "such as that people who have bought scuba gear are good candidates for taking Australian vacations."[26]

Some Pitfalls in Data Mining

Data mining is not infallible. Indeed, it can lead to costly misinterpretations. For instance, it was once common for medical researchers to mine health records looking for parts of the country—"hot spots"—with higher than average cancer rates. Though the findings were vastly upsetting to residents, often the higher cancer rates were simply a statistical fluke.

Business Week economics editor Peter Coy points out that there are four pitfalls in data mining:[27]

- **Some oddities are pure chance:** A formula that happens to fit the data of the past won't necessarily predict the future. Some odd correlations of data are simply the result of pure chance.

- **Evidence can be found to support any preconception:** Says Andrew Lo, a finance professor at Massachusetts Institute of Technology: "Given enough time, enough attempts, and enough imagination, almost any pattern can be teased out of any data set."[28]

- **Too many factors can produce invalid results:** The more factors or variables the data-mining analysis tools consider, the more likely the program will find relationships, valid or not.

- **Explanations should be plausible:** "A finding makes more sense if there's a plausible theory for it," says Coy. "But a beguiling story can disguise weaknesses in the data."

Is data about you finding its way into data warehouses? No doubt it is. Gathering data isn't difficult. You participate in probably hundreds of trans-actions a year, recorded in point-of-sale terminals (✔ p. 235), teller machines, credit-card files, and 1-800 telemarketing responses. Sooner or later, some of the records of your past activities will be used, most likely by marketing companies, to try to influence you.

9.9 The Ethics of Using Databases: Concerns About Accuracy & Privacy

KEY QUESTION

What are some ethical concerns about the uses of databases?

Preview & Review: Databases may contain inaccuracies or be incomplete. They also may endanger privacy—in the areas of finances, health, employment, and commerce. There may be a danger of certain information becoming monopolized that perhaps should not be.

Ethics

"The corrections move by bicycle while the stories move at the speed of light," says Richard Lamm, a former governor of Colorado.

Lamm was lamenting that he was quoted out of context by a Denver newspaper in a speech he made in 1984. Yet several years afterward—long after the paper had run a correction—he still saw the error repeated in later newspaper articles.[29]

How do such mistakes get perpetuated? The answer, suggests journalist Christopher Feola, is the Misinformation Explosion. "Fueled by the growing popularity of both commercial and in-house computerized news databases," he says, "journalists have found it that much easier to repeat errors or rely on the same tired anecdotes and experts."[30]

If news reporters—who are supposed to be trained in careful handling of the facts—can continue to repeat inaccuracies found in databases, what about those without training who have access to computerized facts? How can you be sure that databases with essential information about you—medical, credit, school, employment, and so on—are accurate and, equally important, are secure in guarding your privacy? We examine the topics of *accuracy and completeness* and of *privacy* in this section.

Matters of Accuracy & Completeness

Databases—including public databases such as Nexis/Lexis, Dialog, and Dow Jones News/Retrieval—can provide you with *more* facts and *faster* facts but not always *better* facts. Penny Williams, professor of broadcast journalism at Buffalo State College in New York and formerly a television anchor and reporter, suggests there are five limitations to bear in mind when using databases for research:[31]

- **You can't get the whole story:** For some purposes, databases are only a foot in the door. There may be many facts or aspects to the topic you are looking into that are not in a database. Reporters, for instance, find a database is a starting point, but it may take old-fashioned shoe leather to get the rest of the story.

- **It's not the gospel:** Just because you see something on a computer screen doesn't mean it's accurate. Numbers, names, and facts may need to be verified in other ways.

- **Know the boundaries:** One database service doesn't have it all. For example, you can find full text articles from the *New York Times* on Lexis/Nexis, from the *Wall Street Journal* on Dow Jones News/ Retrieval, and from the *San Jose Mercury News* on America Online, but no service carries all three.

- **Find the right words:** You have to know which keywords (search words) to use when searching a database for a topic. As Lynn Davis, a professional researcher with ABC News, points out, in searching for stories on guns, the keyword "can be guns, it can be firearms, it can be handguns, it can be pistols, it can be assault weapons. If you don't cover your bases, you might miss something."[32]

• **History is limited:** Most public databases, Davis says, have information going back to 1980, and a few into the 1970s, but this poses problems if you're trying to research something that happened or was written about earlier.

Matters of Privacy

Privacy **is the right of people to not reveal information about themselves.** Who you vote for in a voting booth and what you say in a letter sent through the U.S. mail are private matters. However, the ease with which databases and communications lines may pull together and disseminate information has put privacy under extreme pressure.

As you've no doubt discovered, it's no trick at all to get your name on all kinds of mailing lists. Theo Theoklitas, for instance, has received applications for credit cards, invitations to join video clubs, and notification of his finalist status in Ed McMahon's $10 million sweepstakes. Theo is a 6-year-old black cat who's been getting mail ever since his owner sent in an application for a rebate on cat food.[33] A whole industry has grown up of professional information gatherers and sellers, who collect personal data and sell it to fund-raisers, direct marketers, and others.

How easy is it to find out information about you or anyone else? Kimberly Dorcik, 26, of Fairfield, California, was turned down for a job with a women's clothing store because the store checked her credit history and found $30,000 in disputed medical bills.[34] Rosana Rivera, a 23-year-old mother, who was a party in a class-action lawsuit against Texaco in conjunction with a refinery explosion near her neighborhood, was frightened to learn that a private investigator hired by the oil company had generated a five-page computer printout about her from just her name alone.[35] A California man, obsessed with a woman he had once known, was able to hatch intricate schemes of harassment—from within a maximum-security prison. He filed post office change-of-address forms so her mail was forwarded to him in prison and obtained a credit report on her. He even sent the IRS forged power-of-attorney forms so he could get her tax returns.[36]

In the 1970s, the Department of Health, Education, and Welfare developed a set of five Fair Information Practices. These rules have since been adopted by a number of public and private organizations. The practices also led to the enactment of a number of laws to protect individuals from invasion of privacy. (■ *See Panel 9.11.*) Perhaps the most important law is the Federal Privacy Act, or Privacy Act of 1974. **The *Privacy Act of 1974* prohibits secret personnel files from being kept on individuals by government agencies or their contractors. It gives individuals the right to see their records, to see how the data is used, and to correct errors.**

Another significant piece of legislation was the Freedom of Information Act, passed in 1970. **The *Freedom of Information Act* allows ordinary citizens to have access to data gathered about them by federal agencies.** Most privacy laws regulate only the behavior of government agencies or government contractors. For example, the *Computer Matching and Privacy Protection Act* of 1988 prevents the government from comparing certain records to try to find a match. This law does not affect most private companies.

Hoping to head off federal regulation they fear might hamstring online commerce, in 1997 major providers of personal information, including Lexis/Nexis, announced the database industry's first set of privacy standards.[37–39] Among other things, they agreed to release private data only to so-called qualified subscribers who promise to use the information appropriately. However, privacy activists remained unimpressed.

Fair Information Practices

1. There must be no personal data record-keeping systems whose existence is a secret from the general public.
2. People have the right to access, inspect, review, and amend data about them that is kept in an information system.
3. There must be no use of personal information for purposes other than those for which it was gathered without prior consent.
4. Managers of systems are responsible and should be held accountable and liable for the reliability and security of the systems under their control, as well as for any damage done by those systems.
5. Governments have the right to intervene in the information relationships among private parties to protect the privacy of individuals.

Important Federal Privacy Laws

Freedom of Information Act (1970): Gives you the right to look at data concerning you that is stored by the federal government. A drawback is that sometimes a lawsuit is necessary to pry it loose.

Fair Credit Reporting Act (1970): Bars credit agencies from sharing credit information with anyone but authorized customers. Gives you the right to review and correct your records and to be notified of credit investigations for insurance or employment. A drawback is that credit agencies may share information with anyone they reasonably believe has a "legitimate business need." Legitimate is not defined.

Privacy Act (1974): Prohibits federal information collected about you for one purpose from being used for a different purpose. Allows you the right to inspect and correct records. A drawback is that exceptions written into the law allow federal agencies to share information anyway.

Family Educational Rights and Privacy Act (1974): Gives students and their parents the right to review, and to challenge and correct, students' school and college records; limits sharing of information in these records.

Right to Financial Privacy Act (1978): Sets strict procedures that federal agencies must follow when seeking to examine customer records in banks; regulates financial industry's use of personal financial records. A drawback is that the law does not cover state and local governments.

Privacy Protection Act (1980): Prohibits agents of federal government from making unannounced searches of press offices if no one there is suspected of a crime.

Cable Communications Policy Act (1984): Restricts cable companies in the collection and sharing of information about their customers.

Computer Fraud and Abuse Act (1986): Allows prosecution for unauthorized access to computers and databases. A drawback is that people with legitimate access can still get into computer systems and create mischief without penalty.

Electronic Communications Privacy Act (1986): Makes eavesdropping on private conversations illegal without a court order.

Computer Security Act (1987): Makes actions that affect the security of computer files and telecommunications illegal.

Computer Matching and Privacy Protection Act (1988): Regulates computer matching of federal data; allows individuals a chance to respond before government takes adverse actions against them. A drawback is that many possible computer matches are not affected, such as those done for law-enforcement or tax reasons.

Video Privacy Protection Act (1988): Prevents retailers from disclosing video-rental records without the customer's consent or a court order.

■ PANEL 9.11

The five Fair Information Practices and important federal privacy laws
The Fair Information Practices were developed by the U.S. Department of Health, Education, and Welfare in the early 1970s. They have been adopted by many public and private organizations since.

Of particular concern for privacy are the areas of finances, health, employment, commerce, and communications:

- **Finances:** Banking and credit are two private industries for which there are federal privacy laws on the books. **The *Fair Credit Reporting Act of 1971* allows you to have access to and gives you the right to challenge your credit records.** If you have been denied credit, this access must be given to you free of charge. **The *Right to Financial Privacy Act* of 1978 sets restrictions on federal agencies that want to search customer records in banks.**

  In the past, credit bureaus have been severely criticized for disseminating errors and for having reports that were difficult for ordinary readers to understand. Recent changes to the Credit Reporting Act hold providers of consumer credit information, such as department stores and mortgage companies, liable for the accuracy of the information they provide. The changes also allow you to correct errors on your credit report, and you can do so via a toll-free number. (See the Experience Box at the end of this chapter.) The three major credit bureaus in the United States are Experian (formerly TRW), Equifax, and TransUnion.

- **Health:** No federal laws protect medical records in the United States (except those related to treatment for drug and alcohol abuse and psychiatric care, or records in the custody of the federal government). Of course, insurance companies can get a look at your medical data, but so can others that you might not suspect.

  Getting a divorce or suing an employer for wrongful dismissal? A lawyer might subpoena your medical records in hopes of using, say, a drinking problem or medical care for depression against you. When employers have information about personal health, they often use it in making employment-related decisions, according to one study.[40] (■ *See Panel 9.12.*)

■ PANEL 9.12

Who can see your medical and drug records?
You have the right to look at your own medical records. Here are some others who may legally see them also.

Medical-related personnel
- Your health plan's administrators
- Your doctor
- Your doctor's office staff
- Hospital employees with access to the hospital's computer system
- Pharmacists
- Health information collectors, as when you fill out applications for health or life insurance
- Public health agencies
- The Centers for Disease Control and Prevention

Employer-related personnel
- People in your employer's human resources department
- Your employer, if the company has its own health-benefits plan

Insurance-claims personnel
- Health insurance claims processors
- Insurance claims processors at firms hired by your employer
- Contractors hired by your health plan or employer to review claims and monitor health-care use

Your best strategy is to not routinely fill out medical questionnaires or histories. You should also not tell any business more than it needs to know about your health. You can ask your doctor to release only the minimum amount of information that can be released. Finally, ask for a copy of your medical records if you have doubts about the information your doctor or hospital has on you.[41]

- **Employment:** Until recently, private employers have been the least regulated by privacy legislation. If you apply for a job, for instance, a background-checking service may verify your educational background and employment history. It may also take a look at your credit, driving violations, workers' compensation claims, and criminal record if any. However, recent changes to the Credit Reporting Act now require employers to tell applicants if credit histories are being used as part of hiring and obtain the job seekers' permission to request their credit history. In addition, the employers must tell applicants that they have the right to dispute any inaccurate information.[42]

  Once you're hired, though, it's another story. Nearly two-thirds of employers spy on their employees, according to a 1997 survey by the American Management Association, and up to a quarter of them don't tell their employees about it.[43,44] Spying includes reviewing employees' computer files; recording their voice mail, e-mail, and phone calls; and videotaping workers. The only federal law hindering employer surveillance is the 1986 Electronic Communications Act, which prohibits companies from eavesdropping on spoken personal conversations. However, employers can listen to business phone calls and monitor all nonspoken personal communications.

  Note that e-mail generated on a company's e-mail system becomes property of the company.

- **Commerce:** As we've seen, marketers of all kinds would like to get to know you. For example, Virginia Sullivan, a retired school teacher, every month weighed the junk mail she received. She found after 11 months that she had received about 98 pounds' worth. Sullivan also noticed that the junk mail companies seemed to know personal details of her private life, such as her age and buying habits.

  "We constantly betray secrets about ourselves," says Erik Larson, author of *The Naked Consumer*, "and these secrets are systematically collected by the marketers' intelligence network."[45] Now these secrets turn up in both regular junk mail and in e-mail solicitations. There are, however, a number of things you can do to avoid putting yourself on mailing lists in the first place.

  With few exceptions, the law does not prohibit companies from gathering information about you for one purpose and using it without your permission for another.[46] This information is culled from both public information sources, such as driver's license records, and commercial transactions, such as warranty cards. One exception is the Video Privacy Protection Act of 1988. **The *Video Privacy Protection Act* prevents retailers from disclosing a person's video rental records without their consent or a court order.**

  "Somewhere along the way," Larson points out, "the data keepers made the arbitrary decision that everyone is automatically on their lists unless they ask to be taken off."[47] Why not have a law that keeps consumers off all lists unless they ask to be included? Congress has considered this approach, but lobbyists for direct marketing companies object that it would put them out of business.

README

Practical Matters: Tactics for Protecting Privacy in the Digital Era

We constantly—and often unknowingly—"shed" information about ourselves, says Erik Larson, author of *The Naked Consumer*. That information then is collected and sold to commercial marketers, which is how we end up on mailing lists.

Here are some tips for keeping what's left of your private life private.[48-50]

1. Don't give out your telephone number. "I'm even reluctant to put my phone number on a check, even if I'm asked to do so," says Larson. "It's none of their business."

2. Just say no to telemarketers. Learn to say "I don't take phone solicitations. Please remove my name from your list, as the law requires." (And if you decide to respond to a telemarketer, don't give out a credit-card number; ask to have a bill sent to you.)

3. Be aware that if you do mail-order shopping, those companies sell their customer lists—which will result in more catalogs in your mailbox.

4. Remove your name from direct-mail and telemarketing lists. You can do this by writing to:
Direct Marketing Association
Mail/Telephone Preference Service
P.O. Box 9008 (for direct-mail lists)
or P.O. Box 9014 (for telemarketing lists)
Farmingdale, NY 11735

5. When filling out warranty cards, don't fill out the lifestyle survey. Marketers "rely on the fact that a lot of people assume their warranty won't be valid unless they do," says Larson.

6. Don't give your Social Security number to anyone unless required to do so by federal law. This is difficult to do, because so many organizations now use this as your ID number. But resist when you can.

7. Think twice about giving out information online. Look for Web sites that publish a privacy notice when they request personal information, which means they won't share your name and data without your permission.

Privacy concerns don't stop with the use or misuse of information in databases. As we have seen (Chapter 8), it also extends to privacy in communications. Although the government is constrained by several laws on acquiring and disseminating information, and listening in on private conversations, privacy advocates still worry. In recent times, the government has tried to impose new technologies that would enable law-enforcement agents to gather a wealth of personal information. Proponents have urged that Americans must be willing to give up some personal privacy in exchange for safety and security. We discuss this matter in Chapter 12.

Monopolizing Information

"We want to capture the entire human experience throughout history," says Corbis Corporation chief executive officer Doug Rowan.[51] Corbis was formed in 1989 by software billionaire Bill Gates to acquire digital rights to fine art and photographic images that can be viewed electronically—in everything from electronic books to computerized wall hangings.[52] In 1995 Corbis acquired the Bettmann Archive of 17 million photographs, for scanning into its digital database.[53] Its founder, Dr. Otto Bettmann, called his famous collection a "visual story of the world," and indeed many of the images are unique. They include tintypes of black Civil War soldiers, the 1937 crash of the *Hindenburg* dirigible, John F. Kennedy Jr. saluting the casket of his assassinated father. In 1996, Corbis also gained exclusive rights to use famous nature photographer Ansel Adams's photographs for CD-ROM disks and online distribution.[54,55]

However, when Rowan says Corbis wants to capture all of human experience, he means not just photos and art works from the likes of the National Gallery in London and the State Hermitage Museum in St. Petersburg, Russia, for which Corbis also owns digital imaging rights. "Film, video, audio," he says. "We are interested in those fields too."

Ethics

Are there any ethical problems with one company having in its database the exclusive digital rights to our visual and audio history? Like many museums and libraries (such as the Library of Congress), Corbis joins a trend toward democratizing art and scholarship by converting the images and texts of the past into digital form and making them available to people who could never travel to, say, London or St. Petersburg.

However, when Gates acquired the Bettmann images, for example, the move put their future use "into the hands of an aggressive businessman who, unlike Dr. Bettmann, is planning his own publishing ventures," points out one reporter. "While Mr. Gates's initial plans will make Bettmann images more widely accessible, this savvy competitor now ultimately controls who can use them—and who can't."[56] Adds Paul Saffo, of the nonprofit Institute for the Future, "The cultural issue raised by the Bettmann purchase is whether we're seeing history sold to the highest bidder or we'll eventually see history made more accessible to the public as a result."[57] Curators of art museums are afraid that the rights to art works will slip away for less than they are worth or that the images will be pirated or used in silly ways.[58]

Onward

When Cynthia Schoenbrun was laid off as a research administrator from a computer software company, she found something even better. She used her own personal computer to link up with people she knew in Russia and became part of a new field known as *information brokering.*

What is an information broker? "Part librarian, part private eye, and part computer nerd," one writer explains, "an information broker searches for everything written and published on a given subject, be it an obscure corner of the biomedical market or the whereabouts of a German engineering expert."[59] Schoenbrun, for instance, searches computer databases and her network of contacts to find business and investment information about Russia and other former countries of the Soviet Union. She then sells this information to clients.

The majority of information brokers, who are mainly in one- or two-person firms, are people who have seen a chance to own a business without making a heavy investment. Among other advantages, the profession gives people a lot of flexibility in setting their own hours.

One need not become an information broker, however, to benefit from being able to search a database. Doctors, lawyers, and other professionals are turning to databases in order to keep up with the information explosion within their fields. In the new world of computers and communications, everyone should at least know the rudiments of this skill.

Preventing Your Identity from Getting Stolen

Kathryn Rambo, 28, of Los Gatos, California, got her first clue that she was a victim of "identity theft" when a bank manager telephoned. He said that a person claiming to be her had just left after filling out an account application. Not long afterward, Rambo learned that she had a new $35,000 sports utility vehicle listed in her name, along with five credit cards, a $3000 loan, and even an apartment—none of which she'd asked for.

"I cannot imagine what would be weirder, or would make you angrier, than having someone pretend to be you, steal all this money, and then leave you to clean up all their mess later," said Rambo, a special-events planner.[60] Added to this was the eerie matter of constantly having to prove that she was, in fact, herself: "I was going around saying, 'I am who I am!'"[61]

Identity Theft: Stealing Your Good Name—and More

Theft of identity (TOI) is a crime in which thieves hijack your very name and identity and use your good credit rating to get cash or to buy things. To start, all they need is your full name or Social Security number. Using these, they tap into Internet databases and come up with other information—your address, phone number, employer, driver's license number, mother's maiden name, and so on. Then they're off to the races, applying for credit everywhere.

For instance, someone might walk into a clothing store, fill out an instant-credit application, open an account, and walk out with thousands of dollars in new merchandise—which doesn't cost them a dime but will cause you hours of grief.[62] Often, however, thieves simply mail-order whatever they want. "No face-to-face contact, no weapons, no fingerprints, no nothing," says one almost-victim, who refused to give her Social Security number to a pleasant-voiced telephone caller pretending to need to update her Visa card account. "Just a phone call to J. Crew and that $180 gray cashmere sweater with the ribbed sleeves is hers, my treat."[63]

In Rambo's case, someone had used information lifted from her employee-benefits form. The spending spree went on for months, unbeknownst to her, before she got the tip-off from the bank. The reason it took so long was that Rambo never saw any bills. They went to the address listed by the impersonator, who made a few payments to keep creditors at bay while she ran up even more bills. For Rambo, straightening out the mess required months of frustrating phone calls, time off from work, court appearances, and legal expenses.

How Does Identity Theft Start?

Identity theft typically starts in one of several ways:[64]

- **Wallet or purse theft:** There was a time when a thief would steal a wallet or purse, take the cash, and toss everything else. No more. Everything from keys to credit cards can be parlayed into further thefts.

  This is a good reason, incidentally, why you shouldn't put your ATM or phone calling-card PIN number or other passwords on a slip of paper in your wallet; thieves know to look for these.

  Even picture IDs on credit cards may not be enough of a deterrent, since many retailers never look at them.

- **Mail theft:** Thieves also consider mailboxes fair game. The mail will yield them bank statements, credit-card statements, new checks, tax forms, and other personal information.

  Some thieves have counterfeit keys that allow them to open up apartment-house mailboxes.

- **Mining the trash:** You might think nothing of throwing away credit-card offers, portions of utility bills, or old canceled checks. But "dumpster diving" can produce gold for thieves. Credit-card offers, for instance, may have limits of $5000 or so. (And if you cancel a credit card, thieves may get hold of any replacement card and activate it by using other information about you.)

  Some thieves target trash cans in certain neighborhoods and pick through the garbage looking for valuable documents. Be sure to completely tear up any discarded mail that might be useful in the wrong hands.

- **Telephone solicitation:** Prospective thieves may call you up and pretend to represent a bank, credit-card company, government agency, or the like in an attempt to pry loose essential data about you. (If, however, you initiate the call and have a trusted business relationship with the organization, you're on safer ground.)

- **Insider access to databases:** You never know who has, or could have, access to databases containing your personnel records, credit records, car-loan applications, bank documents, and so on. A co-worker obtained San Franciscan Leonard Dudin's Social Security number and used it to open American Express, Chase Bankcard, Cellular One, and many other accounts, racking up more than $200,000 in bills.[65] This is one of the harder TOI methods to guard against, although there are things you can do, as we will describe.

The Internet can also be used to find out more information about you. An experienced Net user may be able to use your name to learn your Social Security number, for instance. "This, the closest thing to a universal identifier that Americans have, is used by banks and other companies as a method of identification," says one account, "even though the numbers are widely available from public sources. It can be used to unearth other, even more private tidbits, which are also stored and traded among myriad computer databases."[66] If you're a student or faculty member at a college that uses Social Security numbers for IDs, your number may well have migrated into the public record.[67]

When a Criminal Has Your Number: What Happens During Identity Theft

"Information is cash," says Ed Howard, executive director of the Center for Law in the Public Interest in Los Angeles.[68] Once criminals have access to your credit, they can use it for both short-term and long-term thievery:[69]

● **Short-term theft:** Usually within the first week of the theft, the wrongdoers quickly attack your credit-card accounts. They may do this both by taking cash advances and by buying things such as appliances or jewelry that can be quickly converted to cash.

● **Long-term theft:** As happened to Kathryn Rambo, long-term theft can go on for months without your knowing it, particularly if the thieves are diverting bills with your name on them to a new address. (If they rent an apartment, they may not actually live there but use it as a mail drop or place from which to operate other illicit schemes.) The criminals may use credit from accounts they've established to make payments on other accounts. This can continue until your credit has been run through and your credit rating is worthless.

While this is going on, it's possible that you could apply for a loan, or even a job, and be turned down. (Ask the organization to put the reason for the rejection in writing.) Most likely, though, the first you will learn of your identity's being stolen is when collection agencies come after you. Verna Willis, a nurse in Brooklyn, only realized she was a TOI victim when the dunning notices started arriving from Macy's and other new credit-card accounts. "It's making me go crazy," she said, "because I feel so completely helpless."[70]

What to Do Once Theft Happens

If you're the victim of a physical theft (or even loss), as when your wallet is snatched, you should immediately contact—first by phone, and then in writing—all your credit-card companies, other financial institutions, the Department of Motor Vehicles, and any other organization whose cards you use that are now compromised. Be sure to call utility companies—telephone, electricity, and gas; identity thieves

can run up enormous phone bills. Also call the local police and your insurance company to report the loss.

It's important to notify financial institutions *within two days of learning of your loss* because then you are legally responsible for only the first $50 of any theft.

If you become aware of fraudulent transactions taking place, immediately contact the fraud units of the three major credit bureaus: Equifax, Experian, and TransUnion. (Experian was formerly called TRW.) (■ *See Panel 9.13, next page.*)

If your Social Security number has been fraudulently used, alert the Social Security Administration (800-772-1213). It's possible, as a last resort, to have your Social Security number changed.

If you have a check guarantee card that was stolen, if your checks have been lost, or if a new checking account has been opened in your name, there are two organizations to notify so that payment on any fraudulent checks will be denied. They are Telecheck (800-366-2424) and National Processing Company (800-526-5380).

If your mail has been used for fraudulent purposes or if an identity thief filed a change of address form, look in the phone directory under U.S. Government Postal Service for the local Postal Inspector's office.

How to Prevent Identity Theft

One of the best ways to keep your finger on the pulse of your financial life is to make a regular request—once a year, say—for a copy of your credit report from one or all three of the main credit bureaus. (■ *Refer back to Panel 9.13.*) This will show you whether there is any unauthorized activity. Reports cost $8.

In addition, there are some specific measures you can take to guard against personal information getting into the public realm.[71,72]

● **Check your credit-card billing statements:** If you see some fraudulent charges, report them immediately. If you don't receive your statement, call the creditor first. Then call the post office to see if a change of address has been filed under your name.

● **Treat credit cards and other important papers with respect:** Make a list of your credit cards and other important documents, and the list of numbers to call if you need to report them lost. (You can photocopy the cards front and back, but make sure the numbers are legible.)

Carry only one or two credit cards at a time. Carry your Social Security card, passport, or birth certificate only when needed.

Don't dispose of credit-card receipts in a public place, such as a public trash can.

Don't give out your credit-card numbers or Social Security number over the phone, unless you have some sort of trusted relationship with the party on the other end.

Tear up credit-card offers before you throw them away.

Keep tax records and other financial documents in a safe place.

- **Treat passwords with respect:** Memorize passwords and PINs. Don't use your birth date, mother's maiden name, or similar common identifiers, which thieves may be able to guess.

- **Treat checks with respect:** Pick up new checks at the bank. Shred canceled checks before throwing them away. Don't let merchants write your credit-card number on the check (illegal in some states).

- **Watch out for "shoulder surfers" when using phones and ATMs:** When using PINs and passwords at public telephones and automated teller machines, shield your hand so that anyone watching through binoculars or using a video camera—"shoulder surfers"—can't read them.

Most TOI victims never learn how identity thieves got hold of their personal IDs. Even if you're careful, there are no guarantees that you won't ever undergo the kind of ghastly experience that Kathryn Rambo did. But the odds are a lot better you won't.

■ PANEL 9.13

The three major credit bureaus

To Report Fradulent Transactions

Call the fraud units at all three credit bureaus. Ask each one to put a "fraud alert" on your report. Also, ask them to attach a statement to your report asking creditors to call you before verifying new credit applications.

Equifax
Consumer Fraud Unit
800-525-6285
M–F 9:00–6:00 EST

Experian
Consumer Fraud Assistance Department
800-301-7195
M–F 7:30–5:30 CT

TransUnion
Fraud Victim Assistance Department
800-680-7289
M–F 7:00–4:15 PST

To Check or Correct Your Credit Report

Equifax, Experian, and TransUnion each charge $8 to supply you with a copy of your credit report.

Federal law guarantees you a free copy of your report if you've been denied credit within the past 60 days.

The law also allows consumers to get a free report each year they certify they are out of work and looking for a job, receiving welfare, or believe the file is wrong because of fraud.

Equifax
Information Service Center
800-685-1111
P.O. Box 740241
Atlanta, GA 30374-0241

Experian
800-682-7654
P.O. Box 8030
Layton, UT 84041-8030

TransUnion
Consumer Disclosure Center
P.O. Box 390
Springfield, PA 19064-0390

Summary

What It Is/What It Does

Why It's Important

batch processing (p. 423, KQ 9.3) Method of processing whereby data is collected over several days or weeks and then processed all at one time, as a "batch" used to update the master file.

With batch processing, if users need to make a request of the system, they must wait until the batch has been processed. Batch processing is less expensive than online processing and is suitable for work in which immediate answers to queries are not needed.

database (p. 419, KQ 9.1, 9.2) Integrated collection of files in a computer system.

Businesses and organizations build databases to help them keep track of and manage their affairs. In addition, users with online connections to database services have enormous research resources at their disposal.

database administrator (DBA) (p. 417, KQ 9.1) Person who coordinates all related activities and needs for a corporation's database, including database design, implementation, and operation; coordination with users; system security; backup and recovery; and performance monitoring.

The DBA determines database access rights and other matters of concern to users.

database management system (DBMS) (p. 425, KQ 9.5) Also called a *database manager;* software that controls the structure of a database and access to the data; allows users to manipulate more than one file at a time (as opposed to file managers).

This software enables: sharing of data (same information is available to different users); economy of files (several departments can use one file instead of each individually maintaining its own files, thus reducing data redundancy, which in turn reduces the expense of storage media and hardware); data integrity (changes made in the files in one department are automatically made in the files in other departments); security (access to specific information can be limited to selected users).

data dictionary (p. 430, KQ 9.7) File that stores data definitions and descriptions of database structure. It may also monitor new entries to the database as well as user access to the database.

The data dictionary monitors the data being entered to make sure it conforms to the rules defined during data definition. The data dictionary may also help protect the security of the database by indicating who has the right to gain access to it.

data files (p. 421, KQ 9.3) Files that contain data, not programs.

Data files contain content that a user has created and stored using applications propgrams. The program may add an extension, such as .DAT.

data mining (DM) (p. 433, KQ 9.8) Also called *knowledge discovery;* computer-assisted process of sifting through and analyzing vast amounts of data in order to extract meaning and discover new knowledge.

The purpose of data mining is to describe past trends and predict future trends.

What It Is/What It Does

Why It's Important

data storage hierarchy (p. 418, KQ 9.2) The levels of data stored in a computer file: bits, bytes (characters), fields, records, files, and databases.

Bits and bytes are what the computer hardware deals with, so users need not be concerned with them. They will, however, deal with characters, fields, records, files, and databases.

data warehouse (p. 434, KQ 9.8) A database containing cleaned-up data and meta-data (information about the data). Stored using high-capacity disk storage.

Data warehouses combine vast amounts of data from many sources in a database form that can be searched, for example, for patterns not recognizable with smaller amounts of data.

DBMS utilities (p. 430, KQ 9.7) Programs that allow the maintenance of databases by creating, editing, and deleting data, records, and files.

DBMS utilities allow people to establish what is acceptable input data, to monitor the types of data being input, and to adjust display screens for data input.

distributed database (p. 416, KQ 9.1) Geographically dispersed database (located in more than one physical location). Users are connected to it through a client/server network.

Data need not be centralized in one location.

Fair Credit Reporting Act of 1971 (p. 440, KQ 9.9) U.S. law that allows people to have access to and gives them the right to challenge their credit records.

If a person has been denied credit, the law allows such access free of charge.

field (p. 418, KQ 9.2) Unit of data consisting of one or more characters (bytes). An example of a field is your name, your address, *or* your Social Security number.

A collection of fields make up a record. *Also see key field.*

file (p. 418, KQ 9.2) In a database, a collection of related records. An example of a file is collected data on everyone employed in the same department of a company, including all names, addresses, and Social Security numbers.

Integrated files make up a database.

file management system (file manager) (p. 424, KQ 9.4) Software for creating, retrieving, and manipulating files, one file at a time.

In the 1950s, magnetic tape was the storage medium and records and files were stored sequentially. File managers were created to work with these files. Today, however, database managers are more common.

Freedom of Information Act (p. 438, KQ 9.9) U.S. law that allows ordinary citizens to have access to data gathered about them by federal agencies.

The Freedom of Information Act helps diminish any tendency for government agencies to exercise the kinds of powers over citizens that are found in dictatorships.

hierarchical database (p. 427, KQ 9.6) One of the four common arrangements for database management systems; fields or records are arranged in related groups resembling a family tree, with "child" records subordinate to "parent" records. A parent may have more than one child, but a child always has only one parent. To find a particular record, one starts at the top with a parent and traces down the chart to the child.

Hierarchical DBMSs work well when the data elements have an intrinsic one-to-many relationship, as might happen with a reservations system. The difficulty, however, is that the structure must be defined in advance and is quite rigid. There may be only one parent per child and no relationships among the child records.

What It Is/What It Does	Why It's Important

key field (p. 419, KQ 9.2) Field that contains unique data used to identify a record so that it can be easily retrieved and processed. The key field is often an identification number, Social Security number, customer account number, or the like. The primary characteristic of the key field is that it is *unique*.

Key fields are needed to identify and retrieve specific records in a database.

master file (p. 422, KQ 9.3) Data file containing relatively permanent records that are generally updated periodically.

Master files contain relatively permanent information used for reference purposes. Master files are updated through the use of transaction files.

network database (p. 428, KQ 9.6) One of the four common arrangements for database management systems; it is similar to a hierarchical DBMS, but each child record can have more than one parent record. Thus, a child record may be reached through more than one parent.

This arrangement is more flexible than the hierarchical one. However, it still requires that the structure be defined in advance. Moreover, there are limits to the number of links that can be made among records.

object (p. 429, KQ 9.6) Software written in small, reusable chunks and consisting of (1) data in the form of text, sound, video, and pictures and (2) instructions on the action to be taken on the data.

Once written, the software chunk can readily be reused.

object-oriented database system (OODBMS) (p. 429, KQ 9.6) One of the four common database structures; uses objects as elements within database files.

In addition to textual data, an object-oriented database can store, for example, a person's photo, "sound bites" of her voice, and a video clip, as well as methods for processing and outputting the data.

offline storage (p. 423, KQ 9.3) Refers to data that is not directly accessible for processing until a tape or disk has been loaded onto an input device.

The storage medium and data are not under the immediate, direct control of the central processing unit.

online processing (p. 423, KQ 9.3) Also called *real-time processing;* means entering transactions into a computer system as they take place and updating the master files as the transactions occur; requires direct access storage.

Online processing gives users accurate information from an ATM machine or an airline reservations system, for example.

online storage (p. 423, KQ 9.3) Refers to stored data that is directly accessible for processing.

Storage is under the immediate, direct control of the central processing unit; users need not wait for a tape or disk to be loaded onto an input device before they can access stored data.

privacy (p. 438, KQ 9.9) Right of people to not reveal information about themselves.

The ease with which databases and communications lines may pull together and disseminate information has put privacy under extreme pressure.

Privacy Act of 1974 (p. 438, KQ 9.9) U.S. law prohibiting government agencies and their contractors from keeping secret files on personnel.

The law gives individuals the right to see their records, see how the data is used, and correct errors.

program files (p. 421, KQ 9.3) Files containing software instructions.

This term is used to differentiate program files from data files.

query by example (QBE) (p. 431, KQ 9.7) Feature of query-language programs whereby the user asks for information in a database by using a sample record to define the qualifications he or she wants for selected records.

QBE further simplifies database use.

query language (p. 430, KQ 9.7) Easy-to-use computer language for making queries to a database and retrieving selected records.

Query languages make it easier for users to deal with databases. To retrieve information from a database, users make queries—that is, they use a query language. These languages have commands such as SELECT, DELETE, and MODIFY.

record (p. 418, KQ 9.2) Collection of related fields. An example of a record would be your name *and* address *and* Social Security number.

Related records make up a file.

relational database (p. 428, KQ 9.6) One of the four common arrangements for database management systems; relates, or connects, data in different files through the use of a key field, or common data element. In this arrangement there are no access paths down through a hierarchy. Instead, data elements are stored in different tables made up of rows and columns. The tables are called *relations*, the rows are called *tuples*, and the columns are called *attributes*. Within a table, a row resembles a record. All related tables must have a key field that uniquely identifies each row.

The relational database is the most flexible arrangement. The advantage of relational databases is that the user does not have to be aware of any "structure." Thus, they can be used with little training. Moreover, entries can easily be added, deleted, or modified. A disadvantage is that some searches can be time consuming. Nevertheless, the relational model has become popular for microcomputer DBMSs.

report generator (p. 431, KQ 9.7) In a database management system, a program users can employ to produce on-screen or printed-out documents from all or part of a database.

Report generators allow users to produce finished-looking reports without much fuss.

Right to Financial Privacy Act of 1978 (p. 440, KQ 9.9) U.S. law that sets restrictions on federal agencies that want to search customer records in banks.

The law protects citizens from unauthorized snooping.

shared database (p. 416, KQ 9.1) Database shared by users in one company or organization in one location.

Shared databases give all users in one organization access to the same information.

transaction file (p. 422, KQ 9.3) Temporary data file that holds all changes to be made to the master file: additions, deletions, revisions.

The transaction file is used to periodically update the master file.

Video Privacy Protection Act (p. 441, KQ 9.9) U.S. law that prevents retailers from disclosing a person's video rental records without their consent or a court order.

Although the prohibition applies to only one kind of industry, it still strengthens the privacy law in the United States.

Exercises

Self-Test Exercises

1. According to the data storage hierarchy, databases are composed of:
   a. _____
   b. _____
   c. _____
2. An individual piece of data within a record is called a(n) _____.
3. A special file in the DBMS called the _____ _____ maintains descriptions of the structure of data used in the database.
4. _____ is the right of people not to reveal information about themselves.
5. A(n) _____ _____ coordinates all related activities and needs for an organization's database.

Short-Answer Questions

1. What is the difference between batch and online processing?
2. What are the main disadvantages of traditional file management systems?
3. What is a query language?
4. What is meant by the term *data mining*?
5. Which law protects you against government agencies keeping copies of your personnel records?

Multiple-Choice Questions

1. Which of the following is a disadvantage of database management systems?
   a. cost issues
   b. data redundancy
   c. lack of program independence
   d. lack of data integrity
   e. All of the above
2. Which of the following database organizations should you choose if you need to store photos?
   a. hierarchy
   b. network
   c. relational
   d. object-oriented
   e. None of the above
3. In the event of a hardware or software failure, which of the following approaches might you use to recover lost data?
   a. mirroring
   b. reprocessing
   c. rollforward
   d. rollback
   e. All of the above
4. Which of the following isn't a feature of a DBMS?
   a. data dictionary
   b. utilities
   c. query language
   d. sequential access storage
   e. report generator
5. Which of the following database structures is similar to the hierarchy structure, except that a child record can have more than one parent record?
   a. network
   b. relational
   c. object-oriented
   d. file-management
   e. None of the above

True/False Questions

T F 1. Ensuring backup and recovery of a database is not one of the functions of a database administrator.

T F 2. Old file-management methods provided the user with an easy way to establish relationships among records in different files.

T F 3. The use of key fields makes it easier to

locate a record in a database.

T F   4.   A transaction file contains relatively permanent records that are periodically updated.

T F   5.   A database is an organized collection of integrated files.

Knowledge in Action

1. Interview someone who works with or manages a database at your school or university. What types of records make up the database, and which departments use it? What types of transactions do these departments enact? Which database structure is used? What are the types and sizes of the storage devices? Was the software custom-written?

2. How do you think you could use a microcomputer DBMS in your profession, job, or other activities? What types of data would be stored? What types of searches would you make, and what kinds of reports would you generate?

3. Describe the characteristics of an object-oriented database that might be useful to a large number of people, including yourself. What type of objects would this database contain? How would this database be used? Who would typically access this database? Does this type of database exist already? Why? Why not?

4. Companies exist today that are in the business of selling information. For example, for a fee you can purchase a list of all the businesses in your area that sell sporting goods, use IBM-compatible computers, and have 10 or more employees. The more specific your information request, the more expensive the information is. What type of database do you think would be especially valuable? What kinds of information would this database contain? What would a record look like? (Give an example.) How big do you think the database would get? Is information of the type you are describing already being sold today? How would you find out?

Information Systems

Information Management & Systems Development

"A ll change is a struggle," says one writer. "Dramatic, across-the-company change is war."[1]

This may be a slight exaggeration. Still, as we stand in the wings of the microchip- and telecommunications-driven New Economy, we can see that many of our organizations will have to undergo radical transformations.[2] The newer business vocabulary reflects this knowledge: downsizing, empowerment, reengineering.

"We're on the verge of what is perhaps the most radical redefinition of the workplace since the Industrial Revolution, with some tremendous benefits involved," says a longtime proponent of flexible work arrangements. "Yet the early signs are that corporations are as likely as not to mess this up." The speaker, a management consultant, was referring to the changes, a trickle now turning into a tidal wave, brought about by the mobile office.[3]

Part of the redefinition of the workplace comes with handing employees laptop computers, portable phones, and beepers and telling them to work from their homes, cars, or customers' offices. Part of it involves the use of a grab bag of electronic information organizers, personal communicators, personal digital assistants, and similar gadgets that help untether people from a fixed office.

"Flex-time" shift hours and voluntary part-time telecommuting programs have been around for a few years. Unlike them, however, the new high-tech tools are forcing some profound changes in the way people work. Many people, of course, like the flexibility of a mobile office. However, others resent having to work at home or being unable to limit their work hours. One computer-company vice president worries about getting her staff to stop sending faxes to each other in the middle of the night. Some employees may work 90 hours a week and still feel like they are falling short. In great part, this is because their managers' skills have not kept pace with the trend.[4] At some point, a constant work lifestyle becomes counterproductive.

Are there ways to prepare ourselves? In this chapter, we describe the traditional shape and functions of an organization. We also discuss the layers of managers and their information needs. Next we show how computer-based information systems—TPSs, MISs, DSSs, and the like—can help managers make decisions. Finally, we describe a strategy for rethinking business and information systems. This is the six-phase problem-solving procedure called systems analysis and design.

10.1 Trends Forcing Change in the Workplace

KEY QUESTION

What are the characteristics of some important trends shaping changes in the workplace?

Preview & Review: The trends of automation, downsizing and outsourcing, total quality management, and employee empowerment, among others, have forced organizations to give considerable thought to reengineering. Reengineering is the search for and implementation of radical change in business processes to achieve breakthrough results.

The *virtual office* is essentially a mobile office, as we have described (✔ p. 355). "Using integrated computer and communications technologies," says one description, "corporations will increasingly be defined not by concrete walls or physical space but by collaborative networks linking hundreds, thousands, even tens of thousands of people together."[5] Widely scattered workers can operate as individuals or as if they were all at company headquarters. Such "road warriors" break the time and space barriers of the organization, operating anytime, anywhere.

The virtual office is only one of several trends in recent years that are affecting the way we work. Others, most of which have been under way for some time, have also had a profound impact. They include, but are not limited to, the following:

- Automation
- Downsizing and outsourcing
- Total quality management
- Employee empowerment
- Reengineering

Automation

When John Diebold wrote his prophetic book *Automation* in the 1950s, the computer was nearly new. Yet Diebold predicted that computers would make many changes. First, he suggested, they would change *how* we do our jobs. Second, he thought, they would change the *kind* of work we do.[6] He was right, of course, on both counts. In the 1950s and 1960s, computers changed how factory work, for instance, was done. In the 1970s and 1980s, factory work itself began to decline as the kind of work was changed.

Diebold's third prediction is that the technologies will change the *world* in which we work. "This is the beginning of the next great development in computers and automation," he says, "which has already begun in the 1990s."

Downsizing & Outsourcing

The word *downsizing* has two meanings. **First, *downsizing* means reducing the size of an organization by eliminating workers. Second, it means the movement from mainframe-based computer systems to systems linking smaller computers in networks.**

As a result of automation, economic considerations, and the drive for increased profitability, in recent years many companies have had to downsize their staffs—lay off employees. In the process, they have, in business jargon, "flattened the hierarchy," reducing the levels and numbers of middle managers. Of course, much of the company's work still remains, forcing the rest of the staff to take up the slack. For instance, the secretary may be gone, but the secretarial work remains. The lower-level or middle-level managers found that with personal computers they could accomplish much of this work. They also found that time no longer permitted them to ask the people with the mainframes in the "glass house"—the Information Systems Department—to do some of their work. They simply had to do it themselves, again using microcomputers and networks.

Downsizing also led to another development, outsourcing. **Outsourcing is the contracting with outside businesses or services to perform the work once done by in-house departments.** The outside specialized contractors, whether janitors or computer-system managers, can often do the work more cheaply and efficiently.[7]

Total Quality Management

***Total quality management (TQM)* is managing with an organization-wide commitment to continuous work improvement and really meeting customer**

needs. The group that probably benefited most from TQM principles was the American automobile makers, who had been devastated by better-made foreign imports. However, much of the rest of U.S. industry would probably also have been shut out of competition in the global economy without the quality strides made in the last few years.

In many cases, unfortunately, the push for quality became principally a matter of pursuing the narrow statistical benchmarks favored by TQM experts. This put considerable stress on employees, with no appreciable payoff in customer satisfaction or profitability. For example, Federal Express originally pursued speed over accuracy in its sorting operation. However, it found the number of misdirected packages soared as workers scrambled to meet deadlines.[8] Now companies are looking for a better return on quality-management efforts.

Employee Empowerment

***Empowerment* means giving others the authority to act and make decisions on their own.** The old style of management was to give lower-level managers and employees only the information they "needed" to know, which minimized their power to make decisions. As a result, truly good work could not be achieved because of the attitude "If it's not part of my job, I don't do it." Today's philosophy is that information should be spread widely, not closely held by top managers, to enable employees lower down in the organization to do their jobs better. Indeed, the availability of networks and groupware (✔ p. 80) has enabled the development of task-oriented teams of workers who no longer depend on individual managers for all decisions in order to achieve company goals.

Reengineering

Trends such as the foregoing force—or should force—organizations to face basic realities. Sometimes the organization has to actually *reengineer*—rethink and redesign itself or key parts of it. **Reengineering is the search for and implementation of radical change in business processes to achieve breakthrough results.** Reengineering, also known as *process innovation* and *core process redesign*, is not just fixing up what already exists. Says one description:

> Reengineers start from the future and work backward, as if unconstrained by existing methods, people, or departments. In effect they ask, "If we were a new company, how would we run this place?" Then, with a meat ax and sandpaper, they conform the company to their vision.[9]

Reengineering works best with big processes that really matter, such as new-product development or customer service. Thus, candidates for this procedure include companies experiencing big shifts in their definition, markets, or competition. Examples are information technology companies—computer makers, cable-TV providers, and local and long-distance phone companies—which are wresting with technological and regulatory change. Expensive software systems are available to help companies reengineer and standardize their information systems to give employees the data they need when they need it.

■ **PANEL 10.1**

Is information really gold?

The only thing you can count on in business today is that you can't count on anything, says Thomas Petzinger Jr. Terrified by a random future, however, people still cling to certain venerated myths. Here's one of them.

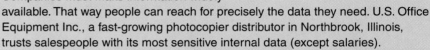

❝❝**Myth #1: Information is gold.**

Wrong. Information has become the lowest commodity. To be sure, information is essential, and like any commodity it varies in quality. But even high-grade information is now in excess supply. The challenge is turning information into action.

The most vital step is paradoxical: Companies must make information widely available. That way people can reach for precisely the data they need. U.S. Office Equipment Inc., a fast-growing photocopier distributor in Northbrook, Illinois, trusts salespeople with its most sensitive internal data (except salaries).

"When all information is available—not just what the company wants them to have—they make much more intelligent decisions," says CEO Mark Challenger. ❞❞

—Thomas Petzinger Jr., *Wall Street Journal*

At the heart of an organization is *information* and how it is used. *(■ See Panel 10.1.)* To understand how to bring about change in an organization, we need to understand how organizations and their managers work—how they need, organize, and use information.

10.2 Organizations: Departments, Tasks, Management Levels, & Types of Information

KEY QUESTIONS

What are the departments, tasks, and levels of managers in an organization, and what types of decisions do they make?

Preview & Review: Common departments in an organization are research and development, production, sales and marketing, accounting and finance, and human resources.

The tasks of managers are planning, organizing, staffing, supervising (leading), and controlling. Managers occupy three levels of responsibility: top, middle, and lower.

Top managers make strategic decisions, using unstructured information. Middle managers make tactical decisions, using semistructured information. Lower managers make operational decisions, using structured information.

Consider any sizable organization you are familiar with. Its purpose is to perform a service or deliver a product. For example, if it's nonprofit, it may deliver the service of educating students or the product of food for famine victims. If it's profit-oriented, it might sell the service of fixing computers or the product of computers themselves.

Information—whether computer-based or not—has to flow within an organization in a way that will help managers, and the organization, achieve their goals. To this end, business organizations are often structured with five departments.

README

Practical Matters: Do You Know What Technology Type You Are? How "Technographics" Targets Potential Information Technology Consumers

Are you a Mouse Potato, Gadget Grabber, Hand Shaker, or some other kind of techno-consumer—or nonconsumer?

In changing organizations, one aspect managers have to consider is how receptive people are to computers, Internet connections, and the like. And, of course, the sellers of such high-tech products and services are even more interested.

To try to predict consumers' behavior regarding technology, Forrester Research, a technology consulting firm, has developed a tool called *Technographics*.[10,11] The classification scheme is based on an annual survey by polling and research firm NPD Group, which asks 131,000 consumers about their motivations, buying habits, and financial ability to purchase technology products. Out of this has come a Technographics chart. (■ *See Panel 10.2, next page.*)

The chart organizes consumers into 10 types according to three criteria: (1) whether they buy/don't buy technology for career, family, or entertainment reasons; (2) whether they are optimistic or pessimistic about new technology; and (3) whether their household incomes are higher or lower (affluent/less affluent).

Career-minded "Fast Forwards," for instance, own an average of 20 technology products per household. Exam-ples of "Fast Forwards" are Carol Linder, 46, a customer-service manager, and husband Robyn, 53, an accountant, who have three school-age children—and two pagers and three PCs. Since they use technology for career, family, and entertainment pursuits, they are inclined to embrace new high-tech products and services. However, though similar in income and family status, Cindy Williams, 46, an administrative secretary, and her husband Gary, 44, a maintenance supervisor, who have two sons, are considered "Traditionalists" because they have a 3-year-old PC and no Internet connection. This suggests they will wait a long time before upgrading.[12] The biggest group is the 70 million "Sidelined Citizens," who are neither interested nor financially able to buy technology.[13]

What technology type do you think you are? To figure this out, ask yourself these questions: (1) In what areas is technology of concern to me—career, family, or entertainment? (2) Am I optimistic or pessimistic about technology in general? (3) How much money do I have to spend on technology? (Single people with incomes above $25,000 and families with incomes above $40,000 have money to spend on technology.)

Departments: R&D, Production, Marketing, Accounting, Human Resources

Depending on the services or products they provide, most organizations have departments that perform five functions: *research and development (R&D)*, *production*, *sales and marketing*, *accounting and finance*, and *human resources (personnel)*.

- **Research and development:** The research and development (R&D) department does two things: (1) It conducts basic research, relating discoveries to the organization's current or new products. (2) It does product development and tests and modifies new products or services created by researchers. Special software programs are available to aid in these functions.

- **Production:** The production department makes the product or provides the service. In a manufacturing company, it takes the raw materials and has people or machinery turn them into finished goods. In many cases, this department uses CAD/CAM software and workstations (✔ p. 23), as well as robotics. (■ *See Panel 10.3.*)

▪ PANEL 10.2

Ten technology types
The types are categorized according to their optimism and pessimism about new technology. Within those categories they are organized according to whether they are affluent or less affluent.

	Career	Family	Entertainment
Affluent optimists	**Fast Forwards** Big-spending, career-oriented technology enthusiasts. Often time-strapped. Business-oriented to improve productivity.	**New Age Nurturers** Also big spenders. Focused on technology for home uses, such as a family PC.	**Mouse Potatoes** High-income, entertainment-focused buyers who like online technology for interactive entertainment and Web surfing.
Less affluent optimists	**Techno-Strivers** Less affluent but up-and-coming professionals or students looking to gain career edge by using pagers, cell phones, online services, etc.	**Digital Hopefuls** Families with limited budgets but interested in new technology. Might buy under-$1000 PC.	**Gadget Grabbers** Lower-income consumers of online entertainment; will buy low-cost high-tech toys.
Affluent pessimists	**Handshakers** Higher income but older consumers, typically managers, who value relationships above technology and don't use their PCs at work.	**Traditionalists** High-income and family-oriented, but cautious about using new technology. Not convinced upgrades and add-ons are worth the investment.	**Media Junkies** High-income, entertainment-driven consumers who prefer TV, VCRs, and older media to PCs.
Less affluent pessimists	**Sidelined Citizens** Low-income technophobes and technology laggards. They are the least likely to puchase any technology.		

▪ PANEL 10.3

Examples of computer-based information systems used in production

Computer-Aided Design
Create, simulate, and evaluate models of products and manufacturing processes.

Computer-Aided Manufacturing
Use computers and robots to fabricate, assemble, and package products.

Factory Management
Plan and control production runs, coordinate incoming orders and raw material requests, oversee cost, and quality assurance programs.

Quality Management
Evaluate product and process specifications, test incoming materials and outgoing products, test production processes in progress, and design quality assurance programs.

Logistics
Purchase and receive materials, control and distribute materials, and control inventory and shipping of products.

Maintenance
Monitor and adjust machinery and processes, perform diagnostics, and do corrective and preventive maintenance.

■ PANEL 10.4

Examples of computer-based information systems used in marketing

Sales Management
Plan, monitor, and support the performance of salespeople and sales of products and services.

Sales Force Automation
Automate the recording and reporting of sales activity by salespeople and the communications and sales support from sales management.

Product Management
Plan, monitor, and support the performance of products, product lines, and brands.

Advertising and Promotion
Help select media and promotional methods and control and evaluate advertising and promotion results.

Sales Forecasting
Produce short- and long-range sales forecasts.

Projected Sales

1999 2000 2001 2002

Market Research
Collect and analyze internal and external data on market variables, developments, and trends.

Marketing Management
Develop marketing strategies and plans based on corporate goals and market research and sales activity data, and monitor and support marketing activities.

- **Sales and marketing:** The marketing department oversees advertising, promotion, and sales. (■ *See Panel 10.4.*) The people in this department plan, price, advertise, promote, package, and distribute the services or goods to customers or clients. The sales reps may use laptop computers, cell phones, wireless e-mail, and faxes in their work while on the road.

- **Accounting and finance:** The accounting and finance department handles all financial matters. It handles cash management, pays bills and taxes, issues paychecks, records payments, makes investments, and compiles financial statements and reports. It also produces financial budgets and forecasts financial performance after receiving information from other departments.

- **Human resources:** The human resources, or personnel, department finds and hires people and administers sick leave and retirement matters. It is also concerned with compensation levels, professional development, employee relations, and government regulations.

Whatever the organization—grocery store, computer maker, law firm, hospital, or university—it is likely to have departments corresponding to these. Although office automation brought about by computers, networks, and groupware has given employees more decision-making power than they used to have, managers in these departments still perform five basic functions.

Management Tasks: Five Functions

Certain specific duties are associated with being a manager. **Management is overseeing the tasks of planning, organizing, staffing, supervising, and controlling business activities.** These five functions, considered the classic tasks of management, are defined as follows:

- **Planning:** Planning is what you do to try to get yourself or your organization from your present position to an even better position. Planning is setting objectives, both long-term and short-term, and developing strategies for achieving them. Whatever you do in planning lays the groundwork for the other four tasks.

- **Organizing:** To achieve your goals, you must organize the parts in a coordinated and integrated effort. Organizing is making orderly arrangements of resources, such as people and materials.

- **Staffing:** Staffing has to do with people. Staffing is selecting, training, and developing people. In some cases, it may be done by specialists, such as those in the human resources department.

- **Supervising:** Supervising is directing, guiding, and motivating employees to work toward achieving the organization's goals. Supervising includes the direct order, of course.

- **Controlling:** Controlling is monitoring the organization's progress and adapting methods toward achieving its goals.

All managers perform all these tasks as part of their jobs. However, the level of responsibility regarding these tasks varies with the level of the manager, as we discuss next.

Management Levels: Three Levels, Three Kinds of Decisions

How do managers carry out the tasks just described? They do it by *making decisions on the basis of the information available to them.* A manager's daily job is to decide on the best course of action, based on the facts known at the time.

For each of the five departments there are three traditional levels of management—top, middle, and lower. These levels are reflected in the organization chart. **An *organization chart* is a schematic drawing showing the hierarchy of formal relationships among an organization's employees. (■** *See Panel 10.5, next page.)*

Managers on each of the three levels have different levels of responsibility and are therefore required to make different kinds of decisions.

- **Top managers—strategic decisions:** The chief executive officer (CEO) or president is the very top manager. However, for our purposes, "top management" refers to the vice presidents, one of whom heads each department.

  ***Top managers* are concerned with long-range planning. Their job is to make strategic decisions. *Strategic decisions* are complex decisions rarely based on predetermined routine procedures, involving the subjective judgment of the decision maker.** *Strategic* means that, of the five management tasks (planning, organizing, staffing, supervising, controlling), top managers are principally concerned with *planning*.

  Besides CEO, president, and vice president, typical titles found at the top management level are treasurer, director, controller (chief accounting officer), and senior partner. Examples of strategic decisions are how growth should be financed and what new markets should be tackled first. Other strategic decisions are deciding the company's 5-year goals, evaluating future financial resources, and deciding how to react to competitors' actions.

  An AT&T vice president of marketing might have to make strategic decisions about promotional campaigns to sell a new paging service. The top manager who runs an electronics store might have to make strategic decisions about stocking a new line of paging devices.

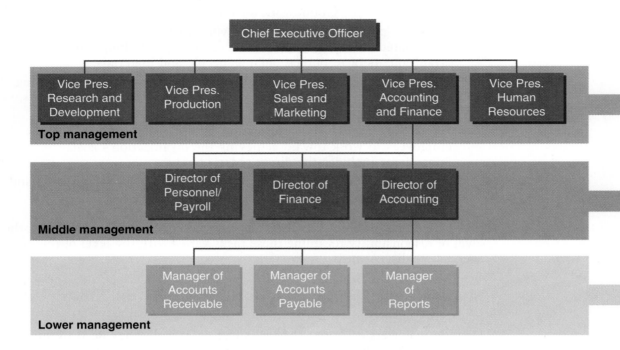

■ PANEL 10.5

Management levels and responsibilities

(Above) An organization generally has five departments: research and development, production, sales & marketing, accounting and finance, and human resources. This organization chart shows the management hierarchy for just one department, Accounting & Finance. Three levels of management are shown—top, middle, and lower. *(Opposite page)* The entire organization can also be represented as a pyramid, with the five departments and three levels of management as shown. Top managers are responsible for strategic decisions, middle managers for tactical decisions, and lower managers for operational decisions. Office automation is changing the flow of information in many organizations, thus "flattening" the pyramid, because not all information continues to flow through traditional hierarchical channels.

- **Middle managers—tactical decisions:** *Middle-level managers* **implement the goals of the organization. Their job is to oversee the supervisors and to make tactical decisions. A *tactical decision* is a decision that must be made without a base of clearly defined informational procedures, perhaps requiring detailed analysis and computations.** *Tactical* means that, of the five management tasks, middle managers deal principally with *organizing* and *staffing*. They also deal with shorter-term goals than top managers do.

  Examples of middle managers are plant manager, division manager, sales manager, branch manager, and director of personnel. An example of a tactical decision is deciding how many units of a specific product (pagers, say) should be kept in inventory. Another is whether or not to purchase a larger computer system.

  The director of sales, who reports to the vice president of sales and marketing for AT&T, sets sales goals for district sales managers

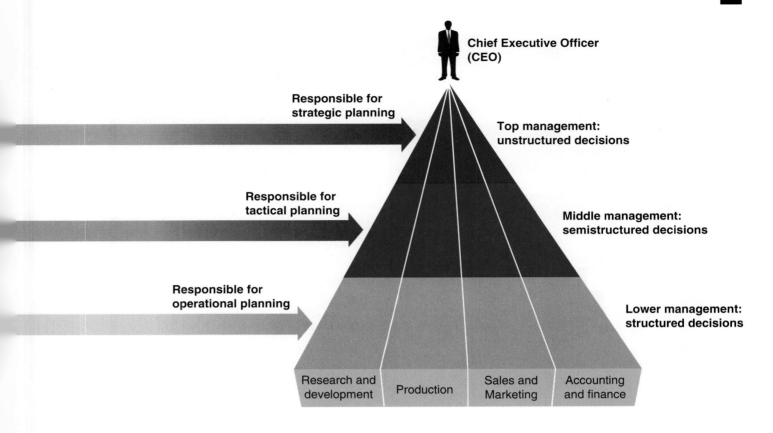

Chief Executive Officer (CEO)

Responsible for strategic planning

Top management: unstructured decisions

Responsible for tactical planning

Middle management: semistructured decisions

Responsible for operational planning

Lower management: structured decisions

| Research and development | Production | Sales and Marketing | Accounting and finance |

throughout the country. They in turn feed him or her weekly and monthly sales reports.

- Lower or supervisory managers—operational decisions: *Lower-level managers,* or *supervisory managers,* **manage or monitor nonmanagement employees. Their job is to make operational decisions. An** *operational decision* **is a predictable decision that can be made by following a well-defined set of routine procedures.** *Operational* means these managers focus principally on *supervising* (leading) and *controlling.* They monitor day-to-day events and, if necessary, take corrective action.

    An example of a supervisory manager is a warehouse manager in charge of inventory restocking. An example of an operational decision is one in which the manager must choose whether or not to restock inventory. (The guideline on when to restock may be determined at the level above.)

    A district sales manager for AT&T would monitor the promised sales and orders for pagers coming in from sales representatives. When sales begin to drop off, the supervisor would need to take immediate action.

Types of Information: Unstructured, Semistructured, & Structured

To make the appropriate decisions—strategic, tactical, operational—the different levels of managers need the right kind of information: unstructured, semistructured, and structured. (■ *See Panel 10.6, next page.)* Examples

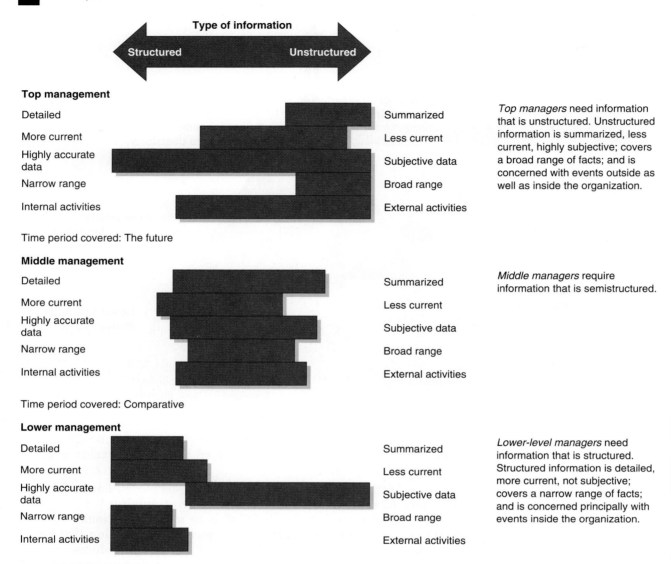

Type of information

Structured ⟷ Unstructured

Top management

Detailed		Summarized
More current		Less current
Highly accurate data		Subjective data
Narrow range		Broad range
Internal activities		External activities

Time period covered: The future

Top managers need information that is unstructured. Unstructured information is summarized, less current, highly subjective; covers a broad range of facts; and is concerned with events outside as well as inside the organization.

Middle management

Detailed		Summarized
More current		Less current
Highly accurate data		Subjective data
Narrow range		Broad range
Internal activities		External activities

Time period covered: Comparative

Middle managers require information that is semistructured.

Lower management

Detailed		Summarized
More current		Less current
Highly accurate data		Subjective data
Narrow range		Broad range
Internal activities		External activities

Time period covered: The past

Lower-level managers need information that is structured. Structured information is detailed, more current, not subjective; covers a narrow range of facts; and is concerned principally with events inside the organization.

■ PANEL 10.6

Types of information: the structured–unstructured continuum

Top managers need information that is unstructured. Unstructured information is summarized, less current, highly subjective; covers a broad range of facts; and is concerned with events outside as well as inside the organization.

   Lower managers need information that is structured. Structured information is detailed, more current, not subjective; covers a narrow range of facts; and is concerned principally with events inside the organization.

   Middle managers require information that is semistructured.

of the three kinds of decision structures appear opposite. (■ *See Panel 10.7.*)

   In general, *all* information to support intelligent decision making at all three levels must be correct—that is, accurate. It must also be complete, including *all* relevant data, yet concise, including *only* relevant data. It must be cost effective, meaning efficiently obtained, yet understandable. It must be current, meaning timely, yet also time sensitive, based on historical, current, or future information needs. Thus, information has three properties:

Decision Structure	Operational Management	Tactical Management	Strategic Management
Unstructured	Cash management	Work group reorganization	New business planning
		Work group performance analysis	Company reorganization
Semistructured	Credit management	Employee performance appraisal	Product planning
	Production scheduling	Capital budgeting	Mergers and aquisitions
	Daily work assignment	Program budgeting	Site location
Structured	Inventory control	Program control	

■ **PANEL 10.7**

Examples of decisions by the type of decision structures and by level of management

1. Level of summarization
2. Degree of accuracy
3. Timeliness

These properties may vary to provide more or less structure, depending on the level of management served and the type of decision making required. **Structured information is detailed, current, and concerned with past events, records a narrow range of facts, and covers an organization's internal activities.** Unstructured information is the opposite. **Unstructured information is summarized, less current, concerned with future events, records a broad range of facts, and covers activities outside as well as inside an organization. Semistructured information includes some structured information and some unstructured information.**

The illustration below shows the information that the three levels of management might deal with in a food-supply business. (■ *See Panel 10.8.*)

Now that we've covered some basic concepts about how organizations are structured and what kinds of information are needed at different levels of management, we need to examine what types of management information systems provide the information.

■ **PANEL 10.8**

Areas covered by the three management levels in a food-supply business

Strategic (unstructured)	Competitive industry statistics		
Tactical (semistructured)	Sales analysis, by customer		
	Reorder analysis of new products		
	Sales analysis, by product line		
	Production forecast		
Operational (structured)	Bill of materials	Order processing	Accounts receivable
	Manufacturing specifications	Online order inquiry	General ledger
	Product specifications	Finished goods inventory	

Food supply business

4th 25,000
1st 3,871
227 63,000
213,000 845

1998 1999
2000 2001

10.3 Management Information Systems

KEY QUESTIONS

What are the six computer-based information systems, and what are their purposes?

Preview & Review: Six basic types of computer-based information systems provide information for decision making.

Transaction processing systems assist lower managers to make operational decisions.

Management information systems help middle managers to make tactical decisions.

Decision support systems and executive support systems support top managers in making strategic decisions.

Office automation systems reduce the manual labor required to operate an efficient office environment.

Expert systems are used at all levels for specific problems.

Top managers make strategic decisions using unstructured information, as we have seen. Middle managers make tactical decisions using semistructured information. Lower-level managers make operational decisions using structured information. The purpose of a computer-based information system is to provide managers (and various categories of employees) with the appropriate kind of information to help them make decisions.

Here we describe the following types of computer-based information systems, corresponding to the three management layers and their requirements.

- **For lower managers:** Transaction processing systems (TPSs)
- **For middle managers:** Management information systems (MISs)
- **For top managers:** Decision support systems (DSSs) and executive information systems (EISs)
- **For all levels, including nonmanagement:** Office automation systems (OASs) and expert systems (ESs)

Let us consider these. (■ *See Panel 10.9.*)

Transaction Processing Systems: For Lower Managers

In most organizations, particularly business organizations, most of what goes on takes the form of transactions. **A *transaction* is a recorded event having to do with routine business activities.** This includes everything concerning the product or service in which the organization is engaged: production, distribution, sales, orders. It also includes materials purchased, employees hired, taxes paid, and so on. Today in most organizations, the bulk of such transactions are recorded in a computer-based information system. These systems tend to have clearly defined inputs and outputs, and there is an emphasis on efficiency and accuracy. Transaction processing systems record data but do little in the way of converting data into information.[14]

A *transaction processing system (TPS)* is a computer-based information system that keeps track of the transactions needed to conduct business. Some features of a TPS are as follows:

- **Input and output:** The inputs to the system are transaction data: bills, orders, inventory levels, and the like. The output consists of processed transactions: bills, paychecks, and so on.
- **For lower managers:** Because the TPS deals with day-to-day matters, it is principally of use to supervisory managers. That is, the TPS helps in making *operational* decisions. Such systems are not usually helpful to middle or top managers.

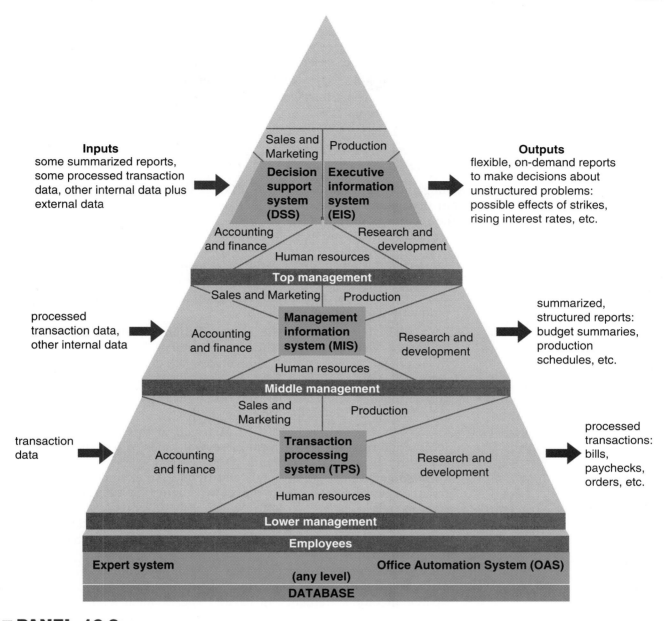

Inputs
some summarized reports, some processed transaction data, other internal data plus external data

Outputs
flexible, on-demand reports to make decisions about unstructured problems: possible effects of strikes, rising interest rates, etc.

Sales and Marketing | Production

Decision support system (DSS) | **Executive information system (EIS)**

Accounting and finance | Research and development

Human resources

Top management

Sales and Marketing | Production

Accounting and finance | **Management information system (MIS)** | Research and development

Human resources

Middle management

processed transaction data, other internal data

summarized, structured reports: budget summaries, production schedules, etc.

Sales and Marketing | Production

Accounting and finance | **Transaction processing system (TPS)** | Research and development

Human resources

Lower management

transaction data

processed transactions: bills, paychecks, orders, etc.

Employees

Expert system | **(any level)** | **Office Automation System (OAS)**

DATABASE

■ PANEL 10.9

Six information systems for three levels of management

The pyramid shows the following: (1) The three levels of management: top, middle, and lower. (2) The five departments for each level: research and development, production, sales and marketing, accounting and finance, and human resources. (3) The kinds of computer-based information systems corresponding to each management level. (4) The kind of data input for each level, and the kind of information output.

- **Produces detail reports:** A manager at this level typically will receive information in the form of detail reports. **A *detail report* contains specific information about routine activities.** An example might be the information needed to decide whether to restock inventory.
- **One TPS for each department:** Each department or functional area of an organization—research and development, production, sales and

marketing, accounting and finance, and human resources—usually has its own TPS. For example, the accounting and finance TPS handles order processing, accounts receivable, inventory and purchasing, accounts payable, order processing, and payroll.

- **Basis for MIS and DSS:** The database of transactions stored in a TPS is used to support management information systems and decision support systems.

Management Information Systems: For Middle Managers

A *management information system (MIS)* is a computer-based information system that uses data recorded by TPS as input into programs that produce routine reports as output.

Features of an MIS are as follows:

- **Input and output:** Inputs consist of processed transaction data, such as bills, orders, and paychecks, plus other internal data. Outputs consist of summarized, structured reports: budget summaries, production schedules, and the like.
- **For middle managers:** An MIS is intended principally to assist middle managers. That is, it helps them with *tactical* decisions. It enables them to spot trends and get an overview of current business activities.
- **Draws from all departments:** The MIS draws from all five departments or functional areas, not just one.
- **Produces several kinds of reports:** Managers at this level usually receive information in the form of several kinds of reports: *summary, exception, periodic, on-demand.*

    Summary reports **show totals and trends.** An example would be a report showing total sales by office, by product, by salesperson, or as total overall sales.

    Exception reports **show out-of-the-ordinary data.** An example would be an inventory report that lists only those items that number fewer than 10 in stock.

    Periodic reports **are produced on a regular schedule.** These may be daily, weekly, monthly, quarterly, or annually. They may contain sales figures, income statements, or balance sheets. Such reports are usually produced on paper, such as computer printouts.

    On-demand reports **produce information in response to an unscheduled demand.** A director of finance might order an on-demand credit-background report on an unknown customer who wants to place a large order. On-demand reports are often produced on a terminal or microcomputer screen rather than on paper.

Decision Support Systems: For Top Managers

A *decision support system (DSS)* is a computer-based information system that provides a flexible tool for analysis and helps managers focus on the future. To reach the DSS level of sophistication in information technology, an organization must have established a transaction processing system and a management information system.

Some features of a DSS are as follows:

- **Inputs and outputs:** Inputs consist of some summarized reports, some processed transaction data, and other internal data. They also include data that is external to that produced by the organization. This external data may be produced by trade associations, marketing research firms, the U.S. Bureau of the Census, and other government agencies.

   The outputs are flexible, on-demand reports on which a top manager can make decisions about unstructured problems.

- **Mainly for top managers:** A DSS is intended principally to assist top managers, although it is now being used by middle managers too. Its purpose is to help them make *strategic* decisions—decisions about unstructured problems, often unexpected and nonrecurring. These problems may involve the effect of events and trends outside the organization. Examples are rising interest rates or a possible strike in an important materials-supplying industry.

- **Produces analytic models:** The key attribute of a DSS is that it uses *models*. **A *model* is a mathematical representation of a real system.** The models use a DSS database, which draws on the TPS and MIS files, as well as external data such as stock reports, government reports, national and international news. The system is accessed through DSS software.

   The model allows the manager to do a simulation—play a "what if" game—to reach decisions. Thus, the manager can simulate an aspect of the organization's environment in order to decide how to react to a change in conditions affecting it. By changing the hypothetical inputs to the model—number of workers available, distance to markets, or whatever—the manager can see how the model's outputs are affected.

Many DSSs are developed to support the types of decisions faced by managers in specific industries, such as airlines or real estate.[15] Curious how airlines decide how many seats to sell on a flight when so many passengers are no-shows? American Airlines developed a DSS, the yield management system, that helps managers decide how much to overbook and how to set prices for each seat so that a plane is filled and profits are maximized. Wonder how owners of those big apartment complexes set rents and lease terms? Investors in commercial real estate use a DSS called RealPlan to forecast property values up to 40 years into the future, based on income, expense, and cash-flow projections. Ever speculate about how insurance carriers set different rates or how Arby's and McDonald's decide where to locate a store? Many companies use DSSs called geographic information systems (GISs), such as MapInfo and Atlas GIS, which integrate geographic databases with other business data and display maps. (■ *See Panel 10.10, next page.*)

Executive Information Systems: Also for Top Managers

An *executive information system (EIS)* is an easy-to-use DSS made especially for top managers; it specifically supports strategic decision making. An EIS is also called an *executive support system (ESS)*. It draws on data not only from systems internal to the organization but also from those outside, such as news services or market-research databases.

■ PANEL 10.10

Geographic DSS for earthquake insurance

Using geographic information systems, such as MapInfo, insurance underwriters can set rates and examine potential liability in the event of a natural disaster. This one presents a visual analysis of policyholders living near earthquake fault lines in California.

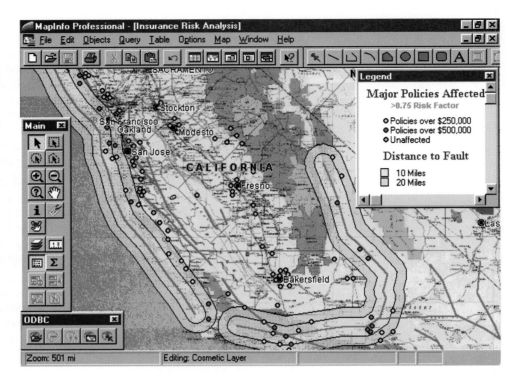

An EIS might allow senior executives to call up predefined reports from their personal computers, whether desktops or laptops. They might, for instance, call up sales figures in many forms—by region, by week, by anticipated year, by projected increases. The EIS includes capabilities for analyzing data and doing "what if" scenarios. An EIS may also allow a manager to browse through information on all aspects of the organization and to focus on areas he or she believes require attention. (■ *See Panel 10.11.*)

■ PANEL 10.11

EIS screen

This Comshare EIS, called Commander Decision, can easily calculate averages, variances, ratios, and summaries.

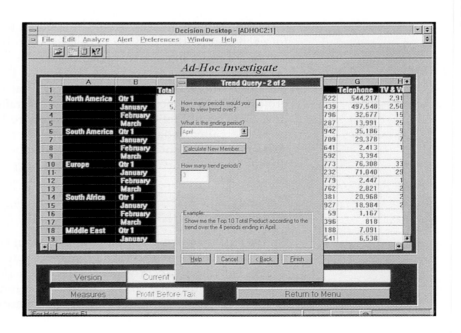

Office Automation & Expert Systems: Information for Everyone

TPSs, MISs, DSSs, EISs—the alphabet soup of information systems discussed so far—are designed for managers of various levels. But there are two types of information systems that are intended for workers of all levels, including those who aren't managers: *office automation systems* and *expert systems*.

- Office automation systems: ***Office automation systems (OASs) are those that combine various technologies to reduce the manual labor required to operate an efficient office environment.*** Used throughout all levels of an organization, OAS technologies include fax, voice mail, e-mail, scheduling software, word processing, and desktop publishing, but there are other technologies as well. (■ *See Panel 10.12.)*

   The backbone of an OAS is a network—LAN, intranet, extranet—that connects everything together. All office functions—dictation, typing, filing, copying, fax, microfilm and records management, telephone calls and switchboard operations—are candidates for integration into the network.

- Expert systems: **An *expert system* is a set of interactive computer programs that helps users solve problems that would otherwise require the assistance of a human expert.** Expert systems are created on the basis of knowledge collected on specific topics from human experts, and they imitate the reasoning process of a human being. Expert systems have emerged from the field of artificial intelligence (discussed in Chapter 12), the branch of computer science that is attempting to create computer systems that simulate human reasoning and sensation.

   Expert systems are used by both management and nonmanagement personnel to solve specific problems, such as how to reduce production costs, improve workers' productivity, or reduce environmental impact. Expert systems are usually run on large computers because of these systems' giant appetite for memory, although some microcomputer expert systems also exist. For example, Negotiator Pro for IBM and Macintosh computers helps executives plan effective negotiations with other people by examining their personality types and recommending negotiating strategies. An expert system screen for a large computer system is shown on the next page. (■ *See Panel 10.13.)*

■ PANEL 10.12

Office automation systems

The backbone is a network linking these technologies.

Office Automation Systems

Electronic Publishing Systems	Electronic Communications Systems	Electronic Collaboration Systems	Image Processing Systems	Office Management Systems
• Word processing • Desktop publishing • Copying systems	• Electronic mail • Voice mail • Facsimile • Desktop videoconferencing	• Electronic meeting systems • Collaborative work systems • Teleconferencing • Telecommuting	• Electronic document management • Other image processing • Presentation graphics • Multimedia Systems	• Electronic office accessories • Electronic scheduling • Task management

■ PANEL 10.13

Expert system screen
This screen is from a United Airlines gate assignment expert system, which analyzes airplane traffic to help workers assign gates to incoming planes.

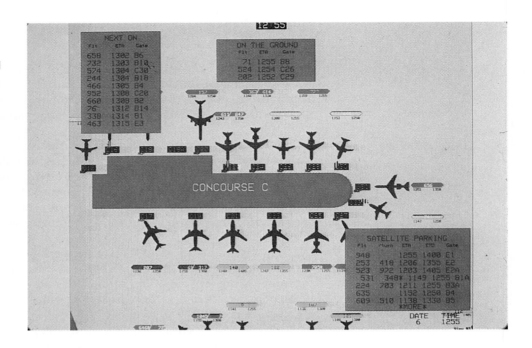

Flattening the Pyramid: The Changing Organization

The old organization chart shaped like a pyramid is a legacy that dates back almost a century, with roots in the military. But it doesn't work as well in today's world.

"If you want your company to be innovative, a rigid organization chart that dictates that all lines of communication is a corporate straitjacket," says Alan Webber, founding editor of *Fast Company* magazine. "If you want your company to be global, you have to be willing to experiment with an organization chart that looks like a spider's web. And if you want your company to be fast and agile, then you might want to try no organization chart at all."[16]

As we indicated earlier in the chapter, networks and new technologies are flattening the traditional pyramid-shaped hierarchical structure of management levels. Now, in addition to being disseminated in a top-down (or bottom-up) manner, information is equally likely to be distributed in a lateral or horizontal manner. (■ *See Panel 10.14.*)

■ PANEL 10.14

The changing hierarchy: information goes sideways
The traditional pyramid-shaped hierarchical structure of management and the resulting distribution of information is changing. In the new, less structured organization, information is also disseminated in a more lateral or horizontal manner.

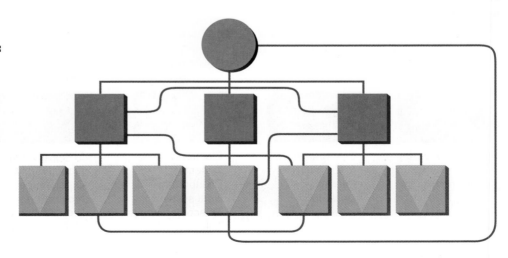

Companies are trying to use new technologies to allow more fluid management structures that release people to be more creative and productive. For instance, at Sun Microsystems in Mountain View, California, the organization chart looks more like Swiss cheese than a pyramid. The company of 15,000 people is tied together by its e-mail system, which generates up to 2 million e-mail messages a day—a clue to the way work really gets done. "Your e-mail flow determines whether you're really part of the organization," says John Gage, the company's chief scientist. The people who get the most messages and participate in the most important exchanges are the ones with the most power—regardless of the picture of the organization chart.[17] An open environment encourages people to innovate.

We have described how an organization and its managers work and what their information needs are. Now let us see how changes can be made to keep up with new demands. We describe the tool of systems analysis and design.

10.4　The Six Phases of Systems Analysis & Design

KEY QUESTION

What are the six phases of the systems development life cycle?

Preview & Review: Knowledge of systems analysis and design helps you explain your present job, improve personal productivity, and lessen risk of a project's failure. The initiative for suggesting a need to analyze and possibly change an information system may come from users, managers, or technical staff.

The six phases of systems design and analysis are known as the systems development life cycle (SDLC). The six phases are (1) preliminary investigation, (2) systems analysis, (3) systems design, (4) systems development, (5) systems implementation, and (6) systems maintenance.

Organizations can make mistakes, of course, and big organizations can make *really big* mistakes.

README

Case Study: Virtual Teams Transcend Time & Space

"If people are more than 50 feet apart, they are not likely to collaborate."

That was what the old so-called 50-Foot Rule stated. Today things have changed. Modern communications technology has now given us global collaboration through what are known as *virtual teams*.

Virtual teams, according to experts Jessica Lipnack and Jeffrey Stamps, are groups of people who work closely together even though they are separated by space, time, and organizational barriers.[18]

Collaborative networks—combinations of local-area and wide-area communications networks—linking hundreds or thousands of people can allow businesses to form and dissolve clusters of workers on a moment's notice.

Forming a small virtual team—a handful of people on the East and West coasts working on a print ad, for example—might seem relatively easy. But NCR Corporation created a virtual team of more than a thousand people spread out at 17 locations to develop a new computer system. "Using a high-speed full-bandwidth audio/video/data link," says one account, "the virtual team completed the project on budget and ahead of schedule."[19]

With communications technology, space and time are no longer the biggest hurdles for virtual teams; the greatest ones are organizational—that is, the pyramid hierarchy and Main Street business norms like caution, continuity, and conservatism.

California's state Department of Motor Vehicles' databases needed to be modernized, and in 1988 Tandem Computers said they could do it. "The fact that the DMV's database system, designed around an old IBM-based platform, and Tandem's new system were as different as night and day seemed insignificant at the time to the experts involved," said one writer investigating the project later.[20] The massive driver's license database, containing the driving records of more than 30 million people, first had to be "scrubbed" of all information that couldn't be translated into the language used by Tandem computers. One such scrub yielded 600,000 errors. Then the DMV had to translate all its IBM programs into the Tandem language. "Worse, DMV really didn't know how its current IBM applications worked anymore," said the writer, "because they'd been custom-made decades before by long-departed programmers and rewritten many times since." Eventually the project became a staggering $44 million loss to California's taxpayers.

In Denver, airport officials weren't trying to upgrade an old system but to do something completely new. At the heart of the Denver International Airport was supposed to be a high-tech system to whisk baggage between terminals so fast that passengers would practically never have to wait for their luggage. As the system failed test after test, airport officials eventually decided they had to *build a manual baggage system*—at an additional cost of $50 million. Spending the money on old technology, it developed, was cheaper than continuing to spend millions paying interest on construction bonds for a nonoperating airport.[21]

Both these examples show how important planning is, especially when an organization is trying to launch a new kind of system. How do you avoid such mistakes? By employing systems analysis and design.

Why Know About Systems Analysis & Design?

But, you may say, you're not going to have to wrestle with problems on the scale of motor-vehicle departments and airports. That's a job for computer professionals. You're mainly interested in using computers and communications to increase your own productivity. Why, then, do you need to know anything about systems analysis and design?

There are several reasons:

- **Explaining your present job:** Especially if you work in a large organization, you may find your department or your job the focus of a study by a systems analyst. Knowing how the procedure works will help you better explain how your job works or what goals your department is supposed to achieve. In progressive companies, management is always interested in suggestions for improving productivity. This is the method for expressing your ideas.

- **Improving personal productivity:** You can use the procedures to solve problems in your own work area, such as buying and implementing new kinds of information technology. Will acquiring handheld pen-based portable computers, such as personal digital assistants, help you in making sales calls, for example? Systems analysis and design can help you make some determinations.

- **Reducing the risks of a project's failure:** You can apply the steps in systems analysis and design to an application of information technology. Is having a voice-mail "telephone tree" better than having calls directed by a receptionist? Or will the system actually backfire by being too impersonal? Systems analysis and design can help you avoid failure.

The Purpose of a System

Suppose you are managing a fleet of delivery trucks for a small family-owned business. When the drivers need to refuel their trucks, they come in to the head office and borrow one of a number of gasoline credit cards. These cards are simply kept in an office desk drawer. You suspect that the reason fuel bills are so high are that drivers are also filling up their personal cars and charging it to the company. (A better idea would be to open an account with one local gas station. You could then direct the gasoline seller to bill you only for filling the company's trucks.)

Is this a system? It certainly is. **A *system* is defined as a collection of related components that interact to perform a task in order to accomplish a goal.** A system may not work very well, but it is nevertheless a system. The point of systems analysis and design is to ascertain how a system works and then take steps to make it better.

An organization's computer-based information system consists of hardware, software, people, procedures, and data, as well as communications setups. These work together to provide people with information for running the organization.

From time to time, organizations need to change their information systems. The reasons may be new marketing opportunities, changes in government regulations, introduction of new technology, a merger with another company, or other changes. The company may be as big as a cable-TV company trying to set up a billing system for movies on-demand. Or it may be as small as a two-person graphic design business trying to change its invoice and payment system. When this happens, the time is ripe for applying systems analysis and design.

Getting the Project Going: How It Starts, Who's Involved

All it takes is a single individual who believes that something badly needs changing to get the project rolling. An employee may influence a supervisor. A customer or supplier may get the attention of someone in higher management. Top management on its own may decide to take a look at a system that looks inefficient. A steering committee may be formed to decide which of many possible projects should be worked on.

Participants in the project are of three types:

- Users: The system under discussion should always be developed in consultation with users, whether floor sweepers, research scientists, or customers. Indeed, inadequate user involvement in analysis and design can be a major cause of a system's failing for lack of acceptance.
- Management: Managers within the organization should also be consulted about the system.
- Technical staff: Members of the company's information systems (IS) department, consisting of systems analysts and programmers, need to be involved. For one thing, they may well have to carry out and execute the project. Even if they don't, they may have to work with outside IS people contracted to do the job.

Complex projects will require one or several systems analysts. **A *systems analyst* is an information specialist who performs systems analysis, design, and implementation.** The job of systems analyst, according to the U.S. Labor Department, promises to be one of the jobs for which there will be the most

■ **PANEL 10.15**

U.S. projected job growth, 1996–2006

Systems analysts fall just behind cashiers—a job that pays far less—among occupations adding the most positions in the next few years.

Occupations adding the most positions	Gain 1996–2006
Cashiers	530,000
Systems analysts	**520,000**
General managers and top executives	467,000
Registered nurses	411,000
Retail salespersons	408,000
Truck drivers	404,000
Home health aides	378,000
Teacher aides	370,000

demand in the coming years, with the position gaining 520,000 jobs in the 1996–2006 decade. (■ *See Panel 10.15.*)

His or her job is to study the information and communications needs of an organization and determine what changes are required to deliver better information to people who need it. "Better" information means information that can be summarized in the acronym "CART"—complete, accurate, relevant, and timely. The systems analyst achieves this goal through the problem-solving method of systems analysis and design—known as "the systems approach." Large and complex projects usually require the services of several systems analysts who are specialists (for example, in databases, client/server programming, or personal computing).

The Six Phases of Systems Analysis & Design

Systems analysis and design **is a six-phase problem-solving procedure for examining an information system and improving it.** The six phases make up what is called the systems development life cycle. **The** *systems development life cycle (SDLC)* **is defined as the step-by-step process that many organizations follow during systems analysis and design.** The number of phases may vary from one company to another, and even the name of the process may differ (*applications development cycle, systems development cycle, structured development life cycle,* for instance). Still, the general objectives remain the same.

Whether applied to a Fortune 500 company or a three-person engineering business, the six phases in systems analysis and design may be said to be as follows. (■ *See Panel 10.16.*)

1. Preliminary investigation: Conduct preliminary analysis, propose alternative solutions, and describe the costs and benefits of each solution. Submit a preliminary plan with recommendations.
2. Systems analysis: Gather data, analyze the data, and make a written report.
3. Systems design: Make a preliminary design and then a detailed design, and write a report.
4. Systems development: Acquire the hardware and software and test the system.
5. Systems implementation: Convert the hardware, software, and files to the new system and train the users.
6. Systems maintenance: Audit the system, request feedback from its users, and evaluate it periodically.

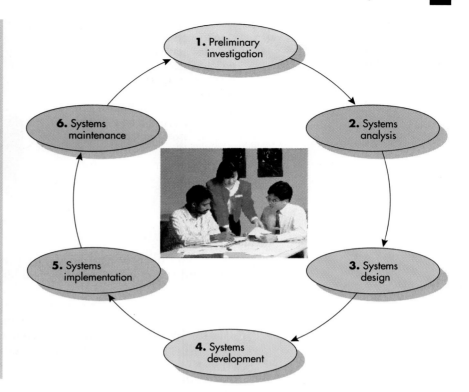

1. *Preliminary investigation:* Conduct preliminary analysis, propose alternative solutions, describe costs and benefits of each solution, and submit a preliminary plan with recommendations.
2. *Systems analysis:* Gather data, analyze the data, and make a written report.
3. *Systems design:* Make a preliminary design and then a detailed design, and write a report.
4. *Systems development:* Acquire the hardware and software and test the system.
5. *Systems implementation:* Convert the hardware, software, and files to the new system and train the users.
6. *Systems maintenance:* Audit the system, and evaluate it periodically.

■ PANEL 10.16

The systems development life cycle
An SDLC typically includes six phases.

Phases often overlap, and a new one may start before the old one is finished. After the first four phases, management must decide whether to proceed to the next phase. *User input and review is a critical part of each phase.* Let us proceed to look at each phase.

10.5 The First Phase: Conduct a Preliminary Investigation

KEY QUESTION

What does a preliminary investigation involve?

Preview & Review: In the first phase, preliminary investigation, a systems analyst conducts a preliminary analysis, determining the organization's objectives and the nature and scope of the problems. The analyst then proposes some possible solutions, comparing costs and benefits. Finally, he or she submits a preliminary plan to top management with recommendations.

The objective of Phase 1, *preliminary investigation*, is to conduct a preliminary analysis, propose alternative solutions, describe costs and benefits, and submit a preliminary plan with recommendations. (■ *See Panel 10.17, next page.*) This phase of preliminary investigation is often called a *feasibility study*.

If you are doing a systems analysis and design, it is safe, even preferable, to assume that you know nothing about the problem at hand. In the first phase, it is your job mainly to ask questions, do research, and try to come up with a preliminary plan. During this process, a systems analyst trained in *JAD (joint applications development)*—which uses highly organized, intensive workshops to bring together system owners, users, analysts, and designers—may run a 3- to 5-day workshop to replace months of traditional interviews and follow-up meetings.

■ PANEL 10.17

First phase: preliminary investigation

1. Conduct preliminary analysis. This includes stating the objectives, defining nature and scope of the problem.
2. Propose alternative solutions: leave system alone, make it more efficient, or build a new system.
3. Describe costs and benefits of each solution.
4. Submit a preliminary plan with recommendations.

1. Conduct the Preliminary Analysis

In this step, you need to find out what the organization's objectives are and the nature and scope of the problem under study.

- **Determine the organization's objectives:** Even if a problem pertains only to a small segment of the organization, you cannot study it in isolation. You need to find out what the objectives of the organization itself are. Then you need to see how the problem being studied fits in with them.

    To define the objectives of the organization, you can do the following:

    (1) *Read internal documents* about the organization. These can include original corporate charters, prospectuses, annual reports, and procedures manuals.

    (2) *Read external documents* about the organization. These can include news articles, accounts in the business press, reports by securities analysts, audits by independent accounting firms, and similar documents. You should also read up on reports on the competition (as in trade magazines, investors services' newsletters, and annual reports).

    (3) *Interview important executives* within the company. Within the particular area you are concerned with, you can also interview key users. Some of this may be done face to face. However, if you're dealing with people over a wide geographical area you may spend a lot of time on the phone or using e-mail.

    From these sources, you can find out what the organization is supposed to be doing and, to some extent, how well it is doing it. Also, you should try to understand the "organizational culture," the set of shared attitudes, values, goals, and practices that characterize the organization.

- **Determine the nature and scope of the problems:** You may already have a sense of the nature and scope of a problem. This may derive from the very fact that you have been asked to do a systems analysis and design project. However, with a fuller understanding of the goals of the organization, you can now take a closer look at the specifics.

    Is too much time being wasted on paperwork? on customers waiting? on employees working on nonessential tasks? How pervasive is the problem within the organization? outside of it? What people are most affected? And so on. Your reading and your interviews should give you a sense of the character of the problem.

2. Propose Alternative Solutions

In delving into the organization's objectives and the specific problem, you may have already discovered some solutions. Other possible solutions can come from interviewing people inside the organization, clients or customers affected by it, suppliers, and consultants. You can also study what competitors are doing. With this data, you then have three choices. You can leave the system as is, improve it, or develop a new system.

- **Leave the system as is:** Perhaps the problem really isn't bad enough to have to take the measures and spend the money required to get rid of it. This is often the case. Some paper-based or nontechnological systems may work best that way.

- **Improve the system:** Maybe changing a few key elements in the system—upgrading to a new computer or new software, or doing a bit of employee retraining, for example—will do the trick. Efficiencies might be introduced over several months, if the problem is not serious.

- **Develop a new system:** If the existing system is truly harmful to the organization, radical changes may be warranted. A new system would not mean just tinkering around the edges, introducing a new piece of hardware or software. It could mean changes in every part and at every level.

3. Describe Costs & Benefits

Whichever of the three alternatives is chosen, it will have costs and benefits. In this step, you need to indicate what these are.

The changes or absence of changes will have a price tag, of course, and you need to indicate what it is. Costs may depend on benefits, which may offer savings. There are all kinds of benefits that may be derived.[22] A process will be speeded up, streamlined through elimination of unnecessary steps, or combined with other processes. Input errors or redundant output may be reduced. Systems and subsystems may be better integrated. Users may be happier with the system. Customers or suppliers may interact better with the system. Security may be improved. Costs may be cut.

4. Submit a Preliminary Plan

Now you need to wrap up all your findings in a written report. The readers of this report will be the executives (probably top managers) who are in a position to decide in which direction to proceed—make no changes, change a little, or change a lot—and how much money to allow the project. You should describe the potential solutions, costs, and benefits and indicate your recommendations. If management approves the feasibility study, then the systems analysis phase can begin.

10.6 The Second Phase: Do an Analysis of the System

KEY QUESTION

How is systems analysis carried out?

Preview & Review: In the second phase, systems analysis, a systems analyst gathers data, using the tools of written documents, interviews, questionnaires, observation, and sampling. Next he or she analyzes the data, using such tools as data flow diagrams, systems flowcharts, and decision tables. Finally, the analyst writes a report.

■ PANEL 10.18

Second phase: systems analysis

1. Gather data, using tools of written documents, interviews, questionnaires, observations, and sampling.
2. Analyze the data, using CASE tools, data flow diagrams, systems flowcharts, connectivity diagrams, grid charts, and decision tables.
3. Write a report.

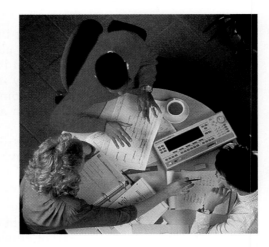

The objective of Phase 2, *systems analysis*, is to gather data, analyze the data, and write a report. (■ *See Panel 10.18.*) Systems *analysis* describes *what a system should do* to satisfy the needs of users. Systems *design*—the next phase—specifies *how the system will do it.*[23]

In this second phase of the SDLC, you will follow the course that management has indicated after having read your Phase 1 feasibility report. We are assuming that they have ordered you to perform Phase 2—to do a careful analysis or study of the existing system in order to understand how the new system you proposed would differ. This analysis will also consider how people's positions and tasks will have to change if the new system is put into effect.

1. Gather Data

In gathering data, there are a handful of tools that systems analysts use, most of them not terribly technical. They include written documents, interviews, questionnaires, observation, and sampling.

- **Written documents:** A great deal of what you need is probably available in written documents: reports, forms, manuals, memos, business plans, policy statements, and so on. Documents are a good place to start because they at least tell you how things are or are supposed to be. These tools will also provide leads on people and areas to pursue further.

  One document of particular value is the organization chart. An *organization chart* shows levels of management and formal lines of authority. We showed an example of an organization chart at the beginning of the chapter. (■ *Refer back to Panel 10.5, p. 462.*)

- **Interviews:** Interviews with managers, workers, clients, suppliers, and competitors will also give you insights. Interviews may be structured or unstructured.

  Structured interviews include only questions you have planned and written out in advance. By sticking with this script and not asking other questions, you can then ask people identical questions and compare their answers.

  Unstructured interviews also include questions prepared in advance, but you can vary from the line of questions and pursue other subjects if it seems productive.

- **Questionnaires:** Questionnaires are useful for getting information from large groups of people when you can't get around to interviewing everyone. Questionnaires may also yield more information because respondents can be anonymous. In addition, this tool is convenient, inexpensive, and yields a lot of data. However, people may not return their forms, results can be ambiguous, and with anonymous questionnaires you can't follow up.

- **Observation:** No doubt you've sat in a coffee shop or on a park bench and just done "people watching." This can be a tool for analysis, too. Through observation you can see people interact with each other and how paper moves through an organization.

  Observation can be nonparticipant or participant. If you are a *nonparticipant observer*, and people know they are being watched, they may falsify their behavior in some way. If you are a *participant observer*, you may get more insights by experiencing the conflicts and responsibilities of the people you are working with.

- **Sampling:** If your data-gathering phase involves a large number of people or a large number of events, it may simplify things just to study a sample. That is, you can do a sampling of the work of 5 people instead of 100, or 20 instances of a particular transaction instead of 500.

2. Analyze the Data

Once the data is gathered, you need to come to grips with it and analyze it. Many analytical tools, or modeling tools, are available. **Modeling tools enable a systems analyst to present graphic, or pictorial, representations of a system.** Examples of modeling tools are *data flow diagrams, systems flowcharts, connectivity diagrams, grid charts, decision tables,* and *object-oriented analysis.*

- **Data flow diagrams:** A *data flow diagram (DFD)* **graphically shows the flow of data through a system**—that is, the essential processes of a system, along with inputs, outputs, and files. (■ *See Panel 10.19, next page.*)

  In analyzing the current system and preparing data flow diagrams, the systems analyst must also prepare a data dictionary, which is then used and expanded during all remaining phases of the systems development life cycle. **A *data dictionary* defines all the elements of data that make up the data flow.** Among other things, it records what each data element is by name, how long it is (how many characters), where it is used (files in which it will be found), as well as any numerical values assigned to it. This information is usually entered into a data dictionary software program.

- **Systems flowcharts:** Another tool is the *systems flowchart*, also called the *system flow diagram.* **A *systems flowchart* diagrams the major inputs, outputs, and processes of a system.** Unlike a data flow diagram, a systems flowchart graphically depicts all aspects of a system. (■ *See Panel 10.20, page 483.*)

  Note: A *systems* flowchart is not the same as a *program* flowchart, which is very detailed. (We describe program flowcharts in Chapter 11.)

- **Connectivity diagrams:** **A *connectivity diagram* is used to map network connections of people, data, and activities at various locations.** (■ *See Panel 10.21, page 484.*) Because connectivity diagrams are concerned with communications networks, we may expect to see these in increasing use.

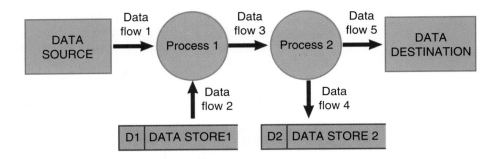

Explanation of standard data flow diagram symbols

Name or Name
Terminator Symbols (entity name)
(person or organization outside the
system boundaries)

File Name or File Name
Data Store Symbol

Name or Name
Process Symbol

Name
→
Data Flow Symbol
(inputs and outputs)

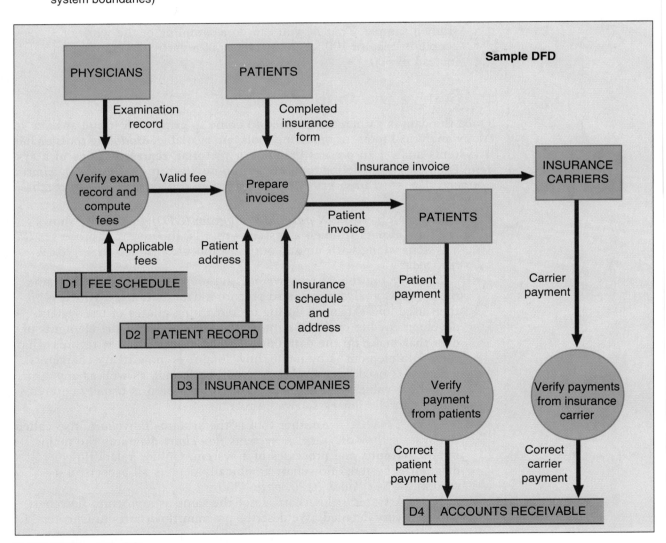

■ PANEL 10.19

Data flow diagram

(Top) Example of data flow diagram and explanation of symbols. *(Bottom)* Sample of data flow diagram of a physician's billing system.

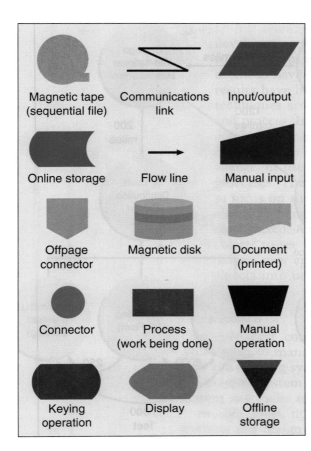

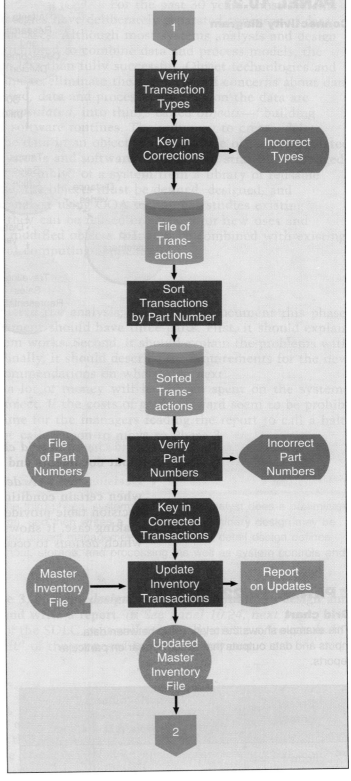

■ PANEL 10.20

Systems flowchart

(Left) Symbols. *(Right)* Example of a flowchart showing how inventory transactions are reflected in an updated master inventory file.

■ PANEL 10.24

**Third phase: systems
design**

1. Do a preliminary design, using
   CASE tools, prototyping tools,
   and project management software, among others.
2. Do a detail design, defining
   requirements for output, input,
   storage, and processing and
   system controls and backup.
3. Write a report.

- **CASE tools:** *CASE* (**for** *computer-aided software engineering*) **tools
  are programs that automate various activities of the SDLC in several
  phases.** (■ *See Panel 10.25.)* This technology is intended to speed up
  the process of developing systems and to improve the quality of the
  resulting systems.[24] Examples of such programs are Excelerator,
  Iconix, System Architect, and Powerbuilder.

  CASE tools may be used at almost any stage of the systems
  development life cycle, not just design. So-called *front-end CASE
  tools* are used during the first three phases—preliminary analysis,
  systems analysis, systems design—to help with the early analysis and
  design. So-called *back-end CASE tools* are used during later stages—
  systems development and implementation—to help in coding and
  testing, for instance.

  CASE tools may be used for many functions. For example, they
  may be used to draw diagrams of system components, which can be
  linked to other components. They can help systems analysts estimate
  and analyze the feasibility of a design. They may be used to do
  prototyping.

  Prototyping **refers to using workstations, CASE tools, and other
  software applications to build working models of system components
  so that they can be quickly tested and evaluated.** Thus, **a** *prototype* **is
  a limited working system developed to test out design concepts.** A

■ PANEL 10.25

CASE tool for banking

This screen is from a banking
system CASE tool by Iconix. It
shows a model for a transaction at an automated teller
machine. Users of the program would enter details
pertaining to their particular
situations.

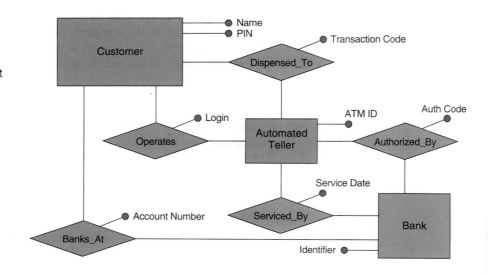

prototype, which may be constructed in just a few days, allows users to find out immediately how a change in the system might benefit them. For example, a systems analyst might develop a menu as a possible screen display, which users could try out. The menu can then be redesigned or fine-tuned, if necessary.

- **Project management software:** As we described in Chapter 2, *project management software* consists of programs used to plan, schedule, and control the people, costs, and resources required to complete a project on time. Project management software often uses Gantt charts and PERT charts.

  A *Gantt chart* uses lines and bars to indicate the duration of a series of tasks. The time scale may range from minutes to years. The Gantt chart allows you to see whether tasks are being completed on schedule.

  A *PERT (Program Evaluation Review Technique) chart* shows not only timing but also relationships among the tasks of a project. The relationships are represented by arrows joining numbered circles that represent events. Elapsed time is indicated alongside the arrows.

2. Do a Detail Design

A *detail design*, also called a *physical design*, describes how a proposed information system will deliver the general capabilities described in the preliminary design. The detail design usually considers the following parts of the system in this order: *output requirements, input requirements, storage requirements, processing requirements,* and *system controls and backup.*

- **Output requirements:** What do you want the system to produce? That is the first requirement to determine. In this first step, the systems analyst determines what media the output will be—whether hardcopy and/or softcopy. He or she will also design the appearance or format of the output, such as headings, columns, menu, and the like.

- **Input requirements:** Once you know the output, you can determine the inputs. Here, too, you must define the type of input, such as keyboard or source data-entry (✔ p. 195). You must determine in what form data will be input and how it will be checked for accuracy. You also need to figure what volume of data the system can be allowed to take in.

- **Storage requirements:** Using the data dictionary as a guide, you need to define the files and databases in the information system. How will the files be organized? What kind of storage devices will be used? How will they interface with other storage devices inside and outside of the organization? What will be the volume of database activity?

- **Processing requirements:** What kind of computer or computers will be used to handle the processing? What kind of operating system will be used? Will the computer or computers be tied to others in a network? Exactly what operations will be performed on the input data to achieve the desired output information? What kinds of user interface are desired?

- **System controls and backup:** Finally, you need to think about matters of security, privacy, and data accuracy. You need to prevent unauthorized users from breaking into the system, for example, and snooping in people's private files. You need to have auditing procedures and set up specifications for testing the new system (Phase

4). You need to institute automatic ways of backing up information and storing it elsewhere in case the system fails or is destroyed.

3. Write a Report

All the work of the preliminary and detail designs will end up in a large, detailed report. When you hand over this report to senior management, you will probably also make some sort of presentation or speech.

10.8 The Fourth Phase: Develop the System

KEY QUESTION

What is done during systems development?

Preview & Review: The fourth phase, systems development, consists of acquiring the hardware and software and then testing the system.

In Phase 4, *systems development*, the systems analyst or others in the organization acquire the software, acquire the hardware, and then test the system. (■ *See Panel 10.26.*)

The fourth phase begins once management has accepted your report containing the design and has "greenlighted" the way to development. Depending on the size of the project, this is the phase that will probably involve the organization in spending substantial sums of money. It could also involve spending a lot of time. However, at the end you should have a workable system.

1. Acquire Software

During the design stage, the systems analyst may have had to address what is called the "make-or-buy" decision, but that decision certainly cannot be avoided now. **In the *make-or-buy decision*, you decide whether you have to create a program—have it custom-written—or buy it, meaning simply purchase an existing software package.** Sometimes programmers decide they can buy an existing program and modify it rather than write it from scratch.

If you decide to create a new program, then the question is whether to use the organization's own staff programmers or hire outside contract programmers (*outsource* it). Whichever way you go, the task could take many months.

Programming is an entire subject unto itself, and we address it in Chapter 11.

■ **PANEL 10.26**

Fourth phase: systems development

1. Acquire software.
2. Acquire hardware.
3. Test the system.

2. Acquire Hardware

Once the software has been chosen, the hardware to run it must be acquired or upgraded. It's possible your new system will not require obtaining any new hardware. It's also possible that the new hardware will cost millions of dollars and involve many items: microcomputers, mainframes, monitors, modems, and many other devices. The organization may find it's better to lease rather than to buy some equipment, especially since chip capability has traditionally doubled every 18 months.

3. Test the System

With the software and hardware acquired, you can now start testing the system. Testing is usually done in two stages: *unit testing,* then *system testing.*

- Unit testing: **In *unit testing,* also called *modular testing,* individual parts of the program are tested, using test (made-up, or sample) data.** If the program is written as a collaborative effort by multiple programmers, each part of the program is tested separately.
- System testing: **In *system testing,* the parts are linked together, and test data is used to see if the parts work together.** At this point, actual organization data may be used to test the system. The system is also tested with erroneous and massive amounts of data to see if the system can be made to fail ("crash").

At the end of this long process, the organization will have a workable information system, one ready for the implementation phase.

10.9 The Fifth Phase: Implement the System

KEY QUESTIONS

How is the system implemented, and what are the four options?

Preview & Review: The fifth phase, systems implementation, consists of converting the hardware, software, and files to the new system and of training the users. Conversion may proceed in four ways: direct, parallel, phased, or pilot.

Whether the new information system involves a few handheld computers, an elaborate telecommunications network, or expensive mainframes, the fifth phase will involve some close coordination in order to make the system not just workable but successful. **Phase 5, *systems implementation,* consists of converting the hardware, software, and files to the new system and training the users.** (■ *See Panel 10.27.*)

■ **PANEL 10.27**

Fifth phase: systems implementation

1. Convert hardware, software, and files through one of four types of conversions: direct, parallel, phased, or pilot.
2. Compile final documentation.
3. Train the users.

1. Convert to the New System

***Conversion*, the process of converting from an old information system to a new one, involves converting hardware, software, and files.**

Hardware conversion may be as simple as taking away an old PC and plunking a new one down in its place. Or it may involve acquiring new buildings and putting in elaborate wiring, climate-control, and security systems.

Software conversion means making sure the applications that worked on the old equipment can be made to work on the new.

File conversion, or *data conversion*, means converting the old files to new ones without loss of accuracy. For example, can the paper contents from the manila folders in the personnel department be input to the system with a scanner? Or do they have to be keyed in manually, with the consequent risk of errors being introduced?

There are four strategies for handling conversion: *direct, parallel, phased,* and *pilot.* (■ *See Panel 10.28.*)

■ **PANEL 10.28**

Converting to a new system: four ways

Four strategies for converting to a new system are direct, parallel, phased, and pilot.

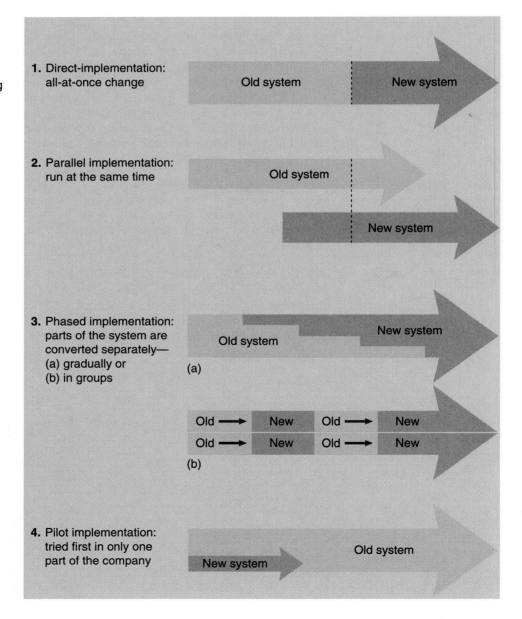

1. Direct-implementation: all-at-once change

   Old system | New system

2. Parallel implementation: run at the same time

   Old system

   New system

3. Phased implementation: parts of the system are converted separately—
   (a) gradually or
   (b) in groups

   (a) Old system | New system

   (b) Old → New Old → New
       Old → New Old → New

4. Pilot implementation: tried first in only one part of the company

   New system | Old system

- Direct approach: ***Direct implementation* mean the user simply stops using the old system and starts using the new one.** The risk of this method is evident: What if the new system doesn't work? If the old system has been discontinued, there is nothing to fall back on.

- Parallel approach: ***Parallel implementation* means that the old and new systems are operated side by side until the new system has shown it is reliable,** at which time the old system is discontinued. Obviously there are benefits in taking this cautious approach. If the new system fails, the organization can switch back to the old one. The difficulty with this method is the expense of paying for the equipment and people to keep two systems going at the same time.

- Phased approach: ***Phased implementation* means that parts of the new system are phased in separately**—either at different times (parallel) or all at once in groups (direct).

- Pilot approach: ***Pilot implementation* means that the entire system is tried out but only by some users.** Once the reliability has been proved, the system is implemented with the rest of the intended users. The pilot approach still has its risks, since *all* the users of a particular group are taken off the old system. However, the risks are confined to only a small part of the organization.

In general, the phased and pilot approaches are the most favored methods. Phased is best for large organizations in which people are performing different jobs. Pilot is best for organizations in which all people are performing the same task (such as order takers at a direct-mail house).

2. Compile Final Documentation

Developing good documentation is an ongoing process during all phases of the SDLC. Examples include manuals of operating procedures and sample data-entry display screens, forms, and reports.[25] If CASE tools have been used, much of the documentation will have been automatically generated and updated during the SDLC.

3. Train the Users

Back in the beginning of this book (Chapter 1), we pointed out that people are one of the important elements in a computer system. You wouldn't know this, however, to see the way some organizations have neglected the role of people when implementing a new computer system. An information system, however, is no better than its users. Hence, involving users from the beginning of the SDLC and ensuring proper training are essential.

Training is done with a variety of tools. They run from documentation (instruction manuals) to videotapes to live classes to one-on-one, side-by-side teacher-student training. Sometimes training is done by the organization's own staffers; at other times it is contracted out.

10.10 The Sixth Phase: Maintain the System

KEY QUESTION

How is system maintenance accomplished?

Preview & Review: The last phase, systems maintenance, adjusts and improves the system through system audits, user feedback, and periodic evaluations.

Phase 6, *systems maintenance,* adjusts and improves the system by having system audits, user feedback, and periodic evaluations and by making changes based on new conditions. (■ *See Panel 10.29.*)

■ PANEL 10.29

Sixth phase: systems maintenance

The sixth phase is to keep the system running through system audits and periodic evaluations.

Even with the conversion accomplished and the users trained, the system won't just run itself. There is a sixth—and never-ending—phase in which the information system must be monitored to ensure that it is successful. Maintenance includes not only keeping the machinery running but also updating and upgrading the system to keep pace with new products, services, customers, government regulations, and other requirements.

Two tools that are sometimes considered part of the maintenance phase are *auditing* and *evaluation.*

- Auditing: ***Auditing* means an independent review of an organization's information system to see if all records and systems are as they should be.** Often a systems analyst will design an *audit trail.* An audit trail helps independent auditors trace the record of a transaction from its output back through all processing and storage to its source.

- Evaluation: Auditing, which is usually done by an accountant, is one form of evaluation. Other evaluations may be done by the head systems analyst or outside systems analysts. Evaluations may also be done by a user or client who is able to compare the workings of the system against some preset criteria.

Every system that has been implemented should have a plan for decommissioning the system—for replacing or discontinuing it. Once the system is old enough to present new problems, the SDLC may be started all over again to design and develop a replacement. In large corporations, a typical SDLC can be measured in years, not in months or weeks.

The six steps in the systems development life cycle are summarized in the table opposite. (■ *See Panel 10.30.)*

Onward

At the Massachusetts Institute of Technology, a robot called the Collection Machine wanders aimlessly through a university lab at night picking up empty aluminum soda cans and disposing of them in a recycling bin. It does its work without a central computer processor or any human intervention.

Abstracting from this experience, MIT developed some guidelines for developing a system of distributed control: (1) Do simple things first. (2) Learn to do them flawlessly. (3) Add new layers of activity over the results of the simple tasks. (4) Don't change the simple things. (5) Make the new layer work as flawlessly as the simple. (6) Repeat, *ad infinitum.*[26]

Simply and flawlessly. Might these also be used as words to live by for programming? We examine this subject in the next chapter.

■ PANEL 10.30

Summary of the systems development life cycle

Phase	Tasks
Phase 1: Preliminary investigation	1. Conduct preliminary analysis. This includes stating the objectives, defining nature and scope of the problem. 2. Propose alternative solutions: leave system alone, make it more efficient, or build a new system. 3. Describe costs and benefits of each solution.
Phase 2: Systems analysis	1. Gather data, using such tools as written documents, interviews, questionnaires, observation, and sampling. 2. Analyze the data, using data flow diagrams, systems flowcharts, connectivity diagrams, grid charts, and decision tables. 3. Write a report.
Phase 3: Systems design	1. Do a preliminary design, using prototyping tools, CASE tools, and project management software. 2. Do a detail design, defining requirements for output, input, storage, and processing and system controls backup. 3. Write a report.
Phase 4: Systems development	1. Acquire software. 2. Acquire hardware. 3. Test the system.
Phase 5: Systems implementation	1. Convert hardware, software, and files through one of four types of conversion: direct, parallel, phased, or pilot. 2. Train the users.
Phase 6: Systems maintenance	1. Audit the system. 2. Evaluate the system periodically.

Experience Box

Critical Thinking Tools

"Clear thinkers aren't born that way," says one writer. "They work at it."[27]

The systems development life cycle is basically an exercise in clear thinking—critical thinking. Critical thinking happens throughout all of systems analysis and design, particularly in the first phase, preliminary analysis. It also is an essential part of the problem-solving procedure called programming.

"We live in a society that is enlarging the boundaries of knowledge at an unprecedented rate," says James Randi, "and we cannot keep up with more than a small portion of what is made available to us."

Randi deplores the uncritical thinking he sees on every hand—people basing their lives on horoscopes, numerology, and similar nonsense. To mix the data available to our senses "with childish notions of magic and fantasy is to cripple our perception of the world around us," he says. "We must reach for the truth, not for the ghosts of dead absurdities."[28]

Reaching for the truth may not come easily; it is a stance toward the world, developed with practice. To achieve this, we have to wrestle with obstacles that are mostly of our own making: mind-sets.

Mind-Sets

By the time we are grown, our minds have become "set" in various patterns of thinking that affect the way we respond to new situations and new ideas. These mind-sets are the result of our personal experiences and the various social environments in which we grew up. Such mind-sets determine what ideas we think are important and, conversely, what ideas we ignore.

"Because we can't pay attention to all the events that occur around us," points out one book on clear thinking, "our minds filter out some observations and facts and let others through to our conscious awareness."[29] Herein lies the danger: "As a result we see and hear what we subconsciously want to, and pay little attention to facts or observations that have already been rejected as unimportant."

Having mind-sets makes life comfortable. However, as the foregoing writers point out, "Familiar relationships and events become so commonplace that we expect them to continue forever. Then we find ourselves completely unprepared to accept changes that are necessary, even when they stare us in the face."

Critical Thinking

To break past mind-sets, we need to learn to think critically. *Critical thinking* is sorting out conflicting claims, weighing the evidence for them, letting go of personal biases, and arriving at reasonable views. Critical thinking means actively seeking to understand, analyze, and evaluate information in order to solve specific problems. It is very much a feature of the problem-solving process of the systems development life cycle in general and of programming in particular.

Critical thinking, we said, is simply clear thinking, an attribute that can be developed. "Before making important choices," says one writer, clear thinkers "try to clear emotion, bias, trivia and preconceived notions out of the way so they can concentrate on the information essential to making the right decision."[30] All it takes is practice.

Critical thinking is the process of sorting out conflicting claims, weighing the evidence for them, letting go of personal biases, and arriving at reasonable views. Critical thinking means actively seeking to understand, analyze, and evaluate information in order to solve specific problems.

The Reasoning Tool: Deductive & Inductive Arguments

The tool for breaking through mind-sets is reasoning. *Reasoning*—giving reasons in favor of this assertion or that—is essential to critical thinking and solving problems. Reasoning is put in the form of *arguments,* which consist of one or more *premises,* or reasons, logically supporting a result or outcome called a *conclusion.*

An example of an argument is as follows:

Premise 1: All students must pass certain courses in order to graduate.

Premise 2: I am a student who wants to graduate.

Conclusion: Therefore, I must pass certain courses.

Note the tip-off word "Therefore," which signals that a conclusion is coming. In real life, such as arguments on radio and TV shows, in books and magazines and newspapers, and the like, the premises and conclusions are not so neatly labeled. Still, there are clues: the words *because, since,* and *for* usually signal premises. The words *therefore, hence,* and *so* signal conclusions. Not all groups of sentences form arguments. Often they may form anecdotes or other types of exposition or explanation.[31]

The two main kinds of correct or valid arguments are inductive and deductive:

- **Deductive argument:** A deductive argument is defined as follows: *If its premises are true, then its conclusions are also true.* In other words, if the premises are true, the conclusions cannot be false.

- **Inductive argument:** An inductive argument is defined as follows: *If the premises are true, the conclusions are PROBABLY true, but the truth is not guaranteed.* An inductive argument is sometimes known as a "probability argument."

An example of a *deductive argument* is as follows:[32]
>*Premise 1:* All students experience stress in their lives.
>
>*Premise 2:* Reuben is a student.
>
>*Conclusion:* Therefore, Reuben experiences stress in his life.

This argument is deductive—the conclusion is *definitely* true if the premises are *definitely* true.

An example of an *inductive argument* is as follows:[33]
>*Premise 1:* Stress can cause illness.
>
>*Premise 2:* Reuben experiences stress in his life.
>
>*Premise 3:* Reuben is ill.
>
>*Conclusion:* Therefore, stress may be the cause of Reuben's illness.

Note the word "may" in the conclusion. This argument is inductive—the conclusion is not stated with absolute certainty; rather, it suggests only that stress *may* be the cause. The link between premises and conclusion is not definite because there may be other reasons for Reuben's illness.

Some Types of Incorrect Reasoning

Patterns of incorrect reasoning are known as *fallacies.* Learning to identify fallacious arguments will help you avoid patterns of faulty thinking in your own writing and thinking and identify it in others'.

Jumping to Conclusions   Also known as *hasty generalization,* the fallacy called *jumping to conclusions* means that a conclusion has been reached when not all the facts are available.

Example: As a new manager coming into a company that instituted the strategy of total quality management (TQM) 12 months earlier, you see that TQM has not improved profitability this year. Thus, you order TQM junked in favor of more traditional business strategies. However, what you don't know is that the traditional business strategies employed prior to TQM had an even *worse* effect on profitability.

Irrelevant Reason or False Cause   The faulty reasoning known as *non sequitur* (Latin for "It does not follow") might be better called *false cause* or *irrelevant reason.* Specifically, it means that the conclusion does not follow logically from the supposed reasons stated earlier. There is no *causal* relationship.

Example: You receive an A on a test. However, because you felt you hadn't been well prepared, you attribute your success to your friendliness with the professor. Or to your horoscope. Or to wearing your "lucky shirt." None of these "reasons" have anything to do with the result.

Irrelevant Attack on a Person or Opponent   Known as an *ad hominem* argument (Latin for "to the person"), the *irrelevant attack on an opponent* attacks a person's reputation or beliefs rather than his or her argument.

Example: Your boss insists you may not hire a certain person as a programmer because he or she has been married and divorced nine times. The fallacy here is thinking that the person's marital history has any bearing on his or her present ability to do programming.

Slippery Slope   The *slippery slope* is a failure to see that the first step in a possible series of steps does not lead inevitably to the rest.

Example: The "Domino theory," under which the United States waged wars against Communism, was a slippery-slope argument that assumed that if Communism triumphed in one country (for example, Nicaragua), then it would inevitably triumph in other regions (the rest of Central America), finally threatening the borders of the United States itself.

Appeal to Authority   The *appeal to authority* argument (known in Latin as *argumentum ad verecundiam*) uses authorities in one area to pretend to validate claims in another area in which the person is not an expert.

Example: You see the appeal to authority argument used all the time in advertising. But how medically qualified is a professional golfer to speak about headache remedies?

Circular Reasoning   The *circular reasoning* argument rephrases a statement to be proven and then uses the new, similar statement as supposed proof that the original statement is in fact true.

Example: You declare that you can drive safely at high speeds with only inches separating you from the car ahead because you have driven this way for years without an accident.

Straw Man Argument   The *straw man* argument is when you misrepresent your opponent's position to make it easier to attack, or when you attack a weaker position while ignoring a stronger one. In other words, you sidetrack the argument from the main discussion.

Example: Politicians use this argument all the time. Attacking a legislator for being "fiscally irresponsible" in supporting funds for a gun-control bill when what you really object to is the fact of gun control is a straw man argument.

Appeal to Pity   The *appeal to pity* argument appeals to mercy rather than makes an argument on the merits of the case itself.

Example: Begging the dean not to expel you for cheating because your parents are poor and made sacrifices to put you through college represents this kind of argument.

Questionable Statistics   Statistics can be misused in many ways as supporting evidence. The statistics may be unknowable, drawn from an unrepresentative sample, or otherwise suspect.

Example: Stating how much money is lost to taxes because of illegal drug transactions is speculation because such transactions are hidden or underground.

The Importance of Having *No* Opinion

It is not necessary to have an opinion pro or con about everything. Indeed, the basis of the scientific method is that a great deal of what is considered to be "the truth" is established *tentatively*. This is why scientists talk in terms of probabilities: nothing is definite or 100% certain, only probable and only for the time being. This means always having an awareness that other evidence may come along at some point to change the existing hypothesis or mode of thinking. If you continually take this attitude, you are indeed a critical thinker.

Suggested Resources

Kahane, Howard. *Logic and Contemporary Rhetoric: The Use of Reason in Everyday Life* (6th ed.). Belmont, CA: Wadsworth, 1992. A comprehensive and entertaining look at the most common fallacies used in politics, the mass media, textbooks, and everyday conversation.

Ruchlis, Hy, and Sandra Oddo. *Clear Thinking: A Practical Introduction.* Buffalo, NY: Prometheus, 1990. Simply and entertainingly describes barriers to clear thought: superstitions, stereotypes, prejudices, and conflicting opinions.

Summary

What It Is/What It Does	Why It's Important
auditing (p. 492, KQ 10.10) Refers to an independent review of an organization's information system to see if all records and systems are as they should be; often done during Phase 6 of the SDLC. Often a systems analyst will design an audit trail.	An audit trail helps independent auditors trace the record of a transaction from its output back through all processing and storage to its source.
computer-aided software engineering (CASE) tools (p. 486, KQ 10.7) Software that provides computer-automated means of designing and changing systems.	CASE tools may be used in almost any phase of the SDLC, not just design. So-called *front-end CASE tools* are used during the first three phases—preliminary analysis, systems analysis, systems design—to help with the early analysis and design. So-called *back-end CASE tools* are used during two later phases—systems development and implementation—to help in coding and testing, for instance.
connectivity diagram (p. 481, KQ 10.6) Modeling tool used to map network connections of people, data, and activities at various locations.	Because connectivity diagrams are concerned with communications networks, we may expect to see these in increasing use.
conversion (p. 490, KQ 10.9) Process of converting from an old information system to a new one; involves converting hardware, software, and files. There are four strategies for handling conversion: *direct, phased, parallel,* and *pilot.*	In order to smoothly switch from an old system to a new one, an orderly plan of conversion must be determined ahead of time.
data dictionary (p. 481, KQ 10.6) Record of the definition of all the elements of data that make up the data flow in a system.	In analyzing a current system and preparing data flow diagrams, systems analysts must prepare a data dictionary, which is used and expanded during subsequent phases of the SDLC.
data flow diagram (DFD) (p. 481, KQ 10.6) Modeling tool that graphically shows the flow of data through a system.	A DFD diagrams the processes that change data into information. DFDs have only four symbols, for source or destination of data, data flow, data processing, and data storage, so they are easy to use.

What It Is/What It Does	Why It's Important

decision support system (DSS) (p. 468, KQ 10.3) Computer-based information system that helps managers with nonroutine decision-making tasks. Inputs consist of some summarized reports, some processed transaction data, and other internal data. They also include data from sources outside the organization, such as trade associations, marketing research firms, and government agencies. The outputs are flexible, on-demand reports from which a top manager can make decisions about unstructured problems.

A DSS is installed to help top managers and middle managers make *strategic* decisions—decisions about unstructured problems, those involving events and trends outside the organization (for example, rising interest rates). The key attribute of a DSS is that it uses *models.* The DSS database, which draws on the TPS and MIS files, as well as outside data, is accessed through DSS software.

decision table (p. 484, KQ 10.6) Modeling tool that shows the decision rules that apply when certain conditions occur and what actions to take.

A decision table provides a model of a simple, structured decision-making case.

detail design (p. 487, KQ 10.7) Second stage of Phase 3 of the SDLC; also called a *physical design.* Describes how a proposed information system will deliver the general capabilities described in the preliminary design phase. The detail design usually considers the following system requirements: *output, input, storage, processing,* and *system controls and backup.*

A new system must be designed in detail before any hardware and software can be developed/purchased.

detail report (p. 467, KQ 10.3) Report that contains specific information about routine activities.

Detail reports are produced by transaction processing systems and are commonly used by lower-level managers.

direct implementation (p. 491, KQ 10.9) Method of system conversion; the users simply stop using the old system and start using the new one.

The risk of this method is that there is nothing to fall back on if the old system has been discontinued.

downsizing (p. 455, KQ 10.1) (1) Reducing the size of an organization by eliminating workers. (2) Moving from mainframe-based computer systems to systems linking smaller computers in networks.

Downsizing of staff is important because as layoffs have shrunk middle and lower management, the remaining people have had to handle more information themselves, leading to greater use of microcomputers and networked systems.

empowerment (p. 456, KQ 10.1) Refers to giving others the authority to act and make decisions on their own.

Old-style management gave lower-level managers and employees only the information they "needed" to know, reducing their power to make decisions. Today information is apt to be spread widely, empowering employees lower down in the organization and enabling them to do their jobs better.

exception reports (p. 468, KQ 10.3) Middle-management reports that show out-of-the-ordinary data—for example, an inventory report listing only those items that number fewer than 10 in stock.

Exception reports highlight matters requiring prompt decisions by management.

What It Is/What It Does

executive information system (EIS) (p. 469, KQ 10.3) Also called an *executive support system (ESS);* easy-to-use DSS made especially for top managers that specifically supports strategic decision making. It draws on data both from inside and outside the organization (for example, news services, market-research databases).

expert system (p. 471, KQ 10.3) Set of interactive computer programs that helps users solve problems that would otherwise require the assistance of a human expert.

grid chart (p. 484, KQ 10.6) Modeling tool that shows the relationship between data on input documents and data on output documents.

lower-level managers (p. 463, KQ 10.2) Also called *supervisory managers;* the lowest level in the hierarchy of the three types of managers. Their job is to make operational decisions, monitoring day-to-day events and, if necessary, taking corrective action.

make-or-buy decision (p. 488, KQ 10.8) Decision made in Phase 4 (programming) of the SDLC concerning whether the organization has to make a program—have it custom-written—or buy it, meaning simply purchase an existing software package.

management (p. 460, KQ 10.2) Level of personnel that oversees the tasks of planning, organizing, staffing, supervising, and controlling business activities.

management information system (MIS) (p. 468, KQ 10.3) Computer-based information system that derives data from all of an organization's departments and produces *summary, exception, periodic,* and *on-demand* reports of the organization's performance.

middle-level managers (p. 462, KQ 10.2) One of the three types of managers; they implement the goals of the organization. Their job is to oversee the supervisors and to make tactical decisions.

model (p. 469, KQ 10.3) Mathematical representation of a real system; models are often used in a DSS.

Why It's Important

The EIS includes capabilities for analyzing data and doing "what if" scenarios.

Expert systems are used by management and nonmanagement personnel to solve sophisticated problems.

Grid charts are used in the systems design phase of the SDLC.

Lower managers need information that is structured—that is, detailed, current, and past-oriented, covering a narrow range of facts and events inside the organization.

The decision taken affects the costs and time required to develop the system.

Different levels of managers need different kinds of information on which to make decisions.

An MIS principally assists middle managers, helping them make *tactical* decisions—spotting trends and getting an overview of current business activities.

Middle managers require information that is both structured and unstructured.

A model allows the manager to do a simulation—play a "what if" game—to reach decisions. By changing the hypothetical inputs to the model, one can see how its outputs are affected.

What It Is/What It Does	Why It's Important
modeling tools (p. 481, KQ 10.6) Analytical tools like charts, tables, and diagrams used by systems analysts. Examples are data flow diagrams, decision tables, systems flowcharts, and object-oriented analysis.	Modeling tools enable a systems analyst to present graphic, or pictorial, representations of a system.
office automation system (OAS) (p. 471, KQ 10.3) Computer information system that combines various technologies to reduce the manual labor needed to operate an office efficiently; used at all levels of an organization.	An OAS uses a network to integrate such technologies as fax, voice mail, e-mail, scheduling software, word processing, and desktop publishing and make them available throughout an organization.
on-demand reports (p. 468, KQ 10.3) Middle-management reports that produce information in response to an unscheduled demand; often produced on screen rather than on paper.	On-demand reports help managers make decisions about nonroutine matters (for example, whether to grant credit to a new customer placing a big order).
operational decision (p. 463, KQ 10.2) Type of decision made by lower-level managers; predictable decision that can be made by following a well-defined set of routine procedures.	*Operational* means focusing principally on supervising and controlling instead of on the other management tasks of planning, organizing, and staffing.
organization chart (p. 461, KQ 10.2) Schematic drawing showing the hierarchy of relationships among an organization's employees.	Organization charts show levels of management and formal lines of authority.
outsourcing (p. 455, KQ 10.1) Contracting with outside businesses or services to perform the work previously done by in-house departments, whether janitorial tasks or systems analysis.	Outside specialized contractors often can do work more cheaply and efficiently.
parallel implementation (p. 491, KQ 10.9) Method of system conversion whereby the old and new systems are operated side by side until the new system has shown it is reliable.	If the new system fails, the organization can switch back to the old one. The difficulty is the expense of paying for equipment and people to operate two systems simultaneously.
periodic reports (p. 468, KQ 10.3) Middle-management reports that are produced on a regular schedule, such as daily, weekly, monthly, quarterly, or annually. They are usually produced on paper, such as computer printouts.	Periodic reports, such as sales figures, income statements, or balance sheets, help managers make routine decisions.
phased implementation (p. 491, KQ 10.9) Method of system conversion whereby parts of the new system are phased in gradually, perhaps over several months, or all at once, in groups.	This conversion strategy is prudent, though it can be expensive.

What It Is/What It Does	**Why It's Important**

pilot implementation (p. 491, KQ 10.9) Method of system conversion whereby the entire system is tried out by only some users. Once the reliability has been proved, the system is implemented with the rest of the intended users.

The pilot approach has risks, since all the users of a particular group are taken off the old system. However, the risks are confined to a small part of the organization.

preliminary design (p. 485, KQ 10.7) Also called a *logical design;* first stage of Phase 3 of the SDLC; describes the general functional capabilities of a proposed information system. Tools that may be used include *CASE tools* and project management software.

During the preliminary design phase, staff reviews the system requirements and then considers major components of the system. Usually several alternative systems (called *candidates*) are considered, and the costs and benefits of each are evaluated.

preliminary investigation (p. 477, KQ 10.5) Phase 1 of the SDLC; the purpose is to conduct a preliminary analysis (determine the organization's objectives, determine the nature and scope of the problem), propose alternative solutions (leave the system as is, improve the efficiency of the system, or develop a new system), describe costs and benefits, and submit a preliminary plan with recommendations.

The preliminary investigation lays the groundwork for the other phases of the SDLC.

prototype (p. 486, KQ 10.7) A limited working system, or part of one. It is developed to test out design concepts.

A prototype, which may be constructed in just a few days, allows users to find out immediately how a change in the system might benefit them.

prototyping (p. 486, KQ 10.7) Involves building a working model of all or part of a system so that it can be quickly tested and evaluated; uses workstations, CASE tools, and other applications software.

Prototyping is part of the preliminary design stage of Phase 3 of the SDLC.

reengineering (p. 456, KQ 10.1) Also known as *process innovation* and *core process redesign;* refers to the search for and implementation of radical change in business processes to achieve breakthrough results.

Reengineering is not just fixing up what already exists; it works best with big processes that really matter (for example, new-product development or customer service). Thus, candidates for this procedure include companies experiencing big shifts in their definition, markets, or competition. Examples are information technology companies—computer makers, cable-TV providers, and local and long-distance phone companies—which are wrestling with technological and regulatory change.

semistructured information (p. 465, KQ 10.2) Information that does not necessarily result from clearly defined, routine procedures. Middle managers need semistructured information that is detailed but more summarized than information for operating managers.

Semistructured information involves review, summarization, and analysis of data to help plan and control operations and implement policy formulated by upper managers.

What It Is/What It Does	Why It's Important

strategic decision (p. 461, KQ 10.2) Type of decision made by top managers; rarely based on predetermined routine procedures but involving the subjective judgment of the decision maker.

Strategic means that of the five management tasks (planning, organizing, staffing, supervising, controlling), top managers are principally concerned with planning.

structured information (p. 465, KQ 10.2) Detailed, current information concerned with past events; it records a narrow range of facts and covers an organization's internal activities.

Lower-level managers need easily defined information that relates to the current status and activities within the basic business functions.

summary reports (p. 468, KQ 10.3) Reports that show totals and trends (for example, a report showing total sales by office, by product, and by salesperson).

Summary reports are used by middle managers to make decisions.

system (p. 475, KQ 10.4) Collection of related components that interact to perform a task in order to accomplish a goal.

Understanding a set of activities as a system allows one for look for better ways to reach the goal.

systems analysis (p. 480, KQ 10.6) Phase 2 of the SDLC; the purpose is to gather data (using written documents, interviews, questionnaires, observation, and sampling), analyze the data, and write a report.

The results of systems analysis will determine whether the system should be redesigned.

systems analysis and design (p. 476, KQ 10.4) Problem-solving procedure for examining an information system and improving it; consists of the six-phase *systems development life cycle.*

The point of systems analysis and design is to ascertain how a system works and then take steps to make it better.

systems analyst (p. 475, KQ 10.4) Information specialist who performs systems analysis, design, and implementation.

The systems analyst studies the information and communications needs of an organization to determine how to deliver information that is more accurate, timely, and useful. The systems analyst achieves this goal through the problem-solving method of systems analysis and design.

systems design (p. 485, KQ 10.7) Phase 3 of the SDLC; the purpose is to do a preliminary design and then a detail design, and write a report.

Systems design is one of the most crucial phases of the SDLC; at the end of this stage executives decide whether to commit the time and money to develop a new system.

systems development (p. 488, KQ 10.8) Phase 4 of the SDLC; hardware and software for the new system are acquired and tested. The fourth phase begins once management has accepted the report containing the design and has approved the way to development.

This phase may involve the organization in investing substantial time and money.

systems development life cycle (SDLC) (p. 476, KQ 10.4) Six-phase process that many organizations follow during systems analysis and design: (1) *preliminary investigation;* (2) *systems analysis;* (3) *systems design;* (4) *systems development;* (5) *systems implementation;* (6) *systems maintenance.* Phases often overlap, and a new one may start before the old one is finished. After the first four phases, management must decide whether to proceed to the next phase. User input and review is a critical part of each phase.

The SDLC is a comprehensive tool for solving organizational problems, particularly those relating to the flow of computer-based information.

systems flowchart (p. 481, KQ 10.6) Also called the *system flow diagram;* modeling tool that uses many symbols to diagram the input, processing, and output of data in a system as well the interaction of all the parts in a system.

A systems flowchart graphically depicts all aspects of a system.

systems implementation (p. 489, KQ 10.9) Phase 5 of the SDLC; consists of converting the hardware, software, and files to the new system and training the users.

This phase is important because it involves putting design ideas into operation.

systems maintenance (p. 491, KQ 10.10) Phase 6 of the SDLC; consists of adjusting and improving the system by having system audits, user feedback, and periodic evaluations and by making changes based on new conditions.

This phase is important for keeping a new system operational and useful.

system testing (p. 489, KQ 10.8) Part of Phase 4 of the SDLC; the parts of a new program are linked together, and test data is used to see if the parts work together.

Test data may consist of actual data used within the organization. Also, erroneous and massive amounts of data may be used to see if the system can be made to fail.

tactical decision (p. 462, KQ 10.2) Type of decision made by middle managers that is without a base of clearly defined informational procedures, perhaps requiring detailed analysis and computations.

Tactical means that, of the five management tasks (planning, organizing, staffing, supervising, controlling), middle managers deal principally with organization and staffing.

top managers (p. 461, KQ 10.2) One of the three types of managers; also called *strategic managers,* they are concerned with long-range planning and strategic decisions.

Top managers need information that is unstructured—that is, summarized, less current, future-oriented, covering a broad range of facts, and concerned with events outside as well as inside the organization.

total quality management (TQM) (p. 455, KQ 10.1) Philosophy of management based on an organization-wide commitment to continuous work improvement and meeting customer needs.

Many industries have benefited from TQM principles, such as American automobile makers.

What It Is/What It Does	**Why It's Important**

transaction (p. 466, KQ 10.3) Recorded event having to do with routine business activities (for example, materials purchased, employees hired, or taxes paid).

Today in most organizations the bulk of transactions are recorded in a computer-based information system.

transaction processing system (TPS) (p. 466, KQ 10.3) Computer-based information system that keeps track of the transactions needed to conduct business. Inputs are transaction data (for example, bills, orders, inventory levels, production output). Outputs are processed transactions (for example, bills, paychecks). Each functional area of an organization—Research and Development, Production, Marketing, and Accounting and Finance—usually has its own TPS.

The TPS helps supervisory managers in making *operational decisions.* The database of transactions stored in a TPS are used to support a management information system and a decision support system.

unit testing (p. 489, KQ 10.8) Part of Phase 4 of the SDLC; also called *modular testing.* In this stage, individual parts of a new program are tested, using test data; precedes system testing.

If the program is written as a collaborative effort by multiple programmers, each part of the program is tested separately.

unstructured information (p. 465, KQ 10.2) Summarized, less current information concerned with future events; it records a broad range of facts and covers activities outside as well as inside an organization.

Top managers need information in the form of highly unstructured reports. The information should cover large time periods and survey activities outside as well as inside the organization.

Self-Test

1. Middle managers make _____ decisions.

2. A _____ is a collection of related components that interact to perform a task in order to accomplish a goal.

3. _____ _____ is when the old system is halted on a given date and the new system is activated.

4. _____ refers to the transition from mainframe-based computer systems to linking smaller computers in networks.

5. Information has three properties that vary in importance depending on the decision and the decision maker. They are:

   a. _____

   b. _____

   c. _____

Short-Answer Questions

1. List the six phases of the SDLC.
2. What is a decision support system?
3. What is the purpose of the systems development phase of the SDLC?
4. What is an executive information system?
5. What five departments exist in most companies?

Multiple-Choice Questions

1. Which of the following isn't a trend that is forcing change in the workplace?
   a. downsizing and outsourcing
   b. total quality management
   c. reengineering
   d. employee empowerment
   e. increased salaries

2. Which of the following is used to support the making of tactical decisions?
   a. transaction processing system
   b. management information system
   c. decision support system
   d. executive information system
   e. All of the above

3. Which of the following describes the method of trying out a new system on a few users?
   a. direct approach
   b. parallel approach
   c. phased approach
   d. pilot approach
   e. None of the above

4. Which of the following is used to support all levels of management (including nonmanagement)?
   a. transaction processing system
   b. management information system
   c. decision support system
   d. expert system
   e. All of the above

5. In which phase of the SDLC are CASE tools, Gantt charts, and PERT charts used?
   a. Phase 1
   b. Phase 2
   c. Phase 3
   d. Phase 4
   e. Phase 5

True/False Questions

T F 1. During the system testing phase of the SDLC, the system is often fed erroneous data.

T F 2. Gantt charts are used to automate various activities in the SDLC.

T F 3. A transaction processing system supports day-to-day business activities.

T F   4.   Decision support systems are used mainly by upper management.

T F   5.   A management information system isn't always computer-based.

Knowledge in Action

1. Design a system that would handle the input, processing, and output of a simple form of your choice. Use a data flow diagram to illustrate the system.

2. Interview a student majoring in computer science who plans to become a systems analyst. Why is this person interested in this field? What does he or she hope to accomplish in it? What courses must be taken to satisfy the requirements for becoming an analyst? What major changes in systems design and analysis does this person forecast for the next five years?

3. Does your university/college have an information systems department that is responsible for developing and supporting all the university information systems? If so, interview a management staff member about the services and functions of the department. Can this person identify the various levels of management within the department? What kinds of user input were requested when the department was being set up? Does it use any sophisticated decision support software? What kinds of services does the department offer to students?

4. In this chapter we describe six trends affecting the reengineering of the workplace. Profile an organization that has been forced to reengineer some of its processes. Where did you find information about this organization? To date, what processes have been reengineered? Why? What additional processes, if any, does the organization plan to reengineer in the future? Which ones? Why? Do you think this organization will be more successful as a result of these reengineering efforts?

5. Suppose you are the owner of a bakery in a medium-sized city. You have three full-time employees and two part-time employees. Customers choose from a variety of baked goods, a selection of gourmet coffees, and have the option of sitting at one of ten tables. The bakery is busiest in the morning hours.

   Because of the growing popularity of your restaurant, you are considering opening a second bakery in town. You would also like to add additional tables to the current bakery, but this would require some remodeling of the existing space. Unfortunately, for now, it is only feasible for you to pursue one of these efforts. What should you do? Open a second bakery or expand the existing one? What information will you need in order to make the best decision? What is your decision?

Software Development

Programming & Languages

THE DAY THE WORLD CRASHES, blared the cover of *Newsweek.* CAN WE FIX THE 2000 BUG BEFORE IT'S TOO LATE?[2]

What is this dark prophecy, and what is the "2000 Bug"? Readers of this book in 1999 may or may not have already felt its effects. Readers in 2000 will know whether civilization has dodged a bullet.

The 2000 Bug—or Millennium Bug, or Year 2000 (abbreviated Y2K) Problem—has come about because in the old mainframe days, when computer memory and storage capacity were limited, programs were designed to save space when recording a year by storing just its last two digits. For example, the year 1999 is stored as "99." When 2000 rolls around, "99" will become "00." To computers, the advent of the millennium will set us *back* a century—to 1900.

Thus, if you were born in 1934 and applied for Social Security benefits in 1999, the computer would subtract "34" from "99" to calculate your age at 65. But if you applied for benefits in 2000, the computer would subtract "34" from "00" to calculate your age at . . . what?

Confusion could reign everywhere. For banks, the best-case scenario is that if computers can't read the date 2000 they just won't work. The worst case is that they will miscalculate all kinds of numbers, from mortgage payments to stock prices.[3] "Traffic signals could go haywire, causing gridlock and car wrecks," suggests one account. "Cash machines might jam up, mighty corporations could go bankrupt, and convicted felons could escape from jail."[4] Says another, "[E]laborate computer systems run everything from hospital life-support systems to air-traffic control to the military's secret weapons programs. So the day after the world's biggest New Year's Eve party on 12/31/99 may be even more interesting than the one before it."[5]

Government agencies, including the Social Security Administration, are spending billions of dollars addressing the Y2K Problem. Estimates on the cost of fixing the Millennium Bug worldwide range from $200 billion to $600 billion. "At the lower figure, the year 2000 problem would still qualify as the most expensive industrial accident of modern times," says one commentator. "As disasters go, it tops Japan's Kobe earthquake ($100 billion) *plus* the Los Angeles earthquake ($40 billion) *plus* Hurricane Andrew ($30 billion)."[6]

The Year 2000 Problem came about because programmers were doing what they are paid to do: solve problems and save money. In the 1960s, at a time when computers were mainframes and the expense of storing data was enormous, programmers made a sensible trade-off. By reducing dates on documents by two digits—two bytes on a computer—programmers could dramatically reduce storage requirements.[7] According to the *Journal of Systems Management,* "The two-digit year format saved the typical organization $1 million per gigabyte of total storage in the 30-year period from 1963–1992."[8] If invested, these savings could have returned $15 million per gigabyte over the same period.

It's not as though civilization has had no warning. Banks first met the Y2K problem back in the 1970s, when computers balked in making calculations past the millennium. Because technology marches forward so rapidly, however, no one expected so many aging computer systems, known as "legacy systems," to still be running as we rounded the corner into the new century. As you might expect, recent systems—including microcomputer systems—contain upgraded software that expands date fields from two digits to four. As for the others, there is much work to be done. The commercial banking system alone, for instance, has 9.3 billion lines of programming code to be proofed.[9] Culling through millions of lines to fix every two-digit date reference is not easy, and one mistake could jeopardize the entire operation. "If routine bug fixes"—the kind of errors that programmers correct every day—"are akin to finding defective seats in a football stadium," says computer writer Gina Smith, "the Year 2000 fix is more like repairing every wobbly chair in America."[10]

As a result, former programmers like Donald Fowler, 52, of Largo, Florida, have been called out of retirement to go back and fix the future. Golfing and tanning are now reserved for weekends, as during the week Fowler pores over lines of computer code in a windowless room. For this he is paid $75,000 a year. "They made an offer I couldn't refuse," he says.[11]

The Year 2000 Problem has shown the world just how important programming and programmers are. Let us see what this subject is about.

11.1 Programming: A Five-Step Procedure

KEY QUESTIONS

What is a program, and what are the five steps in producing it?

Preview & Review: Programming is a five-step procedure for producing a program—a list of instructions—for the computer.

Many people assume that the numbers that appear in a printout of, say, a spreadsheet are probably correct. There is something about the look of a finished product that inspires faith in the result.

However, the numbers are correct only if the data and processing procedures are correct. People often think of programming as simply typing words and numbers into a computer. This is part of it, but only a small part. Basically, programming is a *method of solving problems.* That is, it uses **algorithms, each a set of ordered steps for solving a problem.** (*Algorithm* essentially means the same thing as *logic.*)

What a Program Is

To see how programming works, consider what a program is. **A *program* is a list of instructions that the computer must follow in order to process data into information.** The instructions consist of *statements* used in a programming language, such as BASIC.

As we said in Chapter 1, *applications software* is defined as software that can perform useful work on general-purpose tasks. Examples are programs that do word processing, desktop publishing, or payroll processing.

Applications software may be *packaged* or *customized.*

- Packaged software: This is the kind that you buy from a computer store or mail-order house or download from a vendor's Web site, which is the kind used by most PC users. *Packaged software* is an "off-the-shelf," prewritten program developed for sale to the general public. (We might also include here shareware and freeware, since they come ready to use right away.)

- Customized software: *Customized software* is software designed for a particular customer. This is an applications program that is created or custom-made. It is usually written by a professional programmer, although you can do this, too, for some kinds of applications. Customized software is written to perform a task that cannot be done with packaged software.

The decision whether to buy or create a program is *one* of the phases in the systems development life cycle, as discussed in Chapter 10. (■ *See Panel 11.1.)* If the decision is made to develop a new system, this requires taking some further steps.

What Programming Is

A program, we said, is a list of instructions that the computer must follow in order to process data into information. **Programming, also called *software engineering,* is a multistep process for creating that list of instructions.** Only one of those steps (the step called *coding*) consists of sitting at the keyboard typing words into a computer.

The five steps are as follows. (■ *Refer again to Panel 11.1.)*

1. Clarify the problem.
2. Design a solution.
3. Code the program.
4. Test the program.
5. Document and maintain the program.

■ **PANEL 11.1**

Where programming fits in the systems development life cycle
The fourth phase of the six-phase systems development life cycle has a five-step procedure of its own. These five steps are the problem-solving process called *programming.*

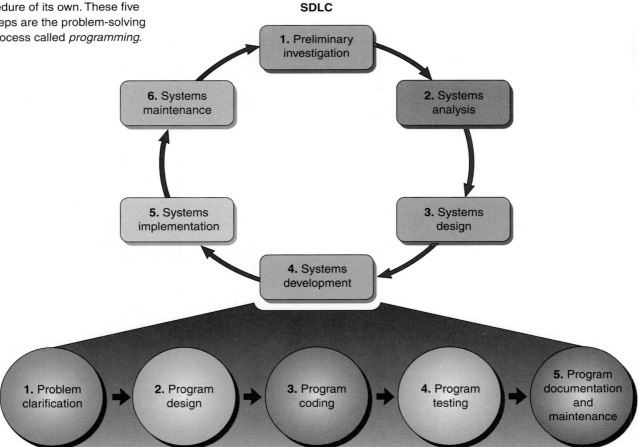

SDLC

1. Preliminary investigation
2. Systems analysis
3. Systems design
4. Systems development
5. Systems implementation
6. Systems maintenance

1. Problem clarification
2. Program design
3. Program coding
4. Program testing
5. Program documentation and maintenance

READ ME

Case Study: Programmers Wanted—*Really* Wanted

Programmers are very well paid.

"A typical programmer's wages, now some $70,000, are jumping 13% a year," says a *Business Week* article, "and far higher in the hottest niches, such as Java Internet software. . . . These days, $20,000 signing bonuses are commonplace and stock options are being handed out with as little fanfare as office supplies."[12]

Demand for software professionals worldwide is growing by more than 500,000 a year.[13] Colleges complain that they can't hire—or keep—enough programmers.[14] Companies regularly steal each others' programming employees, using $30,000 bonuses or 30% raises.[15] Mainframe programmers in their 60s and 70s are being lured out of retirement.[16] Programming work is being shipped offshore—mostly to India—and recruiters scour India, Brazil, Russia, and elsewhere to lure trained employees to the U.S.[17] University of Toledo sophomore Suleyman Gokyigit, 18, who is blind, earns $12,500 a year working part time at programming.[18] For software stars—the most talented programmers—there are even "talent agents" who place them with firms for temporary assignments, sometimes paying as much as $200 an hour. "The best of these hired code slingers," says one report, "can command incomes of $300,000 or more a year."[19]

What's responsible for making programming such a hot career? There are a whole bunch of reasons.

First, there are two short-range emergencies.[20] (1) The Year 2000 Problem described in the text requires that billions of lines of code be reviewed and reworked in order to stave off chaos come January 1, 2000. (2) The European Union's decision to convert to a common currency, the euro, in 1999 is requiring a massive amount of programming.

Then there are important long-range trends:

- **More products operate with software:** Software has become a vital part of new products, from manufacturing equipment to microwave ovens. "There are, for example, 300,000 lines of computer code in a cellular telephone," says one article, "and new cars can contain millions of lines of code."[21]

- **The Internet boom:** The popularity of the Internet has so speeded up software development cycles that it has created a voracious industry demand for programming talent.[22]

- **Technological change is slowing programmer productivity:** Programmer output is down in the U.S. A study of 46 countries showed the United States is at the *bottom* of the global ranking. (Canada is first.) U.S. developers produced an average of 354,000 lines of software code in 1996, down almost 50% from 690,000 in 1995.

  Why the decline? There is a "learning curve effect" of technological change, suggests Howard Rubin, who conducted the study. By staying away from the leading edge in software development, other countries aren't experiencing as big a drop in productivity.[23]

- **The number of computer-science graduates has declined:** Finally, the number of students majoring in computer science is far short of the demand. In fact, the problem has gotten worse: In 1986, according to the Department of Education, about 42,000 people graduated with computer science degrees in the United States. In 1995, it was only 24,404.[24] As a result, one out of 10 computing jobs at information technology companies is going unfilled.

Not everyone has the temperament to be a programmer, which can require long hours of exacting work in front of a computer. But many have found it fulfilling. Robert Silvers, for instance, was able to combine programming with his interest in photography.

"In 1991, I saw a portrait made of seashells arranged on a board," says Silvers, 29, a graduate of MIT. "When you saw it from a distance, it made a portrait of a man's face. But up close, the portrait disappeared."[25]

If it worked for seashells, he thought, why not for photographs? Silvers wrote a program that would create so-called photomosaics, portraits and other images that are made up of what he calls "tiles"—small photos, stamps, paper currency, or whatever. Often he starts with a large collection of "tiles," close to 1000. Then he scans in a larger photo, such as a picture of Abraham Lincoln.

"The computer tries to pick photos [tiles] that look like that part of the image it's trying to represent," says Silvers. If the tiles are made up of Civil War–era photographs, for example, a glint in Lincoln's left eye may turn out to be the white hair of a Civil War soldier in that tile.

Silvers has published a book, *Photomosaics,* and makes signed posters of his images available through his Web site (*http://www. photomosaic.com*).

Clearly, programming appeals to the puzzle-solving kind of mind.

11.2 The First Step: Clarify the Programming Needs

KEY QUESTION

How are programming needs clarified?

Preview & Review: Programmers break the definition of the problem down into six mini-steps. They include (1) clarifying program objectives and users; specifying (2) output, (3) input, and (4) processing requirements; (5) studying the feasibility of implementing the program; and (6) documenting the analysis. An important part of the process is the make-or-buy decision—whether to buy off-the-shelf software or custom-make a program.

The *problem clarification* step requires performing six mini-steps. They include clarifying program objectives and users, output, input, and processing tasks, then studying their feasibility and documenting them. (■ *See Panel 11.2.)* Let us consider these six mini-steps.

1. Clarify Objectives & Users

You solve problems all the time. A problem might be deciding whether to take a required science course this term or next. Or you might try to solve the problem of grouping classes so you can fit in a job. In such cases, you are specifying your *objectives.* Programming works the same way. You need to write a statement of the objectives you are trying to accomplish—the problem you are trying to solve. If the problem is that your company's systems analysts have designed a new computer-based payroll processing program and brought it to you as the programmer, you need to clarify the programming needs.

You also need to make sure you know who the users of the program will be. Will they be people inside the company, outside, or both? What kind of skills will they bring?

2. Clarify Desired Outputs

Make sure you understand the outputs—what the system designers want to get out of the system—before you specify the inputs. For example, what kind of hardcopy is wanted? What information should the outputs include? This step may require several meetings with systems designers and users to make sure you're creating what they want.

3. Clarify Desired Inputs

Once you know the kind of outputs required, you can then think about input. What kind of input data is needed? What form should it appear in? What is its source? What type or style of user interface is required by the users?

■ PANEL 11.2

First step: clarify programming needs

1. Specify program objectives and program users.
2. Specify output requirements.
3. Specify input requirements.
4. Specify processing requirements.
5. Study feasibility of implementing program.
6. Document the analysis.

4. Clarify the Desired Processing

Here you make sure you understand the processing tasks that must occur in order for input data to be processed into output data.

5. Double-Check the Feasibility of Implementing the Program

Is the kind of program you're supposed to create feasible within the present budget and timeline? Will it require hiring a lot more staff? Will it take too long to accomplish?

Sometimes programmers decide they can buy an existing program and modify it rather than write it from scratch (that is, if copyright isn't an issue).

6. Document the Analysis

Throughout this first step on program clarification, programmers must document everything they do. This includes writing objective specifications of the entire process being described.

11.3 The Second Step: Design the Program

KEY QUESTION

How is a program designed?

Preview & Review: In the second step, programmers design a solution. This consists of three mini-steps. (1) The program logic is determined through a top-down approach and modularization, using a hierarchy chart. (2) Next the program is designed with certain tools: pseudocode and/or flowcharts with control structures. (3) Finally, the design is tested with a structured walkthrough.

Assuming the decision is to make, or custom-write, the program, you then move on to design the solution specified by the systems analysts. In the *program design* step, the software is designed in three mini-steps. First, the program logic is determined through a top-down approach and modularization, using a hierarchy chart. Then it is designed in detailed form, either in narrative form, using pseudocode, or graphically, using flowcharts—or both. Finally, the design is tested with a structured walkthrough. (■ *See Panel 11.3.)*

It used to be that programmers took a kind of a seat-of-the-pants approach to programming. Programming was considered an art, not a science. Today, however, most programmers use a design approach called structured programming. **Structured programming takes a top-down approach that breaks**

■ **PANEL 11.3**

Second step: program design

1. Determine program logic through top-down approach and modularization, using a hierarchy chart.
2. Design details using pseudocode and/or flowcharts, preferably involving control structures.
3. Test design with structured walkthroughs.

programs into modular forms. It also uses standard logic tools called control structures (sequential, selection, and iteration).

The point of structured programming is to make programs more efficient (with fewer lines of code) and better organized (more readable), and to have better notations so that they have clear and correct descriptions.

The three mini-steps of program design are as follows.

1. Determine the Program Logic, Using Top-Down Approach

Logically laying out the program is like outlining a long term paper before you proceed to write it. **Top-down program design proceeds by identifying the top element, or module, of a program and then breaking it down in hierarchical fashion to the lowest level of detail. The top-down program design is used to identify the program's processing steps, or modules.** After the program is designed, the actual coding proceeds from the bottom up, using the modular approach.

The concept of modularization is important. *The beauty of modularization is that an entire program can be more easily developed because the parts can be developed and tested separately.*

A *module* is a processing step of a program. Each module is made up of logically related program statements. (Sometimes a module is called a *sub-procedure* or *subroutine*.) An example of a module might be a programming instruction that simply says "Open a file, find a record, and show it on the display screen." It is best if each module has only a single function, just as an English paragraph should have a single, complete thought. This rule limits the module's size and complexity, which facilitates testing and debugging.

Top-down program design can be represented graphically in a hierarchy chart. **A *hierarchy chart*, or *structure chart*, illustrates the overall purpose of the program, identifying all the modules needed to achieve that purpose and the relationships among them.** (■ *See Panel 11.4.*) The program must move in sequence from one module to the next until all have been processed. There must be three principal modules corresponding to the three principal computing operations—input, processing, and output. (In Panel 11.4 they are "Read input," "Calculate pay," and "Generate output.")

2. Design Details, Using Pseudocode and/or Flowcharts & Preferably Control Structures

Once the essential logic of the program has been determined, through the use of top-down programming and hierarchy charts, you can then go to work on the details.

There are two ways to show details—write them or draw them; that is, use *pseudocode* or use *flowcharts*. Most projects use both methods.

- Pseudocode: **Pseudocode is a method of designing a program using normal human-language statements to describe the logic and processing flow.** (■ *See Panel 11.5.*) Pseudocode is like an outline or summary form of the program you will write.

   Sometimes pseudocode is used simply to express the purpose of a particular programming module in somewhat general terms. With the use of such terms as IF, THEN, or ELSE, however, the pseudocode follows the rules of *control structures*, an important aspect of structured programming, as we shall explain.

- Program flowcharts: We described system flowcharts in the previous chapter. Here we consider program flowcharts. **A *program flowchart***

■ PANEL 11.4

A hierarchy chart
This represents a top-down design for a payroll program. Here the modules, or processing steps, are represented from the highest level of the program down to details. The three principal processing operations—input, processing, and output—are represented by the modules in the second layer: "Read input," "Calculate pay," and "Generate output." Before tasks at the top of the chart can be performed, all the ones below must be performed. Each module represents a logical processing step.

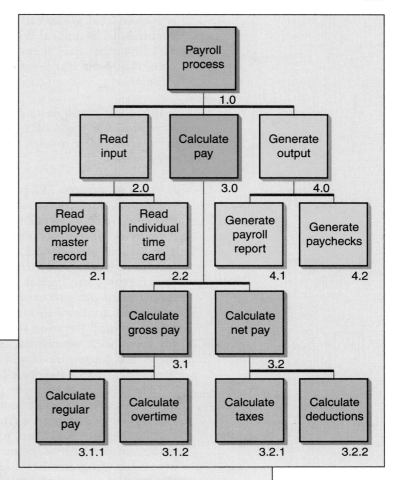

1. Each module must be of manageable size.

2. Each module should be independent and have a single function.

3. The functions of input and output are clearly defined in separate modules.

4. Each module has a single entry point (execution of the program module always starts at the same place) and a single exit point (control always leaves the module at the same place).

5. If one module refers to or transfers control to another module, the latter module returns control to the point from which it was "called" by the first module.

■ PANEL 11.5

Pseudocode

```
START
DO WHILE (so long as) there are records
      Read a customer billing account record
      IF today's date is greater than 30 days from
      date of last customer payment
            Calculate total amount due
            Calculate 5% interest on amount due
            Add interest to total amount due to calculate
            grand total
            Print on invoice overdue amount
      ELSE
            Calculate total amount due
      ENDIF
      Print out invoice
END DO
END
```

is a chart that graphically presents the detailed series of steps (algorithms, or logical flow) needed to solve a programming problem. The flowchart uses standard symbols—called *ANSI symbols,* after the American National Standards Institute, which developed them. (■ *See Panel 11.6.)*

The flowchart symbols might seem clear enough. But how do you know how to express the *logic* of a program? How do you know how to reason it out so it will really work? The answer is to use control structures, as explained next.

- **Control structures:** When you're trying to determine the logic behind something, you use words like "if" and "then" and "else." (For example, without actually using these exact words, you might reason something like this: "*If* she comes over, *then* we'll go out to a movie, *else* I'll just stay in and watch TV.") Control structures make use of the same words. **A *control structure,* or *logic structure,* is a structure that controls the logical sequence in which computer program instructions are executed. In structured program design, three control structures are used to form the logic of a program: sequence, selection, and iteration (or loop).** (■ *See Panel 11.7, p. 518.)* These are the tools with which you can write structured programs and take a lot of the guesswork out of programming. (Additional variations of these three basic structures are also used.)

  One thing that all three control structures have in common is *one entry* and *one exit.* The control structure is entered at a single point and exited at another single point. This helps simplify the logic so that it is easier for others following in a programmer's footsteps to make sense of the program. (In the days before this requirement was instituted, programmers could have all kinds of variations, leading to the kind of incomprehensible program known as *spaghetti code.*)

  Let us consider the three control structures:

  (1) **In the *sequence control structure,* one program statement follows another in *logical order.*** In the example shown in Panel 11.7, there are two boxes ("statement" and "statement"). One box could say "Open file," the other "Read a record." There are no decisions to make, no choices between "yes" or "no." The boxes logically follow one another in sequential order.

  (2) **The *selection control structure*—also known as an *IF-THEN-ELSE* structure—is a structure that represents choice. It offers two paths to follow when a decision must be made by a program.** An example of a selection structure is as follows:

  IF a worker's hours in a week exceed 40
  THEN overtime hours equal the number of hours exceeding 40
  ELSE the worker has no overtime hours.

  A variation on the usual selection control structure is the *case control structure.* This offers more than a single yes-or-no decision. The case structure allows several alternatives, or "cases," to be presented. "IF Case 1 occurs, THEN do thus-and-so. IF Case 2 occurs, THEN follow an alternative course . . ." and so on. The case control structure saves the programmer the trouble of having to indicate a lot of separate IF-THEN-ELSE conditions.

  (3) **The *iteration,* or *loop,* control structure is a structure in which a process may be repeated as long as a certain condition remains true.** There are two types of iteration structures—*DO UNTIL* and *DO WHILE.*

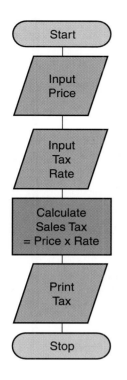

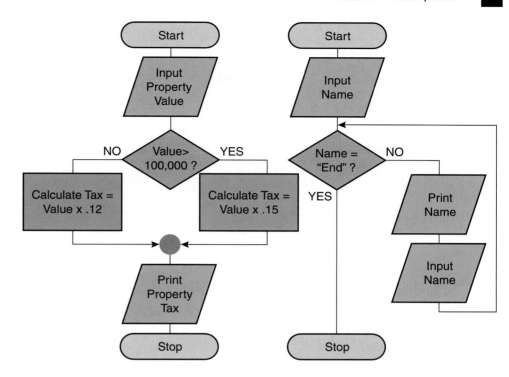

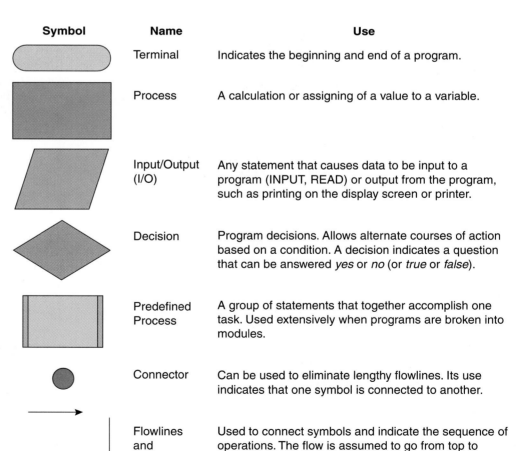

■ PANEL 11.6

Examples of program flowcharts and explanation of flowchart symbols
(Top) Three sample flowcharts. *(Bottom)* ANSI symbols for program flowcharts.

Symbol	Name	Use
Terminal	Terminal	Indicates the beginning and end of a program.
Process	Process	A calculation or assigning of a value to a variable.
Input/Output	Input/Output (I/O)	Any statement that causes data to be input to a program (INPUT, READ) or output from the program, such as printing on the display screen or printer.
Decision	Decision	Program decisions. Allows alternate courses of action based on a condition. A decision indicates a question that can be answered *yes* or *no* (or *true* or *false*).
Predefined Process	Predefined Process	A group of statements that together accomplish one task. Used extensively when programs are broken into modules.
Connector	Connector	Can be used to eliminate lengthy flowlines. Its use indicates that one symbol is connected to another.
Flowlines and Arrowheads	Flowlines and Arrowheads	Used to connect symbols and indicate the sequence of operations. The flow is assumed to go from top to bottom and from left to right. Arrowheads are only required when the flow violates the standard direction.

Sequence control structure
(one program statement follows another
in logical order)

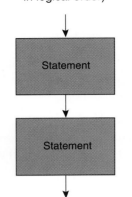

**Iteration control structures:
DO UNTIL and DO WHILE**

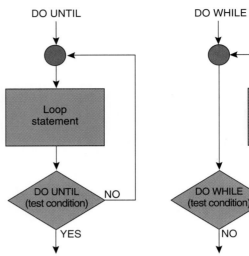

Selection control structure
(IF-THEN-ELSE)

■ **PANEL 11.7**

The three control structures
The three structures used in structured program design to form the
logic of a program are *sequence*,
selection, and *iteration*.

Variation on selection: the case control structure
(more than a single yes-or-no decision)

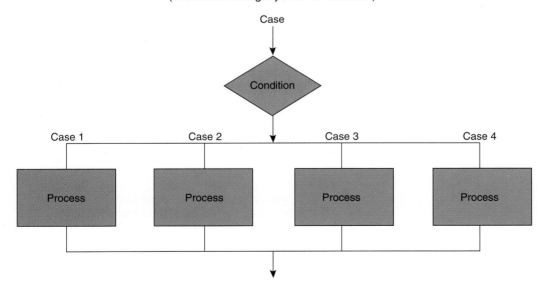

An example of a DO UNTIL structure is as follows:
DO read in employee records UNTIL there are no more employee records.
An example of a DO WHILE structure is as follows:
DO read in employee records WHILE—that is, as long as—there continue to be employee records.

What seems to be the difference between the two iteration structures? It is simply this: If there are several statements that need to be repeated, you need to decide when to *stop* repeating them. You can decide to stop them at the *beginning* of the loop, using the DO WHILE structure. Or you can decide to stop them at the *end* of the loop, using the DO UNTIL structure. The DO UNTIL iteration means that the loop statements will be executed at least once. This is because the iteration statements are executed *before* the program checks whether to stop.

3. Do a Structured Walkthrough

No doubt you've had the experience, after reading over a research paper or project you've done, of being surprised when a friend (or instructor) pointed out some things you missed. The same thing happens to programmers.

In the ***structured walkthrough,* a programmer leads other people in the development team through a segment of code.** The structured walkthrough is actually an established part of the design phase. It consists of a formal review process in which others—fellow programmers, systems analysts, and perhaps users—scrutinize ("walk through") the programmer's work. They review the parts of the program for errors, omissions, and duplications in processing tasks. Because the whole program is still on paper at this point, these matters are easier to correct now than they are later. Some programmers get very nervous before a structured walkthrough, treating it as some sort of test of their competence. Others see it merely as a cooperative endeavor.

11.4 The Third Step: Code the Program

KEY QUESTION

What is involved in coding a program?

Preview & Review: Coding the program is actually writing the program, translating the logic of the design into a programming language. It consists of choosing the appropriate programming language and following its rules, or syntax, exactly.

Once the design has been developed and reviewed in a walkthrough, the actual writing of the program begins. Writing the program is called *coding.* (■ *See Panel 11.8, next page.)* Coding is what many people think of when they think of programming, although it is only one of the five steps. Coding consists of translating the logic requirements from pseudocode or flowcharts into a programming language—the letters, numbers, and symbols that make up the program.

1. Select the Appropriate Programming Language

A *programming language* is a set of rules that tells the computer what operations to do. Examples of well-known programming languages are BASIC, COBOL, and C. These languages are called "high-level languages," as we explain in a few pages.

Not all languages are appropriate for all uses. Some, for example, have strengths in mathematical and statistical processing. Others are more appropriate for database management. Some are for mainframes; others are for

■PANEL 11.8

Third step: program coding
The third step in programming is to translate the logic of the program worked out from pseudocode or flowcharts into a high-level programming language, following its grammatical rules.

1. Select the appropriate high-level programming language.
2. Code the program in that language, following the syntax carefully.

microcomputers. Thus, the language needs to be chosen based on such considerations as what purpose the program is designed to serve and what languages are already being used in the organization or field you are in. We consider these matters in the second half of this chapter.

2. Follow the Syntax

In order for a program to work, you have to follow **the *syntax*, the rules of a programming language.** Programming languages have their own grammar just as human languages do. But computers are probably a lot less forgiving if you use these rules incorrectly.

11.5 The Fourth Step: Test the Program

KEY QUESTION

How is a program tested?

Preview & Review: Testing the program consists of desk-checking, debugging the program of errors, and running real-world data to make sure the program works.

Program testing involves running various tests, such as desk-checking and debugging, and then running real-world data to make sure the program works. (■ *See Panel 11.9.*) Two principal activities are *desk-checking* and *debugging.* These steps are called *alpha testing.*

1. Perform Desk-Checking

***Desk-checking* is simply reading through, or checking, the program to make sure that it's free of errors and that the logic works.** In other words, desk-checking is sort of like proofreading. This step should be taken before the program is actually run on a computer.

2. Debug the Program

Once the program has been desk-checked, further errors, or "bugs," will doubtless surface. **To *debug* means to detect, locate, and remove all errors in a computer program.** Mistakes may be syntax errors or logical errors. ***Syntax errors* are caused by typographical errors and incorrect use of the programming language. *Logic errors* are caused by incorrect use of control structures.** Programs called *diagnostics* exist to check program syntax and display syntax-error messages. Diagnostic programs thus help identify and solve problems. The debugging step is sometimes called *alpha testing.*

■ PANEL 11.9

Fourth step: program testing

The fourth step is to test the program and "debug" it of errors so it will work properly. The word "bug" dates from 1945, when a moth was discovered lodged in the wiring of the Mark I computer. The moth disrupted the execution of the program.

1. Desk-check the program to discover errors.
2. Run the program and debug it (alpha testing).
3. Run real-world data (beta testing).

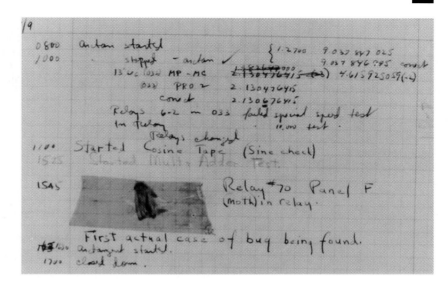

3. Run Real-World Data

After desk-checking and debugging, the program may run fine—in the laboratory. However, it then needs to be tested with real data and real users, called *beta testing*. Indeed, it is even advisable to test it with *bad data*—data that is faulty, incomplete, or in overwhelming quantities—to see if you can make the system crash. Many users, after all, may be far more heavy-handed, ignorant, and careless than programmers have anticipated.

The testing process may take several trials using different test data before the programming team is satisfied the program can be released. Even then, some bugs may still remain, because there comes a point where the pursuit of errors becomes uneconomical. This is one reason why many users are nervous about using the first version of a commercial software package.

11.6 The Fifth Step: Document & Maintain the Program

KEY QUESTION

What is involved in documenting and maintaining a program?

Preview & Review: Documenting the program consists of writing a description of the purpose and process of the program. Documentation should be prepared for users, operators, and programmers. Maintaining the program includes keeping it in working condition, error-free, and up to date.

Writing the program documentation is the fifth step in programming. The resulting **documentation consists of written descriptions of what a program is and how to use it.** Documentation is not just an end-stage process of programming. It has been (or should have been) going on throughout all programming steps. Documentation is needed for people who will be using or involved with the program in the future. (■ *See Panel 11.10, next page.*)

Documentation should be prepared for several different kinds of readers—users, operators, and programmers.

1. Prepare User Documentation

When you buy a commercial software package, such as a spreadsheet, you normally get a manual with it, as well as an online help system. This is *user documentation.* Programmers need to write documentation to help nonprogrammers use the programs.

■ PANEL 11.10

Fifth step: program documentation and maintenance
The fifth step is really the culmination of activity that has been going on through all the programming steps—documentation. Developing written and electronic descriptions of a program and how to use it needs to be done for different people— users, operators, and programmers. Maintenance is an ongoing process.

1. Write user documentation.
2. Write operator documentation.
3. Write programmer documentation.
4. Maintain the program.

2. Prepare Operator Documentation

The people who run large computers are called *computer operators*. Because they are not always programmers, they need to be told what to do when the program malfunctions. The *operator documentation* gives them this information.

3. Write Programmer Documentation

Long after the original programming team has disbanded, the program may still be in use. If, as is often customary, a fifth of the programming staff leaves every year, after 5 years there could be a whole new bunch of programmers who know nothing about the software. *Program documentation* helps train these newcomers and enables them to maintain the existing system.

4. Maintain the Program

A word about maintenance: *Maintenance* is any activity designed to keep programs in working condition, error-free, and up to date. Maintenance includes adjustments, replacements, repairs, measurements, tests, and so on. Modern organizations are changing so rapidly—in products, marketing strategies, accounting systems, and so on—that these changes are bound to be reflected in their computer systems. Thus, maintenance is an important matter, and documentation must be available to help programmers make adjustments in existing systems.

The five steps of the programming process are summarized in the table opposite. (■ *See Panel 11.11.*)

11.7 Five Generations of Programming Languages

KEY QUESTION

What are the five generations of programming languages?

Preview & Review: Languages are said to have evolved in "generations," from machine language to natural languages. The five generations are machine language, assembly language, high-level languages, very-high-level languages, and natural languages.

As we've said, a *programming language* is a set of rules that tells the computer what operations to do. Programmers, in fact, use these languages to create other kinds of software. Many programming languages have been written, some with colorful names (SNOBOL, HEARSAY, DOCTOR, ACTORS, JOVIAL). Each is suited to solving particular kinds of problems. What is it that all these languages have in common? Simply this: ultimately they must be reduced to digital form—a 1 or 0, electricity on or off—because that is all the computer can work with.

■ PANEL 11.11

Summary of the five programming steps

Step	Activities
Step 1: Problem definition	1. Specify program objectives and program users. 2. Specify output requirements. 3. Specify input requirements. 4. Specify processing requirements. 5. Study feasibility of implementing program. 6. Document the analysis.
Step 2: Program design	1. Determine program logic through top-down approach and modularization, using a hierarchy chart. 2. Design details using pseudocode and/or using flowcharts, preferably using control structures. 3. Test design with structured walkthrough.
Step 3: Program coding	1. Select the appropriate high-level programming language. 2. Code the program in that language, following the syntax carefully.
Step 4: Program testing	1. Desk-check the program to discover errors. 2. Run the program and debug it (alpha testing). 3. Run real-world data (beta testing).
Step 5: Program documentation and maintenance	1. Convert hardware, software, and files. 2. Write operator documentation. 3. Write programmer documentation. 4. Maintain the program.

To begin to see how this works, it's important to understand that there are five *levels* or *generations* of programming languages, ranging from low-level to high-level. **The five *generations of programming languages* start at the lowest level with (1) machine language. They then range up through (2) assembly language, (3) high-level languages, and (4) very-high-level languages. At the highest level are (5) natural languages.** Programming languages are said to be *lower level* when they are closer to the language that the computer itself uses—the 1s and 0s. They are called *higher level* when they are closer to the language people use—more like English, for example.

Beginning in 1945, the five levels or generations have evolved over the years, with later generations gradually coming into greater use with programmers. The births of the generations are as follows.

- First generation, 1945—*Machine language*
- Second generation, mid-1950s—*Assembly language*
- Third generation, early 1960s—*High-level languages:* FORTRAN, COBOL, BASIC, C, Ada
- Fourth generation, early 1970s—*Very-high-level languages:* SQL, Intellect, NOMAD, FOCUS
- Fifth generation, early 1980s—*Natural languages*

Let us consider these five generations. (■ *See Panel 11.12, next page.*)

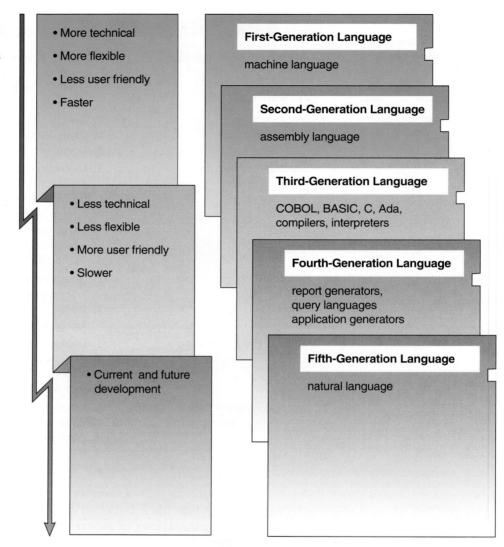

- More technical
- More flexible
- Less user friendly
- Faster

- Less technical
- Less flexible
- More user friendly
- Slower

- Current and future development

First-Generation Language

machine language

Second-Generation Language

assembly language

Third-Generation Language

COBOL, BASIC, C, Ada, compilers, interpreters

Fourth-Generation Language

report generators, query languages application generators

Fifth-Generation Language

natural language

First Generation: Machine Language

The lowest level of language, **machine language is the basic language of the computer, representing data as 1s and 0s.** (■ *See Panel 11.13.)* Machine-language programs vary from computer to computer; that is, they are *machine dependent.*

These binary digits, which correspond to the on and off electrical states of the computer, are clearly not convenient for people to read and use. Believe it or not, though, programmers *did* work with these mind-numbing digits. When the next generation of programming languages came along, assembly language, there must have been great sighs of relief.

Second Generation: Assembly Language

Assembly language is a low-level programming language that allows a computer user to write a program using abbreviations or more easily remembered words instead of numbers. (■ *Refer to Panel 11.13 again.)* For example, the letters MP could be used to represent the instruction MULTIPLY and STO to represent STORE.

■ PANEL 11.13

Three generations of programming languages.

(Top) Machine language is all binary 0s and 1s—difficult for people to work with. *(Middle)* Assembly language uses abbreviations for major instructions (such as MP for MULTIPLY). This is easier for people to use, but still not easy. *(Bottom)* COBOL, a third-generation language, uses English words that can be understood by people.

First generation
Machine language

```
11110010 01110011 1101 001000010000 0111 000000101011
11110010 01110011 1101 001000011000 0111 000000101111
11111100 01010010 1101 001000010010 1101 001000011101
11110000 01000101 1101 001000010011 0000 000000111110
11110011 01000011 0111 000001010000 1101 001000010100
10010110 11110000 0111 000001010100
```

Second generation
Assembly language

```
PACK  210(8,13),02B(4,7)
PACK  218(8,13),02F(4,7)
MP    212(6,13),21D(3,13)
SRP   213(5,13),03E(0),5
UNPK  050(5,7),214(4,13)
OI    054(7),X FO
```

Third generation
COBOL

```
MULTIPLY HOURS-WORKED BY PAY-RATE GIVING GROSS-PAY ROUNDED
```

As you might expect, a programmer can write instructions in assembly language faster than in machine language. Nevertheless, it is still not an easy language to learn, and it is so tedious to use that mistakes are frequent. Moreover, assembly language has the same drawback as machine language in that it varies from computer to computer—it is machine dependent.

We now need to introduce the concept of a *language translator*. Because a computer can execute programs only in machine language, a translator or converter is needed if the program is written in any other language. **A *language translator* is a type of system software (✔ p. 104) that translates a program written in a second-, third-, or higher-generation language into machine language.**

Language translators are of three types:

* Assemblers   * Compilers   * Interpreters

An *assembler,* or *assembler program,* is a program that translates the assembly-language program into machine language. We describe compilers and interpreters in the next section.

Third Generation: High-Level Languages

A *high-level language* resembles some human language such as English; an example is COBOL, which is used for business applications. (**■** *Refer again to Panel 11.13.*) A high-level language allows users to write in a familiar notation, rather than numbers or abbreviations. Most also are not machine dependent—they can be used on more than one kind of computer. Examples of familiar languages of this sort are FORTRAN, BASIC, Pascal, and C.

Assembly language needs an assembler as a language translator. The translator for high-level languages is, depending on the language, either a *compiler* or an *interpreter.*

- **Compiler—execute later: A *compiler* is a language translator that converts the entire program of a high-level language into machine language BEFORE the computer executes the program.** The programming instructions of a high-level language are called the *source code.* The compiler translates it into machine language, which in this case is called the *object code.* The significance of this is that the object code *can be saved.* Thus, it can be executed later (as many times as desired) rather than run right away. (■ *See Panel 11.14.*)

  Examples of high-level languages using compilers are COBOL, FORTRAN, Pascal, and C.

- **Interpreter—execute immediately: An *interpreter* is a language translator that converts each high-level language statement into machine language and executes it IMMEDIATELY, statement by statement.** No object code is saved, as with the compiler. Therefore, interpreted code generally runs more slowly than compiled code. However, code can be tested line by line.

  An example of a high-level language using an interpreter is BASIC.

  Who cares, you might ask, whether you can run a program now or later. (After all, "later" could be only a matter of seconds or minutes.) Here's the significance: When a compiler is used, it requires *two* steps before the program can be executed. These two steps are the source code and the object code. The interpreter, on the other hand, requires only *one* step. However, the advantage of a compiler language is that once you have obtained the object code, *the program executes faster.* The advantage of an interpreter language, on the other hand, is that *programs are easier to develop.*

■ PANEL 11.14

Compiler

This language translator converts the high-level language (source code) into machine language (object code) before the computer can execute the program.

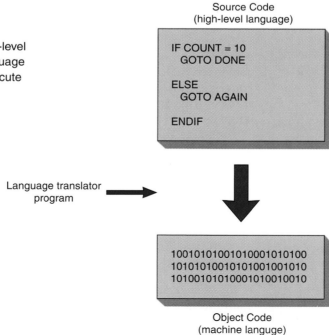

Source Code
(high-level language)

```
IF COUNT = 10
   GOTO DONE

ELSE
   GOTO AGAIN

ENDIF
```

Language translator
program

```
100101010010100001010100
101010100010101001001010
101001010100010100100010
```

Object Code
(machine langue)

Third-generation, high-level languages are also known as *procedural languages.* That is, programs set forth precise procedures, or series of instructions, meaning that the programmer has to follow a proper order of actions to solve a problem. To do that, the programmer has to have a detailed knowledge of programming and the computer it would run on.

For example, say you want to take a taxi to a particular restaurant. If you tell the taxi driver precisely how to get to the restaurant, that's procedural; you have to know how to get there yourself, and you will probably get there efficiently. However, if you simply tell the taxi driver to "Take me to the Doggy Diner restaurant," then you're saying only what you want, which is nonprocedural; you may not get to the restaurant in an efficient manner, but it's easier for you.

The fourth generation of languages represented a step up because they are *nonprocedural,* as we shall explain.

Fourth Generation: Very-High-Level Languages

A *very-high-level language* is often called a *4GL,* for *4th-generation language.* 4GLs are much more user-oriented and allow users to develop programs with fewer commands compared with third-generation languages, although 4GLs require more computing power. 4GLs are also called *rapid application development (RAD) tools.*

4GLs are called *nonprocedural* because programmers—and even users— can write programs that need only tell the computer what they want done, not all the procedures for doing it. That is, they do not have to specify all the programming logic or otherwise tell the computer *how* the task should be carried out. This saves programmers a lot of time because they do not need to write as many lines of code as they do with procedural languages.

Fourth-generation languages consist of report generators, query languages, application generators, and interactive database management system (DBMS) programs. Some 4GLs are tools for end-users, some are tools for programmers.

- **Report generators:** A *report generator,* also called a report writer, is a program for end-users that is used to produce a report. The report may be a printout or a screen display. It may show all or part of a database file. You can specify the format in advance—columns, headings, and so on—and the report generator will then produce data in that format.

  Report generators (an example is RPG III) were the precursor to today's query languages.

- **Query languages:** A *query language* is an easy-to-use language for retrieving data from a database management system. The query may be expressed in the form of a sentence or near-English command. Or the query may be obtained from choices on a menu.

  Examples of query languages are SQL (for Structured Query Language) and Intellect. For example, with Intellect, which is used with IBM mainframes, you can ask an English-language question such as "Tell me the number of employees in the sales department."

- **Application generators:** An *application generator* is a programmer's tool that generates applications programs from descriptions of the problem rather than by traditional programming. The benefit is that the programmer does not need to specify *how* the data should be processed. The application generator is able to do this because it consists of modules that have been preprogrammed to accomplish various tasks.

Programmers use application generators to help them create parts of other programs. For example, the software is used to construct on-screen menus or types of input and output screen formats. NOMAD and FOCUS, two database management systems, include application generators.

4GLs may not entirely replace third-generation languages because they are usually focused on specific tasks and hence offer fewer options. Still, they improve productivity because programs are easy to write.

Fifth Generation: Natural Languages

***Natural languages* are of two types. The first are ordinary human languages: English, Spanish, and so on. The second are programming languages that use human language to give people a more natural connection with computers.** Some of the query languages mentioned above under 4GLs might seem pretty close to human communication, but natural languages try to be even closer. With 4GLs, you can type in some rather routine inquiries. An example of a request in FOCUS might be:

SUM SHIPMENTS BY STATE BY DATE.

Natural languages allow questions or commands to be framed in a more conversational way or in alternative forms. For example, with a natural language, you might be able to state:

I WANT THE SHIPMENTS OF PERSONAL DIGITAL ASSISTANTS FOR ALABAMA AND MISSISSIPPI BROKEN DOWN BY CITY FOR JANUARY AND FEBRUARY. ALSO, MAY I HAVE JANUARY AND FEBRUARY SHIPMENTS LISTED BY CITIES FOR PERSONAL COMMUNICATORS SHIPPED TO WISCONSIN AND MINNESOTA.

Natural languages are part of the field of study known as *artificial intelligence* (discussed in detail in Chapter 12). Artificial intelligence (AI) is a group of related technologies that attempt to develop machines to emulate human-like qualities, such as learning, reasoning, communicating, seeing, and hearing.

The dates of the principal programming languages are shown in the timeline running along the bottom of these facing pages. (■ *See Panel 11.15.*)

■ **PANEL 11.15**

Timeline for development of programming languages and formatting tools

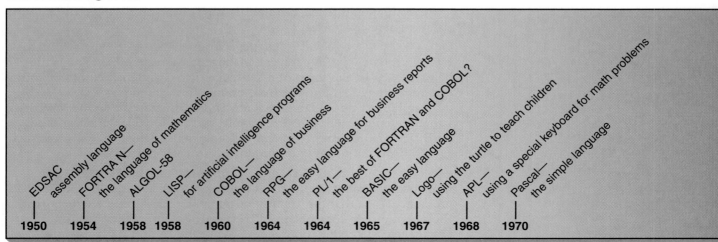

EDSAC — 1950
assembly language — 1954
FORTRAN — the language of mathematics — 1958
ALGOL-58 — 1958
LISP — for artificial intelligence programs — 1960
COBOL — the language of business — 1964
RPG — the easy language for business reports — 1964
PL/1 — the best of FORTRAN and COBOL? — 1965
BASIC — the easy language — 1967
Logo — using the turtle to teach children — 1968
APL — using a special keyboard for math problems
Pascal — the simple language — 1970

11.8 Programming Languages Used Today

KEY QUESTIONS

What are some third-generation languages, and what are they used for?

Preview & Review: Third-generation programming languages used today are FORTRAN, COBOL, BASIC, Pascal, C, LISP, PL/1, RPG, Logo, APL, FORTH, PROLOG, Ada, and the database programming languages.

Let us now turn back and consider some of the third-generation, or high-level, languages in use today.

FORTRAN: The Language of Mathematics & the First High-Level Language

Developed in 1954 by IBM, **FORTRAN (for *FORmula TRANslator*) was the first high-level language.** (■ *See Panel 11.16, next page.*) Originally designed to express mathematical formulas, it is still the most widely used language for mathematical, scientific, and engineering problems. It is also useful for complex business applications, such as forecasting and modeling. However, because it cannot handle a large volume of input/output operations or file processing, it is not used for more typical business problems. The newest version of FORTRAN is FORTRAN 90.

FORTRAN has both advantages and disadvantages:

- Advantages: (1) FORTRAN can handle complex mathematical and logical expressions. (2) Its statements are relatively short and simple. (3) FORTRAN programs developed on one type of computer can often be easily modified to work on other types.
- Disadvantages: (1) FORTRAN does not handle input and output operations to storage devices as efficiently as some other higher-level languages. (2) It has only a limited ability to express and process nonnumeric data. (3) It is not as easy to read and understand as some other high-level languages.

COBOL: The Language of Business

Formally adopted in 1960, **COBOL (for *COmmon Business-Oriented Language*) is the most frequently used business programming language for large computers.** (■ *Refer again to Panel 11.16.*) Its most significant attribute is that it is extremely readable. For example, a COBOL line might read:

MULTIPLY HOURLY-RATE BY HOURS-WORKED GIVING GROSS-PAY

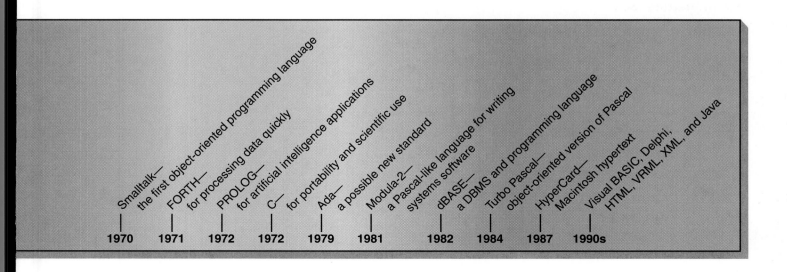

Smalltalk— the first object-oriented programming language — 1970

FORTH— for processing data quickly — 1971

PROLOG— for artificial intelligence applications — 1972

C— for portability and scientific use — 1972

Ada— a possible new standard — 1979

Modula-2— a Pascal-like language for writing systems software — 1981

dBASE— a DBMS and programming language — 1982

Turbo Pascal— object-oriented version of Pascal — 1984

HyperCard— Macintosh hypertext — 1987

Visual BASIC, Delphi, HTML, VRML, XML, and Java — 1990s

FORTRAN

```
               IF (XINVO .GT. 500.00) THEN

                   DISCNT = 0.07 * XINVO

               ELSE

                   DISCNT = 0.0

               ENDIF

               XINVO = XINVO – DISCNT
```

COBOL

```
OPEN-INVOICE-FILE.
        OPEN I-O INVOICE FILE.

READ-INVOICE-PROCESS.
        PERFORM READ-NEXT-REC THROUGH READ-NEXT-REC-EXIT UNTIL END-OF-FILE.
        STOP RUN.

READ-NEXT-REC.
        READ INVOICE-REC
            INVALID KEY
                    DISPLAY 'ERROR READING INVOICE FILE'
                    MOVE 'Y' TO EOF-FLAG
                    GOTO READ-NEXT-REC-EXIT.
        IF INVOICE-AMT > 500
                COMPUTE INVOICE-AMT = INVOICE-AMT – (INVOICE-AMT * .07)
                REWRITE INVOICE-REC.

READ-NEXT-REC-EXIT.
        EXIT.
```

BASIC

```
10  REM       This Program Calculates a Discount Based on the Invoice Amount
20  REM              If Invoice Amount is Greater Than 500, Discount is 7%
30  REM              Otherwise Discount is 0
40  REM
50  INPUT "What is the Invoice Amount"; INV.AMT
60  IF INV.AMT A> 500 THEN LET DISCOUNT = .07 ELSE LET DISCOUNT = 0
70  REM              Display results
80  PRINT "Original Amt", "Discount", "Amt after Discount"
90  PRINT INV.AMT, INV.AMT * DISCOUNT, INV.AMT – INV.AMT * DISCOUNT
100 END
```

Pascal

```
    if INVOICEAMOUNT > 500.00 then

        DISCOUNT := 0.07 * INVOICEAMOUNT

    else

        DISCOUNT := 0.0;

    INVOICEAMOUNT := INVOICEAMOUNT – DISCOUNT
```

C

```
    if (invoice_amount > 500.00)

        DISCOUNT = 0.07 * invoice_amount;

    else

        discount = 0.00;

    invoice_amount = invoice_amount – discount;
```

■ PANEL 11.16

Third-generation languages compared: five examples
This shows how five languages handle the same statement. The statement specifies that a customer gets a discount of 7% of the invoice amount if the invoice is greater than $500; if the invoice is lower, there is no discount.

Writing a COBOL program resembles writing an outline for a research paper. The program is divided into four divisions—Identification, Environment, Data, and Procedure. The divisions in turn are divided into sections, which are divided into paragraphs, which are further divided into sections. The Identification Division identifies the name of the program and the author (programmer) and perhaps some other helpful comments. The Environment Division describes the computer on which the program will be compiled and executed. The Data Division describes what data will be processed. The Procedure Division describes the actual processing procedures.

COBOL, too, has both advantages and disadvantages.

- Advantages: (1) It is machine independent. (2) Its English-like statements are easy to understand, even for a nonprogrammer. (3) It can handle many files, records, and fields. (4) It easily handles input/output operations.

- Disadvantages: (1) Because it is so readable, it is wordy. Thus, even simple programs are lengthy, and programmer productivity is slowed. (2) It cannot handle mathematical processing as well as FORTRAN.

BASIC: The Easy Language

BASIC was developed by John Kemeny and Thomas Kurtch in 1965 for use in training their students at Dartmouth College. By the late 1960s, it was widely used in academic settings on all kinds of computers, from mainframes to PCs.

BASIC (for *Beginner's All-purpose Symbolic Instruction Code*) used to be the most popular microcomputer language and is considered the easiest programming language to learn. (■ *Refer back to Panel 11.16.)* Although it is available in compiler form, the interpreter form is more popular with first-time and casual users. This is because it is interactive, meaning that user and computer can communicate with each other during the writing and running of the program. Today there is no one version of BASIC. One of the popular current evolutions is Visual BASIC, discussed shortly.

The advantage and disadvantages of BASIC are as follows:

- Advantage: The primary advantage of BASIC is its ease of use.

- Disadvantages: (1) Its processing speed is relatively slow, although compiler versions are faster than interpreter versions. (2) There is no one version of BASIC, although in 1987 ANSI adopted a new standard that eliminated portability problems—that is, problems with running it on different computers.

Pascal: The Simple Language

Named after the 17th-century French mathematician Blaise Pascal, **Pascal is an alternative to BASIC as a language for teaching purposes and is relatively easy to learn. (■** *Refer back to Panel 11.16.)* A difference from BASIC is that Pascal uses structured programming.

A compiled language, Pascal offers these advantages and disadvantages:

- Advantages: (1) Pascal is easy to learn. (2) It has extensive capabilities for graphics programming. (3) It is excellent for scientific use.

- Disadvantage: Pascal has limited input/output programming capabilities, which limits its business applications.

C: For Portability & Scientific Use

"C" is the language's entire name, and it does not "stand" for anything. Developed at Bell Laboratories in the early 1970s, **C is a general-purpose, compiled language that works well for microcomputers and is portable among many computers.** (■ *Refer back to Panel 11.16.*) It was originally developed for writing system software. (Most of the UNIX operating system [✔ p. 122] was written using C.) Today it is widely used for writing applications including word processing, spreadsheets, games, robotics, and graphics programs. It is now considered a necessary language for programmers to know.

Here are the advantages and disadvantages of C:

- Advantages: (1) C works well with microcomputers. (2) It has a high degree of portability—it can be run without change on a variety of computers. (3) It is fast and efficient. (4) It enables the programmer to manipulate individual bits in main memory.
- Disadvantages: (1) C is considered difficult to learn. (2) Because of its conciseness, the code can be difficult to follow. (3) It is not suited to applications that require a lot of report formatting and data file manipulation.

Several other high-level languages exist that, though not as popular or as famous as the foregoing, are well-known enough that you may encounter them. Some of them are special-purpose languages. We will introduce them in order of their appearance.

LISP: For Artificial Intelligence Programs

LISP (for LISt Processor) is a third-generation language used principally to construct artificial intelligence programs. Developed at the Massachusetts Institute of Technology in 1958 by mathematician John McCarthy, LISP is used to write expert systems and natural-language programs. *Expert systems* (✔ p. 471) are programs that are imbued with knowledge by a human expert; the programs can walk you through a problem and help solve it.

PL/1: The Best of FORTRAN & COBOL?

PL/1 was introduced in 1964 by IBM. **PL/1 (for *Programming Language 1*) is a third-generation language designed to process both business and scientific applications.** It contains many of the best features of FORTRAN and COBOL and is quite flexible and easy to learn. However, it is also considered to have so many options as to diminish its usefulness. As a result, it has not given FORTRAN and COBOL much competition.

RPG: The Easy Language for Business Reports

RPG was also introduced in 1964 by IBM and has evolved through several important versions since. **RPG (for *Report Program Generator*) was a highly structured and relatively easy-to-learn third-generation language designed to help generate business reports.** The user filled out a special form specifying what information the report should include and in what kind of format.

In 1970, improvements were introduced in RPG II. A successor, RPG III, an interactive fourth-generation language, has since appeared that uses menus to give programmers choices.

Logo: Using the Turtle to Teach Children

Logo was developed at the Massachusetts Institute of Technology in 1967 by Seymour Papert, using a dialect of LISP. **Logo is a third-generation language designed primarily to teach children problem-solving and programming skills.** At the basis of Logo is a triangular pointer, called a "turtle," which responds to a few simple commands such as forward, left, and right. The pointer produces similar movements on the screen, enabling users to draw geometric patterns and pictures on screen. Because of its highly interactive nature, Logo is used not only by children but also to produce graphic reports in business.

APL: Using a Special Keyboard for Math Problems

APL was designed in 1962 by Kenneth Iverson for use on IBM mainframes. **APL (for _A Programming Language_) is a third-generation language that uses a special keyboard with special symbols to enable users to solve complex mathematical problems in a single step.** The special keyboard is required because the APL symbols are not part of the familiar ASCII (✔ p. 156) character set. Though hard to read, this mathematically oriented and scientific language is still found on a variety of computers.

FORTH: For Processing Data Quickly

FORTH was created in 1971 by Charles Moore. **FORTH (for _FOuRTH-generation language_) is a third-generation language designed for real-time control tasks, as well as business and graphics applications.** The program is used on all kinds of computers, from PCs to mainframes, and runs very fast because it requires less memory than other programs. Because it runs so fast, it is used in applications that must process data quickly. Thus, it is used to process data acquired from sensors and instruments, as well as in arcade game programs and robotics.

PROLOG: For Artificial Intelligence Applications

Invented in 1972 by Alan Colmerauer of France, PROLOG did not receive much attention until 1979, when a new version appeared. **PROLOG (for _PROgramming LOGic_) is used for developing artificial intelligence applications,** such as natural language programs and expert systems.

Ada: A Possible New Standard

Ada is an extremely powerful structured programming language designed by the U.S. Department of Defense to ensure portability of programs from one application to another. Ada was named for Countess Ada Lovelace, considered the world's "first programmer." Based on Pascal, Ada was originally intended to be a standard language for weapons systems. However, it has been used in successful commercial applications.

An advantage of Ada is that it is a structured language, with a modular design, so that pieces of a large program can be written and tested separately. Moreover, because it has features that permit the compiler to check it for errors before the program is run, programmers are more apt to write error-free programs. However, Ada has a high level of complexity and difficulty. Moreover, business users already have so much invested in COBOL, FORTRAN, and C that they have little motivation to switch over to this new language.

Database Programming Languages: dBASE, Access, FoxPro, & Paradox

Created by Wayne Ratliff to manage a company football pool, dBASE was originally named Vulcan. It has since gone through a number of versions and companies; it became dBASE II when the program was acquired by Ashton-Tate in 1982. (Ashton-Tate in turn was acquired by Borland International in 1991, and the most current version is dBASE V.)

Most users are familiar with dBASE as a database management system (DBMS), for controlling the structure of a database and access to the data. However, dBASE is also a Pascal-like, fourth-generation programming language. Other DBMS programs, such as Access, FoxPro, and Paradox, are also used as programming languages for end-user development.

11.9 Object-Oriented & Visual Programming

KEY QUESTIONS

How does OOP work, and what are some of the OOP-based languages?

Preview & Review: Object-oriented programming (OOP) is a programming method that combines data with instructions for processing that data to create a self-sufficient "object," or block of preassembled programming code, that can be used in other programs.

Three concepts of OOP are encapsulation, inheritance, and polymorphism.

Some examples of OOP languages are Smalltalk, C++, Turbo Pascal, and Hypertalk.

Visual programming is a method of creating programs in which the programmer makes connections between objects by drawing, pointing, and clicking on diagrams and icons.

Consider how it was for the computer pioneers, programming in machine language or assembly language. Novices putting together programs in BASIC or Pascal can breathe a collective sigh of relief that they weren't around at the dawn of the Computer Age. Even some of the simpler third-generation languages represent a challenge because they are procedure oriented, forcing the programmer to follow a predetermined path.

Fortunately, two new developments have made things somewhat easier—*object-oriented programming* and *visual programming.*

Object-Oriented Programming: Block by Block

Imagine you're programming in a traditional third-generation language, such as BASIC, creating your coded instructions one line at a time. As you work on some segment of the program (such as how to compute overtime pay), you may think, "I'll bet some other programmer has already written something like this. Wish I had it. It would save a lot of time."

Fortunately, a kind of recycling technique now exists. This is object-oriented programming. Let us explain this in four steps:

1. **What OOP is:** *Object-oriented programming* (*OOP,* pronounced "oop") is a programming method that combines data with the instructions for processing that data, resulting in a self-sufficient "object" that can be used in other programs. The important thing here is the object.

2. **What an "object" is:** An *object* is a block of preassembled programming code that is a self-contained module. The module contains, or encapsulates, both (1) a chunk of data and (2) the processing instructions that may be called on to be performed on that data.

3. **When an object's data is to be processed—sending the "message":** Once the object becomes part of a program, the processing instructions may or may not be activated. That only happens when a "message" is sent. A *message* is an alert sent to the object when an operation involving that object needs to be performed.

4. **How the object's data is processed—the "methods":** The message need only identify the operation. How it is actually to be performed is embedded within the processing instructions that are part of the object. These instructions about the operations to be performed on data within the object are called the *methods.*

Once you've written a block of program code (that computes overtime pay, for example), it can be reused in any number of programs. Thus, unlike traditional programming, with OOP you don't have to start from scratch—that is, reinvent the wheel—each time. (■ *See Panel 11.17.*)

Object-oriented programming takes longer to learn than traditional programming because it means training oneself to a new way of thinking. Once learned, however, the beauty of OOP is that an object can be used repeatedly

■ PANEL 11.17

Conventional versus object-oriented programs

Conventional Programs

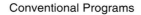

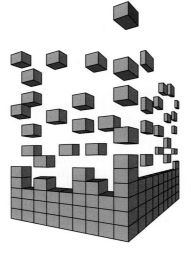

When building conventional programs, programmers write every line of code from scratch.

Object-Oriented Programs

With object-oriented programs, programmers can use blocks, or "objects," of pre-assembled modules containing data and the associated processing instructions.

in different applications and by different programmers, speeding up development time and lowering costs.

Three Important Concepts of OOP

Object-oriented programming has three important concepts, which go under the jaw-breaking names of *encapsulation, inheritance,* and *polymorphism.*[26] Actually, these are not as fearsome as they look:

- **Encapsulation:** *Encapsulation* means an object contains (encapsulates) both (1) data and (2) the processing instructions about it, as we have seen. Once an object has been created, it can be reused in other programs. An object's uses can also be extended through concepts of *class* and *inheritance.*

- **Inheritance:** Once you have created an object, you can use it as the foundation for similar objects that have the same behavior and characteristics. All objects that are derived from or related to one another are said to form a *class.* Each class contains specific instructions (methods) that are unique to that group.

  Classes can be arranged in hierarchies—classes and subclasses. *Inheritance is* the method of passing down traits of an object from classes to subclasses in the hierarchy. Thus, new objects can be created by *inheriting* traits from existing classes.

  Writer Alan Freedman gives this example: "The object MACINTOSH could be one instance of the class PERSONAL COMPUTER, which could inherit properties from the class COMPUTER SYSTEMS."[27] If you were to add a new computer, such as COMPAQ, you would need to enter only what makes it *different* from other computers. The *general* characteristics of personal computers could be inherited.

- **Polymorphism:** Polymorphism means "many shapes." In object-oriented programming, *polymorphism* means that a message (generalized request) produces different results based on the object that it is sent to.

  Polymorphism has important uses. It allows a programmer to create procedures about objects whose exact type is not known in advance but will be known at the time the program is actually run on the computer. Freedman gives this example: "A screen cursor may change its shape from an arrow to a line depending on the program mode." The processing instructions "to move the cursor on screen in response to mouse movement would be written for 'cursor,' and polymorphism would allow that cursor to be whatever shape is required at runtime." It would also allow a new cursor shape to be easily integrated into the program.[28]

Examples of OOP Programming Languages

Some examples of OOP programming languages are Smalltalk, C++, and Hypertalk.

- **Smalltalk—the first OOP language:** Smalltalk was invented by computer scientist Alan Kay in 1970 at Xerox Corporation's Palo Alto Research Center in California. **Smalltalk, the first OOP language, uses a keyboard for entering text, but all other tasks are performed with a mouse.**

- **C++—more than C:** *C++* ("C plus plus")—**the plus signs stand for "more than C"—combines the traditional C programming language with object-oriented capability.** C++ was created by Bjarne Stroustrup. With C++, programmers can write standard code in C without the object-oriented features, use object-oriented features, or do a mixture of both. C++ is used for developing applications software.

- **Hypertalk—the language for HyperCard:** HyperCard, software introduced for the Apple Macintosh in 1987, is based on the concept of *cards* and *stacks* of cards—just like notecards, only they are electronic. A card is a screenful of data that makes up a single record; cards are organized into related files called stacks. Using a mouse, you can make your way through the cards and stacks to find information or discover connections between ideas.

  HyperCard is not precisely an object-oriented programming language, but a language called Hypertalk is. *Hypertalk,* **which uses OOP principles, is the language used in the HyperCard program to manipulate HyperCard stacks.**

Visual Programming

Essentially, visual programming takes OOP to the next level. The goal of visual programming is to make programming easier for programmers and more accessible to nonprogrammers by borrowing the object orientation of OOP languages but exercising it in a graphical or visual way. Visual programming enables users to think more about the problem solving than about handling the programming language. There is no learning of syntax.

Visual programming **is a method of creating programs in which the programmer makes connections between objects by drawing, pointing, and clicking on diagrams and icons and by interacting with flowcharts.** Thus, the programmer can create programs by clicking on icons that represent common programming routines. This type of programming became necessary to develop GUI-based (✔ p. 108) applications because earlier text-based languages are not efficient for this.

Visual BASIC, developed by Microsoft in the early 1990s, is the most popular visual programming language. It offers a visual environment for program construction, allowing users to build various application components using drag-and-drop tools, buttons, scroll bars, and menus. Visual BASIC, a fourth-generation OOP-based language, works with Microsoft Windows to create other Windows-compatible applications.

11.10 Internet Programming: HTML, XML, VRML, Java, & ActiveX

KEY QUESTION

What are the features of HTML, XML, VRML, Java, and ActiveX?

Preview & Review: Programming languages used to build linked multimedia sites on the World Wide Web include HTML, XML, VRML, Java, and the set of controls called ActiveX.

Many of the thousands of Internet data and information sites around the world are text-based only; that is, the user sees no graphics, animation, or video and hears no sound. The World Wide Web, however, permits all of this.

One way to build such multimedia sites on the Web is to use some fairly recently developed programming languages and standards: HTML, XML, VRML, Java, and ActiveX.

- **HTML—for creating 2-D Web documents and links:** *HTML (hypertext markup language, ✔ p. 326)* is a markup language that lets people create on-screen documents for the Internet that can easily be linked by words and pictures to other documents. HTML is a type of code that embeds simple codes within standard ASCII (✔ p. 156) text documents to provide an integrated, two-dimensional display of text and graphics. In other words, a document created with any word processor and stored in ASCII format can become a Web page with the addition of a few HTML codes.

  One of the main features of HTML is the ability to insert hypertext links (✔ p. 326) into a document. Hypertext links enable you to display another Web document simply by clicking on a link area—usually underlined or highlighted—on your current screen. One document may contain links to many other related documents. The related documents may be on the same server (✔ p. 24) as the first document, or they may be on a computer halfway around the world. A link may be a word, a group of words, or a picture.

  Most commercial applications software packages, such as Microsoft Word, can save documents in HTML format. In addition, various HTML editors—commercial HTML packages—also exist to help people who don't want to learn everything about HTML but who want to create their own Web pages. You will, however, need a server that can store your Web pages once they are created. (For a fee, Internet service providers offer server space.)

- **XML—for making the Web work better:** The chief characteristics of HTML are its simplicity and its ease in combining plain text and pictures. But, in the words of journalist Michael Krantz, "HTML simply lacks the software muscle to handle the business world's endless and complex transactions."[29]

  Enter XML. Whereas HTML makes it easy for humans to read Web sites, **XML (extensible markup language) makes it easy for machines to read Web sites by enabling Web developers to add more "tags" to a Web page.** At present, when you use your browser to click on a Web site, search engines can turn up too much, so that it's difficult to pinpoint the specific site—such as one with a recipe for a low-calorie chicken dish for 12—for you. Says Krantz, "XML makes Web sites smart enough to tell other machines whether they're looking at a recipe, an airline ticket, or a pair of easy-fit blue jeans with a 34-inch waist." XML lets Web site developers put "tags" on their Web pages that describe information in, for example, a food recipe as "ingredients," "calories," "cooking time," and "number of portions." Thus, your browser no longer has to search the entire Web for a low-calorie poultry recipe for 12.

- **VRML—for creating 3-D Web pages:** VRML rhymes with "thermal." **VRML (virtual reality modeling language) is a type of programming language used to create three-dimensional Web pages.** For example, there are 3-D cities that you can tour online. One firm called BigBook *(http://www.bigbook.com)* has used VRML to create improved telephone-book Yellow Pages. On screen you seem to fly through a three-dimensional rendering of San Francisco, for example, that has the precision of a two-dimensional street map. When you pass your cursor over a building, its address pops up. When you search for the locations of businesses, the buildings housing those businesses are emphasized.[30]

Practical Matters: Web-Authoring Tools—How to Create Your Own Simple Web Site, Easily & for Free

"The full power of the Web will only be unleashed," says technology writer Stephen Wildstrom, "when it's as easy to post information online as it is to write a memo."[31]

We may be getting closer. The World Wide Web is a great way to get information about yourself and your work to co-workers or to potential customers. To do so, however, you need to create a Web page and put it online.

Fortunately, some free, easy-to-use Web-authoring tools for building simple Web sites are available. All of the following let you create Web pages using icons and menus; you don't need to know HTML to get the job done.[32]

- **Online services:** If you're a member of America Online, you can use AOL's Personal Publisher (keyword: *personal publisher*).

- **Browsers:** Netscape's Communicator comes with a Web-building program called Composer. The recent releases of Microsoft Explorer also offer Web-authoring tools.

- **Word processing software:** Microsoft offers Internet Assistants for all its Office applications, including Word, its word processing program. (You can download Internet Assistant at *http://www.microsoft.com/word/internet/ia/*)

- **Desktop publishing software:** If you've taken up desktop publishing, you should know that Adobe PageMaker and Microsoft Publisher come with simple Web page editors.

Once you've created your Web site, you'll need to "publish" it—upload (transfer) it to a Web server in order for it to be viewed on the Internet. You can get upload instructions from your online service or Internet service provider (ISP). Some ISPs will give you a certain amount of space on their servers at no additional charge.

Other specific packages for building Web sites are Adobe's PageMill, Soft Quad's Hot Metal Pro, and Microsoft's FrontPage Editor. If you want to create more sophisticated sites, you can try using tools such as Home Page from Claris or Trellix from Trellix Corporation. (A trial version of Trellix is downloadable from *http://www.trellix.com*)

Incidentally, Internet services are competing ferociously for talented, experienced Web page designers, according to Thomas Madzy, president of Sundoro Development, an Itasca, Illinois, Internet consultant. "With 200,000 openings nationwide for Web experts and other techies," reports one news item, "signing bonuses of 50% to 75% of the typical $60,000 to $80,000 annual salary are common, [Madzy] says."[33]

VRML is not an extension of HTML, and so HTML Web browsers cannot interpret it. Thus, users need a VRML add-on (plug-in), such as Netscape's Live3D, to receive VRML Web pages. Users who are not on a large computer system also need a high-end microcomputer such as a Power Macintosh or Pentium-based PC. Like HTML, VRML is a document-centered ASCII language. Unlike HTML, it tells the computer how to create 3-D worlds. VRML pages can also be linked to other VRML pages.

Even though VRML's designers wanted to let nonprogrammers create their own virtual spaces quickly and painlessly, it's not as simple to describe a three-dimensional scene as it is to describe a page in HTML. However, many existing modeling and CAD (✔ p. 78) tools now offer VRML support, and new VRML-centered software tools are arriving.

- Java—for creating interactive Web pages: Available from Sun Microsystems and derived from C++, Java is a major departure from the HTML coding that makes up most Web pages. Sitting atop markup languages such as HTML and XML, *Java* **is an object-oriented programming language that allows programmers to build applications that can run on any operating system.** With Java, big application programs can be broken into mini-applications, or "applets," that can be downloaded off the Internet and run on any computer. Moreover, Java enables a Web page to deliver, along with visual content, applets that when downloaded can make Web pages interactive.

If the use of Java becomes widespread, the Web will be transformed from the information-delivering medium it is today into a completely interactive computing environment. You will be able to treat the Web as a giant hard disk loaded with a never-ending supply of software applications.

Microcomputers are becoming available with special Java microprocessors designed to run Java software directly. However, Java is not compatible with many existing microprocessors, such as those from Intel and Motorola. For this reason, users need to use a small "interpreter" program, called a Java Virtual Machine, which translates a Java program into a language that any computer or operating system can understand. They also need a Java-capable browser in order to view Java special effects.

Java development programs are available for programmers. In addition, Java software packages—such as ActionLine, Activator Pro, AppletAce, and Mojo—are available to give nonprogrammers the ability to add multimedia effects to their Web pages, producing applets that can be viewed by any Java-equipped browser. Such packages can be used by anyone who understands multimedia file formats and is willing to experiment with menu options.

● ActiveX—also for creating interactive Web pages: ActiveX was developed by Microsoft as an alternative to Java for creating interactivity on Web pages. Indeed, Java and ActiveX are the two major contenders in the Web-applet war for transforming the Web into a complete interactive environment.

ActiveX **is a set of controls, or reusable components, that enables programs or content of almost any type to be embedded within a Web page.** Whereas Java requires you to download an applet each time you visit a Web site, with ActiveX the component is downloaded only once, then stored on your hard drive for later and repeated use.

Thus, the chief characteristic of ActiveX is that it features *reusable* components—small modules of software code that perform specific tasks (such as a spelling checker), which may be plugged seamlessly into other applications. With ActiveX you can obtain from your hard disk any file that is suitable for the Web—such as a Java applet, animation, or pop-up menu—and insert it directly into an HTML document.

Programmers can create ActiveX controls or components in a variety of programming languages, including C, C++, Visual Basic, and Java. Thousands of ready-made ActiveX components are now commercially available from numerous software development companies.

Onward

What do object-oriented, visual, and Internet programming imply for the future? Will tomorrow's programmer look less like a writer typing out words and more like an electrician wiring together circuit components, as one magazine suggests?[34] What does this mean for the five-step programming model we have described?

Some institutions are now teaching only object-oriented design techniques, which allow the design and ongoing improvement of working program models. Here programming stages overlap, and users repeatedly flow through analysis, design, coding, and testing stages. Thus, users can test out new parts of programs and even entire programs as they go along. They need not wait until the end of the process to find out if what they said they wanted is what they really wanted.

This new approach to programming is not yet in place in business, but in a few years it may be. If you're interested in being able to communicate with programmers in the future, or in becoming one yourself, the new approaches are worth your attention.

How to Buy Software

You can buy music audiotapes at the 7-Eleven. Sometimes you can buy or rent videocassette tapes at the gas station minimart. Computer software is not quite there yet, but it may be getting close. You can buy software at Sears, Circuit City, and Sam's Club. But should you? Let's consider the various sales outlets.[35-37]

Getting Ready to Buy

Whether shopping for system software or applications software, you need to be clear on a few things before you buy.

Do You Know Your Needs?

Before talking to anyone about software you should have a clear idea of what you want your computer to do for you.

Are you mainly writing research papers? keeping track of performance of employees reporting to you? projecting sales figures? building a mailing list and launching a fund-raising campaign? publishing a newsletter? teaching children about computers? You want the machine to serve you, not vice versa. Do you need to share documents that you create with other users? If so, what programs are they using?

Do You Know What Software You Want?

The safest course is to pick software used successfully by people you know. Or look for ratings in the leading computer magazines. Brands that consistently get high ratings in magazine reviews are generally likely to be reliable.

Do You Know the Latest Version & Release?

If you know the name of the software you want to use, or at least inquire about, so much the better. If you have a particular brand and type in mind, make sure it's the most recent version and release.

A new *version* of a software package resembles a model change on a car. It adds all kinds of new features, generally making the software more powerful and versatile. Versions are usually numbered in ascending order. MS-DOS 6.22 is a later version of Microsoft's Disk Operating System software than DOS 4.1 or DOS 3.3.

A *release number* identifies a specific version of a program. A program labeled 5.2, for example, is the third release of the fifth version. (The first and second releases were 5.0 and 5.1.) New releases generally incorporate routine enhancements and correct the annoying errors called software bugs.

Experienced users have a horror of using the very first version, version 1.0, of anything because the software generally still contains numerous bugs.

Most software manufacturers are continually upgrading their products. Often the upgrades are made available—usually online, over the Internet—at considerably less expense to purchasers of earlier versions.

Do You Know If an Upgrade Is Coming Out?

Find out if a new version of the software you're interested in is just around the corner. You may want to hold off buying until it's available.

Upgrades are to software as new models are to the auto industry. Every year or so, software manufacturers bring out a new version or release featuring incremental improvements, just like the car makers do.

Will the Salespeople Speak Your Language?

Selling ice cream does not require a lot of product knowledge. Selling software does (or should). Some salespeople know their wares but talk down to newcomers to try to impress them with their knowledge. Others have only the scantest familiarity with their products, although they may be patient with novices' questions. You want someone who is both knowledgeable and helpful. That may require a little investigation on your part.

Software Sellers: The Range of Outlets

The types of software sellers are as follows:

- Small hardware and software retail stores
- Small software-only retail stores
- Computer superstores
- Electronics, office, and department stores and warehouse clubs
- Mail-order companies and direct mail
- Software manufacturers (vendors) via their Web sites

Small Retail Stores for Both Hardware & Software

A small retail store selling both computer hardware and software may be a good place to go if you need to buy a PC as well as software. Prices won't beat those of discounters and superstores, but you may be able to find a well-informed, knowledgeable staff. Examples of small retail stores are those found in chains of dealers such as CompuAdd, MicroAge, and Software City. Many office-products superstores also have software.

Small Retail Stores for Software Only

If you already have a microcomputer, you may find the salespeople at software-only stores even better able to talk to you about the nuances of programs than those in the previous group. Stores that belong to national chains, such as Software Etc. and Babbage's, may carry a couple of thousand software titles. Generally such stores have a number of computers on the premises so you can try out software before buying. Prices are often discounted and may match those of computer superstores.

Computer Superstores

Computer superstores range up to 30,000 square feet in size, versus, say, 2000 square feet for small retail stores. Not only do they carry all kinds of hardware, including computer furniture, but also upwards of 2000–3000 software titles. Like smaller stores, they offer computers for trying out software. Unlike some smaller stores, they also offer classes to train you in the use of particular software packages. Finally, they have extensive technical departments for installing software and readying and repairing hardware. Examples of chains with superstores are CompuAdd, CompUSA, and Computer City SuperCenters.

Electronics, Office, & Department Stores & Warehouse Clubs

You can tell that computer programs have reached the mainstream when they are no longer sold as specialty items in specialty stores. Walk into electronics stores like Radio Shack, Circuit City, The Good Guys, and Best Buy and you'll find software there, too. Large office-supply stores often have software departments, too. Some department stores such as Sears, Montgomery Ward, and Dayton Hudson carry both hardware and software. It's doubtful, however, that the selection and prices are as competitive as those in the other types of stores we've described.

Discount warehouse clubs include stores like Costco, Office Club, Office Depot, and Sam's Club. These are membership operations that provide consumers with merchandise—often including computer hardware and software—at severely discounted prices. Some warehouse clubs are oriented toward business owners and managers, and so most of their software offerings are most appropriate for business. Some drawbacks are that these stores may not have repair services, customer support, or salespeople with deep product knowledge.

Mail-Order Companies & Direct Mail

With mail order, you dial a toll-free number, tell the order taker what you want, and charge the purchase to your credit card. The product is delivered to you by delivery service, often overnight. As long as you know what you want and pick a reputable company, mail order is an effective—and cost-effective—way of getting software. Indeed, mail order is probably the cheapest way to get software.

There are two ways to order software by mail order.

- **Mail-order companies:** These companies sell all kinds of software (and perhaps some hardware "peripherals"—

equipment like printers and monitors). Examples are CompuAdd Express, Software Unlimited, MicroWarehouse and PC Connection for IBM-compatibles, and MacWarehouse and Mac Connection for Macintoshes.

- **Direct mail:** Some software manufacturers make their products available through direct mail. Examples are Microsoft and IBM.

Vendor's Web Sites

Many software vendors offer software and software upgrades for sale via their Web sites. Usually you can download the software (after paying by credit card) or ask that a CD be sent to you.

Sensible Software Shopping: Some Tips

In software shopping, you're concerned not only with getting a good price but also protecting yourself if things go wrong. Here are a few tips.

Ask What Follow-Up Help Is Available

It's worth asking what kind of follow-up help the software seller offers. Actually, most retailers don't provide any such assistance, although they may provide classes for an extra charge. Many software packages come with some sort of tutorial to help you get started. The tutorial—an instructional book, videotape, or floppy disk—will lead you through a prescribed sequence of steps to learn the product.

Technical support is generally offered by the software manufacturer through a telephone number. Some manufacturers provide toll-free numbers, some do not.

Confirm the Price

Catalogs and ads are frequently revised. Sometimes there are hidden extra fees, such as for shipping charges, "restocking" charges if you return merchandise, or extra fees for credit-card purchases.

Ask About Money-Back Guarantees

If you open a software package, try it out, and find you don't like it, can you return it? Be sure to ask. Some sellers will make refunds, others will accept only unopened software. Some will make money-back guarantees only for 14–30 days after purchase, some have no time limits on software returns. In any event, save your receipts and be sure to try out the purchase within the allotted time.

Pay by Credit Card

Whether buying in a store or over the phone by mail order, use a credit card. That gives you some leverage to cancel the sale. Some credit-card companies offer added protection in the form of warranties of their own.

Summary

Summary

What It Is/What It Does	Why It's Important
ActiveX (p. 540, KQ 11.10) A set of controls, or reusable components, that enables programs or content of almost any type to be embedded within a Web page.	A principal contender with Java for transforming the Web into a complete interactive computing environment.
Ada (p. 533, KQ 11.8) Powerful third-generation, structured programming language designed by U.S. Defense Department to ensure portability of programs from one application to another.	Based on Pascal and originally intended to be a standard language for weapons systems, Ada has been used successfully in commercial applications.
algorithm (p. 509, KQ 11.1) Set of ordered, logical steps for solving a problem.	Algorithms are the basis of computer programming.
APL (p. 533, KQ 11.8) A Programming Language; third-generation mathematically oriented and scientific language designed in 1962 for use on IBM mainframes.	APL uses a special keyboard with special symbols to enable users to solve complex mathematical problems in a single step.
assembler (p. 525, KQ 11.7) Also called *assembler program;* language translator program that translates assembly-language programs into machine language.	Computers cannot run assembly language; it must first be translated.
assembly language (p. 524, KQ 11.7) Second-generation programming language; it allows a programmer to write a program using more easily remembered abbreviations or words instead of machine language.	A programmer can write instructions in assembly language faster than in machine language.
BASIC (p. 531, KQ 11.8) Beginner's All-purpose Symbolic Instruction Code; developed in 1965, the most popular microcomputer language and the easiest programming language to learn. Most popular versions: True BASIC, QuickBASIC, Visual BASIC for Windows. BASIC was widely used in the late 1960s in academic settings on all kinds of computers and is now used in business.	The interpreter form of BASIC is popular with first-time and casual users because it is interactive—user and computer can communicate during writing and running of a program.
C (p. 532, KQ 11.8) High-level general-purpose programming language that works well for microcomputers and is portable among many computers.	C is useful for writing operating systems, database management software, and some scientific applications. C is widely used in commercial software development, including games, robotics, and graphics.
C++ (p. 537, KQ 11.9) High-level programming language that combines the traditional C programming language with object-oriented capability (the plus signs mean "more than C").	With C++, programmers can write standard code in C without the object-oriented features, use object-oriented features, or do a mixture of both.

What It Is/What It Does

COBOL (p. 529, KQ 11.8) COmmon Business-Oriented Language; high-level programming language of business. First standardized in 1968, the language has been revised three times, most recently as COBOL-85.

compiler (p. 526, KQ 11.7) Language translator that converts the entire program of a high-level language (called source code) into machine language (called object code) for execution later. Examples of compiler languages: COBOL, FORTRAN, Pascal, C.

control structure (p. 516, KQ 11.3) Also called *logic structure;* in structured program design, the programming structure that controls the logical sequence in which computer program instructions are executed. Three control structures are used to form the logic of a program: *sequence, selection,* and *iteration* (or loop).

debug (p. 520, KQ 11.5) Part of program testing; the detection, location, and removal of syntax and logic errors in a program.

desk-checking (p. 520, KQ 11.5) Form of program testing; programmers read through a program to ensure it's error-free and logical.

documentation (p. 521, KQ 11.6) Written descriptions of a program and how to use it; supposed to be done during all programming steps.

FORTH (p. 533, KQ 11.8) FOuRTH-generation language; created in 1971, a third-generation language designed for real-time control tasks as well as business and graphics applications. Used on all kinds of computers, it runs very fast because it requires less memory than other languages.

FORTRAN (p. 529, KQ 11.8) FORmula TRANslator; developed in 1954, it was the first high-level language and was designed to express mathematical formulas. The newest version is FORTRAN 90.

generations of programming languages (p. 523, KQ 11.7) Five increasingly sophisticated levels (generations) of programming languages: (1) machine language, (2) assembly language, (3) high-level languages, (4) very high-level languages, (5) natural languages.

Why It's Important

The most significant attribute of COBOL is that it is extremely readable. COBOL is the language used by the majority of mainframe users, although it will also run on microcomputers.

Unlike other language translators (assemblers and interpreters), with a compiler program the object code can be saved and executed later rather than run right away. The advantage of a compiler is that, once the object code has been obtained, the program executes faster.

One thing that all three control structures have in common is one entry and one exit. The control structure is entered at a single point and exited at another single point. This helps simplify the logic so that it is easier for others following in a programmer's footsteps to make sense of the program.

Debugging may take several trials using different data before the programming team is satisfied the program can be released. Even then, some errors may remain, because trying to remove all of them may be uneconomical.

Desk-checking should be done before the program is actually run on a computer.

Documentation is needed for all people who will be using or involved with the program in the future—users, operators, and programmers.

Because FORTH runs so fast, it is used in applications that must process data quickly—for example, sensors, arcade games, robotics.

The most widely used language for mathematical, scientific, and engineering problems, FORTRAN is also useful for complex business applications, such as forecasting and modeling. Because it cannot handle a large volume of input/output operations or file processing, it is not used for more typical business problems.

Programming languages are said to be *lower level* when they are closer to the language used by the computer (0s and 1s) and *higher level* when closer to the language used by people (for example, English).

What It Is/What It Does	**Why It's Important**

hierarchy chart (p. 514, KQ 11.3) Also called *structure chart;* a diagram used in programming to illustrate the overall purpose of a program, identifying all the modules needed to achieve that purpose and the relationships among them.

In a hierarchy chart, the program must move in sequence from one module to the next until all have been processed. There must be three principal modules corresponding to the three principal computing operations—input, processing, and output.

high-level language (p. 525, KQ 11.7) Also known as *third-generation languages;* they somewhat resemble human languages. Examples: FORTRAN, COBOL, BASIC, Pascal.

High-level languages allow programmers to write in a familiar notation rather than numbers or abbreviations. Most can also be used on more than one kind of computer.

Hypertalk (p. 537, KQ 11.9) Language that uses principles of object-oriented programming in the Apple Macintosh HyperCard program to manipulate related files called *stacks.* Using a mouse, users can go through the stacks to find information or discover connections between ideas.

Though HyperCard is not precisely an object-oriented language, Hypertalk is.

interpreter (p. 526, KQ 11.7) Language translator that converts each high-level language statement into machine language and executes it immediately, statement by statement. An example of a high-level language using an interpreter is BASIC.

Unlike with the language translator called the compiler, no object code is saved. The advantage of an interpreter is that programs are easier to develop.

iteration control structure (p. 516, KQ 11.3) Also known as *loop structure;* one of the control structures used in structured programming. A process is repeated as long as a certain condition remains true. The programmer can stop the repetition at the *beginning* of the loop, using the DO WHILE iteration structure, or at the *end* of the loop, using the DO UNTIL iteration structure (which means the loop statements will be executed at least once).

Iteration control structures help programmers write better organized programs.

Java (p. 540, KQ 11.10) Object-oriented programming language used to create applications that will run on any operating system. Big applications can be broken into mini-applications, "applets," that can be downloaded from the Internet. For example, some applets enable Web pages to be interactive.

Java may be able to transform the Internet from just an information-delivering medium into a completely interactive computing environment.

language translator (p. 525, KQ 11.7) Type of system software that translates a program written in a second-, third-, or higher-generation language into machine language. Language translators are of three types: *assemblers, compilers,* and *interpreters.*

Because a computer can execute programs only in machine language, a translator is needed if the program is written in any other language.

LISP (p. 532, KQ 11.8) LIST Processor; developed in 1958, a third-generation language used principally to construct artificial intelligence programs.

LISP is used to write expert systems and natural language programs. Expert systems are programs imbued with knowledge by a human expert; such programs can walk users through a problem and help them solve it.

What It Is/What It Does	Why It's Important

logic errors (p. 520, KQ 11.5) Programming errors caused by not using control structures correctly.

If a program has logic errors, it will not run correctly or perhaps not run at all.

Logo (p. 533, KQ 11.8) Developed in 1967, a third-generation language designed primarily to teach children problem-solving and programming skills, though it is also used to produce graphic reports in business.

Logo uses a triangular pointer, called a *turtle,* which responds to a few simple commands (such as forward, left, right) and produces similar movements on screen, enabling users to draw geometric patterns and pictures.

machine language (p. 524, KQ 11.7) Lowest level of programming language; the language of the computer, representing data as 1s and 0s. Most machine-language programs vary from computer to computer—they are machine dependent.

Machine language, which corresponds to the on and off electrical states of the computer, is not convenient for people to use. Assembly language and higher-level languages were developed to make programming easier.

module (p. 514, KQ 11.3) A processing step of a program; sometimes called a *subprocedure* or *subroutine.* Each module is made up of logically related program statements.

Each module has only a single function, which limits the module's size and complexity.

natural languages (p. 528, KQ 11.7) (1) Ordinary human languages (for instance, English, Spanish); (2) fifth-generation programming languages that use human language to give people a more natural connection with computers.

Though still in their infancy, natural languages are getting close to human communication.

object-oriented programming (OOP) (p. 534, KQ 11.9) Programming method in which data and the instructions for processing that data are combined into a self-sufficient object—piece of software—that can be used in other programs. Examples of OOP languages: Smalltalk, C++, Turbo Pascal, Hypertalk.

Objects can be reused and interchanged among programs, producing greater flexibility and efficiency than is possible with traditional programming methods.

Pascal (p. 531, KQ 11.8) High-level programming language; an alternative to BASIC as a language for teaching purposes that is relatively easy to learn.

Pascal has extensive capabilities for graphics programming and is excellent for scientific use.

PL/1 (p. 532, KQ 11.8) Programming Language 1; introduced in 1964, a third-generation language designed to process both business and scientific applications. It is quite flexible and easy to learn.

PL/1 contains many of the best features of FORTRAN and COBOL; however, it also has so many options as to diminish its usefulness and so has not given them much competition.

program (p. 509, KQ 11.1) List of instructions the computer follows to process data into information. The instructions consist of statements written in a programming language (for example, BASIC).

Without programs, data could not be processed into information by a computer.

program flowchart (p. 514, KQ 11.3) Chart that graphically presents the detailed, logical steps (algorithms) needed to solve a programming problem. It uses standard symbols called ANSI symbols.

The program flowchart presents the detailed series of steps needed to solve a programming problem.

What It Is/What It Does	Why It's Important

programming (p. 510, KQ 11.1) Also called *software engineering;* five-step process for creating software instructions: (1) clarify the problem; (2) design a solution; (3) write (code) the program; (4) test the program; (5) document and maintain the program.

Programming is one step in the systems development life cycle (*covered in Chapter 10*).

programming language (p. 519, KQ 11.4) Set of rules that allow programmers to tell the computer what operations to follow. The five levels (generations) of programming languages are (1) machine language, (2) assembly language, (3) high-level (procedural) languages (FORTRAN, COBOL, BASIC, Pascal, C, RPG, etc.), (4) very high-level (nonprocedural) languages (RPG III, SQL, Intellect, NOMAD, FOCUS, etc.), and (5) natural languages.

Not all programming languages are appropriate for all uses. Thus, a language must be chosen to suit the purpose of the program and to be compatible with other languages being used by users.

PROLOG (p. 533, KQ 11.8) PROgramming LOGic; invented in 1972, a third-generation language used for developing artificial intelligence applications.

PROLOG is used to develop natural language programs and expert systems.

pseudocode (p. 514, KQ 11.3) Method of designing a program using normal human-language statements to describe the logic and processing flow. Pseudocode is like doing an outline or summary form of the program to be written.

By using such terms as IF, THEN, or ELSE, pseudocode follows the rules of control structures, an important aspect of structured programming.

RPG (p. 532, KQ 11.8) Report Program Generator; introduced in 1964, a third-generation language that has evolved through several versions (RPG, RPG II, RPG III) and is designed to help generate business reports. Users fill out a special form specifying the information to be included and its format.

RPG is highly structured and relatively easy to use.

selection control structure (p. 516, KQ 11.3) Also known as an *IF-THEN-ELSE structure;* one of the control structures used in structured programming. It offers two paths to follow when a decision must be made by a program.

Selection control structures help programmers write better organized programs.

sequence control structure (p. 516, KQ 11.3) One of the control structures used in structured programming; each program statement follows another in logical order. There are no decisions to make.

Sequence control structures help programmers write better organized programs.

Smalltalk (p. 536, KQ 11.9) Object-oriented programming (OOP) language, invented in 1970; a keyboard is used to enter text, but all other tasks are performed with a mouse.

Smalltalk was the first OOP language.

What It Is/What It Does	**Why It's Important**

structured programming (p. 513, KQ 11.3) Method of programming that takes a top-down approach, breaking programs into modular forms and using standard logic tools called control structures (sequence, selection, iteration).

Structured programming techniques help programmers write better organized programs, using standard notations with clear, correct descriptions.

structured walkthrough (p. 519, KQ 11.3) Program review process that is part of the design phase of the programming process; a programmer leads other development team members in reviewing a segment of code to scrutinize the programmer's work.

The structured walkthrough helps programmers find errors, omissions, and duplications, which are easy to correct because the program is still on paper.

syntax (p. 520, KQ 11.4) "Grammar" rules of a programming language that specify how words and symbols are put together.

Each programming language has its own syntax, just as human languages do.

syntax errors (p. 520, KQ 11.5) Programming errors caused by typographical errors and incorrect use of the programming language.

If a program has syntax errors, it will not run correctly or perhaps not run at all.

top-down program design (p. 514, KQ 11.3) Method of program design; a programmer identifies the top or principal processing step, or module, of a program and then breaks it down in hierarchical fashion into smaller processing steps. The design can be represented in a top-down hierarchy chart.

Top-down program design enables an entire program to be more easily developed because the parts can be developed and tested separately.

very-high-level language (p. 527, KQ 11.7) Also known as *fourth-generation language (4GL);* more user-oriented than third-generation languages, 4GLs require fewer commands. 4GLs consist of report generators, query languages, application generators, and interactive database management system programs. Some 4GLs are tools for end-users, some are tools for programmers.

Programmers can write programs that need only tell the computer what they want done, not all the procedures for doing it, which saves them the time and labor of having to write many lines of code.

visual programming (p. 537, KQ 11.9) Method of creating programs; the programmer makes connections between objects by drawing, pointing, and clicking on diagrams and icons. Programming is made easier because the object orientation of object-oriented programming is used in a graphical or visual way.

Visual programming enables users to think more about the problem solving than about handling the programming language.

VRML (virtual reality modeling language) (p. 538, KQ 11.10) Type of programming language used to create three-dimensional (3-D) Web pages.

VRML expands the information-delivering capabilities of the Web.

XML (extensible markup language) (p. 538, KQ 11.10) Programming language used to make it easy for machines to read Web sites by allowing Web developers to add more "tags" to a Web page.

XML is more powerful software than HTML, allowing information on a Web site to be described by general tags—for example, identifying one piece of information in a recipe as "cooking time" and others as "ingredients."

Exercises

Self-Test

1. Machine language is a _____ - generation language.
2. Fifth-generation languages are often called _____ _____ languages.
3. _____ is used for creating 2-D Web documents and links.
4. In the _____ control structure, one program statement follows another in logical order.
5. _____ is used for creating 3-D Web pages.

Short-Answer Questions

1. What is natural language?
2. How do third-generation languages differ from first- and second-generation languages?
3. What is visual programming?
4. What were the reasons behind the development of high-level programming languages?
5. Why is documentation important during program development?

Multiple-Choice Questions

1. _____ is a relatively simple high-level language that was developed to help students learn programming.
   a. BASIC
   b. COBOL
   c. C
   d. ADA
   e. All of the above

2. Query languages, report generators, and applications generators are examples of _____ _____ -generation languages.
   a. first
   b. second
   c. third
   d. fourth
   e. fifth

3. Which of the following is used to design a program using English-like statements?
   a. pseudocode
   b. program flowcharts
   c. control structures
   d. structured walkthroughs
   e. None of the above

4. Which of the following control structures use DO UNTIL?
   a. sequence
   b. selection
   c. iteration (loop)
   d. pseudocode
   e. None of the above

5. Assemblers, compilers, and interpreters are types of _____ _____.
   a. programming languages
   b. language translators
   c. alpha testers
   d. application generators
   e. All of the above

True/False Questions

T F 1. The rules for using a programming language are called *syntax*.

T F 2. A query language allows the user to easily retrieve information from a database management system.

T F 3. Objects found in an object-oriented program can be reused in other programs.

T F 4. It is correct to refer to programming as *software engineering*.

T F 5. A syntax error can be caused by a simple typographical error.

Knowledge in Action

1. Suppose you're in charge of creating a Web site that will be used by the other students in your class to access assignment information, a list of suggested reading materials, relevant Web addresses, and other course information. How might you implement HTML, XML, VRML, and JAVA to create this site?

2. Some experts think that, before long, we will have only one superapplication to run on our computers instead of several separate applications packages. However, people in the computer industry are commonly overoptimistic about the speed at which new developments will occur. What do you think will be the obstacles to achieving a superapplication?

3. Visit the computer laboratory at your school.

   a. Identify which high-level languages are available.

   b. Determine if each language processor identified is a compiler or an interpreter.

   c. Determine if the language processors are available for microcomputers, larger computers, or both.

   d. Identify any microcomputer-based electronic spreadsheet software and database management systems software. Have any applications been created with these tools that are used in the lab or by the lab staff?

4. Interview several students who are majoring in computer science and studying to become programmers. What languages do they plan to master? Why? What kinds of jobs do they expect to get? What kinds of future developments do they anticipate in the field of software programming?

5. Check the yellow pages in your phone book, and contact a company that develops custom-designed software. What languages do they use to write the software? Does this company follow the five stages of software development described in this book, or does it use another set of stages? If another set of software-development stages is used, what are its characteristics?

Society & the Digital Age

Challenges & Promises

key questions

You should be able to answer the following questions:

Clearly, information technology is driving the new world of jobs, leisure, and services, and nothing is going to stop it. Indeed, predicts one futurist, by 2010 probably 90% of the workforce will be affected by the four principal information technologies—computer networks, imaging technology, massive data storage, and (as we discuss in this chapter) artificial intelligence.[1]

Where will you be in all this? People pursuing careers find the rules are changing very rapidly. Up-to-date skills are becoming ever more crucial. Job descriptions of all kinds are being redefined. Even familiar jobs are becoming more demanding. Today, experts advise, you need to prepare to continually upgrade your skills, prepare for specialization, and prepare to market yourself.

"Computer technology is the most powerful and the most flexible technology ever developed," says Terry Bynum, who chairs the American Philosophical Association's Committee on Philosophy and Computing. "Even though it's called a technical revolution, at heart it's a social and ethical revolution because it changes everything we value."[2]

In this chapter, we consider both the challenges and the promises of computers and communications in relation to society. First let us consider the following *challenges* of the Digital Age:

- The blueprint for the Information Superhighway
- Security issues—accidents, hazards, crime, viruses—and security safeguards
- Quality-of-life issues—environment, mental health, the workplace
- Economic issues—employment and the haves/have-nots

We will then consider the following *promises:*

- The roles of intelligent agents and avatars
- Artificial intelligence
- The promised benefits of the Information Revolution

12.1 The Information Superhighway: Is There a Grand Design?

KEY QUESTION

What are the NII, the new Internet, the Telecommunications Act, and the 1997 White House plan?

Preview & Review: The Information Superhighway envisions using wired and wireless capabilities of telephones and networked computers with cable TV. It may evolve following a model backed by the federal government called the National Information Infrastructure, along with newer versions of the Internet—VBNS, Internet 2, and NGI. Or it may evolve out of competition brought on by the deregulation of long-distance and local-telephone companies, cable companies, and television broadcasters created by the 1996 Telecommunications Act. A 1997 White House plan suggests the government should stay out of Internet commerce.

As we said in Chapter 1, the *Information Superhighway* is a vision or a metaphor. It envisions a fusion of the two-way wired and wireless capabilities of telephones and networked computers with cable and satellite TV's capacity to transmit thousands of programs, generally based on the evolving system of the Internet. When completed, it is hoped that the I-way will give us lightning-fast (high-bandwidth) voice and data exchange, multimedia, interactivity, and nearly universal and low-cost access—and that it will do so reliably and securely. Whether you're a Russian astronaut aloft in the spacecraft Mir, a Bedouin tribesman in the desert with a PDA/cell phone, or a Canadian work-at-home mother with her office in a spare bedroom, you'll be able to connect with nearly anything or anybody anywhere. You'll be able to participate in telephony, teleconferencing, telecommuting, teleshopping, telemedicine, tele-education, tele-voting, and even tele-psychotherapy, to name a few possibilities.

What shape will the Information Superhighway take? Some government officials hope it will follow a somewhat orderly model, such as that envisioned in the National Information Infrastructure (of which new versions of the Internet are a part). Others hope it will evolve out of competition intended by the passage of the 1996 Telecommunications Act. Still others hope that a White House document, *A Framework for Global Electronic Commerce*, offers a realistic policy. Let us look at these.

The National Information Infrastructure

As portrayed by U.S. government officials, **the *National Information Infrastructure (NII)* is a kind of grand vision for today's existing networks and technologies as well as technologies yet to be deployed. Services would be delivered by telecommunications companies, cable-television companies, and the Internet.** Applications would be varied—education, health care, information access, electronic commerce, and entertainment.[3] (■ *See Panel 12.1, next page.*)

Who would put the pieces of the NII together? The national policy is to let private industry do it, with the government trying to ensure fair competition among phone and cable companies and compatibility among various technological systems.[4] In addition, NII envisions open access to people of all income levels.

The New Internet: VBNS, Internet 2, & NGI

The National Information Infrastructure is, as mentioned, more of a national vision than an actual blueprint. And lately we have been hearing less about NII and more about new Internet networks: *Internet 2*, the *Next Generation Internet*, and *VBNS*.

What are these, and does this mean that *three* new networks will be built? Actually, all three names refer to the same network. This will be a new high-speed Internet that will unclog the clogged electronic highway that the present Net has become. All three will use the same fiber-optic cable, high-speed switching devices, and software. The reason for the differences, as one report explains, is that "government officials and university administrators want to emphasize their individual roles in reaching the common goal."[5]

Here's what the three efforts represent:[6–11]

- **VBNS: Linking supercomputers and other banks of computers across the nation, *VBNS* (for *Very-high-speed Backbone Network Service*) is the main government component to upgrade the "backbone," or primary hubs of data transmission.** Speeds would be at 1000 times current Internet speeds.

Commerce

With inexpensive access charges, small companies could afford to act like big ones. Boundaries would be erased between company departments, suppliers, and customers. Designs for new products could be tested and exchanged with factories in remote locations. With information flowing faster, goods could be sent to market faster and inventories kept low.

Government

An information highway could extend electronic democracy through electronic voting, allow interactive local-government meetings between electors and elected, and help deliver government services such as administering Social Security forms.

Education

"Virtual" classrooms and distance learning would replace lecture halls and scheduled class times. Students could take video field trips to distant places and get information from remote museums and libraries (such as the Library of Congress).

Home Services

Consumers would be able to receive movies on demand, home shopping, and videogames; do electronic bill paying; and tap into libraries and schools.

Health Care

Through telemedicine, health-care providers and researchers could share medical images, patient records, and research and perform long-distance patient examinations. Interactive, multimedia materials directed to the public would outline health-care options.

Information

Government records, patents, contracts, legal documents, and satellite maps could be made available to the public online. Libraries could also be digitized, with documents available for downloading.

Mobile Communications

Users with handheld personal communicators or personal digital assistants would be able to send and receive voice, fax, text, and video messages anywhere.

■ PANEL 12.1

The National Information Infrastructure: the grand vision

These are some of the services envisioned by supporters of the National Information Infrastructure, a kind of grand national vision. The NII would rely on existing networks and technologies and technologies yet to be deployed, such as faster versions of the Internet.

Financed by the National Science Foundation and managed by telecommunications giant MCI, VBNS will involve only the top 100 research universities, whereas Internet 2 would eventually touch everyone on the line. VBNS has already been underway since 1996, and most of the present members are also members of Internet 2.

- Internet 2: **Internet 2 is a cooperative university-business program to enable high-end users to quickly and reliably move huge amounts of data, using VBNS as the official backbone.** Whereas VBNS would provide data transfer at 1000 times present speeds, Internet 2 would operate at 100 times current Internet speeds.

  In effect, Internet 2 will add "toll lanes" to the Internet that already exists today to speed things up. The purpose is to advance videoconferencing, research, and academic collaboration—to enable a kind of "virtual university." Presently more than 117 universities and about 25 companies are participants.

- Next Generation Internet: **The *Next Generation Internet (NGI)* is the U.S. government's broad new program to parallel the university-and-business-sponsored effort of Internet 2, and is designed to provide money to six government agencies to be used to help tie the campus high-performance backbones into the broader federal infrastructure.** The technical goals for NGI include connecting at least 100 sites, including universities, federal national laboratories, and other research organizations. Speeds, as mentioned, would be 100 times those of today's Internet, and 10 sites would be connected at speeds that are 1000 times as fast.

  NGI would work to make a new generation of networks that is faster, more reliable, and more secure than today's Internet. It would also promote applications such as distance education and distance research, including experiments involving simultaneous "real-time" use of scientific instruments and databases at different sites.

In general, all of these are modeled after the original Internet, except that they will use high-speed fiber-optic circuits and more sophisticated software. Internet 2 and NGI should be available to the public by 2003.[12]

So imagine yourself doing the following: One NGI project would allow a company's executives in different parts of the world to meet via "tele-immersion"—that is, "a 3-D setting with stereo-quality sound and dazzling graphics presentations," as one writer describes it. "Participants would don 3-D goggles at their desktop computer—much as game players do with virtual reality technology—and literally be able to share the same pen to sketch out ideas on an easel."[13]

And a few years down the road, there'll be something else. Douglas Van Houweling is a scientist who helped create the original Internet and is now leading research on Internet 2. "Five years from now," he said in early 1998, "we'll be talking about Internet 3."[14]

The 1996 Telecommunications Act

"Let the telecom wars begin," proclaimed the *Wall Street Journal* in February 1996.[15]

"Let the telecom dogfight begin," announced *Business Week*—14 months later.[16]

Has the battle started yet? What's taking so long? This requires some explanation.

After years of legislative attempts to overhaul the 1934 Communications Act, in February 1996 President Clinton signed into law (using a high-tech stylus and an electronic tablet) a new telecommunications law. The *Telecommunications Act of 1996* undoes 60 years of federal and state communications regulations and is designed to let phone, cable, and TV businesses compete and combine more freely.

Supporters compared the law's significance to the fall of Communism, saying it would create jobs, expand consumer choices, and lower phone and cable rates.[17] Opponents—mostly consumer groups—said consolidation of businesses would cost jobs and probably raise rates. The law, predicted one article, "is certain to accelerate the convergence of local and long-distance phone businesses with cable operators, cellular companies, broadcast concerns, computer makers, and others."[18]

In general, the law allowed the following:[19-22] It permitted greater competition between local and long-distance telephone companies, as well as between the telephone and cable industries. In particular, the legislation allowed cable companies (such as TCI) and long-distance carriers (AT&T, Sprint, MCI) to offer local telephone service. It also allowed the regional Bell operating companies (at that time seven, now probably four) to offer not just local phone service but also long-distance services. Both local and long-distance phone companies are permitted to enter the cable-TV business, offering video via wired or wireless means. Cable services are allowed to go into cross-ownership with telephone companies in small communities. Cable services are also permitted to boost their rates. Television stations received a new broadcast spectrum to be used for digital TV.

So, did we get the "telecom wars" we were led to expect—a mighty struggle to offer lower rates, better service, more competition? Not exactly. Or at least not yet.

"The only point on which all parties agree," says Laurence Tribe, Harvard professor of constitutional law, "is that the law isn't working as intended, and that American consumers are still waiting for free and healthy competition in communications services."[23]

The 1997 White House Plan for Internet Commerce

In July 1997, the Clinton administration unveiled a document that is significant because it endorses a governmental hands-off approach to the Internet—or, as it more grandly calls it, "the Global Information Infrastructure."[24] Behind the title *A Framework for Global Electronic Commerce*, authored by a White House group, was a plan whose gist is this: Government should stay out of the way of Internet commerce.

"Where government is needed," it states, "its aim should be to support and enforce a predictable, minimalist, consistent, and legal environment for commerce."[25] Thus, government involvement might be necessary for protecting intellectual property against electronic piracy, for example, or negotiating international agreements on tariffs and taxation. It also might have to see that law enforcement has some access to encrypted (coded) data, an idea the computer industry strongly opposes. One thing government should not do, the report emphasized, is tax the Internet, although some states and cities would like to do just that.

Otherwise, the plan states that private companies, not government, should take the lead in promoting the Internet as an electronic marketplace, in adopting self-regulation, and in devising ratings systems to help parents guide their children away from objectionable online content. The policy "literally follows the first rule of the Hippocratic oath: Do no harm," says one observer. "It lets businesses and consumers determine the Internet's growth."[26]

Eli Noam, professor of finance and economics at the Columbia Business School, remains a skeptic, believing that cyberspace will be regulated no matter what the White House says.

> For all the rhetoric of an Internet "free trade zone," will the United States readily accept an Internet that includes Thai child pornographers, Albanian tele-doctors, Cayman Island tax dodges, Monaco gambling, Nigerian blue sky stock schemes, Cuban mail-order catalogues? Or, for that matter, American violators of privacy, purveyors of junk E-mail, or "self-regulating" price-fixers? Unlikely. And other countries will feel the same on matters they care about.[27]

We consider some of these concerns in the following sections.

12.2 Security Issues: Threats to Computers & Communications Systems

KEY QUESTION

What are some characteristics of the six key security issues for information technology?

Preview & Review: Information technology can be disabled by a number of occurrences. It may be harmed by human, procedural, and software errors; by electromechanical problems; and by "dirty data." It may be threatened by natural hazards and by civil strife and terrorism. Criminal acts perpetrated against computers include theft of hardware, software, time and services, and information; and crimes of malice and destruction. Computers may be harmed by viruses. Computers can also be used as instruments of crime. Criminals may be employees, outside users, hackers, crackers, and professional criminals.

Security issues go right to the heart of the workability of computer and communications systems. Here we discuss the following threats to computers and communications systems:

- Errors and accidents
- Natural and other hazards
- Crime against computers and communications
- Crime using computers and communications
- Worms and viruses
- Computer criminals

Errors & Accidents

We frequently hear about "computer errors," but often what seems to be the computer's fault is human indifference or bad management.

For instance, Brian McConnell of Roanoke, Virginia, found that he couldn't get past a bank's automated telephone system to talk to a real person. This was not the fault of the system so much as of the people at the bank. McConnell, president of a software firm, thereupon wrote a program that automatically phoned eight different numbers at the bank. Employees picking up the phone heard the recording, "This is an automated customer complaint. To hear a live complaint, press . . ."[28]

In general, errors and accidents in computer systems may be classified as human errors, procedural errors, software errors, electromechanical problems, and "dirty data" problems.

- **Human errors:** Quite often, when experts speak of the "unintended effects of technology," what they are referring to are the unexpected things people do with it. People can complicate the workings of a system in three ways:[29]

  (1) Humans often are not good at assessing their own information needs. Thus, for example, many users will acquire a computer and communications system that either is not sophisticated enough or is far more complex than they need.

  (2) Human emotions affect performance. For example, one frustrating experience with a computer is enough to make some people abandon the whole system. But throwing your computer out the window, of course, isn't going to get you any closer to learning how to use it better.

  (3) Humans act on their perceptions, which in modern information environments are often too slow to keep up with the equipment. You can be so overwhelmed by information overload, for example, that decision making may be just as faulty as if you had too little information.

- **Procedural errors:** Some spectacular computer failures have occurred because someone didn't follow procedures. Consider the 2½-hour shutdown of Nasdaq, the nation's second largest stock market. Nasdaq is so automated that it likes to call itself "the stock market for the next 100 years." A few years ago, Nasdaq was shut down by an effort, ironically, to make the computer system more user-friendly. Technicians were phasing in new software, adding technical improvements a day at a time. A few days into this process, the technicians tried to add more features to the software, flooding the data-storage capability of the computer system. The result was a delay in opening the stock market that shortened the trading day.[30]

- **Software errors:** We are forever hearing about "software glitches" or "software bugs" (✔ p. 520). Recall that a *software bug* is an error in a program that causes it not to work properly.

  An example of a somewhat small error was when a school employee in Newark, New Jersey, made a mistake in coding the school system's master scheduling program. When 1000 students and 90 teachers showed up for the start of school at Central High School, half the students had incomplete or no schedules for classes. Some classrooms had no teachers while others had four instead of one.[31]

- **Electromechanical problems:** Mechanical systems, such as printers, and electrical systems, such as circuit boards, don't always work. They may be improperly constructed, get dirty or overheated, wear out, or become damaged in some other way. Power failures (brownouts and blackouts) can shut a system down. Power surges can burn out equipment.

  Modern systems, argues Yale University sociologist Charles Perrow, are made up of thousands of parts, all of which interrelate in ways that are impossible to anticipate. Because of that complexity, he says, what he calls "normal accidents" are inevitable. That is, it is almost certain that some combinations of minor failures will eventually amount to something catastrophic. Indeed, it is just such collections of small failures that led to catastrophes such as the blowing up of the Challenger space shuttle in 1986 and the near-meltdown of the Three Mile Island nuclear-power plant in 1979.[32] In the Digital Age, "normal accidents" will not be rarities but are to be expected.

- **"Dirty data" problems:** When keyboarding a research paper, you undoubtedly make a few typing errors (which, hopefully, you clean up). So do all the data-entry people around the world who feed a continual stream of raw data into computer systems. A lot of problems are caused by this kind of "dirty data." *Dirty data* is data that is incomplete, outdated, or otherwise inaccurate.

  A good reason for having a look at your records—credit, medical, school—is so that you can make any corrections to them before they cause you complications. As the president of a firm specializing in business intelligence writes, "Electronic databases, while a time-saving resource for the information seeker, can also act as catalysts, speeding up and magnifying bad data."[33]

Natural & Other Hazards

Some disasters do not merely lead to temporary system downtime; they can wreck the entire system. Examples are natural hazards, and civil strife and terrorism.

- **Natural hazards:** Whatever is harmful to property (and people) is harmful to computers and communications systems. This certainly includes natural disasters: fires, floods, earthquakes, tornadoes, hurricanes, blizzards, and the like. If they inflict damage over a wide area, as have ice storms in eastern Canada or hurricanes in Florida, natural hazards can disable all the electronic systems we take for granted. Without power and communications connections, automated teller machines, credit-card verifiers, and bank computers are useless.

- **Civil strife and terrorism:** We may take comfort in the fact that wars and insurrections seem to take place in other parts of the world. Yet we are not immune to civil unrest, such as the riots that wracked Los Angeles following the 1993 trial of police officers for the beating of Rodney King. Nor are we immune, apparently, to acts of terrorism, such as the 1993 bombing of New York's World Trade Center or the 1995 blowing up of the Oklahoma City federal building. In the New York case, companies found themselves frantically having to move equipment to new offices and reestablishing their computer networks.

  In recent times, magazines have run stories on Department of Defense plans for "cyberwar," a nonbloody kind of information warfare in which the computer systems of adversaries could be disabled through the use of viruses, phony radio messages, and electronic jamming.[34,35] The Pentagon itself (which has 650,000 terminals and workstations, 100 WANs, and 10,000 LANs) has been taking steps to reduce its own systems' vulnerability to intruders.[36]

Ethics

Crimes Against Computers & Communications

An *information-technology crime* can be of two types. It can be an illegal act perpetrated *against* computers or telecommunications. Or it can be the *use* of computers or telecommunications to accomplish an illegal act. Here we discuss the first type.

Crimes against information technology include theft—of hardware, of software, of computer time, of cable or telephone services, of information. Other illegal acts are crimes of malice and destruction. Some examples are as follows:

- **Theft of hardware:** Stealing of hardware can range from shoplifting an accessory in a computer store to removing a laptop or cellular phone from someone's car. Professional criminals may steal shipments of microprocessor chips off a loading dock or even pry cash machines out of shopping-center walls.

  Eric Avila, 26, a history student at the University of California at Berkeley, had his doctoral dissertation—involving six years of painstaking research—stored on the hard drive of his Macintosh PowerBook when a thief stole the machine out of his apartment. Although he had copied an earlier version of his dissertation (70 pages entitled "Paradise Lost: Politics and Culture in Post-War Los Angeles") onto a diskette, the thief stole that, too. "I'm devastated," Avila said. "Now it's gone, and there is no way I can recover it other than what I have in my head." To make matters worse, he had no choice but to pay off the $2000 loan for a computer he did not have anymore.[37] The moral, as we've said repeatedly in this book: *Always make backup copies of your important data, and store them in a safe place—away from your computer.*

- **Theft of software:** Stealing software can take the form of physically making off with someone's diskettes, but it is more likely to be the copying of programs. Software makers secretly prowl electronic bulletin boards in search of purloined products, then try to get a court order to shut down the bulletin boards. They also look for companies that "softlift"—buying one copy of a program and making copies for as many computers as they have.

  Many pirates are reported by co-workers or fellow students to the "software police," the Software Publishers Association. The SPA has a toll-free number (800-388-7478) for anyone to report illegal copying and initiate antipiracy actions. In mid-1994, two New England college students were indicted for allegedly using the Internet to encourage the exchange of copyrighted software.[38]

  Another type of software theft is copying or counterfeiting of well-known software programs. These pirates often operate in China, Taiwan, Mexico, Russia, and various parts of Asia and Latin America. In some countries, more than 90% of U.S. microcomputer software in use is thought to be illegally copied.[39]

- **Theft of time and services:** The theft of computer time is more common than you might think. Probably the biggest use of it is people using their employer's computer time to play games. Some people also may run sideline businesses.

  Theft of cable and telephone services has increased over the years. One cable company reported it lost $12 million a year to pirates using illegal set-top converter boxes. Recently, thieves have been able to crack the codes of the fast-growing digital satellite industry, using illegal decoders.[40] Under federal law, a viewer with an illegal decoder box can face up to 6 months in jail and a $1000 fine.

  For years "phone phreaks" have bedeviled the telephone companies. They have also found ways to get into company voice-mail systems, then use an extension to make long-distance calls at the company's expense. In addition, they have also found ways to tap into cellular phone networks and dial for free.

- **Theft of information:** "Information thieves" have been caught infiltrating the files of the Social Security Administration, stealing confidential personal records and selling the information. Thieves have also broken into computers of the major credit bureaus and have stolen credit information. They have then used the information to charge purchases or have resold it to other people. On college campuses, thieves have snooped into or stolen private information such as grades.

The recipients of huge sums of illicit profits, such as drug traffickers, are also going high-tech, doing their money laundering by using home-banking software, for example, to zip money across borders. Authorities fear that the rise of cybercash—for instance, use of smart cards containing memory chips that can be filled or emptied with the equivalent of cash—will turn money laundering into a financial crime that will be harder than ever to track. Says a U.S. Treasury Department official, "That's the drug kingpin of the future: The guy walking around with a chip in his pocket worth a few million."[41]

- **Crimes of malice and destruction:** Sometimes criminals are more interested in abusing or vandalizing computers and telecommunications systems than in profiting from them. For example, a student at a Wisconsin campus deliberately and repeatedly shut down a university computer system, destroying final projects for dozens of students. A judge sentenced him to a year's probation, and he left the campus.[42]

Ethics

Crimes Using Computers & Communications

Just as a car can be used to assist in a crime, so can a computer or communications system. For example, four college students on New York's Long Island who met via the Internet used a specialized computer program to steal credit-card numbers, then, according to police, went on a one-year, $100,000 shopping spree. When arrested, they were charged with grand larceny, forgery, and scheming to defraud.[43]

In addition, investment fraud has come to cyberspace. Many people now use online services to manage their stock portfolios through brokerages hooked into the services. Scam artists have followed, offering nonexistent investment deals and phony solicitations and manipulating stock prices.

Information technology has also been used simply to perpetrate mischief. For example, three students at a Wisconsin campus faced disciplinary measures after distributing bogus e-mail messages, one of which pretended to be a message of resignation sent by the university's chancellor.[44]

Worms & Viruses

Worms and viruses are forms of high-tech maliciousness. A *worm* is a program that copies itself repeatedly into memory or onto a disk drive until no more space is left. An example is the worm program unleashed by a student at Cornell University that traveled through an e-mail network and shut down thousands of computers around the country.

A *virus* is a "deviant" program that attaches itself to computer systems and destroys or corrupts data (✔ p. 128). Viruses are passed in two ways:

- **By diskette:** The first way is via an infected diskette, such as one you might get from a friend or a repair person. It's also possible to get a virus from a sales demo disk or even (in 3% of cases) from a shrink-wrapped commercial disk.

- **By network:** The second way is via a network, as from e-mail or an electronic bulletin board. This is why, when you're looking into all the freebie games and other software available online, you should use virus-scanning software to check downloaded files.

The virus usually attaches itself to your hard disk. It might then display annoying messages ("Your PC is stoned—legalize marijuana") or cause Ping-Pong balls to bounce around your screen and knock away text. More seriously, it might add garbage to or erase your files or destroy your system software. It may evade your detection and spread its havoc elsewhere.

Viruses may take several forms, the two main traditional ones being boot-sector viruses and file viruses. A more recent one is the *macro virus.* (■ *See Panel 12.2.)* There have been many strains of viruses in recent years, some of them quite well known (Stoned, Jerusalem B, Lehigh, Pakistani Brain, Michaelangelo). Some 6000 viruses have been identified, but only a few hundred of them have been found "in the wild," or in general circulation. Although most are benign, some are intended to be destructive. Some virus writers do it for the intellectual challenge or to relieve boredom, but others do it for revenge, typically against an employer.[45] One virus writer calling

■ PANEL 12.2 Types of viruses

- **Boot-sector virus:** The boot sector is that part of the system software containing most of the instructions for booting, or powering up, the system. The boot sector virus replaces these boot instructions with some of its own. Once the system is turned on, the virus is loaded into main memory before the operating system. From there it is in a position to infect other files. Any diskette that is used in the drive of the computer then becomes infected. When that diskette is moved to another computer, the contagion continues. Examples of boot-sector viruses: AntCMOS, AntiEXE, Form.A, NYB (New York Boot), Ripper, Stoned.Empire.Monkey.

- **File virus:** File viruses attach themselves to executable files—those that actually begin a program. (In DOS these files have the extensions .com and .exe.) When the program is run, the virus starts working, trying to get into main memory and infecting other files.

- **Multipartite virus:** A hybrid of the file and boot-sector types, the multipartite virus infects both files and boot sectors, which makes it better at spreading and more difficult to detect. Examples of multipartite viruses are Junkie and Parity Boot.

   A type of multipartite virus is the *polymorphic virus,* which can mutate and change form just as human viruses can. Such viruses are especially troublesome because they can change their profile, making existing antiviral technology ineffective.

   A particularly sneaky multipartite virus is the *stealth virus,* which can temporarily remove itself from memory to elude capture. An example of a multipartite, polymorphic stealth virus is One Half.

- **Macro virus:** Macro viruses take advantage of a procedure in which miniature programs, known as macros, are embedded inside common data files, such as those created by e-mail or spreadsheets, which are sent over computer networks. Until recently, such documents have typically been ignored by antivirus software. Examples of macro viruses are Concept, which attaches to Word documents and e-mail attachments, and Laroux, which attaches to Excel spreadsheet files. Fortunately, the latest versions of Word and Excel come with built-in macro virus protection.

- **Logic bomb:** Logic bombs, or simply bombs, differ from other viruses in that they are set to go off at a certain date and time. A disgruntled programmer for a defense contractor created a bomb in a program that was supposed to go off two months after he left. Designed to erase an inventory tracking system, the bomb was discovered only by chance.

- **Trojan horse:** The Trojan horse covertly places illegal, destructive instructions in the middle of a legitimate program, such as a computer game. Once you run the program, the Trojan horse goes to work, doing its damage while you are blissfully unaware. An example of a Trojan horse is FormatC.

himself Hellraiser, who in his pre-computer youth used to roam New York City streets with a can of spray paint, says "Viruses are the electronic form of graffiti."[46]

The fastest-growing virus in history, many experts say, was the *Word Concept* virus (or simply *Concept* virus), which worried people because it was able to sneak past security devices by hitching rides on e-mail and other common Internet files.[47] Concept attached itself to documents created by Microsoft's popular word processing program, Word 6.0 or higher. The virus doesn't cause damage, but some destructive relatives continue to surface. For example, Wazzu rearranges words in your document and inserts the word "wazzu" at random.[48]

A variety of virus-fighting programs are available at stores, although you should be sure to specify the viruses you want to protect against. *Antivirus software* scans a computer's hard disk, floppy disks, and main memory to detect viruses and, sometimes, to destroy them. We described some antivirus programs in Chapter 3 and antivirus measures in the Experience Box at the end of Chapter 5.

Ethics

Computer Criminals

What kind of people are perpetrating most of the information-technology crime? Over 80% may be employees; the rest are outside users, hackers and crackers, and professional criminals.

- **Employees:** "Employees are the ones with the skill, the knowledge, and the access to do bad things," says Donn Parker, an expert on computer security at SRI International in Menlo Park, California. "They're the ones, for example, who can most easily plant a 'logic bomb.'" Dishonest or disgruntled employees, he says, pose "a far greater problem than most people realize."[49] Says Michigan State University criminal justice professor David Carter, who surveyed companies about computer crime, "Seventy-five to 80% of everything happens from inside."[50]

    Most common frauds, Carter found, involved credit cards, telecommunications, employees' personal use of computers, unauthorized access to confidential files, and unlawful copying of copyrighted or licensed software. In addition, the increasing use of laptops off the premises, away from the eyes of supervisors, concerns some security experts. They worry that dishonest employees or outsiders can more easily intercept communications or steal company trade secrets.

    Workers may use information technology for personal profit or steal hardware or information to sell. They may also use it to seek revenge for real or imagined wrongs, such as being passed over for promotion. Sometimes they may use the technology simply to demonstrate to themselves that they have power over people. This may have been the case with a Georgia printing-company employee convicted of sabotaging the firm's computer system. As files mysteriously disappeared and the system randomly crashed, other workers became so frustrated and enraged that they quit.

- **Outside users:** Suppliers and clients may also gain access to a company's information technology and use it to commit crimes. Both suppliers and clients have more access as electronic connections such as electronic data interchange (✔ p. 333) systems become more commonplace.

- **Hackers and crackers:** *Hackers* are people who gain unauthorized access to computer or telecommunications systems for the challenge or even the principle of it. For example, Eric Corley, publisher of a magazine called *2600: The Hackers' Quarterly,* believes that hackers are merely engaging in "healthy exploration." In fact, by breaking into corporate computer systems and revealing their flaws, he says, they are performing a favor and a public service. Such unauthorized entries show the corporations involved the leaks in their security systems.[51]

  Crackers also gain unauthorized access to information technology but do so for malicious purposes. (Some observers think the term *hacker* covers malicious intent, also.) Crackers attempt to break into computers and deliberately obtain information for financial gain, shut down hardware, pirate software, or destroy data.

  The tolerance for "benign explorers"—hackers—has waned. Most communications systems administrators view any kind of unauthorized access as a threat, and they pursue the offenders vigorously. Educators try to point out to students that universities can't provide an education for everybody if hacking continues. The most flagrant cases of hacking are met with federal prosecution. A famous instance involved five young New York–area men—with pseudonyms such as Acid Phreak, Phiber Optik, and Scorpion— calling themselves the Masters of Deception who had used cheap computers to break into some of the nation's most powerful information systems.[52] They were indicted for computer trespass, with Phiber Optik receiving the longest sentence, a year in federal prison.[53]

- **Professional criminals:** Members of organized crime rings don't just steal information technology. They also use it the way that legal businesses do—as a business tool, though for illegal purposes. For instance, databases can be used to keep track of illegal gambling debts and stolen goods. Not surprisingly, the old-fashioned illegal bookmaking operation has gone high-tech, with bookies using computers and fax machines in place of betting slips and paper tally sheets.

  Microcomputers, scanners, and printers can be used to forge checks, immigration papers, passports, and driver's licenses. Telecommunications can be used to transfer funds illegally. For instance, Russian computer hackers were able to break into a Citibank electronic money transfer system and steal more than $10 million before they were caught.[54]

As information-technology crime has become more sophisticated, so have the people charged with preventing it and disciplining its outlaws. Campus administrators are no longer being quite as easy on offenders and are turning them over to police. Industry organizations such as the Software Publishers Association are going after software pirates large and small. (Commercial software piracy is now a felony, punishable by up to 5 years in prison and fines of up to $250,000 for anyone convicted of stealing at least 10 copies of a program, or more than $2500 worth of software.) Police departments in cities as far apart as Medford, Massachusetts, and San Jose, California, now have police patrolling a "cyber beat." That is, they cruise online bulletin boards looking for pirated software, stolen trade secrets, child molesters, and child pornography.

In 1988, after the last widespread Internet break-in, the U.S. Defense Department created the Computer Emergency Response Team (CERT). Although it has no power to arrest or prosecute, CERT provides round-the-clock international information and security-related support services to users of the Internet. Whenever it gets a report of an electronic snooper, whether on the Internet or on a corporate e-mail system, CERT stands ready to lend assistance. It counsels the party under attack, helps them thwart the intruder, and evaluates the system afterward to protect against future break-ins.

12.3 Security: Safeguarding Computers & Communications

KEY QUESTION

What are the characteristics of the four components of security?

Preview & Review: Information technology requires vigilance in security. Four areas of concern are identification and access, encryption, protection of software and data, and disaster-recovery planning.

The ongoing dilemma of the Digital Age is balancing convenience against security. *Security* **is a system of safeguards for protecting information technology against disasters, systems failure, and unauthorized access that can result in damage or loss.** We consider four components of security:

- Identification and access
- Encryption
- Protection of software and data
- Disaster-recovery planning

Identification & Access

Are you who you say you are? The computer wants to know.

There are three ways a computer system can verify that you have legitimate right of access. Some security systems use a mix of these techniques. The systems try to authenticate your identity by determining (1) what you have, (2) what you know, or (3) who you are.

- **What you have—cards, keys, signatures, badges:** Credit cards, debit cards, and cash-machine cards all have magnetic strips or built-in computer chips that identify you to the machine. Many require you to display your signature, which someone may compare as you write it. Computer rooms are always kept locked, requiring a key. Many people also keep a lock on their personal computers. In addition, a computer room may be guarded by security officers, who may need to see an authorized signature or a badge with your photograph before letting you in.

  Of course, credit cards, keys, and badges can be lost or stolen. Signatures can be forged. Badges can be counterfeited.

- **What you know—PINs, passwords, and digital signatures:** To gain access to your bank account through an automated teller machine (ATM), you key in your PIN. **A** *PIN,* **or** *personal identification number,* **is the security number known only to you that is required to access the system.** Telephone credit cards also use a PIN. If you carry either an ATM or a phone card, *never* carry the PIN written down elsewhere in your wallet (even disguised).

  A *password* **is a special word, code, or symbol that is required to access a computer system.** Passwords are one of the weakest security links, says AT&T security expert Steven Bellovin. Passwords can be guessed, forgotten, or stolen. To reduce a stranger's guessing, Bellovin recommends never choosing a real word or variations of your name, birth date, or those of your friends or family. Instead you should mix letters, numbers, and punctuation marks in an oddball sequence of no fewer than eight characters.[55]

The advice is sound, but the problem today is that many people have to remember several passwords. "Now password overload has become a plague, with every computer online service, voice-mail box, burglar-alarm disarmer, and office computer network demanding a unique string of code," says technology writer William Bulkeley.[56] Still, in line with Bellovin's suggestions, he offers these possibilities as good passwords: *2b/orNOT2b%. Alfred!E!Newman7.* He also reports the strategy of Glenn Maxwell, a computer instructor from Farmington, Michigan, who uses an obvious and memorable password, but shifts the position of his hands on the keyboard, creating a meaningless string of characters. (Thus, *ELVIS* becomes *R;BOD* when you move your fingers one position right on the keyboard.)

Skilled hackers may break into national computer networks and detect passwords as they are being used. Or, using the telephone, they pretend to be computer technicians to cajole passwords out of employees. They may even find access codes in discarded technical manuals in trash bins.

A relatively new technology is the digital signature, which security experts hope will lead to a world of paperless commerce. A *digital signature* is a string of characters and numbers that a user signs to an electronic document being sent by his or her computer. The receiving computer performs mathematical operations on the alphanumeric string to verify its validity. The system works by using a *public-private key system*. That is, the system involves a pair of numbers called a private key and a public key. One person creates the signature with a secret private key, and the recipient reads it with a second, public key. "This process in effect notarizes the document and ensures its integrity," says one writer.[57]

For example, when you write your boss an electronic note, you sign it with your secret private key. (This could be some bizarre string beginning 479XY283 and continuing on for 25 characters.) When your boss receives the note, he or she looks up your public key. Your public key is available from a source such as an electronic bulletin board, the Postal Service, or a corporate computer department. If the document is altered in any way, it will no longer produce the same signature sequence.

- **Who you are—physical traits:** Some forms of identification can't be easily faked—such as your physical traits. Biometrics tries to use these in security devices. **Biometrics is the science of measuring individual body characteristics.**

For example, before a number of University of Georgia students can use the all-you-can-eat plan at the campus cafeteria, they must have their hands read. As one writer describes the system, "a camera automatically compares the shape of a student's hand with an image of the same hand pulled from the magnetic strip of an ID card. If the patterns match, the cafeteria turnstile automatically clicks open. If not, the would-be moocher eats elsewhere."[58]

Besides handprints, other biological characteristics read by biometric devices are fingerprints (computerized "finger imaging"), voices, the blood vessels in the back of the eyeball, the lips, and even one's entire face.

Some computer security systems have a "call-back" provision. In a *call-back system*, the user calls the computer system, punches in the password, and hangs up. The computer then calls back a certain preauthorized number. This measure will block anyone who has somehow got hold of a password but is calling from an unauthorized telephone.

Ethics

Encryption

PGP is a computer program written for encrypting computer messages—putting them into secret code. **Encryption, or enciphering, is the altering of data so that it is not usable unless the changes are undone.** PGP (for Pretty Good Privacy) is so good that it is practically unbreakable; even government experts can't crack it. (This is because it uses a two-keys method similar to that described above under digital signatures.)

Encryption is clearly useful for some organizations, especially those concerned with trade secrets, military matters, and other sensitive data. Some maintain that the future of Internet commerce is at stake, because transactions cannot flourish over the Net unless they are secure.[59] However, from the standpoint of our society, encryption is a two-edged sword. For instance, police in Sacramento, California, found that PGP blocked them from reading the computer diary of a convicted child molester and finding links to a suspected child pornography ring. *Should* the government be allowed to read the coded e-mail of its citizens? What about its being blocked from surveillance of overseas terrorists, drug dealers, and other enemies? The government maintains that it needs access to scrambled data for national security and law enforcement.

At present, there are limitations on encryption technology governing the export of hardware and software containing data-scrambling features. U.S. technology firms complain they are losing the export market because of these controls. In addition, however, the FBI and the National Security Agency have been pushing to have all encryption products sold in the United States include a "back door" allowing the government to be able to peep at any message. Ironically, the government's encryption policy seems to violate the very spirit of openness expressed in the White House's *A Framework for Global Electronic Commerce*, discussed above. Encryption is opposed not only by civil libertarians who fear "back door" access could lead to government snooping but also by most information technology companies (Microsoft, Sun, Apple, IBM) and the majority of the regional telephone companies.[60–62]

Protection of Software & Data

Organizations go to tremendous lengths to protect their programs and data. As might be expected, this includes educating employees about making backup disks, protecting against viruses, and so on. (We discussed these matters in detail elsewhere, especially in the Experience Box at the end of Chapter 5.)

Other security procedures include the following:

- **Control of access:** Access to online files is restricted only to those who have a legitimate right to access—because they need them to do their jobs. Many organizations have a transaction log that notes all accesses or attempted accesses to data.

- **Audit controls:** Many networks have *audit controls* that track which programs and servers were used, which files opened, and so on. This creates an *audit trail*, a record of how a transaction was handled from input through processing and output.

- **People controls:** Because people are the greatest threat to a computer system, security precautions begin with the screening of job applicants. That is, résumés are checked to see if people did what they said they did. Another control is to separate employee functions, so that people are not allowed to wander freely into areas not essential to their jobs. Manual and automated controls—input controls, processing controls, and output controls—are used to check that data is handled accurately and completely during the processing cycle. Printouts, printer ribbons, and other waste that may yield passwords and trade secrets to outsiders is disposed of through shredders or locked trash barrels.

Disaster-Recovery Plans

A *disaster-recovery plan* **is a method of restoring information processing operations that have been halted by destruction or accident.** "Among the countless lessons that computer users have absorbed in the hours, days, and weeks after the [1993 New York City] World Trade Center bombing," wrote one reporter, "the most enduring may be the need to have a disaster-recovery plan. The second most enduring lesson may be this: Even a well-practiced plan will quickly reveal its flaws."[63]

Mainframe computer systems are operated in separate departments by professionals, who tend to have disaster plans. Mainframes are usually backed up. However, many personal computers, and even entire local area networks, are not backed up. The consequences of this lapse can be great. It has been reported that on average, a company loses as much as 3% of its gross sales within 8 days of a sustained computer failure. In addition, the average company struck by a computer failure lasting more than 10 days never fully recovers.[64]

A disaster-recovery plan is more than a big fire drill. It includes a list of all business functions and the hardware, software, data, and people to support those functions. It includes arrangements for alternate locations, either hot sites or cold sites. A *hot site* is a fully equipped computer center, with everything needed to resume functions. A *cold site* is a building or other suitable environment in which a company can install its own computer system. The disaster-recovery plan includes ways for backing up and storing programs and data in another location, ways of alerting necessary personnel, and training for those personnel.

12.4 Quality-of-Life Issues: The Environment, Mental Health, & the Workplace

KEY QUESTION

How does information technology create environmental, mental-health, and workplace problems?

Preview & Review: Information technology can create problems for the environment, people's mental health (isolation, gambling, Net addiction, and stress), and the workplace (misuse of technology and information overload).

Earlier in this book, we pointed out the worrisome effects of technology on intellectual property rights and truth in art and journalism (✔ pp. 84, 294), on censorship (✔ p. 338), on health matters and ergonomics (✔ p. 243), on environmental matters (✔ p. 176) and on privacy (✔ p. 437). Here we discuss some other quality-of-life issues related to information technology:

- Environmental problems
- Mental-health problems
- Workplace problems

Environmental Problems

"This county will do peachy fine without computers," says Micki Haverland, who has lived in rural Hancock County, Tennessee, for 20 years.[65] Telecommunications could bring jobs to an area that badly needs them, but several people moved there precisely because they like things the way they are—pristine rivers, unspoiled forests, and mountain views.

But it isn't just people in rural areas who are concerned. Suburbanites in Idaho and Utah, for example, worry that lofty metal poles topped by cellular-transmitting equipment will be eyesores that will destroy views and property values.[66] City dwellers everywhere are concerned that the federal government's 1996 decision to deregulate the telecommunications industry will lead to a rat's nest of roof antennas, satellite dishes, and above-ground transmission stations. As a result, telecommunications companies are now experimenting with hiding transmitters in the "foliage" of fake trees made of metal.

Political scientist James Snider, of Northwestern University, points out that the problems of the cities could expand well beyond the cities if telecommuting triggers a massive movement of people to rural areas. "If all Americans succeed in getting their dream homes with several acres of land," he writes, "the forests and open lands across the entire continental United States will be destroyed" as they become carved up with subdivisions and roads.[67]

Mental-Health Problems: Isolation, Gambling, Net-Addiction, Stress

Monica Ainsworth of Princeton, New Jersey, communicates every day with her therapist, although she never sees him. Even though she travels a lot, she's able to connect with him because the therapeutic sessions take place in cyberspace.[68]

But is it really therapy, and does it work? Definitely not, says clinical psychologist Leonard Holmes of Newport News, Virginia. Although he does one to three "consultations" a week by e-mail, he tells clients beforehand that the arrangement is not suitable for complex problems. "You can't diagnose and treat disorders by computer contact," he says. "Too many powerful nonverbal cues are missing."[69] Still, an online consultation may be beneficial if the problem is a simple one, he thinks. For example, he has counseled single people who tend to abandon relationships after the initial romantic stage passes. Similarly, clinical psychologist David Sommers of Kensington, Maryland, offers psychological advice—but not traditional therapy—on line. "A lot of what I do is just provide information and help people feel less alone," he says.[70]

Cyberspace therapy has caused a lot of concern among practitioners in the American Psychological Association. Indeed, the APA offers a Web site *(http://www.apa.org)* that directs users to resources for mental health and to a "help center" for suggestions on dealing with a variety of problems. But one should be aware that some supposed cybertherapists don't even have a license to practice, that hackers may eavesdrop on therapy, and that Net counselors aren't bound by the same kinds of laws that protect in-person therapists.

Still, online therapy represents an attempt to deal with some of the problems that people bring to the Internet. Consider:

- **Isolation:** Automation allows us to go days without actually speaking with or touching another person, from buying gas to playing games. Even the friendships we make online in cyberspace, some believe, "are likely to be trivial, short lived, and disposable—junk friends." Says one writer, "We may be overwhelmed by a continuous static of information and casual acquaintance, so that finding true soul mates will be even harder than it is today."[71]

- **Gambling:** Gambling is already widespread in North America, but information technology could make it almost unavoidable. Although gambling by wire is illegal in the U.S., all kinds of moves are afoot to get around it. For example, host computers for Internet casinos and sports books have been established in Caribbean tax havens, and satellites, decoders, and remote-control devices are being used so TV viewers can do racetrack wagering from home.

  Some mental-health professionals are concerned with the long-range effects. "About 5% [of the users or viewers] will be compulsive gamblers and another 10% to 15% will be problem gamblers," says Kevin O'Neill of New Jersey's Council on Compulsive Gambling. "Compulsive gamblers want action, which is what interactive television [or computers] can give you."[72] In Congress, legislation has been introduced making it a crime to transmit money wagers or gambling information over the Internet.

- **Net addiction:** Don't let this happen to you: "A student e-mails friends, browses the World Wide Web, blows off homework, botches exams, flunks out of school."[73] This is a description of the downward spiral of the "Net addict," often a college student—because schools give students no-cost/low-cost linkage to the Internet—though it can be anyone. Some become addicted (although until recently some professionals felt "addiction" was too strong a word) to chat groups, some to online pornography, some simply to the escape from real life.[74,75] Indeed, sometimes the computer replaces one's spouse or boyfriend/girlfriend in the user's affections. In one instance, a man sued his wife for divorce for having an "online affair" with a partner who called himself The Weasel.[76,77]

- **Stress:** In a 1995 survey of 2802 American PC users, three-quarters of the respondents (ranging in age from children to retirees) said personal computers had increased their job satisfaction and were a key to success and learning. However, many found PCs stressful: 59% admitted getting angry at them within the previous year. And 41% said they thought computers have reduced job opportunities rather than increased them.[78]

ARE YOU AN INTERNET ADDICT?

- Do you stay on line longer than you intended?
- Has tolerance developed so that longer periods of time are needed on line?
- Do you call in sick to work, skip classes, go to bed late or wake up early to use the Internet?
- Do you experience withdrawal symptoms (increased depression, moodiness, or anxiety) when you are off line?
- Have you given up recreational, social, or occupational activities because of the Internet?
- Do you continue to use the Internet despite recurrent problems it creates in your real life (work, school, financial, or family problems)?
- Have you made several unsuccessful attempts to cut down the amount of time you use the Internet?

Psychologist and sociologist Sherry Turkle of MIT believes that when it comes to mental health, information technology is neither a blessing nor a curse. In fact, she holds, "The Internet is not a drug." Rather, people who seem addicted may be "working through important personal issues in the safety of life on the screen."[79]

In her book *Life on the Screen: Identity in the Age of the Internet*, she suggests that cyberspace can make people's lives more fulfilling by getting them to face issues of identity and relationships that they have never had to confront before.[80-82] In particular, people exploring the computer-based fantasy worlds known as *MUDs* (for *Multi-User Domains*) on the Net can communicate anonymously with others and try on different roles, genders, or animals or even beings from other planets. "The way we used to think about identity is that people had a core self, a one," she says. Now, in the behavior of people playing Internet role-playing games, she sees evidence that identity itself consists of different constructions or personae.

Ethics

Workplace Problems

First the mainframe computer, then the desktop stand-alone PC, and most recently the networked computer were all brought into the workplace for one reason only: to improve productivity. How is it working out? Let's consider two aspects: the misuse of technology, and information overload.

- **Misuse of technology:** "For all their power," says an economics writer, "computers may be costing U.S. companies tens of billions of dollars a year in downtime, maintenance, and training costs, useless game playing, and information overload."[83]

    Consider games. Employees may look busy, staring into their computer screens with brows crinkled. But it could be they're just hard at work playing Doom or surfing the Net. Workers with Internet access average 10 hours a week online.[84] However, fully 23% of computer game players use their office PCs for their fun, according to one survey.[85] A study of employee online use at one major company concluded that the average worker wastes 1½ hours each day.[86]

    Another reason for so much wasted time is all the fussing that employees do with hardware and software. Says one editor, "Back in the old days, when I toiled on a typewriter, I never spent a whole morning installing a new ribbon. . . . I did not scan the stores for the proper cables to affix to my typewriter or purchase books that instructed me on how to get more use from my liquid white-out."[87] One study estimated that microcomputer users waste 5 billion hours a year waiting for programs to run, checking computer output for accuracy, helping co-workers use their applications, organizing cluttered disk storage, and calling for technical support.[88]

    Many companies don't even know what kind of microcomputers they have, who's running them, or where they are. The corporate customer of one computer consultant, for instance, swore it had 700 PCs and 15 users per printer. An audit showed it had 1200 PCs with one printer each.[89]

    A particularly interesting misuse is the continual upgrade. Ask yourself, Do I really need that slick new product? Ron Erickson, former chairman of the Egghead Software stores, says he uses an old version of a word processing program. His advice to consumers: "Don't get the new version if the old one is working O.K."[90] As for many businesses, he says, the rule should be: "Stop buying new software, and train employees on what you have."

- **Information overload:** "My boss basically said, 'Carry this pager seven days a week, 24 hours a day, or find another job,'" says the chief architect for a New Jersey school system. (He complied, but pointedly notes that the pager's "batteries run out all the time.")[91] "It used to be considered a status symbol to carry a laptop computer on a plane," says futurist Paul Saffo. "Now anyone who has one is clearly a working dweeb who can't get the time to relax. Carrying one means you're on someone's electronic leash."[92]

   The new technology is definitely a two-edged sword. Cellular phones, pagers, fax machines, and modems may untether employees from the office. But they tend to work longer hours under more severe deadline pressure than do their tethered counterparts who stay at the office, according to one study.[93] Moreover, the gadgets that once promised to do away with irksome business travel by ushering in a new era of communications have done the opposite—created the office-in-a-bag that allows business travelers to continue to work from airplane seats and hotel room desks.

   What does being overwhelmed with information do to you, besides inducing stress and burnout? One result is that because we have so many choices to entice and confuse us we may become more reluctant to make decisions. Home buyers, for instance, now take twice as long as a decade earlier to sign a contract on a new house, organizations take months longer to hire top executives, and managers tend to consider worst-case scenarios rather than benefits when considering investing in a new venture.[94]

   "The volume of information available is so great that I think people generally are suffering from a lack of meaning in their lives," says Neil Postman, chair of the department of culture and communication at New York University. "People are just adrift in the sea of information, and they don't know what the information is about or why they need it."[95]

People and businesses are beginning to realize the importance of coming to grips with these problems. Some companies are employing GameCop, a software program that catches unsuspecting employees playing computer games on company time.[96] Some are installing special software (asset-management programs) that tell them how many PCs are on their networks and what they run. Some are imposing strict hardware and software standards to reduce the number of different products they support.[97] To avoid information overload, some people—those who have a choice—no longer carry cell phones or even look at their e-mail. Others are installing so-called *Bozo filters*, software that screens out trivial e-mail messages and cellular calls and assigns priorities to the remaining files. Still others are beginning to employ programs called *intelligent agents* (discussed shortly) to help them make decisions.

But the real change may come as people realize that they need not be tied to the technological world in order to be themselves, that solitude is a scarce resource, and that seeking serenity means streamlining the clutter and reaching for simpler things.

12.5   Economic Issues: Employment & the Haves/Have-Nots

KEY QUESTION

How may technology affect the unemployment rate and the gap between rich and poor?

Ethics

Preview & Review: Many people worry that jobs are being reduced by the effects of information technology. They also worry that it is widening the gap between the haves and have-nots.

"If you'd had any brains," Yale University professor David Gelernter read in the letter from the Unabomber (later identified as Theodore Kaczynski), whose mail bomb had savagely disfigured the computer scientist, "you would realize that there are a lot of people out there who resent bitterly the way techno-nerds like you are changing the world."[98]

People who don't like technology in general, and today's information technology in particular, have been called *neo-Luddites*. The original Luddites were a group of weavers in northern England who, while proclaiming their allegiance to a mythical King Ludd, in 1812–1814 went about smashing modern looms that moved cloth production out of the hands of peasant weavers and into inhumane factories. Although the term now seems to connote a knee-jerk antagonism to technology, in actuality the 19th-century Luddites were desperate, brave people, with no other means of employment and no future after being stripped of their livelihoods.

Nevertheless, in recent times a number of books (such as Clifford Stoll's *Silicon Snake Oil*, Stephen Talbott's *The Future Does Not Compute*, and Mark Slouka's *War of the Worlds*) have appeared that have tried to provide a counterpoint to the hype and overselling of information technology. Some of these strike a sensible balance, but some make the alarming case that technological progress is actually no progress at all—indeed, it is a curse. The two biggest charges (which are related) are, first, that information technology is killing jobs and, second, that it is widening the gap between the rich and the poor.

Technology, the Job Killer?

Certainly ATMs do replace bank tellers, fast-pass electronic systems do replace turnpike-toll takers, and Internet travel agents do lure customers away from small travel agencies. There's no question that technological advances play an ambiguous role in social progress.

But is it true, as technology critic Jeremy Rifkin says, that intelligent machines are replacing humans in countless tasks, "forcing millions of blue-collar and white-collar workers into temporary, contingent, and part-time employment and, worse, unemployment"?[99]

This is too large a question to be fully considered in this book. The economy of North America is undergoing powerful structural changes, brought on not only by the widespread diffusion of technology but also by greater competition, increased global trade, the shift from manufacturing to service employment, the weakening of labor unions, more flexible labor markets, more rapid immigration, partial deregulation, and other factors.[100–102]

A counterargument is that jobs don't disappear, they just change. Or the jobs that do disappear represent drudgery. "If your job has been replaced by a computer," says Stewart Brand, "that may have been a job that was not worthy of a human."[103]

Gap Between Rich & Poor

"In the long run," says MIT economist Paul Krugman, "improvements in technology are good for almost everyone. . . . Unfortunately, what is true in the long run need not be true over shorter periods."[104] We are now, he believes, living through one of those difficult periods in which technology doesn't produce widely shared economic gains but instead widens the gap between those who have the right skills and those who don't.

A U.S. Department of Commerce survey of "information have-nots" reveals that about 20% of the poorest households in the U.S. do not have telephones. Moreover, only a fraction of those poor homes that do have phones will be able to afford the information technology that most economists agree is the key to a comfortable future.[105] The richer the family, the more likely it is to have and use a computer.

Schooling—especially college—makes a great difference. Every year of formal schooling after high school adds 5–15% to annual earnings later in life.[106] Being well educated is only part of it, however; one should also be technologically literate. Employees who are skilled at technology "earn roughly 10–15% higher pay," according to the chief economist for the U.S. Labor Department.[107]

Advocates of information access for all find hope in the promises of NII proponents for "universal service" and the wiring of every school to the Net. But this won't happen automatically. Ultimately we must become concerned with the effects of growing economic disparities on our social and political health.

Now that we've considered the challenges, let us discuss some of the promises of information technology not described so far. Some of these are truly impressive. They include *intelligent agents* and *avatars*, *artificial intelligence*, and the *promises of the Information Revolution*.

12.6 Intelligent Agents & Avatars

KEY QUESTION

What are some characteristics of intelligent agents and avatars?

Preview & Review: Intelligent agents are computer programs that act as electronic secretaries, e-mail filters, and electronic news clipping services. Internet agents called spiders roam the Web assembling page information to put into databases for later searching by search engines. Avatars are either (1) a graphical image of you or someone else on a computer screen or (2) a graphical personification of a computer or process that's running on a computer.

What the online world really needs is a terrific librarian. "What bothers me most," says Christine Borgman, chair of the UCLA Department of Library and Information Science, "is that computer people seem to think that if you have access to the Web, you don't need libraries."[108] But what's in a library is standardized and well organized and what's on the Web is overwhelming, unstandardized, and chaotic.

As a solution, scientists have been developing so-called *intelligent agents* to send out on computer networks to find and filter information. And to make them more friendly they are inventing graphical on-screen personifications called *avatars*. Let's consider both of these.

Intelligent Agents

An *intelligent agent* is a computer program that performs work tasks on your behalf, including roaming networks and compiling data. A software agent acts as an electronic assistant that will perform, in the user's stead, such tasks as filtering messages, scanning news services, and similar secretarial chores. It will also travel over communications lines to nearly any kind of computer database, collecting files to add to a database.

Examples of agents are the following:[109–115]

- **Digital secretaries:** Wildfire is a speech-recognition system, a "digital secretary" that will answer the phone, take messages, track you down on your cell phone and announce the caller, place calls for you, and remind you of appointments. Other electronic systems are available from Webley Systems and Access Point. Some of these systems can handle e-mail and faxes in much the same way as phone calls.

- **E-mail filters:** BeyondMail will filter your e-mail, alerting you to urgent messages, telling you which require follow-up, and sorting everything according to priorities. For people whose e-mail threatens to overwhelm them with "cyberglut," such an agent is a godsend.

- **Electronic clipping services:** Several companies offer customized electronic news services (Heads Up, I-News, Journalist, Personal Journal, News Hound, the Personal Internet Newspaper) that will scan online news sources and publications looking for information that contains keywords you have previously specified. Some will rank a selected article according to how closely it fits your request. Others will pull together articles in the form of a condensed electronic newspaper. These are all forms of push technology, as we discussed in Chapter 7 (✔ p. 331).

- **Internet agents: spiders, crawlers, and robots:** We have already discussed (✔ pp. 133, 330) search tools, such as search engines and directories, to help you find topics on the Internet.

  What interests us here, however, are the intelligent agents used to assemble the database that a search engine searches. Most such databases are created by *spiders*. *Spiders*—also known as *crawlers* or *robots*—are software programs that roam the Web, looking for new Web sites by following links from page to page. When a spider finds a new page, it adds information about it—its title, address, and perhaps summary of contents—to the search engine's database.

Avatars

Want to see yourself—or a stand-in for yourself—on your computer screen? Then try using a kind of cyberpersona called an *avatar*. An *avatar* is either (1) a graphical image of you or someone else on a computer screen or (2) a graphical personification of a computer or process that's running on a computer.

- **Avatar as yourself or others:** The on-screen version of yourself could be "anything from a human form to a pair of cowboy boots with lips," writes technology columnist Denise Caruso. "Users move them around while talking (via keyboard) with other avatars on the same screen."[116] In CompuServe's Worlds Chat, subscribers participating in online "chat rooms," which are furnished like cartoon stage sets, can get together with other users, each of whom can construct an avatar from a variety of heads, clothing, shoes, and even animal identities.

- **Avatars representing a process:** "The driving force behind avatars is the ongoing search for an interface that's easier and more comfortable to use," says one writer, "especially for the millions of people who are still computerphobic."[117] One difficulty with designing computer-controlled avatars—called *agents, characters,* and *bots*—is making sure that they don't make people react negatively to them. Thus, instead of faces or personifications, it may be better to use pictures of notepads, checkbooks, and similar objects.

Agents and avatars are still in their infancy. In time, however, the promise is that they will make information technology much easier by helping to tame the "cyberglut" and by helping us deal more effectively with all the on-screen choices available to us. No doubt their improved versions will draw on the field of artificial intelligence, as we discuss next.

12.7 Artificial Intelligence

KEY QUESTION

What are some characteristics of the seven key areas of artificial intelligence?

Preview & Review: Artificial intelligence (AI) is a research and applications discipline that includes the areas of robotics, perception systems, expert systems, natural-language processing, fuzzy logic, neural networks, and genetic algorithms. Another area, artificial life, is the study of computer instructions that act like living organisms. The Turing test has long been used as a standard to determine whether a computer possesses "intelligence." Behind all aspects of AI are ethical questions.

You're having trouble with your new software program. You call the customer "help desk" at the software maker. Do you get a busy signal or get put on hold to listen to music (or, worse, advertising) for several minutes? Technical support lines are often swamped, and waiting is commonplace. Or, to deal with your software difficulty, do you find yourself dealing with . . . other software?

This event is not unlikely. Programs that can walk you through a problem and help solve it are called *expert systems* (✔ p. 471). As the name suggests, these are systems that are imbued with knowledge by a human expert. Expert systems are one of the most useful applications of an area known as *artificial intelligence.*

Artificial intelligence (AI) is a group of related technologies that attempt to develop machines to emulate human-like qualities, such as learning, reasoning, communicating, seeing, and hearing. Today the main areas of AI are:

- Robotics
- Perception systems
- Expert systems
- Natural language processing
- Fuzzy logic
- Neural networks
- Genetic algorithms

We will consider these and also an area known as *artificial life.*

Robotics

Robotics is a field that attempts to develop machines that can perform work normally done by people. The machines themselves, of course, are called *robots.* As we mentioned in Chapter 5 (✔ p. 234), a robot is an automatic device that performs functions ordinarily ascribed to human beings or that operates with what appears to be almost human intelligence. Dante II, for

■ PANEL 12.3

Robotics

(Top left) Dante II volcano explorer. *(Top right)* The BOA (Big-On-Asbestos) is used to strip and bag asbestos-containing insulation materials from pipes. *(Bottom left)* Testing Nomad in a desert in Chile, before using it to explore another planet's surface. *(Bottom right)* NASA's Mars Pathfinder, used to explore part of the surface of Mars in 1997.

instance, is an eight-legged, 10-foot-high, satellite-linked robot used by scientists to explore the inside of Mount Spurr, an active volcano in Alaska. *(See ■ Panel 12.3.)* Robots may be controlled from afar, as in an experiment at the University of Southern California in which Internet users thousands of miles away were invited to manipulate a robotic arm to uncover objects in a sandbox.[118]

Perception Systems

Perception systems are sensing devices that emulate the human capabilities of sight, hearing, touch, and smell. Clearly, perception systems are related to robotics, since robots need to have at least some sensing capabilities. Examples of perception systems are vision systems, used for pattern recognition. Vision systems are used, for example, to inspect products for quality control in factory assembly lines. To discriminate among parts or shapes, a robot measures the varying intensities of light of the parts. Each intensity has a numbered value that is compared to a similar palette of intensities stored in the system's memory. If the intensity is not recognized, the part is rejected. An example of this kind of perception system is the Bin Vision Systems used by General Electric to pick up specific parts.

Expert Systems

An *expert system,* you'll recall, is an interactive computer program that helps users solve problems that would otherwise require the assistance of a human expert. We briefly mentioned expert systems in Chapter 10 in the context of management information systems, but they are used for many other purposes as well.

The expert system MYCIN helps diagnose infectious diseases. PROSPECTOR assesses geological data to locate mineral deposits. DENDRAL identifies chemical compounds. Home-Safe-Home evaluates the residential environment of an elderly person. Business Insight helps businesses find the best strategies for marketing a product. REBES (Residential Burglary Expert System) helps detectives investigate crime scenes. CARES (Computer Assisted Risk Evaluation System) helps social workers assess families for risks of child abuse. CLUES (Countrywide Loan Underwriting Expert System) evaluates home-mortgage-loan applications. Crush takes a body of expert advice and combines it with worksheets reflecting a user's business situation to come up with a customized strategy to "crush competitors."

All these programs simulate the reasoning process of experts in certain well-defined areas. That is, professionals called knowledge engineers interview the expert or experts and determine the rules and knowledge that must go into the system. For example, to develop Muckraker, an expert system to assist with investigative reporting, the knowledge engineers interviewed journalists. (■ *See Panel 12.4.*)

■ PANEL 12.4   Developing the expert system Muckraker

Steve Weinberg, former executive director of Investigative Reporters and Editors, agreed to provide his knowledge of investigative journalism as the basis for Muckraker, an expert system for journalists. Here he describes how he was interviewed by a knowledge engineer, Louanna Furbee.

❝It was Louanna Furbee . . . who worked hardest in the early stages to puzzle out the underlying logic (if any, I worried) of how I worked on an investigation. She explained that, after interviewing me, she would try to reduce what I had said into concepts. She would write each concept separately on an index card, then ask me to sort the cards into groupings. From those groupings, she hoped to sketch a tree of knowledge, which the computer programmers could then translate into electronic impulses.

When I viewed the tree a week later, I was amazed at how Furbee had managed to translate my words into a graphic that would be the basis of a computer program. She had sketched 57 connected branches. The two main trunks were 'paper trails' and 'people trails,' a distinction I had made when she interviewed me. (When conducting an investigation, I almost always consult paper first, then find the people to help explain the paper.)

On the 'paper trail' trunk, Furbee sketched my distinctions between primary-source documents and secondary-source accounts. She also captured my thinking about how the type of subject (Is the story primarily about an individual, an institution, or an issue?) determines which documents I will seek first. . . .

On the 'people' trunk, Furbee focused on the two main problems I had found of most concern to journalists: getting in the door and, once inside, conducting the interview successfully. She worked in branches reflecting my thinking on when to request an interview by letter or telegram rather than by telephone, on dealing with secretaries and other potential bars to access, on how to bring an off-the-record source back on the record.

Using her tree, the rest of the Expert Systems team began to imprint my thinking onto a computer disk. . . .❞

—Steve Weinberg, "Steve's Brain," *Columbia Journalism Review*

When the system was unveiled, the first screen after the title read: "Muckraker's purpose is to provide advice on following the paper trail and interviewing sources. After a series of questions, Muckraker will make a recommendation. Use Muckraker to help plan your investigation."

Programs incorporate not only the experts' surface knowledge ("textbook knowledge") but also deep knowledge ("tricks of the trade"). What, exactly, is this latter kind of knowledge? "An expert in some activity has by definition reduced the world's complexity by his or her specialization," say some authorities. One result is that "much of the knowledge lies outside direct conscious awareness. . . ."[119]

An expert system consists of three components: *knowledge base, inference engine,* and *user interface.*

- **Knowledge base:** A *knowledge base* is an expert system's database of knowledge about a particular subject. This includes relevant facts, information, beliefs, assumptions, and procedures for solving problems. The basic unit of knowledge is expressed as an IF-THEN-ELSE rule ("IF this happens, THEN do this, ELSE do that"). Programs can have as many as 10,000 rules. A system called ExpertTAX, for example, which helps accountants figure out a client's tax options, consists of over 2000 rules.

- **Inference engine:** The *inference engine* is the software that controls the search of the expert system's knowledge base and produces conclusions. It takes the problem posed by the user of the system and fits it into the rules in the knowledge base. It then derives a conclusion from the facts and rules contained in the knowledge base.

  Reasoning may be by a forward chain or backward chain. In the forward chain of reasoning, the inference engine begins with a statement of the problem from the user. It then proceeds to apply any rule that fits the problem. In the backward chain of reasoning, the system works backward from a question to produce an answer.

- **User interface:** The user interface is the display screen that the user deals with. It gives the user the ability to ask questions and get answers. It also explains the reasoning behind the answer.

Natural Language Processing

Natural languages are ordinary human languages, such as English. (A second definition is that they are programming languages, called fifth-generation languages, that give people a more natural connection with computers—✔ p. 528.) **Natural-language processing is the study of ways for computers to recognize and understand human language,** whether in spoken or written form.

Think how challenging it is to make a computer translate English into another language. In one instance, the English sentence "The spirit is willing, but the flesh is weak" came out in Russian as "The wine is agreeable, but the meat is spoiled." The problem with human language is that it is often ambiguous and often interpreted differently by different listeners.

Today you can buy a handheld computer that will translate a number of English sentences—principally travelers' phrases ("Please take me to the airport")—into another language. This trick is similar to teaching an English-speaking child to sing "Frère Jacques." More complex is the work being done by AI scientists trying to discover ways to endow the computer with an "understanding" of how human language works. This means working with ideas about the instinctual instructions or genetic code that babies are born with for understanding language.

Still, some natural-language systems are already in use. Intellect is a product that uses a limited English vocabulary to help users orally query databases. LUNAR, developed to help analyze moon rocks, answers questions about geology from an extensive database. Verbex, used by the U.S. Postal

Service, lets mail sorters read aloud an incomplete address and replies with the correct zip code.

In the future, natural-language comprehension may be applied to incoming e-mail messages so that such messages can be filed automatically. However, this would require that the program understand the text rather than just look for certain words.[120]

Fuzzy Logic

A relatively new concept being used in the development of natural languages is fuzzy logic. The traditional logic behind computers is based on either/or, yes/no, true/false reasoning. Such computers make "crisp" distinctions, leading to precise decision making. **Fuzzy logic is a method of dealing with imprecise data and uncertainty, with problems that have many answers rather than one.** Unlike classical logic, fuzzy logic is more like human reasoning: It deals with probability and credibility. That is, instead of being simply true or false, a proposition is *mostly* true or *mostly* false, or *more* true or *more* false.

A frequently given example of an application of fuzzy logic is in running elevators. How long will most people wait in an elevator before getting antsy? About a minute and a half, say researchers at the Otis Elevator Company. The Otis artificial intelligence division has thus done considerable research into how elevators may be programmed to reduce waiting time.[121] Ordinarily when someone on a floor in the middle of a building pushes the call button, the system will send whichever elevator is closest. However, that car might be filled with passengers, who will be delayed by the new stop, whereas another car that is farther away might be empty. In a fuzzy-logic system, the computer assesses not only which car is nearest but also how full the cars are before deciding which one to send.

Neural Networks

Fuzzy-logic principles are being applied in another area of AI, neural networks. **Neural networks use physical electronic devices or software to mimic the neurological structure of the human brain.** Because they are structured to mimic the rudimentary circuitry of the cells in the human brain, they learn from example and don't require detailed instructions.

To understand how neural networks operate, let us compare them to the operation of the human brain.

- **The human neural network:** The word *neural* comes from neurons, or nerve cells. The neurons are connected by a three-dimensional lattice called *axons*. Electrical connections between neurons are activated by *synapses.*

  The human brain is made up of about 100 billion neurons. However, these cells do not act like "computer memory" sites. No cell holds a picture of your dog or the idea of happiness. You could eliminate any cell—or even a few million—in your brain and not alter your "mind." Where do memory and learning lie? In the electrical connections between cells, the *synapses.* Using electrical pulses, the neurons send "on/off" messages along the synapses.

- **The computer neural network:** In a hardware neural network, the nerve cell is replaced by a transistor, which acts as a switch. Wires connect the cells (transistors) with each other. The synapse is replaced by an electronic component called a resistor, which determines whether a cell should activate the electricity to other cells. A software neural network emulates a hardware neural network, although it doesn't work as fast.

Computer-based neural networks use special AI software and complicated fuzzy-logic processor chips to take inputs and convert them to outputs with a kind of logic similar to human logic.

Ordinary computers mechanically obey instructions according to set rules. However, neural-network computers, like children, learn by example, problem solving, and memory by association. The network "learns" by fine-tuning its connections in response to each situation it encounters. (In the brain, learning takes place through changes in the synapses.) If you're teaching a neural network to speak, for example, you train it by giving it sample words and sentences as well as the pronunciations. The connections between the electronic "neurons" gradually change, allowing more or less current to pass.

Using software from a neural-network producer, Intel has developed a neural-network chip that contains many more transistors than the Pentium. Other chip makers are also working on neural network chips. Over the next few years, these chips will begin to bring the power of these silicon "brains" not only to your PC but also to such tasks as automatically balancing shifting laundry loads in washing machines.

Neural networks are already being used in a variety of situations. One such program learned to pronounce a 20,000-word vocabulary overnight.[122] Another helped a mutual-fund manager to outperform the stock market by 2.3–5.6 percentage points over three years.[123] At a San Diego hospital emergency room in which patients complained of chest pains, a neural-network program was given the same information given doctors. It correctly diagnosed patients with heart attacks 97% of the time, compared to 78% for the human physicians.[124] In Chicago, a neural net system has also been used to evaluate patient X-rays to look for signs of breast cancer. It outperformed most doctors in distinguishing malignant tumors from benign ones.[125] Banks use neural-network software to spot irregularities in purchasing patterns associated with individual accounts, thus often noticing when a credit card is stolen before its owner does.[126]

Genetic Algorithms

A *genetic algorithm* is a program that uses Darwinian principles of random mutation to improve itself. The algorithms are lines of computer code that act like living organisms. Different sections of code haphazardly come together, producing programs. Like Darwin's rules of evolution, many chunks of code compete with each other to see which can best perform the desired solution—the aim of the program. Some chunks will even become extinct. Those that survive will combine with other survivors and will produce offspring programs.[127–130]

Expert systems can capture and preserve the knowledge of expert specialists, but they may be slow to adapt to change. Neural networks can sift through mountains of data and discover obscure relationships, but if there is too much or too little data they may be ineffective—garbage in, garbage out. Genetic algorithms, by contrast, use endless trial and error to learn from experience—to discard unworkable approaches and grind away at promising approaches with the kind of tireless energy of which humans are incapable.

The awesome power of genetic algorithms has already found applications. Organizers of the Paralympic Games used it to schedule events. LBS Capital Management Fund of Clearwater, Florida, uses it to help pick stocks for a pension fund it manages. In something called the FacePrints project, witnesses use a genetic algorithm to describe and identify criminal suspects. Texas Instruments is drawing on the skills that salmon use to find spawning grounds to produce a genetic algorithm that shipping companies can use to let packages "seek" their own best routes to their destinations. A hybrid expert system–genetic algorithm called Engeneous was used to boost performance in the Boeing 777 jet engine, a feat that involved billions of mind-boggling calculations.

Computer scientists still don't know what kinds of problems genetic algorithms work best on. Still, as one article pointed out, "genetic algorithms have going for them something that no other computer technique does: they have been field-tested, by nature, for 3.5 billion years."[131]

Ethics

Artificial Life, the Turing Test, & AI Ethics

Genetic algorithms would seem to lead us away from mechanistic ideas of artificial intelligence and into more fundamental questions: "What is life, and how can we replicate it out of silicon chips, networks, and software?" We are dealing now not with artificial intelligence but with artificial life. *Artificial life,* or *A-life,* is a field of study concerned with "creatures"—computer instructions, or pure information—that are created, replicate, evolve, and die as if they were living organisms.[132]

Of course, "silicon life" does not have two principal attributes associated with true living things—it is not water- and carbon-based. Yet in another respect such "creatures" mimic life: If they cannot learn or adapt, they perish.

How can we know when we have reached the point where computers have achieved human intelligence? How will you always know, say, if you're on the phone, whether you're talking to a human being or to a computer? Clearly, with the strides made in the fields of artificial intelligence and artificial life, this question is no longer just academic.

Interestingly, this matter was addressed back in 1950 by Alan Turing, an English mathematician and computer pioneer. Turing predicted that by the end of the century computers would be able to mimic human thinking and converse so naturally that their communications would be indistinguishable from a person's. Out of these observations came the Turing test. The *Turing test* is a test or game for determining whether a computer is considered to possess "intelligence" or "self-awareness."

In the Turing test, a human judge converses by means of a computer terminal with two entities hidden in another location. One entity is a person typing on a keyboard. The other is a software program. As the judge types in and receives messages on the terminal, he or she must decide whether the entity is human. In this test, intelligence, the ability to think, is demonstrated by the computer's success in fooling the judge.

Judith Anne Gunther participated as one of eight judges in the third annual Loebner Prize Competition, which is based on Turing's ideas.[133] (There have been other competitions since.) The "conversations" are restricted to predetermined topics, such as baseball. This is because today's best programs have neither the databases nor the syntactical ability to handle an unlimited number of subjects. Conversations with each entity are limited to 15 minutes. At the end of the contest, the program that fools the judges most is the one that wins.

Gunther found that she wasn't fooled by any of the computer programs. The winning program, for example, relied as much on deflection and wit as it did on responding logically and conversationally. (For example, to a judge trying to discuss a federally funded program, the computer said: "You want logic? I'll give you logic: shut up, shut up, shut up, shut up, shut up, now go away! How's that for logic?") However, Gunther *was* fooled by one of the five humans, a real person discussing abortion. "He was so uncommunicative," wrote Gunther, "that I pegged him for a computer."

Behind everything to do with artificial intelligence and artificial life—just as it underlies everything we do—is the whole matter of *ethics*. In his book *Ethics in Modeling*, William A. Wallace, professor of decision sciences at Rensselaer Polytechnic Institute, points out that many users are not aware that computer software, such as expert systems, is often subtly shaped by the ethical judgments and assumptions of the people who create them. In one instance, he points out, a bank had to modify its loan-evaluation software after it discovered that it tended to reject some applications because it unduly emphasized old age as a negative factor. Another expert system, used by health maintenance organizations (HMOs), instructs doctors on when they should opt for expensive medical procedures, such as magnetic resonance imaging tests. HMOs like expert systems because they help control expenses, but critics are concerned that doctors will have to base decisions not on the best medicine but simply on "satisfactory" medicine combined with cost cutting.[134]

Clearly, there is no such thing as completely "value-free" technology. Human beings build it, use it, and have to live with the results.

12.8 The Promised Benefits of the Information Revolution

KEY QUESTION

What are the expected future benefits of the Information Revolution?

Preview & Review: The Information Revolution promises great benefits in the areas of education and information, health, commerce and electronic money, entertainment, and government and electronic democracy

Paul Saffo, called "the sage of cyberspace," is a techno-forecaster whose gadget-filled office is located at the Institute for the Future in Menlo Park, California. "He is smart, quotable, and knows just about everyone worth knowing in the wired world," says one article about him.[135]

Some of the industries that are big today, he predicts, will disappear: network television, telephone companies, personal computers, telecommunications as we presently know it. But enterprises that now appear at the margins—home shopping is one example—are the centers of what will become new industries.

Looking ahead to a moment when you are holding in your hand your "information appliance," the gadget that will help you access anybody and anything anywhere, how might this connection affect your life? Let's look at some areas of promise.

Education & Information

The government is interested in reforming education, and technology can assist that effort. Presently the United States has more computers in its classrooms than other countries, but the machines are older and teachers aren't as computer-literate. A recent study shows that 61.2% of urban schools have phone lines they could use for Internet access, while 42% own modems. The poorer the school district, the less likely it is to have modems.[136]

Computers can be used to create "virtual" classrooms not limited by scheduled class time. Several institutions (Stanford, MIT) have been replacing the lecture hall with forms of learning featuring multimedia programs, workstations, and television courses at remote sites. The Internet could be used to enable students to take video field trips to distant places and to pull information from remote museums and libraries.

As we have seen, making information available—and having it make sense—is one of our greatest challenges. Can everything in the Library of Congress be made available online to citizens and companies? What about government records, patents, contracts, and other legal documents? Or satellite-taken geographical maps? Indeed, satellite imaging, based on technology from Cold War spy satellites, is now so good that companies, cities, and other buyers can use it to get views of land use, traffic patterns—and even of your backyard. (The technology also has some people worried about privacy invasion and a free-for-all expansion of espionage.)

Of particular interest is distance learning, or the "virtual university." *Distance learning* is the use of computer and/or video networks to teach courses to students outside the conventional classroom. Until recently, distance learning has been largely outside the mainstream of campus life. That is, it concentrates principally on part-time students, those who cannot easily travel to campus, those interested in noncredit classes, or those seeking special courses in business or engineering. However, part-timers presently make up about 45% of all college enrollments. This, says one writer, is "a group for whom 'anytime, anywhere' education holds special appeal."[137]

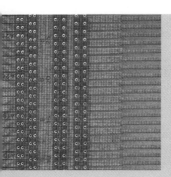

READ ME

Practical Matters: Distance Learning

Distance learning, one writer points out, is "a contemporary version in some ways of the correspondence schools of the 1950s."[138] Students take courses given at a distance, using television or a computer network, and the teacher may be close by in another room or on the other side of the world.

Of course, there are drawbacks to this process. "Some courses just don't translate well at a distance," says Pam Dixon, author of *Virtual College.* "Sometimes it's better to get your hands dirty and experience the course in a physical setting. Would you have wanted your dentist to have learned his or her craft entirely by videoconferencing?"[139] But, Dixon points out, business, writing, computers, mathematics, and library science, to name just a few courses and career paths, are well suited to teaching from afar. Indeed, students may get more attention from teachers online than they would from teachers in a large classroom with hundreds of students. In addition, says tele-education entrepreneur Glenn Jones, distance learning has the capability of offering computer access to virtual libraries of print, photos, recordings, and movies.[140]

To be sure, distance learning may take a bit of getting used to, as Joseph Walter, professor of communication studies at Northwestern University, found in studying the subject. "We learned a lot," says Walter, "including such things as the fact that the computer changes the dynamics of communication. It takes four or five times as long as speaking when you have to type out what you want to say."[141] Moreover, as Jones points out, it is still the enthusiasm and wit and skill of caring teachers that make the difference in the learning experience.

But in return for giving up being in a live classroom, students gain the convenience of being able to take courses not offered locally, often at times that suit their own schedules, not the space needs of an educational institution. This works best for the growing numbers of part-time and nontraditional students, those not among the full-time campus-bound 18- to 24-year-olds. In fact, this is the premise of one of the most ambitious plans for distance learning—the Western Governors University, a "virtual university" that will use technology to deliver courses from colleges and corporations in 16 western states and Guam.[142] WGU will offer a "Smart Catalogue/Advisor" that students can use to see if course offerings meet their schedules, interests, and budgets.

If you can't get the educational offerings you want locally, where do you start your search for the distance-learning equivalent? You might begin by looking at Distance Learning on the Net *(http://www.hoyle.com/distance.htm)* and the Comprehensive Distance Education List of Resources *(http://talon.extramural.uiuc.edu/ramage/disted.html).*

Health

The government is calling for an expansion of "telemedicine," the use of telecommunications to link health-care providers and researchers, enabling them to share medical images, patient records, and research. Of particular interest would be the use of networks for "teleradiology" (the exchange of X-rays, CAT scans, and the like), so that specialists could easily confer. Telemedicine would also allow long-distance patient examinations, using video cameras and, perhaps, virtual-reality kinds of gloves that would transmit and receive tactile sensations.

Commerce & Electronic Money

Businesses clearly see the Internet as a way to enhance productivity and competitiveness. However, the changes will probably go well beyond this.

The thrust of the original Industrial Revolution was separation—to break work up into its component parts to permit mass production. The effect of computer networks in the Digital Revolution, however, is unification—to erase boundaries between company departments, suppliers, and customers.[143]

Indeed, the parts of a company can now as easily be global as down the hall from one another. Thus, designs for a new product can be tested and exchanged with factories in remote locations. With information flowing faster, goods can be sent to market faster and inventories kept reduced. Says an officer of the Internet Society, "Increasingly you have people in a wide variety of professions collaborating in diverse ways in other places. The whole notion of 'the organization' becomes a blurry boundary around a set of people and information systems and enterprises."[144]

The electronic mall, in which people make purchases online, is already here. Record companies, for instance, are making sound excerpts and videos of new albums available on Web sites; you can sample the album and then order it sent as a cassette or CD. Banks in cyberspace are allowing customers to adopt avatars or personas of themselves and then meet in three-dimensional virtual space on the World Wide Web where they can query bank tellers and officers and make transactions. Wal-Mart Stores and Microsoft have developed a joint online shopping venture that allows shoppers to browse online and buy merchandise.

Cybercash or E-cash will change the future of money. Whether they take the form of smart cards or of electronic blips online, cybercash will probably begin to displace (though not completely supplant) checks and paper currency. This would change the nature of how money is regulated as well as the way we spend and sell.

Entertainment

Among the future entertainment offerings could be movies on-demand, video games, and gaming ("telegambling"). *Video on-demand* would allow viewers to browse through a menu of hundreds of movies, select one, and start it when they wanted. This definition is for true video on-demand, which is like having a complete video library in your house. (An alternative, simpler form could consist of running the same movie on multiple channels, with staggered starting times.) True video on-demand will require a server, a storage system with the power of a supercomputer that would deliver movies and other data to thousands of customers at once.

Government & Electronic Democracy

Will information technology help the democratic process? There seem to be two parts to this. The first is its use as a campaign tool, which may, in fact, skew the democratic process in some ways. The second is its use in governing and in delivering government services.

Santa Monica, California, established a computer system, called Public Electronic Network (PEN), which residents may hook into free of charge. PEN gives Santa Monica residents access to city council agendas, staff reports, public safety tips, and the public library's online catalog. Citizens may also enter into electronic conferences on topics both political and non-political; this has been by far the most popular attraction.

PEN could be the basis for wider forms of electronic democracy. For example, electronic voting might raise the percentage of people who vote. Interactive local-government meetings could enable constituents and town council members to discuss proposals.

The Internet could also deliver federal services and benefits. A few years ago, the government unveiled a program in which Social Security pensioners and other recipients of federal aid without bank accounts could use a plastic automated-teller-machine card to walk up to any ATM and withdraw the funds due them.

Onward

How do most of us view the way change takes place?

We believe it occurs slowly and predictably, like the process of water boiling, says economics writer Robert Samuelson. That is, the water warms gradually, getting hotter until it boils.

Quite often, however, change occurs suddenly, wrenchingly, dramatically—and is quite different from what came before. "Life and history aren't always water coming slowly to a boil," he says. "Sometimes they're a critical mass triggering radical change."[145]

This is what makes predictions difficult. "Hardly anyone foresaw . . . the explosion of the Internet," Samuelson points out, just as hardly anyone foresaw the collapse of Communism or the advent of AIDS. These were all major, world-shaking events, yet they were not on the radar screens of most futurists and planners.

Thus, if change often occurs in abrupt, surprising fashion, how can we really make predictions about the future?

Still, as we said at the beginning of the chapter, in a world of breakneck change, you can still thrive. The most critical knowledge, however, may turn out to be self-knowledge.

This is not the end. It is the beginning.

Experience Box

Job Searching on the Internet & World Wide Web

"If you haven't done a job search in a while, you will find many changes in a modern-day, high-quality search for a new position," says Mary Anne Buckman, consultant at Career Directions Inc.[146] Indeed, technological change has so affected the whole field of job hunting that futurists refer to it as a *paradigm shift*. This means that the change is of such magnitude that the "prevailing structure is radically, rapidly, and unalterably transformed by new circumstances."[147]

In the Experience Box at the end of Chapter 8 we described how to prepare a computer-friendly résumé. Here let's go a step further and describe how you can use the Internet and the World Wide Web to help you search for jobs. Online areas of interest for the job seeker include:

- Resources for career advice
- Ways for you to find employers
- Ways for employers to find you

Resources for Career Advice

It's 3 A.M. Still, if you're up at this hour (or indeed at any other time) you can still find job-search advice, tips on interviewing and résumé writing, and postings of employment opportunities around the world. One means for doing so is to use your Web browser to use a directory such as Yahoo! *(http://www.yahoo.com)* to obtain a list of popular Web sites. In the menu, you can click on Business and Economy, then Employment, then Jobs. This will bring up a list of sites that offer career advice, résumé postings, job listings, research about specific companies, and other services. (Caution: As might be expected, there is also a fair amount of junk out there: get-rich-quick offers, résumé-preparation firms, and other attempts to separate you from your money.)

Advice about careers, occupational trends, employment laws, and job hunting is also available through online chat groups and bulletin boards, such as those on the online services—America Online, CompuServe, Microsoft Network, and Prodigy. For instance, CompuServe offers career-specific discussion groups, such as the PR Marketing Forum. Through these groups you can get tips on job searching, interviewing, and salary negotiations. In addition, you might wish to check the U.S. Bureau of Labor Statistics Web site *(http://stats.bls.gov/eophome.htm),* which contains employment projections and a list of fastest-growing occupations; Career Magazine *(http://www.careermag.com);* Job Search Advice for College Grads *(http://www.collegegrad.com);* and JobSmart Salary Survey Links for all fields *(http://www.jobsmart.org/tools/salary/sal-prof.htm).*

Ways for You to Find Employers

As you might expect, companies seeking people with technical backgrounds and technical people seeking employment pioneered the use of cyberspace as a job bazaar. However, as the public's interest in commercial services and the Internet has exploded, the technical orientation of online job exchanges has changed. Now, says one writer, "interspersed among all the ads for programmers on the Internet are openings for English teachers in China, forest rangers in New York, physical therapists in Atlanta, and models in Florida."[148] Most Web sites are free to job seekers, although some may require you to fill out an online registration form.

Some jobs are posted on Usenets by individuals, companies, and universities or colleges, such as computer networking company Cisco Systems of San Jose, California, and the University of Utah in Salt Lake City. Others are posted by professional or other organizations, such as the American Astronomical Society, Jobs Online New Zealand, and Volunteers in Service to America (VISTA). Some of the principal organizations posting job listings are listed in the box on the next page. (■ *See Panel 12.5.)*

The difficulty with searching through these resources is that it can mean wading through thousands of entries in numerous databanks, with many of them not being suitable for or interesting to you. An alternative to trying to find an employer is to have employers find you.

Ways for Employers to Find You

Because of the Internet's low (or zero) cost and wide reach, do you have anything to lose by posting your résumé online for prospective employers to view? Certainly you might if the employer happens to be the one you're already working for. In addition, you have to be aware that you lose control over anything broadcast into cyberspace—you're putting your credentials out there for the whole world to see, and you need to be somewhat concerned about who might gain access to them.

Posting your résumé with an electronic jobs registry is certainly worth doing if you have a technical background, since technology companies in particular find this an efficient way of screening and hiring. However, it may also benefit people with less technical backgrounds. Online recruitment "is popular with companies because it pre-screens applicants for at least basic computer skills," says one writer. "Anyone who can master the Internet is likely to know something about word processing, spreadsheets, or database searches, knowledge required in most good jobs these days."[149]

Résumés may be prepared as we described in the Experience Box at the end of Chapter 8. The latest variant, however, is to produce a résumé with hypertext links and/or clever graphics and multimedia effects, then put it on a Web site to entice employers to chase after you. If you don't know how to do this, there are many companies that—for a fee—

can convert your résumé to HTML (✔ p. 326) and publish it on their own Web sites. Some of these services can't dress it up with fancy graphics or multimedia, but since complex pages take longer for employers to download anyway, the extra pizzazz is probably not worth the effort. In any case, for you the bottom line is how much you're willing to pay for these services. For instance, Résumé Innovations *(http://www.resume-innovations.com)* charges $85 to write a résumé and nothing to post it on a Web site. There are a number of Web sites on which you can post your résumé, sometimes for free.

Companies are also beginning to replace their recruiters' campus visits with online interviewing. For example, the firm VIEWnet of Madison, Wisconsin, offers first-round screenings or interviews for summer internships through its teleconferencing "InterVIEW" technology, which allows video signals to be transmitted (at 17 frames per second) via phone lines.

■ PANEL 12.5 Organizations posting job listings on the Web

- **America's Job Bank:** A joint venture of the New York State Department of Labor and the Federal Employment and Training Administration, America's Job Bank *(http://www.ajb.dni.us/index.html)* advertises more than 100,000 jobs of all types. There are links to each state's employment office. More than a quarter of the jobs posted are sales, service, or clerical. Another quarter are managerial, professional, and technical. Other major types are construction, trucking, and manufacturing.
- **Career Mosaic:** A service run by Bernard Hodes Advertising, Career Mosaic *(http://www.careermosaic.com)* offers links to nearly 200 major corporations, most of them high-technology companies. One section is aimed at college students and offers tips on résumés and networking. A major strength is the JOBS database, which lets you fill out forms to narrow your search, then presents you with a list of jobs meeting your criteria.
- **Career Path:** Career Path *(http://www.careerpath.com/)* is a classified-ad employment listing from numerous American newspapers, which you can search either individually or all at once. Major papers include the *Boston Globe,* the *Chicago Tribune,* the *Los Angeles Times,* the *New York Times,* the *San Jose Mercury News,* and the *Washington Post.*
- **E-Span:** One of the oldest and biggest services, the E-Span Interactive Employment Network *(http://www.espan.com)* features all-paid ads from employers.
- **FedWorld:** This bulletin board *(http://www.fedworld.gov)* offers job postings from the U.S. Goverment.
- **Internet Job Locator:** Combining all major job-search engines on one page, the Internet Job Locator *(http://www.joblocator.com/jobs)* lets you do a search of all of them at once.

- **JobHunt:** Started by Stanford University geologist Dane Spearing, JobHunt *(http://www.job-hunt.org)* contains a list of more than 700 sites related to online recruiting.
- **JobTrak:** The nation's leading online job listing service, JobTrak *(http://www.jobtrak.com)* claims to have 35,000 students and alumni visiting the site each day, with more than 300,000 employers and 750 college career centers posting 3000 new jobs daily.
- **JobWeb:** Operated by the National Association of colleges and Employers, Job Web *(http://www.jobweb.org/)* is a college placement service with 1600 U.S. member universities and colleges and 1400 employer organizations. It claims to have served over 1 million college students and alumni.
- **Monster Board:** Not just for computer techies, the Monster Board *(http://www.monster.com)* offers real jobs for real people, although a lot of the companies listed are in the computer industry.
- **NationJob Network:** Based in Des Moines, Iowa, NationJob Network *(http://www.nationjob.com)* lists job opportunities primarily in the Midwest. A free feature called P.J. Scout sends job seekers news of new jobs.
- **Online Career Center:** Based in Indianapolis, Online Career Center *(http:www.occ.com/occ/)* is a nonprofit national recruiting service listing jobs at more than 3000 companies. About 30% of the jobs are nontechnical, with many in sales and marketing and in health care.
- **Workplace:** An employment resource offering staff and administrative positions in colleges and universities, government, and the arts *(http://galaxy.einet.net.galaxy/Community/Workplace.html).*

Summary

What It Is/What It Does	Why It's Important
artificial intelligence (p. 578, KQ 12.7) Group of related technologies that attempt to develop machines to emulate human-like qualities, such as learning, reasoning, communicating, seeing, and hearing.	AI is important for enabling machines to do things formerly possible only with human effort.
biometrics (p. 568, KQ 12.3) Science of measuring individual body characteristics.	Biometrics is used in some computer security systems—for example, to verify individual's fingerprints before allowing access.
disaster-recovery plan (p. 570, KQ 12.3) Method of restoring information processing operations that have been halted by destruction or accident.	Disaster recovery plans are important for companies desiring to resume computer and business operations in short order.
encryption (p. 569, KQ 12.3) Also called *enciphering;* the altering of data so that it is not usable unless the changes are undone.	Encryption is useful for users transmitting trade or military secrets or other sensitive data.
fuzzy logic (p. 582, KQ 12.7) Method of dealing with imprecise data and uncertainty, with problems that have many answers rather than one.	Unlike traditional "crisp," yes/no digital logic, fuzzy logic deals with probability and credibility.
genetic algorithm (p. 583, KQ 12.7) Program that uses Darwinian principles of random mutation to improve itself.	Genetic algorithms use trial and error to learn from experience, thus constantly improving themselves.
information-technology crime (p. 561, KQ 12.2) Crime of one of two types: an illegal act perpetrated against computers or telecommunications; or the use of computers or telecommunications to accomplish an illegal act.	Information-technology crimes cost billions of dollars every year.
intelligent agent (p. 576, KQ 12.6) Computer program that performs work tasks on a user's behalf, including roaming networks and compiling data.	Agents scan databases and electronic mail; clerical agents answer telephones and send faxes; search engines use spider agents to find new Web sites.
Internet 2 (p. 557, KQ 12.1) Cooperative university-business program to upgrade the Internet, allowing high-end users to quickly and reliably move huge amounts of data using "toll lanes" provided by VBNS, the high-speed data transmission system connecting major research centers.	Internet 2 would provide speeds 100 times that of today's Internet. The improvement could advance videoconferencing, research, and academic collaboration among the members—currently more than 117 universities and about 25 companies.

What It Is/What It Does	Why It's Important
National Information Infrastructure (NII) (p. 555, KQ 12.1) U.S. government vision for the Information Superhighway; services wil be delivered via the networks and technologies of several information providers—the telecommunications companies, cable-TV companies, and the Internet.	Services could include education, health care, information, commerce, and entertainment.
natural language processing (p. 581, KQ 12.7) Study of ways for computers to recognize and understand human language, whether in spoken or written form.	Natural language processing could further reduce the barriers to human/computer communications.
neural networks (p. 582, KQ 12.7) Field of artificial intelligence; networks that use physical electronic devices or software to mimic the neurological structure of the human brain, with, for instance, transistors for nerve cells and resistors for synapses.	Neural networks are able to mimic human learning behavior and pattern recognition.
Next Generation Internet (p. 557, KQ 12.1) U.S. government's broad program to parallel the university/business effort of Internet-2, helping to tie that high-performance network into the broader federal infrastructure. NGI funds six government agencies' programs directed at the effort.	Using high-speed fiber-optic circuits and sophisticated software, NGI, Internet 2, and VBNS all aim to improve on the original Internet. NGI and Internet 2 are planned to be available by 2003.
password (p. 567, KQ 12.3) Special word, code, or symbol that is required to access a computer system.	One of the weakest links in computer security, passwords can be guessed, forgotten, or stolen.
personal identification number (PIN) (p. 567, KQ 12.3) Security number known only to an individual user, who cannot access the system without it.	PINs are required to access many computer systems and automated teller machines.
robotics (p. 578, KQ 12.7) Field of artificial intelligence that attempts to develop robots, machines that can perform work normally done by people.	Robots are performing more and more functions in business and the professions.
security (p. 567, KQ 12.3) System of safeguards for protecting information technology against disasters, systems failure, and unauthorized access, all of which can result in damage or loss.	With proper security, organizations and individuals can minimize losses caused to information technology from disasters, systems failures, and unauthorized access.
VBNS (Very-High-Speed Backbone Network Service) (p. 555, KQ 12.1) Part of the effort to upgrade the Internet; VBNS is the U.S. government's project to link supercomputers and other banks of computers across the nation at speeds 1000 times faster than the current Internet. Begun in 1996, VBNS will involve only the top 100 research universities in the United States, but it will also have "toll lanes" for other users.	VBNS is the main U.S. government component to upgrade primary hubs of data transmission (the Internet's "backbone"). Most of the present members of VBNS will also be part of Internet 2.

Self-Test

1. The purpose of _____ is to scan a computer's disk devices and memory to detect viruses and, sometimes, to destroy them.

2. Data that is incomplete, outdated, or otherwise inaccurate is referred to as _____ _____.

3. List four areas in which the Information Superhighway promises great benefits.

   a _____
   b. _____
   c. _____
   d. _____

4. _____ use physical electronic devices or software to mimic the structure of the human brain.

5. So that information processing operations can be restored after destruction or accident, a company should adopt a _____.

Short-Answer Questions

1. What is an intelligent agent?
2. What is the significance of the 1996 Telecommunications Act?
3. How would you define *information-technology crime*?
4. What is the difference between a hacker and a cracker?

Multiple-Choice Questions

1. What are the terms *spiders* and *crawlers* most closely related to?
   a. Internet
   b. modems
   c. electronic news services
   d. artificial intelligence
   e. None of the above

2. Which of the following is an interactive computer program that helps users solve problems?
   a. robot
   b. perception system
   c. expert system
   d. natural language
   e. All of the above

3. Which of the following *isn't* necessarily a component of an expert system?
   a. natural language
   b. knowledge base
   c. inference engine
   d. user interface
   e. All of the above

4. Which of the following involves the study of ways for computers to understand human language?
   a. natural language
   b. knowledge base
   c. inference engine
   d. user interface
   e. All of the above

5. Which of the following groups perpetrate over 80% of information technology crime?
   a. hackers
   b. crackers
   c. professional criminals
   d. employees
   e. None of the above

True/False Questions

T F   1. A digital signature looks the same as your signature on a check.

T F   2. Encrypted data isn't directly usable.

T F   3. An avatar is a computer program that can roam networks and compile data.

T F   4. The 1996 Telecommunications Act permits greater competition between local and long-distance telephone companies.

T F   5. Viruses can be passed to another computer by a diskette or through a network.

Knowledge in Action

1. In addition to *2600: The Hacker's Quarterly*, where do hackers find new information about their field? Are support groups available? In what ways do hackers help companies? Does a hacker underground exist? Research your answers using current computer periodicals and/or the Internet.

2. Assuming you have a microcomputer in your home that includes a modem, what security threats, if any, should you be concerned with? List as many ways as you can think of to ensure that your computer is protected.

3. Explore the National Information Infrastructure (NII) in more detail. Create an executive report describing the objectives for the NII, its guiding principles, and its agenda for action. Does the NII exist today or is it a plan for the future? Or both? Research your answers using current periodicals and/or the Internet.

4. What's your opinion about the issue of free speech on an electronic network? Research some recent legal decisions in various countries, as well as some articles on the topic, and then give a short report about what you think. Should the contents of messages be censored? If so, under what conditions?

5. Artificial intelligence professional societies, such as the American Association for Artificial Intelligence (contact info: 415-328-3123 or *http://www.aaai.org*), provide a variety of published material as well as symposia, workshops, conferences, and related services and activities for those involved in various AI fields. These societies can be easily located by using a Web browser and then searching for the phrase "artificial intelligence." Contact one or more societies and obtain information on activities, services, and fees.

Answers

Self-Test Questions
1. antivirus software   2. dirty data   3. education and information, health, commerce and electronic money, entertainment, government and electronic democracy   4. neural networks   5. disaster-recovery plan

Short-Answer Questions
1. A computer program that can perform work on your behalf such as roaming networks and compiling data.   2. It undoes 60 years of federal and state communications regulations enabling phone, cable, and TV businesses to compete and combine more freely.   3. An illegal act perpetrated against computers or telecommunications, an illegal act involving the use of computers or telecommunications.   4. A hacker gains unauthorized access to a computer or telecommunications system for the challenge of it, whereas a cracker may do the same for malicious reasons.

Multiple-Choice Questions
1. a   2. c   3. a   4. a   5. d

True/False Questions
1. F   2. T   3. F   4. T   5. T

Notes

Chapter 1

1. Thomas A. Stewart, "The Information Age in Charts," *Fortune*, April 4, 1994, pp. 75–79.
2. Tom Mandel, in "Talking About Portables," *Wall Street Journal*, November 16, 1992, p. R18–R19.
3. Field Institute survey, reported in Jonathan Marshall, "High Tech Often Equals Higher Pay," *San Francisco Chronicle*, September 3, 1996, p. A5.
4. Alan Krueger, Princeton University, cited in Marshall, 1996.
5. Donald Spencer, *Webster's New World Dictionary of Computer Terms*, 4th ed. (New York: Prentice Hall, 1992), p. 206.
6. Annie D. Johnson, quoted in Jared Sandberg and Thomas E. Weber, "WebTV Finds a Following Among Net-Surfing Seniors," *Wall Street Journal*, April 8, 1997, pp. B1, B4.
7. "What Does 'Digital' Mean in Regard to Electronics?" *Popular Science*, August 1997, pp. 91–94.
8. Charles McCoy, "Meet the Jetsons," *Wall Street Journal*, June 16, 1997, p. R6.
9. Amy Saltzman, "Making It in a Sizzling Economy," *U.S. News & World Report*, June 23, 1997, pp. 50–58.
10. Julie Schmit, "90 Hours a Week and Loving It," *USA Today*, April 3, 1996, pp. 1A, 2A.
11. Saltzman, 1997.
12. Debby Atkins, quoted in Paul Davidson, "High-Tech Help Wanted: No Experience Necessary," *USA Today*, December 23, 1997, p. 1B.
13. Information Technology Association of America, reported in Stephen Baker, Ann Barrett, and Linda Himelstein, "Calling All Nerds," *Business Week*, March 10, 1997, pp. 36–37.
14. Davidson, 1997.
15. Marshall, 1996.
16. Gene Koretz, "The Payoff from Computer Skills," *Business Week*, November 3, 1997, p. 30.
17. Mary Beth Marklein, "Dream Jobs Abound for High-Tech Grads," *USA Today*, April 7, 1997, p. 8D.
18. Nick Wingfield, "The Webmaster," *Wall Street Journal*, December 8, 1997, pp. R20, R22.
19. Buck Consultants, reported in Barbara B. Buchholz, "Casting a Wider Net for Web Site Workers," *New York Times*, February 2, 1997, sec. 3, p. 11.
20. Baker, Barrett, and Himelstein, 1997.
21. McCoy, 1997.
22. McCoy, 1997.
23. Barbara Simmons and Gary Chapman, "Information Highway Has Many Potholes," *San Francisco Chronicle*, January 17, 1994, p. B3.
24. Lawrence M. Fisher, "Intel Wins Contract to Develop World's Fastest Supercomputer," *New York Times*, September 8, 1995, p. C2.
25. Gautem Naik, "In Digital Dorm, Click on Return for Soda," *Wall Street Journal*, January 23, 1997, pp. B1, B6.
26. Blanton Fortson, in "Talking About Portables," 1992.
27. Elizabeth Fernandez, "Homeless but Wired," *San Francisco Examiner*, January 28, 1996, pp. A-1, A-14.
28. Link Resources, reported in Ilana DeBare, "Telecommuting Sparks Debate Over Safety," *San Francisco Chronicle*, August 4, 1997, pp. B1, B2.
29. IDC/Link, reported in Susan J. Wells, "For Stay-Home Workers, Speed Bumps on the Telecommute," *New York Times*, August 17, 1997, sec. 3, pp. 1, 14.
30. Michael Capochiano, commenting on 1996 USA Today/IntelliQuest Technology Monitor study, reported in Leslie Miller, "Most Users See Internet as Happy Medium," *USA Today*, June 18, 1996, p. 4E.
31. Mike Snider, "Fewer Homes on Line, Study Shows," *USA Today*, September 3, 1997, p. 1D.
32. Study by Dataquest, reported in Jon Swartz, "A Rush to Plug PCs into the Internet," *San Francisco Chronicle*, August 21, 1997, pp. A1, A15.
33. Walter S. Mossberg, "The Marriage of TV and Home Computer May Last This Time," *Wall Street Journal*, August 7, 1997, p. B1.
34. Katie Hafner and Jennifer Tanaka, "Info Hits Home," *Business Week*, November 11, 1996, pp. 84–88.
35. Gina Smith, "Info Appliances," *Popular Science*, November 1996, pp. 26–27.
36. James Gleick, "Addicted to Speed," *New York Times Magazine*, September 28, 1997, pp. 54–61.
37. John D. Dvorak, "Avoiding Information Overload," *PC Magazine*, December 17, 1996, p. 87.
38. Tom Forester and Perry Morrison, *Computer Ethics: Cautionary Tales and Ethical Dilemmas in Computing* (Cambridge, MA: The MIT Press, 1990), pp. 1–2.
39. Interview with Ted Selker, "More Power to Go," *Byte*, July 1997, p. 32.
40. Anonymous. (1990, January). Der Mensch wird immer dümmer. *Der Spiegel*, pp. 98–103.
41. Library of Congress, http://lcweb.loc.gov.
42. Debbie G. Longman and Rhonda H. Atkinson, *College Learning and Study Skills*, 2nd ed. (St. Paul, MN: West, 1992), p. 4.
43. Mervill Douglass and Donna Douglass, *Manage Your Time, Manage Your Work, Manage Yourself* (New York: American Management Association, 1980).
44. W. M. Beneke and M. B. Harris, "Teaching Self-Control of Study Behavior," *Behavior Research and Therapy*, 1972, *10*, 35–41.
45. E. B. Zechmeister and S. E. Nyberg, *Human Memory: An Introduction to Research and Theory* (Pacific Grove, CA: Brooks/Cole, 1982).
46. B. K. Bromage and R. E. Mayer, "Quantitative and Qualitative Effects of Repetition on Learning from Technical Text," *Journal of Educational Psychology*, 1982, *78*, 271–78.
47. Longman and Atkinson, 1992, pp. 148–53.
48. H. C. Lindgren, *The Psychology of College Success: A Dynamic Approach* (New York: Wiley, 1969).
49. R. J. Palkovitz and R. K. Lore, "Note Taking and Note Review: Why Students Fail Questions Based on Lecture Material," *Teaching of Psychology*, 1980, *7*, 159–61.
50. Palkovitz and Lore, 1980, pp. 159–61.
51. F. P. Robinson, *Effective Study*, 4th ed. (New York: Harper & Row, 1970).
52. J. Langan and J. Nadell, *Doing Well in College: A Concise Guide to Reading, Writing, and Study Skills* (New York: McGraw-Hill, 1980), pp. 93–110.
53. Langan and Nadell, 1980, p. 104.

Chapter 2

1. John Markoff, "A Free and Simple Computer Link," *New York Times*, December 8, 1993, p. C1.
2. Software Publishers Association, reported in Alan Deutschman, "Mac vs. Windows: Who Cares?" *Fortune*, October 4, 1993, p. 114.
3. Claris survey of small businesses, reported in USA Snapshots, "Small-Business Software," *USA Today*, July 17, 1996, p. 1B.
4. Deutschman, 1993.
5. Knight-Ridder News Service, "Atomic Bomberman Is a Real Blast," *San Jose Mercury News*, August 31, 1997, p. 7F.
6. Joan Indiana Rigdon, "Nintendo Catches Up to Sony in Market for Most-Advanced Video-Game Players," *New York Times*, February 3, 1997, p. B3.
7. Jim Erickson, "New Video College Offers Games Degree," *San Francisco Examiner*, September 8, 1996, p. C-10; reprinted from Seattle Post-Intelligencer.
8. Neil Gross, "Zap! Splat! Smarts? Why Video Games May Actually Help Your Children Learn," *Business Week*, December 23, 1996, pp. 64–71.
9. Seymour Papert, quoted in Gross, 1996.
10. Steve G. Steinberg, "Back in Your Court, Software Designers," *Los Angeles Times*, December 7, 1995, pp. D2, D11.
11. Rick Tetzeli, "Videogames: Serious Fun," *Fortune*, December 27, 1993, pp. 110–116.
12. Dean Takahashi, "Internet Transforms the Way PC Games Are Developed," *Wall Street Journal*, June 19, 1997, p. B4.
13. Jon Swartz, "AOL Enters the Game," *San Francisco Chronicle*, June 17, 1997, pp. C1, C2.
14. Carey Goldberg, "Game Centers Lure Computer Loners to High-Tech Team Activities," *New York Times*, August 4, 1997, p. A7.
15. David Elrich, "New Video Games: Despite Promises, Violence Rules," *New York Times*, July 3, 1997, p. B13.
16. Debra Jo Immergut, "Blood on the Net: Computer Gamers See Red," *Wall Street Journal*, August 7, 1997, p. A12.
17. Edward Rothstein, "Girl Software: A Fantasy World Stressing Advice and the Anxiety of Romance," *New York Times*, February 17, 1997, p. 29.
18. Karen De Witt, "Girl Games on Computers, Where Shoot 'em Up Simply Won't Do," New York Times, June 23, 1997, p. C3.
19. Rebecca L. Eisenberg, "The Barbie Syndrome," *San Francisco Examiner*, May 4, 1997, pp. D-5, D-12.
20. Susan Gregory Thomas, "Great Games for Girls," *U.S. News & World Report*, November 25, 1996, pp. 108–110.
21. International Network of Women in Technology 1995 survey, reported in Janet Rae-Dupree and Dean Takahashi, "High-Tech Women Still Struggle," *San Jose Mercury News*, June 2, 1996, pp. 1A, 24A.
22. Dan Gillmor, "Technology Ignores Half a World of Talent," *San Jose Mercury News*, June 2, 1996, p. 1F.
23. American Association of Engineering Societies, reported in Rae-Dupree and Takahashi, 1996.
24. Kathy Wheeler, quoted in in Rae-Dupree and Takahashi, 1996.
25. NetSmart Inc., reported in Paul M. Eng, "What Do Women Want Online?" *Business Week*, October 20, 1997, p. 142C.
26. "A Woman's Perspective," *Wall Street Journal*, March 13, 1997, p. B6.
27. Robin Frost, "Women On-line: Cybergrrl Aims to Show the Way," *Wall Street Journal*, May 30, 1996, p. B8.
28. Stacey J. Miller, "The Galloping Letdown," *ComputerLife*, August 1997, p. 109.
29. Herb Brody, "Video Games That Teach?" *Technology Review*, November/December 1993, pp. 50–57.
30. Jay Sivin-Kachala, Interactive Educational Systems Design, quoted in Nicole Carroll, "How Computers Can Help Low-Achieving Students," *USA Today*, November 20, 1995, p. 5D.
31. Wayne Kawamoto, "Make It So! Automating Your Business Is Easier Than You Think," *Bay Area Computer Currents*, May 6/May 19, 1997, pp. 39–48.
32. Barbara Kantrowitz, "In Quicken They Trust," *Newsweek*, May 2, 1994, pp. 65–66.
33. Edward Rothstein, "Between the Dream and the Reality Lies the Shadow. Or Is It the Interface?" *New York Times*, December 11, 1995, p. C3.
34. Randall Stross, "Netscape: Inside the Big Software Giveaway," *Fortune*, March 30, 1998, pp. 150–152.
35. Baruch College–Harris Poll, commissioned by *Business Week*; reported in Amy Cortese, "A Census in Cyberspace," *Business Week*, May 5, 1997, pp. 84–85.
36. Stephen H. Wildstrom, "Let Your Laptop Do the Navigating," *Business Week*, October 6, 1997, p. 23.
37. David Einstein, "Maps You Don't Have to Fold," *San Francisco Chronicle*, September 9, 1997, p. C4.
38. Michael J. Himowitz, "Software to Get You from Hither to Yon," *Fortune*, March 3, 1997 p. 188.
39. "A Better Street Guide," *Newsweek*, March 17, 1997, p. 10.
40. Phillip Robinson, "Street-Wise," *San Jose Mercury News*, April 20, 1997, pp. 1E, 4E.
41. Jan Norman, "Office Suite Office," *San Francisco Examiner*, November 17, 1996, pp. B-5, B-7.
42. Michael Finley, "Lotus Gives Its All for the Team," *San Jose Mercury News*, March 9, 1997, p. 7F.
43. Margaret Trejo, quoted in Richard Atcheson, "A Woman for *Lear's*," *Lear's*, November 1993, p. 87.
44. Stacey Richardson, quoted in Peter H. Lewis, "Pairing People Management with Project Management," *New York Times*, April 11, 1993, sec. 3, p. 12.
45. Glenn Rifkin, "Designing Tools for the Designers," *New York Times*, June 18, 1992, p. C6.
46. Bernie Ward, "Computer Chic," *Sky*, April 1993, pp. 84–90.
47. Claudia H. Deutsch, "Not Making Them Like They Used To," *New York Times*, March 31, 1997, pp. C1, C2.
48. John Ennis, quoted in Peter Plagens and Ray Sawhill, "Throw Out the Brushes," *Newsweek*, September 1, 1997, pp. 76–77.
49. Plagens and Sawhill, 1997.
50. Jim Creighton, quoted in "Meetings Will Gain Importance," *The Futurist*, November-December 1996, pp. 45–46.
51. David Kirkpatrick, "Groupware Goes Boom," *Fortune*, December 27, 1993, p. 100.
52. Associated Press, quoted in Ellen Goodman, "Computercide on My Mind," *San Francisco Chronicle*, July 31, 1997, p. A21; reprinted from Boston Globe.
53. Phillip Robinson, "End the Hype: Computers Must Work Better," *San Jose Mercury News*, August 24, 1997, p. 3E.
54. Paul Krugman, "The Paper-Bag Revolution," *New York Times Magazine*, September 28, 1997, pp. 52–53.
55. Bill Husted, "Sometimes More Is Really Less," *San Jose Mercury News*, June 22, 1997, p. 7F.
56. Don Clark and David Bank, "Microsoft May Face a Backlash Against 'Bloatware,' " *Wall Street Journal*, November 18, 1996, pp. B1, B4.
57. Stephen Manes, "The Life of a Computer User: One Frustration After Another," *San Jose Mercury News*, November 24, 1996, p. 5F; reprinted from *New York Times*.
58. John Merchant, "Help Is Just a Keystroke Away! Sure," *New York Times*, April 21, 1996, sec. 3, p. 13.
59. Eric Rocco, Dataquest, reported in Jared Sandberg, "PC Users Battle Help Lines, and Pay for the Privilege," *Wall Street Journal*, June 27, 1996, p. B1.
60. Denise K. Magner, "Verdict in a Plagiarism Case," *Chronicle of Higher Education*, January 5, 1994, pp. A17, A20.
61. William Grimes, "A Question of Ownership of Images," *New York Times*, August 20, 1993, p. B7.
62. Andy Ihnatko, "Right-Protected Software," *MacUser*, March 1993, pp. 29–30.

63. Business Software Alliance, reported in Deborah Shapley, "Corporate Web Police Hunt Down E-Pirates," *New York Times*, May 19, 1997, p. C5.
64. Brian Rust, University of Wisconsin, reported in Karla Haworth, "Publishers Press Colleges to Stop Software Piracy by Their Students," *Chronicle of Higher Education*, July 11, 1997, pp. A19-A20.
65. Bruce Haring, "Sound Advances Open Doors to Bootleggers," *USA Today*, May 27, 1997, p. 8D.
66. Goldie Blumenstyk, "Comics and Centerfolds on Web Pages Pose a Copyright Problem for Colleges," *Chronicle of Higher Education*, September 27, 1996, pp. A29–A30.
67. Haworth, 1997.

Chapter 3

1. Alan Robbins, "Why There's Egg on Your Interface," *New York Times*, December 1, 1996, sec. 3, p. 12.
2. Raj Reddy, reported in Dean Takahashi, "While You Wait, Why Not Make a List, Get Voice Mail, File a Nail?" *Wall Street Journal*, March 11, 1997, p. B1.
3. Rory J. O'Connor, "Microsoft Talks Fail; Suit Likely," *San Jose Mercury News*, May 17, 1998, p. 24A.
4. David Bank, "Microsoft's Gates Urges Clients Focus on Its Windows NT," *Wall Street Journal*, October 3, 1997, p. B5.
5. Stuart Card, quoted in Kevin Maney, "Computer Windows May Be Obsolete," *USA Today*, August 25, 1995, p. 2B.
6. Peter H. Lewis, "Champion of MS-DOS, Admirer of Windows," *New York Times*, April 4, 1993, sec. 3, p. 11.
7. Paul Krugman, "The Paper-Bag Revolution," *New York Times Magazine*, September 28, 1997, pp. 52–53.
8. Ken Wasch, quoted in Lisa Green, "Windows 95 Drops the Curtain on DOS," *USA Today*, August 22, 1995, p. 2B.
9. Cathy Booth, "Steve's Job: Restart Apple," *Time*, August 18, 1997, pp. 28–34.
10. Lee Gomes, "Apple Seeks to Ensure Programs for Rhapsody Will Run on Mac," *Wall Street Journal*, May 13, 1997, p. B6.
11. Jim Carlton, "Apple to Roll Out New Operating System Gradually, Scrapping Quick Transition," *Wall Street Journal*, October 3, 1997, p. B5.
12. Amy Cortese, "IBM Rides into Microsoft Country," *Business Week*, June 6, 1994, pp. 111–112.
13. James Gleick, "Making Microsoft Safe for Capitalism," *New York Times Magazine*, November 5, 1995, pp. 50–57, 64.
14. Don Clark, "Windows 95 Birthday Isn't Gala for All," *Wall Street Journal*, August 22, 1996, pp. B1, B3.
15. Don Clark, "Microsoft Corp. Delays Release of Windows 95 System Upgrade," *Wall Street Journal*, September 16, 1997, p. B10.
16. International Data Corp., reported in Tom Abate, "Novell's Comeback Chief," *San Francisco Chronicle*, September 12, 1997, pp. B1, B2.
17. Axxel Knutson, reported in Julia Angwin, "Stock of the Week: Novell," *San Francisco Chronicle*, September 8, 1997, p. B3.
18. The Gartner Group, cited in Abate, 1997.
19. Richard Scocozza, reported in Angwin, 1997.
20. Ben Smith, quoted in John Montgomery, "Putting Unix in All the Right Places," *Byte*, January 1998, pp. 96I-96N.
21. Denise Caruso, "Netscape's Decision to Give Away Code Could Alter the Software Industry," *New York Times*, February 2, 1998, p. C3.
22. Steve Lohr, "Microsoft Sets Its Sights on Corporate Computing," *New York Times*, May 19, 1997, pp. C1, C7.
23. David E. Kalish, Associated Press, "Microsoft's Windows NT Takes on the Mainframes," *San Francisco Chronicle*, May 21, 1997, p. B2.
24. David Kirkpatrick, "He Wants *All* Your Business—And He's Starting to Get It," *Fortune*, May 26, 1997, pp. 58–68.
25. Walter S. Mossberg, "New Windows Desktop Fails to Show the Way to Easier PC Use," *Wall Street Journal*, October 9, 1997, p. B1.
26. Reuters, "Windows 98 Opens a New Channel," *San Francisco Chronicle*, October 9, 1997, p. D3.
27. Mossberg, October 9, 1997.
28. David Bank, "Microsoft's Gates Urges Clients Focus on Its Windows NT," *Wall Street Journal*, October 3, 1997, p. B5.
29. Tosca Moon Lee, "Utility Software: Your PC's Life Preserver," *PC Novice*, March 1993, pp. 68–73.
30. "Under the Hood," *Fortune Technology Buyer's Guide*, Winter 1997, pp. 118–119.
31. Janet Rae-Dupree, "Help for Hard Drives," *San Jose Mercury News*, April 28, 1997, pp. 1E, 3E.
32. Harry Goldblatt, "Calling Doctor Hard Drive," *Fortune*, June 9, 1997, p. 140.
33. Soo-Yin Jue, quoted in Rae-Dupree, 1997.
34. Nikki Stange, quoted in Goldblatt, 1997.
35. Gillian Coolidge, "Investigating the Lost Files of Peter Norton, PC Pioneer," *PC Novice*, May 1992, pp. 14–18.
36. "Under the Hood," 1997.
37. "A New Model for Personal Computing," *San Jose Mercury News*, August 13, 1995, p. 27A. 6
38. Lee Gomes, "Hollow Dreams," *San Jose Mercury News*, November 12, 1995, pp. 1D, 3D.
39. Joseph Jennings, "The End of Wintel?" *San Francisco Examiner*, December 17, 1995, pp. B-5, B-7.
40. Mark Fleming [letter] and Mike McGowan [letter], "Present at the Creation of the Net," *Business Week*, December 25, 1995, p. 12.
41. Jim Carlton, "Apple to Roll Out New Operating System Gradually, Scrapping Quick Transition," *Wall Street Journal*, October 3, 1997, p. B5.
42. Michael H. Martin, "Digging Data Out of Cyberspace," *Fortune*, April 1, 1996, p. 147.
43. Richard Scoville, "Find It on the Net," *PC World*, January 1996, pp. 125-130.
44. David Haskin, "Power Search," *Internet World*, December 1997, pp. 78-92.
45. Martin, 1996.
46. Scoville, 1996.

Chapter 4

1. Michael S. Malone, "The Tiniest Transformer," *San Jose Mercury News*, September 10, 1995, pp. 1D–2D; excerpted from *The Microprocessor: A Biography* (New York: Telos/Springer Verlag, 1995).
2. Hadas Dembo, "The Way Things Were," *Wall Street Journal*, November 16, 1992, pp. R16–R17.
3. James J. Mitchell, "Chip Makers Step Up to the 12-inch Plate," *San Jose Mercury News*, March 2, 1997, pp. 1E, 3E.
4. Laurence Hooper, "No Compromises," *Wall Street Journal*, November 16, 1992, p. R8.
5. Phillip Robinson, "When the Power Fails," *San Jose Mercury News*, December 17, 1995, pp. 1F, 6F.
6. Robinson, 1995.
7. Harry Somerfield, "Surge Protectors Vital to Shield Electronics," *San Francisco Chronicle*, November 22, 1995, sec. Z-1, p. 7.
8. Suzanne Weixel, *Easy PCs*, 2nd ed. (Indianapolis: Que Corp., 1993).
9. Bruce Haring, "Power Outages Give PC Owners an Unpleasant Jolt," *USA Today*, August 19, 1996, p. 3D.
10. Andy Reinhardt, "Pentium: The Next Generation," *Business Week*, May 12, 1997, pp. 42–43.
11. Susan Moran, "PowerPC's Waning Market," *San Francisco Chronicle*, November 29, 1997, pp. B1, B2.
12. Kevin Maney, "Moore's Law Still Intact," *USA Today*, September 25, 1997, p. 2B.
13. Dean Takahashi, "Intel's Top Chip Architect to Unveil His Latest Creation," *Wall Street Journal*, October 10, 1997, pp. B1, B9.
14. John Markoff, "Intel Is Gambling with a New and More Powerful Set of Chips," *New York Times*, August 27, 1997, p. C8.
15. Otis Port, "Gordon Moore's Crystal Ball," *Business Week*, June 23, 1997, p. 120.
16. John Markoff, "New Chip May Make Today's Computer Passé," *New York Times*, September 17, 1997, pp. A1, C5.
17. Dean Takahashi, "Intel Sets Price Cuts of 13% to 40% on Computer Chips," *Wall Street Journal*, October 29, 1997, p. B11.
18. Craig Barrett, reported in Bloomberg News, "The Intel View of Future PCs," *New York Times*, April 23, 1997, p. C2.
19. Andy Reinhardt, "Intel's New Chip Has Real Flash," *Business Week*, September 29, 1997, p. 40.
20. James Kim, "Intel Doubles the Memory in a Flash," *USA Today*, September 18, 1997, p. 4B.
21. James L. Pappas, quoted in Laurie Flynn, "Which Peripheral Plug Goes Where? Help Is on the Drawing Board," *New York Times*, March 27, 1995, p. C9.
22. Flynn, 1995.
23. Cary Lu, "Power Struggle," *Inc. Tech*, 1997, No. 3, p. 35.
24. Donna Rosato, "Airlines Empower Fliers with Laptops," *USA Today*, September 30, 1996, p. 1B.
25. Stuart F. Brown, "Batteries Are Getting Charged Up," *Fortune*, October 27, 1997, pp. 240[C]–240[T].
26. Microsoft Press, *Microsoft Press Computer Dictionary*, 3rd ed. (Redmond, WA: Microsoft Press, 1997).
27. Alan Freedman, *The Computer Desktop Encyclopedia* (New York: AMACOM, 1996).
28. Lu, 1997.
29. Stephen H. Wildstrom, "Laptop Helpers," *Business Week*, June 16, 1997, p. 21.
30. Merritt Jones, quoted in "FYI," *Popular Science*, November 1995, pp. 88–89.
31. Jonathan Marshall, "TI Is Betting on DSP," *San Francisco Chronicle*, July 9, 1997, pp. B1, B4.
32. Michael Malone, "Microprocessor at 25," *San Jose Mercury News*, October 20, 1996, pp. 1E, 3E.
33. Robert D. Hof and Otis Port, "Silicon Dreamers vs. the PC," *Business Week*, May 13, 1996, pp. 78–80.
34. Stephen H. Wildstrom, "A Multimedia Power Surge," *Business Week*, December 30, 1996, p. 22.
35. Walter Mossberg, "MMX Has Much to Offer, But Less Than Hype Suggests," *Wall Street Journal*, February 13, 1997, p. B1.
36. Jennifer Tanaka, "A New Chip Off the Block," *Newsweek*, February 17, 1997, p. 70.
37. Dean Takahashi, "Intel's Chip Innovations Could Scramble PC Industry," *Wall Street Journal*, October 21, 1997, p. B6.
38. Laurence Zuckerman, "IBM to Make Smaller and Faster Chips," *New York Times*, September 22, 1997, pp. C1, C6.
39. Raju Narisetti, "IBM to Announce Computer-Chip Breakthrough," *Wall Street Journal*, September 22, 1997, p. B6.
40. Reuters, "Intel, HP Unveil New Computer Architecture," *San Francisco Chronicle*, October 15, 1997, pp. B1, B2.
41. Markoff, August 27, 1997.
42. Brent Schlender, "Killer Chip," *Fortune*, November 10, 1997, pp. 70–80.
43. Larry Armstrong, Otis Port, and Steven V. Brull, "GaAs Guzzlers on the Info Highway?" *Business Week*, August 19, 1996, pp. 78–79.
44. Bernard Meyerson, quoted in F.V. Frank Vizard, "Cheaper, Faster Chips," *Popular Science*, November 1997, p. 48.
45. Andrew Pollack, "A Japan Offer of Billion-Bit Chips for '98," *New York Times*, February 14, 1995, p. C4.
46. Molly Williams, Bloomberg Business News, "Developing the 1-Gigabit Chip," *San Francisco Chronicle*, October 26, 1995, pp. D1, D2.
47. George Johnson, "Giant Computer Virtually Conquers Space and Time," *New York Times*, Septmber 2, 1997, pp. B7, B10.
48. Gyan Bhanot, quoted in Johnson, 1997.
49. David Einstein, "Linking PCs Will Create Powerful Computer," *San Francisco Chronicle*, August 14, 1997, pp. D1, D4.
50. David Stipp, "Scientists Boost Superconductor Current Density," *Wall Street Journal*, April 19, 1995, p. B8.
51. Otis Port, "How to Soup Up Superconductors," *Business Week*, May 1, 1995, p. 128.
52. David P. Hamilton and Dean Takahashi, "Scientists Are Battling to Surmount Barriers in Microchip Advances," *Wall Street Journal*, December 10, 1996, pp. A1, A17.
53. Jane E. Allen, Associated Press, "New Computers May Use DNA Instead of Chips," *San Francisco Chronicle*, May 3, 1995, p. B2.
54. Steven Levy, "Computers Go Bio," *Newsweek*, May 1, 1995, p. 63.
55. Gina Kolata, "A Vat of DNA May Become Fast Computer of the Future," *New York Times*, April 11, 1995, pp. B5, B8.
56. Sharon Begley and Gregory Beals, "Computing Is in Their Genes," *Newsweek*, April 29, 1995, p. 5.
57. Vincent Kiernan, "DNA-Based Computers Could Race Past Supercomputers, Researchers Predict," *Chronicle of Higher Education*, November 28, 1997, pp. A23–A24.
58. Steve Mann, quoted in Judith Gaines, "MIT Graduate's Clothes Make for a Truly Personal Computer," *San Francisco Chronicle*, October 1, 1997, p. A7; reprinted from *Boston Globe*.
59. Stewart Alsop, "Computers in Your Clothes," *Fortune*, October 27, 1997, pp. 269–70.
60. Elizabeth Weise, "Try These Computers for Size," *San Francisco Chronicle*, August 22, 1996, p. B3.
61. John W. Verity and Paul C. Judge, "Making Computers Disappear," *Business Week*, June 24, 1996, pp. 118–19.
62. Mariette Dichristina and Suzanne Kantra Kirschner, "What's New," *Popular Science*, May 1996, p. 10.
63. Susan Gregory Thomas, "The Networked Family," *U.S. News & World Report*, December 1, 1997, pp. 66–80.
64. Stephen Manes, "New Stuff: Let the Buyer Be Aware," *New York Times*, April 22, 1997, p. B13.
65. G. Pascal Zachary, "Why We Can't Part with Those Vintage PCs," *Wall Street Journal*, July 2, 1997, pp. B1, B4.
66. Dan Gillmor, "Old Computer Will Mean a Lot to Those in Need," *San Jose Mercury News*, December 24, 1995, p. 1F.
67. Lawrence J. Magid, "Computer Users Can Do Their Part in Recycling," *San Jose Mercury News*, May 26, 1996, pp. 1F, 2F.
68. Kevin Maney, "Technology Moving Too Fast? Be Glad It's Not the 1840s," *USA Today*, January 30, 1997, p. 2B.
69. Stephen Ambrose, reported in Maney, January 30, 1997.
70. Robert C. Post, reported in Steve Lohr, "The Future Came Faster in the Old Days," *New York Times*, October 5, 1997, sec. 4, pp. 1, 4.
71. Paul Saffo, reported in Lohr, 1997.
72. Meg Greenfield, "Back to the Future," *Newsweek*, January 27, 1997, p. 96.
73. Jeffrey R. Young, "Invasion of the Laptops: More Colleges Adopt Mandatory Computing Programs," *Chronicle of Higher Education*, December 5, 1997, pp. A33–35.
74. Amy Grenier, quoted in Thomas J. DeLoughry, "Portable Computers, Light and Powerful, Gain Popularity on College Campuses," *Chronicle of Higher Education*, October 6, 1993, pp. A21, A24.
75. David E. Kalish, "Laptop Computers Go Mainstream," *San Francisco Chronicle*, April 21, 1997, p. C5.
76. Kalish, 1997.
77. "Hardtops," *Newsweek*, February 24, 1997, p. 10.
78. Todd Copilevitz, "Some Laptops Are Built RAM Tough," *San Francisco Examiner*, January 19, 1997, pp. D-5, D-6; reprinted from *Dallas Morning News*.
79. Stephen Wildstrom and Neil Gross, "Laptop Lowdown," *Business Week*, November 24, 1997, pp. 114–18.
80. Martin Reynolds, Dataquest Corp., cited in Lee Gomes and Evan Ramstad, "Bargain PCs Thrill Buyers, Worry Makers," *Wall Street Journal*, February 24, 1997, pp. B1, B8.
81. Copilevitz, 1997.
82. "Small Wonders," *Technology Buyer's Guide*, *Fortune*, Winter 1997, pp. 44–56.

83. Stratford Sherman, "The Pleasures of Light Laptops . . . ," *Fortune,* August 4, 1997, pp. 216–17.
84. Reported in "Small Wonders," 1997.
85. Ted C. Fishman, "Lapping the Field," *Worth,* September 1997, pp. 131–34.
86. David M. Deal, "Gadgets Solve Travel Problems," *USA Today,* June 17, 1997, p. 14E.
87. David A Kaplan, "Palm Reading," *Newsweek,* July 21, 1997, pp. 46–47.
88. "On the Go," *Technology Buyer's Guide, Fortune,* Winter 1997, pp. 64–72.
89. Wendy Taylor and Marty Jerome, "Handheld Computers Power Up," *San Francisco Chronicle,* November 27, 1997, p. E5.
90. "On the Road," *Technology Buyer's Guide, Fortune,* Winter 1998, pp. 96–106.
91. Paul C. Judge, "Small Is Dandy . . . Up to a Point," *Business Week,* November 24, 1997, p. 124.

Chapter 5

1. Heather Fisher, quoted in Carol Jouzaitis, "Step Right Up, and Pay Your Taxes and Tickets," *USA Today,* October 2, 1997, p. 4A.
2. Jouzaitis, 1997.
3. David M. Halbfinger, "Coming Soon: City Hall at Your Fingertips," *New York Times,* September 9, 1997, pp. A1, A16.
4. Connie Guglielmo, "Here Come the Super-ATMs," *Fortune,* October 14, 1996, pp. 232–34.
5. Betsy Wade, "E-Tickets Begin to Catch On," *New York Times,* August 10, 1997, sec. 4, p. 4.
6. Jouzaitis, 1997.
7. David Gelernter, quoted in Associated Press, "Bombing Victim Says He's Lucky to Be Alive," *San Francisco Chronicle,* January 28, 1994, p. A15.
8. Kevin Maney, "Computers Link with Chinese Language," *USA Today,* November 6, 1997, pp. 1B, 2B.
9. Don Norman, quoted in Cynthia Crossen, "Print Scrn, Num Lock and Other Mysteries of the Keyboard," *Wall Street Journal,* October 22, 1996, pp. B1, B8.
10. David R. Miller, "Imagining the Perfect Keyboard" [letter], *Wall Street Journal,* November 14, 1996, p. A23.
11. David Lieberman, "Do-it-all Box Could Start a Cable Revolution," *USA Today,* December 16, 1997, p. 6B.
12. David Bank, "TCI Uses Hi-Tech 'Layer Cake' to Ward Off Microsoft," *Wall Street Journal,* December 16, 1997, p. B4.
13. Leslie Cauley, "TCI, Others in Pact with NextLevel to Buy Digital-TV Set-Top Devices," *Wall Street Journal,* December 18, 1997, p. B10.
14. Stephen H. Wildstrom, "Eeeek! A Better Mouse," *Business Week,* December 1, 1997, p. 20.
15. Dean Takahashi, "New Mice Let You Shimmy and Shake Across Your Screen," *Wall Street Journal,* November 20, 1997, p. B1.
16. Amy Harmon, "The Computer Mouse: Where Art and Science Meet," *New York Times,* December 13, 1997, pp. A11, A12.
17. David Berquel, quoted in Mary Geraghty, "Pen-Based Computer Seen as Tool to Ease Burden of Note Taking," *Chronicle of Higher Education,* November 9, 1994, p. A22.
18. Jeff Pelline, "CellNet Sees a Big Business in Meter Reading," *San Francisco Chronicle,* September 28, 1995, pp. D1, D2.
19. Jonathan Marshall, "Requiem for Meter Readers?" *San Francisco Chronicle,* April 18, 1997, pp. C1, C4.
20. Ken Butcher, cited in Catherine Young and Willy Stern, "Maybe They Should Call Them 'Scammers,'" *Business Week,* January 16, 1995, pp. 32–33.
21. California Public Interest Research Group, "Scanners or Scammers," reported in Peter Sinton, "Checkout Mistakes Prevalent," *San Francisco Chronicle,* October 30, 1997, pp. C1, C2.
22. Claudia H. Deutsch, "There's Gold in Those Old Photos in the Attic," *New York Times,* June 30, 1997, p. C6.
23. George James, "Agents Raid Brand-Name Counterfeiters," *New York Times,* September 28, 1995, p. C5.
24. "Will Ben's New Look Stop Counterfeits?" *New York Times,* September 28, 1995, p. C5.
25. "Student Charged with Counterfeiting," *Chronicle of Higher Education,* December 12, 1997, p. A8.
26. Barbara Rudolph, "Some Like Them Hot," *Time,* November 14, 1994, p. 76.
27. "New Bill to Beat Counterfeiters," *San Francisco Chronicle,* September 28, 1995, p. A1.
28. Jonathan Rabinovitz, "Internet Becomes Day Care Centers' Big Mother," *San Francisco Chronicle,* December 16, 1997, p. A5; reprinted from New York Times.
29. Lawrence A. Armour, "The Big Picture, and How to Get It," *Fortune,* December 8, 1997, pp. 265–66.
30. Kim Komando, "The Komputer Klinic," *America West,* June 1997, p. 28.
31. Kevin Maney, "Kodak CEO Faces Critical Moment," *USA Today,* September 17, 1997, pp. 1B–2B.
32. Mike Langberg, "Kodak's Online Photograph Venture Bold but Still Fuzzy," *San Jose Mercury News,* September 7, 1997, pp. 1E, 2E.
33. Carol J. Castaneda, "Red-Light Runners Caught in the Act," *USA Today,* June 17, 1997, p. 4A.
34. Faith Bremner, "Sensors Make Snowy Roads Safer," *Reno Gazette-Journal,* December 6, 1996, pp. 1A, 10A.
35. Kate Murphy, "Get Along Little Dogie #384-591E," *New York Times,* July 21, 1997, p. C4.
36. Peter Sinton, "The E-Nose Knows," *San Francisco Chronicle,* December 2, 1997, pp. C1, C14.
37. Del Jones, "Mobil Speedpass Makes Gas a Point-at-Pump Purchase," *USA Today,* February 20, 1997, p. 1B.
38. Skip Wollenberg, "Mobil Gas Gadget Feeds Your Need for Speed," *San Francisco Chronicle,* February 20, 1997, p. B3.
39. Matthew L. Wald, "E-Z Pass to Cross George Washington Monday, Heading to Washington," *New York Times,* July 25, 1997, p. A13.
40. Laurie J. Flynn, "High-Technology Dog Tags for More Than Just Dogs," *New York Times,* August 12, 1996, p. C5.
41. Michael Lawrence Ellis III, reported in Joe Sharkey, "In Plainspoken English, America's Still Whopperjawed," *New York Times,* November 16, 1997, sec. 3, p. 7.
42. Speech Recognition Facts, www.iglou.com/vrsky/spfacts.htm, March 1998.
43. Tom Abate, "New Chip Verifies Fingerprints," *San Francisco Chronicle,* May 22, 1997, p. B3.
44. "Fingers Hold Key to Your ID—Most of the Time" [editorial], *USA Today,* July 22, 1997, p. 10A.
45. Saul Hansell, "Is This an Honest Face?" *New York Times,* August 20, 1997, pp. C1, C6.
46. Peter Sinton, "Soon, ATM Users Could Get Cash with Just a Glance," *San Francisco Chronicle,* April 11, 1997, pp. A1, A17.
47. Gordon Faircough, "Futuristic Identification Systems Hit the Here and Now," *Wall Street Journal,* December 1, 1997, p. B2.
48. Bill Hendrix, "New Theory About Music: It Tunes Brain for Learning," *San Jose Mercury News,* June 25, 1996, pp. 1E, 11E.
49. Jeffrey Biegel, quoted in Ira Rosenblum, "Live Video Is Joining Sound on the Web," *New York Times,* July 8, 1997, p. B5.
50. "Flat-Panel TV," *Popular Science,* December 1997, p. 34.
51. Paul M. Eng, "A Cyberscreen So Tiny It Fits on a Dime," *Business Week,* April 21, 1997, p. 126C.
52. Chris O'Malley, "Large, Lush LCDs," *Popular Science,* March 1997, p. 31.
53. *Microsoft Press Computer Dictionary,* 3rd ed. (Redmond, WA: *Microsoft Press,* 1977), p. 257.
54. Bruce Brown, "Portable Printers: Fewer Trade-offs," *PC Magazine,* August 1995, pp. 215–25.
55. Jim Seymour, *On the Road: The Portable Computing Bible* (New York: Brady, 1992), pp. 184–85.
56. David L. Wheeler, "Recreating the Human Voice," *Chronicle of Higher Education,* January 19, 1996, pp. A8–A9.
57. Sreenath Sreenivasan, "Blind Users Add Access on the Web," *New York Times,* December 2, 1996, p. C7.
58. Diana Berti, quoted in Timothy L. O'Brien, "Aided by Computers, Many of the Disabled Form Own Businesses," *Wall Street Journal,* October 8, 1993, pp. A1, A5.
59. Jonathan Marshall, "Videophone Finds Niche," *San Francisco Chronicle,* November 6, 1997, pp. D1, D6.
60. Joel Brinkley, "Getting the Picture," *New York Times,* November 24, 1997, pp. C1, C12.
61. Jon Healey, "Dawn of Digital Television Arrives," *San Jose Mercury News,* November 15, 1998, pp. 1A, 26A.
62. Osvaldo Arias, quoted in N. R. Kleinfield, "Stepping into Computer, Disabled Savor Freedom," *New York Times,* March 12, 1995, sec. 1, p. 17.
63. Peter Sinton, "ATMs Wherever You Go," *San Francisco Chronicle,* November 8, 1997, pp. D1, D2.
64. Randall E. Stross, "A Cannes for Nerds," *U.S. News & World Report,* February 24, 1997, pp. 51–52.
65. Peter Sinton, "Smart Card Technology Heads for the Internet," *San Francisco Chronicle,* May 16, 1997, pp. B1, B2.
66. Frank Vizard, "Home Withdrawals," *Popular Science,* January 1998, p. 38.
67. Peter Sinton, "7-Eleven Machines Check IDs," *San Francisco Chronicle,* December 9, 1997, pp. C1, C7.
68. Janet Guyon, "Smart Plastic," *Fortune,* October 13, 1997, p. 56.
69. George Church, "Leave Your Cash at Home," *Time,* October 13, 1997, p. 64.
70. James Kim, "'Smart' Card Blitz Begins in Manhattan," *USA Today,* October 6, 1997, p. 1B.
71. "Food Stamps Are Replaced by Plastic Cards in Texas," *New York Times,* November 27, 1995, p. A14.
72. Carl Pascarella, quoted in Marc Levinson and Adam Rogers, "The End of Money?" *Newsweek,* October 30, 1995, pp. 62–65.
73. Steve Kaye, quoted in Elisa Williams, "Making Your Home Work," *San Francisco Examiner,* October 8, 1995, pp. B-1, B-8; reprinted from *Orange County Register.*
74. Phat X. Chiem, "Image Is Everything with Digital X-Rays," *San Francisco Chronicle,* August 22, 1997, pp. B1, B2.
75. Philip J. Hilts, "Digital X-Ray Systems to Replace Old Films with Electronic Images," *New York Times,* September 30, 1997, p. B15.
76. Peter Sinton, "Smart Card Technology Heads for the Internet," *San Francisco Chronicle,* May 16, 1997, pp. B1, B2.
77. "Shopping on Your Dashboard," *The Futurist,* January-February 1994, p. 8.
78. Steven Findlay, "A 'Smart Card' for Medical Bills?" *USA Today,* December 10, 1997, p. 1B.
79. Neil Gross, "Power-Assisted Squirming," *Business Week,* September 29, 1997, p. 106.
80. Otis Port, "A Needle with Supersharp Sensors," *Business Week,* July 22, 1996, p. 59.
81. Steve Sternberg, "Device that Helps Quadriplegics Use Hands Wins Approval," *USA Today,* August 19, 1997, p. 1A.
82. Daniel Goleman, "Laugh and Your Computer Will Laugh with You, Someday," *New York Times,* January 7, 1997, pp. B9, B14.
83. Malcolm W. Browne, "How Brain Waves Can Fly a Plane," *New York Times,* March 7, 1995, pp. B1, B10.
84. Don Clark, "Mind Games: Soon You'll Be Zapping Bad Guys Without Lifting a Finger," *Wall Street Journal,* June 16, 1995, p. B12.
85. Tom Abate, "Marin Investor Bets on an Impulse," *San Francisco Examiner,* July 2, 1995, pp. B-1, B-7.
86. Malcolm W. Browne, "Neuron Talks to Chip and Chip to Nerve Cell," *New York Times,* August 22, 1995, pp. B1, B9.
87. Mary Madison, "Mind Control for Computers," *San Francisco Chronicle,* December 2, 1995, pp. A1, A13.
88. Jennifer Benjamin, quoted in Robert D. Hershey Jr., "Graduating with Credit Problems," *New York Times,* November 10, 1996, sec. p. 15.
89. Joshua Wolf Shenk, "In Debt All the Way Up to Their Nose Rings," *U.S. News & World Report,* June 9, 1997, pp. 38–39.
90. Roper College Track Financial Services Study, reported in Judy Lowe, "Credit Cards 101: Helping Students Manage Easy Plastic," *Christian Science Monitor,* September 30, 1996, p. 13.
91. Survey by Tahira K. Hira, Iowa State University, reported in Hershey, 1996.
92. Timothy L. O'Brien, "Giving Credit Where Debt Is Due," *New York Times,* December 14, 1997, sec. 4, p. 14.
93. J. Foren, "College Students Piling on Credit Card Debt," *San Francisco Examiner,* December 1, 1991, p. E-9.
94. L. Kutner, "College Students with Big Credit Card Bills May Be Learning an Economics Lesson the Hard Way," *New York Times,* August 19, 1993, p. B4.
95. Halimah Abdullah, "Easy Credit Lures Students Down Path to Onerous Debt," *New York Times,* August 24, 1997, sec. 1, p. 18.
96. Shenk, 1997.
97. Suzanne Boas, Consumer Credit Counseling Service, Atlanta, cited in Lowe, 1996.
98. "Check Cards: Should You Replace Your ATM Card?" *Consumer Reports,* October 1997, pp. 68–69.
99. David J. Morrow, "Handy? Surely, but Debit Card Has Risks, Too," *New York Times,* July 13, 1997, sec. 1, pp. 1, 9.
100. Shenk, 1997.
101. "Fidler to Kent State," *Quill,* December 1996, p. 8.
102. Kevin Maney, "High-Tech Tablets: Next Step for Newspapers?" *USA Today,* June 5, 1997, p. 6B.
103. Paul M. Eng, "Web Surfing That'll Give You Whiplash," *Business Week,* April 7, 1997, p. 136C.
104. Phillip Robinson, "Home Theater No Longer Just for the Rich and Famous," *San Jose Mercury News,* November 30, 1997, p. 9S.
105. Diana Hembree and Ricardo Sandoval, "The Lady and the Dragon," *Columbia Journalism Review,* August 1991, pp. 44–45.
106. Bureau of Labor Statistics, cited in Ellen Neuborne, "Workers in Pain; Employers Up in Arms," *USA Today,* January 9, 1997, pp. 1B, 2B.
107. Edward Felsenthal, "An Epidemic or a Fad? The Debate Heats Up Over Repetitive Stress," *Wall Street Journal,* July 14, 1994, pp. A1, A4.
108. Felsenthal, 1994.
109. Ilana DeBare, "Eyestrain a Bulging Problem," *San Francisco Chronicle,* July 14, 1997, pp. B1, B2.
110. Jane E. Brody, "Reading a Computer Screen Is Different from Reading a Book, and Has Different Effects on the Eyes," *New York Times,* August 7, 1997, p. B6.
111. "New Suspect in Alzheimer's Risk," *San Jose Mercury News,* July 31, 1994, p. 23A; reprinted from *Los Angeles Times.*
112. "Mobile Phones Safe, but They Heat Brain," *San Francisco Chronicle,* May 23, 1997, p. B1.
113. Study by Michael Repacholia and others for World Health Organization, reported in Jonathan Marshall, "Cell Phones Linked to Cancer in Mice," *San Francisco Chronicle,* May 9, 1997, pp. A1, A11; reported in *Radiation Research.*
114. Marty Jerome, "Boot Up or Die," *PC Computing,* April 1998, pp. 172–86.

Chapter 6

1. Keith McCurdy, "'Killer Apps' of the '90s," *San Francisco Examiner,* January 19, 1997, pp. D-5, D-6.
2. Lawrence M. Fisher, "IBM Plans to Announce Leap in Disk-Drive Capacity," *New York Times,* December 30, 1997, p. C2.
3. Lawrence J. Magid, "It's Easy to Add Disk Space or a Little Zip," *Los Angeles Times,* April 27, 1998, p. D4.

4. "IBM Unveils a Disk That Can Hold Twice the Current Capacity," *Wall Street Journal,* December 30, 1997, p. B2.
5. Gina Smith, "On Sound, Screens and SCSI vs. EIDE," *San Francisco Examiner,* May 26, 1996, pp. B-5, B-6.
6. Craig Clarke, quoted in Leslie Miller, "Zippy New Disk Drive Pumps Up Storage," *USA Today,* May 4, 1995, p. 6D.
7. Stephen H. Wildstrom, "Cheap and Easy Storage Space," *Business Week,* May 22, 1995, p. 26.
8. Evan I. Schwartz, "CD-ROM: A Mass Medium at Last," *Business Week,* July 19, 1993, pp. 82–83.
9. Joel Smith, "108 Years in a Box," *San Jose Mercury News,* October 19, 1997, p. 3F; reprinted from *Detroit News.*
10. William M. Bulkeley, "Publishers Deliver Reams of Data on CDs," *Wall Street Journal,* February 22, 1993, p. B6.
11. Special advertising section, "The New Presentation Technology: How Business Puts It to Work," *Business Week,* November 10, 1997, pp. 23–34.
12. Henry M. Levin, quoted in Kate Murphy, "Pitfalls vs. Promise in Training by CD-ROM," *New York Times,* May 6, 1996, p. C3.
13. Bridget O'Connor, quoted in Murphy, 1996.
14. Richard C. Hsu and William E. Mitchell, "Books Have Endured for a Reason . . . ," *New York Times,* May 25, 1997, sec. 3, p. 12.
15. Richard Clark, quoted in Murphy, 1996.
16. Jack McGarvey, ". . . But Computers Are Clearly the Future," *New York Times,* May 25, 1997, sec. 3, p. 12.
17. Edward Baig, "Be Happy, Film Freaks," *Business Week,* May 26, 1997, pp. 172–73.
18. Dennis Normile, "Get Set for the Super Disc," *Popular Science,* February 1996, pp. 55–58.
19. "DVD Stands for DiVideD," *Byte,* January 1998, p. 77.
20. David Lieberman, "Some Studios Fear Films Will Be Illegally Copied," *USA Today,* September 2, 1997, pp. 1B, 2B.
21. Bruce Orwall, "Disney Overcomes DVD Reservations, Will Issue Titles on Disk by Year End," *Wall Street Journal,* September 5, 1997, p. B4.
22. Bruce Orwall, "A 'Disposable' Video Disk Threatens to Undercut Nascent Market for DVDs," *Wall Street Journal,* September 9, 1997, p. B8.
23. Leonard Wiener, "New TVs and DVD: Wait or Buy Now?" *U.S. News & World Report,* November 24, 1997, p. 96.
24. Paul M. Eng, Robert D. Hof, and Hiromi Uchida, "It's a Whole New Game: The Hards vs. the Cards," *Business Week,* June 8, 1992, pp. 101–103.
25. Janet L. Fix, "Deal Secures First Data's Credit Lead," *USA Today,* September 18, 1995, pp. 1B, 2B.
26. Stephen Manes, "Time and Technology Threaten Digital Archives . . . ," *New York Times,* April 7, 1998, p. B15.
27. Mike Snider, "Obsolescence: The No. 1 Built-In Feature," *USA Today,* October 29, 1997, p. 6D.
28. Laura Tangley, "Whoops, There Goes Another CD-ROM," *U.S. News & World Report,* February 16, 1998, pp. 67–68.
29. Marcia Stepanek, "From Digits to Dust," *Business Week,* April 20, 1998, pp. 128-130.
30. Manes, 1998.
31. Stephen Manes, ". . . But with Luck and Diligence, Treasure-Troves of Data Can Be Preserved," *New York Times,* April 7, 1998, p. B15.
32. Dan Gillmor, "A Computer Whose Time Has Come—Internet Appliance," *San Jose Mercury News,* January 28, 1996, p. 1F.
33. Lawrence Magid, "Need Backup? Try the Internet," *San Francisco Examiner,* July 27, 1997, p. B-7.
34. Phillip Robinson, "Online Data Backup Has Its Pros, Cons," *San Jose Mercury News,* August 31, 1997, pp. 1F, 4F.
35. "Gargantua's 'Lossless' Compression," *The Australian,* March 22, 1994, p. 32; reprinted from *The Economist.*
36. Peter Coy, "Invasion of the Data Shrinkers," *Business Week,* February 14, 1994, pp. 115–16.
37. "Gargantua's 'Lossless' Compression," 1994.
38. Coy, 1994.
39. "IBM Unveils a Disk . . . ," 1997.
40. Fisher, 1997.
41. Charles Petit, "Tomorrow's Disk to Store Even More," *San Francisco Chronicle,* April 18, 1995, pp. B1, B4.
42. Tom Dellecave Jr., "Jukeboxes: High, Low Notes," *InformationWeek,* March 6, 1995, pp. 40–44.
43. Mark Maremont, "A Magnetic Mug Shot on Your Credit Card?" *Business Week,* April 24, 1995, p. 58.
44. Otis Port, "Sifting Through Data with a Neural Net," *Business Week,* October 30, 1995, p. 70.
45. Matthew May, "Digital Videos on the Way," *The Times* (London), April 15, 1994, p. 34.
46. Richard Brandt, "Can Larry Beat Bill?" *Business Week,* May 15, 1995, pp. 88–96.
47. Larry Armstrong, "Holographic Memories Get Easier to Write," *Business Week,* October 13, 1997, p. 126.
48. John Markoff, "Tiny Magnets May Form Basis for Computing Breakthrough," *New York Times,* January 27, 1997, p. C2.
49. Robert Birge, quoted in Amai Kumar Naj, "Researchers Isolate Bacteria Protein That Can Store Data in 3 Dimensions," *Wall Street Journal,* September 4, 1991, p. B4.
50. Peter H. Lewis, "Besides Storing 1,000 Words, Why Not Store a Picture Too?" *New York Times,* October 11, 1992, sec. 3, p. 8.
51. William Safire, "Art vs. Artifice," *New York Times,* January 3, 1994, p. A11.
52. Hans Fantel, "Sinatra's 'Duets,' Music Recording or Wizardry?" *New York Times,* January 1, 1994, p. 13.
53. Cover, *Newsweek,* June 27, 1994.
54. Cover, *Time,* June 27, 1994.
55. Jonathan Alter, "When Photographs Lie," *Newsweek,* July 30, 1990, pp. 44–45.
56. Fred Ritchin, quoted in Alter, 1990.
57. Bill Allen, quoted in Michael Satchell, "Antartica and the Polar Bear," *U.S. News & World Report,* January 12, 1998, pp. 48–51.
58. Robert Zemeckis, cited in Laurence Hooper, "Digital Hollywood: How Computers Are Remaking Movie Making," *Rolling Stone,* August 11, 1994, pp. 55–58, 75.
59. Woody Hochswender, "When Seeing Cannot Be Believing," *New York Times,* June 23, 1992, pp. B1, B3.
60. Kathleen O'Toole, "High-Tech TVs, Computers Blur Line Between Artificial, Real," *Stanford Observer,* November-December 1992, p. 8.
61. Michael Rogers, quoted in Tod Oppenheimer, "*Newsweek*'s Voyage Through Cyberspace," *Columbia Journalism Review,* December 1993, pp. 34–37.
62. Bronwyn Fryer, "Smile! You're on My Computer," *Business Week Enterprise,* June 9, 1997, pp. ENT 20–23.
63. Phillip Robinson, "Organizing and Editing Digital Photos Can Be a Snap," *San Jose Mercury News,* November 23, 1997, pp. 1F, 5F.
64. Richard Folkers, "Pixelated Photography," *U.S. News & World Report,* May 12, 1997, pp. 77–78.
65. Mike Langberg, "HP a Digital Photography Hot-Shot," *San Jose Mercury News,* May 25, 1997, p.4E.
66. "Instant Images," *Fortune, Technology Buyer's Guide,* Winter 1997, pp. 185–87.
67. Phillip Robinson, "What to Shoot for in a Digital Camera," *San Jose Mercury News,* November 16, 1997, pp. 1F, 4F.
68. Susan Gregory Thomas, "A Photo Lab on Your Desk," *U.S. News & World Report,* November 25, 1996, pp. 104–106.
69. Stewart Alsop, "Digital Photography Is the Next Big Thing," *Fortune,* August 4, 1997, pp. 220–21.
70. Mike Langberg, "Kodak's Online Photograph Venture Bold but Still Fuzzy," *San Jose Mercury News,* September 7, 1997, pp. 1E, 2E.

Chapter 7

1. Andy Reinhardt, Peter Elstrom, and Paul Judge, "Zooming Down the I-Way," *Business Week,* April 7, 1997, pp. 76–87.
2. Virginia Brooks, quoted in Reinhardt et al., 1997.
3. Rebecca Quick, "The Crusaders," *Wall Street Journal,* December 8, 1997, pp. R6, R8.
4. Jon Swartz, "SFO's Latest Arrival—the Internet," *San Francisco Chronicle,* February 15, 1997, pp. A1, A13.
5. Jonathan Marshall, "Silicon Valley Complex Offers High-Speed Link," *San Francisco Chronicle,* September 17, 1997, pp. A1, A13.
6. David B. Whittle, *Cyberspace: The Human Dimension* (New York: W.H. Freeman, 1997).
7. "Living Online," *The Futurist,* July-August 1997, p. 54.
8. "Send that Fax Via Internet," *Business Week,* October 27, 1997, p. 18E2.
9. Jonathan Marshall, "New Device Eliminates Fax Machine," *San Francisco Chronicle,* June 24, 1997, p. C4.
10. N. R. Kleinfield, "For Homeless, Free Voice Mail," *New York Times,* January 30, 1995, p. A12.
11. Mark Dillard, quoted in Marcia Vickers, "Don't Touch That Dial: Why Should I Hire You?" *New York Times,* April 13, 1997, sec. 3, p. 11.
12. Cowles/Simba, reported in "Online," *Popular Science,* March 1998, p. 29.
13. Jesse Kornbluth, "The Truth About the Web," *San Francisco Chronicle,* January 23, 1996, p. C4.
14. David Einstein, "Internet Surfing Is a Snap," *San Francisco Chronicle,* September 23, 1997, p. C4.
15. *The Digital Economy,* U.S. Department of Commerce, Washington, DC, April 16, 1998.
16. U.S. Department of Commerce.
17. Nielsen Media Research and CommerceNet survey, reported in Leslie Miller, "1 in 4 Now Using the Net," *USA Today,* December 11, 1997, p. 1A.
18. Dataquest Inc. study, reported in Jon Swartz, "A Rush to Plug PCs into the Internet," *San Francisco Chronicle,* August 21, 1997, pp. A1, A15.
19. David Landis, "Exploring the Online Universe," *USA Today,* October 7, 1993, p. 4D.
20. Arthur M. Louis, "What You Should Look for When Shopping on an ISP," *San Francisco Chronicle,* April 1, 1997, p. D11.
21. Jared Sandberg, "What Do They Do On-line?" *Wall Street Journal,* December 9, 1996, p. R8.
22. David Einstein, "What They Want Is E-mail," *San Francisco Chronicle,* February 20, 1996, pp. B1, B6.
23. Peter H. Lewis, "The Good, the Bad and the Truly Ugly Faces of Electronic Mail," *New York Times,* September 6, 1994, p. B7.
24. Robert Rossney, "E-Mail's Best Asset—Time to Think," *San Francisco Chronicle,* October 5, 1995, p. E7.
25. Walter S. Mossberg, "With E-Mail, You'll Get the Message Without the Hang-ups," *Wall Street Journal,* April 20, 1995, p. B1.
26. Lewis, 1994.
27. Michelle Quinn, "E-Mail Is Popular—but Far from Perfect," *San Francisco Chronicle,* April 21, 1995, pp. B1, B8.
28. David Cay Johnson, "Not So Fast: E-Mail Sometimes Slows to a Crawl," *New York Times,* January 7, 1996, sec. 4, p. 5.
29. David W. De Long, quoted in Judith H. Dobrzynski, "@ Wit's End: Coping with E-Mail Overload," *New York Times,* April 28, 1996, sec. 4, p. 2.
30. Elizabeth P. Crowe, "The News on Usenet," *Bay Area Computer Currents,* August 8–21, 1995, pp. 94–95.
31. Anibal Torres, quoted in Associated Press, "Internet Helps Trace Missing Baby," *San Francisco Chronicle,* November 20, 1997, p. A17, A21.
32. "On-Line Stores," *Internet World,* October 1997.
33. Walter Mossberg, "'Push' Technology Sometimes Pushes News You Can't Use," *Wall Street Journal,* March 27, 1997, p. B1.
34. Amy Cortese, "A Way Out of the Web Maze," *Business Week,* February 24, 1997, pp. 94–104.
35. Susan Gregory Thomas, "The News Pushers," *U.S. News & World Report,* March 17, 1997, p. 76.
36. Edward Rothstein, "Making the Internet Come to You, Through 'Push' Technology," *New York Times,* January 20, 1997, p. C5.
37. J. D. Lasica, "When Push Comes to News," *American Journalism Review,* May 1997, pp. 32–39.
38. Vin Crosbie, quoted in Lasica, 1997.
39. Walter Mossberg, "Average Home Users May Find Push Service Slow and Distracting," *Wall Street Journal,* October 16, 1997, p. B1.
40. David Bank, "New Web Browsers Play Down TV-Channel Approach," *Wall Street Journal,* September 30, 1997, pp. B1, B12.
41. Amy Cortese, "Here Comes the Intranet," *Business Week,* February 26, 1996, pp. 76–84.
42. Thomas E. Weber, "The Web's Little Secret: It's Staler Than You Think," *Wall Street Journal,* May 22, 1997, p. B6.
43. Julia Angwin, "The Net's Latest Catch," *San Francisco Chronicle,* February 4, 1997, pp. C1, C2.
44. Michael J. Himowitz, "Web Phone Calls Made Easy," *Fortune,* November 10, 1997, p. 230.
45. Jonathan Marshall, "Internet Saves Callers $$$," *San Francisco Chronicle,* August 26, 1997, p. C4.
46. Steve Rosenbush, "Internet Could Revolutionize Phone Service," *USA Today,* February 10, 1998, pp. 1B, 2B.
47. Eric J. Savitz, "'Net Threat," *Barron's,* October 13, 1997, pp. 37–43.
48. David Einstein, "Tuning in to the Dead via the Web," *San Francisco Chronicle,* February 10, 1998, p. C3.
49. David Einstein, "New Era Dawns as Radio Comes to the Internet," *San Francisco Chronicle,* May 18, 1995, pp. B1, B2.
50. Thomas E. Weber, "Will Video Be an Internet Star?" *Wall Street Journal,* July 17, 1997, p. B6.
51. Joan E. Rigon, "Coming Soon to the Internet: Tools to Add Glitz to the Web's Offerings," *Wall Street Journal,* August 16, 1995, pp. B1, B2.
52. Dean Takahashi, "Microsoft, Netscape Agree on Standard for 3-D on Internet," *Wall Street Journal,* August 4, 1997, p. B5.
53. Frank Vizard, "Touch the Internet," *Popular Science,* September 1997, p. 62.
54. John Seabrook, "My First Flame," *The New Yorker,* June 6, 1994, pp. 70–79.
55. Virgina Shea, quoted in Ramon G. McLeaod, "netiquette-Cyberspace's Cryptic Social Code," *San Francisco Chronicle,* March 6, 1996, pp. A1, A10.
56. Yahoo!, cited in Del Jones, "Cyber-porn Poses Workplace Threat," *USA Today,* November 27, 1995, p. 1B.
57. Lawrence J. Magid, "Be Wary, Stay Safe in the On-line World," *San Jose Mercury News,* May 15, 1994, p. 1F.
58. Peter H. Lewis, "Limiting a Medium without Boundaries," *New York Times,* January 15, 1996, pp. C1, C4.
59. John M. Broder, "Making America Safe for Electronic Commerce," *New York Times,* June 22, 1997, sec. 4, p. 4.
60. Margaret Mannix and Susan Gregory Thomas, "Exposed Online," *U.S. News & World Report,* June 23, 1997, pp. 59-61.
61. Noah Matthews, "Shareware," *San Jose Mercury News,* October 12, 1997, p. 4F.
62. David Post, quoted in Rebecca Quick, "Don't Expect Your Secrets to Get Kept on the Internet," *Wall Street Journal,* February 6, 1998, p. B5.
63. Amitai Etzioni, "Some Privacy, Please, for E-Mail," *New York Times,* November 23, 1997, sec. 3, p. 12.
64. Gregory L. Vistica and Evan Thomas, "Backlash in the Ranks," *Newsweek,* April 20, 1998, p. 27.
65. Survey by Equifax and Louis Harris & Associates, cited in Bruce Horovitz, "80% Fear Loss of Privacy to Computers," *USA Today,* October 31, 1995, p. 1A.
66. Kurt Andersen, "The Age of Unreason," *The New Yorker,* February 3, 1997, pp. 40–43.
67. Peter Applebome, "On the Internet, Term Papers Are Hot Items," *New York Times,* June 8, 1997, sec. 1, pp. 1, 20.
68. Lee De Cesare, "Virtual Term Papers" [letters], *New York Times,* June 10, 1997, p. A20.

69. Bruce Leland, quoted in Applebome, 1997.
70. Eugene Dwyer, "Virtual Term Papers" [letters], *New York Times*, June 10, 1997, p. A20.
71. Mitchell Zimmerman, "How to Track Down Collegiate Cyber-Cheaters" [letters], *New York Times*, June 15, 1997, sec. 4, p. 14.
72. William L. Rukeyser, "How to Track Down Collegiate Cyber-Cheaters" [letters], *New York Times*, June 15, 1997, sec. 4, p. 14.
73. David Rothenberg, "How the Web Destroys the Quality of Students' Research Papers," *Chronicle of Higher Education*, August 15, 1997, p. A44.
74. Brian Hecht, "Net Loss," *The New Republic*, February 17, 1997, pp. 15–18.

Chapter 8

1. Anthony Ramirez, "Why Phone Numbers Don't Add Up," *New York Times*, August 10, 1997, sec. 4, p. 4.
2. Patrick J. Lyons, "The Trauma of the 90's: Adding New Area Codes," *New York Times*, March 10, 1997, p. C5.
3. The Wahlstrom Report, Stamford, CN, reported in "More Area Codes Needed," *The Futurist*, September-October 1997, p. 5.
4. Bruce Bennett, quoted in Jonathan Marshall, "Ready for Longer Phone Numbers?" *San Francisco Chronicle*, July 26, 1997, pp. D1, D2.
5. Marshall, 1997.
6. Alvin Toffler, quoted in Marianne Roberts, "Computers Replace Commuters," *PC Novice*, September 1992, p. 27.
7. Scott Bowles, "Sharing a Ride a Luxury to Some," *USA Today*, January 29, 1998, pp. 1A, 2A.
8. FIND/SVP survey, reported in Patricia Commins, "Telecommuting Accelerates with Tight Job Market," *San Jose Mercury News*, October 12, 1997, pp. 1PC, 2PC.
9. June Langhoff, "Telecommuting Makes Sense (and Cents)," *San Francisco Examiner*, October 19, 1997, pp. D-5, D-7.
10. Carolyn McCann, quoted in Mary Beth Marklein, "Telecommuters Gain Momentum," *USA Today*, June 18, 1996, p. 6E.
11. Alison L. Sprout, "Moving Into the Virtual Office," *Fortune*, May 2, 1994, p. 103.
12. Leon Jaroff, "Age of the Road Warrior," *Time*, Spring 1995, pp. 38–40.
13. James J. Mitchell, "Office Sharing," *San Jose Mercury News*, April 20, 1997, pp. 1D, 2D.
14. Alice Bredin, "Summertime, and Working in a Home Office Isn't Easy," *New York Times*, June 9, 1996, sec. 3, p. 10.
15. Melanie Warner, "Working at Home—the Right Way to Be a Star in Your Bunny Slippers," *Fortune*, March 3, 1997, pp. 165–66.
16. Jim Koerlin, "Even BART Can't Make Me Telecommute," *San Francisco Examiner*, September 21, 1997, pp. B-9, B-10.
17. Catherine Rossbach, quoted in Susan J. Wells, "Stay-Home Workers, Speed Bumps on the Telecommute," *New York Times*, August 17, 1997, sec. 3, pp. 1, 14.
18. Odette Pollar, "Home May Be Where the Business Is," *San Francisco Chronicle*, September 28, 1997, p. J3.
19. Jan Boyd and Stan Bunger, "The Show Must Go On—Gates Interview Will Air," *San Francisco Chronicle*, January 29, 1998, p. E3.
20. Stuart Weiss, "Will Working at Home Work for You?" *Business Week*, August 19, 1996, p. 84E8.
21. Laurie M. Grossman, "Truck Cabs Turn into Mobile Offices as Drivers Take on White-Collar Tasks," *Wall Street Journal*, August 3, 1993, pp. B1, B5.
22. Bob Spoer, quoted in Marilyn Lewis, "Tethered to Work," *San Jose Mercury News*, October 1, 1995, pp. 1A, 18A.
23. Jacob M. Schlesinger, "Get Smart," *Wall Street Journal*, October 21, 1991, p. R18.
24. Peter Hart, quoted in Lewis, 1995.
25. Walter S. Mossberg, "Attempts to Speed Up Modem Connections Are Off to a Slow Start," *Wall Street Journal*, March 13, 1997, p. B1.
26. Erik Nachbar, quoted in Dean Takahashi, "In Playing Games on the Internet, Connections Count," *Wall Street Journal*, September 11, 1997, pp. B1, B8.
27. John Marshall, "New ISDN Has E-mail, Voice, Data," *San Francisco Chronicle*, June 14, 1997, pp. D1, D2.
28. Leslie Cauley, "Baby Bells Rediscover Fast ISDN Service," *Wall Street Journal*, January 22, 1996, p. B5.
29. Kevin Maney, "Moving to Fast Lanes on the Net," *USA Today*, October 31, 1996, p. 1B.
30. Gregg Keizer, "Screaming for Bandwidth," *Computerlife*, January 1997, pp. 58–64.
31. Jonathan Marshall, "Reducing the World Wide Wait," *San Francisco Chronicle*, June 4, 1997, pp. D1, D4.
32. Andrew J. Kessler, "Cancel That ISDN Order," *Forbes*, January 26, 1998, p. 88.
33. "Just How Fast?" *Time*, September 23, 1996, p. 55.
34. Grant Balkema, quoted in Mark Robichaux, "Cable Modems Are Tested and Found to Be Addictive," *Wall Street Journal*, December 27, 1995, pp. 13, 17.
35. Lucien Rhodes, "The Race for More Bandwidth," *Wired*, January 1996, pp. 140–145.
36. Forrester Research Inc., cited in Peter Coy, "The Big Daddy of Data Haulers?" *Business Week*, January 29, 1996, pp. 74–76.
37. Walter Mossberg, "New Satellite Network for Internet Access Isn't Up to Speed," *Wall Street Journal*, September 18, 1997, p. B1.
38. Kevin Maney, "'Megahertz' Remains a Mega-Mystery to Most," *USA Today*, February 13, 1997, p. 4B.
39. Susan Horner, quoted in James Brooke, "Telemarketing Finds a Ready Labor Market in Hard-Pressed North Dakota," *New York Times*, February 3, 1997, p. A8.
40. Hank Kahrs, quoted in Evan Ramstad, "Works in Progress," *Wall Street Journal*, September 11, 1997, pp. R6, R8.
41. Phillip Robinson, "Finding Yourself: Global Positioning Technology Helps Pinpoint Your Location," *San Jose Mercury News*, June 1, 1997, pp. 1E, 4E.
42. James Gleick, "Lost in Space," *New York Times Magazine*, October 26, 1997, pp. 24, 26.
43. Ross Kerber and Dean Takahashi, "Location, Location, Location," *Wall Street Journal*, September 11, 1997, p. R18.
44. Carolyn Nielson, "GPS Ready to Take Off," *San Francisco Examiner*, June 4, 1995, pp. B-5, B-6.
45. David Perlman, "Satellite Network Captures Volcano Drama in Hawaii," *San Francisco Chronicle*, February 17, 1997, p. A4.
46. Anthony Ramirez, "Cheap Beeps: Across Nation, Electronic Pagers Proliferate," *New York Times*, July 19, 1993, pp. A1, C2.
47. Oscar Suris, "Still One Way," *Wall Street Journal*, September 11, 1997, pp. R21, R25.
48. Ramstad, 1997.
49. Ramstad, 1997.
50. Robert Berner, "The Missing Link," *Wall Street Journal*, September 11, 1997, pp. R23, R25.
51. Eric Schine, Peter Elstrom, Amy Barrett et al., "The Satellite Biz Blasts Off," *Business Week*, January 27, 1997, pp. 62–70.
52. G. Christian Hill, "The Spoils of War," *Wall Street Journal*, September 11, 1997, pp. R1, R4.
53. Hill, 1997.
54. C. Michael Armstrong, quoted in Jeff Cole, "In New Space Race, Companies Are Seeking Dollars from Heaven," *Wall Street Journal*, October 10, 1995, pp. A1, A10.
55. Rebecca Quick, "Talk Isn't Cheap," *Wall Street Journal*, September 11, 1997, p. R10.
56. Mike Langberg, "Hard Cell," *San Jose Mercury News*, May 18, 1997, pp. 1F, 2F.
57. Phillip Robinson, "The Cellular Pitch," *San Jose Mercury News*, March 30, 1997, pp. 1F, 6F.
58. Jon Healey, "Celler's Market," *San Jose Mercury News*, October 26, 1997, pp. 1F, 4F.
59. Chris O'Malley, "Sorting Out Cellphones," *Popular Science*, February 1998, pp. 54–59.
60. National Highway Traffic Safety Administration study, reported in Associated Press, "Cell Phones One of Many Driver Distractions," *Reno Gazette-Journal*, January 8, 1998, p. 3A.
61. "Who Needs a Cell Phone?" *Consumer Reports*, February 1997, pp. 10–15.
62. John W. Verity, "This Year, Servers Will Be King," *Business Week*, January 8, 1996, p. 94.
63. Scott McCartney, "PC Servers Making Inroads as Their Power Accelerates," *Wall Street Journal*, January 17, 1995, p. B3.
64. Bill Gates, cited in Susan Gregory Thomas, "Making a Fast Connection," *U.S. News & World Report*, March 24, 1997, pp. 67–68.
65. David L. Wilson, "Students' Popular Internet Sites Slow Campus Networks to a Crawl," *Chronicle of Higher Education*, February 28, 1997, pp. A26, A28.
66. Bruce Brown, "U.S. Mail Beats E-Mail in a Classic Tortoise-Hare Contest," *San Jose Mercury News*, May 19, 1996, pp. 1F, 6F.
67. Network Technology, Santa Clara, study reported in David S. Hizenrath, "Ballooning Volume of E-Mail Slows Delivery Times to a Crawl," *San Jose Mercury News*, May 18, 1997, p. 2F; reprinted from Washington Post.
68. Bob Quillin, "Strike Up the Bandwidth," *San Francisco Examiner*, September 14, 1997, pp. B-5, B-8.
69. Bill Kennard, quoted in Associated Press, "FCC Chairman Wants Faster Net Access," *San Francisco Chronicle*, January 31, 1998, p. D2.
70. Edmund L. Andrews, "67 Nations Agree to Freer Markets in Communications," *New York Times*, February 16, 1997, sec. 1, pp. 1, 6.
71. Marc Newman, quoted in Robert S. Boyd, "Commercial Satellites Launch 'Unwired Planet,'" *San Jose Mercury News*, May 18, 1997, pp. 1E, 14E.
72. Robert Kinzie, cited in Boyd, 1997.
73. Boyd, 1997.
74. Schine, Elstrom, Barrett et al., 1997.
75. William J. Cook, "1997, A New Space Odyssey," *U.S. News & World Report*, March 3, 1997, pp. 45–52.
76. Quein Hard, "Iridium Creates New Plan for Global Cellular Service," *Wall Street Journal*, August 18, 1997, p. B2.
77. Elizabeth Weise, "Company's Bold $9 Billion Plan to Bring the World Online," *San Francisco Chronicle*, March 22, 1997, p. A7.
78. Kevin Maney, "Internet in the Sky," *USA Today*, April 30, 1997, pp. 1B, 2B.
79. James Kim, "Scrambling for the Sky," *USA Today*, June 18, 1997, pp. 1B, 2B.
80. Kim, 1997.
81. Jonathan Marshall, "Speedy New Technology Begins Race to Market," *San Francisco Chronicle*, February 18, 1997, pp. B1, B2.
82. Howard Banks, "The Law of the Photon," *Forbes*, October 6, 1997, pp. 66–73.
83. Jonathan Marshall, "Photonics Industry Is Changing Communications," *San Francisco Chronicle*, July 29, 1997, pp. C1, C7.
84. Otis Port, "A Lens That Tricks the Light Fantastic," *Business Week*, July 21, 1997, pp. 118–19.
85. Marshall, "Photonics Industry Is Changing Communications," 1997.
86. David Payne, quoted in Banks, 1997.
87. William Gartner, quoted in Banks, 1997.
88. Ross Kerber, "Utilities Reach Out to Add Phone, Cable Service," *Wall Street Journal*, January 27, 1997, pp. B1, B14.
89. Richard Woodbury, "Power Player," *Time*, June 9, 1997, p. P8.
90. Gautam Naik, "Electric Outlets Could Be Link to the Internet," *Wall Street Journal*, October 7, 1997, p. B6.
91. Chris Burke, quoted in Gautam Naik, "Putting Telecom Services on Power Lines Could Spark Internet Usage in Europe," *Wall Street Journal*, October 9, 1997, p. B8.
92. Frank Vizard, "Turning Phone Lines into TV Lines," *Popular Science*, August 1997, p. 50.
93. Bloomberg Business News, "Starved for Phone Service, Latin America Gobbles Up Cellular," *San Jose Mercury News*, August 4, 1996, pp. 1E, 6E.
94. Ricardo Sandoval, "Cell Phones Booming in Latin America," *San Jose Mercury News*, September 21, 1997, pp. 1E, 5E.
95. Michael H. Martin, "When Info Worlds Collide," *Fortune*, October 28, 1996, pp. 130–33.
96. Saundra Banks Loggins, quoted in Kenneth Howe, "Firm Turns Hiring into a Science," *San Francisco Chronicle*, September 19, 1992, pp. B1, B2.
97. William M. Bulkely, "Employees Use Software to Track Résumés," *Wall Street Journal*, June 23, 1992, p. B6.
98. Resumix, cited in Margaret Mannix, "Writing a Computer-Friendly Résumé," *U.S. News & World Report*, October 26, 1992, pp. 90–93.
99. David Rampe, "Cyberspace Résumés Fit the Modern Job Hunt," *New York Times*, Feburary 3, 1997, p. C6.
100. Joyce Lain Kennedy and Thomas J. Morrow, *Electronic Résumé Revolution: Creating a Winning Résumé for the New World of Job Seeking* (New York: Wiley, 1994).
101. Kathleen Pender, "Jobseekers Urged to Pack Lots of 'Keywords' into Résumés," *San Francisco Chronicle*, May 16, 1994, pp. B1, B4.
102. Howard Bennett and Chuck McFadden, "How to Stand Out in a Crowd," *San Francisco Examiner & Chronicle*, October 17, 1993, help wanted section, p. 29.
103. Richard Bolles, quoted in Sylvia Rubin, "How to Open Your Job 'Parachute' Afer College," *San Francisco Chronicle*, February 24, 1994, p. E9.

Chapter 9

1. Carole A. Lane, quoted in Leslie Miller, "You Are a Database and Access Abounds," *USA Today*, June 9, 1997, p. 6D.
2. S. C. Gwynne, "How Casinos Hook You," *Time*, November 17, 1997, pp. 68–69.
3. Christina Binkley, "Harrah's Builds Database About Patrons," *Wall Street Journal*, September 2, 1997, pp. B1, B10.
4. Mike Snider, "Putting the Nation's Past On Line," *USA Today*, April 17, 1996, pp. 1D, 2D.
5. Scott Berinato, "Founding Fathers Find New Life On-line," *San Francisco Examiner*, May 5, 1996, pp. B-5, B-6.
6. James H. Billington, quoted in Mike Snider, "Research Archives in Cyberspace," *USA Today*, April 10, 1997, p. 6D.
7. Amanda Kell, "Modern Monks—Holy, High Tech," *San Francisco Chronicle*, January 9, 1995, p. B3.
8. Jeffrey R. Young, "Modern-Day Monastery," *Chronicle of Higher Education*, January 19, 1996, p. A21.
9. Joseph B. White, Don Clark, and Silvia Ascarelli, "This German Software Is Complex, Expensive—and Wildly Popular," *Wall Street Journal*, March 14, 1997, pp. A1, A8.
10. William J. Cook, "What Makes IBM Run?" *U.S. News & World Report*, June 9, 1997, pp. 48–49.
11. Cook, 1997.
12. David M. Kroenke, *Database Processing*, 5th ed. (Upper Saddle River, NJ: Prentice Hall, 1995), p. 271.
13. James A. Larson, *Database Directories* (Upper Saddle River, NJ: Prentice Hall PTR, 1995).
14. Larson, 1995.
15. Kroenke, 1995, p. 467.
16. Jonathan Berry, John Verity, Kathleen Kerwin, and Gail DeGeorge, "Database Marketing," *Business Week*, September 5, 1994, pp. 56–62.
17. Sara Reese Hedberg, "The Data Gold Rush," *Byte*, October 1995, pp. 83–88.
18. John W. Verity, "Coaxing Meaning Out of Raw Data," *Business Week*, February 3, 1997, pp. 134–38.
19. Hedberg, 1995.
20. Cheryl D. Krivda, "Data-Mining Dynamite," *Byte*, October 1995, pp. 97–103.
21. Edmund X. DeJesus, "Data Mining," *Byte*, October 1995, p. 81.
22. Krivda, 1995.
23. Karen Watterson, "A Data Miner's Tools," *Byte*, October 1995, pp. 91–96.
24. Watterson, 1995.
25. Watterson, 1995.

26. Watterson, 1995.
27. Peter Coy, "He Who Mines Data May Strike Fools Gold," *Business Week*, June 16, 1997, p. 40.
28. Andrew W. Lo, quoted in Coy, 1997.
29. Richard Lamm, quoted in Christopher J. Feola, "The Nexis Nightmare," *American Journalism Review*, July/August 1994, pp. 39–42.
30. Feola, 1994.
31. Penny Williams, "Database Dangers," *Quill*, July/August 1994, pp. 37–38.
32. Lynn Davis, quoted in Williams, 1994.
33. Associated Press, "Many Companies Are Willing to Give a Cat a Little Credit," *San Francisco Chronicle*, January 8, 1994, p. C1.
34. Anthony Ramirez, "Name, Résumé, References. And How's Your Credit?" *New York Times*, August 31, 1997, sec. 3, p. 8.
35. Nina Bernstein, "On Line, High-Tech Sleuths Find Private Facts," *New York Times*, September 15, 1997, pp. A1, A12.
36. Ken Hoover, "Prisoner's Long-Distance Victims," *San Francisco Chronicle*, June 1, 1993, pp. A1, A6.
37. Steve Lohr, "Rare Alliance on Privacy for Software," *New York Times*, June 12, 1997, pp. C1, C7.
38. "Database Firms to Curb What They Sell," *San Francisco Chronicle*, June 11, 1997, pp. B1, B2.
39. Leslie Miller, "Guidelines for Privacy On Line Set," *USA Today*, June 10, 1997, p. 1D.
40. David Linowes, cited in Joseph Anthony, "Who's Reading Your Medical Records?" *American Health*, November 1993, pp. 54–58.
41. Steven Findlay, "Prescription for Patient Privacy," *USA Today*, September 11, 1997, pp. 1A, 2A.
42. Ramirez, 1997.
43. American Management Association, reported in Maggie Jackson, "Most Firms Spy on Employees, Survey Finds," *San Francisco Chronicle*, May 23, 1997, pp. B1, B4.
44. Dana Hawkins, "Who's Watching Now?" *U.S. News & World Report*, September 15, 1997, pp. 55–58.
45. Erik Larson, quoted in Martin J. Smith, "Marketers Want to Know Your Secrets," *San Francisco Examiner*, November 21, 1993, pp. E-3, E-8.
46. Larson, reported in Smith, 1993.
47. CNET.COM, "Some Tips to Protect Privacy," *San Francisco Chronicle*, June 11, 1996, p. C4.
48. Joshua Quittner, "Invasion of Privacy," *Time*, August 25, 1997, pp. 28–35.
49. Deborah L. Jacobs, "They've Got Your Name. You've Got Their Junk," *New York Times*, March 13, 1994, sec. 3, p. 5.
50. Larson, reported in Smith, 1993.
51. Doug Rowan, quoted in Ronald B. Lieber, "Picture This: Bill Gates Dominating the Wide World of Digital Content," *Fortune*, December 11, 1995, p. 38.
52. Kathy Rebello, "The Ultimate Photo Op?" *Business Week*, October 23, 1995, p. 40.
53. Steve Lohr, "Huge Photo Archive Bought by Software Billionaire Gates," *New York Times*, October 11, 1995, pp. A1, C5.
54. Don Clark, "Bill Gates's Corbis Gains Sole Rights to Artist's Works," *Wall Street Journal*, April 2, 1996, p. B9.
55. Steve Lohr, "Gates Buys Rights to Adams Photo Images," *New York Times*, April 2, 1996, p. C13.
56. Wendy Bounds, "Bill Gates Owns Otto Bettmann's Lifework," *The Wall Street Journal*, January 17, 1996, pp. B1, B2.
57. Paul Saffo, quoted in Lohr, 1995.
58. Phil Patton, "The Pixels and Perils of Getting Art On Line," *New York Times*, August 7, 1994, sec. 2, pp. 1, 31.
59. John R. Emshwiller, "Firms Finds Profits Searching Databases," *Wall Street Journal*, January 25, 1993, pp. B1, B2.
60. Kathryn Rambo, quoted in Ramon G. McLeod, "New Thieves Prey on Your Very Name," *San Francisco Chronicle*, April 7, 1997, pp. A1, A6.
61. Rambo, quoted in T. Trent Gegax, "Stick 'Em Up? Not Anymore. Now It's Crime by Keyboard," *Newsweek*, July 21, 1997, p. 14.
62. Kathryn Kranhold, "'Identity Theft' Bill Leaves Credit Bureaus in the Cold," *Wall Street Journal*, June 4, 1997, pp. CA1, CA4.
63. Wendy Lichman, "Say Goodbye to Your Credit Card," *San Francisco Chronicle*, October 30, 1997, p. E10.
64. McLeod, 1997.
65. Emerald Yeh and Christine McMurry, "Having Your Identity Stolen Out from Under You," *San Francisco Chronicle*, May 26, 1996, "Sunday" section, p. 6.
66. Saul Hansell, "Identity Crisis: When a Criminal's Got Your Number," *New York Times*, June 16, 1996, sec. 4, pp. 1, 5.
67. Joshua Quittner, "No Privacy on the Web," *Time*, June 2, 1997, pp. 64–64.
68. Ed Howard, quoted in Peter Sinton, "ID Theft a Growing Problem," *San Francisco Chronicle*, October 20, 1997, p. B3.
69. McLeod, 1997.
70. Verna Willis, quoted in Hansell, 1996.
71. McLeod, 1997.
72. Peter Sinton, "How to Protect Personal Information," *San Francisco Chronicle*, October 20, 1997, p. B3.

Chapter 10

1. Thomas A. Stewart, "Reengineering: The Hot New Managing Tool," *Fortune*, August 23, 1993, pp. 40-48.
2. John Huey, "Waking Up to the New Economy," *Fortune*, June 27, 1994, pp. 36-46.
3. Gil Gordon, cited in Sue Shellenbarger, "Overwork, Low Morale Vex the Mobile Office," *Wall Street Journal*, August 17, 1994, pp. B1, B4.
4. Shellenbarger, 1994.
5. Samuel E. Bleecker, "The Virtual Organization," *The Futurist*, March-April 1994, pp. 9–14.
6. John Diebold, "The Next Revolution in Computers," *The Futurist*, May–June 1994, pp. 34–37.
7. Jonathan Marshall, "Contracting Out Catching On," *San Francisco Chronicle*, August 22, 1994, pp. D1, D3.
8. David Greising, "Quality: How to Make It Pay," *Business Week*, August 8, 1994, pp. 54–59.
9. Stewart, 1993.
10. "Do You Know Your Technology Type?" *The Futurist*, September-October 1997, pp. 10–11.
11. Paul C. Judge, "Are Tech Buyers Different?" *Business Week*, January 26, 1998, pp. 64–68.
12. Judge, 1998.
13. "Do You Know Your Technology Type?" 1997.
14. William Cats-Baril and Ronald Thompson, *Information Technology and Management* (Burr Ridge, IL: Irwin/McGraw-Hill, 1997), P. 159.
15. James O'Brien, *Management Information Systems* (Burr Ridge, IL: Irwin/McGraw-Hill, 1996), p. 375.
16. Alan Webber, "The Best Organization Is No Organization," *USA Today*, March 6, 1997, p. 13A.
17. Webber, 1997.
18. Jessica Lipnack and Jeffrey Stamps, *Virtual Teams: Reaching Across Space, Time, and Organization with Technology* (New York: John Wiley & Sons, 1997).
19. "Virtual Teams Transcend Space and Time," *The Futurist*, September-October 1997, p. 59.
20. Gary Webb, "Potholes, Not 'Smooth Transition,' Mark Project," *San Jose Mercury News*, July 3, 1994, p. 18A.
21. Dirk Johnson, "Denver May Open Airport in Spite of Glitches," *New York Times*, July 17, 1994, p. A12.
22. K. Kendall and J. Kendall, *Systems Analysis and Design* (Englewood Cliffs, NJ: Prentice Hall, 1992), p. 39.
23. James O'Brien, *Introduction to Information Systems*, 8th ed. (Burr Ridge, IL: Irwin/McGraw-Hill, 1997), p. 385.
24. Jeffrey L. Whitten and Lonnie Bentley, *Systems Analysis and Design Methods*, 4th ed. (Burr Ridge, IL: Irwin/McGraw-Hill, 1997), p. 104.
25. O'Brien, 1997, p. 410.
26. Kevin Kelly, *Out of Control*; cited in Rick Tetzeli, "Managing in a World Out of Control," *Fortune*, September 5, 1994, p. 111.
27. R. Wild, "Maximize Your Brain Power," *Men's Health*, April 1992, pp. 44–49.
28. James Randi, "Help Stamp Out Absurd Beliefs," *Time*, April 13, 1992, p. 80.
29. Hy Ruchlis and Sandra Oddo, *Clear Thinking: A Practical Introduction* (Buffalo, NY: Prometheus, 1990), p. 109.
30. Wild, 1992.
31. Howard Kahane, *Logic and Contemporary Rhetoric: The Uses of Reason in Everyday Life*, 5th ed. (Belmont, CA: Wadsworth, 1988).
32. J. Rasool, C. Banks, and M.-J. McCarthy, *Critical Thinking: Reading and Writing in a Diverse World* (Belmont, CA: Wadsworth, 1993), p. 132.
33. Rasool, Banks, and McCarthy, 1993, p. 132.

Chapter 11

1. Jeffery L. Sheler, "Dark Prophecies," *U.S. News & World Report*, December 15, 1997, pp. 63–71.
2. *Newsweek*, June 2, 1997, cover.
3. Debra Sparks, "Will Your Bank Live to See the Millennium?" *Business Week*, January 26, 1998, pp. 74–75.
4. Paul Sweeney, "End of the World as We Know It," *Inc.*, August 1997, p. 29.
5. Marilyn Vos Savant, "Ask Marilyn," *Parade Magazine*, October 26, 1997, p. 15.
6. Holman W. Jenkins Jr., "Turns Out the Year 2000 Problem Is Just the Beginning," *Wall Street Journal*, May 6, 1997, p. A23.
7. Kevin Maney, "Year 2000: Small Oversight Is Big Problem," *USA Today*, March 15, 1996, p. 2B.
8. *Journal of Systems Management*, quoted in Jenkins, 1997.
9. Jenkins, 1997.
10. Gina Smith, "The Millennial Mess," *Popular Science*, February 1997, p. 26.
11. Donald Fowler, quoted in Rajiv Chandrasekaran, "Older Computer Programmers Go Back to Fix the Future," *San Francisco Chronicle*, March 6, 1997, p. A4.
12. Stephen Baker, Gary McWilliams, and Manjeet Kripalani, "Forget the Huddled Masses: Send Nerds," *Business Week*, July 21, 1997, pp. 110–16.
13. James J. Mitchell, "Software Expert Shortage Growing More Acute," *San Jose Mercury News*, May 25, 1997, pp. 1D, 2D.
14. Jeffrey R. Young, "Colleges Complain They Can't Hire—or Keep—Enough Computer Specialists," *Chronicle of Higher Education*, September 5, 1997, pp. A36–37.
15. Ilana DeBare, "Programmers in the Driver's Seat," *San Francisco Chronicle*, January 1, 1998, p. D3.
16. Ilana DeBare, "Computer Old-Timers Recruited to Get the Jump on the Year 2000," *San Francisco Chronicle*, March 24, 1997, pp. A1, A15.
17. Jonathan Kaufman, "A U.S. Recruiter Goes Far Afield to Bring in High-Tech Workers," *Wall Street Journal*, January 8, 1998, pp. A1, A2.
18. Lee Berton, "A Young, Blind Whiz on Computers Makes a Name in Industry," *Wall Street Journal*, August 15, 1997, p. B1.
19. Laurence Zuckerman, "Agent to the Software Stars," *New York Times*, September 8, 1997, pp. C1, C6.
20. Mitchell, 1997.
21. Mitchell, 1997.
22. Zuckerman, 1997.
23. Howard Rubin, study for META Group, Stamford, Conn., reported in Ira Sager, "U.S. Programmers Are Falling Off the Pace," *Business Week*, May 5, 1997, p. 126C.
24. Young, 1997.
25. Robert Silvers, quoted in Cathy Hainer, "Programming a Photo-Art Form," *USA Today*, December 22, 1997, p. 6D.
26. Alan Freedman, *The Computer Glossary*, 6th ed. (New York: AMACOM, 1993), p. 370.
27. Freedman, 1993.
28. Freedman, 1993.
29. Michael Krantz, "Keeping Tabs Online," *Time*, November 10, 1997, pp. 81–82.
30. "Big City, BigBook," Newsweek, August 1996, p. 13.
31. Stephen H. Wildstrom, "Web Sites Made Simpler," *Business Week*, January 26, 1998, p. 15.
32. David Einstein, "Web Site Creation Is Free, Easy," *San Francisco Chronicle*, January 20, 1998, p. B3.
33. "Work Week," Wall Street Journal, November 18, 1997, p. A1.
34. Peter D. Varhol, "Visual Programming's Many Faces," *Byte*, July 1994, pp. 187–88.
35. Gretchen Boehr, "Where to Buy Software," *PC Novice*, June 1992, pp. 20–25.
36. Walter S. Mossberg, "For PC Shoppers Who Know What They Want, Mail Order Delivers," *Wall Street Journal*, March 25, 1993, p. B1.
37. Walter S. Mossberg, "Talk Is Cheap? Not If You're Calling for Software Support," *Wall Street Journal*, October 14, 1993, p. B1.

Chapter 12

1. Andy Hines, "Jobs and Infotech," *The Futurist*, January-February 1994, pp. 9–13.
2. Terry Bynum, quoted in Lawrence Hardy, "Tapping into New Ethical Quandaries," *USA Today*, August 1, 1995, p. 6D.
3. Patricia Schnaidt, "The Electronic Superhighway," *LAN Magazine*, October 1993, pp. 6–8.
4. Al Gore, reported in "Toward a Free Market in Telecommunications," *Wall Street Journal*, April 19, 1994, p. A18.
5. Jeffrey R. Young, "Confusion Abounds About 3 Net Internet Efforts and Their Names," *Chronicle of Higher Education*, July 25, 1997, pp. A22, A23.
6. Young, 1997.
7. Robyn Meredith, "Building 'Internet 2,'" *New York Times*, February 2, 1998, p. C3.
8. Jon Swartz, "Need for Speed Spawns 2 Internetlets," *San Francisco Chronicle*, July 28, 1997, pp. A1, A13.
9. Mike Snider, "Envisioning the Internet's Next Step," *USA Today*, May 15, 1997, p. 6D.
10. Jeffrey R. Young, "Searching for 'Killer Applications,'" *Chronicle of Higher Education*, August 8, 1997, pp. A22, A23.
11. Colleen Cordes, "Federal Support for New Version of Internet Hinges on 5 Spending Bills," *Chronicle of Higher Education*, September 5, 1997, p. A38.
12. Meredith, 1998.
13. Swartz, 1997.
14. Douglas E. Van Houweling, quoted in Meredith, 1998.
15. Wall Street Journal News Roundup, "Likely Mergers Herald an Era of Megacarriers," *Wall Street Journal*, February 2, 1996, pp. B1, B3.
16. Peter Elstrom, "Let the Telecom Dogfight Begin," *Business Week*, April 7, 1997, p. 42.
17. Steven Levy, "Now for the Free-for-All," *Newsweek*, February 12, 1996, pp. 42–44.
18. Bryan Gruley and Albert R. Karr, "Bill's Passage Represents Will of Both Parties," *Wall Street Journal*, February 2, 1996, pp. B1, B3.
19. *Wall Street Journal News Roundup*, 1996.
20. Levy, "Now for the Free-for-All," 1996.
21. Richard Zoglin, "We're All Connected," *Time*, February 12, 1996, p. 52.
22. Gautam Naik, "Bell Companies Ready to Charge into Long-Distance," *Wall Street Journal*, February 5, 1996, p. B4.
23. Laurence H. Tribe, "The FCC vs. the Constitution," *Wall Street Journal*, September 9, 1997, p. A8.
24. John M. Broder, "Let It Be," *New York Times*, June 30, 1997, pp. C1, C9.
25. *A Framework for Global Electronic Commerce*, quoted in Steven Levy, "Bill and Al Get It Right," *Newsweek*, July 7, 1997, p. 80.
26. Jim Hornthal, quoted in Jon Swartz, "Clinton Advocates Net Self-Rule," *San Francisco Chronicle*, July 2, 1997, pp. B1, B2.
27. Eli M. Noam, "An Unfettered Internet? Keep Dreaming," *New York Times*, July 11, 1997, p. A2.

28. "Frustrated Bank Customer Lets His Computer Make Complaint," *Los Angeles Times*, October 20, 1993, p. A28.
29. We are grateful to Prof. John Durham for contributing these ideas.
30. Arthur M. Louis, "Nasdaq's Computer Crashes," *San Francisco Chronicle*, July 16, 1994, pp. D1, D3.
31. Joseph F. Sullivan, "A Computer Glitch Causes Bumpy Start in a Newark School," *New York Times*, September 18, 1991, p. A25.
32. Malcolm Gladwell, "Blowup," *The New Yorker*, January 22, 1996, pp. 32–36.
33. Leonard M. Fuld, "Bad Data You Can't Blame on Intel," *Wall Street Journal*, January 9, 1995, p. A12.
34. Douglas Waller, "Onward Cyber Soldiers," *Time*, August 21, 1995, pp. 38–44.
35. Mark Thompson, "If War Comes Home," *Time*, August 21, 1995, pp. 44–46.
36. John J. Fialka, "Pentagon Studies Art of 'Information Warfare,' to Reduce Its Systems' Vulnerability to Hackers," *Wall Street Journal*, July 3, 1995, p. A10.
37. Henry K. Lee, "UC Student's Dissertation Stolen with Computer," *San Francisco Chronicle*, January 27, 1994, p. A15.
38. Thomas J. DeLoughry, "2 Students Are Arrested for Software Piracy," *The Chronicle of Higher Education*, April 20, 1994, p. A32.
39. Suzanne P. Weisband and Seymour E. Goodman, "Subduing Software Pirates," *Technology Review*, October 1993, pp. 31–33.
40. Jeffrey A. Trachtenberg and Mark Robichaux, "Crooks Crack Digital Codes of Satellite TV," *Wall Street Journal*, January 12, 1996, pp. B1, B9.
41. Stanley E. Morris, quoted in David E. Sanger, "Money Laundering, New and Improved," *New York Times*, December 24, 1995, sec. 4, p. 4.
42. David L. Wilson, "Gate Crashers," *The Chronicle of Higher Education*, October 20, 1993, pp. A22–A23.
43. John T. McQuiston, "4 College Students Charged with Theft Via Computer," *New York Times*, March 18, 1995, p. 38.
44. Jeremy L. Milk, "3 U. of Wisconsin Students Face Punishment for Bogus E-Mail Messages," *The Chronicle of Higher Education*, October 20, 1993, p. A25.
45. Peter H. Lewis, "Cybervirus Whodunit: Who Creates This Stuff?" *New York Times*, September 4, 1995, p. 20.
46. Julian Dibbell, "Viruses Are Good for You," *Wired*, February 1995, pp. 126–135, 175–180.
47. James Kim, "Understanding Concept Virus Is Key to Prevention," *USA Today*, March 11, 1996, p. 5B.
48. Stan Miastkowski, "Virus Killers," *PC World*, March 1997, pp. 181–90.
49. Donald Parker, quoted in William M. Carley, "Rigging Computers for Fraud or Malice Is Often an Inside Job," *Wall Street Journal*, August 27, 1992, pp. A1, A5.
50. David Carter, quoted in Associated Press, "Computer Crime Usually Inside Job," *USA Today*, October 25, 1995, p. 1B.
51. Eric Corley, cited in Kenneth R. Clark, "Hacker Says It's Harmless, Bellcore Calls It Data Rape," *San Francisco Examiner*, September 13, 1992, p. B-9; reprinted from Chicago Tribune.
52. Michelle Slatalla and Joshua Quittner, *Masters of Deception: The Gang that Rules Cyberspace* (New York: HarperCollins, 1995).
53. Joshua Quittner, "Hacker Homecoming," *Time*, January 23, 1995, p. 61.
54. Associated Press, "Russian Hackers Caught," *San Francisco Chronicle*, August 19, 1995, p. D1.
55. Steven Bellovin, cited in Jane Bird, "More than a Nuisance," *The Times* (London), April 22, 1994, p. 31.
56. William M. Bulkeley, "To Read This, Give Us the Password . . . Ooops! Try It Again," *Wall Street Journal*, April 19, 1995, pp. A1, A8.
57. Robert Lee Hotz,"Sign on the Electronic Dotted Line," *Los Angeles Times*, October 19, 1993, pp. A1, A16.
58. Eugene Carlson, "Some Forms of Identification Can't Be Handily Faked," *Wall Street Journal*, September 14, 1993, p. B2.
59. Justin Matlkick, "Security of Online Markets Could Well Be at Stake," *San Francisco Chronicle*, September 16, 1997, A21.
60. Jon Swartz, "New Strict Encryption Controls Rejected," *San Francisco Chronicle*, September 25, 1997, pp. D1, D10.
61. Jonathan Weber, "Encryption Bares High Tech's Weakness," *San Jose Mercury News*, September 21, 1997, p. 3E.
62. John Markoff, "White House Wants Control of High-Tech Scrambling," *San Francisco Examiner*, September 7, 1997, pp. B-1, B-3; reprinted from New York Times.
63. John Holusha, "The Painful Lessons of Disruption," *New York Times*, March 17, 1993, pp. C1, C5.
64. The Enterprise Technology Center, cited in "Disaster Avoidance and Recovery Is Growing Business Priority," special advertising supplement in *LAN Magazine*, November 1992, p. SS3.
65. Micki Haverland, quoted in Fred R. Bleakley, "Rural County Balks at Joining Global Village," *Wall Street Journal*, January 4, 1996, pp. B1, B2.
66. David Ensunsa, "Proposed Cell-Phone Pole Faces Challenge," *The Idaho Statesman*, June 23, 1995, p. 4B.
67. James H. Snider, "The Information Superhighway as Environmental Menace," *The Futurist*, March-April 1995, pp. 16–21.
68. Marilyn Elias, "Concern Over On-line Counseling," *USA Today*, August 14, 1997, p. 8D.
69. Leonard Holmes, quoted in Elias, 1997.
70. David Sommers, quoted in Diana McKeon Charkalis, "Web Can Be a Link to Your Shrink," *USA Today*, February 26, 1997, p. 4D.
71. Andrew Kupfer, "Alone Together," *Fortune*, March 20, 1995, pp. 94–104.
72. Kevin O'Neill, quoted in David Lieberman, "Racetracks Are Betting on Interactive TV," *USA Today*, October 25, 1995, p. 1B.
73. Marco R. della Cava, "Are Heavy Users Hooked or Just On-line Fanatics?" *USA Today*, January 16, 1996, pp. 1A, 2A.
74. Kenneth Howe, "Diary of an AOL Addict," *San Francisco Chronicle*, April 5, 1995, pp. D1, D3.
75. Kendall Hamilton and Claudia Kalb, "They Log On, but They Can't Log Off," *Newsweek*, December 18, 1995, pp. 60–61.
76. Associated Press, "Husband Accuses Wife of Having Online Affair," *San Francisco Chronicle*, February 2, 1996, p. A3.
77. Karen S. Peterson and Leslie Miller, "Cyberflings Are Heating Up the Internet," *USA Today*, February 6, 1996, pp. 1D, 2D.
78. Survey by Microsoft Corporation, reported in Don Clark and Kyle Pope, "Poll Finds Americans Like Using PCs but May Find Them to Be Stressful," *Wall Street Journal*, April 10, 1995, p. B3.
79. Sherry Turkle, quoted in Susan Wloszczyna, "MIT Prof Taps into Culture of Computers," *USA Today*, November 27, 1995, pp. 1D, 2D.
80. Sherry Turkle, *Life on the Screen: Identity in the Age of the Internet* (New York: Simon & Schuster, 1995).
81. Tom Abate, "Exploring the World of Cyber-Psychology," *San Francisco Examiner*, December 3, 1995, pp. B-1, B-3.
82. Christopher Lehmann-Haupt, "The Self in Cyberspace, Decentered and Opaque," *New York Times*, December 7, 1995, p. B4.
83. Jonathan Marshall, "Some Say High-Tech Boom Is Actually a Bust," *San Francisco Chronicle*, July 10, 1995, pp. A1, A4.
84. Yahoo!/Jupiter Communications survey, reported in Del Jones, "On-line Surfing Costs Firms Time and Money," *USA Today*, December 8, 1995, pp. 1A, 2A.
85. Coleman & Associates survey, reported in Julie Tilsner, "Meet the New Office Party Pooper," *Business Week*, January 29, 1996, p. 6.
86. Webster Network Strategies survey, reported in Jones, 1995.
87. Steven Levy, quoted in Marshall, 1995.
88. STB Accounting Systems 1992 survey, reported in Jones, 1995.
89. Ira Sager and Gary McWilliams, "Do You Know Where Your PCs Are?" *Business Week*, March 6, 1995, pp. 73–74.
90. Ron Erickson, "More Software. Gee." *New York Times*, August 4, 1995, p. A19.
91. Alex Markels, "Words of Advice for Vacation-Bound Workers: Get Lost," *Wall Street Journal*, July 3, 1995, pp. B1, B5.
92. Paul Saffo, quoted in Laura Evenson, "Pulling the Plug," *San Francisco Chronicle*, December 18, 1994, "Sunday" section, p. 53.
93. Daniel Yankelovich Group report, cited in Barbara Presley Noble, "Electronic Liberation or Entrapment," *New York Times*, June 15, 1994, p. C4.
94. Annette Kornblum, "Maybe. Maybe Not." *San Jose Mercury News*, March 10, 1996, pp. 1C, 5C; reprinted from the Washington Post.
95. Neil Postman, quoted in Evenson, 1994.
96. Mike Snider, "Keeping PC Play Out of the Office," *USA Today*, January 26, 1995, p. 3D.
97. Sager & McWilliams, 1995.
98. Unabomber letter to David Gelernter, quoted in Bob Ickes, "Die, Computer, Die!" *New York*, July 24, 1995, pp. 22–26.
99. Jeremy Rifkin, "Technology's Curse: Fewer Jobs, Fewer Buyers," *San Francisco Examiner*, December 3, 1995, p. C-19.
100. Michael J. Mandel, "Economic Anxiety," *Business Week*, March 11, 1996, pp. 50–56.
101. Bob Herbert, "A Job Myth Downsized," *New York Times*, March 8, 1996, p. A19.
102. Robert Kuttner, "The Myth of a Natural Jobless Rate," *Business Week*, October 20, 1997, p. 26.
103. Stewart Brand, in "Boon or Bane for Jobs?" *The Futurist*, January-February 1997, pp. 13–14.
104. Paul Krugman, "Long-Term Riches, Short-Term Pain," *New York Times*, September 25, 1994, sec. 3, p. 9.
105. Department of Commerce survey, cited in "The Information 'Have Nots'" [editorial], *New York Times*, September 5, 1995, p. A12.
106. Beth Belton, "Degree-based Earnings Gap Grows Quickly," *USA Today*, February 16, 1996, p. 1B.
107. Alan Krueger, quoted in LynNell Hancock, Pat Wingert, Patricia King, Debra Rosenberg, and Allison Samuels, "The Haves and the Have-Nots," *Newsweek*, February 27, 1995, pp. 50–52.
108. Christine Borgman, quoted in Gary Chapman, "What the On-line World Really Needs Is an Old-Fashioned Librarian," *San Jose Mercury News*, August 21, 1995, p. 3D; reprinted from Los Angeles Times.
109. Jared Sandberg, "Digital Secretaries Can Help Free You from Phone Tangles," *Wall Street Journal*, February 5, 1998, p. B1.
110. John W. Verity and Richard Brandt, "Robo-Software Reports for Duty," *Business Week*, February 14, 1994, pp. 110–113.
111. Katie Hafner, "Have Your Agent Call My Agent," *Newsweek*, February 27, 1995, pp. 76–77.
112. Laurie Flynn, "Electronic Clipping Services Cull Cyberspace and Fetch Data for You. But They Can't Think for You," *New York Times*, May 8, 1995, p. C6.
113. Peter Wayner and Alan Joch, "Agents of Change," *Byte*, March 1995, pp. 94–95.
114. Kurt Indermaur, "Baby Steps," *Byte*, March 1995, pp. 97–104.
115. Peter Wayner, "Free Agents," *Byte*, March 1995, pp. 105–114.
116. Denise Caruso, "Virtual-World Users Put Themselves in a Sort of Electronic Puppet Show," *New York Times*, July 10, 1995, p. C5.
117. Tom R. Halfhill, "Agents and Avatars," *Byte*, February 1996, pp. 69–72.
118. David L. Wilson, "On-Line Treasure Hunt," *The Chronicle of Higher Education*, March 17, 1995, pp. A19–A20.
119. Robert Benfer Jr., Louanna Furbee, and Edward Brent Jr., quoted in Steve Weinberg, "Steve's Brain," *Columbia Journalism Review*, February 1991, pp. 50–52.
120. Tom Foremski, "Read It and Weep: E-Mail Overload Ahead," *San Francisco Examiner*, April 9, 1995, pp. B-5, B-6.
121. Jeanne B. Pinder, "Fuzzy Thinking Has Merits When It Comes to Elevators," *New York Times*, September 22, 1993, pp. C1, C7.
122. Gene Bylinsky, "Computers That Learn by Doing," *Fortune*, September 6, 1993, pp. 96–102.
123. Robert McGrough, "Fidelity's Bradford Lewis Takes Aim at Indexes with His 'Neural Network' Computer Program," *Wall Street Journal*, October 27, 1992, pp. C1, C21.
124. Michael Waldholz, "Computer 'Brain' Outperforms Doctors Diagnosing Heart Attack Patients," *Wall Street Journal*, December 2, 1991, p. B78.
125. Otis Port, "A Neural Net to Snag Breast Cancer," *Business Week*, March 13, 1995, p. 95.
126. Otis Port, "Computers That Think Are Almost Here," *Business Week*, July 17, 1995, pp. 68–72.
127. Port, July 17, 1995.
128. Sharon Begley and Gregory Beals, "Software au Naturel," *Newsweek*, May 8, 1995, pp. 70–71.
129. Gautam Naik, "In Sunlight and Cells, Science Seeks Answers to High-Tech Puzzles," *Wall Street Journal*, January 16, 1996, pp. A1, A5.
130. Begley and Beals, 1995.
131. Begley and Beals, 1995.
132. Peter H. Lewis, "'Creatures' Get a Life, Although It Is an Artificial One," *New York Times*, October 13, 1993, p. B7.
133. Judith Anne Gunther, "An Encounter With A.I.," *Popular Science*, June 1994, pp. 90–93.
134. Laura Johannes, "Meet the Doctor: A Computer That Knows a Few Things," *Wall Street Journal*, December 18, 1995, p. B1.
135. Jerry Carroll and Laura Evenson, "The Sage of Cyberspace," *San Francisco Chronicle*, June 6, 1996, "Sunday," pp. 1, 4.
136. Tamara Henry, "Many Schools Can't Access Net Offer," *USA Today*, November 3, 1995, p. 7D.
137. Robert L. Johnson, "Extending the Reach of 'Virtual' Classrooms," *The Chronicle of Higher Education*, July 6, 1994, pp. A19–A23.
138. Shelly Freierman, "For Lifelong Learning, Click Here," *New York Times*, August 25, 1997, p. C6.
139. Pam Dixon, *Virtual College* (Princeton, NJ: Peterson's, 1996).
140. Glenn R. Jones, *Cyberschools: An Education Renaissance* (Jones Digital Century, 1997).
141. Joseph B. Walther, quoted in William H. Honon, "Northwestern University Takes a Lead in Using the Internet to Add Sound and Sight to Courses," *New York Times*, May 28, 1997, p. A17.
142. Goldie Blumenstyk, "Western Governors U. Takes Shape as a New Model for Higher Education," *The Chronicle of Higher Education*, February 6, 1998, p. A21–A24.
143. Myron Magnet, "Who's Winning the Information Revolution," *Fortune*, November 30, 1992, pp. 110–117.
144. Tony Rutkowski, quoted in Schnaidt, 1993.
145. Robert J. Samuelson, "The Way the World Works," *Newsweek*, January 12, 1998, p. 52.
146. Mary Anne Buckman, quoted in Carol Kleiman, "Tailor Your Resume for Inclusion in a Company Database," *San Jose Mercury News*, April 14, 1996, pp. PC1–PC2.
147. David Borchard, "Planning for Career and Life," *The Futurist*, January-February 1995, pp. 8–12.
148. Jonathan Marshall, "Surfing the Internet Can Land You a Job," *San Francisco Chronicle*, July 17, 1995, pp. D1, D3.
149. Marshall, 1995.

Index

Boldface page numbers indicate pages on which key terms are defined.

Photo & Art Credits

Page 1 © Photo Disc; **6** © Rainbow; **15** (top, bottom) © PhotoDisc; **16** (top) Intel; **16** (middle) Sony; **16** (bottom) Hewlett-Packard; **17** (top) © PhotoDisc; **Panel 1.5** (left) NEC, (middle) Seagate, (right) Iomega; **18** © PhotoDisc; **19** (top) Adobe, (bottom) Microsoft; **Panel 1/6** clockwise from top left © Karen Kosmanski / Photo Net / PNI, © Tom Tracy / Photo Network / PNI, IBM, IBM, IBM, Adastra Systems, (middle) Intel; **27** (top) Motorola, (bottom) Sharp Electronics Corp.; **Panel 1.8** Reprinted from the October 6, 1997, issue of *Business Week* by special permission, copyright © 1997 by The McGraw-Hill Companies, Inc.; **30** "Riven" by Cyan Productions; **Panel 1.9** data from H. C. Lindgren, *The Psychology of College Success: A Dynamic Approach* (New York: Wiley, 1969); **47** © Luciano Galiardi / The Stock Market; **53** (top) *Williams-Sonoma Guide to Good Cooking*, Broderbund; **72** Door-to-Door Copilot, NavROute Software; **Panel 2.10** (both) IBM; **81** Lincoln Premier, Ford Motors; **85** Elastic Reality, Inc.; **87** Microsoft Office 97 Professional Edition; **Panel 2.11** copyright © 1998 *San Jose Mercury News*, all rights reserved.Reproduced with permission. Use of this material does not imply endorsement of the *San Jose Mercury News*; **103** © PhotoDisc; **145** © PhotoDisc; **Panel 4.1** text: Springer-Verlag, 1995, reprinted with permission of Telos/Springer Verlag, New York; photo: © Dan McCoy; **Panel 4.2** IBM Archives; **Panel 4.3** both photos © Dan McCoy; **150** IBM; **160** both photos by Brian K. Williams; **162** Intel; **Panel 4.16** Hewlett-Packard; **171** Brian K. Williams; **175** from Steve Man's Web site; **Panel 4.17** photo courtesy of IBM; **Panel 4.18** (left) Brian K. Williams, (right) Tandy; **Panel 4.19** IBM; **193** © PhotoDisc; **194** IBM; **199** Mouseman by Logitech; **201** (top) Marble FX by Logitech, (middle) IBM, (bottom) Brian K. Williams; **Panel 5.5** adapted from *Macworld*, May 1996, p. 101; **Panel 5.6** FTG Data Systems; **Panel 5.7** Calcomp Ultraslate; **Panel 5.8** both from Stylistic 1000; **Panel 5.9** both courtesy of NCR; **Panel 5.10** (left) CheckMate Electronics; **207** (top) © PhotoDisc, (bottom) IBM; **Panel 5.12** Psion; **Panel 5.13** © Arnold Zann / Black, Star; **Panel 5.14** AP Worldwide Photos; **Panel 5.16** photo courtesy of Casio; **213** Tom Burdete, U.S. Geological Service; **Panel 5.18** (left) © Liz Hafalia, *San Francisco Chronicle*,(right) Reuters / Archive Photos; **Panel 5.21** (top right, bottom left) Planar, (bottom right) Fujitsu; **Panel 5.22** (left) Citizen, (right) Hewlett-Packard; **Panel 5.23** Hewlett-Packard; **Panel 5.24** Hewlett-Packard; **Panel 5.26** adapted from Phillip Robinson, "Color Comes In," *San Jose Mercury News*, October 1, 1995, pp. 1F, 6F; **Panel 5.27** (both) courtesy of Calcomp; **Panel 5.28** Hewlett-Packard; **Panel 5.29** (top left) Autodesk, (top right) UPL Research, Inc., (middle right) Autodesk, (bottom right) © Peter Menzel; (bottom left) © Charles Gupton; **235** (top and bottom) © Wernher Krutein / Photovault; **236** (top) IBM; **236** (bottom) United Telecom; **Panel 5.30** AT&T Global Info / Solution; **239** Robert A. Flynn, Inc.; **Panel 5.31** Dr. Irfan Essa, Georgia Tech; **Panel 5.33** (left) Kinesis, (right) Microsoft; **Panel 5.34** © Frank Bevans; **263** © PhotoDisc; **267** (top) IBM, (bottom) © PhotoDisc; **Panel 6.5** (left) Iomega, (right) Syquest; **276** APS; **Panel 6.8** (left) Toshiba, (right) IBM: **279** (bottom) © Frank Bevans; **Panel 6.9** Kodak; **283** Toshiba / The Benjamin Group; **287** (all) IBM; **Panel 6.13** © Paul Higdon, New York Times; **307** © PhotoDisc; **Panel 7.2** Hewlett-Packard; **Panel 7.9** Relevant Knowledge Web Report, September 1997; **Panel 7.10** adapted from Howard Bryant (art by Wes Killingbeck), "Netphones Hit Sound Barriers," *San Jose Mercury News*, August 3, 1997, p. 1D, copyright © 1998 *San Jose Mercury News*, all rights reserved, reproduced with permission. Use of this material does not imply endorsement of the *San Jose Mercury News*; **337** (left) Sony Electronics, Inc.; **Panel 7.11** December 13, 1993, p. 72, *PC Novice*, Peed Corp., 120 W. Harvest Drive, Lincoln NB 68521; **353** © PhotoDisc; **Panel 8.4** (top) Hayes, (bottom) Multitech Systems; **Panel 8.7** (top left and top right) AT&T, (bottom left) U.S. Sprint; **Panel 8.8** (photo) Brian K. Williams; **Panel 8.9** (photo) UPI / Corbis-Bettmann; **Panel 8.10** adapted from Tom Curley, (art by Kevin Rechin), "GPS: High-Tech How to Get There," *USA Today*, May 7, 1996, p. 8A, © 1996 *USA Today*, reprinted with permission; **Panel 8.11** adapted from Jared Schneidman, "How It Works," *Wall Street Journal*, February 11, 1994, p. R5, and *Popular Science* © March 3, 1997; **375** © Peter Townes / Photo Network / PNI; **Panel 8.26** adapted from Alan Freedman, *The Computer Desktop Encyclopedia* (New York:AMA-COM, 1996), p. 617; **413** © PhotoDisc; **Panel 9.2** (photo) AP Wideworld Photos, (text) copyright 1996, *The Chronicle of Higher Education*, reprinted with permission; **Panel 9.10** adapted from "Data Mining Process," illustrated by Victor Grad, © 1995, p. 84, in Sara Reese Hedberg, "The Data Gold Rush," *Byte*, October 1995, pp. 83–88; **Panel 9.12** © Michael P. Manheim / Photo Network / PNI; **453** © PhotoDisc; **Panel 10.2** © Jon Feingarten / Stock, Boston / PNI; (text) "Are You Still Clinging to Those Chestnuts of Business? Read On," May 9, 1997, p. 81; **Panels 10.3 and 10.4** adapted from *Management Information Systems* (Burr Ridge, IL: Irwin/McGraw-Hill, 1996), pp. 314, 319; **Panel 10.13** United Airlines Gate Assignment Display System; **Panel 10.14** William Cats-Baril, *Information Technology and Management* (Burr Ridge: IL: Irwin/McGraw-Hill, 1997), p. 109; **Panel 10.15** U.S. Labor Department, adapted from *Business Week*, January 1, 1998, p. 22; **Panel 10.16** (photo) © PhotoDisc; **Panel 10.17** (photo) © PhotoDisc; **Panel 10.18** © PhotoDisc; **Panel 10.24** © (photo) PhotoDisc;**Panel 10.26** (photo) © PhotoDisc; **Panel 10.27** (photo) © PhotoDisc; **Panel 10.29** (photo) © PhotoDisc; **507** © PhotoDisc; **Panel 11.2** © PhotoDisc; **Panel 11.3** © PhotoDisc; **Panel 11.6** (top) Julia Case Bradley, *QuickBASIC Using Modular Structure* (Burr Ridge: IL: Irwin/McGraw-Hill, 1996), p. 15; **Panel 11.8** ©PhotoDisc; **Panel 11.9** Naval Surface Warfare Center, Dahlgren, VA; **Panel 11.10** © PhotoDisc; **572** reprinted with permission of Kimberly S. Young; **Panel 12.3** all photos courtesy of the Robotics Institute, Carnegie Melon; **Panel 12.4** reprinted from *Columbia Journalism Review*, January/February 1991, © 1991 by *Columbia Journalism Review*; photo decoration for all Readme boxes © PhotoDisc.

README Boxes
Page 325 *San Jose Mercury News*, July 17,1994, p. 1E. From Simson L. Garfinkel. © 1994 Simon Garfinkel. **343** Ramon G. McLeod, "Netiquette—Cyberspace's Cryptic Social Code," *San Francisco Chronicle*, March 6, 1996, pp. A1, A10. © SAN FRANCISCO CHRONICLE. Reprinted by permission. **442** *San Francisco Examiner*, November 21, 1993, p. E-3. From Martin J. Smith; reprinted from *Orange County Register*. From *The Naked Consumer* by Erik Larson (New York: Viking/Penguin USA).